Handbook of
MINERALOGY

VOLUME II

Silica, Silicates

Part 1

John W. Anthony
Richard A. Bideaux
Kenneth W. Bladh
Monte C. Nichols

To Marty Scott,
with thanks and the
compliments of the authors

Richard G Bideaux

MINERAL DATA PUBLISHING
Tucson, Arizona

Credit: Crystal drawing of epitaxial overgrowth of amphibole enclosing
pyroxene from Russell, New York, USA. See G.H. Williams (1889),
Amer. J. Sci., 39, 352–358.

IBM-PC is a trademark of International Business Machines.

TEX is a trademark of the American Mathematical Society.

Library of Congress Cataloging-in-Publication Data

Handbook of mineralogy.

Includes bibliographical references.
Contents: v. 1. Elements, sulfides, sulfosalts – v. 2. Silica,
silicates.
1. Mineralogy–Handbooks, manuals, etc.   I. Anthony, John W.
(John Williams), 1920–
QE366.8.H36 1990   549   89-13673
ISBN 0-9622097-0-8 (v. 1)   CIP

   9   8   7   6   5   4   3   2   1

Note: It is not practical for a book to provide complete cross-referencing between
mineral names and elemental constituents. Accordingly, a computer program, **SEARCH,** for searching the chemical formulae of all mineral species,
with its associated up-to-date file of mineral formulae and names, has been
written. This program runs on any IBM-PC or compatible computer having
640K memory and MS-DOS 2.2 or above. It is supplied on a $5\frac{1}{4}$ inch, 1.2
MB or $3\frac{1}{2}$ inch, 1.44MB floppy disk. Further information is available from
Mineral Data Publishing at the address below.

Mineral Data Publishing
P.O. Box 37072
Tucson, Arizona 85740

ACKNOWLEDGEMENTS

Invaluable assistance was rendered through review of early drafts of selected species descriptions by the following mineralogists: Petr Černý, Pete Dunn, Carl Francis, Edward Grew, Frank Hawthorne, Cornelius Klein, Bernard Leake, Paul Moore, and William Wise. Locality information was also reviewed by John White, Jr. and William Pinch. Our thanks especially to Hans-Jürgen Wilke, who consulted his extensive records of world-wide locality data on our behalf. Remaining errors of course are ours.

Steven Steinke and Edward Hasenor provided initial summaries of some prominent silicate species, giving us the extra impetus needed to reach the end of such a massive literature search and abstraction task. We are pleased to acknowledge both their help and good mineralogical judgement.

The *Mineral Powder Diffraction File Data Book* of the International Centre for Diffraction Data was consulted for comparison to our original literature sources and to insure the completeness of our data. In those cases where a mineral's pattern was available only from the ICDD, they have generously granted us permission for reproduction.

The relative importance of localities for inclusion in the "Distribution" section was judged in part through examination of the following extensive collections, by courtesy of their curatorial staffs: the National Museum of Natural History, Washington, D.C., through Jeffery Post; Harvard University, Cambridge, Massachusetts, through Carl Francis; in London, the Natural History Museum [formerly the British Museum (Natural History)], through Peter Embrey and Robert Symes; and in Paris, at the School of Mines, Claude Guillemin and Jean-François Poullen; at the Sorbonne, Pierre Bariand; and at the Natural History Museum, Henri-Jean Schubnel.

Type specimen locations were augmented from the International Mineralogical Association *Reference Sample Catalog,* through the courtesy of its present keepers, Anthony Stalder and Anthony Kampf.

Martha Scott, Karen Schwartz, and Olga Graham patiently deciphered and entered the information for many species from our handwritten forms into the initial computer files. This potential source of errors we are happy to report is now in the past, with direct computer entry by the authors now routine. Proofreading by Barbara Nichols and Linda Dufek was essential to finding our errors of both commission and omission. Curtis Schuh provided computer support, prepared Figures 1–3, and assisted with other arcane typographical matters.

Typesetting for draft and final book pages of this volume was accomplished using the TEX computerized typesetting system developed at Stanford University by Donald Knuth. Use of this system greatly eased our burden of producing typographically elegant printed pages directly from a database.

We have been able to examine original literature in nearly all cases, through the Science Library of the University of Arizona, the Orton Memorial Library of Geology at the Ohio State University, the Branner Geological Library at Stanford University, and the USGS libraries at Menlo Park and Denver. Use of a specialized library on gemology was provided by Eric Van Valkenburg.

Access to obscure references was facilitated by use of the abstracts provided by the *American Mineralogist, Mineralogical Abstracts,* and *Chemical Abstracts.* These were especially useful for their coverage and translation of the Russian and Chinese literature. Accuracy in transliteration of Russian locality names was assisted by Hans-Jürgen Wilke and Bill Smith.

For this second volume, the major international mineralogical journals were examined through 1992. New species descriptions and crystal structure determinations abstracted in the New Mineral Names section of the *American Mineralogist* during 1993 and 1994, and for which complete data was published, are also included. Numerous less-accessible references were called to our attention through use of the USGS Ford-Fleischer File, the latest edition of which encompasses the mineralogical literature through 1987. For selected species, the GEO-REF database of the American Geological Institute was utilized to broaden our search of the more recent literature. Still, we recognize that our coverage of the literature is perforce less than perfect, and we solicit pertinent information for inclusion of corrections and additions in future volumes and reissues.

Finally, we thank the mineralogists, mineral collectors, and librarians who ordered copies of our first volume in numbers such that the first printing has been exhausted. This is important to our continued ability to complete the series. The first volume has been reprinted in time to accompany this second volume; we will keep all volumes in print through the appearance of the fifth and last.

John W. Anthony
(1920 – 1992)
Richard A. Bideaux
Kenneth W. Bladh
Monte C. Nichols

January, 1995

INTRODUCTION

The four decades following the appearance of the first two volumes of the seventh edition of *Dana's System of Mineralogy* have seen the number of known mineral species more than double — from about 1500 species at mid-century to around 3600 at this writing.

This burgeoning is nowhere more pronounced than within the silicate minerals (the subject of this volume II of the *Handbook of Mineralogy*) whose representatives have increased about three-fold. While over 900 silicate species are included here, fewer than 300 of these received mention in the sixth edition of *Dana's System* (published in 1892, with later appendices to 1915.)

This proliferation stems from several factors: the continuing development of a firm base of physical and chemical theory for the crystalline state, the appearance of a panoply of remarkably sophisticated instrumental methods nicely suited to the study of minerals, and the timely blossoming of computer science, principal progeny of the transistor.

The digital computer and its associated software have not only eased the burden of that computationally most arduous of crystallographic tasks, the determination of atomic sites within the unit cell, they have markedly accelerated the collection and interpretation of all manner of mineralogical data. Indeed, their presence is now central to the operation of essentially all analytical instrumentation having any degree of complexity.

This *Handbook of Mineralogy* series was conceived in order to gather in convenient form the data crucial to identification of all mineral species and to provide relatively up-to-date references containing information central to the definition of each species. Our intent is to provide data sufficient to distinguish a species from all others. If additional information seems desirable, primary or secondary references are given for each mineral in which the species is further discussed.

Today's fundamental mineralogical knowledge is effectively presented in a number of popular introductory textbooks, for example: Berry, Mason, and Dietrich's (1983) *Mineralogy*; Klein and Hurlbut's (1993) *Manual of Mineralogy*; and Blackburn and Dennen's (1994) *Principles of Mineralogy*. We remind those who peruse these volumes that mastery of that information is essential to use of the *Handbook of Mineralogy*. Our volumes present data, not the background necessary to use or understand it.

A cursory glance will reveal what may appear to be a somewhat unconventional organizational scheme. The well-known chemical-structural classification of the *System* has been adopted for the pattern of the overall work because, from a geological point of view, this classification has the advantage of tending to cluster together minerals of similar provenance. However, within each volume of this work, the minerals have been arranged alphabetically. We have selected this scheme to facilitate the location of individual minerals.

As an auxiliary aid, we incorporate a diagrammatic classification of the amphiboles. Application of the labyrinthine nomenclature of this complex group of minerals has been significantly influenced by the aforementioned advances in crystal chemistry, electron microprobe analysis, and computer science.

We describe structural and chemical relations with other species under "Group." We have generally followed Fleischer and Mandarino's (1995) *Glossary of Mineral Species* as a guide to those minerals having enough properties in common and numerous enough to warrant designation together as a group.

In the species descriptions, we distinguish between data extracted from the literature and information inferred from group or other related species knowledge. This latter category is enclosed in [square brackets] with attribution to its source by analogy.

Group names of long standing are used under "Association" when individual species names within the group are not accurately known. Also in this category are some species names which are used without suffixes, notably

Table 1. Distribution of mineral categories among the *Handbook of Mineralogy* volumes. Major categories are listed in bold type followed by minor categories in alphabetical order.

VOLUME I	VOLUME II	VOLUME III	VOLUME IV	VOLUME V
Elements	**Silica**	**Halides**	**Arsenates**	**Borates**
Sulfides	**Silicates**	**Hydroxides**	**Phosphates**	**Carbonates**
Sulfosalts		**Oxides**	**Uranates**	**Sulfates**
Alloys		Antimonates	**Vanadates**	Chromates
Antimonides		Antimonites		Germanates
Arsenides		Arsenites		Iodates
Bismuthinides		Carbides		Molybdates
Intermetallics		Nitrides		Nitrates
Selenides		Phosphides		Organics
Sulfhalides		Silicides		Selenates
Sulfoxides		V-oxysalts		Selenites
Tellurides				Sulfites
				Tellurates
				Tellurites
				Tungstates

those containing rare earth elements when the predominant element is unknown.

The distribution of minerals among the five volumes is indicated in Table 1. Major categories are listed in bold type in the table and on the spines of the corresponding volumes as a quick guide to the contents. Minor categories are listed in alphabetical order following the major categories.

We have distilled the pertinent facts defining each mineral to one page. Although this concentration may result in seemingly callous winnowing of sacred classical descriptive matter, such scholarly facts, peripheral to our present objectives, can always be consulted in the literature referenced for each species. The hallowed angle table and crystal drawings were early victims of this parsimony.

For species names and chemical formulae, principal guides are the *American Mineralogist*; the recommendations of the International Mineralogical Association's Commission on New Minerals and Mineral Names; Fleischer and Mandarino's (1995) *Glossary of Mineral Species*; Nickel and Nichols' (1991) *Mineral Reference Manual*; and Clark's (1993) *Hey's Mineral Index*. Where the philosophies of these are in conflict, we attempt to steer a prudent course.

We have tended to incorporate data from the naturally occurring mineral, preferably type material where available, rather than from its synthetic equivalent. Where data are missing and should be determined, the designation "n.d." (not determined) appears, so highlighting species for which essential observations are lacking.

Our order of presentation of mineral properties should prove an expeditious alternative to conventional formats. In our departure from those conventions we have tried to establish logical groupings of related properties. While we believe these arrangements to be consistent and self-explanatory, we present an explanatory page at the end of this introduction as a further guide to both the order and content of the mineral entries presented in this volume.

When lacking from our referenced literature sources, yet possible to define, we have calculated densities in many cases. These are from data given in the *Handbook*; the unit cell data and contents, and either the simplified chemical formula or an actual analysis. We report the results in [square brackets]. The reader may compare these to the measured densities as a check on internal consistency.

X-ray powder diffraction lines and their intensities are listed for all those minerals for which they were available, preference again being given to patterns from natural material whenever possible. The origin of the X-ray powder data is included in the references, and the locality or source of the sample precedes each pattern listing.

Up to four chemical analyses have been selected from the literature to illustrate the range of substitution in a mineral's structure, usually for major elements only. We often include an idealized analysis for comparison, calculated especially for this work using the atomic weights given in Table 2.

Localities given under "Distribution" are limited to about a dozen worldwide. This seems not unduly restrictive since about half of the minerals are known from only a single locality, and a further quarter from no more than six. For the less common and rare species we have attempted to provide comprehensive locality information.

For species having numerous localities, criteria for inclusion of a locality are the occurrence of especially well crystallized, pure, or abundant material. We have tended to drop occurrences stated as "reported from" in the older literature; these often require authentication by modern methods. We have also tried to avoid incompletely given locality data unless they are the only data available for a species. Locality information is not referenced.

The whereabouts of type material is noted if mentioned in one of the references, if cited in one of the various compilations made by major museums, or in the IMA's *Reference Sample Catalog* of type specimens. If unknown to us, this is noted as "n.d." only if the mineral was described after 1900. We specify neither the nature of type material, whether holotype, cotype, etc., nor the type locality.

As to our definition of what constitutes a silicate mineral, an obvious initial requirement was that its structure contain mutually bonded silicon and oxygen atoms.

The majority of silicon- and oxygen-bearing minerals presented few problems, as their structures consist either of independent SiO_4 tetrahedra, groups thereof scattered throughout, or cross-linked in geometrically fascinating configurations such as rings, chains, sheets, and frameworks. Priority as a silicate was accorded even if the SiO_4 is subordinate in amount to other cations, so long as it occupies an independent structural position.

But there are, indeed, minerals which contain silicon in other and more unusual forms. Certain phosphates in which SiO_4 substitutes diadochically for PO_4 were treated as phosphates, not silicates, if the SiO_4 is subordinate in atomic ratio to PO_4; but if the SiO_4 exceeds PO_4, the mineral was termed a silicate.

Another problem, rarely encountered, was silicon present in the mineral composition but with no independent oxygen to which it can bond (for example, thaumasite, containing silicon, but no oxygen other than that in hydroxyl). These few minerals we still term as silicates.

In our reviews of the literature we have tried to minimize distortions of the facts we have selected. Whenever possible, we have consulted the original literature, as well as secondary sources. In general, we have not listed those references from which we have extracted no data, the principal exception being papers on crystal structures which we mention whether or not we actually quote from them. We always provide the full reference, even for those important references which occur repeatedly. Abbreviations of mineralogical terms largely conform to the practice of the *American Mineralogist*. The explanatory page at the end of this introduction presents a number of these.

Table 2. Listing of element names, element symbols, atomic numbers and atomic weights used for calculations of density, Z, and idealized chemical analyses made especially for these volumes. The values for the atomic weights are those given in: Commission on Atomic Weights and Isotopic Abundances (1986) Atomic weights of the elements 1985. *Pure and Applied Chemistry*, 58, 1677–1692. (© 1986 IUPAC).

Element	Sym.	At. No.	At. Wt.	Element	Sym.	At. No.	At. Wt.
Actinium†	Ac	89	227.028	Mendelevium†	Md	101	258.10
Aluminum	Al	13	26.982	Mercury	Hg	80	200.59
Americium†	Am	95	243.061	Molybdenum	Mo	42	95.94
Antimony	Sb	51	121.75	Neodymium	Nd	60	144.24
Argon	Ar	18	39.948	Neon	Ne	10	20.180
Arsenic	As	33	74.922	Neptunium†	Np	93	237.048
Astatine†	At	85	209.987	Nickel	Ni	28	58.69
Barium	Ba	56	137.327	Niobium	Nb	41	92.906
Berkelium†	Bk	97	249.075	Nitrogen	N	7	14.007
Beryllium	Be	4	9.012	Nobelium†	No	102	259.101
Bismuth	Bi	83	208.980	Osmium	Os	76	190.2
Boron	B	5	10.811	Oxygen	O	8	15.999
Bromine	Br	35	79.904	Palladium	Pd	46	106.42
Cadmium	Cd	48	112.411	Phosphorus	P	15	30.974
Calcium	Ca	20	40.078	Platinum	Pt	78	195.08
Californium†	Cf	98	242.059	Plutonium†	Pu	94	244.064
Carbon	C	6	12.011	Polonium†	Po	84	208.982
Cerium	Ce	58	140.115	Potassium	K	19	39.098
Cesium	Cs	55	132.905	Praseodymium	Pr	59	140.908
Chlorine	Cl	17	35.453	Promethium†	Pm	61	144.913
Chromium	Cr	24	51.996	Protactinium†	Pa	91	231.036
Cobalt	Co	27	58.933	Radium†	Ra	88	226.025
Copper	Cu	29	63.546	Radon†	Rn	86	222.018
Curium†	Cm	96	247.070	Rhenium	Re	75	186.207
Dysprosium	Dy	66	162.50	Rhodium	Rh	45	102.906
Einsteinium†	Es	99	252.083	Rubidium	Rb	37	85.468
Erbium	Er	68	167.26	Ruthenium	Ru	44	101.07
Europium	Eu	63	151.965	Samarium	Sm	62	150.36
Fermium†	Fm	100	257.095	Scandium	Sc	21	44.956
Fluorine	F	9	18.998	Selenium	Se	34	78.96
Francium†	Fr	87	223.020	Silicon	Si	14	28.086
Gadolinium	Gd	64	157.25	Silver	Ag	47	107.868
Gallium	Ga	31	69.723	Sodium	Na	11	22.990
Germanium	Ge	32	72.61	Strontium	Sr	38	87.62
Gold	Au	79	196.966	Sulfur	S	16	32.066
Hafnium	Hf	72	178.49	Tantalum	Ta	73	180.948
Helium	He	2	4.003	Technetium†	Tc	43	97.907
Holmium	Ho	67	164.930	Tellurium	Te	52	127.60
Hydrogen	H	1	1.008	Terbium	Tb	65	158.925
Indium	In	49	114.82	Thallium	Tl	81	204.383
Iodine	I	53	126.904	Thorium†	Th	90	232.038
Iridium	Ir	77	192.22	Thulium	Tm	69	168.934
Iron	Fe	26	55.847	Tin	Sn	50	118.710
Krypton	Kr	36	83.80	Titanium	Ti	22	47.88
Kurchatovium	Ku	104	261.11	Tungsten	W	74	183.85
Lanthanum	La	57	138.906	Uranium†	U	92	238.029
Lawrencium†	Lr	103	260.105	Vanadium	V	23	50.942
Lead	Pb	82	207.2	Xenon	Xe	54	131.29
Lithium	Li	3	6.941	Ytterbium	Yb	70	173.04
Lutetium	Lu	71	174.967	Yttrium	Y	39	88.906
Magnesium	Mg	12	24.305	Zinc	Zn	30	65.39
Manganese	Mn	25	54.938	Zirconium	Zr	40	91.224

† These elements have no stable isotopes; the atomic weight of the most stable isotope is listed.

Crystal Data: Crystal system and point group in Hermann-Mauguin symbols. Description of the visual appearance of single crystals and aggregates. Common crystal forms and twin laws are indicated.

Physical Properties: Megascopic and simple measurable properties of the pure mineral, especially those depending on cohesion. Density is in g/cm^3. Reactions to excitation by X-rays, electrons, and ultraviolet (UV) light are given: SW = short-wave; LW = long-wave. Radioactivity is noted.

Optical Properties: Characteristics depending on the interaction with white light, first with a hand sample (transparency, color, and streak); secondly in polished or thin section, viewed through a reflecting or polarizing light microscope (color, pleochroism, and anisotropism; optical class, sign, indices of refraction, etc.). Spectral reflectance data (R or R_1–R_2) are given as available, with each wavelength (in nanometers) listed in parentheses followed by the corresponding reflectance value in percent.

Cell Data: Space group symbol and unit cell parameters in Ångstrom units.

X-ray Powder Pattern: Source of material, followed by up to seven most intense maxima, presented as d-spacing in Å followed by relative intensity in parentheses. Maxima known to be unresolved doublets have a "d" appended to their respective intensity values. A "b" is appended to the intensity of those broadened for other reasons.

Chemistry: Determinations of major and minor elements in the natural mineral, reported as weight percentages. Non-idealized empirical formulae are from the references; calculated analyses are based on idealized chemical formulae. Sources of the materials analyzed are given. Abbreviations used for analytical instrumentation: AA = atomic absorption; DTA = differential thermal analysis; EGA = evolved gas analysis; ICP = inductively-coupled plasma; IR = infrared; TGA = thermogravimetric analysis; XRF = X-ray fluorescence.

Polymorphism & Series: Polymorphs and polytypes are noted as appropriate. Series are isomorphous with complete substitution and continuous variation of properties with composition.

Mineral Group: Notes membership in a group of structurally similar species.

Occurrence: A description of the geological processes and physicochemical environments inferred for the species' origin.

Association: Minerals in close spatial and inferred genetic relationship with the species. In order of closest association (equilibrium) to farthest, or most abundant to least, where known. Multiple lists indicate species with multiple parageneses, and their sources.

Distribution: The most important geographic localities for occurrences of the species, for up to a dozen localities.

Name: Significance and origin of the name.

Type Material: Museum holdings of type material.

References: Citations used to define the data summary for the species. We are the source for [calc. dens.] or [Z] values in square brackets and for all idealized chemical analyses.

AMPHIBOLES

The general amphibole structural formula is: $A_{0-1}B_2C_5^{vi}T_8^{iv}O_{22}(OH,F,Cl)_2$, where

A = A sites: Ca, K, Na, Pb
B = M4 sites: Ca, Fe^{2+}, Li, Mg, Mn^{2+}, Na
C = M1+M2+M3 octahedral sites: Al, Fe^{2+}, Fe^{3+}, Li, Mg, Mn^{2+}, Mn^{3+}, Ti
T = tetrahedral sites: Al, Be, Fe^{3+}, Si, Ti.

Other elements which may occur in notable amounts include Ba, Cr^{3+}, Ni, Sr, Zn.

We use [square brackets] on the amphibole chemical formulae to emphasize these structural sites. This diverges from Fleischer and Mandarino's (1995) *Glossary of Mineral Species*, our primary source for mineral formulae, but it is preferable to explicitly show the structural sites due to their importance in classification.

Determination of the name of an amphibole, according to the IMA nomenclature adopted in this volume, requires a complete chemical analysis. Wet chemical analyses provide all of the chemical data necessary; electron microprobe analyses however are more prevalent today. Their advantage is more likely homogeneity of analyzed material, but with less analytical sensitivity for Li, Be, Cl, F, and an inability to determine the oxidation states of especially Fe and Mn, or the amount of H_2O.

Before naming, the elements determined in a chemical analysis must be systematically allocated among the structural sites in the general amphibole formula. Full discussion of these allocation schemes is to be found in the general amphibole references.

Ambiguities in allocation and sometimes the resulting name may arise due to the choice of analytical method, undetermined elements and oxidation states, method of normalizing the structural formula, or compositions falling close to classification boundaries. The minor element content of some structural sites may dictate that one or more prefixes or adjectival modifiers be applied to the fundamental name.

The rules for fundamental names, prefixes, and adjectival modifiers reflect elemental ratios in the structural sites and are explained in Leake (1978). These complex relations are usually simplified on diagrams as slices through multidimensional compositional space. We provide three such diagrams, Figures 1–3. Note that names alternative to those which the overall scheme would supply are in use in a few special situations.

End member and intermediate species, descriptions of which can be found in this volume, are in **boldface Roman type** in these diagrams. Some end member species, not yet found in nature, are in ***slanted boldface Roman type***. Minor intermediate species, which may or may not occur in nature, but are not described in the text, are in *slanted Roman type*. Additional end member species, necessary to complete the classification, are listed on the page following.

General amphibole references

(1) Leake, B.E. (1968) A catalog of analyzed calciferous and subcalciferous amphiboles together with their nomenclature and associated minerals. Geol. Soc. Amer. Special Paper 98, 210 p. (2) Leake, B.E., Compiler (1978) Nomenclature of amphiboles. Mineral. Mag., 42, 533–563; Amer. Mineral., 63, 1023–1052. (3) Hawthorne, F.C. (1983) The crystal chemistry of the amphiboles. Can. Mineral., 21, 173–480. (4) Rock, N.M.S. and B.E. Leake (1984) The International Mineralogical Association amphibole nomenclature scheme: computerization and its consequences. Mineral. Mag., 48, 211–227. (5) Fleischer, M. and R.E. Wilcox (1984) Microscopic determination of the nonopaque minerals. U.S. Geol. Survey Bull. 1627, 453 p. (6) Richard, L.R. and D.B. Clarke (1990) AMPHIBOL: a program for calculating structural formulae and for classifying and plotting chemical analyses of amphiboles. Amer. Mineral., 75, 421–423. (7) Tindle, A.G. and P.C. Webb (1994) PROBE-AMPH – a spreadsheet program to classify microprobe-derived amphibole analyses. Computers & Geosciences, 20, 1201–1228.

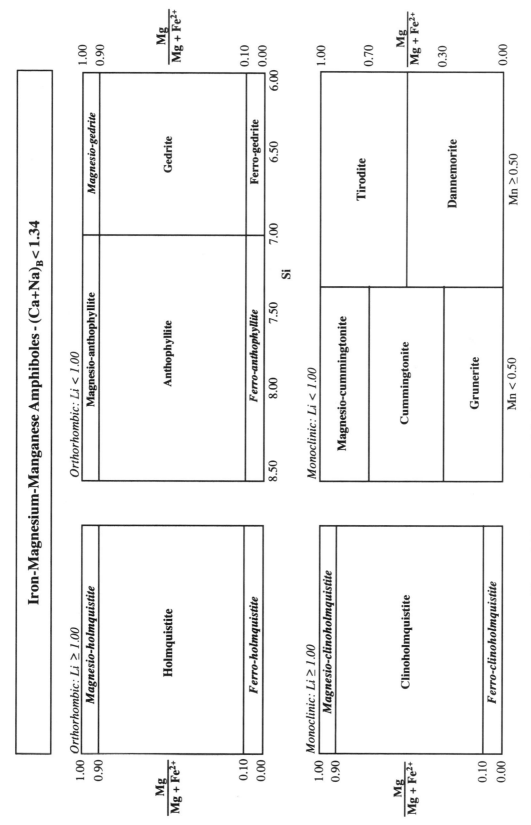

Figure 1. The nomenclature of the iron-magnesium-manganese amphiboles.

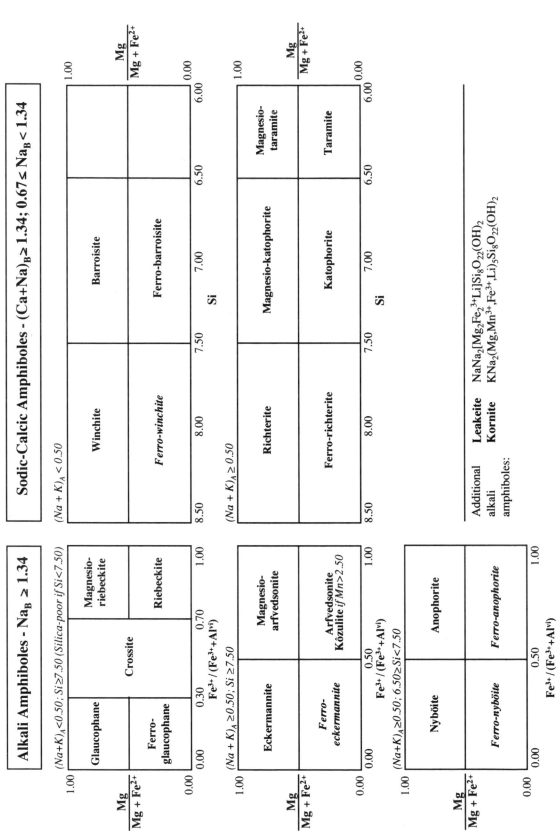

Figure 2. The nomenclature of the alkali and sodic-calcic amphiboles.

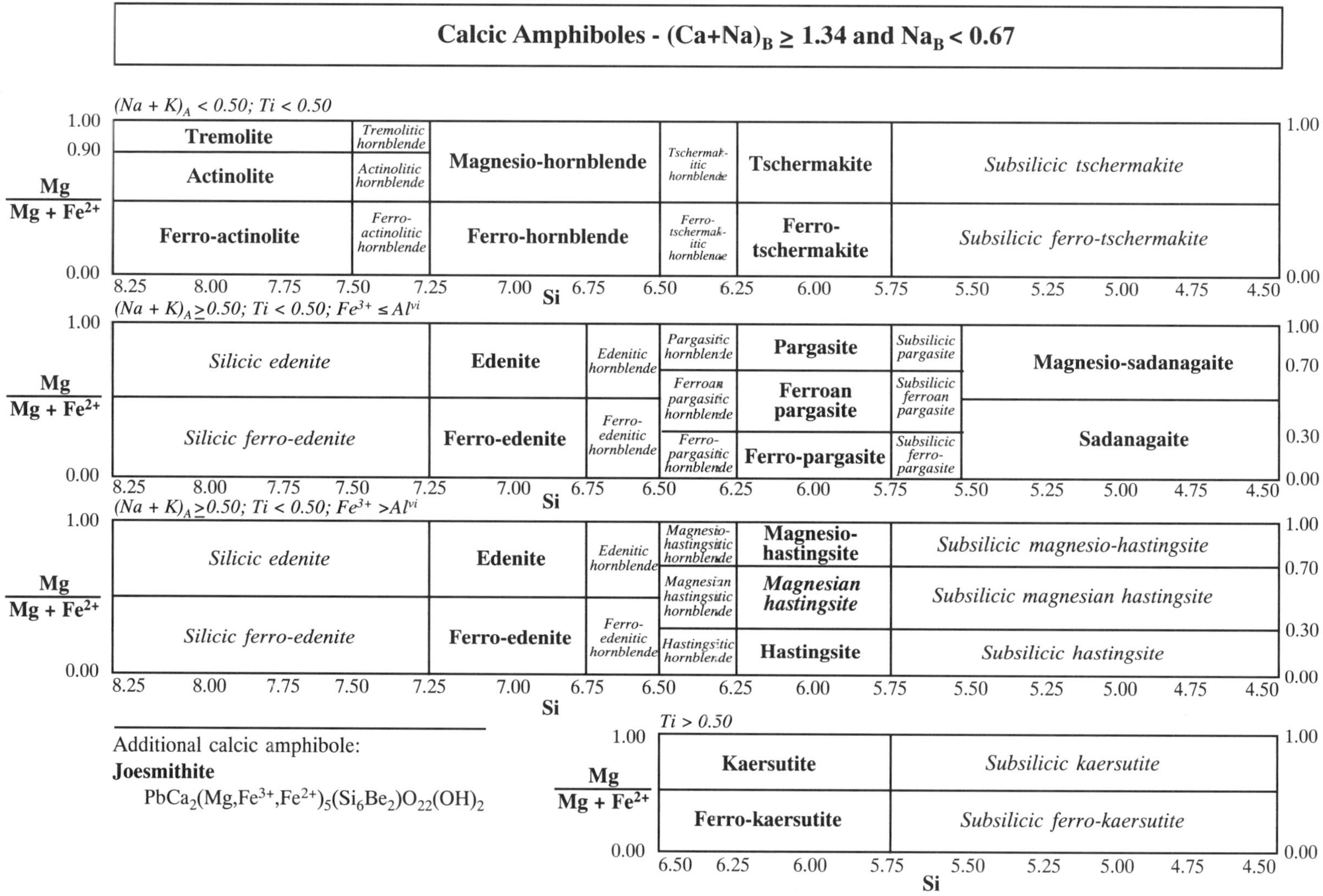

Calcic Amphiboles - $(Ca+Na)_B \geq 1.34$ and $Na_B < 0.67$

$(Na + K)_A < 0.50; Ti < 0.50$

Tremolite	*Tremolitic hornblende*	**Magnesio-hornblende**	*Tschermakitic hornblende*	**Tschermakite**	*Subsilicic tschermakite*
Actinolite	*Actinolitic hornblende*				
Ferro-actinolite	*Ferro-actinolitic hornblende*	**Ferro-hornblende**	*Ferro-tschermakitic hornblende*	**Ferro-tschermakite**	*Subsilicic ferro-tschermakite*

$\dfrac{Mg}{Mg + Fe^{2+}}$ axis from 1.00, 0.90, 0.00

Si axis: 8.25 8.00 7.75 7.50 7.25 7.00 6.75 6.50 6.25 6.00 5.75 5.50 5.25 5.00 4.75 4.50

$(Na + K)_A \geq 0.50; Ti < 0.50; Fe^{3+} \leq Al^{vi}$

Silicic edenite	**Edenite**	*Edenitic hornblende*	*Pargasitic hornblende*	**Pargasite** / *Subsilicic pargasite*	**Magnesio-sadanagaite**
			Ferroan pargasitic hornblende	**Ferroan pargasite** / *Subsilicic ferroan pargasite*	
Silicic ferro-edenite	**Ferro-edenite**	*Ferro-edenitic hornblende*	*Ferro-pargasitic hornblende*	**Ferro-pargasite** / *Subsilicic ferro-pargasite*	**Sadanagaite**

$\dfrac{Mg}{Mg + Fe^{2+}}$ axis right: 1.00, 0.70, 0.30, 0.00

$(Na + K)_A \geq 0.50; Ti < 0.50; Fe^{3+} > Al^{vi}$

Silicic edenite	**Edenite**	*Edenitic hornblende*	*Magnesio-hastingsitic hornblende*	**Magnesio-hastingsite**	*Subsilicic magnesio-hastingsite*
			Magnesian hastingsitic hornblende	***Magnesian hastingsite***	*Subsilicic magnesian hastingsite*
Silicic ferro-edenite	**Ferro-edenite**	*Ferro-edenitic hornblende*	*Hastingsitic hornblende*	**Hastingsite**	*Subsilicic hastingsite*

$\dfrac{Mg}{Mg + Fe^{2+}}$ axis right: 1.00, 0.70, 0.30, 0.00

Additional calcic amphibole:

Joesmithite

$PbCa_2(Mg,Fe^{3+},Fe^{2+})_5(Si_6Be_2)O_{22}(OH)_2$

$Ti > 0.50$

Kaersutite	*Subsilicic kaersutite*
Ferro-kaersutite	*Subsilicic ferro-kaersutite*

Si axis: 6.50 6.25 6.00 5.75 5.50 5.25 5.00 4.75 4.50

Figure 3. The nomenclature of the calcic amphiboles.

xiii

Additional amphibole end members necessary to complete the classification:

Iron-Magnesium-Manganese amphiboles:

Sodium anthophyllite, $Na(Mg, Fe^{2+})_2(Mg, Fe^{2+})_5(Si_7Al)O_{22}(OH)_2$

Sodium gedrite, $Na(Mg, Fe^{2+})_2[(Mg, Fe^{2+})_4Al](Si_6Al_2)O_{22}(OH)_2$

Sodic-Calcic amphiboles:

Alumino-barroisite, $[NaCa][Mg_3Al_2](Si_7Al)O_{22}(OH)_2$

Alumino-katophorite, $Na[NaCa][Fe_4^{2+}Al](Si_7Al)O_{22}(OH)_2$

Alumino-taramite, $Na[NaCa][Fe_3^{2+}Al_2](Si_6Al_2)O_{22}(OH)_2$

Alumino-winchite, $[NaCa][Mg_4Al]Si_8O_{22}(OH)_2$

Ferri-barroisite, $[NaCa][Mg_3Fe_2^{3+}](Si_7Al)O_{22}(OH)_2$

Ferri-katophorite, $Na[NaCa][Fe_4^{2+}Fe^{3+}](Si_7Al)O_{22}(OH)_2$

Ferri-taramite, $Na[NaCa][Fe_3^{2+}Fe_2^{3+}](Si_6Al_2)O_{22}(OH)_2$

Ferri-winchite, $[NaCa][Mg_4Fe^{3+}]Si_8O_{22}(OH)_2$

Ferro-alumino-barroisite, $[NaCa][Fe_3^{2+}Al_2](Si_7Al)O_{22}(OH)_2$

Ferro-alumino-winchite, $[NaCa][Fe_4^{2+}Al]Si_8O_{22}(OH)_2$

Ferro-ferri-barroisite, $[NaCa][Fe_3^{2+}Fe_2^{3+}](Si_7Al)O_{22}(OH)_2$

Ferro-ferri-winchite, $[NaCa][Fe_4^{2+}Fe^{3+}]Si_8O_{22}(OH)_2$

Magnesio-alumino-katophorite, $Na[NaCa][Mg_4Al](Si_7Al)O_{22}(OH)_2$

Magnesio-alumino-taramite, $Na[NaCa][Mg_3Al_2](Si_6Al_2)O_{22}(OH)_2$

Magnesio-ferri-katophorite, $Na[NaCa][Mg_4Fe^{3+}](Si_7Al)O_{22}(OH)_2$

Magnesio-ferri-taramite, $Na[NaCa][Mg_3Fe_2^{3+}](Si_6Al_2)O_{22}(OH)_2$

Calcic amphiboles:

Alumino-ferro-hornblende, $Ca_2[Fe_4^{2+}Al](Si_7Al)O_{22}(OH)_2$

Alumino-magnesio-hornblende, $Ca_2[Mg_4Al](Si_7Al)O_{22}(OH)_2$

Alumino-tschermakite, $Ca_2[Mg_3Al_2](Si_6Al_2)O_{22}(OH)_2$

Ferri-tschermakite, $Ca_2[Mg_3Fe_2^{3+}](Si_6Al_2)O_{22}(OH)_2$

Ferro-alumino-tschermakite, $Ca_2[Fe_3^{2+}Al_2](Si_6Al_2)O_{22}(OH)_2$

Ferro-ferri-tschermakite, $Ca_2[Fe_3^{2+}Fe_2^{3+}](Si_6Al_2)O_{22}(OH)_2$

VOLUME II
Silica, Silicates

Part 1

Crystal Data: [Tetragonal] (by analogy to braunite). *Point Group:* $[4/m \ 2/m \ 2/m.]$ Anhedral grains, to 50 μm.

Physical Properties: Hardness = n.d. VHN = 870–950, average 920 (25 g load). D(meas.) = n.d. D(calc.) = 4.96 (synthetic).

Optical Properties: Opaque. *Color:* Black; gray in reflected light. *Streak:* Brownish black. *Luster:* Metallic. *Anisotropism:* Weak.

Cell Data: *Space Group:* $[I4_1/acd.]$ $a = 9.406(1)$ $c = 18.546(3)$ Z = [8]

X-ray Powder Pattern: Synthetic $(\text{Cu}_{0.98}\text{Mn}_{0.02}^{2+})_{\Sigma=1.00}\text{Mn}_6^{3+}\text{SiO}_{12}$. 2.702 (100), 1.6507 (30), 2.350 (15), 2.133 (15), 1.459 (14), 1.4016 (11), 1.6627 (10)

Chemistry:

	(1)
SiO_2	10.1
TiO_2	0.23
Al_2O_3	0.58
Fe_2O_3	4.7
Mn_2O_3	72.1
CuO	11.6
MgO	< 0.04
CaO	0.16
Total	99.47

(1) Evvia Island, Greece; by electron microprobe, corresponds to $(\text{Cu}_{0.88}\text{Mn}_{0.10}^{2+}\text{Ca}_{0.02})_{\Sigma=1.00}$ $(\text{Mn}_{5.51}^{3+}\text{Fe}_{0.36}^{3+}\text{Al}_{0.07}\text{Ti}_{0.02}\text{Cu}_{0.02})_{\Sigma=5.98}\text{Si}_{1.03}\text{O}_{12}$.

Polymorphism & Series: Forms a series with braunite.

Occurrence: In very low-grade, high-pressure metamorphic Mn, Al-rich quartzites.

Association: Quartz, shattuckite, tenorite, sursassite, piemontite, ardennite, rutile, hollandite, clinochlore.

Distribution: At Mili, Evvia Island, and Apikia, Andros Island, Cyclades Islands, Greece.

Name: For Dr. Irmgard Abs-Wurmbach (1938–), German mineralogist.

Type Material: Institute for Mineralogy, Ruhr University, Bochum, Germany; National Museum of Natural History, Washington, D.C., USA.

References: (1) Reinecke, T., E. Tillmanns, and H.-J. Bernhardt (1991) Abswurmbachite, $\text{Cu}^{2+}\text{Mn}_6^{3+}[\text{O}_8/\text{SiO}_4]$, a new mineral of the braunite group: natural occurrence, synthesis, and crystal structure. Neues Jahrb. Mineral., Abh., 163, 117–143. (2) (1992) Amer. Mineral., 77, 670 (abs. ref. 1).

Crystal Data: Monoclinic. *Point Group:* $2/m$. As bladed crystals, to 15 cm; columnar, may be kinked or bent; radiating fibrous to asbestiform; granular to massive. *Twinning:* Simple or lamellar, common $\|$ {100}; lamellar, less common $\|$ {001}.

Physical Properties: *Cleavage:* Good on {110}, with intersections of 56° and 124°; parting on {100}. *Tenacity:* Brittle, tough in fibrous aggregates ("nephrite jade"). Hardness = 5–6 D(meas.) = 3.03–3.24 D(calc.) = [3.07]

Optical Properties: Transparent to translucent. *Color:* Bright green to grayish green; in thin section, colorless, pale green to deep green. *Streak:* White. *Luster:* Vitreous, silky. *Optical Class:* Biaxial (–). *Pleochroism:* Weak; X = pale yellow, yellowish green; Y = pale yellow-green, green; Z = pale green, deep greenish blue. *Orientation:* $Y = b$; $Z \wedge c =$ 14°–18.5°. *Dispersion:* $r < v$, weak. $\alpha = 1.613$–1.646 $\beta = 1.624$–1.656 $\gamma = 1.636$–1.666 2V(meas.) = 79°–86°

Cell Data: *Space Group:* $C2/m$. a = 9.891(1)   b = 18.200(1)   c = 5.305(1) $\beta = 104.64(1)°$ Z = 2

X-ray Powder Pattern: Sobotin, Czech Republic. (ICDD 25-157). 2.719 (100), 2.543 (100), 3.401 (80), 8.47 (70), 4.91 (70), 3.143 (70), 2.959 (70)

Chemistry:

	(1)		(1)
SiO_2	55.17	MnO	0.18
TiO_2	0.17	MgO	16.21
Al_2O_3	2.69	CaO	12.08
FeO	11.07	Na_2O	0.82
		Total	98.39

(1) Southeast Anglesey, Wales; by electron microprobe, corresponding to $(Ca_{1.83}Na_{0.22})_{\Sigma=2.05}$ $(Mg_{3.41}Fe^{2+}_{1.26}Al_{0.23}Fe^{3+}_{0.05}Mn_{0.02}Ti_{0.02})_{\Sigma=4.99}(Si_{7.79}Al_{0.21})_{\Sigma=8.00}O_{22}(OH)_2$.

Polymorphism & Series: Forms a series with tremolite and ferro-actinolite.

Mineral Group: Amphibole (calcic) group: $0.5 \leq Mg/(Mg + Fe^{2+}) \leq 0.89$; $(Na + K)_A < 0.5$; $Na_B < 0.67$; $(Ca + Na)_B \geq 1.34$; $Si \geq 7.5$.

Occurrence: Produced by low-grade regional or contact metamorphism of magnesium carbonate, mafic, or ultramafic rocks; also in glaucophane-bearing blueschists.

Association: Talc, epidote, chlorite, glaucophane, pumpellyite, lawsonite, albite.

Distribution: Many localities. In Austria, on Mt. Greiner, Zillertal, and at Untersulzbachtal. From Zermatt, Valais, Switzerland. At Snarum and Arendal, Norway. From the Ural Mountains, Russia. In the USA, from Gouverneur, St. Lawrence Co., New York; Franklin and Newton, Sussex Co., New Jersey; Chester, Windsor Co., Vermont; in the Fairfax quarry, Centreville, Fairfax Co., Virginia; Crestmore, Riverside Co., California; at Salida, Chaffee Co., Colorado. "Nephrite jade" occurs, in the USA, south and east of Lander, Fremont Co., Wyoming; north from Cape San Martin, Monterey Co., California; and around Jade Mountain, near the Kobuk River, Alaska. Along the Fraser River, British Columbia, Canada. Around Mt. Cook, South Island, New Zealand. Fine material from the Kunlun Mountains, Sinkiang Uighur Autonomous Region, China.

Name: From the Greek for *ray*, in allusion to the mineral's common radiating fibrous habit.

References: (1) Dana, E.S. (1892) Dana's system of mineralogy, (6th edition), 389–391. (2) Deer, W.A., R.A. Howie, and J. Zussman (1963) Rock-forming minerals, v. 2, chain silicates, 249–262. (3) Mitchell, J.T., F.D. Bloss, and G.V. Gibbs (1971) Examination of the actinolite structure and four other C2/m amphiboles in terms of double bonding. Zeits. Krist., 133, 273–300. (4) Gibbons, W. and M. Gyopari (1986) A greenschist protolith for blueschist in Anglesey, U.K. In: B.W. Evans and E.H. Brown, Eds., Blueschists and eclogites, Geol. Soc. Amer. Memoir 164, 217–228. (5) Phillips, W.R. and D.T. Griffen (1981) Optical mineralogy, 229–231.

Crystal Data: Monoclinic. *Point Group:* $2/m$. Prismatic crystals with blunt to steep terminations, to 35 cm, striated lengthwise, can be bent or twisted. In sprays of acicular crystals, fibrous, in radial concretions. *Twinning:* Simple and lamellar twinning common on {100}.

Physical Properties: *Cleavage:* Good on {110}, (110) $\wedge$ (1$\bar{1}$0) $\sim$87°; parting on {100}. *Fracture:* Uneven. *Tenacity:* Brittle. Hardness = 6   D(meas.) = 3.50–3.60   D(calc.) = 3.576

Optical Properties: Translucent to opaque. *Color:* Dark green to greenish black, reddish brown, black; bright green to yellow-green in thin section. *Streak:* Pale yellowish gray. *Luster:* Vitreous to slightly resinous.
Optical Class: Biaxial (–); varieties rich in Ca, Mg, Fe^{2+} are (+). *Pleochroism:* X = emerald-green, deep green; Y = grass-green, deep green, yellow; Z = brownish green, green, yellowish brown, yellow. *Orientation:* $Y = b$; $X \wedge c = -10°$ to 12°; $Z \wedge a = 6°$–28°. *Dispersion:* $r > v$, moderate to strong. *Absorption:* $X > Y > Z$. $\alpha = 1.722$–1.776 $\beta = 1.780$–1.820 $\gamma = 1.795$–1.836   2V(meas.) = 60°–70°

Cell Data: *Space Group:* $C2/c$ (synthetic).   a = 9.658   b = 8.795   c = 5.294 $\beta = 107.42°$ Z = 4

X-ray Powder Pattern: Narssârssuk, Greenland.
2.900 (100), 6.369 (90), 4.416 (80), 2.983 (70), 2.4701 (60), 1.7293 (60), 1.3975 (60)

Chemistry:

	(1)		(1)
SiO$_2$	51.35	CaO	1.25
TiO$_2$	1.10	Na$_2$O	12.66
Al$_2$O$_3$	2.15	K$_2$O	0.15
Fe$_2$O$_3$	28.66	H$_2$O$^+$	0.12
FeO	2.24	H$_2$O$^-$	0.17
MgO	0.10	Total	99.95

(1) Itapirapurã, São Paulo, Brazil; corresponds to $(Na_{0.94}Ca_{0.02}K_{0.01})_{\Sigma=0.97}(Fe^{3+}_{0.83}Fe^{2+}_{0.07}Al_{0.07}$ $Ti_{0.03}Mg_{0.01})_{\Sigma=1.01}(Si_{1.98}Al_{0.02})_{\Sigma=2.00}O_6$.

Mineral Group: Pyroxene group.

Occurrence: Common in alkalic igneous rocks, carbonatites, and pegmatites. From regionally metamorphosed schists, gneisses, and iron formations; in blueschist facies rocks, and from sodium metasomatism in granulites. An authigenic mineral in some shales and marls.

Association: Potassic feldspar, nepheline, riebeckite, arfvedsonite, aenigmatite, astrophyllite, catapleiite, eudialyte, sérandite, apophyllite.

Distribution: Some localities for good crystals are: in Norway, at Rundemyr, near Kongsberg; from Skaadoe, near Brevik; on Låven and other islands, Langesundsfjord; and at Drammen and Arendal. In the Alnö complex, Sweden. At Huesca, Huesca Province, Spain. From the Lovozero and Khibiny massifs, Kola Peninsula, Russia. At Narssârssuk, Greenland. From Oldonyo Dili, Tanzania. Large crystals from Mt. Malosa, Zomba district, Malawi. In the USA, from Pitcairn, Russell, and LaSalle, St. Lawrence Co., and at Natural Bridge, Jefferson Co., New York; at Magnet Cove, Hot Springs Co., Arkansas; in the Bear Paw Mountains, Hill Co., Montana. Large crystals from Mont Saint-Hilaire, Quebec, Canada.

Name: From Ægir, the Scandinavian sea-god, as first described from Norway.

References: (1) Dana, E.S. (1892) Dana's system of mineralogy, (6th edition), 364–366. (2) Deer, W.A., R.A. Howie, and J. Zussman (1978) Rock-forming minerals, (2nd edition), v. 2A, single-chain silicates, 482–519. (3) Frondel, C. and C. Klein (1965) Ureyite [kosmochlor], NaCrSi$_2$O$_6$: a new meteoritic pyroxene. Science, 149, 742–744.

Aenigmatite $\quad\quad\quad$ Na$_2$Fe$_5^{2+}$TiSi$_6$O$_{20}$

Crystal Data: Triclinic, pseudomonoclinic. *Point Group:* $\bar{1}$. Crystals poorly developed, prismatic, to 8 cm; as irregular segregations. *Twinning:* By rotation $\perp$ (01$\bar{1}$) or about [010] of the pseudomonoclinic cell; polysynthetic.

Physical Properties: *Cleavage:* Good on {010}, {100}. *Fracture:* Uneven. *Tenacity:* Brittle. Hardness = 5.5 D(meas.) = 3.81 D(calc.) = [3.84]

Optical Properties: Translucent to opaque. *Color:* Velvet-black; in thin section, reddish brown to black. *Streak:* Reddish brown. *Luster:* Vitreous to greasy.
Optical Class: Biaxial (+). *Pleochroism:* X = yellow brown; Y = red-brown; Z = dark brown to black. *Orientation:* $Y = b$; $Z \wedge c = 45°$. *Dispersion:* $r < v$, very strong. $\alpha = 1.793–1.81$ $\beta = 1.804–1.82$ $\gamma = 1.87–1.90$ $2V$(meas.) = 30°–50°

Cell Data: *Space Group:* $P\bar{1}$. $a = 10.406(13)$ $b = 10.813(14)$ $c = 8.926(6)$ $\alpha = 104°56(9)'$ $\beta = 96°52(11)'$ $\gamma = 125°19(6)'$ $Z = 2$

X-ray Powder Pattern: Kola Peninsula, Russia.
8.09 (vs), 3.145 (vs), 2.706 (s), 2.547 (s), 2.937 (ms), 2.119 (ms), 2.414 (m)

Chemistry:

	(1)	(2)	(3)		(1)	(2)	(3)
SiO$_2$	39.62	38.3	41.84	MgO	1.65	0.82	
TiO$_2$	9.66	8.37	9.27	CaO	0.44	1.42	
Al$_2$O$_3$	0.64	2.56		Na$_2$O	7.20	6.35	7.19
Fe$_2$O$_3$	4.64			K$_2$O	0.04	0.02	
FeO	33.92	40.2	41.70	Cl	0.02		
MnO	2.46	1.93		H$_2$O$^+$	0.05		
				Total	100.34	99.97	100.00

(1) Khibinite quarry, Kirovsk, Kola Peninsula, Russia. (2) Motzfeldt centre, Greenland; by electron microprobe. (3) Na$_2$Fe$_5$TiSi$_6$O$_{20}$.

Polymorphism & Series: Forms a series with wilkinsonite.

Mineral Group: Aenigmatite group.

Occurrence: A primary constituent in sodium-rich alkalic volcanics, pegmatites, and other silica-poor igneous rocks.

Association: Aegirine, augite, riebeckite, arfvedsonite, hedenbergite, fayalite, ilmenite.

Distribution: Notable studied occurrences include: at Naujakasik, near the Tunugdliarfik Fjord, and on the Kangerdluarssuk Plateau, in the Ilímaussaq intrusion; from Narssârssuk; and elsewhere in Greenland. In the Khibiny and Lovozero massifs, Kola Peninsula, Russia. At Sandefjord, Norway. From Cuddua Mida, Sicily, Italy. In the USA, from the Fourche Mountains, Hot Springs Co., Arkansas, and Santa Rosa, Sonoma Co., California. In Australia, from Warrumbungle volcano, Nandewar volcano, and the Mt. Warning complex, New South Wales; and the Peak Range Province, Queensland. From Logan Point quarry, Dunedin volcano, New Zealand.

Name: From the Greek for *riddle*, apparently an allusion to its (formerly) uncertain chemical composition.

Type Material: Mining Academy, Freiberg, Germany, 29166.

References: (1) Dana, E.S. (1892) Dana's system of mineralogy, (6th edition), 403–404. (2) Deer, W.A., R.A. Howie, and J. Zussman (1978) Rock-forming minerals, (2nd edition), v. 2A, single-chain silicates, 640–654. (3) Thompson, R.N. and J.E. Chisholm (1969) Synthesis of aenigmatite. Mineral. Mag., 37, 253–255. (4) Jones, A.P. (1984) Mafic silicates from the nepheline syenites of the Motzfeldt centre, South Greenland. Mineral. Mag., 48, 1–12.

$$Ca_4(Al, Fe^{3+}, Mg, Fe^{2+})_{10}Si_{12}O_{35}(CO_3)(OH)_{12} \cdot 12H_2O \qquad \textbf{Aërinite}$$

Crystal Data: Monoclinic. *Point Group:* n.d. As masses of cryptocrystalline fibers; earthy, compact.

Physical Properties: Hardness = ~3   D(meas.) = 2.48(2)   D(calc.) = [2.46]

Optical Properties: Semitransparent. *Color:* Blue to blue-green.
Optical Class: Biaxial (−). *Pleochroism:* Intense; X = bright blue; $Y = Z$ = pale beige.
$\alpha = 1.510(5)$ $\beta = 1.560(5)$ $\gamma = 1.580(5)$ 2V(meas.) = n.d.   2V(calc.) = 63°

Cell Data: *Space Group:* n.d. $a = 14.690(15)$ $b = 16.872(15)$ $c = 5.170(15)$ $\beta = 94°45'$
$Z = [2]$

X-ray Powder Pattern: Saint-Pandelon, France.
14.65 (100), 4.050 (80), 2.721 (75), 2.812 (50), 3.798 (35), 3.651 (35), 2.135 (35)

Chemistry:

	(1)	(2)
SiO_2	36.91	37.87
TiO_2	trace	
Al_2O_3	17.20	18.64
Fe_2O_3	7.31	
FeO	3.47	7.00
MnO	0.10	
MgO	3.15	6.
CaO	12.03	11.93
Na_2O	0.27	2.70
K_2O	0.04	0.12
H_2O	17.79	15.45
CO_2	2.30	
P_2O_5	0.71	
Total	101.28	99.71

(1) Saint-Pandelon, France; H_2O by TGA, CO_2 by chromatography; corresponds to
$(Ca_{4.04}Na_{0.16}K_{0.02})_{\Sigma=4.22}(Al_{6.36}Fe^{3+}_{1.72}Mg_{1.47}Fe^{2+}_{0.91}Mn_{0.03})_{\Sigma=10.49}(Si_{11.59}P_{0.19})_{\Sigma=11.78}O_{35.21}$
$(OH)_{12.11}(CO_3)_{0.99} \cdot 12.54H_2O$. (2) Morocco.

Occurrence: A hydrothermal mineral of the zeolite facies, formed at relatively low temperature, in fractures cutting mafic igneous rocks (Saint-Pandelon, France).

Association: Prehnite, scolecite, mesolite (Saint-Pandelon, France).

Distribution: At Caserras, Juseu, and Estopiñan, Huesca Province; Tartaren, Lerida Province; and other less-well-defined localities in Spain. From Saint-Pandelon, Landes, France. In Morocco, at Ourika. From the Gunsight Mountains, Pima Co., Arizona, USA.

Name: From the Greek for *sky blue* for its distinctive color.

Type Material: National School of Mines, Paris, France.

References: (1) Dana, E.S. (1892) Dana's system of mineralogy, (6th edition), 1025.
(2) Termier, H. and G. Termier (1946) Sur la presence d'aérinite au Maroc. Compt. Rend. Soc. Géol. France, 78–80 (in French). (3) (1948) Chem. Abs., 42, 7204 (abs. ref. 2). (4) Azambre, B. and P. Monchoux (1988) Précisions minéralogiques sur l'aérinite: nouvelle occurrence à Saint-Pandelon (Landes, France). Bull. Minéral., 111, 39–47 (in French with English abs.).

Afghanite $(Na, Ca, K)_8(Si, Al)_{12}O_{24}(SO_4, Cl, CO_3)_3 \cdot H_2O$

Crystal Data: Hexagonal. *Point Group:* $6/m\ 2/m\ 2/m$, $6mm$, or $\overline{6}\ m2$. As stout to slender, thin tabular laths and as rounded grains.

Physical Properties: *Cleavage:* Perfect on $\{10\overline{1}0\}$. *Fracture:* Conchoidal. Hardness = 5.5–6 D(meas.) = 2.55 D(calc.) = 2.65

Optical Properties: Transparent. *Color:* Blue; in thin section, colorless. *Luster:* Vitreous. *Optical Class:* Uniaxial (+). $\omega = 1.523(2)$ $\epsilon = 1.529(2)$

Cell Data: *Space Group:* $P6_3/mmc$, $P6_3mc$, or $P\overline{6}2c$. a = 12.77(3) c = 21.35(4) Z = 3

X-ray Powder Pattern: Sar-e-Sang, Afghanistan.
3.688 (100), 3.298 (100), 4.82 (80), 3.997 (60), 2.865 (60), 2.130 (60), 1.792 (60)

Chemistry:

	(1)	(2)
SiO_2	30.8	32.96
Al_2O_3	25.	25.45
CaO	16.5	11.98
Na_2O	12.6	12.52
K_2O	2.7	3.07
Cl	4.6	3.98
H_2O	0.7	
CO_2	0.4	
SO_3	8.5	10.82
$-O = Cl_2$	1.0	0.9
Total	100.8	99.88

(1) Sar-e-Sang, Afghanistan; corresponds to $(Na_{4.86}Ca_{3.52}K_{0.69})_{\Sigma=9.07}(Si_{6.13}Al_{5.87})_{\Sigma=12.00}O_{25.21}$ $[Cl_{1.55}(SO_4)_{1.27}(CO_3)_{0.11}]_{\Sigma=2.93} \cdot 0.46H_2O$. (2) Edwards, New York, USA; by electron microprobe, corresponds to $(Na_{4.63}Ca_{2.45}K_{0.75})_{\Sigma=7.83}(Si_{6.28}Al_{5.72})_{\Sigma=12.00}O_{24.08}[(SO_4)_{1.55}Cl_{1.29}]_{\Sigma=2.84}$.

Mineral Group: Cancrinite group.

Occurrence: In thin veinlets cutting lazurite crystals (Sar-e-Sang, Afghanistan); in silicified limestone xenoliths in pumice (Pitigliano quarry, Italy).

Association: Lazurite, sodalite, nepheline, phlogopite, olivine, diopside, vesuvianite, calcite, pyrite.

Distribution: At Sar-e-Sang, Badakhshan Province, Afghanistan. In Russia, in the Malaya Bystraya and Tultuy lazurite deposits, near Lake Baikal, and in other unspecified deposits in eastern Siberia. From Lyadzhuar-Darinsk, Pamir Mountains, Tadzhikistan. In the Pitigliano quarry, near Grosseto, Tuscany, Italy. From the Edwards mine, Edwards, St. Lawrence Co., New York, USA. At Lake Harbour, Baffin Island, Newfoundland, Canada.

Name: For the country of first occurrence, AFGHANistan.

Type Material: Mineralogy-Crystallography Laboratory, University P. and M. Curie, Paris; National School of Mines, Paris, France; The Natural History Museum, London, England, 1969,72.

References: (1) Bariand, P., F. Cesbron, and R. Giraud (1968) Une nouvelle espèce minérale: l'afghanite de Sar-e-Sang, Badakhshan, Afghanistan. Comparaison avec les minéraux du groupe de la cancrinite. Bull. Soc. fr. Minéral., 91, 34–42 (in French with English abs.). (2) (1968) Amer. Mineral., 53, 2105 (abs. ref. 1). (3) Hogarth, D. (1979) Afghanite: new occurrences and chemical composition. Can. Mineral., 17, 47–52.

Crystal Data: Monoclinic. *Point Group:* m. Crystals prismatic, elongated and striated $\parallel$ [010], to 11 cm, many tabular on {100} and {201}; in radial fibrous spherulites; massive.

Physical Properties: *Cleavage:* Perfect on {10$\bar{1}$}; good on {100}. *Fracture:* Conchoidal. *Tenacity:* Brittle. Hardness = 3–4   D(meas.) = 2.630   D(calc.) = 2.643(5)   Piezoelectric.

Optical Properties: Transparent. *Color:* Colorless or white. *Streak:* White. *Luster:* Vitreous.
Optical Class: Biaxial (+). *Orientation:* $Y = b$; $X \wedge c = 31.2°$ *Dispersion:* $r < v$.
$\alpha = 1.616–1.618$ $\beta = 1.619–1.621$ $\gamma = 1.631–1.634$ 2V(meas.) = $50°–56°$

Cell Data: *Space Group:* Cc. $a = 16.278(1)$ $b = 5.6321(4)$ $c = 13.236(1)$ $\beta = 134.898°$
$Z = 4$

X-ray Powder Pattern: Kimberley, South Africa.
2.83 (100), 6.61 (90), 3.18 (90), 2.73 (90), 3.28 (70), 2.15 (70), 2.34 (60)

Chemistry:

	(1)	(2)	(3)
SiO_2	33.96	34.65	35.09
Al_2O_3, Fe_2O_3	0.60		
MgO		trace	
CaO	49.28	48.94	49.13
F		0.18	
H_2O^+	15.89	16.08	15.78
H_2O^-	0.12	0.01	
$-O = F_2$		0.10	
Total	99.85	99.76	100.00

(1) Kimberley, South Africa. (2) Crestmore, California, USA. (3) $Ca_3(SiO_3OH)_2 \cdot 2H_2O$.

Occurrence: A product of contact metamorphism of limestones.

Association: Apophyllite, natrolite, thaumasite, merwinite, spurrite, gehlenite, ettringite, portlandite, hillebrandite, foshagite, brucite, calcite.

Distribution: From the Dutoitspan diamond mine, Kimberley, and the Wessels mine, near Kuruman, Cape Province, South Africa. In Ireland, at Scawt Hill, near Larne, Co. Antrim. From the Schellkopf, near Brenk, Eifel district, and on the Zeilberg, near Maroldsweisach, Bavaria, Germany. At Campomorto, Montalto di Castro, Lazio, Italy. From Nahal Ayalon, Israel. In the USA, at Crestmore, Riverside Co., California. In the Mihara mine and at Fuka, near Bicchu, Okayama Prefecture, Japan.

Name: For Alpheus Fuller Williams (1874–1953), General Manager, DeBeers Consolidated Mines, Kimberley, South Africa.

Type Material: The Natural History Museum, London, England, 1925,80; National Museum of Natural History, Washington, D.C., USA, 95237.

References: (1) Parry, J. and F.E. Wright (1925) Afwillite, a new hydrous calcium silicate, from Dutoitspan Mine, Kimberley, South Africa. Mineral. Mag., 20, 277–285. (2) (1925) Amer. Mineral., 10, 447 (abs. ref. 1). (3) Switzer, G. and E.H. Bailey (1953) Afwillite from Crestmore, California. Amer. Mineral., 38, 629–633. (4) Heller, L. and H.F.W. Taylor (1956) Crystallographic data for the calcium silicates. H.M. Stationary Office, London, 50–53. (5) Malik, K.M.A. and J.W. Jeffrey (1976) A re-investigation of the structure of afwillite. Acta Cryst., 32, 475–480. (6) Kusachi, I., C. Henmi, and K. Henmi (1989) Afwillite and jennite from Fuka, Okayama Prefecture, Japan. Mineral. J. (Japan), 14, 279–292.

Crystal Data: Triclinic. *Point Group:* $\bar{1}$. Crystals elongated ∥ [001], to 10 cm; platy aggregates.

Physical Properties: *Cleavage:* Excellent on {110} and {1$\bar{1}$0}, poor on {010}.
Hardness = 5.5   D(meas.) = 2.902   D(calc.) = 2.887   Fluoresces bright pink under LW UV and duller pink under SW UV.

Optical Properties: Transparent to translucent. *Color:* White to grayish or greenish white.
Luster: Pearly on cleavages.
Optical Class: Biaxial (–). $\alpha = 1.567$ $\beta = 1.579$ $\gamma = 1.581$ 2V(meas.) = n.d.
2V(calc.) = 47°

Cell Data: *Space Group:* $P\bar{1}$. a = 7.759(2)   b = 18.946(3)   c = 6.986(1) $\alpha = 89.88(2)°$
$\beta = 116.65(2)°$ $\gamma = 94.32(2)°$ Z = 4

X-ray Powder Pattern: Villedieu Township, Canada.
3.19 (vs), 3.14 (vs), 3.44 (s), 3.33 (s), 2.58 (s), 2.31 (s), 2.04 (ms)

Chemistry:

	(1)	(2)		(1)	(2)
SiO_2	57.79	60.92	CaO	25.70	28.43
TiO_2	0.01		SrO	0.16	
ZrO_2	0.18		BaO	0.06	
Al_2O_3	1.32		Na_2O	7.90	7.86
RE	2.57		K_2O	0.22	
Fe_2O_3	0.11		F	4.45	4.82
MnO	0.25		H_2O^+	0.4	
MgO	0.02		$-O = F_2$	1.87	2.03
			Total	99.27	100.00

(1) Villedieu Township, Canada; by wet chemical analysis, spectrophotometry, and flame photometry, total Fe as Fe_2O_3, H_2O by the Penfield method. (2) $NaCa_2Si_4O_{10}F$.

Occurrence: In pegmatite lenses and pods and in mafic gneisses in a regionally metamorphosed agpaitic alkalic rock complex (Villedieu Township, Canada).

Association: Hiortdahlite, mosandrite, miserite, britholite, vlasovite, calcite, fluorite, clinohumite, gittinsite, norbergite, zircon, biotite, phlogopite, galena (Villedieu Township, Canada); miserite, aegirine, eudialyte, quartz (Wausau, Wisconsin, USA).

Distribution: In Canada, from the Sheffield Lake complex, Kipawa River, Villedieu Township, Quebec. In the USA, in the Wausau complex, Marathon Co., Wisconsin. In the Dara-i-Pioz massif, Alai Range, Tien Shan, Tadzhikistan. From the Murun massif, southwest of Olekminsk, Yakutia, Russia.

Name: For Professor Stuart O. Agrell, petrologist, Cambridge University, Cambridge, England.

Type Material: Canadian Museum of Nature, Ottawa; Royal Ontario Museum, Toronto, Canada, M34496; Harvard University, Cambridge, Massachusetts, 117050; National Museum of Natural History, Washington, D.C., USA, 127007; Cambridge University, Cambridge; The Natural History Museum, London, England, 1979,431.

References: (1) Gittins, J., M.G. Brown, and B.D. Sturman (1976) Agrellite, a new rock-forming mineral in regionally metamorphosed agpaitic alkalic rocks. Can. Mineral., 14, 120–126. (2) (1977) Amer. Mineral., 62, 173–174 (abs. ref. 1). (3) Ghose, S. and C. Wan (1979) Agrellite, Na(Ca, RE)$_2$Si$_4$O$_{10}$F: a layer structure with silicate tubes. Amer. Mineral., 64, 563–572.

Crystal Data: Triclinic.   *Point Group:* $\bar{1}$ or 1.   As sprays of bladed prismatic crystals, to 0.4 mm; commonly fibrous.

Physical Properties: *Cleavage:* Perfect on {010}.   Hardness = n.d. D(meas.) = 2.96   D(calc.) = 2.951

Optical Properties: Translucent. *Color:* Bluish green.
Optical Class: Biaxial (+). *Pleochroism:* X = very light bluish green; $Y = Z$ = brilliant bluish green. *Orientation:* $X = b$; $Z \wedge c = 15°$. $\alpha = 1.550(1)$ $\beta = 1.583(1)$ $\gamma = 1.641(1)$ 2V(meas.) = 80(1)°   2V(calc.) = 76.4°

Cell Data: *Space Group:* $P1$ or $P\bar{1}$.   a = 13.637(5)   b = 14.507(4)   c = 13.620(2) $\alpha = 107.16(2)°$ $\beta = 105.45(2)°$ $\gamma = 110.57(2)°$ Z = 3

X-ray Powder Pattern: Ajo, Arizona, USA.
12.25 (100), 2.455 (12), 4.08 (10), 3.061 (10), 3.381 (8), 2.832 (8), 2.258 (8)

Chemistry:

	(1)
SiO$_2$	41.2
Al$_2$O$_3$	3.81
FeO	0.11
MnO	0.02
CuO	42.2
CaO	0.04
Na$_2$O	0.84
K$_2$O	2.50
H$_2$O	8.35
Total	99.07

(1) Ajo, Arizona, USA; Si, Al, and Cu by electron microprobe, corresponding to $(K_{0.70}Na_{0.36}Ca_{0.01})_{\Sigma=1.07}(Cu_{6.97}Fe_{0.02})_{\Sigma=6.99}Al_{0.98}Si_{9.00}O_{24}(OH)_{6.00} \cdot 3.09H_2O$.

Occurrence: In oxidized copper-rich base-metal deposits.

Association: Shattuckite, conichalcite, quartz, muscovite, pyrite (Ajo, Arizona, USA); creaseyite, fluorite (Potter-Cramer property, Arizona, USA); shattuckite, duhamelite, sillénite (Munihuaza, Mexico); quartz, papagoite (Messina, South Africa).

Distribution: In the USA, in Arizona, from the New Cornelia mine, Ajo, Pima Co.; at the Moon Anchor mine and Potter-Cramer property, near Wickenburg, Maricopa Co.; and a prospect in Copper Creek, Pinal Co. From Munihuaza, near Alamos, Sonora, Mexico. At Messina, Transvaal, South Africa.

Name: For the type occurrence at Ajo, Arizona, USA.

Type Material: National Museum of Natural History, Washington, D.C., USA, 113220.

References: (1) Schaller, W.T. and A.C. Vlisidis (1958) Ajoite, a new hydrous aluminum copper silicate. Amer. Mineral., 43, 1107–1111. (2) Kato, T. and Y. Miúra (1976) Cell dimension of ajoite. Mineral. J. (Japan), 8, 234–239. (3) (1980) Mineral. Abs., 31, 415 (abs. ref. 2). (4) Chao, G.Y. (1981) Ajoite: new data. Amer. Mineral., 66, 201–203.

Crystal Data: Triclinic. *Point Group:* $\bar{1}$. Rarely as sheaves, to 1 cm, of radiating prisms, elongated and striated $\|$ [100]; fibrous, fine granular, massive. *Twinning:* On $\{0\bar{2}1\}$; twin axis $\perp \{0\bar{2}1\}$, lamellar in thin section.

Physical Properties: *Cleavage:* Good on $\{010\}$, poor on $\{0\bar{1}2\}$. Hardness = 6
D(meas.) = 3.48   D(calc.) = 3.47

Optical Properties: Transparent. *Color:* Yellow-orange to orange-brown. *Luster:* Vitreous. *Optical Class:* Biaxial (+). *Pleochroism:* X = colorless; Y = pale yellow; Z = light canary-yellow. *Orientation:* $X \wedge \{010\} = 58°$; $Y \wedge \{010\} = 30°$. $\alpha = 1.698(1)$ $\beta = 1.704(1)$
$\gamma = 1.720(1)$ 2V(meas.) = 65.5°

Cell Data: *Space Group:* $P\bar{1}$. a = 8.337(2)   b = 10.367(2)   c = 7.629(1) $\alpha = 104.46(1)°$
$\beta = 93.81(2)°$ $\gamma = 104.18(1)°$ Z = 1

X-ray Powder Pattern: Akatore Creek, New Zealand.
4.665 (100), 3.310 (90), 2.214 (80), 9.681 (60), 3.466 (50), 3.063 (50), 2.866 (50)

Chemistry:

	(1)	(2)
SiO_2	36.4	36.9
TiO_2	0.03	0.1
Al_2O_3	8.3	6.2
FeO	1.0	0.9
MnO	47.7	44.2
MgO	0.3	0.8
CaO	0.2	0.2
H_2O	6.21	[10.7]
Total	100.14	[100.0]

(1) Akatore Creek, New Zealand; by electron microprobe, corresponding to $(\mathrm{Mn_{8.61}Fe_{0.19}}$ $\mathrm{Mg_{0.09}Ca_{0.05})_{\Sigma=8.94}Al_{2.09}Si_{7.75}O_{23.17}(OH)_{8.83}}$. (2) Norberg, Sweden; by electron microprobe, average of eight analyses, H_2O by difference; corresponds to $(\mathrm{Mn_{8.29}Mg_{0.27}Fe_{0.19}Ca_{0.05})_{\Sigma=8.80}}$ $\mathrm{Al_{1.64}Si_{8.16}O_{23.17}(OH)_8}$.

Occurrence: In a manganiferous metachert and carbonate lens in schists (Akatore Creek, New Zealand); in manganiferous potassium-rich felsic metavolcanics (Norberg, Sweden).

Association: Rhodochrosite, pyroxmangite, rhodonite, spessartine, quartz, tinzenite, apatite, todorokite, alabandite, hübnerite (Akatore Creek, New Zealand); ganophyllite, rhodochrosite, pyrolusite (Norberg, Sweden).

Distribution: In New Zealand, three km south of Akatore Creek, east Otago, South Island. From Norberg, Sweden.

Name: For the locality near Akatore Creek, New Zealand.

Type Material: University of Otago, Dunedin; Geological Survey of New Zealand, Lower Hutt, New Zealand; National Museum of Natural History, Washington, D.C., USA, 137285, 142541.

References: (1) Read, P.B. and A. Reay (1971) Akatoreite, a new manganese silicate from Eastern Otago, New Zealand. Amer. Mineral., 56, 416–426. (2) Ounchanum, P. and S. Morad (1987) Paragenesis of akatoreite and ganophyllite in the manganiferous rocks of the Häste field, Norberg ore district, central Sweden. Neues Jahrb. Mineral., Abh., 157, 225–244. (3) Burns, P.C. and F.C. Hawthorne (1993) Edge-sharing $\mathrm{Mn^{2+}O_4}$ tetrahedra in the structure of akatoreite, $\mathrm{Mn_9^{2+}Al_2Si_8O_{24}(OH)_8}$. Can. Mineral., 31, 321–329.

Crystal Data: Tetragonal. *Point Group:* $\overline{4}2m$. Crystals short prismatic to thin tabular, which may resemble squashed, octahedrally-modified cubes; granular, massive. *Twinning:* On {100} and {001}, cruciform.

Physical Properties: *Cleavage:* Distinct on {001}, poor on {110}. *Fracture:* Uneven to conchoidal. *Tenacity:* Brittle. Hardness = 5–6   D(meas.) = 2.944   D(calc.) = 2.922

Optical Properties: Transparent to translucent. *Color:* Colorless, yellowish gray, green, brown; in thin section, colorless to yellow. *Luster:* Vitreous to resinous.
Optical Class: Uniaxial (+). *Absorption:* Weak; $E > O$. $\omega = 1.630–1.632$ $\epsilon = 1.639–1.648$

Cell Data: *Space Group:* $P\overline{4}2_1m$ (synthetic). $a = 7.8288(8)$ $c = 5.0052(5)$ $Z = 2$

X-ray Powder Pattern: Synthetic.
2.87 (100), 3.09 (30), 1.764 (30), 2.039 (20), 2.488 (18), 3.73 (14), 5.55 (12)

Chemistry:

	(1)	(2)	(3)
SiO₂	46.55	44.07	44.08
TiO₂		0.05	
Al₂O₃	0.96	0.33	
Fe₂O₃		0.00	
FeO	0.12	0.47	
MnO	0.00	0.11	
ZnO		1.56	
MgO	13.30	12.87	14.78
CaO	39.30	40.09	41.14
Na₂O		< 0.05	
Total	100.23	[99.55]	100.00

(1) Vesuvius, Italy. (2) Cascade Mountain, New York, USA; by electron microprobe; original total given as 99.52%. (3) Ca₂MgSi₂O₇.

Polymorphism & Series: Forms a series with gehlenite.

Mineral Group: Melilite group.

Occurrence: A product of contact metamorphism of siliceous limestones and dolostones, and in rocks of the sanidinite facies. Also forms from alkalic magmas rich in calcium.

Association: Monticellite, wollastonite, spurrite, larnite, merwinite, "fassaite," grossular, diopside, forsterite.

Distribution: Occurrences are typically of intermediate series members; relatively pure examples have been found at Monte Somma and Vesuvius, Campania, and Monte Cavalluchio, Sacrofano, near Rome, Lazio, Italy. From Crestmore, Riverside Co., California; on Cascade Mountain, Adirondack Mountains, Essex Co., New York; and from the Tres Hermanas district, Luna Co., New Mexico, USA. On Dufresne Hill, Oka, Quebec, Canada. From near Kilchoan, Ardnamurchan, Argyllshire, Scotland. At Scawt Hill, near Larne, Co. Antrim, Ireland. In the Wessels mine, near Kuruman, Cape Province, South Africa.

Name: For Anders Richard Åkerman (1837–1922), Swedish metallurgist.

References: (1) Dana, E.S. (1892) Dana's system of mineralogy, (6th edition), 474–477. (2) Deer, W.A., R.A. Howie, and J. Zussman (1986) Rock-forming minerals, (2nd edition), v. 1B, disilicates and ring silicates, 285–334. (3) Ervin, G. and E.F. Osborn (1949) X-ray data on synthetic melilites. Amer. Mineral., 34, 717–722. (4) Valley, J.W. and E.J. Essene (1980) Åkermanite in the Cascade Slide xenolith and its significance for regional metamorphism in the Adirondacks. Contr. Mineral. Petrol., 74, 143–152. (5) Swainson, I.P., M.T. Dove, W.S. Schmahl, and A. Putnis (1992) Neutron powder diffraction study of the åkermanite–gehlenite solid solution series. Phys. Chem. Minerals, 19, 185–195.

Crystal Data: Monoclinic. *Point Group: 2/m.* Crystals fibrous ∥ [010]; as radiating aggregates and balls, to 7.5 cm.

Physical Properties: *Cleavage:* Perfect on {010}. Hardness = 4.5 D(meas.) = 6.488(3) D(calc.) = [6.30]

Optical Properties: Transparent to translucent. *Color:* Colorless to white, cream, or light gray. *Luster:* Adamantine.
Optical Class: Biaxial (–). *Orientation:* Y = b. *Dispersion:* $r < v$, strong, weak inclined.
$\alpha = 1.945$–1.947 $\beta = 1.955$–1.961 $\gamma = 1.959$–1.968 2V(meas.) = 65°

Cell Data: *Space Group: P2/n.* a = 12.247 b = 7.059 c = 11.236 $\beta = 113.12°$ Z = 12

X-ray Powder Pattern: Tsumeb, Namibia.
3.34 (100), 3.56 (95), 3.53 (75), 2.300 (75), 3.23 (70), 2.987 (70), 3.25 (60)

Chemistry:

	(1)	(2)	(3)
SiO_2	21.11	20.01	21.21
Al_2O_3		0.09	
FeO	0.09	0.18	
MnO		0.02	
PbO	78.13	78.95	78.79
CaO	trace	0.09	
insol.	0.61		
Total	99.94	99.34	100.00

(1) Alamos, Mexico. (2) Tsumeb, Namibia; by electron microprobe. (3) $PbSiO_3$.

Occurrence: As a rare secondary mineral in the oxidized zone of lead-bearing base metal deposits.

Association: Wulfenite, leadhillite, cerussite (Alamos, Mexico); leadhillite, anglesite, melanotekite, fleischerite, kegelite, hematite (Tsumeb, Namibia); diaboleite, phosgenite, cerussite, wulfenite, willemite (Tiger, Arizona, USA); melanotekite, shattuckite, wickenburgite (Rawhide mine, Arizona, USA).

Distribution: From Mexico, in Sonora, at Alamos, and the San Pascual mine, Zimapan, Hidalgo. In the USA, from Arizona, in the Mammoth-St. Anthony mine, at Tiger, Pinal Co., the Lucky Cuss mine, Tombstone, Cochise Co., and the Rawhide mine, Artillery Mountains, Maricopa Co. As exceptional crystal groups from Tsumeb, Namibia.

Name: For the Mexican type locality at Alamos.

Type Material: Harvard University, Cambridge, Massachusetts, USA, 85509, 85510.

References: (1) Palache, C. and H.E. Merwin (1909) Alamosite, a new lead silicate from Mexico. Amer. J. Sci., 27, 399–401. (2) Mackay, A.L. (1952) The unit cell and space-group of alamosite ($PbSiO_3$). Mineral. Mag., 29, 933–945. (3) Boucher, M.L. and D.R. Peacor (1968) The crystal structure of alamosite, $PbSiO_3$. Zeits. Krist., 126, 98–111. (4) Medenbach, O. and K. Schmetzer (1975) Alamosit ($PbSiO_3$) von Tsumeb — ein neuer Fundpunkt. Neues Jahrb. Mineral., Abh., 123, 138-147 (in German).

Crystal Data: Triclinic. *Point Group:* $\bar{1}$. Crystals commonly tabular ∥ {010}, may be curved, to 3 cm; divergent aggregates, granular, cleavable massive. *Twinning:* Common around [010] or ⊥ {010}, giving polysynthetic striae on {001} or {010}; many other laws, contact, simple and multiple.

Physical Properties: *Cleavage:* Perfect on {001}, very good on {010}, imperfect on {110}. *Fracture:* Uneven to conchoidal. *Tenacity:* Brittle. Hardness = 6–6.5 D(meas.) = 2.60–2.65 D(calc.) = 2.609–2.621

Optical Properties: Transparent to translucent. *Color:* White to gray, bluish, greenish, reddish; may be chatoyant. *Streak:* White. *Luster:* Vitreous, typically pearly on cleavages. *Optical Class:* Biaxial (+) (low); (–) (high). *Dispersion:* $r < v$, weak (low). $\alpha = 1.526$–1.530 $\beta = 1.531$–1.533 $\gamma = 1.534$–1.541 2V(meas.) = 85°–90° (low); 52°–54° (high).

Cell Data: *Space Group:* $C\bar{1}$ (low). a = 8.137(1) b = 12.785(1) c = 7.1583(4) $\alpha = 94.26(1)°$ $\beta = 116.60(1)°$ $\gamma = 87.71(1)°$ Z = 4, or *Space Group:* $C\bar{1}$ (high). a = 8.149 b = 12.880 c = 7.106 $\alpha = 93.37°$ $\beta = 116.30°$ $\gamma = 90.28°$ Z = 4

X-ray Powder Pattern: Amelia, Virginia, USA (low).
3.196 (100), 3.780 (25), 6.39 (20), 3.684 (20), 4.030 (16), 3.663 (16), 2.933 (16)

Chemistry:

	(1)	(2)	(3)
SiO_2	68.71	68.74	66.04
Al_2O_3	19.63	19.44	21.26
CaO	0.22		2.13
Na_2O	11.72	11.82	10.57
K_2O	0.03		
Total	100.31	100.00	100.00

(1) Alp Rischuna, Switzerland. (2) $NaAlSi_3O_8$. (3) $Na_{0.90}Ca_{0.10}Al_{1.10}Si_{2.90}O_8$.

Polymorphism & Series: Low- and high-temperature structural modifications are recognized.

Mineral Group: Feldspar group, plagioclase series.

Occurrence: A major constituent of granites and granite pegmatites, alkalic diorites, basalts, and in hydrothermal and alpine veins. A product of potassium metasomatism and in low-temperature and low-pressure metamorphic facies and in some schists. Detrital and authigenic in sedimentary rocks.

Association: Quartz, orthoclase, muscovite, biotite, "hornblende."

Distribution: Widespread; a few localities for good crystals are: in Switzerland, from St. Gotthard, Ticino and Tavetsch, Graubünden. From Roc Tourné, near Modane, Savoie, France. On Mt. Greiner, Zillertal, Tirol, Austria. At Baveno, Piedmont, and in the Pfitschtal, Trentino-Alto Adige, Italy. From Mursinka, Ural Mountains, and Miass, Ilmen Mountains, Southern Ural Mountains, Russia. In the USA, at Haddam and Middletown, Middlesex Co., Connecticut; Amelia, Amelia Co., Virginia; from Diana, Lewis Co., and Dekalb, Macomb, and Pierrepont, St. Lawrence Co., New York. On Prince of Wales Island, Alaska; in the Pala and Mesa Grande districts, San Diego Co., California. At Bathurst, and Wicklow Township, Hastings Co., Ontario, Canada. From Virgem da Lapa and Morro Velho, Minas Gerais, Brazil.

Name: From the Latin, *albus*, for *white*, its characteristic color.

References: (1) Dana, E.S. (1892) Dana's system of mineralogy, (6th edition), 325–333. (2) Deer, W.A., R.A. Howie, and J. Zussman (1963) Rock-forming minerals, v. 4, framework silicates, 94–165. (3) Phillips, W.R. and D.T. Griffen (1981) Optical mineralogy, 352–360. (4) Armbruster, T., H.B. Bürgi, M. Kunz, E. Gnos, S. Brönnimann, and C. Lienert (1990) Variation of displacement parameters in structure refinements of low albite. Amer. Mineral., 75, 135–140.

Crystal Data: n.d. *Point Group:* n.d. Irregular, tabular to platy crystals, $< 2 \ \mu m$, with curled edges when viewed under the electron microscope; as microscopic rounded aggregates.

Physical Properties: Hardness = Soft. D(meas.) = n.d. D(calc.) = n.d. Positive identification of minerals in the smectite group may need data from DTA curves, dehydration curves, and X-ray powder patterns before and after treatment by heating and with organic liquids.

Optical Properties: Translucent. *Color:* Colorless, pale yellow or green.
Optical Class: n.d. $n = 1.558-1.567$

Cell Data: *Space Group:* n.d. Z = n.d.

X-ray Powder Pattern: Taro Valley, Italy.
24.8 (100), 12.4 (67), 3.10 (12), 8.27 (6), 4.96 (2), 3.54 (2), 2.48 (0.5)

Chemistry:

	(1)
SiO_2	53.15
Al_2O_3	3.48
Fe_2O_3	3.48
MnO	0.03
MgO	27.40
CaO	1.10
Na_2O	1.18
H_2O	10.18
Total	100.00

(1) Taro Valley, Italy; by AA and XRF.

Polymorphism & Series: Talc-saponite mixed-layer mineral.

Mineral Group: Smectite group.

Occurrence: An alteration product in serpentinized ophiolites and residual in soils derived therefrom; in altered dolostones.

Association: Talc, chlorite, serpentine, calcite.

Distribution: In Italy, in Emilia-Romagna, at Monte Chiaro, Taro Valley; Frassinoro, Modena; and Ferriere, Nure Valley. From Kinshasa, Shaba Province, Zaire.

Name: To honor Professor Andrea Alietti (1923–), of the University of Modena, Modena, Italy, who first studied the structure of the mineral.

Type Material: n.d.

References: (1) Alietti, A. (1958) Some interstratified clay minerals of the Taro Valley. Clay Minerals Bull., 3, 207–211. (2) Veniale, F. and H.W. van der Marel (1969) Identification of some 1:1 regular interstratified trioctahedral clay minerals. Proc. Int. Clay Conf., Tokyo, 1, 233–244. (3) (1972) Amer. Mineral., 57, 598 (abs. ref. 2). (4) Alietti, A. and J. Mejsner (1980) Structure of a talc/saponite mixed-layer mineral. Clays and Clay Minerals, 28, 388–390. (5) Bailey, S.W. (1982) Nomenclature for regular interstratifications. Amer. Mineral., 67, 394–398.

$(Ca, Ce)_2(Al, Fe^{2+}, Fe^{3+})_3(SiO_4)(Si_2O_7)O(OH)$ Allanite-(Ce)

Crystal Data: Monoclinic; may be metamict. *Point Group: 2/m.* Crystals tabular ∥ {100}, or prismatic to acicular ∥ [010], to 90 cm; granular, massive. *Twinning:* Polysynthetic, common on {100}.

Physical Properties: *Cleavage:* Imperfect on {001}; poor on {100} and {110}. *Fracture:* Conchoidal to uneven. *Tenacity:* Brittle. Hardness = 5.5–6 D(meas.) = 3.5–4.2 D(calc.) = 4.11 May be radioactive.

Optical Properties: Translucent to opaque. *Color:* Brown to black; brown or green in thin section. *Streak:* Gray. *Luster:* Vitreous, resinous to submetallic. *Optical Class:* Biaxial (+) or (−); isotropic when metamict. *Pleochroism: X =* pale olive-green, reddish brown; *Y =* dark brown, brownish yellow; *Z =* dark reddish brown, greenish brown. *Orientation: Y = b; X ∧ c =* 1°–47°; *Z ∧ a =* 26°–72°. *Dispersion: r > v,* strong. n = 1.54–1.72, metamict. α = 1.690–1.791 β = 1.700–1.815 γ = 1.706–1.828 2V(meas.) = 40°–123°

Cell Data: *Space Group: P2_1/m.* a = 8.932   b = 5.770   c = 10.1575 β = 114.69°   Z = 2

X-ray Powder Pattern: Luangwe Bridge area, Zambia; non-metamict.
2.920 (100), 2.714 (66), 3.53 (43), 2.627 (41), 2.182 (36), 2.886 (28), 2.158 (23)

Chemistry:

	(1)		(1)		(1)
SiO_2	29.79	Ce_2O_3	12.8	MgO	0.64
TiO_2	0.02	RE_2O_3	4.78	CaO	10.10
Al_2O_3	12.75	Fe_2O_3	10.64	H_2O^+	1.64
Y_2O_3	0.04	FeO	9.02	H_2O^-	0.10
La_2O_3	7.2	MnO	0.85	Total	100.37

(1) Luangwe Bridge area, Zambia; wet chemical and spectrographic analysis; $RE_2O_3 = Pr_2O_3$ 1.0%, Nd_2O_3 3.10%, Sm_2O_3 0.60%, Gd_2O_3 0.06%, Er_2O_3 0.02%. Analyses have been made of materials rich in La and Y but they are not otherwise described.

Mineral Group: Epidote group.

Occurrence: An accessory in some granites and granite pegmatites, syenites, more rarely in gabbroic pegmatites. Rarely in schists, gneisses, and some contact metamorphosed limestones; a clastic component of sediments.

Association: Epidote, muscovite, fluorite.

Distribution: Widespread; some localities affording good crystals or rich material follow. From Qáqarssuatsiaq, Aluk, south Greenland. At Ytterby and Finbo, Sweden. From Kragerö, Arendal, and on Hitterö Island, Flekkefjord, Norway. From Sillböle, Stansvik, and Vaarala, Finland. At Miass, Ilmen Mountains, Southern Ural Mountains, Russia. From the Trimouns mine, near Luzenac, Ariège, France. In the USA, at Pacoima Canyon, Los Angeles Co., California; in the Baringer Hill pegmatite and on the Rode Ranch, Llano Co., Texas. In Canada, from Olden Township, Frontenac Co., Quebec; in the McDonald mine, Hybla, Ontario; and at Hoidas Lake, near Nisikkatch Lake, Saskatchewan. From near Telixtlahuaca, Oaxaca, Mexico. In the Mary Kathleen mine, Queensland, and at Broken Hill, New South Wales, Australia. From Ribuae, Mozambique.

Name: For Thomas Allan (1777–1833), Scottish mineralogist who discovered the species.

Type Material: University of Copenhagen, Copenhagen, Denmark, 5, 6; The Natural History Museum, London, England, 94377.

References: (1) Dana, E.S. (1892) Dana's system of mineralogy, (6th edition), 522–526. (2) Deer, W.A., R.A. Howie, and J. Zussman (1986) Rock-forming minerals, (2nd edition), v. 1B, disilicates and ring silicates, 151–179. (3) Vlasov, K.A., Ed. (1966) Mineralogy of rare elements, v. II, 302–308. (4) Dollase, W.A. (1971) Refinement of the crystal structures of epidote, allanite and hancockite. Amer. Mineral., 56, 447–464.

Crystal Data: Monoclinic. *Point Group:* $2/m$. Crystals rounded, slender; in stout plates, some deeply striated. As fan-shaped aggregates; commonly irregular, massive. *Twinning:* On {001}, lamellar, common; also on {105}, {305}.

Physical Properties: *Fracture:* Conchoidal. *Tenacity:* Brittle. Hardness = 5.5
D(meas.) = 4.020   D(calc.) = [4.20]

Optical Properties: Transparent to translucent. *Color:* Pinkish to reddish brown, deep pink, grayish pink; in thin section, pink or buff in thick plates. *Luster:* Vitreous to dull.
Optical Class: Biaxial (−). *Pleochroism:* Pinks in thicker sections. *Orientation:* $Z = b$.
Dispersion: $r > v$. $\alpha = 1.756$ $\beta = 1.780$ $\gamma = 1.792$ 2V(meas.) = 72°

Cell Data: *Space Group:* $P2_1/c$. $a = 8.275$ $b = 4.850$ $c = 10.720$ $\beta = 104.64°$ Z = 2

X-ray Powder Pattern: Franklin, New Jersey, USA.
1.799 (10), 2.860 (8), 2.598 (6), 3.127 (5), 2.725 (5), 2.425 (5), 2.357 (5)

Chemistry:

	(1)	(2)	(3)		(1)	(2)	(3)
SiO_2	24.90	24.65	24.38	MgO	2.16	0.78	
TiO_2	0.00	0.14		CaO	0.74	0.11	
Al_2O_3	trace	0.00		F		2.20	
Fe_2O_3	0.00			H_2O		[2.60]	3.66
FeO	1.40	0.84		$-O = F_2$		0.93	
MnO	70.35	70.15	71.96	Total	99.55	[100.54]	100.00

(1) Bald Knob, North Carolina, USA. (2) Do.; by electron microprobe; H_2O calculated from stoichiometry. (3) $Mn_5(SiO_4)_2(OH)_2$.

Polymorphism & Series: Dimorphous with ribbeite; forms a series with chondrodite.

Mineral Group: Humite group.

Occurrence: Hydrothermally deposited, in lenses in a manganese-bearing vein (Bald Knob, North Carolina, USA); in veins crosscutting franklinite ore near pegmatites in a metamorphosed stratiform Zn-Mn deposit (Franklin, New Jersey, USA).

Association: Spessartine, rhodonite, galaxite (Bald Knob, North Carolina, USA); kolicite, holdenite, magnussonite, adelite, kraisslite, chlorophoenicite, franklinite, willemite, barite, calcite (Sterling Hill, New Jersey, USA).

Distribution: In the USA, from Bald Knob, near Sparta, Alleghany Co., North Carolina; crystals from Franklin and Sterling Hill, Ogdensburg, Sussex Co., New Jersey; in the Germolis prospect, Fiddleton, Amador Co., and at Alum Rock Park, Santa Clara Co., California; at Eureka, Hinsdale Co., and the Sunnyside mine, San Juan Co., Colorado. In the Benallt mine, Rhiw, Lleyn Peninsula, Wales. From Långban, and in the Brattfors mine, Nordmark, Värmland, Sweden. In the Hanawa mine, Iwate Prefecture; the Kaso mine, Tochigi Prefecture; the Taguchi mine, Shidara, Aichi Prefecture; and the Rito mine and Hamayokokawa mine, Nagano Prefecture, Japan.

Name: For the occurrence in Alleghany Co., North Carolina, USA.

Type Material: n.d.

References: (1) Ross, C.S. (1932) The manganese minerals of a vein near Bald Knob, North Carolina. Amer. Mineral., 17, 1–18. (2) Rogers, A.F. (1935) The chemical formula and crystal system of alleghanyite. Amer. Mineral., 20, 25–35. (3) Rentzeperis, P.J. (1970) The crystal structure of alleghanyite, $Mn_5[(OH)_2|(SiO_4)_2]$. Zeits. Krist., 132, 1–18. (4) Winter, G.A., E.J. Essene, and D.R. Peacor (1983) Mn-humites from Bald Knob, North Carolina: mineralogy and phase equilibria. Amer. Mineral., 68, 951–959. (5) Dunn, P.J. (1985) Manganese humites and leucophoenicites from Franklin and Sterling Hill, New Jersey: parageneses, compositions, and implications for solid solution limits. Amer. Mineral., 70, 379–387.

Crystal Data: Amorphous (?). *Point Group:* n.d. Rarely observed as ring-shaped particles, with diameters of 50 Å, which in three dimensions may represent sections through hollow spherules or polyhedra. As hyaline crusts and masses; stalactites and flowstones.

Physical Properties: *Fracture:* Conchoidal to earthy. *Tenacity:* Brittle. Hardness = 3 D(meas.) = 2.75 D(calc.) = n.d.

Optical Properties: Translucent. *Color:* White, pale blue to sky-blue, green, brown. *Luster:* Vitreous to earthy, waxy. *Optical Class:* Isotropic. $n = 1.468$–1.512 (air-dried).

Cell Data: *Space Group:* n.d. Z = n.d.

X-ray Powder Pattern: n.d.

Chemistry:

	(1)	(2)	(3)		(1)	(2)	(3)
SiO_2	21.39	28.31	29.17	CaO	1.96	trace	trace
TiO_2		0.40	0.42	Na_2O		1.91	1.48
Al_2O_3	35.20	34.41	33.81	K_2O		0.29	0.16
Fe_2O_3		0.56	0.56	H_2O^+	40.86	10.60	11.20
MgO		0.08	0.04	H_2O^-		23.20	23.00
				Total	99.41	99.76	[99.84]

(1) Allentown, Lehigh Co., Pennsylvania, USA; CaO as $(Ca, Mg)CO_3$. (2) Iijima, Japan. (3) Kanuma, Japan; original total given as 99.81%.

Occurrence: A weathering product of volcanic ash. In hydrothermally altered igneous rocks, from the breakdown of feldspars, and in hydrothermal veins, typically related to copper deposits. In sedimentary rocks, including chalk and coal beds.

Association: Quartz, cristobalite, imogolite, gibbsite, vermiculite, chrysocolla, limonite.

Distribution: Of widespread occurrence. From Gräfenthal, near Saalfeld, Thuringia, and at Schneeberg and Schwarzenberg, Saxony, Germany. From Jáchymov (Joachimsthal), Czech Republic. In the Calabona mine, Alghero, and the Rosas mine, Sulsis, Sardinia, Italy. At Laurium, Greece. From near Woolwich, Kent, and Wheal Hamblyn, Devon, England. At the Chessy copper mine, near Lyons, Rhône, France. In the USA, from Bisbee, and the Maid of Sunshine mine, Gleeson, Cochise Co., Arizona; from Kelly, Socorro Co., New Mexico; at Cerro Gordo, Inyo Co., California; and at Friedensville, Lehigh Co. and Cornwall, Lebanon Co., Pennsylvania. In the Mbobo Mkulu Cave, Transvaal, South Africa. From Japan, in the Misotsuchi volcanic ash bed at Iijima, Nagano Prefecture, and the Kanumatsuchi bed at Kanuma, Tochigi Prefecture. At Mt. Shank, South Australia.

Name: From the Greek for *other* and *to appear*, referring to a change of appearance under the blowpipe.

Type Material: Mining Academy, Freiberg, Germany, 26104.

References: (1) Dana, E.S. (1892) Dana's system of mineralogy, (6th edition), 693–694. (2) Kitagawa, Y. (1974) Dehydration of allophane and its structural formula. Amer. Mineral., 59, 1094–1098. (3) Henmi, T. and K. Wada (1976) Morphology and composition of allophane. Amer. Mineral., 61, 379–390. (4) Wada, S.I. and K. Wada (1977) Density and structure of allophane. Clay Minerals Bull., 12, 289–298. (5) Bailey, S.W. (1980) Summary of recommendations of AIPEA nomenclature committee on clay minerals. Amer. Mineral., 65, 1–7.

Crystal Data: Hexagonal. *Point Group:* $\bar{3}\,2/m$. Irregular accumulations, to 1 mm.

Physical Properties: *Fracture:* Conchoidal. *Tenacity:* Brittle. Hardness = 5–6 D(meas.) = 2.76(5) D(calc.) = 2.78 Vivid orange-red fluorescence in UV.

Optical Properties: Transparent. *Color:* Colorless to weak brownish pink. *Luster:* Vitreous. *Optical Class:* Uniaxial (+). $\omega = 1.618(2)$ $\epsilon = 1.626(2)$

Cell Data: *Space Group:* $R\bar{3}m$. $a = 14.046(2)$ $c = 60.60(2)$ $Z = 6$

X-ray Powder Pattern: Mt. Alluaiv, Russia; could be mistaken for eudialyte. 2.960 (10), 2.825 (10), 7.14 (8), 1.762 (8), 4.30 (7), 2.148 (7), 3.36 (5)

Chemistry:

	(1)		(1)
SiO_2	53.3	SrO	1.0
TiO_2	6.0	BaO	0.6
ZrO_2	0.2	Na_2O	18.6
La_2O_3	0.2	K_2O	0.2
Ce_2O_3	0.8	Cl	0.8
Nb_2O_5	3.9	H_2O	1.7
MnO	3.6	$-O = Cl_2$	0.2
CaO	8.6	Total	99.3

(1) Mt. Alluaiv, Russia; by electron microprobe, average of three grains, H_2O by coulometry; corresponding to $(Na_{17.47}Sr_{0.28}Ce_{0.14}K_{0.12}Ba_{0.11}La_{0.03})_{\Sigma=18.15}(Ca_{4.46}Mn_{1.47})_{\Sigma=5.93}$ $(Ti_{2.18}Nb_{0.85}Zr_{0.05})_{\Sigma=3.08}Si_{25.82}O_{73.26}Cl_{0.66} \cdot 2.75H_2O$.

Occurrence: In ultra-agpaitic pegmatites in a differentiated alkalic massif.

Association: Eudialyte, nepheline, sodalite, potassic feldspar, arfvedsonite, aegirine.

Distribution: On Mt. Alluaiv, in the Lovozero massif, Kola Peninsula, Russia.

Name: For its occurrence on Mt. Alluaiv, Kola Peninsula, Russia.

Type Material: A.E. Fersman Mineralogical Museum, Academy of Sciences, Moscow, Russia; The Natural History Museum, London, England, 1994,1.

References: (1) Khomyakov, A.P., G.N. Netschelyustov, and R.K. Rastsvetaeva (1990) Alluaivite $Na_{19}(Ca, Mn)_6(Ti, Nb)_3Si_{26}O_{74}Cl \cdot 2H_2O$ – a new titanosilicate of eudialyte-like structure. Zap. Vses. Mineral. Obshch., 119(1), 117–120 (in Russian). (2) (1991) Amer. Mineral., 76, 1728 (abs. ref. 1).

Crystal Data: Cubic. *Point Group:* $4/m\,\bar{3}\,2/m$. Typically well-formed dodecahedra or trapezohedra, to 1 m; also in rounded grains and massive.

Physical Properties: *Cleavage:* Parting on {110}. *Fracture:* Subconchoidal. *Tenacity:* Brittle. Hardness = 7–7.5 D(meas.) = 4.318 D(calc.) = 4.313

Optical Properties: Transparent to translucent. *Color:* Deep red, brownish red, red-violet, black; may be sectored. *Streak:* White. *Luster:* Vitreous to resinous. *Optical Class:* Isotropic; anomalously biaxial. *Dispersion:* Weak. $n = 1.830$

Cell Data: *Space Group:* $Ia3d$. a = 11.526 Z = 8

X-ray Powder Pattern: Cavendish Township, Peterborough Co., Ontario, Canada. (ICDD 9-427).
2.569 (100), 1.540 (50), 2.873 (40), 1.599 (40), 4.04 (30), 1.866 (30), 1.660 (30)

Chemistry:

	(1)	(2)
SiO_2	37.39	36.21
TiO_2	0.16	
Al_2O_3	20.72	20.49
Fe_2O_3	0.83	
FeO	36.37	43.30
MnO	0.86	
MgO	3.85	
CaO	0.41	
Total	100.59	100.00

(1) Falun, Sweden; corresponds to $(Fe_{2.49}^{2+}Mn_{0.14}Ca_{0.05})_{\Sigma=2.68}(Al_{1.94}Fe_{0.05}^{3+}Ti_{0.01})_{\Sigma=2.00}$ $(Si_{2.99}Al_{0.01})_{\Sigma=3.00}O_{12}$. (2) $Fe_3Al_2(SiO_4)_3$.

Polymorphism & Series: Forms two series, with pyrope, and with spessartine.

Mineral Group: Garnet group.

Occurrence: The most common garnet, typically in mica schists and gneisses, from regionally metamorphosed argillaceous sediments and pelites; also in contact metamorphic hornfels. In granites and eclogites; in sedimentary rocks; as a detrital mineral.

Association: Biotite, cordierite, chlorite, staurolite, andalusite, kyanite, sillimanite, hematite, plagioclase, amphibole, pyroxene.

Distribution: Widespread. Some localities for fine crystals include: in the Zillertal, Tirol, Austria. At Falun, Sweden. In Norway, from near Bodø. From the Akhmatovsk deposit, near Zlatoust, Ural Mountains, Russia. In the USA, from Roxbury, Litchfield Co., and Southbury, New Haven Co., Connecticut; at Auburn, Androscoggin Co., and Topsham, Sagadahoc Co., Maine; from Westfield and Russell, Hampden Co., Massachusetts; and at Hanover, Grafton Co., New Hampshire. From Avondale, Chester Co., Pennsylvania; a large commercial deposit at the Barton mine, North Creek, Warren Co., New York; at Michigamme, Marquette Co., Michigan; from Salida, Chaffee Co., Colorado; and at Ft. Wrangell, Stikine River, Alaska. In Brazil, at Pernambuco, Bahia. From the Miami district, Zimbabwe. At Broken Hill and Thackaringa, New South Wales, and in the Harts Range, Northern Territory, Australia. From Yamanoo, Ibaragi Prefecture, and the Ishikawa district, Fukushima Prefecture, Japan.

Name: For Alabanda in Turkey, a cutting center in antiquity.

References: (1) Dana, E.S. (1892) Dana's system of mineralogy, (6th edition), 437–447. (2) Deer, W.A., R.A. Howie, and J. Zussman (1982) Rock-forming minerals, (2nd edition), v. 1A, orthosilicates, 468–698, esp. 537–589. (3) Novak, G.A. and G.V Gibbs (1971) The crystal chemistry of the silicate garnets. Amer. Mineral., 56, 791–825.

Crystal Data: Triclinic. *Point Group:* 1. As tapering pseudohexagonal prisms elongated along [001], to 2 mm; also tabular. *Twinning:* Common as six-fold sector twins on {001} and polysynthetic twins ∥ {010} prism edges.

Physical Properties: *Cleavage:* Perfect on {001}. *Tenacity:* Brittle. Hardness = 2.5–3 D(meas.) = 2.78 D(calc.) = 2.70

Optical Properties: Transparent. *Color:* Colorless, white, pink to lilac, or pale green. *Streak:* White with pale green tint. *Luster:* Pearly to somewhat metallic on cleavage surfaces. *Optical Class:* Biaxial (+). *Dispersion:* $r < v$. $\alpha = 1.5967(5)$ $\beta = 1.5986(5)$ $\gamma = 1.615(1)$ 2V(meas.) = 18°

Cell Data: *Space Group:* $C1$. $a = 5.307(1)$ $b = 9.195(2)$ $c = 14.068(3)$ $\alpha = 90.09(2)°$ $\beta = 90.25(2)°$ $\gamma = 89.96(2)°$ $Z = [4]$

X-ray Powder Pattern: Saranovskoye deposits, Russia; $2H_2$.
7.0 (100), 3.51 (100), 2.476 (80), 1.925 (70), 1.528 (60), 2.600 (40), 1.462 (35)

Chemistry:

	(1)	(2)	(3)
SiO_2	20.95	22.39	21.56
Al_2O_3	35.21	33.73	36.58
FeO	8.28	7.49	
MnO	trace	0.34	
MgO	22.88	23.69	28.93
CaO	0.58		
H_2O	13.02	[12.36]	12.93
Total	100.92	[100.00]	100.00

(1) Chester, Massachusetts, USA. (2) Pensacola Mountains, Antarctica; by electron microprobe, H_2O by difference. (3) $Mg_2Al(SiAl)O_5(OH)_4$.

Polymorphism & Series: $2H_1$, $2H_2$, 6R polytypes.

Mineral Group: Kaolinite-serpentine group.

Occurrence: A product of low-grade metamorphism of Al, Mg-rich rocks.

Association: Vesuvianite, chlorite (Pensacola Mountains, Antarctica); magnetite, rutile, diaspore (Chester, Massachusetts, USA); grossular, calcite, diopside, clinozoisite (Black Lake, Canada).

Distribution: At the Emery mine, Chester, Hampden Co., Massachusetts, USA. From the Lake Asbestos mine, Black Lake, Quebec, Canada. In Antarctica, in the Dufek mafic massif, Pensacola Mountains. From the Postmasburg manganese deposits, Cape Province, South Africa. From Russia, at the Saranovskoye chromite deposits, Northern Ural Mountains. At Hällefors, Sweden. On Mt. Sobotka, Silesia, Poland.

Name: Honors James Ames, a mine owner.

Type Material: National Museum of Natural History, Washington, D.C., USA, 80715.

References: (1) Dana, E.S. (1892) Dana's system of mineralogy, (6th edition), 655. (2) Deer, W.A., R.A. Howie, and J. Zussman (1963) Rock-forming minerals, v. 3, sheet silicates, 164–169. (3) Steadman, R. and P. M. Nuttall (1962) The crystal structure of amesite. Acta Cryst., 15, 510. (4) Hall, S.H. and S.W. Bailey (1976) Amesite from Antarctica. Amer. Mineral., 61, 497–499. (5) Hall, S.H. and S.W. Bailey (1979) Cation ordering pattern in amesite. Clays and Clay Minerals, 27, 241–247. (6) Anderson, C.S. and S.W. Bailey (1981) A new cation ordering pattern in amesite–$2H_2$. Amer. Mineral., 66, 185–195. (7) Wiewióra, A., J.A. Rausell-Colom, and T. García-González (1991) The crystal structure of amesite from Mount Sobotka: a nonstandard polytype. Amer. Mineral., 76, 647–652.

Crystal Data: Monoclinic, pseudotetragonal. *Point Group:* 2. As well-formed pseudotetragonal pyramidal crystals, to 5 mm, showing {011} and {110}, or more rarely {111} and {110}.

Physical Properties: Hardness = ~4.5 D(meas.) = 2.06–2.23 D(calc.) = 2.146–2.178

Optical Properties: Transparent. *Color:* Colorless. *Luster:* Vitreous. *Streak:* White. *Optical Class:* Biaxial (–). *Orientation:* X = b; Z ∧ c = 12°. α = 1.485 β = 1.490 γ = 1.494 2V(meas.) = 82°

Cell Data: *Space Group:* I2. a = 10.226–10.26 b = 10.422–10.44 c = 9.884–9.92 β = 88°19′–91°30′ Z = [2]

X-ray Powder Pattern: Hegau, Germany.
2.722 (100), 4.220 (90), 3.141 (80), 7.295 (55), 2.704 (50), 3.238 (45), 5.108 (40)

Chemistry:

	(1)	(2)
SiO$_2$	36.38	34.81
Al$_2$O$_3$	29.46	29.53
Fe$_2$O$_3$	trace	
MgO	trace	
CaO	0.22	
SrO	0.03	
BaO	trace	
Na$_2$O	8.22	8.98
K$_2$O	12.96	13.64
H$_2$O	12.80	13.04
Total	100.07	100.00

(1) Hegau, Germany; by electron microprobe; corresponds to K$_{1.88}$Na$_{1.80}$Ca$_{0.02}$ Al$_{3.93}$Si$_{4.12}$O$_{16}$·4.84H$_2$O. (2) K$_2$Na$_2$Al$_4$Si$_4$O$_{16}$·5H$_2$O.

Mineral Group: Zeolite group.

Occurrence: In veinlets cutting melilite-nephelinite volcanic rocks and pyroclastics (Hegau, Germany); in natrolite veinlets cutting ijolite-urtite pegmatites and apatite-nepheline rocks (Kola Peninsula, Russia).

Association: Merlinoite, aragonite, calcite (Hegau, Germany); natrolite (Kola Peninsula, Russia).

Distribution: In the Höwenegg quarry, Hegau, Baden-Württemberg, Germany. From the Kukisvumchorr apatite deposit, Khibiny massif, Kola Peninsula, Russia.

Name: To honor Giovan Battista Amici (1786–1863), physicist, optician, and inventor of microscope optical elements.

Type Material: University of Modena, Modena, Italy; The Natural History Museum, London, England; National Museum of Natural History, Washington, D.C., USA, 145843.

References: (1) Alberti, A., G. Hentschel, and G. Vezzalini (1979) Amicite, a new zeolite. Neues Jahrb. Mineral., Abh., 11, 481–488. (2) Alberti, A. and G. Vezzalini (1979) The crystal structure of amicite, a zeolite. Acta Cryst., 35, 2866–2869. (3) (1980) Amer. Mineral., 65, 808 (abs. refs. 1 and 2). (4) Khomyakov, A.P., G.E. Cherepivskaya, T.A. Kurova, and V.V. Kaptsov (1982) First occurrence of amicite (K$_2$Na$_2$Al$_4$Si$_4$O$_{16}$·5H$_2$O) in the USSR. Doklady Acad. Nauk SSSR, 263, 978–980 (in Russian). (5) (1982) Chem. Abs., 97, 9281 (abs. ref. 4).

Aminoffite $Ca_3Be_2Si_3O_{10}(OH)_2$

Crystal Data: Tetragonal. *Point Group:* $4/m$. As small, euhedral, pyramidal crystals. *Twinning:* Commonly twinned, polysynthetically on {001}.

Physical Properties: *Cleavage:* Poor on {001}. *Fracture:* Conchoidal. *Tenacity:* Brittle. Hardness = 5.5–6   D(meas.) = 2.94   D(calc.) = 2.86   Strongly piezoelectric.

Optical Properties: Transparent. *Color:* Colorless to light yellowish. *Luster:* Vitreous. *Optical Class:* Uniaxial (–); may be abnormally biaxial. $\omega = 1.647$ $\epsilon = 1.637$

Cell Data: *Space Group:* $P4_2/n$. $a = 9.865(2)$ $c = 9.930(2)$ $Z = 4$

X-ray Powder Pattern: Långban, Sweden.
2.614 (100), 2.84 (90), 4.02 (80), 2.141 (80), 6.97 (70), 4.40 (70), 3.48 (70)

Chemistry:

	(1)	(2)
SiO_2	42.49	43.28
Al_2O_3	4.41	
Fe_2O_3	0.31	
MnO	0.19	
BeO	6.20	12.01
CaO	40.27	40.39
H_2O	6.45	4.32
Total	[100.32]	100.00

(1) Långban, Sweden; original total given as 100.33%. (2) $Ca_3Be_2Si_3O_{10}(OH)_2$.

Occurrence: In cavities in massive magnetite (Långban, Sweden); in fluorite veins at the contact between marbles and "hornblende" granites associated with hastingsite nepheline syenites (Dugdinsk massif, Russia); in fluorite veins in tinguaite and nephelinite dikes (Bayankolsk dike field, Russia).

Association: Magnetite, goethite (Långban, Sweden); fluorite (Russia).

Distribution: From Långban, Värmland, Sweden. In Russia, found in the Bayankolsk dike field, Tuva, and in the Dugdinsk massif; also in the Lake Baikal area, eastern Siberia.

Name: For Dr. Gregori Aminoff (1883–1947), Swedish mineralogist and expert on Långban mineralogy, associated with the Riksmuseum, Stockholm, Sweden.

Type Material: Harvard University, Cambridge, Massachusetts, 106917; National Museum of Natural History, Washington, D.C., USA, 137291, R7823; The Natural History Museum, London, England.

References: (1) Hurlbut, C.S. (1937) Aminoffite, a new mineral from Långban, Sweden. Geol. Fören. Förhandl. Stockholm, 59, 290–292. (2) (1938) Amer. Mineral., 23, 293 (abs. ref. 1). (3) Mandarino, J.A. (1964) X-ray powder data for aminoffite. Amer. Mineral., 49, 212–214. (4) Moore, P.B. (1968) Relation of the manganese-calcium silicates, gageite and harstigite: a correction. Amer. Mineral., 53, 1418–1420. (5) Coda, A., G. Rossi, and L. Ungaretti (1967) The crystal structure of aminoffite. Atti Rend. Accad. Lincei, 43(3–4), 225–232. (6) (1968) Chem. Abs., 69, 71315 (abs. ref. 5). (7) (1967) Str. Rep., 32, 466–467 (abs. ref. 5).

Crystal Data: Tetragonal. *Point Group:* $4/m$. As minute aggregates replacing analcime crystals. *Twinning:* Common, repeated on $\{110\}$.

Physical Properties: Hardness = n.d. D(meas.) = 2.29(5)   D(calc.) = 2.24

Optical Properties: Translucent. *Color:* White. *Luster:* Resinous to vitreous. *Optical Class:* Uniaxial (+). $n = 1.518(2)$

Cell Data: *Space Group:* $I4_1/a$. $a = 13.214(1)$ $c = 13.713(2)$ $Z = 16$

X-ray Powder Pattern: Fujioka, Japan.
5.43 (100), 3.30 (80), 5.53 (50), 3.43 (40), 2.955 (20), 2.859(20), 2.839 (10)

Chemistry:

	(1)
SiO_2	62.67
Al_2O_3	22.43
K_2O	4.43
$(NH_4)_2O$	8.70
H_2O	1.77
Total	[100.00]

(1) Fujioka, Japan; by electron microprobe; after subtraction of elements attributed to dolomite and analcime contamination, recalculated to 100.00%; corresponds to $[(NH_4)_{0.68}K_{0.19}]_{\Sigma=0.87}$ $Al_{0.89}Si_{2.12}O_6$.

Occurrence: In veinlets, fractures, and cavities in hydrothermally altered crystalline schist, as powdcry pscudomorphous replacements of analcime crystals.

Association: Analcime, dolomite.

Distribution: In the Tatarazawa quarry, Fujioka, Gumma Prefecture, Japan.

Name: For *ammonia* in its chemical composition and its relation to *leucite*.

Type Material: National Science Museum, Tokyo, Japan; National Museum of Natural History, Washington, D.C., USA, 165991.

References: (1) Hori, H., K. Nagashima, M. Yamada, R. Miyawaki, and T. Marubashi (1986) Ammonioleucite, a new mineral from Tatarazawa, Fujioka, Japan. Amer. Mineral., 71, 1022–1027.

Crystal Data: Monoclinic. *Point Group:* $2/m$. As acicular prismatic crystals, to 1 cm, elongated and commonly striated $\parallel$ [001], with cross sections appearing rhombohedral or hexagonal.

Physical Properties: *Cleavage:* Good on {100}. *Fracture:* Conchoidal. *Tenacity:* Brittle.
Hardness = 3–5 D(meas.) = 2.40(5) D(calc.) = 2.38

Optical Properties: Transparent to translucent. *Color:* Colorless. *Luster:* Vitreous.
Optical Class: Biaxial (+). *Orientation:* $Z = b$; $Y \wedge c = 10°$. *Dispersion:* $r < v$, weak.
$\alpha = 1.533$ $\beta = 1.534$ $\gamma = 1.538$ 2V(meas.) = 57(2)° 2V(calc.) = 59°

Cell Data: *Space Group:* $C2/c$. $a = 18.830(2)$ $b = 11.517(2)$ $c = 5.190(1)$
$\beta = 100.86(1)°$ $Z = 4$

X-ray Powder Pattern: Amstall, Austria.
9.75 (100), 3.603 (100), 3.816 (90), 5.43 (70), 4.714 (60), 3.175 (60), 4.069 (40)

Chemistry:

	(1)
SiO_2	49.41
Al_2O_3	22.84
CaO	13.80
Cl	1.75
H_2O	12.40
$-O = Cl_2$	0.39
Total	99.81

(1) Amstall, Austria; by electron microprobe, corresponds to $Ca_{0.98}Al_{1.78}Si_{3.26}O_{12.80}Cl_{0.20}H_{5.46}$.

Occurrence: In open fissures cutting pegmatitic schlieren, in hydrothermally altered graphite-bearing metamorphic rocks.

Association: Apatite, rutile, siderite, albite, laumontite, calcite, vivianite.

Distribution: In the Amstall graphite quarry, Amstall, Austria.

Name: For the type locality at Amstall, Austria.

Type Material: Institute of Mineralogy and Crystallography, Vienna University; Natural History Museum, Vienna, Austria.

References: (1) Quint, R. (1987) Description and crystal structure of amstallite, $CaAl(OH)_2[Al_{0.8}Si_{3.2}O_8(OH)_2] \cdot [(H_2O)_{0.8}Cl_{0.2}]$, a new mineral from Amstall, Austria. Neues Jahrb. Mineral., Monatsh., 253–262. (2) (1988) Amer. Mineral., 73, 1492–1493 (abs. ref. 1).

Crystal Data: Cubic; tetragonal, orthorhombic, or monoclinic, pseudocubic, with degree of ordering. *Point Group:* $4/m\ \bar{3}\ 2/m$; $4/m\ 2/m\ 2/m$, $2/m\ 2/m\ 2/m$, or $2/m$. Crystals commonly trapezohedra {211}, to 25 cm. Also granular, compact, massive, typically showing concentric structure. *Twinning:* Polysynthetic on {001}, {110}.

Physical Properties: *Cleavage:* Very poor on {100}. *Fracture:* Subconchoidal. *Tenacity:* Brittle. Hardness = 5–5.5   D(meas.) = 2.24–2.29   D(calc.) = 2.271   Weakly piezoelectric; weakly electrostatic when rubbed or heated.

Optical Properties: Transparent to translucent. *Color:* White, colorless, gray, pink, greenish, yellowish; in thin section, colorless. *Luster:* Vitreous. *Optical Class:* Isotropic; anomalously biaxial (–). $n = 1.479$–1.493 2V(meas.) = 0°–85°

Cell Data: *Space Group:* $Ia3d$; $I4_1/acd$, $Ibca$, or $I2/a$.   a = 13.723–13.733   Z = 16 (cubic).

X-ray Powder Pattern: Låven Island, Langesundsfjord, Norway.
3.43 (10), 5.61 (8), 2.925 (8), 1.743 (6), 2.693 (5), 2.505 (5), 1.903 (5)

Chemistry:

	(1)	(2)	(3)		(1)	(2)	(3)
SiO$_2$	54.19	54.58	54.58	Na$_2$O	11.08	13.50	14.08
Al$_2$O$_3$	23.12	23.05	23.16	K$_2$O	1.62	0.00	
MgO	0.10	0.10		H$_2$O$^+$	8.20	8.70	8.18
CaO	1.54	0.45		Total	99.85	100.38	100.00

(1) Cyclopean Islands, Italy. (2) Mazé, Niigata Prefecture, Japan. (3) NaAlSi$_2$O$_6$·H$_2$O.

Polymorphism & Series: Forms a series with pollucite.

Mineral Group: Zeolite group.

Occurrence: In the groundmass or vesicles of silica-poor intermediate and mafic igneous rocks, typically basalts and phonolites, from late-stage hydrothermal solutions, or disseminated due to deuteric alteration. In lake beds, altered from pyroclastics or clays, or as a primary precipitate; authigenic in sandstones and siltstones.

Association: Zeolites, prehnite, calcite, quartz, glauconite.

Distribution: Some localities for outstanding specimens follow. In Italy, on the Cyclopean Islands, and from Val di Fassa and Alpe di Siusi, Trentino-Alto Aldige. Large crystals from Kotchechovmo, Krasnoyarski Krai, Russia. At Breidhdalsheidhi, Iceland. In the Dean quarry, St. Keverne, Lizard Peninsula, Cornwall, England. Around Glasgow, Dumbartonshire, Scotland. In the USA, from the Lake Superior district, Houghton Co., Michigan; at Cornwall, Lebanon Co., Pennsylvania; from Bergen Hill, Hudson Co., and West Paterson, Passaic Co., New Jersey; on Table Mountain, Jefferson Co., Colorado; in the Price Creek quarry, Benton Co., Oregon. From the Bay of Fundy district, Nova Scotia; large crystals from Mont Saint-Hilaire, Quebec, Canada. From Flinders, Victoria, Australia.

Name: From the Greek for *weak*, alluding to the weak electrostatic charge developed when heated or rubbed.

References: (1) Dana, E.S. (1892) Dana's system of mineralogy, (6th edition), 595–598. (2) Deer, W.A., R.A. Howie, and J. Zussman (1963) Rock-forming minerals, v. 4, framework silicates, 338–350. (3) Saha, P. (1959) Geochemical and X-ray investigations of natural and synthetic analcites. Amer. Mineral., 44, 300–313. (4) Černý, P. (1974) The present status of the analcime-pollucite series. Can. Mineral., 12, 334–341. (5) Coombs, D.S. (1955) X-ray investigations on wairakite and non-cubic analcime. Mineral. Mag., 30, 699. (6) Mazzi, F. and E. Galli (1978) Is each analcime different? Amer. Mineral., 63, 448–460. (7) Pechar, F. (1988) The crystal structure of natural monoclinic analcime (NaAlSi$_2$O$_6$·H$_2$O). Zeits. Krist., 184, 63–69.

Anandite

$$(Ba, K)(Fe^{2+}, Mg)_3(Si, Al, Fe)_4O_{10}(S, OH)_2$$

Crystal Data: Monoclinic or orthorhombic. *Point Group:* $2/m$, m, or $2/m\ 2/m\ 2/m$. Poorly developed prism faces give a hexagonal outline to cleavage flakes.

Physical Properties: *Cleavage:* {001}, perfect. Hardness = 3–4 D(meas.) = 3.94 D(calc.) = 3.94

Optical Properties: Nearly opaque. *Color:* Black. *Luster:* Vitreous.
Optical Class: Biaxial (+). *Pleochroism:* Y = green; Z = brown. *Orientation:* $Y = b$; $Z \wedge a = $ 8°–16°. *Dispersion:* Strong. $\alpha = 1.85(1)$ β = n.d. $\gamma = > 1.88$ 2V(meas.) = n.d.

Cell Data: *Space Group:* $C2/c$ or Cc. a = 5.412(5) b = 9.434(5) c = 19.953(10) $\beta = 94°52(10)'$ Z = 2, or *Space Group:* $Pnmn$. a = 5.439(1) b = 9.509(2) c = 19.878(6) Z = 2

X-ray Powder Pattern: Wilagedera iron prospect, Sri Lanka.
3.320 (100), 4.995 (85), 2.490 (80), 9.92 (60), 2.716 (50), 2.681 (45), 3.430 (40)

Chemistry:

	(1)	(2)		(1)	(2)
SiO_2	25.20	24.59	Na_2O	0.10	0.06
TiO_2	0.28	0.08	K_2O	0.93	0.24
Al_2O_3	4.85	0.80	F		0.12
Fe_2O_3	6.98		Cl		0.91
FeO	33.10	41.73	H_2O^+	1.98	
MnO	0.66	0.92	H_2O^-	0.12	
MgO	3.39	2.89	S	2.96	4.25
CaO	0.16	0.00	$-O = S, (F, Cl)_2$	1.48	2.38
BaO	20.35	23.05	Total	99.58	97.26

(1) Wilagedera iron prospect, Sri Lanka. (2) Do.; by electron microprobe; Mössbauer spectroscopy indicates Fe^{2+}:Fe^{3+} = 1.19.

Polymorphism & Series: 2O, $2M_1$, 1M polytypes.

Mineral Group: Mica group.

Occurrence: As monomineralic veinlets and lenses in a banded magnetite deposit capped by banded magnetite-barite rock. The ore zone and associated interbedded calc-schists and gneisses are of metasedimentary origin, of approximately granulite facies grade.

Association: Magnetite, chalcopyrite, pyrite, pyrrhotite, barite.

Distribution: At the Wilagedera iron prospect, North Western Province, Sri Lanka.

Name: For Dr. Ananda Kentush Coomaraswamy (1877–1947), the first Director of the Mineral Survey of Ceylon (Sri Lanka).

Type Material: n.d.

References: (1) Pattiaratchi, D.B., E. Saari, and T.G. Sahama (1967) Anandite, a new barium iron silicate from Wilagedera, North Western Province, Ceylon. Mineral. Mag., 36, 1–4. (2) Filut, M.A., A.C. Rule, and S.W. Bailey (1985) Crystal structure refinement of anandite-2Or, a barium- and sulfur-bearing trioctahedral mica. Amer. Mineral., 70, 1298–1308.

Crystal Data: Orthorhombic. *Point Group:* $2/m\ 2/m\ 2/m$. As euhedral crystals or columnar aggregates having nearly square cross sections, commonly elongated ‖ [001], to 20 cm. Also fibrous, compact, massive. *Twinning:* On {101}, rare.

Physical Properties: *Cleavage:* Good on {110}, poor on {100}. *Fracture:* Uneven to subconchoidal. *Tenacity:* Brittle. Hardness = 6.5–7.5 D(meas.) = 3.13–3.16 D(calc.) = 3.149

Optical Properties: Transparent to nearly opaque with inclusions. *Color:* Pink, violet, yellow, green, white, gray; in thin section, colorless to pink or green. *Streak:* White. *Luster:* Vitreous to subvitreous.
Optical Class: Biaxial (–) or (+). *Pleochroism:* Weak; X = reddish pink; $Y = Z$ = greenish yellow. *Orientation:* $X = c$; $Y = b$; $Z = a$. *Dispersion:* $r < v$, strong. $\alpha = 1.632$ $\beta = 1.636$ $\gamma = 1.643$ 2V(meas.) = 71°–86°

Cell Data: *Space Group:* $Pnnm$. a = 7.7980(7) b = 7.9031(10) c = 5.5566(5) Z = 4

X-ray Powder Pattern: Minas Gerais, Brazil.
5.542 (100), 4.527 (90), 2.773 (90), 2.170 (90), 3.924 (60), 3.524 (60), 2.466 (50)

Chemistry:

	(1)	(2)
SiO$_2$	36.74	37.08
TiO$_2$	0.01	
Al$_2$O$_3$	62.70	62.92
Fe$_2$O$_3$	0.36	
FeO	0.05	
CuO	0.02	
MgO	0.03	
K$_2$O	0.07	
H$_2$O$^+$	0.15	
H$_2$O$^-$	0.01	
Total	100.14	100.00

(1) Goat Mountain, Shoshone Co., Idaho, USA. (2) Al$_2$SiO$_5$.

Polymorphism & Series: Trimorphous with kyanite and sillimanite; forms a series with kanonaite.

Occurrence: A result of contact metamorphism of argillaceous sediments, also in regionally metamorphosed schists. Rare in granites and pegmatites, which however afford the largest crystals. Detrital in some sandstones.

Association: Kyanite, sillimanite, cordierite, corundum, garnet, tourmaline, mica.

Distribution: Widespread; some localities for good crystals follow. From Hornachuelos, near Cordoba, Spain. On the Lisens Alp, Selraintal, Tirol, Austria. At Gefrees, Bodenmais, and elsewhere in Bavaria, Germany. From the Claggau quarry, Co. Galway, Ireland. In the USA, from Lancaster, Worcester Co., Massachusetts; Leiperville, Delaware Co., Pennsylvania; around Custer, Custer Co., South Dakota; in California, from Fresno, Fresno Co., near Ogilby, Cargo Muchacho Mountains, Imperial Co., and near Daltons Ranch, Madera Co. From Mt. Howden, Bimbowrie, South Australia. Gem crystals from the Santa Teresa district, Espírito Santo, Brazil.

Name: For an occurrence in the Andalusia region, Spain.

References: (1) Dana, E.S. (1892) Dana's system of mineralogy, (6th edition), 496–498. (2) Deer, W.A., R.A. Howie, and J. Zussman (1982) Rock-forming minerals, (2nd edition), v. 1A, orthosilicates, 759–779. (3) Skinner, B.J., S.P. Clark, Jr., and D.E. Appleman (1961) Molar volumes and thermal expansions of andalusite, kyanite, and sillimanite. Amer. J. Sci., 259, 651–668. (4) Winter, J.K. and S. Ghose (1979) Thermal expansion and high-temperature crystal chemistry of the Al$_2$SiO$_5$ polymorphs. Amer. Mineral., 64, 573–586. (5) Gunter, M. and F.D. Bloss (1982) Andalusite-kanonaite series: lattice and optical parameters. Amer. Mineral., 67, 1218–1228.

Crystal Data: Triclinic. *Point Group:* $\bar{1}$. Crystals rare, to 2 cm; commonly massive, cleavable, or granular. *Twinning:* Commonly twinned after Albite, Pericline, and Carlsbad laws. Also twinned after a law similar to the Manebach law.

Physical Properties: *Cleavage:* Perfect on {001}, less perfect on {010}; {110} also observed. *Fracture:* Uneven to conchoidal. *Tenacity:* Brittle. Hardness = 6–6.5 D(meas.) = 2.66–2.68 D(calc.) = 2.68

Optical Properties: Transparent to translucent. *Color:* White, gray, green, yellow, flesh-red. *Streak:* White. *Luster:* Subvitreous to pearly.
Optical Class: Biaxial (+). *Dispersion:* $r > v$, weak. $\alpha = 1.545$–1.556 $\beta = 1.548$–1.558 $\gamma = 1.552$–1.563 2V(meas.) = 78°–90° (low); 73°–80° (high).

Cell Data: *Space Group:* $P\bar{1}$. $a = 8.151(3)$ $b = 12.829(5)$ $c = 14.206(7)$ $\alpha = 93.62(3)°$ $\beta = 116.21(2)°$ $\gamma = 89.70(2)°$ $Z = 8$

X-ray Powder Pattern: Crestmore, Riverside Co., California, USA (low).
3.21 (100), 3.18 (90), 4.04 (80), 3.76 (70), 3.65 (70), 3.14 (70), 2.93 (70b)

Chemistry:

	(1)	(2)	(3)		(1)	(2)	(3)
SiO_2	57.59	60.76	55.59	Na_2O	6.39	8.12	5.73
Al_2O_3	25.84	24.82	28.30	K_2O	0.55		
Fe_2O_3	0.92			H_2O^+	0.32		
MgO	trace			H_2O^-	0.05		
CaO	8.45	6.30	10.38	Total	100.11	100.00	100.00

(1) Beaver Bay, Lake Co., Minnesota, USA. (2) $Na_{0.70}Ca_{0.30}Al_{1.30}Si_{2.70}O_8$. (3) $Na_{0.50}Ca_{0.50}Al_{1.50}Si_{2.50}O_8$.

Polymorphism & Series: Low and high temperature variants differing in structural detail are recognized.

Mineral Group: Feldspar group, plagioclase series.

Occurrence: Widespread in igneous rocks of intermediate silica content, as syenites and andesites. Characteristic of granulite to amphibolite facies metamorphism, commonly as antiperthite; as detrital grains in sedimentary rocks.

Association: Quartz, potassic feldspar, biotite, "hornblende," magnetite.

Distribution: A widespread mineral, although crystals are not common. In the andesite lavas of the Andes Mountains, South America, as at Marmato, Columbia. Around St. Raphaël, Estérel Mountains, Var, and near Chagey, Haute-Saône, France. At Bodenmais, Bavaria, Germany. On Mt. Arcuentu, Sardinia, Italy. From Vapnefjord, Iceland. At Sannidal and Arendal, Norway. In the USA, at Sanford, York Co., Maine. In Japan, at Kaneda, Miyagi Prefecture; Naka, Iwojima Island; Kuzuhara, Toyama Prefecture; and a number of other places. From Minsen, Korea. At Cape Grant, Victoria, Australia.

Name: For the Andes Mountains of South America where it is abundant in the lavas.

References: (1) Dana, E.S. (1892) Dana's system of mineralogy, (6th edition), 325–327, 333–334. (2) Deer, W.A., R.A. Howie, and J. Zussman (1963) Rock-forming minerals, v. 4, framework silicates, 94–165. (3) Phillips, W.R. and D.T. Griffen (1981) Optical mineralogy, 352–360. (4) Goodyear, J. and W.J. Duffin (1954) The identification and determination of plagioclase feldspars by the X-ray powder method. Mineral. Mag., 30, 306–326. (5) Steurer, W. and H. Jagodzinski (1988) The incommensurately modulated structure of an andesine. Acta Cryst., 44, 344–351.

Crystal Data: Cubic. *Point Group:* $4/m\,\overline{3}\,2/m$. Commonly well-crystallized dodecahedra, trapezohedra, or combinations, to 5 cm. Also granular to massive.

Physical Properties: *Fracture:* Uneven to conchoidal. *Tenacity:* Brittle. Hardness = 6.5–7 D(meas.) = 3.8–3.9   D(calc.) = 3.859

Optical Properties: Transparent to translucent. *Color:* Yellow, greenish yellow to emerald-green, dark green; brown, brownish red, brownish yellow; grayish black, black; may be sectored. *Streak:* White. *Luster:* Adamantine to resinous, dull.
Optical Class: Isotropic; typically weakly anisotropic. $n = 1.887$

Cell Data: *Space Group:* $Ia3d$.   a = 12.056   Z = 8

X-ray Powder Pattern: Synthetic.
2.696 (100), 3.015 (60), 1.6112 (60), 2.462 (45), 1.9564 (25), 1.6728 (25), 1.1195 (25)

Chemistry:

	(1)	(2)
SiO_2	34.91	35.47
TiO_2	trace	
Al_2O_3	0.69	
Fe_2O_3	30.40	31.42
MgO	0.58	
CaO	33.20	33.11
H_2O^-	0.19	
Total	99.97	100.00

(1) Reškovic stream, Serbia, Yugoslavia; corresponds to $(Ca_{3.01}Mg_{0.07})_{\Sigma=3.08}(Fe_{1.94}^{3+}Al_{0.02})_{\Sigma=1.96}$ $(Si_{2.95}Al_{0.05})_{\Sigma=3.00}O_{12}$. (2) $Ca_3Fe_2(SiO_4)_3$.

Polymorphism & Series: Forms two series, with grossular, and with schorlomite.

Mineral Group: Garnet group.

Occurrence: In skarns from contact metamorphosed impure limestones or calcic igneous rocks; in chlorite schists and serpentinites; in alkalic igneous rocks, then typically titaniferous.

Association: Vesuvianite, chlorite, epidote, spinel, calcite, dolomite, magnetite.

Distribution: Widespread; fine examples from; in Italy, at Frascati, Alban Hills, Lazio; the Val Malenco, Lombardy; the Ala Valley, Piedmont; and Larcinaz, Val d'Aosta. At Dognecea (Dognaczka) and Oraviţa (Oravicza), Banat, Romania. From Vaskö, Hungary. At Zermatt, Valais, Switzerland. From Arendal, Norway. In the Wessels mine, near Kuruman, Cape Province, South Africa. In Russia, gem crystals from the Bobrovka River, Nizhni Tagil district, Ural Mountains; at Sineretschenskoje, north of Vladivostock. In the USA, from Stanley Butte, Graham Co., Arizona; on Garnet Hill, Calaveras Co., and around the Gem mine, San Benito Co., California; at Franklin and Sterling Hill, Sussex Co., New Jersey; from Magnet Cove, Hot Springs Co., Arkansas; and on Prince of Wales Island, Alaska. In Mexico, found near Cárdenas, Chihuahua.

Name: After J.B. d'Andrada e Silva (1763–1838), Brazilian mineralogist who described a variety.

References: (1) Dana, E.S. (1892) Dana's system of mineralogy, (6th edition), 437–447.
(2) Deer, W.A., R.A. Howie, and J. Zussman (1982) Rock-forming minerals, (2nd edition),
v. 1A, orthosilicates, 468–698, esp. 617–641. (3) Novak, G.A. and G.V Gibbs (1971) The crystal chemistry of the silicate garnets. Amer. Mineral., 56, 791–825. (4) (1960) NBS Circ. 539, 9, 22.

Andremeyerite

$BaFe_2^{2+}Si_2O_7$

Crystal Data: Monoclinic, pseudo-orthorhombic. *Point Group:* $2/m$. As crystals, < 0.2 mm, with $\{100\}$ and $\{010\}$ dominant, also $\{011\}$ and $\{120\}$. *Twinning:* Multiple twinning on $\{100\}$.

Physical Properties: *Cleavage:* Perfect on $\{100\}$ and $\{010\}$. Hardness $= 5.5$ VHN $= 440$
D(meas.) $= 4.15$ D(calc.) $= 4.14$

Optical Properties: Transparent to translucent. *Color:* Pale emerald-green.
Optical Class: Biaxial $(+)$. *Pleochroism:* $X =$ pale bluish green; $Y = Z =$ colorless with faint brownish tint. *Orientation:* $Z = b$; $X \wedge c = 2°$ at 670 nm to 61° at 470 nm. *Dispersion:* Very strong. *Absorption:* $X > Y \simeq Z$. $\alpha = 1.740(5)$ $\beta = 1.740(5)$ $\gamma = 1.760(5)$
2V(meas.) $= 0°-80°$

Cell Data: *Space Group:* $P2_1/c$. $a = 7.488(1)$ $b = 13.785(1)$ $c = 7.085(1)$
$\beta = 118.23(1)°$ $Z = 4$

X-ray Powder Pattern: Mt. Nyiragongo, Zaire.
3.055 (100), 3.122 (80), 3.288 (60), 2.472 (55), 4.63 (40), 2.811 (40), 3.198 (20)

Chemistry:

	(1)	(2)
SiO_2	32.46	28.81
Al_2O_3	1.00	
FeO	31.55	34.44
MnO	1.33	
MgO	0.75	
CaO	0.52	
BaO	32.55	36.75
Na_2O	0.10	
K_2O	0.65	
Total	100.91	100.00

(1) Mt. Nyiragongo, Zaire; by electron microprobe. (2) $BaFe_2Si_2O_7$.

Occurrence: In vesicles of melilite-leucite-nephelinite; may represent crystallization of an associated green glass.

Association: Nepheline, leucite, clinopyroxene, kirschsteinite, melilite, apatite, magnetite, götzenite, troilite, glass.

Distribution: From the rim of Mt. Nyiragongo volcano, Kivu Province, Zaire.

Name: For André Marie Meyer (1890–?), Belgian geologist with the Geological Survey of the Belgian Congo, who first collected the mineral.

Type Material: University of Helsinki, Helsinki, Finland.

References: (1) Sahama, T.G., J. Siivola, and P. Rehtijärvi (1973) Andremeyerite, a new barium iron silicate from Nyiragongo, Zaire. Bull. Geol. Soc. Finland, 45, 1–8. (2) (1974) Amer. Mineral., 59, 381 (abs. ref. 1). (3) Cannillo, E.C., F. Mazzi, and G. Rossi (1988) Crystal structure of andremeyerite, $BaFe(Fe, Mn, Mg)Si_2O_7$. Amer. Mineral., 73, 608–612.

Crystal Data: Monoclinic. *Point Group:* $2/m$. Tabular crystals and cleavage fragments with pseudohexagonal outlines, to 15 cm. Commonly as foliated masses. *Twinning:* Contact twins with composition surface {001} and twin axis [310].

Physical Properties: *Cleavage:* Perfect on {001}. *Tenacity:* Flexible. Hardness = 2.5–3 D(meas.) = 3.3　D(calc.) = [3.36]

Optical Properties: Translucent to transparent. *Color:* Brown, black. *Luster:* Vitreous. *Optical Class:* Biaxial (–). *Pleochroism:* X = brown; $Y = Z$ = dark brown. *Orientation:* $Y = b$. *Dispersion:* $r < v$, weak.　$\alpha = 1.624$　$\beta = 1.672$　$\gamma = 1.672$　2V(meas.) = 0°–5°

Cell Data: *Space Group:* $C2/m$.　a = 5.3860(9)　b = 9.3241(7)　c = 10.2683(9) $\beta = 100.63(1)°$　Z = 2

X-ray Powder Pattern: Synthetic; 1M.
10.264 (100), 3.380 (80), 2.654 (70), 2.465 (40), 1.556 (40), 2.199 (20), 1.692 (20)

Chemistry:

	(1)	(2)	(3)
SiO_2	34.90	32.56	35.21
TiO_2	0.00	1.48	
Al_2O_3	10.90	17.24	9.96
FeO	36.19	34.25	42.11
MnO	1.77	0.03	
MgO	3.40	0.86	
CaO	0.15	0.0	
Na_2O	0.00	0.28	
K_2O	8.10	8.20	9.20
F	2.22		
Cl	0.24		
H_2O			3.52
Total	[97.87]	94.90	100.00

(1) Moina skarn, Tasmania; by electron microprobe, original total given as 97.78%. (2) Kawai mine, Japan; by electron microprobe. (3) $KFe_3AlSi_3O_{10}(OH)_2$.

Polymorphism & Series: 1M polytype.

Mineral Group: Mica group.

Occurrence: In magnesium-poor igneous and metamorphic rocks.

Association: Fluorite, zircon (Rockport, Massachusetts, USA).

Distribution: In the USA, from Cape Ann, east of Rockport, Essex Co., Massachusetts, and on Pikes Peak, El Paso Co., Colorado. From Mont Saint-Hilaire, Quebec, Canada. In the Moina tungsten skarn deposit, 40 km southwest of Devonport, Tasmania. From the Kawai mine, Ena, Gifu Prefecture, Japan. In Russia, at Katugin, Siberia, and Sludorudnik, Ural Mountains. From Flowerdale, near Gairloch, Scotland.

Name: For the first noted occurrence at Cape Ann, Massachusetts, USA.

References: (1) Dana, E.S. (1892) Dana's system of mineralogy, (6th edition), 634. (2) Deer, W.A., R.A. Howie, and J. Zussman (1963) Rock-forming minerals, v. 3, sheet silicates, 55–84. (3) Eugster, H.P. and D.R. Wones (1962) Stability relations of the ferruginous biotite, annite. J. Petrol., 3, 82–125. (4) Hazen, R.M. and C.W. Burnham (1973) The crystal structures of one-layer phlogopite and annite. Amer. Mineral., 58, 889–900. (5) Kwak, T.A.P. and P.W. Askins (1981) Geology and genesis of the F-Sn-W(-Be-Zn) skarn (wrigglite) at Moina, Tasmania. Econ. Geol., 76, 439–467.

Anophorite $NaNa_2[Mg, Fe^{2+}, Fe^{3+}]_5(Si_7Al)O_{22}(OH)_2$

Crystal Data: Monoclinic. *Point Group:* $[2/m.]$ As crystals, prismatic to acicular, to 3 cm. *Twinning:* On {100}.

Physical Properties: *Cleavage:* Very good on {110}, intersecting at 56° and 124°. *Tenacity:* [Brittle.] Hardness = [5–6] D(meas.) = 3.166 D(calc.) = n.d.

Optical Properties: Semitransparent. *Color:* Velvet-black; red-brown to greenish in thin section. *Streak:* Black. *Luster:* Vitreous.
Optical Class: [Biaxial.] *Pleochroism:* X = pale brownish yellow; Y = yellowish green to olive-green; Z = chocolate-brown to violet-brown. *Dispersion:* $r > v$. *Absorption:* $Z > Y > X$.
α = n.d. β = n.d. γ = n.d. 2V(meas.) = 44°20′

Cell Data: *Space Group:* n.d. Z = n.d.

X-ray Powder Pattern: n.d.

Chemistry:

	(1)
SiO_2	49.79
TiO_2	5.37
Al_2O_3	1.98
Fe_2O_3	7.54
FeO	9.18
MnO	0.36
MgO	11.59
CaO	3.16
Na_2O	7.92
K_2O	1.85
H_2O	1.52
Total	100.26

(1) Katzenbuckel, Germany; corresponds to $(Na_{0.77}K_{0.35})_{\Sigma=1.12}(Na_{1.50}Ca_{0.50})_{\Sigma=2.00}(Mg_{2.55}Fe^{2+}_{1.14}$ $Fe^{3+}_{0.47}Ti_{0.60}Mn_{0.05})_{\Sigma=4.81}(Si_{7.29}Fe^{3+}_{0.37}Al_{0.34})_{\Sigma=8.00}O_{22}(OH)_2$.

Mineral Group: Amphibole (alkali) group: $Fe^{3+}/(Fe^{3+} + Al^{vi}) > 0.5$; $Mg/(Mg + Fe^{2+}) \geq 0.5$; $(Na + K)_A \geq 0.5$; $Na_B \geq 1.34$; $(Ca + Na)_B \geq 1.34$; $6.5 < Si < 7.5$.

Occurrence: In shonkinite.

Association: Aegirine, nepheline, sanidine, mica, natrolite.

Distribution: In Germany, on the Katzenbuckel, Odenwald, Baden-Würtemberg.

Name: From the Greek for *ascending*, as its extinction angle is less than that of katophorite.

Type Material: n.d.

References: (1) Ford, W.E. (1915) Dana's system of mineralogy, (6th edition), app. III, 4. (2) (1910) Neues Jahrb. Mineral., Monatsh., 34 (abs., in German). (3) Hintze, C. (1938) Handbuch der Mineralogie. Gruyter & Co., Berlin, 30–31 (in German). (4) Rock, N.M.S. and B.E. Leake (1984) The International Mineralogical Association amphibole nomenclature scheme: computerization and its consequences. Mineral. Mag., 48, 211–227, esp. 212.

Crystal Data: Triclinic. *Point Group:* $\bar{1}$. Crystals commonly short, prismatic on [001], rarely on [010], to 2 cm; lamellar, coarse granular, massive. *Twinning:* Commonly polysynthetic on the Albite law; also after the Pericline, Carlsbad, Manebach, and Baveno laws.

Physical Properties: *Cleavage:* Perfect on {001}, less so on {010}, imperfect on {110}. *Fracture:* Conchoidal to uneven. *Tenacity:* Brittle. Hardness = 6–6.5   D(meas.) = 2.74–2.76 D(calc.) = 2.760

Optical Properties: Transparent to translucent. *Color:* White, grayish, reddish; colorless in thin section. *Streak:* White. *Luster:* Vitreous.
Optical Class: Biaxial (–). *Dispersion:* $r < v$, weak. $\alpha = 1.573–1.577$ $\beta = 1.580–1.585$ $\gamma = 1.585–1.590$ 2V(meas.) = 78°–83°

Cell Data: *Space Group:* $P\bar{1}$ (low).   a = 8.1768   b = 12.8768   c = 14.1690 $\alpha = 93.17°$ $\beta = 115.85°$ $\gamma = 92.22°$ Z = 8

X-ray Powder Pattern: Monte Somma, Italy (low).
3.19 (100), 3.18 (91), 3.21 (63), 3.26 (52), 4.04 (48), 3.12 (39), 3.62 (33)

Chemistry:

	(1)	(2)	(3)
SiO_2	45.88	45.62	43.19
TiO_2	0.04		
Al_2O_3	34.31	35.02	36.65
Fe_2O_3	0.83		
CaO	18.28	18.24	20.16
Na_2O	0.82	1.12	
K_2O	0.11		
H_2O^+	0.14		
Total	100.41	100.00	100.00

(1) Crookdene, Northumberland, England. (2) $Na_{0.10}Ca_{0.90}Al_{1.90}Si_{2.10}O_8$. (3) $CaAl_2Si_2O_8$.

Polymorphism & Series: Trimorphous with dmisteinbergite and svyatoslavite; low- and high-temperature structural modifications are recognized.

Mineral Group: Feldspar group, plagioclase series.

Occurrence: A rare constituent of mafic plutonic and volcanic rocks. In some granulite facies metamorphic rocks; in metamorphosed carbonate rocks; with corundum deposits. Known from meteorites.

Association: Olivine, pyroxene, corundum.

Distribution: A widely distributed rock-forming mineral. Classic occurrences include: from Monte Somma and Vesuvius, Campania; on Mt. Monzoni, Val di Fassa, Trentino-Alto Adige; and from the Cyclopean Islands, Italy. At Tunaberg, Södermanland, Sweden. From near Lojo, Finland. At Bogoslovsk and Barsowka, Ural Mountains, Russia. On Miyakejima Island, Tokyo Prefecture; at Toshinyama, Tochigi Prefecture; the Zao volcano, Yamagata Prefecture; Otaru, Hokkaido; and other places in Japan. In the USA, on Great Sitkin Island, Aleutian Islands, Alaska; from Grass Valley, Nevada Co., California. On Amitok Island, Labrador, Newfoundland, Canada.

Name: From the Greek for *oblique*, for its triclinicity.

References: (1) Dana, E.S. (1892) Dana's system of mineralogy, (6th edition), 325–327, 337–341. (2) Deer, W.A., R.A. Howie, and J. Zussman (1963) Rock-forming minerals, v. 4, framework silicates, 94–165. (3) Phillips, W.R. and D.T. Griffen (1981) Optical mineralogy, 352–360. (4) Kempster, C.J.E., H.D. Megaw, and E.W. Radoslovich (1962) The structure of anorthite, $CaAl_2Si_2O_8$. I. Structure analysis. Acta Cryst., 15, 1005–117. (5) Megaw, H.D., C.J.E. Kempster, and E.W. Radoslovich (1962) The structure of anorthite, $CaAl_2Si_3O_8$. II. Description and discussion. Acta Cryst., 15, 117–135.

Crystal Data: Triclinic. *Point Group:* $\bar{1}$. Short prismatic crystals; also tabular, rhombic, flattened along [010], to 5 cm. *Twinning:* Baveno, Carlsbad, and Manebach laws; polysynthetic albite and pericline law twinning produce a grid pattern on {100}.

Physical Properties: *Cleavage:* Perfect on {001}, less perfect on {010}; partings on {100}, {110}, {$\bar{1}$10}, and {$\bar{2}$01}. *Fracture:* Uneven. *Tenacity:* Brittle. Hardness = 6 D(meas.) = 2.57–2.60 D(calc.) = 2.57

Optical Properties: Transparent. *Color:* Colorless, also white, pale creamy yellow, red, green. *Streak:* White. *Luster:* Vitreous, may be pearly on cleavages.
Optical Class: Biaxial (–). *Orientation:* $Z \wedge b \simeq 5°$. *Dispersion:* $r > v$, weak.
$\alpha = 1.524–1.526$ $\beta = 1.529–1.532$ $\gamma = 1.530–1.534$ 2V(meas.) = $42°–52°$

Cell Data: *Space Group:* $C\bar{1}$. $a = 8.287$ $b = 12.972$ $c = 7.156$ $\alpha = 91.05°$ $\beta = 116.26°$
$\gamma = 90.15°$ Z = 4

X-ray Powder Pattern: Grande Caldeira, Azores. (ICDD 9-478).
3.211 (100), 3.243 (90), 4.106 (16), 2.162 (16), 6.49 (14), 3.768 (14), 3.726 (14)

Chemistry:

	(1)
SiO_2	62.79
Al_2O_3	22.12
Fe_2O_3	0.36
FeO	0.41
CaO	3.76
Na_2O	7.35
K_2O	2.98
H_2O^+	0.19
H_2O^-	0.07
Total	100.03

(1) Mt. Erebus, Ross Island, Antarctica; corresponds to $(Na_{0.64}Ca_{0.18}K_{0.17})_{\Sigma=0.99}$
$(Al_{0.98}Fe^{2+}_{0.02}Fe^{3+}_{0.01})_{\Sigma=1.01}(Si_{2.81}Al_{0.19})_{\Sigma=3.00}O_8$.

Mineral Group: Feldspar (alkali) group; intermediate between low sanidine and high albite.

Occurrence: In high-temperature sodic volcanic and hypabyssal rocks.

Association: Typically in a fine-grained groundmass or weathered out as loose crystals.

Distribution: Rather abundant worldwide. Some localities for well-characterized material include: on Pantelleria and Ustica Islands, Italy. At Larvik, Norway. From Berkum, North Rhine-Westphalia, Germany. On Grande Caldeira Island, Azores. At Ropp, Nigeria. On Mt. Kenya, Kenya. From Kilimanjaro, Tanzania. At Chilposan, near Minchon, North Korea. From Ogaya, Toyama Prefecture, and Madarajima, Saga Prefecture, Japan. At Kakanui, New Zealand. From Mt. Anakie and Mt. Franklin, Daylesford, Victoria, Australia. Large crystals from Mt. Erebus, Ross Island, Antarctica. At Boron, Kern Co., California, USA.

Name: From the Greek for *oblique* and *fracture*, descriptive of the cleavage.

References: (1) Dana, E.S. (1892) Dana's system of mineralogy, (6th edition), 324–325.
(2) Deer, W.A., R.A. Howie, and J. Zussman (1963) Rock-forming minerals, v. 4, framework silicates, 6–93. (3) Phillips, W.R. and D.T. Griffen (1981) Optical mineralogy, 342–344.
(4) Carmichael, I.S.E. and W.S. MacKenzie (1964) The lattice parameters of high-temperature triclinic sodic feldspars. Mineral. Mag., 33, 949–962. (5) Harlow, G.E. (1982) The anorthoclase structures: the effects of temperature and composition. Amer. Mineral., 67, 975–996.

Crystal Data: Orthorhombic. *Point Group:* $2/m\ 2/m\ 2/m$. Crystals rare, to 25 cm; as bladed aggregates of unterminated prismatic crystals. Commonly lamellar or fibrous, asbestiform.

Physical Properties: *Cleavage:* Perfect on {210}, intersecting at 54.5° and 125.5°; distinct on {010} and {100}. *Tenacity:* Brittle; fibers are elastic. Hardness = 5.5–6 D(meas.) = ~2.9–3.5 D(calc.) = 3.09

Optical Properties: Transparent to translucent. *Color:* Gray, brownish gray, yellowish brown, clove-brown, brownish green, emerald-green; in thin section, colorless to pale green or yellow. *Streak:* White or grayish. *Luster:* Vitreous, pearly on cleavage.
Optical Class: Biaxial (+) or (–). *Pleochroism:* When Fe-rich, moderate; X = clove-brown, yellowish brown, grayish brown; Y = clove-brown, brown-gray, brownish; Z = clove-brown to dark brown, grayish blue to green, lilac. *Orientation:* $X = a$; $Y = b$; $Z = c$. *Dispersion:* $r > v$ or $r < v$, weak to moderate. *Absorption:* $Z > Y = X$ or $Z = Y > X$. $\alpha = 1.603$–1.679 $\beta = 1.617$–1.685 $\gamma = 1.627$–1.690 2V(meas.) = ~80°

Cell Data: *Space Group:* $Pnma$. a = 18.544(2) b = 18.026(2) c = 5.282(1) Z = 4

X-ray Powder Pattern: Georgia, USA.
3.05 (100), 3.24 (60), 8.26 (55), 2.84 (40), 2.54 (40), 3.65 (35), 8.9 (30)

Chemistry:

	(1)		(1)		(1)
SiO_2	58.08	FeO	10.18	Na_2O	0.05
TiO_2	0.04	MnO	0.20	K_2O	0.01
Al_2O_3	0.30	MgO	27.99	H_2O	[2.20]
Fe_2O_3	0.65	CaO	0.17	Total	[99.87]

(1) Ochsenkogel, Gleinalpe, Austria; by electron microprobe, Fe^{2+}:Fe^{3+} by wet chemical analysis, H_2O calculated from stoichiometry; corresponding to $(Mg_{5.71}Fe^{2+}_{1.17}Fe^{3+}_{0.07}Ca_{0.02}Mn_{0.02}Na_{0.01})_{\Sigma=7.00}$ $(Si_{7.95}Al_{0.05})_{\Sigma=8.00}O_{22}(OH)_{2.00}$.

Polymorphism & Series: Forms a series with magnesio-anthophyllite and ferro-anthophyllite.

Mineral Group: Amphibole (Fe–Mn–Mg) group: $0.1 \le Mg/(Mg + Fe^{2+}) \le 0.89$; $(Ca + Na)_B < 1.34$; Li < 1.0; Si ≥ 7.0.

Occurrence: From medium- or high-grade metamorphism, in amphibolites, gneisses, metaquartzites, iron formations, granulites, and schists derived from argillaceous sediments, ultramafic, or mafic igneous rocks; a retrograde reaction product.

Association: Cordierite, talc, chlorite, sillimanite, mica, olivine, "hornblende," gedrite, magnesio-cummingtonite, garnet, staurolite, plagioclase.

Distribution: From Kongsberg and Snarum, Norway. At Schneeberg, Saxony, Germany. From Norberg, Sweden. At Heřmanov, Czech Republic. In Greenland, from Fiskenæsset. In the USA, from Chesterfield, Hampshire Co., Massachusetts; the Carleton talc mine, near Chester, Windsor Co., Vermont; near Media, Delaware Co., Pennsylvania; the Day Book deposit, near Spruce Pine, Mitchell Co., North Carolina; in California, at the Winchester quarry, Riverside Co., and near Coffee Creek, Carrville, Trinity Co.; in the Copper Queen mine, Prairie Divide, Park Co., Colorado. From Munglinup, Western Australia.

Name: From the Latin *anthophyllum*, meaning *clove*, in allusion to the mineral's color.

References: (1) Dana, E.S. (1892) Dana's system of mineralogy, (6th edition), 384–385. (2) Deer, W.A., R.A. Howie, and J. Zussman (1963) Rock-forming minerals, v. 2, chain silicates, 211–229. (3) Rabbitt, J.C. (1948) A new study of the anthophyllite series. Amer. Mineral., 33, 263–323. (4) Beatty, S. (1950) X-ray diffraction patterns of asbestos. Amer. Mineral., 35, 579–589. (5) Walitzi, E.M., F. Walter, and K. Ettinger (1989) Verfeinerung der Kristallstruktur von Anthophyllit vom Ochsenkogel/Gleinalpe, Österreich. Zeits. Krist., 188, 237–244 (in German). (6) Phillips, W.R. and D.T. Griffen (1981) Optical mineralogy, 223–225.

Crystal Data: Monoclinic. *Point Group: m.* Minute crystals, typically platy along [001], rarely elongated along [010]; some plates have rectangular outline due to cleavage; commonly bladed or fibrous. *Twinning:* May exhibit two- or three-fold twins rotated 60° about an axis ⊥ [001].

Physical Properties: *Cleavage:* Perfect on {001}, observed on {100} and {010}. *Fracture:* Conchoidal or splintery. Hardness = 2.5–3.5 D(meas.) = 2.65 D(calc.) = 2.61

Optical Properties: Translucent to opaque. *Color:* Green, blue-green, white; colorless to pale green in thin section. *Streak:* White. *Luster:* Resinous, greasy, silky, waxy, earthy. *Optical Class:* Biaxial (–). *Orientation:* $X = c.$ *Dispersion:* $r > v.$ $\alpha = 1.558–1.567$ $\beta = 1.565$ $\gamma = 1.562–1.574$ 2V(meas.) = 37°–61°

Cell Data: *Space Group: Cm.* $a = 43.53(1)$ $b = 9.259(3)$ $c = 7.263(7)$ $\beta = 91°8.4(1.6)'$ $Z = 16$

X-ray Powder Pattern: Griffen Range, Westland, New Zealand.
7.29 (100), 2.525 (100), 3.61 (80), 2.458 (60), 2.172 (60), 1.694 (60), 6.43 (40)

Chemistry:

	(1)	(2)	(3)
SiO_2	41.65	44.50	43.37
Al_2O_3	0.10	1.41	
Fe_2O_3	2.88		
FeO	0.16	0.35	
MnO	0.05		
MgO	41.06	41.56	43.63
H_2O	14.22	12.36	13.00
Total	100.12	100.18	100.00

(1) Nikka Vord quarries, Unst, Shetland Islands, Scotland. (2) State Line pits, Rock Springs, Cecil Co., Maryland, USA; Cr_2O_3 0.06% and NiO 0.095% additionally. (3) $Mg_3Si_2O_5(OH)_4$.

Polymorphism & Series: Polymorphous with clinochrysotile, lizardite, orthochrysotile, and parachrysotile.

Mineral Group: Kaolinite-serpentine group.

Occurrence: Commonly replaces ultramafic rocks, pervasively or in crosscutting veinlets. As a replacement of siliceous dolostone along contacts with diabase sills.

Association: Chromite, magnetite, chrysotile, olivine.

Distribution: Widespread; less common however than lizardite. Some localities for well-studied material include: in the Val Antigorio, Piedmont, Italy. From Prägraten, Tirol, Austria. At Glen Urquhart, Inverness-shire, Scotland. From Hsiu-Yen Hsien, Liaoning Province, China. In the Nishisonnogi area, Nagasaki Prefecture, and the Sasagure area, Fukuoka Prefecture, Japan. From Woodsreef, New South Wales, Australia. In the USA, at Texas, Lancaster Co., Pennsylvania; around Baltimore, Baltimore Co., Maryland; from Brewster, Putnam Co., New York; and at Buck Creek, Clay Co., North Carolina. At Asbestos, Quebec, and Timmins, Ontario, Canada.

Name: For the occurrence at Val Antigorio, Italy.

References: (1) Dana, E.S. (1892) Dana's system of mineralogy, (6th edition), 669–674. (2) Deer, W.A., R.A. Howie, and J. Zussman (1963) Rock-forming minerals, v. 3, sheet silicates, 171–190. (3) Brindley, G.W. and O. von Knorring (1954) A new variety of antigorite (ortho-antigorite) from Unst, Shetland Islands. Amer. Mineral., 39, 794–804. (4) Page, N.J. and R.G. Coleman (1967) Serpentine-mineral analyses and physical properties. U.S. Geol. Sur. Prof. Paper 575-B, B103–B107. (5) Yada, K. (1979) Microstructures of chrysotile and antigorite by high-resolution electron microscopy. Can. Mineral., 17, 679–691. (6) Wicks, F.J. and D.S. O'Hanley (1988) Serpentine minerals: structures and petrology. In: S.W. Bailey, Ed., Hydrous phyllosilicates. Rev. Mineral. 19, MSA, 91–167.

Crystal Data: Monoclinic, probable. *Point Group:* n.d. As minute, twisted or curved beaded spherules of radial fibers, composed of crystals; in seams, exhibiting a silky, fibrous structure on broken surfaces.

Physical Properties: Hardness = 2   D(meas.) = 2.80(2)   D(calc.) = [3.26]

Optical Properties: Transparent to translucent. *Color:* Faïence-blue; in thin section, rich blue. *Luster:* Nonmetallic, silky.
Optical Class: Biaxial (–). *Orientation:* $Y \simeq$ length. $\alpha = 1.610$ $\beta = 1.650$ $\gamma = 1.650$
2V(meas.) = Small.

Cell Data: *Space Group:* n.d. $a = 12.89$ $b = 6.055$ $c = 19.11$ $\beta = 90.42°$ $Z = [2]$

X-ray Powder Pattern: Christmas, Arizona, USA.
12.89 (100), 3.168 (70), 7.663 (50), 10.62 (40), 9.556 (40), 4.491 (40), 4.174 (40)

Chemistry:

	(1)	(2)
SiO$_2$	40.8	39.66
FeO	0.3	
CuO	43.6	47.26
MgO	1.7	
CaO	1.8	
H$_2$O	13.8	13.08
Total	102.0	100.00

(1) Christmas, Arizona, USA; average of two analyses; after recalculation to 100%, corresponds to (Cu$_{7.87}$Mg$_{0.61}$Ca$_{0.46}$Fe$_{0.06}$)$_{\Sigma=9.00}$Si$_{9.75}$O$_{28.5}$·11H$_2$O. (2) Cu$_9$Si$_{10}$O$_{29}$·11H$_2$O.

Occurrence: A retrograde metamorphic or mesogene mineral, formed at the expense of a prograde calc-silicate and sulfide assemblage in tactites; typically in fractures or crackled zones cutting garnet-diopside rock, replacing both these silicates and calcite.

Association: Kinoite, gilalite, stringhamite, junitoite, clinohedrite, xonotlite, apophyllite, calcite, tobermorite.

Distribution: In the Christmas copper mine, Gila Co., Arizona, USA.

Name: For the Apache Indians who inhabit the region in Arizona, USA, in which it occurs.

Type Material: University of Arizona, Tucson, Arizona, USA; The Natural History Museum, London, England, 1980,532.

References: (1) Cesbron, F.P. and S.A. Williams (1980) Apachite and gilalite, two new copper silicates from Christmas, Arizona. Mineral. Mag., 43, 639–641. (2) (1980) Amer. Mineral., 65, 1065 (abs. ref. 1).

Ardennite — $Mn_4^{2+}(Al, Mg)_6(SiO_4)_2(Si_3O_{10})[(As, V)O_4](OH)_6$

Crystal Data: Orthorhombic. *Point Group:* $2/m\ 2/m\ 2/m$. Individual crystals rare, prismatic, with prism faces strongly striated and pyramidal faces smooth; as radiating fibrous groups, to 4 cm.

Physical Properties: *Cleavage:* Perfect on {010}, distinct on {110}; parting on {001}. *Fracture:* Subconchoidal to uneven. *Tenacity:* Brittle. Hardness = 6–7   D(meas.) = 3.69–3.75   D(calc.) = 3.74

Optical Properties: Nearly opaque, translucent in thin splinters. *Color:* Yellow to yellowish brown; in thin section, yellow to brown. *Luster:* Subadamantine. *Optical Class:* Biaxial (+). *Pleochroism:* Strong; X = dark brownish yellow; Y = golden yellow; Z = pale yellow. *Orientation:* $Z = b$ or c. α = n.d. β = 1.74–1.78; birefringence = 0.15–0.20. γ = n.d.   2V(meas.) = 0°–70°

Cell Data: *Space Group:* $Pnmm$.   a = 8.7126(8)   b = 18.5124(11)   c = 5.8108(8)   Z = 2

X-ray Powder Pattern: Salmchâteau, Belgium.
2.574 (100), 2.911 (70), 4.21 (60), 3.15 (60), 2.871 (60), 1.448 (60), 3.76 (50)

Chemistry:

	(1)	(2)		(1)	(2)
SiO_2	27.85	29.70	MnO	25.70	18.93
TiO_2		0.04	CuO	0.26	0.01
Al_2O_3	23.55	22.99	MgO	2.30	5.49
Fe_2O_3	0.86	2.77	CaO	1.17	4.41
As_2O_5	13.25	10.11	H_2O^+	5.10	5.44
V_2O_5	0.89	0.04	H_2O^-	0.02	
FeO	0.00		Total	100.95	99.93

(1) Salmchâteau, Belgium. (2) Haute-Maurienne, France; by electron microprobe, confirmed by XRF and wet analysis, H_2O by coulometry; corresponds to $(Mn_{2.75}^{2+}Ca_{0.81}Mg_{0.43})_{\Sigma=3.99}$ $(Al_{4.65}Mg_{0.98}Fe_{0.36}^{3+})_{\Sigma=5.99}(As_{0.91}O_4)Si_{5.10}O_{18}(OH)_6$.

Occurrence: In pegmatites and quartz veins in schist (Salmchâteau, Belgium); in highly oxidized Mn, Al-rich metasediments (Andros Island, Greece).

Association: Quartz, albite, pyrolusite, piemontite, spessartine, braunite, hematite.

Distribution: At Salmchâteau, near Ottré, and at Bihain, Belgium. From Ala, Piedmont, Italy. At Bonneval-sur-Arc, Haute-Maurienne, Isère, France. In the Merehead quarry, Shepton Mallet, Somerset, England. On Evvia and Andros Islands, Cyclades Islands, Greece. At the Kajlidongri manganese mine, Jhabua district, Madhya Pradesh, India. Along the Asemi-gawa River, Kochi Prefecture, Japan.

Name: For the initial occurrence at Salmchâteau, in the Ardennes Mountains, Belgium.

References: (1) Dana, E.S. (1892) Dana's system of mineralogy, (6th edition), 542–543. (2) Moore, P.B. (1965) Cell data of orientite and its relation to ardennite and zoisite. Can. Mineral., 8, 262–265. (3) Donnay, G. and R. Allmann (1968) Si_3O_{10} groups in the crystal structure of ardennite. Acta Cryst., 24, 845–855. (4) Allman, R. and G. Donnay (1971) Structural relations between pumpellyite and ardennite. Acta Cryst., 27, 1871–1875. (5) Pasero, M. and T. Reinecke (1991) Crystal chemistry, HRTEM analysis and polytypic behavior of ardennite. Eur. J. Mineral., 3, 819–830.

$NaNa_2[(Fe^{2+}, Mg)_4Fe^{3+}]Si_8O_{22}(OH)_2$ — Arfvedsonite

Crystal Data: Monoclinic. *Point Group:* $2/m$. Commonly as elongated prisms, unterminated, to 0.60 m, tabular on {010}. As prismatic aggregates and radiating fibrous clusters. *Twinning:* Simple or lamellar twinning || {100}.

Physical Properties: *Cleavage:* Perfect on {110}, with intersections of 56° and 124°; parting on {010}. *Fracture:* Uneven. *Tenacity:* Brittle. Hardness = 5–6 D(meas.) = $\sim$3.3–3.5 D(calc.) = [3.33]

Optical Properties: Translucent to opaque. *Color:* Black, deep green on thin edges; strongly colored in thin section. *Streak:* Deep bluish gray, gray-green. *Luster:* Vitreous. *Optical Class:* Biaxial (–). *Pleochroism:* Strong, in blue-greens, yellow-browns, or gray-violets. *Orientation:* $Z = b$; $X \wedge c = 5°$–30°; $Y \wedge a = 20°$–45°. *Dispersion:* $r > v$, very strong. *Absorption:* $X > Y > Z$. $\alpha = \sim$1.67–1.700 $\beta = \sim$1.68–1.710 $\gamma = \sim$1.69–1.715 2V(meas.) = $0°$–90°

Cell Data: *Space Group:* $C2/m$. $a = 10.007(2)$ $b = 18.077(2)$ $c = 5.332(1)$ $\beta = 104.101(7)°$ $Z = 2$

X-ray Powder Pattern: Nunarsuatsiak, Tunugdliarfik, Greenland. (ICDD 14-633). 3.161 (100), 2.732 (80), 8.51 (70), 3.423 (45), 2.604 (35), 2.185 (35), 2.550 (25)

Chemistry:

	(1)	(2)		(1)	(2)
SiO_2	48.99	51.34	CaO	0.93	1.02
TiO_2	0.77	1.34	Na_2O	6.94	8.68
Al_2O_3	1.66	0.45	K_2O	3.67	1.60
Fe_2O_3	7.52	3.70	F	0.21	
FeO	26.56	26.74	H_2O	1.64	[1.89]
MnO	0.94	1.78	$-O = F_2$	[0.09]	
MgO	0.45	1.77	Total	[100.19]	[100.31]

(1) Red Wine complex, Labrador, Newfoundland, Canada; by electron microprobe, averaged with wet chemical analysis, original total given as 100.18%; corresponding to $(K_{0.71}Na_{0.29})_{\Sigma=1.00}$ $(Na_{1.84}Ca_{0.16})_{\Sigma=2.00}(Fe^{2+}_{3.60}Fe^{3+}_{0.92}Al_{0.15}Mn_{0.13}Mg_{0.11}Ti_{0.09})_{\Sigma=5.00}(Si_{7.83}Al_{0.17})_{\Sigma=8.00}O_{22}$
(2) Kangerdlugssuaq Fjord, Greenland; by electron microprobe, H_2O calculated from stoichiometry.

Polymorphism & Series: Forms a series with magnesio-arfvedsonite.

Mineral Group: Amphibole (alkali) group: $Fe^{2+}/(Fe^{2+} + Mg) \geq 0.5$; $Fe^{3+}/(Fe^{3+} + Al^{vi}) \geq 0.5$; $(Na + K)_A \geq 0.5$; $Na_B \geq 1.34$; $Mn_C < 2.5$.

Occurrence: Common in alkalic granites and other alkalic plutonic rocks and pegmatites.

Association: Nepheline, albite, aegirine, riebeckite, katophorite, magnesio-katophorite, quartz.

Distribution: Notable occurrences are: in Greenland, in the Ilímaussaq intrusion, from Kangerdlugssuaq Fjord, at Narssârssuk, and at Ivigtut. Around the Langesundsfjord, Norway. At Buchans, Newfoundland, and Mont Saint-Hilaire, Quebec, Canada. In the USA, from St. Peter's Dome, El Paso Co., Colorado; on Hurricane Mountain, Intervale, Carroll Co., New Hampshire; at Magnet Cove, Hot Springs Co., Arkansas; near Burnsville, Yancey Co., North Carolina. From the Lovozero massif, Kola Peninsula, Russia. At Bitola, Macedonia. On Mbolwe Hill, Mkushi River area, Central Province, Zambia. In Australia, from Mittagong, New South Wales.

Name: To honor the Swedish chemist, Johan A. Arfvedson (1792–1841).

References: (1) Dana, E.S. (1892) Dana's system of mineralogy, (6th edition), 401–403. (2) Deer, W.A., R.A. Howie, and J. Zussman (1963) Rock-forming minerals, v. 2, chain silicates, 364–374. (3) Hawthorne, F.C. (1976) The crystal chemistry of the amphiboles: V. The structure and chemistry of arfvedsonite. Can. Mineral., 14, 346–356. (4) Layne, G.D. and J.C. Rucklidge (1982) Astrophyllite from Kangerdlugssuaq, East Greenland. Mineral. Mag., 45, 149–156. (5) Phillips, W.R. and D.T. Griffen (1981) Optical mineralogy, 243–245.

Crystal Data: Orthorhombic, pseudohexagonal. *Point Group:* $2/m\ 2/m\ 2/m$. Sheaves of prismatic crystals, with individuals to 2.5 cm. *Twinning:* Complex, showing sectored triplets; may be lamellar.

Physical Properties: *Cleavage:* Perfect prismatic, one pinacoidal distinct; partings along sector boundaries. Hardness = 7–8   D(meas.) = 2.74–2.76   D(calc.) = 2.787

Optical Properties: Translucent. *Color:* Colorless, fawn, greyish green. *Luster:* Vitreous. *Optical Class:* Biaxial (–), with uniaxial (–) domains. *Orientation:* $X = c$; $Z = b$. $\alpha = 1.550–1.551$ $\beta = 1.557–1.559$ $\gamma = 1.559–1.562$ 2V(meas.) = 60°–65°   2V(calc.) = 57.8°

Cell Data: *Space Group: Pnna.* $a = 13.874(2)$ $b = 18.660(2)$ $c = 10.698(1)$ $Z = 4$

X-ray Powder Pattern: Kongsberg, Norway. (ICDD 20-112).
3.86 (100), 3.41 (90), 2.91 (90), 6.94 (80), 2.78 (80), 4.25 (65), 9.31 (55)

Chemistry:

	(1)	(2)	(3)
SiO_2	46.18	47.6	47.10
Al_2O_3	27.52	27.0	26.64
CaO	9.99	9.50	9.77
SrO	0.04	0.62	
BaO	12.37	12.05	13.35
Na_2O	0.16	0.36	
K_2O	0.13	0.15	
H_2O^+	3.41		
H_2O^-	0.11		
H_2O		2.90	3.14
CO_2		0.31	
Total	99.91	100.49	100.00

(1) Kongsberg, Norway. (2) Rémigny, Canada; by electron microprobe; average of 38 analyses on six samples; corresponds to $(Ba_{0.89}Na_{0.10}K_{0.04})_{\Sigma=1.03}(Ca_{1.92}Sr_{0.07}Na_{0.03})_{\Sigma=2.02}$ $Al_{3.02}(Al_3Si_9O_{30}) \cdot 2[(H_2O)_{0.91}(CO_2)_{0.04}]$. (3) $BaCa_2Al_6Si_9O_{30} \cdot 2H_2O$.

Mineral Group: Milarite group (?).

Occurrence: In calcite veins (Kongsberg, Norway); in manganese-metasomatized diorite included in later granodiorite (Rémigny, Canada).

Association: Axinite, pyrrhotite, quartz (Kongsberg, Norway); albite, manganiferous zoisite, piemontite, prehnite, epidote, muscovite (Rémigny, Canada); celsian, bytownite (Broken Hill, Australia).

Distribution: From the Armen mine, near Kongsberg, Norway. In the Su Zurfuru mine, Fluminese, Sardinia, Italy. Large crystals from the Wasenalp, near Simplon, Valais, Switzerland. At Coire Loch Kander, 11 km south-southeast of Braemar, Scotland. From Chvaletice, Czech Republic. At Rémigny, Quebec, Canada. From 16 km northeast of North Bend, King Co., Washington, USA. At Broken Hill and Pernamoota, 30 km north of Broken Hill, New South Wales, Australia.

Name: For the Armen mine, Kongsberg, Norway.

Type Material: National School of Mines, Paris, France; The Natural History Museum, London, England, 1947,290.

References: (1) Neumann, H. (1939) Armenite, a new mineral. Preliminary note. Norsk. Geol. Tidsskr., 19, 312–313. (2) (1941) Amer. Mineral., 26, 235 (abs. ref. 1). (3) Puoliot, G., P. Trudel, G. Valiquette, and P. Samson (1984) Armenite-thulite-albite veins at Rémigny, Quebec: the second occurrence of armenite. Can. Mineral., 22, 453–464. (4) Armbruster, T. and M. Czank (1992) H_2O ordering and superstructures in armenite, $BaCa_2Al_6Si_9O_{30} \cdot 2H_2O$: a single-crystal X-ray and TEM study. Amer. Mineral., 77, 422–430.

Crystal Data: Monoclinic. *Point Group:* $2/m$, m, or 2. As crystals, to 2 cm, and aggregates. *Twinning:* Polysynthetic, by rotation about [100].

Physical Properties: *Cleavage:* Perfect on {001}, good on {100}. *Tenacity:* Very brittle. Hardness = 4.6   VHN = 310–330   D(meas.) = 2.562–2.593   D(calc.) = 2.71

Optical Properties: Semitransparent. *Color:* Dark to pale brown. *Luster:* Vitreous. *Optical Class:* Biaxial (–). *Orientation:* $Z = b$; $Y \wedge c = 5°$–$7°$. *Dispersion:* $r < v$. $\alpha = 1.563$ $\beta = 1.569$ $\gamma = 1.573$ 2V(meas.) = n.d.

Cell Data: *Space Group:* $C2/m$, Cm, or $C2$.   a = 14.04   b = 14.16   c = 7.81 $\beta = 109°33'$ Z = 4

X-ray Powder Pattern: Khan-Bogdinskii massif, Mongolia.
4.26 (100), 3.05 (100), 6.60 (90), 3.80 (90), 7.05 (50), 2.995 (50), 1.947 (50)

Chemistry:

	(1)	(2)
SiO$_2$	60.12	61.64
TiO$_2$	0.12	
ZrO$_2$	19.80	21.07
Al$_2$O$_3$	0.60	
(Y, RE)$_2$O$_3$	0.55	
Fe$_2$O$_3$	1.31	
MgO	0.19	
CaO	9.15	9.59
Na$_2$O	0.18	
K$_2$O	0.14	
H$_2$O	7.90	7.70
P$_2$O$_5$	0.20	
Total	100.26	100.00

(1) Khan-Bogdinskii massif, Mongolia. (2) CaZrSi$_6$O$_{15}$·2.5H$_2$O.

Occurrence: In schlieren of alkalic granite pegmatite, at the contact of arfvedsonite granite with xenoliths of felsic volcanic rocks.

Association: Quartz, microcline, albite, aegirine, arfvedsonite, monazite, synchesite, titanite, other titanosilicates.

Distribution: In the Khan-Bogdinskii granitic massif, Gobi, Mongolia. In the Strange Lake complex, southeast of Lac Brisson, Quebec and Labrador, Newfoundland, Canada.

Name: For Neil Alden Armstrong (1930–), American astronaut, first man to walk on the Moon.

Type Material: A.E. Fersman Mineralogical Museum, Academy of Sciences, Moscow, Russia.

References: (1) Vladykin, N.V., V.I. Kovalenko, A.A. Kashaev, A.N. Sapozhnikov, and V.A. Pisarskaya (1973) A new mineral of calcium and zirconium, armstrongite. Doklady Acad. Nauk SSSR, 209, 1185–1188 (in Russian). (2) (1974) Amer. Mineral., 59, 208 (abs. ref. 1). (3) Kashaev, A.A. and A.N. Sapozhnikov (1978) Crystal structure of armstrongite. Kristallografiya (Sov. Phys. Crystal.), 23, 956–961 (in Russian).

Crystal Data: Hexagonal. *Point Group: 3m.* As rhombohedral crystals, up to 5 mm.

Physical Properties: *Cleavage:* Rhombohedral. Hardness = 6.5–7 D(meas.) = 3.70 D(calc.) = 3.71

Optical Properties: Transparent. *Color:* Lemon-yellow or citron-yellow. *Luster:* Vitreous. *Optical Class:* Uniaxial (–); anomalously biaxial. $\omega = 1.86$ $\epsilon = 1.83$ 2V(meas.) = 0°–17°

Cell Data: *Space Group: P3c1.* a = 8.36(2) c = 15.30(3) Z = 2

X-ray Powder Pattern: Cherbadung, Switzerland.
3.23 (100), 1.570 (70), 1.153 (70), 2.41 (60), 1.746 (60), 1.315 (60), 4.04 (50)

Chemistry:

	(1)
SiO_2	12.3
TiO_2	6.3
SnO_2	2.1
Al_2O_3	1.2
As_2O_3	58.7
Tl_2O_3	1.0
BeO	2.3
CaO	15.4
Total	99.3

(1) Cherbadung, Switzerland; purity confirmed by electron microprobe; corresponds approximately to $Ca_{2.67}(Ti_{0.67}Sn_{0.13}Tl_{0.04})_{\Sigma=0.84}As_{6.67}^{3+}Si_{2.00}Al_{0.27}Be_{1.00}O_{20}$.

Occurrence: On cleft faces in orthogneiss (Cherbadung, Switzerland).

Association: Magnetite, hematite, titanite, apatite, anatase, malachite, azurite, tennantite, molybdenite (Cherbadung, Switzerland).

Distribution: On the east flank of Pizzo Cervandone, Alpe Devero, Val d'Aosta, Italy. On the west flank of Cherbadung [Pizzo Cervandone], Binntal, Valais, Switzerland.

Name: For elements in the composition, *As, Be, Ca, Si.*

Type Material: Natural History Museum, Basel, Switzerland, SG750; National Museum of Natural History, Washington, D.C., USA, 143117.

References: (1) Graeser, S. (1966) Asbecasit und cafarsit, zwei neue Mineralien aus dem Binnatal (Kt. Wallis). Schweiz. Mineral. Petrog. Mitt., 46, 367–375 (in German with English abs.). (2) (1967) Amer. Mineral., 52, 1583–1584 (abs. ref. 1). (3) Cannillo, E., G. Giuseppetti, and C. Taadini (1970) The crystal structure of asbecasite. Atti Rend. Accad. Lincei, 46, 457–467. (4) (1970) Amer. Mineral., 55, 1818 (abs. ref. 3). (5) Downs, J.W. and G.V. Gibbs (1981) The role of the BeOSi bond in the structures of beryllosilicate minerals. Amer. Mineral., 66, 819–826.

Crystal Data: Tetragonal. *Point Group:* $4/m$. Prismatic along [001], to 0.4 mm, showing {110}, {100}, {001}, and {301}.

Physical Properties: *Fracture:* Conchoidal. *Tenacity:* Brittle. Hardness = n.d. D(meas.) = > 4.07   D(calc.) = 4.69

Optical Properties: Transparent. *Color:* Blue. *Streak:* Light blue. *Luster:* Vitreous. *Optical Class:* Uniaxial (+). $\omega = 1.786(3)$ $\epsilon = 1.800(4)$

Cell Data: *Space Group:* $I4/m$.   a = 14.1852(8)   c = 6.0759(8)   Z = 2

X-ray Powder Pattern: Anticline prospect, Western Australia.
10.2 (100), 4.495 (100), 3.333 (100), 3.013 (90), 5.644 (70), 2.611 (50), 2.805 (30)

Chemistry:

	(1)
SiO_2	14.07
CuO	18.66
PbO	52.17
Cl	2.28
H_2O	[4.22]
CO_2	[10.31]
$-O = Cl_2$	0.51
Total	[101.20]

(1) Anticline prospect, Western Australia; by electron microprobe, average of four analyses, OH and CO_2 confirmed present by infrared spectroscopy, H_2O and CO_2 calculated from stoichiometry; corresponds to $Pb_{3.99}Cu_{4.01}HSi_{4.00}O_{12.03}(HCO_3)_{4.00}(OH)_{4.00}Cl_{1.10}$.

Occurrence: In a weathered shear zone cutting shales and graywackes, as an alteration of galena and probably chalcopyrite.

Association: Diaboleite, duftite, beudantite, caledonite, plattnerite, cerussite, malachite, brochantite.

Distribution: From the Anticline prospect, 11 km west-southwest of Ashburton Downs homestead, Capricorn Range, Western Australia.

Name: Presumably for its occurrence near the Ashburton Downs pastoral lease and homestead, Western Australia.

Type Material: Canadian Museum of Nature, Ottawa, Canada, 58391; Museum of Victoria, Melbourne, Australia, M40712.

References: (1) Grice, J.D., E.H. Nickel, and R.A. Gault (1991) Ashburtonite, a new bicarbonate-silicate mineral from Ashburton Downs, Western Australia: description and structure determination. Amer. Mineral., 76, 1701–1707.

Ashcroftine-(Y)　　　　$K_5Na_5(Y, Ca)_{12}Si_{28}O_{70}(CO_3)_8(OH)_2 \cdot 8H_2O$

Crystal Data: Tetragonal. *Point Group:* $4/m\ 2/m\ 2/m$. As fibrous prismatic crystals, to 2 cm, striated $\|$ [001]; in divergent sprays; as a fine, crystalline powder.

Physical Properties: *Cleavage:* Perfect on {100}, distinct on {001}. *Tenacity:* Flexible. Hardness = 5　D(meas.) = 2.61　D(calc.) = [2.59]

Optical Properties: Transparent to translucent. *Color:* Pale pink, white, pinkish brown, deep violet-brown. *Streak:* White. *Luster:* Vitreous, may be silky.
Optical Class: Uniaxial (+). $\omega = 1.536{-}1.537$　$\epsilon = 1.545{-}1.549$

Cell Data: *Space Group:* $I4/mmm$.　$a = 24.01(1)$　$c = 17.52(1)$　$Z = 4$

X-ray Powder Pattern: Narssârssuk, Greenland.
17.027 (100), 12.043 (90), 7.615 (60), 3.114 (50), 2.687 (50), 5.379 (35), 6.014 (30)

Chemistry:

	(1)	(2)
SiO_2	38.09	42.69
Y_2O_3	26.61	34.37
MnO	0.79	
MgO	0.87	
CaO	5.72	
Na_2O	3.62	3.93
K_2O	5.65	5.97
H_2O^+	4.8	4.11
H_2O^-	6.40	
CO_2	7.2	8.93
Total	99.75	100.00

(1) Narssârssuk, Greenland; Y_2O_3 originally determined as Al_2O_3. (2) $K_5Na_5Y_{12}Si_{28}O_{70}$ $(CO_3)_8(OH)_2 \cdot 8H_2O$.

Occurrence: In cavities in augite syenite (Narssârssuk, Greenland); in igneous breccia and pegmatites in an intrusive alkalic gabbro-syenite complex (Mont Saint-Hilaire, Canada).

Association: Calcite, elpidite, albite, quartz, aegirine, orthoclase, zinnwaldite, graphite (Narssârssuk, Greenland); aegirine, bastnäsite, lorenzenite, brookite, elpidite, leucosphenite, cordylite, narsarsukite (Mont Saint-Hilaire, Canada).

Distribution: From Narssârssuk, Greenland. At Mont Saint-Hilaire, Quebec, Canada. From Cava del 'Osa, near Rome, Lazio, Italy.

Name: For Frederick Noel Ashcroft (1878–1949), benefactor to the British Museum (Natural History), London, England.

Type Material: The Natural History Museum, London, England, 1924,867; University of Copenhagen, Copenhagen, Denmark; Harvard University, Cambridge, Massachusetts, 110266; National Museum of Natural History, Washington, D.C., USA, 95320, R4333.

References: (1) Hey, M.H. and F.A. Bannister (1933) Studies on the zeolites. Part IV. Ashcroftine (kalithomsonite of S.G. Gordon). Mineral. Mag., 23, 305–308. (2) Moore, P.B., J.M. Bennett, and S.J. Louisnathan (1969) Ashcroftine is not a zeolite! Mineral. Mag., 37, 515–517. (3) Moore, P.B., P.K. Sen Gupta, E.O. Schlemper, and S. Merlino (1987) Ashcroftine, ca. $K_{10}Na_{10}(Y, Ca)_{24}(OH)_4(CO_3)_{16}(Si_{56}O_{140}) \cdot 16H_2O$, a structure with enormous polyanions. Amer. Mineral., 72, 1176–1189. (4) Mandarino, J.A. and V. Anderson (1989) Monteregian Treasures. Cambridge Univ. Press, 30.

44

Crystal Data: Tetragonal. *Point Group:* $4/m\ 2/m\ 2/m$. Platy, to about 0.5 mm.

Physical Properties: *Cleavage:* Perfect on {001}. Hardness = n.d. VHN = 146–186 (50 g load). D(meas.) = n.d. D(calc.) = 8.041

Optical Properties: Transparent. *Color:* Yellow to yellow-green. *Streak:* Light yellow to white. *Luster:* Adamantine.
Optical Class: Uniaxial (–). $\omega = 2.39$ $\epsilon = 2.32$ *Anisotropism:* Weak in oil, gray; not noticeable in air.
R_1–R_2: (400) 20.6–20.7, (420) 19.8–19.8, (440) 19.0–19.15, (460) 18.5–18.65, (480) 18.1–18.2, (500) 17.7–17.9, (520) 17.4–17.55, (540) 17.15–17.3, (560) 16.9–17.0, (580) 16.7–16.8, (600) 16.6–16.7, (620) 16.5–16.5, (640) 16.3–16.4, (660) 16.2–16.3, (680) 16.1–16.2, (700) 16.0–16.1

Cell Data: *Space Group:* $I4/mmm$. a = 3.897(2) c = 22.81(2) Z = 1

X-ray Powder Pattern: Kombat mine, Namibia.
2.957 (100), 2.752 (60), 1.627 (60), 1.602 (50), 3.46 (40), 1.984 (40), 1.747 (40)

Chemistry:

	(1)
SiO$_2$	2.0
PbO	93.7
Cl	4.4
$-$O $=$ Cl$_2$	1.0
Total	99.1

(1) Kombat mine, Namibia; by electron microprobe, average of three analyses, corresponding to Pb$_7$Si$_{0.55}$O$_{7.07}$Cl$_{2.02}$.

Occurrence: In manganese silicate lenses nearby a sulfide ore body.

Association: Hematophanite, barite, jacobsite, hematite, copper, molybdophyllite, chlorite.

Distribution: In the Kombat mine, 49 km south of Tsumeb, Namibia.

Name: For the Asis Farm, location of the Kombat mine, which in the Nama language means *drinking place*.

Type Material: Canadian Geological Survey, Ottawa, Canada, 64565; The Natural History Museum, London, England, 1986,304.

References: (1) Rouse, R.C., D.R. Peacor, P.J. Dunn, A.J. Criddle, C.J. Stanley, and J. Innes (1988) Asisite, a silicon-bearing lead oxychloride from the Kombat mine, South West Africa (Namibia). Amer. Mineral., 73, 643–650.

Astrophyllite $(K, Na)_3(Fe^{2+}, Mn)_7Ti_2Si_8O_{24}(O, OH)_7$

Crystal Data: Triclinic. *Point Group:* $\bar{1}$. Crystals tabular $\perp$ {001}, bladed or acicular, some faces commonly striated, to 10 cm; groups may be stellate.

Physical Properties: *Cleavage:* Perfect on {001}, poor on {100}. *Tenacity:* Brittle. Hardness = 3 D(meas.) = 3.2–3.4 D(calc.) = n.d.

Optical Properties: Opaque to translucent in thin leaves. *Color:* Bronze-yellow to gold-yellow, brown to reddish brown. *Streak:* Golden. *Luster:* Submetallic, pearly, greasy. *Optical Class:* Biaxial (+). *Pleochroism:* Strong; X = deep orange-red; Y = orange-yellow; Z = lemon-yellow. *Orientation:* $X \simeq c$; $Y \simeq b$; $Z \simeq a$. *Dispersion:* $r > v$, strong. *Absorption:* $X > Y > Z$. $\alpha = 1.678$–1.695 $\beta = 1.703$–1.726 $\gamma = 1.733$–1.758 2V(meas.) = 66°–84°

Cell Data: *Space Group:* $A\bar{1}$. $a = 5.36$ $b = 11.76$ $c = 21.08$ $\alpha = 85°8'$ $\beta = 90°0'$ $\gamma = 103°13'$ $Z = 2$

X-ray Powder Pattern: Låven Island, Norway; nearly identical with kupletskite. (ICDD 14-194).
10.6 (100), 3.51 (80), 2.77 (60), 2.64 (60), 2.57 (60), 9.82 (30), 1.76 (30)

Chemistry:

	(1)		(1)
SiO_2	36.64	MnO	5.68
TiO_2	10.76	CaO	0.98
ZrO_2	0.66	Na_2O	2.52
Nb_2O_5	0.77	K_2O	5.13
FeO	31.50	H_2O	[4.92]
		Total	[99.56]

(1) Kangerdlugssuaq Fjord, Greenland; by electron microprobe, H_2O calculated from stoichiometry; corresponds to $(K_{1.40}Na_{1.04}Ca_{0.22})_{\Sigma=2.66}(Fe_{5.62}Mn_{1.03})_{\Sigma=6.65}$ $(Ti_{1.73}Nb_{0.07}Zr_{0.07})_{\Sigma=1.87}Si_{7.82}O_{24}(OH)_{7.00}$.

Polymorphism & Series: Forms a series with kupletskite.

Mineral Group: Astrophyllite group.

Occurrence: In nepheline syenites, alkali granites, and their pegmatites; less commonly in fenites and other metasomatized rocks; in nepheline syenite gneisses and paragneisses.

Association: Albite, aegirine, arfvedsonite, nepheline, natrolite, zircon, biotite, leucophanite, eudialyte, catapleiite, djerfisherite, rasvumite, delhayelite, kupletskite.

Distribution: Some prominent localities include: on Låven Island and near Brevik, Langesundsfjord, Norway. In Greenland, at Narssârssuk, in the Ilímaussaq intrusion, and elsewhere. In Spain, south of La Guia, and at Monte Galiñeiro, Vigo, Pontevedra Province. In the USA, in Colorado, at St. Peter's Dome, El Paso Co., and in Maine, near Biddeford, York Co. From Mont Saint-Hilaire, Quebec, and the Red Wine complex, Labrador, Newfoundland, Canada. On Rouma Isle, Los Islands, Guinea. From Pilansberg, Transvaal, South Africa. In the Khan-Bogdinskii granitic massif, Gobi, Mongolia. From Mts. Yukspor and Eveslogchorr, Khibiny massif; in the Lovozero massif, Kola Peninsula; and other less-well-defined localities in Russia.

Name: From the Greek for *star* and *leaf*, for the stellate and foliated micaceous habit.

Type Material: Mining Academy, Freiberg, Germany, 26244.

References: (1) Dana, E.S. (1892) Dana's system of mineralogy, (6th edition), 719–720. (2) Woodrow, P.J. (1967) The crystal structure of astrophyllite. Acta Cryst., 22, 673–678. (3) Layne, G.D., J.C. Rucklidge, and C.K. Brooks (1982) Astrophyllite from Kangerdlugssuaq, East Greenland. Mineral. Mag., 45, 149–156. (4) Mandarino, J.A. and V. Anderson (1989) Monteregian Treasures. Cambridge Univ. Press, 31–32. (5) Phillips, W.R. and D.T. Griffen (1981) Optical mineralogy, 530.

$(Ca, Sr)Mn^{2+}(Al, Fe^{3+})_4[H(Si, P)O_4](PO_4)_3(OH)_4.$ Attakolite

Crystal Data: Monoclinic. *Point Group: 2/m.* Indistinctly crystalline, massive. *Twinning:* Possible twin lamellae are observed in thin section.

Physical Properties: Hardness = 5   D(meas.) = 3.09–3.23   D(calc.) = [3.35]

Optical Properties: Transparent. *Color:* White to pale red; colorless in thin section. *Optical Class:* Biaxial (+). *Orientation:* $Y = b$; $Z = c$; $X \wedge a = 24°$. *Dispersion:* $r < v$ to $r \gg v$, strong. $\alpha = 1.650$–1.655 $\beta = 1.654$–1.664 $\gamma = 1.661$–1.675 2V(meas.) = $75°$–$84°$ 2V(calc.) = $74°$

Cell Data: *Space Group: C2/m.* $a = 17.188(4)$ $b = 11.477(8)$ $c = 7.322(5)$ $\beta = 113.83(4)°$ Z = 4

X-ray Powder Pattern: Västanå mine, Sweden.
3.09 (10), 3.13 (8), 4.34 (7), 2.97 (6), 6.61 (4), 5.68 (4), 3.51 (4)

Chemistry:

	(1)	(2)		(1)	(2)
SiO_2	9.35	5.9	Na_2O	0.03	
Al_2O_3	26.97	26.8	F	0.10	
Fe_2O_3	0.60	3.9	Cl	0.06	
FeO	1.31		H_2O^+	5.92	[10.5]
MnO	7.10	9.7	H_2O^-	0.20	
PbO	0.03		CO_2	0.57	
MgO	0.29		P_2O_5	32.59	33.6
CaO	11.40	6.6	SO_3	0.13	
SrO	3.30	3.0	$-O = (F, Cl)_2$	0.06	
			Total	99.89	[100.0]

(1) Västanå mine, Sweden; contaminated with calcite 1.30% and svanbergite 0.76%. (2) Do.; by electron microprobe, total Fe as Fe_2O_3 on crystallo-chemical grounds, H_2O by difference; corresponds to $(Ca_{0.81}Sr_{0.20})_{\Sigma=1.01}Mn_{0.95}(Al_{3.65}Fe^{3+}_{0.34})_{\Sigma=3.99}[H(Si_{0.68}P_{0.28})_{\Sigma=0.96}O_4](PO_4)_3(OH)_4$.

Occurrence: In an iron deposit.

Association: Berlinite, lazulite, apatite, svanbergite, pyrophyllite, hematite, calcite, quartz.

Distribution: From the Västanå mine, near Näsum, Kristianstad, Sweden.

Name: From the Greek for *salmon*, for the pale red color.

References: (1) Dana, E.S. (1892) Dana's system of mineralogy, (6th edition), 847. (2) Palache, C., H. Berman, and C. Frondel (1951) Dana's system of mineralogy, (7th edition), v. II, 845. (3) Gabrielson, O. and P. Gejer. (1966) The mineral attakolite. Arkiv Mineral. Geol., 3(30), 537–543. (4) (1966) Amer. Mineral., 51, 534 (abs. ref. 3). (5) Grice, J.D. and P.J. Dunn (1992) Attakolite: new data and crystal-structure determination. Amer. Mineral., 77, 1285–1291.

Crystal Data: Monoclinic. *Point Group:* $2/m$. Stubby prismatic crystals, square or octagonal in section, to 10 cm; elongated $\parallel$ [001], acicular, skeletal, dendritic. Exsolution lamellæ and overgrowths common. *Twinning:* Simple or multiple on {100}, common; also on {001}.

Physical Properties: *Cleavage:* Good on {110}, (110) $\wedge$ (1$\bar{1}$0) $\sim$87°; partings on {100} and {010}. *Fracture:* Uneven to conchoidal. *Tenacity:* Brittle. Hardness = 5.5–6 D(meas.) = 3.19–3.56   D(calc.) = 3.31

Optical Properties: Transparent to opaque. *Color:* Black, brown, greenish, violet-brown; in thin section, colorless to gray; clear cores zoned to dark rims common, also oscillatory and sector zoning. *Streak:* Gray-green. *Luster:* Vitreous, resinous to dull.
Optical Class: Biaxial (+). *Pleochroism:* X = pale green, pale brown, green, greenish yellow; Y = pale brown, pale yellow-green, violet; Z = pale green, grayish green, violet. *Orientation:* Y = b; Z $\wedge$ c = $-35°$ to $-50°$; X $\wedge$ a = $-20°$ to $-35°$. *Dispersion:* $r > v$, weak to moderate. *Absorption:* Weak; $Y > Z > X$. $\alpha = 1.671$–1.735 $\beta = 1.672$–1.741 $\gamma = 1.703$–1.774 2V(meas.) = 25°–61°

Cell Data: *Space Group:* $C2/c$. a = 9.699   b = 8.844   c = 5.272 $\beta = 106.97°$ Z = 4

X-ray Powder Pattern: Locality unknown. (ICDD 24-203).
2.994 (100), 3.324 (75), 2.949 (65), 2.516 (65), 2.566 (55), 2.134 (35), 2.895 (34)

Chemistry:

	(1)	(2)		(1)	(2)
SiO_2	51.83	45.30	MgO	16.00	11.20
TiO_2	0.49	4.33	CaO	19.21	22.24
Al_2O_3	3.07	6.72	Na_2O	0.27	0.78
Fe_2O_3	1.38	3.98	K_2O	0.02	0.18
FeO	7.21	5.15	H_2O^+	0.47	0.15
MnO	0.17	0.19	H_2O^-	0.11	0.05
			Total	100.23	100.27

(1) Stillwater complex, Montana, USA; corresponds to $(Ca_{0.76}Fe^{2+}_{0.22}Na_{0.02})_{\Sigma=1.00}(Mg_{0.88}Al_{0.05}$ $Fe^{3+}_{0.04}Ti_{0.01}Mn_{0.01})_{\Sigma=0.99}(Si_{1.92}Al_{0.08})_{\Sigma=2.00}O_6$. (2) Stöffel, Germany; corresponds to $(Ca_{0.91}Na_{0.07}Fe^{2+}_{0.04})_{\Sigma=1.02}(Mg_{0.66}Fe^{3+}_{0.13}Fe^{2+}_{0.09}Ti_{0.08}Al_{0.04})_{\Sigma=1.00}(Si_{1.69}Al_{0.31})_{\Sigma=2.00}O_6$.

Mineral Group: Pyroxene group.

Occurrence: Essential in mafic igneous rocks, basalt, gabbro; common in ultramafic rocks; in some high-grade metamorphic rocks and metamorphosed iron formations.

Association: Orthoclase, sanidine, labradorite, olivine, leucite, amphiboles, pyroxenes.

Distribution: Widespread; only a few classic localities, much studied or providing fine examples, are listed. From Arendal, Norway. In Italy, from Vesuvius, Campania; around Frascati, Alban Hills, Lazio; on Mt. Monzoni, Val di Fassa, Trentino-Alto Adige; at Traversella, Piedmont; and on Mt. Etna, Sicily. Around the Laacher See, Eifel district, Germany. At Ústí nad Lábem (Aussig), Bílina, and Vlčí Hora, near Černošín, Czech Republic. On the Azores and Cape Verde Islands. In Canada, from Renfrew and Haliburton Cos., Ontario; at Otter Lake, Pontiac Co., Quebec; and many other localities. In the USA, from Franklin and Sterling Hill, Ogdensburg, Sussex Co., New Jersey; and at Diana, Lewis Co., and Fine, St. Lawrence Co., New York. From Tomik, Gilgit district, Pakistan. At Kangan, Andhra Pradesh, India.

Name: From the Greek for *luster*, apparently based on the appearance of its cleavage surface.

References: (1) Dana, E.S. (1892) Dana's system of mineralogy, (6th edition), 352–364, esp. 360–362. (2) Deer, W.A., R.A. Howie, and J. Zussman (1978) Rock-forming minerals, (2nd edition), v. 2A, single-chain silicates, 294–398. (3) Clark, J.R., D.E. Appleman, and J.J. Papike (1969) Crystal-chemical characterization of clinopyroxenes based on eight new structure refinements. MSA Spec. Paper 2, 31–50. (4) Phillips, W.R. and D.T. Griffen (1981) Optical mineralogy, 196–198.

Crystal Data: Triclinic. *Point Group:* $\bar{1}$. Stubby, prismatic, striated crystals, to 3 cm; also platy and as subparallel aggregates.

Physical Properties: *Cleavage:* Perfect on {001}, good on {010} and {100}. *Fracture:* Uneven. *Tenacity:* Brittle. Hardness = 5.5–6 D(meas.) = 3.34–3.37 D(calc.) = 3.26

Optical Properties: Opaque to faintly translucent. *Color:* Dark greenish black. *Luster:* Vitreous. *Optical Class:* Biaxial (–). *Pleochroism:* X = deep green; Y = lilac-brown; Z = pale to deep brown. *Dispersion:* $r > v$, strong. $\alpha = 1.700(3)$ $\beta = 1.710(3)$ $\gamma = 1.725(3)$ 2V(meas.) = 50(2)°

Cell Data: *Space Group:* $P\bar{1}$. $a = 7.50(1)$ $b = 12.18(1)$ $c = 6.68(1)$ $\alpha = 86°3(5)'$ $\beta = 93°59(5)'$ $\gamma = 112°19(5)'$ $Z = 2$

X-ray Powder Pattern: Baveno, Italy.
2.75 (100), 2.87 (80), 3.12 (70), 2.17 (60), 2.95 (60b), 2.47 (50), 6.69 (40)

Chemistry:

	(1)	(2)	(3)
SiO_2	52.80	52.44	52.41
TiO_2	0.29	0.00	
Al_2O_3	2.41	0.23	
Fe_2O_3	13.17	14.43	13.93
FeO	7.68	10.12	12.53
MnO	1.89	1.03	
MgO	0.92	0.65	
CaO	19.19	19.67	19.56
Na_2O	0.48	0.08	
H_2O	0.91	1.27	1.57
Total	99.74	[99.92]	100.00

(1) Arendal, Norway. (2) Holyoke, Hampden Co., Massachusetts, USA; original total given as 99.82%. (3) $Ca_2Fe^{2+}Fe^{3+}Si_5O_{14}(OH)$.

Polymorphism & Series: Forms a series with manganbabingtonite.

Occurrence: In veins cutting granite pegmatite and diorite; in cavities and vugs in mafic volcanic rocks and gneisses; in skarns.

Association: Prehnite, calcite, epidote, albite, orthoclase, garnet, quartz, "hornblende," zeolites.

Distribution: Known from a number of localities, most minor; some providing good crystals are: in Norway, at the Brastad mine, Oyestad, near Arendal. In Germany, at Herbornseelbach, Hesse. From Baveno, Piedmont, Italy. In the Khandivali quarry, near Bombay, Maharashtra, India. At Mitani, Kochi Prefecture, Japan. From Noril'sk, western Siberia, Russia. In the USA, in Massachusetts, exceptional crystals from Lane's quarry, Westfield, Hampden Co.; in the Cheapside and Deerfield quarries, East Deerfield, Franklin Co.; also at Winchester Highlands, Uxbridge, Norfolk Co. At Paterson, Passaic Co., and Mine Hill, Morris Co., New Jersey; from the Goose Creek quarry, Leesburg, Loudoun Co., Virginia; and at Lenoir, Caldwell Co., North Carolina.

Name: For Dr. William Babington (1757–1833), Irish physician and mineralogist.

References: (1) Dana, E.S. (1892) Dana's system of mineralogy, (6th edition), 381–382. (2) Washington, H.S. and H.E. Merwin (1923) On babingtonite. Amer. Mineral., 8, 215–223. (3) Palache, C. and F.A. Gonyer (1932) On babingtonite. Amer. Mineral., 17, 295–299. (4) Kosoi, A.L. (1975) The structure of babingtonite. Kristallografiya (Sov. Phys. Crystal.), 20, 730–739 (in Russian). (5) Traill, R.J. and A.P. Sabina (1960) Catalogue of X-ray diffraction patterns and specimen mounts on file at the Geological Survey of Canada. Geol. Survey of Canada, Paper 60-4, 11.

Bafertisite $Ba(Fe^{2+}, Mn^{2+})_2 TiOSi_2O_7(OH, F)_2$

Crystal Data: Monoclinic. *Point Group: m.* Crystals of simple rhombic habit, elongated along [100] and flattened on {001}. As aggregates of acicular crystals, to 1.5 cm. *Twinning:* Ubiquitous on {001}, polysynthetic.

Physical Properties: *Cleavage:* Perfect on {001}. Hardness = ~5   D(meas.) = 3.96–4.25   D(calc.) = 3.8

Optical Properties: Semitransparent. *Color:* Bright red, yellowish red to light brown, straw-brown to tan.
Optical Class: Biaxial (–). *Pleochroism:* X = yellow-red to reddish brown; Y = yellow; Z = pale yellow to greenish yellow. *Orientation:* Y = b; X ∧ a = 5.5°.   *Absorption:* X > Z.
$\alpha = 1.805–1.808$ $\beta = 1.835$ $\gamma = 1.860–1.862$ 2V(meas.) = 54°–86°

Cell Data: *Space Group: Cm.* a = 10.60   b = 13.64   c = 12.47 $\beta = 119°30'$ Z = 8

X-ray Powder Pattern: Baiyun-Obo deposit, China.
2.65 (100), 2.11 (40), 1.72 (40), 2.52 (30), 2.23 (30), 2.07 (30), 1.75 (30)

Chemistry:

	(1)	(2)		(1)	(2)
SiO_2	23.68	25.18	BaO	29.98	26.59
TiO_2	15.39	14.27	K_2O	0.12	0.33
Al_2O_3	0.29	1.00	Na_2O	0.49	0.38
Fe_2O_3	1.08	3.67	F		3.50
Nb_2O_5	0.84	0.32	Cl	0.63	
FeO	22.56	10.82	H_2O^+	1.65	2.16
MnO	1.62	12.77	H_2O^-	1.14	
MgO	0.50		$-O = (F, Cl)_2$	0.14	1.47
CaO	0.37	0.30	Total	100.20	99.82

(1) Baiyun-Obo deposit, China. (2) Burpala massif, Russia; corresponding to $Ba(Mn_{0.90}Fe^{2+}_{0.76} Fe^{3+}_{0.33})_{\Sigma=1.99} TiOSi_2O_7(OH, F)_2$.

Polymorphism & Series: Forms a series with hejtmanite.

Occurrence: In hydrothermal veins (Baiyun-Obo deposit, China); in a dike of microcline granite-aplite (Burpala massif, Russia).

Association: Aegirine, fluorite, barite, bastnäsite (Baiyun-Obo deposit, China).

Distribution: In China, from the Baiyun-Obo mine, near Pao-t'ou, Inner Mongolia. From the Burpala massif, about 120 km north of Lake Baikal, eastern Siberia, Russia. At an undefined locality in Kazakhstan. In the Fountain quarry, near Fountain, Pitt Co., North Carolina, USA.

Name: For elements in the composition, BArium, iron, FERrum, TItanium, and SIlicon.

Type Material: n.d.

References: (1) Peng Ch'i-Jui (1959) The discovery of several new minerals of rare elements. Ti Chih K'o Hsueh, 10, 289 (in Chinese). (2) (1960) Amer. Mineral., 45, 754 (abs. ref. 1) (3) Semenov, E.I. and P'ei-Shan Chang (1959) New mineral bafertisite. Sci. Record (Peking), 3, 652–655 (in Russian). (4) (1960) Amer. Mineral., 45, 1317 (abs. ref. 3). (5) Kuan Ya-Hsien, V.I. Simonov, and N.V. Belov (1963) Crystal structure of bafertisite $BaFe_2TiO[Si_2O_7](OH)_2$. Doklady Acad. Nauk SSSR, 149, 1416–1419 (in Russian). (6) (1963) Chem. Abs., 59, 3386 (abs. ref. 5). (7) Yakovlevskaya, T.A. and D.A. Mineev (1965) Crystals and the optical orientation of bafertisite. Trudy Mineral. Muzeya Akad. Nauk SSSR, 16, 293–294 (in Russian). (8) (1965) Chem. Abs., 63, 9158 (abs. ref. 7). (9) Ganzeev, A.A., A.F. Efimov, and G.V. Lyubomilova (1971) Manganiferous bafertisite from the Burpala massif (northern Baikal) Trudy Mineral. Muzeya Akad. Nauk SSSR, 20, 195-197 (in Russian). (10) (1972) Amer. Mineral., 57, 1005 (abs. ref. 9).

Crystal Data: Monoclinic. *Point Group:* 2/m. Crystals stumpy prismatic, to 250 μm. *Twinning:* Contact twins, individuals having [010] in common.

Physical Properties: *Fracture:* Conchoidal. Hardness = ~6 VHN = 725–783, 750 average (50 g load). D(meas.) = n.d. D(calc.) = 3.48 Cathodoluminescence, dull gray with greenish tint.

Optical Properties: Transparent. *Color:* Colorless. *Luster:* Vitreous. *Optical Class:* Biaxial (+). *Orientation:* $X = c$; $Y = b$; $Z = a$. $\alpha = 1.652$ $\beta = 1.658$ $\gamma = 1.670$ 2V(meas.) = ~72°

Cell Data: *Space Group:* $P2_1/a$. $a = 10.42$ $b = 10.16$ $c = 7.36$ $\beta = 91.1°$ $Z = 4$

X-ray Powder Pattern: Dupezeh Mountain, Iraq.
2.84 (100), 2.98 (85), 3.23 (80), 3.04 (75), 2.88 (70), 7.30 (45), 1.702 (40)

Chemistry:

	(1)
SiO$_2$	29.26
TiO$_2$	2.11
ZrO$_2$	27.00
Al$_2$O$_3$	0.03
Fe$_2$O$_3$	0.11
MgO	0.05
CaO	41.44
Na$_2$O	0.02
Total	100.02

(1) Dupezeh Mountain, Iraq; by electron microprobe, average of 26 analyses of four crystals; estimated to contain HfO$_2$ ~0.16%; corresponds to Ca$_{3.00}$(Zr$_{0.89}$Ti$_{0.11}$)$_{\Sigma=1.00}$ (Si$_{1.98}$Fe$_{0.01}$)$_{\Sigma=1.99}$O$_9$.

Occurrence: In a melilite skarn in contact with banded diorite, in roof pendant xenoliths of calc-silicate marbles and hornfels.

Association: Åkermanite, perovskite, schorlomite, monticellite, wollastonite, foshagite, calcite, phlogopite, spinel, cuspidine, baddeleyite, pyrrhotite, djerfisherite, valleriite (melilite skarn); andesine, kaersutite, titanian augite, ilmenite, titanian magnetite, xonotlite (banded diorite).

Distribution: On Dupezeh Mountain, near Hero Town, Qala-Diza region, northeastern Iraq.

Name: For Baghdad, the capital of Iraq.

Type Material: n.d.

References: (1) Al-Hermezi, H.M., D.M. McKie, and A.J. Hall (1986) Baghdadite, a new calcium zirconium silicate mineral from Iraq. Mineral. Mag., 50, 119–123. (2) (1987) Amer. Mineral., 72, 222 (abs. ref. 1).

Baileychlore $(Zn, Fe^{2+}, Al, Mg)_6(Si, Al)_4O_{10}(OH)_8$

Crystal Data: Triclinic. *Point Group:* $\bar{1}$ or 1. As fine-grained transverse fibers of very small size.

Physical Properties: *Cleavage:* Perfect micaceous. Hardness = < 5 D(meas.) = 3.18(2) D(calc.) = 3.195

Optical Properties: Transparent. *Color:* Green, zoned.
Optical Class: Biaxial. *Pleochroism:* Weak; green to yellow-green. $\alpha = 1.582$ β = n.d.
$\gamma = 1.614$ 2V(meas.) = n.d.

Cell Data: *Space Group:* $C\bar{1}$ or $C1$ (uncertainty due to random stacking of chlorite layers).
$a = 5.346(3)$ $b = 9.257(4)$ $c = 14.40(7)$ $\beta = 97.12(5)°$ Z = [2]

X-ray Powder Pattern: Red Dome deposit, Australia.
7.14 (100), 14.3 (90), 1.542 (60), 2.660 (50), 3.573 (40), 2.450 (35b), 4.600 (30)

Chemistry:

	(1)
SiO_2	32.0
Al_2O_3	12.4
FeO	12.9
MnO	0.15
ZnO	30.5
MgO	4.6
CaO	1.0
H_2O	n.d.

(1) Red Dome deposit, Australia; by electron microprobe; excluding CaO, corresponds to
$(Zn_{2.50}Fe_{1.20}Al_{1.17}Mg_{0.76}Mn_{0.01})_{\Sigma=5.64}(Si_{3.55}Al_{0.45})_{\Sigma=4.00}O_{10}(OH)_8$.

Mineral Group: Chlorite group.

Occurrence: As rims on colloform calcite veins, within a strongly oxidized collapse karst-breccia containing skarn clasts.

Association: Andesine, garnet, vesuvianite, zincian chamosite, goethite, hematite, chalcocite, copper, malachite, calcite.

Distribution: From the Red Dome deposit, 15 km west-northwest of Chillagoe, Queensland, Australia.

Name: For Professor Sturges W. Bailey, Department of Geology and Geophysics, University of Wisconsin, Madison, Wisconsin, USA.

Type Material: Geological Museum, University of Wisconsin, Madison, Wisconsin, 6000/1; National Museum of Natural History, Washington, D.C., USA, 164430; South Australian Museum, Adelaide, Australia, 13592.

References: (1) Rule, A.C. and F. Radke (1988) Baileychlore, the Zn end member of the trioctahedral chlorite series. Amer. Mineral., 73, 135–139.

Crystal Data: Monoclinic. *Point Group:* $2/m$.   Rarely as single crystals, to 0.2 mm, stout, rhombic prisms having oblique terminations or as thin diamond-shaped tablets. As botryoidal and drusy crusts and nodules; most commonly as dense fine-grained aggregates or massive.

Physical Properties: Hardness = 4.5   D(meas.) = 2.88   D(calc.) = 2.94

Optical Properties: Translucent to opaque from clay inclusions. *Color:* Colorless to white. *Luster:* Vitreous to porcelaneous.
Optical Class: Biaxial (–). *Orientation:* $Y = b$; $Z \wedge c = 44°$. $\alpha = 1.624$ $\beta = 1.635$ $\gamma = 1.654$
2V(meas.) = 87°–88°

Cell Data: *Space Group:* $P2_1/c$.   a = 4.83   b = 7.60   c = 9.60 $\beta = 90°2'$ Z = [1]

X-ray Powder Pattern: Sterling Borax mine, California, USA.
3.11 (100), 2.85 (60), 2.236 (60), 3.74 (50), 2.51 (50), 2.18 (50), 2.99 (40)

Chemistry:

	(1)	(2)
SiO_2	27.90	28.90
B_2O_3	27.58	27.91
CaO	35.14	35.97
H_2O^+	8.60	7.22
H_2O^-	0.30	
rem.	0.65	
Total	100.17	100.00

(1) Corkscrew Canyon, Death Valley, California, USA; remainder is oxides of trivalent elements.
(2) $Ca_4B_4(BO_4)(SiO_4)_3(OH)_3 \cdot H_2O$.

Mineral Group: Gadolinite group.

Occurrence: As irregular veins in altered volcanic rock (Baker Canyon, Death Valley, California, USA).

Association: Natrolite, thomsonite (Baker Canyon, Death Valley, California, USA); danburite, calcite, stilbite, datolite, quartz (Charcas, Mexico).

Distribution: In the USA, in California, the mineral probably never occurred at the type locality given as "16 miles northeast of Daggett" which is Borate, 14 km east of Yermo, San Bernardino Co.; it more likely was from later authenticated localities in the Black Mountains, west of Furnace Creek, Death Valley, Inyo Co.; also in the Sterling Borax mine, Tick Canyon, Los Angeles Co. Found near Tory Hill, Bancroft, Ontario, Canada. At Charcas, San Luis Potosí, Mexico. From Sivas, Turkey. In Italy, from near Sestri Levante, Liguria.

Name: For Richard C. Baker, of Nutfield, Surrey, England, Mining Director of the Pacific Coast Borax Co., who discovered the mineral.

Type Material: The Natural History Museum, London, England, 86511; Harvard University, Cambridge, Massachusetts, 92735; National Museum of Natural History, Washington, D.C., USA, 94820.

References: (1) Palache, C., H. Berman, and C. Frondel (1951) Dana's system of mineralogy, (7th edition), v. II, 363. (2) Kramer, H. and R.D. Allen (1956) A restudy of bakerite, priceite, and veatchite. Amer. Mineral., 41, 689–700. (3) Murdoch, J. (1962) Bakerite crystals. Amer. Mineral., 47, 919–923. (4) Pemberton, H.E. (1971) Type locality for bakerite. Amer. Mineral., 56, 1109–1110.

Balangeroite

$(Mg, Fe^{2+}, Fe^{3+}, Mn^{2+})_{42}Si_{16}O_{54}(OH)_{40}$

Crystal Data: Monoclinic. *Point Group:* $2/m$. Fibrous, both loose and compact; asbestiform, elongated along [001], to several cm.

Physical Properties: *Cleavage:* One or more on {$hk0$} are very good. *Tenacity:* Brittle. Hardness = n.d. D(meas.) = 2.96–3.10 D(calc.) = 3.098

Optical Properties: Transparent only in thin section. *Color:* Brown. *Luster:* Vitreous, greasy.
Optical Class: Biaxial (–). *Pleochroism:* Distinct; dark brown ∥ [001]; yellow-brown ⊥ [001]. $\alpha = 1.680(5)$ ⊥ [001]. β = n.d. $\gamma = \sim 1.680$ ∥ [001]. 2V(meas.) = n.d.

Cell Data: *Space Group:* $P2/n$. a = 19.40 b = 9.65 c = 19.40 $\beta = 91.1°$ Z = [2]

X-ray Powder Pattern: Balangero mine, Italy.
2.714 (100), 6.77 (80), 2.674 (75), 3.378 (45), 9.59 (40), 3.278 (40), 2.516 (40)

Chemistry:

	(1)
SiO_2	28.37
TiO_2	0.03
Al_2O_3	0.27
Fe_2O_3	8.89
Cr_2O_3	0.03
FeO	16.95
MnO	3.59
MgO	31.81
CaO	0.13
H_2O	[9.93]
Total	[100.00]

(1) Balangero mine, Italy; by electron microprobe, average of 16 analyses; $Fe^{2+}:Fe^{3+} = 2.12$ was deduced from wet chemical analysis, H_2O by difference; corresponds to $(Mg_{26.74}Fe^{2+}_{7.99}Fe^{3+}_{3.77}Mn^{2+}_{1.71}$ $Al_{0.18}Ca_{0.08}Cr_{0.01}Ti_{0.01})_{\Sigma=40.49}Si_{16}O_{55.81}(OH)_{37.35}$.

Occurrence: In schistose serpentinite in proximity to a large ultramafic massif.

Association: Chrysotile, magnetite, Fe–Ni alloy, olivine, chlorite, titanian clinohumite, diopside, antigorite, calcite, aragonite, "opal," "chalcedony," clay minerals.

Distribution: In the Balangero (San Vittore) mine and at Ponte del Diavolo, Lanzo Valley, and from Santa Maria della Neve, Fiano, Piedmont, Italy.

Name: For the Balangero mine, Piedmont, Italy.

Type Material: Mineralogy Institute, Turin University, Turin, Italy, 14873.

References: (1) Compagnoni, R., G. Ferraris, and L. Fiori (1983) Balangeroite, a new fibrous silicate related to gageite from Balangero, Italy. Amer. Mineral., 68, 214–219. (2) Ferraris, G., M. Mellini, and S. Merlino (1987) Electron-diffraction and electron-microscopy study of balangeroite and gageite: crystal structures, polytypism, and fiber texture. Amer. Mineral., 72, 382–391. (3) Belluso, E. and G. Ferraris (1991) New data on balangeroite and carlosturanite from alpine serpentinites. Eur. J. Mineral., 3, 559–566.

Crystal Data: Orthorhombic. *Point Group:* $2/m\ 2/m\ 2/m$. Crystals needlelike and fibrous, elongated ‖ [001], up to 1 mm; as radiating or parallel sheaflike aggregates.

Physical Properties: *Cleavage:* Perfect on {010}, distinct on {100} and {110}. Hardness = n.d. D(meas.) = 3.32–3.35 D(calc.) = [3.32]

Optical Properties: Transparent to translucent. *Color:* Pale yellowish white; in thin section, colorless. *Luster:* Silky.
Optical Class: Biaxial (–). *Orientation:* $X = b$; $Y = c$; $Z = a$. $\alpha = 1.5807–1.5810$
$\beta = 1.5954–1.5958$ $\gamma = 1.5984–1.6008$ 2V(meas.) = 68°–72°

Cell Data: *Space Group:* $Ccca$. a = 13.60 b = 20.24 c = 5.16 Z = [4]

X-ray Powder Pattern: Hsianghualing area, China.
10.12 (100), 3.39 (91), 4.05 (78), 2.605 (31), 2.390 (28)

Chemistry:

	(1)		(1)
SiO$_2$	33.44	CaO	0.28
TiO$_2$	0.11	BaO	19.07
Al$_2$O$_3$	23.43	Li$_2$O	2.00
Fe$_2$O$_3$	0.32	Na$_2$O	0.26
FeO	0.58	K$_2$O	0.47
MnO	0.05	H$_2$O$^+$	10.30
BeO	0.01	P$_2$O$_5$	0.03
MgO	9.68	Total	100.03

(1) Hsianghualing area, China; corresponding to $(Ba_{0.88}K_{0.07}Na_{0.06}Ca_{0.03})_{\Sigma=1.04}$ $(Mg_{1.69}Al_{0.17}Fe^{2+}_{0.06}Fe^{3+}_{0.03})_{\Sigma=1.95}Li_{0.95}Al_{3.00}(Si_{3.93}Al_{0.07})_{\Sigma=4.00}O_{11.96}(OH)_{8.07}$.

Occurrence: In a miarolitic cavity.

Association: Zinnwaldite, quartz.

Distribution: From the Hsianghualing area, Linwu, Hunan Province, China.

Name: For BArium and LIthium in the composition, and the Chinese for *fibrous*.

Type Material: n.d.

References: (1) X-ray Laboratory, Wuhan Geologic College, Geology Team 654, Hunan Geology Bureau, and Geology Laboratory, Hunan Geology Bureau (1975) A new lithium-bearing mineral in China: balipholite, BaMg$_2$LiAl$_3$[Si$_2$O$_6$]$_2$(OH)$_8$. Scientia Geologia Sinica, 1, 100 (in Chinese with English title). (2) (1976) Amer. Mineral., 61, 338 (abs. ref. 1). (3) (1975) Mineral. Abs., 26, 325 (abs. ref. 1). (4) Wuhan Geologic College; Hunan Geologic Bureau (1977) Balipholite BaMg$_2$LiAl$_3$[Si$_2$O$_6$]$_2$(OH, F)$_8$ and its crystal structure. Ti Chih K'o Hsueh, 65–82 (in Chinese). (5) (1977) Chem. Abs., 87, 120592 (abs. ref. 4).

Crystal Data: Orthorhombic. *Point Group:* $2/m \, 2/m \, 2/m$. Very rarely showing traces of crystal faces; coarsely crystalline to compact, massive.

Physical Properties: *Cleavage:* Good on {110} and {001}. Hardness = 6.5
D(meas.) = 3.065   D(calc.) = 3.073

Optical Properties: Translucent to transparent in thin flakes. *Color:* White.
Luster: Vitreous, pearly on cleavage.
Optical Class: Biaxial (+). *Orientation:* $X = c$; $Y = a$; $Z = b$. $\alpha = 1.5695$ $\beta = 1.5710$
$\gamma = 1.5775$ 2V(meas.) = 41(3)°

Cell Data: *Space Group: Ibam.* $a = 8.496(2)$ $b = 9.983(2)$ $c = 16.755(3)$ Z = 4

X-ray Powder Pattern: Benallt mine, Wales; close to stronalsite.
3.53 (100), 5.20 (90), 8.50 (80), 3.21 (80), 2.90 (80), 2.09 (80), 3.77 (70)

Chemistry:

	(1)	(2)
SiO_2	34.74	36.44
Al_2O_3	31.20	30.92
MnO	0.03	
MgO	1.00	
CaO	0.81	
BaO	21.99	23.25
Na_2O	8.43	9.40
K_2O	0.66	
H_2O	1.08	
Total	99.94	100.00

(1) Benallt mine, Wales. (2) $BaNa_2Al_4Si_4O_{16}$.

Mineral Group: Feldspar group.

Occurrence: In veinlets through manganese ore and in lenses in metamorphosed mudstone (Benallt mine, Wales).

Association: Tephroite, alleghanyite, jacobsite, barite, calcite (Benallt mine, Wales).

Distribution: In the Benallt mine, Rhiw, Lleyn Peninsula, Wales. At Långban, Värmland, Sweden. From the Kalahari manganese field, Cape Province, South Africa.

Name: Derived from the chemical symbols for the major constituents, *Ba, Na, Al, Si.*

Type Material: National School of Mines, Paris, France; The Natural History Museum, London, England, 1944,420–423; Harvard University, Cambridge, Massachusetts, 108656; National Museum of Natural History, Washington, D.C., USA, 105854.

References: (1) Campbell Smith, W., F.A. Bannister, and M.H. Hey (1944) A new barium-feldspar from Wales. Nature, 154, 336–337. (2) (1945) Amer. Mineral., 30, 85 (abs. ref. 1). (3) Campbell Smith, W., F.A. Bannister, and M.H. Hey (1944) Banalsite, a new barium-feldspar from Wales. Mineral. Mag., 27, 33–47. (4) Haga, N. (1973) The crystal structure of banalsite, $BaNa_2Al_4Si_4O_{16}$, and its relation to the feldspar structure. Mineral. J. (Japan), 7, 262–281. (5) Welin, E. (1968) X-ray powder data for minerals from Långban and the related mineral deposits of Central Sweden. Arkiv Mineral. Geol., 4, 499–541.

KCa(Mn^{2+}, Fe^{2+}, Zn, Mg)$_{20}$
(Si, Al)$_{32}$O$_{76}$(OH)$_{16}$·4−12H$_2$O
Bannisterite

Crystal Data: Monoclinic. *Point Group:* 2/m. As prismatic bladed crystals and aggregates, to 20 cm; as anhedral plates bounded by cleavage surfaces.

Physical Properties: *Cleavage:* Perfect on {001}; a second prismatic cleavage is noted microscopically. Hardness = 4 D(meas.) = 2.83–2.84 D(calc.) = 2.84

Optical Properties: Translucent. *Color:* Light to dark brown, black. *Luster:* Resinous. *Optical Class:* Biaxial (−). *Pleochroism:* X = nearly colorless; Y = Z = pale yellow to brown. *Orientation:* Y = b; Z ≃ a. *Dispersion:* r < v, weak to moderate. α = 1.544–1.574 β = 1.586–1.611 γ = 1.589–1.612 2V(meas.) = Small to medium.

Cell Data: *Space Group:* A2/a. a = 22.20–22.32 b = 16.32–16.40 c = 24.69–24.70 β = 94°17′–94°21′ Z = 4

X-ray Powder Pattern: Franklin, New Jersey, USA.
12.33 (100), 3.436 (20), 4.103 (15b), 2.638 (15), 3.077 (12), 2.606 (11), 4.593 (10)

Chemistry:

	(1)	(2)
SiO$_2$	45.4	46.3
Al$_2$O$_3$	3.84	4.2
Fe$_2$O$_3$	2.1	1.44
FeO	16.5	5.67
MnO	19.9	23.4
ZnO	0.0	4.6
MgO	0.47	3.1
CaO	1.55	1.3
Na$_2$O	0.51	0.09
K$_2$O	0.92	1.04
H$_2$O	7.82	9.3
Total	99.0	100.4

(1) Broken Hill, Australia; determinations by a combination of electron microprobe, AA, and flame photometry. (2) Franklin, New Jersey, USA; by electron microprobe; Fe^{2+} and Fe^{3+} by wet methods, K and Na by flame photometry, H$_2$O by DTA-TGA; corresponds to (K$_{0.82}$Na$_{0.11}$)$_{\Sigma=0.93}$ Ca$_{0.86}$(Mn$_{12.21}$Fe$^{2+}_{2.92}$Mg$_{2.85}$Zn$_{2.09}$Fe$^{3+}_{0.66}$)$_{\Sigma=20.73}$(Si$_{28.53}$Al$_{3.05}$)$_{\Sigma=31.58}$O$_{76}$(OH)$_{16}$·12.2H$_2$O.

Occurrence: In metamorphosed Mn-Zn orebodies.

Association: Manganoan and zincian amphiboles, rhodonite, sphalerite, quartz, calcite, barite (Franklin, New Jersey, USA); rhodonite, sphalerite, fluorite, galena, apophyllite, quartz (Broken Hill, Australia).

Distribution: In the Benallt mine, Rhiw, Lleyn Peninsula, Wales. From Franklin, Sussex Co., New Jersey, USA. At Nyberget, Sweden. In the Ananai mine, Kochi Prefecture, Japan. From Broken Hill, New South Wales, Australia.

Name: For Dr. Frederick Allen Bannister (1901–), formerly Keeper of Minerals, British Museum (Natural History), London, England.

Type Material: The Natural History Museum, London, England, 1967,321; Harvard University, Cambridge, Massachusetts, 91862, 91863, 108571; National Museum of Natural History, Washington, D.C., USA, 145728–145730.

References: (1) Smith, M.L. and C. Frondel (1968) The related layered minerals ganophyllite, bannisterite, and stilpnomelane. Mineral. Mag., 36, 893–913. (2) (1969) Amer. Mineral., 54, 577 (abs. ref. 1). (3) Plimer, I.R. (1977) Bannisterite from Broken Hill, Australia. Neues Jahrb. Mineral., Monatsh., 504–508. (4) Dunn, P.J., P.B. Leavens, J.A. Norberg, and R.A. Ramik (1981) Bannisterite: new chemical data and empirical formulae. Amer. Mineral., 66, 1063–1067. (5) Heaney, P.J., J.E. Post, and H.T. Evans, Jr. (1992) The crystal structure of bannisterite. Clays and Clay Minerals, 40, 129–144.

Crystal Data: Tetragonal. *Point Group:* $4/m$ As crystals, to 10 cm; as subhedral crystals and grains.

Physical Properties: *Cleavage:* Fair on $\{110\}$. *Fracture:* Hackly. *Tenacity:* Brittle.
Hardness = 6 VHN = 769 D(meas.) = 4.42–4.71 D(calc.) = 4.69

Optical Properties: Translucent to transparent. *Color:* Light brown to black. *Luster:* Nearly semimetallic to vitreous.
Optical Class: Uniaxial (+). *Pleochroism:* Strong; O = colorless; E = dark brownish to light greenish yellow. *Absorption:* $E > O$. $\omega = 1.94$ $\epsilon = 2.16$

Cell Data: *Space Group:* $I4_1/a$. a = 19.2–20.022 c = 5.908–6.006 Z = 16

X-ray Powder Pattern: Ravalli Co., Montana, USA.
1.335 (100), 1.422 (25), 3.139 (10), 1.763 (9.5), 3.516 (9), 2.234 (9), 1.351 (9)

Chemistry:

	(1)	(2)		(1)	(2)
SiO_2	14.17	15.74	CaO	0.41	0.22
TiO_2	29.33	41.48	BaO	37.55	40.21
Al_2O_3	1.62	0.10	Na_2O	0.20	
Fe_2O_3	3.07	0.22	K_2O	0.13	
Cr_2O_3	0.05		Cl	2.01	1.69
Nb_2O_5	11.50	0.07	H_2O	0.52	
MgO	0.20		$-O = Cl_2$	0.45	
			Total	100.31	99.73

(1) Baiyun-Obo mine, China; corresponds to $(Ba_{3.82}Ca_{0.11})_{\Sigma=3.93}(Ti_{5.84}Nb_{1.35}Fe_{0.60}Al_{0.18})_{\Sigma=7.97}$ $(Si_{3.68}Al_{0.32})_{\Sigma=4.00}O_{28}Cl_{0.88}$. (2) Šebkovice, Czech Republic; by electron microprobe, corresponding to $(Ba_{4.00}Ca_{0.06})_{\Sigma=4.06}(Ti_{7.91}Fe_{0.04}Al_{0.02}Nb_{0.01})_{\Sigma=7.98}(Si_{3.99}Al_{0.01})_{\Sigma=4.00}O_{28}Cl_{0.73}$.

Occurrence: In quartz veins cutting quartzite near alkalic granite and syenite (Baiyun-Obo mine, China); in a carbonate vein cutting "hornblende" gneiss intruded by diabase sills and small pegmatites (Ravalli Co., Montana, USA).

Association: Quartz, calcite, aegirine, galena, pyrite, alkalic amphiboles (Baiyun-Obo mine, China); calcite, dolomite, ancylite, aeschynite, monazite, sodic amphiboles, barite (Ravalli Co., Montana, USA); benitoite, bario-orthojoaquinite, fresnoite, natrolite (Gem mine, California, USA).

Distribution: In China, from the Baiyun-Obo mine, near Pao-t'ou, Inner Mongolia. In the USA, on Sheep Creek, Ravalli Co., Montana, and at the Gem mine, San Benito Co., California. From Šebkovice, Czech Republic. At Karlstein and Jarolden, Waldviertel region, Austria. In the Garaoulère mine, Pierrefitte, Hautes-Pyrénées, France.

Name: For the locality, Pao-t'ou (Baotou in Russian), China.

Type Material: National School of Mines, Paris, France.

References: (1) Peng Ch'i-Jui (1959) The discovery of several new minerals of rare elements. Ti Chih K'o Hsueh, 10, 289 (in Chinese). (2) (1960) Amer. Mineral., 45, 754 (abs. ref. 1). (3) Semonov, V.I. (1960) Baotite, a mineral with a metasilicate ring (Si_4O_{12}). Kristallografiya (Sov. Phys. Crystal.), 5, 544–546 (in Russian). (4) Heinrich, E.W., W.H. Boyer, and F.A. Crowley (1962) Baotite (pao-t'ou-k'uang) from Ravalli County, Montana. Amer. Mineral., 47, 987–993. (5) Nekrasov, Y.K., V.I. Ponomarev, V.I. Simonov, and D.M. Kheiker (1969) Refinement of the atomic structure of baotite and the isomorphic relationships in this mineral. Kristallografiya (Sov. Phys. Crystal.), 14, 602–609 (in Russian). (6) Vlasov, K.A., Ed. (1966) Mineralogy of rare elements, v. II, 566–567. (7) Němec, D. (1987) Baotite – a rock-forming mineral of Ba-rich hyperpotassic dyke rocks. Neues Jahrb. Mineral., Monatsh., 31–42.

Crystal Data: Monoclinic. *Point Group:* $2/m$ or m. As platy deposits, to 5 cm; in patchy granular aggregates. *Twinning:* On {001}, common.

Physical Properties: *Cleavage:* Perfect on {001}. *Tenacity:* Brittle. Hardness = 3.5–4 D(meas.) = 2.92 D(calc.) = 2.91 Brilliant bluish white fluorescence under SW UV.

Optical Properties: Transparent to translucent. *Color:* White. *Luster:* Vitreous, pearly in part.
Optical Class: Biaxial (+). *Orientation:* $X \wedge \perp$ {001} $\simeq 50°$. *Dispersion:* $r > v$, strong.
$\alpha = 1.670$ $\beta = 1.670$–1.671 $\gamma = 1.673$–1.677 2V(meas.) = 32°–60°

Cell Data: *Space Group:* $C2/c$ or Cc. a = 16.941(3) b = 9.746(2) c = 20.907(3)
$\beta = 112.50(10)°$ Z = 4

X-ray Powder Pattern: Dara-i-Pioz massif, Tadzhikistan.
3.22 (100), 2.41 (20), 1.92 (17), 3.02 (5), 3.54 (4), 1.60 (4), 1.49 (4)

Chemistry:

	(1)	(2)		(1)	(2)
SiO_2	50.46	52.31	Li_2O	2.05	3.25
TiO_2	9.55	10.99	Na_2O	0.70	0.22
ZrO_2	2.28		K_2O	2.96	2.89
Fe_2O_3	0.50	0.29	F	1.05	0.34
Nb_2O_5	0.72		H_2O		1.21
MnO	0.12	0.22	$-O = F_2$	0.44	0.14
CaO	30.36	28.25	Total	100.31	99.83

(1) Dara-i-Pioz massif, Tadzhikistan; corresponds to $(K_{0.90}Na_{0.28})_{\Sigma=1.18}Li_{1.96}Ca_{7.75}$ $(Ti_{1.71}Zr_{0.26})_{\Sigma=1.97}Si_{12}O_{37.00}F_{0.79}$. (2) Iwagi Islet, Japan; by electron microprobe, Li by flame photometry, H_2O by gravimetry, and F by specific ion electrode; corresponding to $(K_{0.85}Na_{0.10})_{\Sigma=0.95}Li_{3.00}(Ca_{6.94}Mn_{0.04})_{\Sigma=6.98}(Ti_{1.90}Fe^{3+}_{0.05})_{\Sigma=1.95}Si_{12}O_{35.78}[(OH)_{1.85}F_{0.25}]_{\Sigma=2.10}$.

Occurrence: An accessory mineral in quartz-albite-aegirine veinlets and in albitites in syenites (Dara-i-Pioz massif, Tadzhikistan); a fine-grained accessory mineral in a small aegirine syenite stock in coarse-grained biotite granite (Iwagi Islet, Japan).

Association: Miserite, ekanite, titanite, quartz, albite, aegirine (Dara-i-Pioz massif, Tadzhikistan); albite, aegirine, pectolite, sugilite, allanite, titanite, andradite, zircon, apatite (Iwagi Islet, Japan).

Distribution: From the Dara-i-Pioz massif, Alai Range, Tien Shan, Tadzhikistan. On Iwagi Islet, Ehime Prefecture, Japan.

Name: For Rauf Baratovich Baratov, Soviet petrographer, of Tadzhikistan.

Type Material: Institute of Mineralogy and Geochemistry of Rare Elements, Moscow; A.E. Fersman Mineralogical Museum, Academy of Sciences, Moscow, Russia.

References: (1) Dusmatov, V.D., E.I. Semenov, A.P. Khomyakov, A.V. Bykova, and N.K. Dzharfarov (1975) Baratovite, a new mineral. Zap. Vses. Mineral. Obshch., 104, 580–582 (in Russian). (2) (1976) Amer. Mineral., 61, 1053 (abs. ref. 1). (3) Menchetti, S. and C. Sabelli (1979) The crystal structure of baratovite. Amer. Mineral., 64, 383–389. (4) Murakami, N., T. Kato, and F. Hirowatari (1983) Katayamalite, a new Ca-Li-Ti silicate mineral from Iwagi Islet, southwest Japan. Mineral. J. (Japan), 11, 261–268. (5) (1984) Amer. Mineral., 69, 811–812 (abs. ref. 4). (6) Baur, W.H. and D. Kassner (1992) Katayamalite and baratovite are structurally identical. Eur. J. Mineral., 839–841.

Bario-orthojoaquinite

$(Ba, Sr)_4Fe_2^{2+}Ti_2Si_8O_{26} \cdot H_2O$

Crystal Data: Orthorhombic. *Point Group:* $2/m\ 2/m\ 2/m$ or $mm2$. As pseudotetragonal crystals, steep bipyramidal {111} and truncated by {001}, to 8 mm; pyramidal faces are curved and striated; as aggregates.

Physical Properties: *Cleavage:* {001}, good. Hardness = 5.5 D(meas.) = 3.96 D(calc.) = 3.96

Optical Properties: Transparent to translucent. *Color:* Yellow-brown. *Streak:* Pale yellow. *Luster:* Vitreous.
Optical Class: Biaxial (+). *Pleochroism:* X = very pale yellow; Y = pale yellow; Z = yellow. *Orientation:* X = a; Y = b; Z = c. *Dispersion:* $r > v$, strong. *Absorption:* $Z \gg Y > X$. $\alpha = 1.735$ $\beta = 1.737$ $\gamma = 1.800$ 2V(meas.) = $10°–15°$

Cell Data: *Space Group:* $Ccmm$, $Cc2m$, or $Ccm2$. a = 10.477(5) b = 9.599(1) c = 22.59(1) Z = [4]

X-ray Powder Pattern: Gem mine, California, USA.
2.997 (100), 2.953 (95), 2.824 (90), 5.64 (70), 2.935 (70), 4.30 (62), 3.203 (50)

Chemistry:

	(1)
SiO_2	35.15
TiO_2	11.33
Al_2O_3	0.57
RE_2O_3	0.00
FeO	9.47
MnO	0.62
CaO	0.17
SrO	3.34
BaO	38.56
Na_2O	0.12
H_2O	1.3
Total	100.63

(1) Gem mine, California, USA; by electron microprobe, corresponds to $(Ba_{3.44}Sr_{0.44}Al_{0.15}Ca_{0.04})_{\Sigma=4.07}(Fe^{2+}_{1.80}Mn_{0.12}Na_{0.05})_{\Sigma=1.97}(Ti_{1.94}Al_{0.06})_{\Sigma=2.00}$ $Si_{8.00}O_{26} \cdot 0.93H_2O$.

Mineral Group: Joaquinite group.

Occurrence: In a block of highly fractured basalt subjected to high-pressure metamorphism and serpentinization.

Association: Benitoite, baotite, fresnoite, natrolite.

Distribution: At the Gem mine, San Benito Co., California, USA.

Name: For its BARIum content, ORTHOrhombic symmetry, and membership in the *joaquinite* group.

Type Material: University of California, Santa Barbara, California; Harvard University, Cambridge, Massachusetts, 11952; National Museum of Natural History, Washington, D.C., USA, 149428.

References: (1) Wise, W.S. (1982) Strontiojoaquinite and bario-orthojoaquinite: two new members of the joaquinite group. Amer. Mineral., 67, 809–816.

Crystal Data: Orthorhombic. *Point Group:* $2/m \; 2/m \; 2/m$. As well-formed, laminar crystals, to 5 mm, with {010}, {011}, and {111}; {001} is common, but of poor quality.

Physical Properties: *Cleavage:* Perfect on {010}. Hardness = n.d. D(meas.) = 2.13(1) D(calc.) = 2.11

Optical Properties: Transparent to translucent. *Color:* White to slightly pink. *Luster:* Vitreous.
Optical Class: Biaxial (–). *Orientation:* X = a; Y = b; Z = c. $\alpha = 1.4791(5)$ $\beta = 1.4851(5)$ $\gamma = 1.4891(5)$ 2V(meas.) = n.d. 2V(calc.) = 78(8)°

Cell Data: *Space Group: Amma.* a = 13.643(2) b = 18.200(3) c = 17.842(2) Z = 8

X-ray Powder Pattern: Capo Pula, Sardinia, Italy; cannot be distinguished from stellerite.
9.10 (> 100), 4.054 (100), 3.028 (80), 3.004 (25), 4.659 (20), 2.773 (20), 1.8192 (17)

Chemistry:

	(1)
SiO_2	58.82
Al_2O_3	14.75
Fe_2O_3	0.04
MnO	trace
MgO	0.24
CaO	1.66
SrO	trace
BaO	trace
Na_2O	5.97
K_2O	1.76
H_2O	16.40
Total	99.64

(1) Capo Pula, Sardinia, Italy; by AA, SiO_2 by gravimetry, Al_2O_3 by complexometric titration, H_2O by TGA; corresponds to $(Na_{1.36}K_{0.26}Ca_{0.21}Mg_{0.04})_{\Sigma=1.87}Al_{2.05}Si_{6.93}O_{18} \cdot 6.44H_2O$.

Mineral Group: Zeolite group.

Occurrence: On the walls of large fractures in deeply weathered andesitic and rhyolitic lavas.

Association: Heulandite.

Distribution: Found near Capo Pula, below South Efisio Tower, Nora, Sardinia, Italy.

Name: For Richard Maling Barrer (1910–), New Zealand-born British teacher and student of the chemistry of zeolites.

Type Material: University of Modena, Modena, Italy; National Museum of Natural History, Washington, D.C., USA, 128521.

References: (1) Passaglia, E. and D. Pongiluppi (1974) Sodian stellerite from Capo Pula, Sardegna. Lithos, 7, 69–73. (2) Passaglia, E. and D. Pongiluppi (1975) Barrerite, a new natural zeolite. Mineral. Mag., 40, 208. (3) (1976) Amer. Mineral., 61, 1053 (abs. refs. 1 and 2). (4) Galli, E. and A. Alberti (1975) The crystal structure of barrerite. Bull. Soc. fr. Minéral., 98, 331–340.

Barroisite $\qquad$ $[NaCa][(Mg, Fe^{2+})_3Al_2](Si_7Al)O_{22}(OH)_2$

Crystal Data: Monoclinic. *Point Group:* $2/m$. As rims on, or cores of, other mixed-species amphibole crystals.

Physical Properties: *Cleavage:* [Perfect on {110}, intersecting at ∼56° and ∼124°; partings on {100}, {001}.] *Tenacity:* [Brittle.] Hardness = [5–6] D(meas.) = n.d. D(calc.) = 3.21

Optical Properties: Semitransparent. *Color:* Green; dark green in thin section.
Luster: [Vitreous.]
Optical Class: [Biaxial.] α = n.d. β = n.d. γ = n.d. 2V(meas.) = n.d.

Cell Data: *Space Group:* [$C2/m$.] a = 9.759(1) b = 17.921(2) c = 5.290(1)
β = 104.49(1)° Z = [2]

X-ray Powder Pattern: n.d.

Chemistry:

	(1)	(2)		(1)	(2)
SiO_2	53.26	49.32	MgO	17.84	7.93
TiO_2	0.24	0.22	CaO	7.41	5.94
Al_2O_3	7.90	9.05	Na_2O	3.84	4.30
Cr_2O_3	0.04		K_2O	0.39	0.21
FeO	5.80	19.97	F	0.04	
MnO	0.03	0.35	Cl	0.03	
NiO	0.11		$-O = (F, Cl)_2$	0.03	
			Total	96.90	97.29

(1) Nordfjord, Norway; by electron microprobe, Fe^{2+}:Fe^{3+} by diffraction; corresponds to $Na_{1.04}Ca_{1.10}(Mg_{3.82}Al_{0.70}Fe^{2+}_{0.42}Fe^{3+}_{0.20}K_{0.07}Ti_{0.03})_{\Sigma=5.24}(Si_{7.40}Al_{0.60})_{\Sigma=8.00}O_{22}(OH)_2$. (2) Anglesey, Wales; by electron microprobe, corresponds to $Na_{1.07}Ca_{0.93}(Mg_{1.73}Fe^{2+}_{1.58}Fe^{3+}_{0.86}Al_{0.77}Mn_{0.04}K_{0.04}Ti_{0.02})_{\Sigma=5.04}(Si_{7.21}Al_{0.79})_{\Sigma=8.00}O_{22}(OH)_2$.

Polymorphism & Series: Forms a series with ferro-barroisite.

Mineral Group: Amphibole (sodic-calcic) group: $Mg/(Mg + Fe^{2+}) \geq 0.5$; $(Na + K)_A < 0.5$; $0.67 \leq Na_B \leq 1.33$; $(Ca + Na)_B \geq 1.34$; $Si < 7.5$.

Occurrence: In blueschist facies metamorphic rocks.

Association: Omphacite, glaucophane, crossite, actinolite, calcite.

Distribution: Along Cleary Creek, 20 km north of Fairbanks, and on the northwest side of Kodiak Island, Alaska, USA. In Wales, on Anglesey, near Llanfairpwllgwyngyll. From the Nybö eclogite pod, Nordfjord, Norway.

Name: n.d.

Type Material: n.d.

References: (1) Murgoci, G. (1922) Sur les propriétés des amphiboles bleues. Compt. Rendus Acad. Sci. Paris, 175, 373 (in French). (2) Murgoci, G. (1922) Sur le classification des amphiboles bleues et de certaines hornblendes. Compt. Rendus Acad. Sci. Paris, 175, 426 (in French). (3) Heritsch, H., P. Paulitsch, and E.M. Walitzi (1957) Die Struktur von Karinthin und einer barroisitischen Hornblende. Tschermaks Mineral. Petrog. Mitt., 6, 215–225 (in German). (4) Ungaretti, L., D.C. Smith, and G. Rossi (1981) Crystal-chemistry by X-ray structure refinement and electron microprobe analysis of a series of sodic-calcic to alkali-amphiboles from the Nybö eclogite pod, Norway. Bull. Minéral., 104, 400–412. (5) Gibbons, W. and M. Gyopari (1986) A greenschist protolith for blueschist in Anglesey, U.K. In: B.W. Evans and E.H. Brown, Eds., Blueschists and eclogites, Geol. Soc. Amer. Memoir 164, 217–228.

Crystal Data: Orthorhombic. *Point Group: mm*2. As thin tabular to prismatic crystals displaying square cross sections, to 2 cm; granular, massive.

Physical Properties: *Cleavage:* Good on {001} and {100}, imperfect on {010}. *Tenacity:* Brittle. Hardness = 7 VHN = 734 D(meas.) = 4.066(2) D(calc.) = [4.00] May fluoresce vivid blue.

Optical Properties: Translucent. *Color:* Colorless, white, blue. *Luster:* Vitreous to greasy. *Optical Class:* Biaxial (–) or (+). *Orientation:* X = b; Y = c; Z = a. *Dispersion:* r > v, moderate. α = 1.695(2) β = 1.702(2) γ = 1.708(2) 2V(meas.) = 60°–70°

Cell Data: *Space Group: Pn*2$_1$*a.* a = 9.835(2) b = 11.654(3) c = 4.673(2) Z = 4

X-ray Powder Pattern: Seal Lake, Canada.
3.35 (vs), 2.93 (ms), 2.45 (ms), 3.04 (m), 2.99 (m), 5.84 (mw), 4.56 (mw)

Chemistry:

	(1)	(2)	(3)		(1)	(2)	(3)
SiO$_2$	35.51	36.42	37.54	BeO	16.01	15.77	[15.46]
Al$_2$O$_3$	0.05			MgO	0.21	0.29	
Fe$_2$O$_3$	0.04			CaO	0.42		
FeO		0.19	trace	BaO	47.43	46.49	46.75
ZnO		trace		H$_2$O$^-$	0.57	0.40	
PbO		0.11		Total	100.24	99.67	[99.75]

(1) Långban, Sweden. (2) Franklin, New Jersey, USA. (3) Narssârssuk, Greenland; by electron microprobe, BeO from stoichiometry; corresponds to Ba$_{0.99}$Be$_{2.00}$Si$_{2.02}$O$_{7.03}$.

Occurrence: In a banded vein in a metamorphosed stratiform zinc orebody (Franklin, New Jersey, USA); in an "amazonite" pocket (Park Co., Colorado, USA); in narrow veins in chertlike silica and albite in fenitized gneisses (Seal Lake, Canada); in calcite veins cutting amphibole gneisses (Vishnevye Mountains, Russia).

Association: Hedyphane (Långban, Sweden); hedyphane, willemite (Franklin, New Jersey, USA); microcline, barite, fluorite (Park Co., Colorado, USA); eudialyte (Seal Lake, Canada); bastnäsite, aeschynite, barite, strontianite, thorogummite, harmotome, fluorite, quartz (Vishnevye Mountains, Russia).

Distribution: At Långban, Värmland, Sweden. In Norway, on Arø Island, Langesundsfjord, and at Bratthagen, 10 km northwest of Larvik, Lågendal. From Narssârssuk, Greenland. In the USA, at Franklin, Sussex Co., New Jersey, with exceptional crystals from an undisclosed locality in Park Co., Colorado. In Canada, at Seal Lake and Letitia Lake, Labrador, Newfoundland, and Mont Saint-Hilaire, Quebec. In Russia, from the Vishnevy-Ilmen Mountains, Southern Ural Mountains. In the Pokrovo-Kireevsk structure, Kazakhstan.

Name: From the Greek for *heavy*, for its high density.

References: (1) Dana, E.S. (1892) Dana's system of mineralogy, (6th edition), 562. (2) Palache, C. (1935) The minerals of Franklin and Sterling Hill, Sussex County, New Jersey. U.S. Geol. Sur. Prof. Paper 180, 92. (3) Heinrich, E.W. and R.W. Deane (1962) An occurrence of barylite near Seal Lake, Labrador. Amer. Mineral., 47, 758–763. (4) Robinson, P.D. and J.H. Fang (1977) Barylite, BaBe$_2$Si$_2$O$_7$: its space group and crystal structure. Amer. Mineral., 62, 167–169. (5) Petersen, O.V. and O. Johnson (1980) First occurrence of the rare mineral barylite in Greenland. Tschermaks Mineral. Petrog. Mitt., 27, 35–39. (6) Vlasov, K.A., Ed. (1966) Mineralogy of rare elements, v. II, 87–89.

Crystal Data: Hexagonal. *Point Group:* $3\bar{2}/m$. As platy, subparallel crystals in lamellar and ball-like aggregates, to 3 mm; granular, massive.

Physical Properties: *Cleavage:* {0001}, perfect. *Fracture:* Uneven. *Tenacity:* Brittle. Hardness = 3   D(meas.) = 6.55–6.71   D(calc.) = [6.54]

Optical Properties: Semitransparent. *Color:* Pink to colorless; tarnishes on exposure. *Luster:* Pearly on cleavage. *Optical Class:* Uniaxial (–). $\omega = 2.033–2.07$ $\epsilon = 2.015–2.05$

Cell Data: *Space Group:* $R\bar{3}c$. a = 9.821   c = 38.38   Z = 6

X-ray Powder Pattern: Långban, Sweden.
2.957 (100), 2.765 (80), 2.675 (80), 3.88 (60), 3.16 (60), 4.57 (50), 3.22 (50)

Chemistry:

	(1)	(2)	(3)
SiO_2	16.84	17.1	16.26
Al_2O_3	0.59		
FeO	0.23		
MnO	3.33	3.7	3.20
ZnO	0.30		
PbO	77.35	78.4	80.54
MgO	0.78		
CaO	0.21	0.6	
H_2O^+	0.07		
Total	99.70	99.8	100.00

(1) Franklin, New Jersey, USA. (2) Do.; by electron microprobe. (3) $Pb_8Mn(Si_2O_7)_3$.

Occurrence: In iron ores (Harstig mine, Sweden); as thin films and veinlets in a metamorphosed zinc orebody (Franklin, New Jersey, USA).

Association: Garnet, willemite, axinite, hardystonite (Franklin, New Jersey, USA).

Distribution: In the Harstig mine, Pajsberg, near Persberg, and Långban and Jakobsberg, Värmland, Sweden. From Franklin, Sussex Co., New Jersey, USA.

Name: From the Greek for *heavy* and for SILicon in its composition.

Type Material: Harvard University, Cambridge, Massachusetts, USA, 114630.

References: (1) Dana, E.S. (1892) Dana's system of mineralogy, (6th edition), 421. (2) Shannon, E.V. and H. Berman (1926) Barysilite from Franklin Furnace, New Jersey. Amer. Mineral., 11, 130–132. (3) Glasser, F.P. (1964) New data on barysilite. Amer. Mineral., 49, 1485–1488. (4) Lajzérowicz, J. (1965) Étude par diffraction des rayons X et absorption infra-rouge de la barysilite, $MnPb_8 \cdot 3Si_2O_7$, et composés isomorphes. Acta Cryst., 20, 357–363 (in French with English abs.). (5) Billhardt, H.W. (1969) Synthesis of lead pyrosilicate and other barysilite-like compounds. Amer. Mineral., 54, 510–521. (6) Dunn, P.J. (1985) The lead silicates from Franklin, New Jersey: occurrence and composition. Mineral. Mag., 49, 721–727.

$(Na, K)_3(Ba, Fe^{2+}, Ca)_2$
$(Ti, Fe^{3+})_3O_2(Si_2O_7)_2(F, OH, O)_2$ **Barytolamprophyllite**

Crystal Data: Monoclinic. *Point Group:* $2/m$ or 2. As foliated aggregates of cleavable pseudorhombohedra.

Physical Properties: *Cleavage:* Perfect on {100}, less so on {011}, imperfect on {010}. *Tenacity:* Brittle. Hardness = 2–3 D(meas.) = 3.62–3.66 D(calc.) = [3.82]

Optical Properties: Semitransparent. *Color:* Dark brown. *Luster:* Vitreous. *Optical Class:* Biaxial (+). *Pleochroism:* Strong; X = light yellow; Z = brown. *Orientation:* $Z \wedge c = 6°$–$7°$. *Dispersion:* $r > v$, distinct. *Absorption:* $Z > Y > X$. $\alpha = 1.742$–1.743 $\beta = [1.754]$ $\gamma = 1.776$–1.778 2V(meas.) = $29°$–$30°$

Cell Data: *Space Group:* $C2/m$ or $C2$. $a = 19.833$ $b = 7.089$ $c = 5.393$ $\beta = 96.66°$ $Z = 2$

X-ray Powder Pattern: Kola Peninsula, Russia. (ICDD 34-313).
2.801 (100), 2.153 (90), 1.482 (90), 1.601 (80), 3.447 (70), 1.790 (70), 3.294 (50)

Chemistry:

	(1)		(1)
SiO_2	28.53	BaO	17.24
TiO_2	26.60	Na_2O	9.52
Al_2O_3	1.12	K_2O	3.10
Fe_2O_3	2.72	F	1.60
FeO	2.63	Cl	0.51
MnO	1.75	H_2O^+	0.70
MgO	1.00	P_2O_5	0.06
CaO	1.70	$-O = (F, Cl)_2$	0.78
SrO	1.47	Total	[99.47]

(1) Kola Peninsula, Russia; original total given as 99.34%; corresponds to $(Na_{2.47}K_{0.53})_{\Sigma=3.00}$ $(Ba_{1.05}Fe^{2+}_{0.28}Ca_{0.24}Mn_{0.20}Mg_{0.20}Sr_{0.12})_{\Sigma=2.09}(Ti_{2.68}Fe^{3+}_{0.28})_{\Sigma=2.96}(Si_{3.82}Al_{0.18})_{\Sigma=4.00}O_{14}$ $[F_{0.68}(OH)_{0.62}O_{0.57}Cl_{0.14}]_{\Sigma=2.01}$.

Occurrence: In ijolite (Lovozero massif, Russia).

Association: Aegirine, nepheline, potassic feldspar, cancrinite, apatite (Lovozero massif, Russia); amphibole, eudialyte, natrolite, pectolite (Gardiner complex, Greenland).

Distribution: From the Lovozero massif and on Mts. Rasvumchorr and Koashva, Khibiny massif, Kola Peninsula, and in the Murun massif, southwest of Olekminsk, Yakutia, Russia. At Kahlenberg, Graulai, and Üdersdorf, Eifel district, Germany. From Coyote Peak, near Orick, Humboldt Co., California, USA. In the Gardiner complex, beyond the head of Kangerdlugssuaq Fjord, Greenland.

Name: For the BARium in its composition and its relation to *lamprophyllite*.

Type Material: n.d.

References: (1) Tze-Chung Peng [Zhizhong Peng] and Chien-hung Chang [Jianhong Zhang] (1965) New varieties of lamprophyllite–barytolamprophyllite and orthorhombic lamprophyllite. Scientia Sinica, 14, 1827–1840 (in English). (2) (1966) Amer. Mineral., 51, 1549 (abs. ref. 1). (3) Zhizhong Peng, Jianhong Zhang, and Jinfu Shu (1984) The crystal structure of barytolamprophyllite. Kexue Tongbao, 29, 237–241 (in English). (4) (1984) Chem. Abs., 101, 26260 (abs. ref. 3).

Crystal Data: Orthorhombic, pseudohexagonal. *Point Group:* $mm2$ or $2/m\ 2/m\ 2/m$. As complex elongated crystals, to 10 cm.

Physical Properties: *Cleavage:* Fair on {100}, {010}, and {001}; parting on {110}. *Tenacity:* Brittle. Hardness = ~6 VHN = 764 D(meas.) = 3.432 D(calc.) = 3.49 Some samples are piezoelectric.

Optical Properties: Transparent in thin slivers. *Color:* Dark brown. *Streak:* Rose-brown. *Luster:* Vitreous to slightly oily.
Optical Class: Biaxial (+). *Pleochroism:* X = colorless; Y = yellowish brown; Z = reddish brown. *Orientation:* X = a; Y = b; Z = c. *Dispersion:* $r < v$, strong. $\alpha = 1.730(1)$ $\beta = 1.735(1)$ $\gamma = 1.791(1)$ 2V(meas.) = 7°

Cell Data: *Space Group: Ima2* or *Imam.* a = 10.40–10.50 b = 13.85–13.91 c = 8.08–8.10 Z = 4

X-ray Powder Pattern: Inagli massif, Russia; nearly identical to shcherbakovite.
2.91 (100), 3.39 (50), 2.16 (50), 1.68 (50), 2.09 (40), 3.20 (30), 2.62 (30)

Chemistry:

	(1)		(1)
SiO_2	39.00	CaO	0.27
TiO_2	22.00	SrO	0.03
ZrO_2	1.90	BaO	22.00
Al_2O_3	0.90	Na_2O	8.40
Fe_2O_3	1.80	K_2O	2.60
Nb_2O_5	0.36	H_2O^+	0.50
MnO	0.09	LOI	0.10
MgO	trace	Total	99.95

(1) Inagli massif, Russia; corresponds to $(Na_{1.66}K_{0.34})_{\Sigma=2.00}(Ba_{0.88}Ca_{0.03}Mn_{0.01})_{\Sigma=0.92}$ $(Ti_{1.68}Fe^{3+}_{0.14}Al_{0.11}Zr_{0.09})_{\Sigma=2.02}Si_{3.97}[O_{13.66}(OH)_{0.34}]_{\Sigma=14.00}$.

Occurrence: In aegirine-arfvedsonite-microcline pegmatites in dunites (Inagli massif, Russia).

Association: Microcline, nepheline, aegirine, arfvedsonite, lorenzenite, uranian thorite, eudialyte, apatite, orthoclase (Inagli massif, Russia).

Distribution: From the Inagli massif, 30 km west of Aldan; the Murun massif, southwest of Olekminsk, Yakutia; and on Mt. Rasvumchorr, Khibiny massif, Kola Peninsula, Russia. At Liley, near Üdersdorf, Graulai, and Altburg, Eifel district, Germany.

Name: For Ba, Ti, Si in the composition.

Type Material: Institute of Mineralogy and Geochemistry of Rare Elements, Moscow; Moscow Geological Survey Institute, Moscow; A.E. Fersman Mineralogical Museum, Academy of Sciences, Moscow, Russia.

References: (1) Kravchenko, S.M. and E.V. Vlasova (1959) Rare-metal mineralization associated with nepheline syenites of the alkalic province of Central Aldan. Doklady Acad. Nauk SSSR, 128, 1046–1049 (in Russian). (2) (1960) Amer. Mineral., 45, 908–909 (abs. ref. 1). (3) Kravchenko, S.M., E.V. Vlasova, and N.G. Pinevich (1960) Batisite, a new mineral. Doklady Acad. Nauk SSSR, 133, 657–660 (in Russian). (4) (1960) Amer. Mineral., 45, 1317 (abs. ref. 3). (5) Nikitin, A.V. and N.V. Belov (1962) Crystal structure of batisite $Na_2BaTi_2Si_4O_{14} = Na_2BaTi_2O_2[Si_4O_{12}]$. Doklady Acad. Nauk SSSR, 146, 1401–1403 (in Russian). (6) (1963) Chem. Abs., 58, 5116 (abs. ref. 5). (7) Schmahl, W., E. Tillmanns, and K. Abraham (1981) Struktur und Kristallchemie von Batisite aus der Westeifel. Fortschr. Mineral., 59(1), 174–176 (in German). (8) Schmahl, W.W. and E. Tillmanns (1987) Isomorphic substitutions, straight Si-O-Si geometry, and disorder of tetrahedral tilting in batisite, $(Ba, K)(K, Na)Na(Ti, Fe, Nb, Zr)Si_4O_{14}$. Neues Jahrb. Mineral., Monatsh., 107–118.

Crystal Data: Orthorhombic. *Point Group:* $mm2$. As radiating crystal groups, to 6 cm; as bladed crystals and rosettes of thin plates; also fine feltlike masses of needles, compact fibrous, chalklike massive. *Twinning:* On {100}.

Physical Properties: *Cleavage:* Perfect on {001}, fair on {100}. Hardness = 5.5–6 D(meas.) = 2.71–2.74   D(calc.) = 2.69–2.79   Weakly piezoelectric.

Optical Properties: Translucent to transparent. *Color:* Colorless to white. *Luster:* Vitreous. *Optical Class:* Biaxial (+). *Orientation:* $X = c$; $Y = b$; $Z = a$. $\alpha = 1.578–1.586$ $\beta = 1.579–1.585$ $\gamma = 1.583–1.593$ 2V(meas.) = 22°–58°

Cell Data: *Space Group:* $Am2a$.   a = 19.39(2)   b = 23.19(2)   c = 5.005(9)   Z = 4

X-ray Powder Pattern: Baveno, Italy.
3.74 (100), 3.35 (90), 3.24 (80), 3.13 (70), 3.05 (70), 2.56 (60), 4.19 (50)

Chemistry:

	(1)	(2)
SiO_2	57.64	57.83
Al_2O_3	6.46	10.90
BeO	7.66	5.35
CaO	23.96	23.99
H_2O^+	3.56	
H_2O^-	0.32	1.93
Total	99.60	100.00

(1) Mesa Grande, California, USA; corresponds to $Ca_{4.11}Be_{2.95}Al_{1.22}Si_{9.22}O_{26} \cdot 1.90H_2O$.
(2) $Ca_4Be_2Al_2Si_9O_{26}(OH)_2$.

Occurrence: As druses in miarolitic cavities in granite and associated pegmatites, formed by alteration of beryl and other beryllium-bearing minerals. Also in hydrothermal veins and skarns.

Association: Beryl, helvite, phenakite, bertrandite, quartz, epidote, stilbite, albite, orthoclase, titanite, chlorite, clinozoisite, tremolite.

Distribution: From Baveno and Val d'Ossola, Piedmont, Italy. At Strzegom (Striegau), Silesia, Poland. From the Czech Republic, at Jeclov, and near Drahonín. At Takelund, southeast of Halden, Østfold, Norway. In Switzerland, from the Muotta Nera, Piz Lai Blau, Val Nalps and at Val Casaccia, Val Cristallina, Graubünden. In the USA, from the Himalaya mine, Mesa Grande, and in the Rincon district, San Diego Co., California; at the Foote mine, Kings Mountain, Cleveland Co., North Carolina; from the Rutherford mine, Amelia, Amelia Co., Virginia; and elsewhere. In the Londonderry quarry, near Coolgardie, Western Australia. From Batystau, central Kazakhstan. In Russia, as large crystals at Akoudertia, Siberia; Malyshero, Ural Mountains; and a number of other less-well-defined localities.

Name: For the original occurrence at Baveno, Italy.

Type Material: Municipal Museum of Natural History, Milan, Italy.

References: (1) Artini, E. (1901) Di una nuova specie minerale trovata nel granito di Baveno. Atti Rend. Accad. Lincei, 10, 139–145 (in Italian). (2) Fleischer, M.F. and G. Switzer (1953) The bavenite problem. Amer. Mineral., 38, 988–993. (3) Switzer, G. and L.E. Reichen (1960) Re-examination of pilinite and its identification with bavenite. Amer. Mineral., 45, 757–762. (4) Berry, L.G. (1963) The composition of bavenite. Amer. Mineral., 48, 1166–1168. (5) Cannillo, E. and A. Coda (1966) The crystal structure of bavenite. Acta Cryst., 20, 301–309. (6) Loiacono, G.M., G. Kostecky, and J.S. White, Jr. (1982) Resolution of space group ambiguities in minerals. Amer. Mineral., 67, 846–847.

Crystal Data: Hexagonal. *Point Group:* $\bar{6}m2$. Crystals rarely entirely euhedral, prismatic with partially developed pyramidal terminations, more commonly anhedral to subhedral, < 1 mm. Also as clusters of crystals.

Physical Properties: *Cleavage:* Pyramidal, probably $\{10\bar{1}4\}$; poor on $\{10\bar{1}0\}$; strong parting $\parallel \{0001\}$. Hardness = n.d. D(meas.) = n.d. D(calc.) = 3.82 Moderately strong pale bluish fluorescence under SW UV; strong blue cathodoluminescence.

Optical Properties: Transparent to translucent. *Color:* Colorless.
Optical Class: Uniaxial (+). $\omega = 1.675$–1.681 $\epsilon = 1.685$–1.688

Cell Data: *Space Group:* $P\bar{6}c2$. $a = 6.737$–6.769 $c = 9.997$–10.020 $Z = 2$

X-ray Powder Pattern: Rockall Island, Scotland.
3.80 (100), 2.800 (100), 5.85 (35), 3.38 (35), 2.93 (30), 2.012 (30), 2.213 (25)

Chemistry:

	(1)	(2)	(3)
SiO$_2$	39.71	39.59	39.46
TiO$_2$	0.17	0.51	
ZrO$_2$	26.38	26.12	26.98
SnO$_2$	0.06	0.11	
BaO	33.69	34.12	33.56
Total	100.01	100.45	100.00

(1–2) Rockall Island, Scotland; by electron microprobe. (3) BaZrSi$_3$O$_9$.

Occurrence: A late-stage interstitial mineral in aegirine-riebeckite granite (Rockall Island, Scotland); in granular celsian-bearing rocks near contacts with granite (Big Creek, California, USA).

Association: Elpidite, aegirine, albite, cristobalite, leucophosphite, monazite, quartz, apatite, eudialyte, magnetite, pyrochlore, xenotime, barite, titanite (Rockall Island, Scotland); taramellite, celsian, diopside, actinolite (Big Creek, California, USA).

Distribution: On Apex Ridge, Rockall Island, in the North Atlantic Ocean, 294 km west of St. Kilda Island, Outer Hebrides, Scotland. At the Esquire No. 8 claim, Big Creek, Fresno Co., California, USA.

Name: For BArium and ZIRconium in the composition.

Type Material: The Natural History Museum, London, England, MI 36445.

References: (1) Young, B.R., J.R. Hawkes, R.J. Morriman, and M.T. Styles (1978) Bazirite, BaZrSi$_3$O$_9$, a new mineral from Rockall Island, Inverness-shire, Scotland. Mineral. Mag., 42, 35–40. (2) (1979) Amer. Mineral., 64, 241 (abs. ref. 1). (3) Alfors, J.T. and A. Pabst (1984) Titanian taramellites in western North America. Amer. Mineral., 69, 358–373. (4) Hawthorne, F.C. (1987) The crystal chemistry of the benitoite group minerals and structural relations in (Si$_3$O$_9$) ring structures. Neues Jahrb. Mineral., Monatsh., 16–30.

Crystal Data: Hexagonal. *Point Group:* $6/m \, 2/m \, 2/m$. Hexagonal prisms, to 1 cm, commonly barrel-like and in subparallel aggregation.

Physical Properties: Hardness = 6.5–7   VHN = 851–897   D(meas.) = 2.77–2.85 D(calc.) = n.d. Electromagnetic.

Optical Properties: Semitransparent. *Color:* Light to dark sky-blue.
Optical Class: Uniaxial (–). *Pleochroism:* O = pale greenish yellow; E = intense sky-blue.
$\omega = 1.622$–1.637 $\epsilon = 1.602$–1.622

Cell Data: *Space Group:* $P6/mcc$.   a = 9.521(5)   c = 9.165(5)   Z = 2

X-ray Powder Pattern: Synthetic Be$_3$(Sc$_{1.75}$Fe$^{3+}_{0.25}$)$_{\Sigma=2.00}$Si$_6$O$_{18}$.
3.306 (100), 8.27 (89), 2.960 (75), 4.006 (48), 4.578 (24), 2.577 (17), 1.761 (15)

Chemistry:

	(1)	(2)		(1)	(2)
SiO$_2$	58.80	58.00	MgO	0.82	0.10
Al$_2$O$_3$	0.25	0.80	Li$_2$O	0.00	0.24
Fe$_2$O$_3$	2.21	5.70	Na$_2$O	2.82	1.60
Sc$_2$O$_3$	14.44	14.50	K$_2$O	0.22	0.13
FeO	3.68		Rb$_2$O	0.037	0.25
MnO	1.58	1.43	Cs$_2$O	0.31	2.93
BeO	12.90	14.50	H$_2$O$^+$	2.60	1.10
			Total	100.67	101.28

(1) Central Kazakhstan; corresponds to Be$_{3.06}$(Sc$_{1.26}$Fe$^{2+}_{0.31}$Fe$^{3+}_{0.17}$Mn$_{0.13}$Mg$_{0.12}$Al$_{0.03}$)$_{\Sigma=2.02}$ (Si$_{5.93}$Be$_{0.07}$)$_{\Sigma=6.00}$O$_{18}$•0.87H$_2$O, with (Na$_{0.55}$K$_{0.03}$Cs$_{0.01}$)$_{\Sigma=0.59}$ in the channels. (2) Heftetjern, Norway; by electron microprobe, average of three analyses; total Fe as Fe$_2$O$_3$, Be and Li by AA on bulk sample; corresponds to Be$_{3.1}$(Sc$_{1.24}$Fe$_{0.42}$Mn$_{0.12}$Al$_{0.10}$Mg$_{0.02}$)$_{\Sigma=1.90}$(Na$_{0.30}$Cs$_{0.12}$ Li$_{0.05}$Rb$_{0.02}$K$_{0.01}$)$_{\Sigma=0.50}$(Si$_{5.7}$Be$_{0.3}$)$_{\Sigma=6.00}$O$_{18}$•0.36H$_2$O.

Occurrence: In miarolitic cavities in granite (Baveno, Italy); in alpine veins (Switzerland); in granite pegmatite (Heftetjern, Norway).

Association: Quartz, orthoclase, muscovite, laumontite, albite (Baveno, Italy); quartz, hematite, calcite, chlorite, albite, fluorite (Switzerland); beryl, bavenite (Heftetjern, Norway).

Distribution: From Baveno, Piedmont, Italy. In Switzerland, at Val Strem and elsewhere in Graubünden; on the Witenalp; at Wiler, Reusstal; and on Stollen Oberaar, Grimsel Pass. At Heiligenblut, Carinthia, Austria. From Heftetjern, Tørdal, Norway. At an undisclosed locality in central Kazakhstan. On Mt. Antero, Chaffee Co., Colorado, USA.

Name: For Alessandro E. Bazzi, discoverer of the material from Baveno, Italy.

Type Material: Municipal Museum of Natural History, Milan; University of Milan, Milan, Italy; The Natural History Museum, London, England, 1925,534.

References: (1) Artini, E. (1915) Due minerali di Baveno contenenti terre rare: weibyeite e bazzite. Atti Rend. Accad. Lincei, 24, 313–319 (in Italian). (2) (1921) Mineral. Abs., 1, 204 (abs. ref. 1). (3) Bergerhoff, G. (1955) Über die Kristallstruktur des Bazzit und ihre Beziehungen zu der des Beryll. Tschermaks Mineral. Petrog. Mitt., 35, 410–421 (in German with English abs.). (4) Peyronel, G. (1956) The crystal structure of Baveno bazzite. Acta Cryst., 9, 181–186. (5) Chistyakova, M.B., V.A. Moleva, and Z.P. Razmanova (1966) Bazzite found for the first time in the USSR. Doklady Acad. Nauk SSSR, 169, 1421–1424 (in Russian). (6) (1967) Amer. Mineral., 52, 563–564 (abs. ref. 5). (7) Frondel, C. and J. Ito (1968) Synthesis of the scandium analogue of beryl. Amer. Mineral., 53, 943–953. (8) Juve, G. and S. Bergstøl (1990) Caesian bazzite in granite pegmatite in Tørdal, Telemark, Norway. Mineral. Petrol., 43, 131–136.

Beidellite $(Ca_{0.5}, Na)_{0.3}Al_2(Si, Al)_4O_{10}(OH)_2 \cdot nH_2O$

Crystal Data: Monoclinic, pseudohexagonal. *Point Group:* $2/m$. As thin plates, laths, and ribbons, to 30 μm; in veinlets filling fractures and as claylike masses.

Physical Properties: *Cleavage:* Perfect on {001}. Hardness = 1–2 D(meas.) = 2–3 depending on hydration. D(calc.) = n.d. Positive identification of minerals in the smectite group may need data from DTA curves, dehydration curves, and X-ray powder patterns before and after treatment by heating and with organic liquids.

Optical Properties: Translucent. *Color:* White, reddish brown, brownish gray. *Luster:* Waxy to vitreous.
Optical Class: Biaxial (–). $\alpha = 1.494$ $\beta = 1.536$ $\gamma = 1.536$ 2V(meas.) = $9°$–$16°$

Cell Data: *Space Group:* $C2/m$. $a = 5.179$ $b = 8.970$ $c = 17.57$ $\beta = [\sim 90°]$
Z = n.d.

X-ray Powder Pattern: Black Jack mine, Idaho, USA; glycolated, diffuse pattern.
17.6 (100), 4.42 (100), 3.95 (100), 3.54 (100), 2.50 (100), 1.498 (100), 2.57 (80)

Chemistry:

	(1)	(2)		(1)	(2)
SiO_2	45.32	45.83	CaO	2.76	1.41
TiO_2		0.46	Na_2O	0.10	0.16
Al_2O_3	27.84	22.79	K_2O	0.12	0.09
Fe_2O_3	0.70	5.71	H_2O^+	14.48	9.79
FeO		0.28	H_2O^-	8.16	12.55
MgO	0.16	0.86	Total	99.64	99.93

(1) Black Jack mine, Idaho, USA; corresponds to $(Ca_{0.23}Na_{0.02}K_{0.01})_{\Sigma=0.26}$ $(Al_{1.96}Fe^{3+}_{0.04}Mg_{0.02})_{\Sigma=2.02}(Si_{3.46}Al_{0.54})_{\Sigma=4.00}O_{10}(OH)_2.$ (2) Velka Kopan, Ukraine; corresponds to $(Ca_{0.12}Na_{0.02}K_{0.01})_{\Sigma=0.15}(Al_{1.62}Fe^{3+}_{0.33}Mg_{0.10}Ti_{0.03}Fe^{2+}_{0.02})_{\Sigma=2.10}(Si_{3.54}Al_{0.46})_{\Sigma=4.00}O_{10}(OH)_2.$

Mineral Group: Smectite group.

Occurrence: A constituent of bentonitic clays; an alteration product in hydrothermal mineral deposits, especially porphyry Cu-Mo systems; in soils derived from mafic rocks.

Association: Plagioclase, quartz, orthoclase, montmorillonite, kaolin, allophane, muscovite.

Distribution: Of worldwide occurrence, but pure and well-characterized materials are uncommon. In the USA, from Beidell, Saguache Co., and at Wagon Wheel Gap, Mineral Co., Colorado; in the Black Jack mine, Carson district, Owyhee Co., Idaho; from Arizona, at Morenci, Greenlee Co., Ajo, Pima Co., in the San Manuel mine, Pinal Co., and in the Globe-Miami district, Gila Co.; at Boron, Kern Co., California. In the Princess mine, Namiquipa, Chihuahua, Mexico. From Sibert, Rhône, France. At Unterrupsroth, Bavaria, Germany. From Velka Kopan, Khust, Ukraine.

Name: For Beidell, Colorado, USA.

Type Material: National Museum of Natural History, Washington, D.C., USA, R4762 (Black Jack mine, Idaho, USA); R4761 (Beidell, Colorado, USA, material determined to be a mixture in part).

References: (1) Larsen, E.S. and E.T. Wherry (1925) Beidellite, a new mineral name. J. Wash. Acad. Sci., 15, 465–466. (2) Ross, C.S. and E.V. Shannon (1925) The chemical and optical properties of beidellite. J. Wash. Acad. Sci., 15, 467–468. (3) (1926) Amer. Mineral., 11, 167 (abs. refs. 1 and 2). (4) Deer, W.A., R.A. Howie, and J. Zussman (1963) Rock-forming minerals, v. 3, sheet silicates, 226–245. (5) Weir, A.H. and R. Greene-Kelly (1962) Beidellite. Amer. Mineral., 47, 137–146.

Crystal Data: Hexagonal. *Point Group:* $\bar{6}m2$. As crystals, prismatic to barrel-shaped, with triangular cross section and stepped faces, showing the forms $\{11\bar{2}0\}$ and $\{0001\}$, to 1 mm.

Physical Properties: *Fracture:* Uneven. *Tenacity:* Moderately brittle. Hardness = 6–7 VHN = 900–1030, average 970 (40 g load). D(meas.) = 4.16(3) D(calc.) = 4.25

Optical Properties: Transparent. *Color:* Brown to brownish red. *Streak:* White. *Luster:* Adamantine.
Optical Class: Uniaxial (+); anomalously biaxial. $\omega = 1.928(2)$ $\epsilon = 2.002(5)$ 2V(meas.) = 0°–10°

Cell Data: *Space Group:* $P\bar{6}2m$. a = 8.966(3) c = 7.799(3) Z = 1

X-ray Powder Pattern: Vuoriyärvi complex, Russia.
2.937 (100), 3.888 (51), 7.81 (35), 1.948 (26), 2.750 (25), 3.481 (24), 2.154 (22)

Chemistry:

	(1)
SiO_2	17.80
TiO_2	5.60
ZrO_2	1.20
Al_2O_3	0.14
Fe_2O_3	1.78
Nb_2O_5	42.20
Ta_2O_5	0.15
CaO	0.05
SrO	0.00
BaO	30.30
Na_2O	0.20
K_2O	0.55
Total	99.97

(1) Vuoriyärvi complex, Russia; by electron microprobe, corresponding to $(Ba_{2.74}K_{0.16}Na_{0.09}Ca_{0.01})_{\Sigma=3.00}(Nb_{4.41}Ti_{0.97}Fe^{3+}_{0.31}Zr_{0.13}Al_{0.04}Ta_{0.01})_{\Sigma=5.87}Si_{4.12}O_{24.90}$.

Occurrence: Of secondary origin, formed by alteration of barium-rich pyrochlore during dolomitization of calcite carbonatites in pyroxenites.

Association: Magnetite, pyrochlore, phlogopite, chlorite, pyrite, pyrrhotite, dolomite, carbonate-apatite, barite, alstonite, nenadkevichite.

Distribution: In the Vuoriyärvi carbonatite complex, Kola Peninsula, Russia.

Name: For I.V. Bel'kov (1917–1989), Soviet mineralogist who explored the Kola Peninsula, Russia.

Type Material: A.E. Fersman Mineralogical Museum, Academy of Sciences, Moscow, Russia.

References: (1) Voloshin, A.V., V.V. Subbotin, Y.A. Pakhomovskii, A.Y. Bakhchisaraitsev, N.A. Yamnova, and D.Y. Pushcharovskii (1990) Belkovite $Ba_3(Nb, Ti)_6(Si_2O_7)_2O_{12}$ new mineral from carbonatite of the Vuoriyarvi massif (Kola Peninsula). Doklady Acad. Nauk SSSR, 315, 1218–1220 (in Russian). (2) Voloshin, A.V., V.V. Subbotin, Y.A. Pakhomovskii, A.Y. Bakhchisaraitsev, N.A. Yamnova, and D.Y. Pushcharovskii (1991) Belkovite — a new barium–niobium silicate from carbonatites of the Vuoriyarvi massif (Kola Peninsula, USSR). Neues Jahrb. Mineral., Monatsh., 23–31. (3) (1991) Amer. Mineral., 76, 1728 (abs. ref. 2).

Bellbergite

$(K, Ba, Sr)_2Sr_2Ca_2(Ca, Na)_4Al_{18}Si_{18}O_{72} \cdot 30H_2O$

Crystal Data: Hexagonal. *Point Group:* $6/m \, 2/m \, 2/m$, $\bar{6}m2$, or $6mm$. Rarely as hexagonal dipyramidal crystals, consisting of $\{10\bar{1}2\}$, to 0.3 mm; commonly subparallel intergrowths elongated along [0001].

Physical Properties: *Fracture:* Conchoidal. Hardness = ~5 D(meas.) = 2.20(2) D(calc.) = 2.19

Optical Properties: Transparent. *Color:* Colorless to white; crystals zoned, with transparent ends and white sides. *Streak:* [White.] *Luster:* Vitreous.
Optical Class: Uniaxial (–). $\omega = 1.522(2)$ $\epsilon = 1.507(2)$

Cell Data: *Space Group:* $P6_3/mmc$, $P6_3mc$, or $P\bar{6}2c$. a = 13.244(1) c = 15.988(2) Z = 1

X-ray Powder Pattern: Bellerberg volcano, Germany.
3.80 (100), 6.58 (80), 2.95 (70), 2.21 (70), 2.70 (50), 2.50 (50), 1.83 (50)

Chemistry:

	(1)
SiO_2	34.41
Al_2O_3	27.90
CaO	9.31
SrO	7.63
BaO	1.22
Na_2O	0.70
K_2O	1.95
H_2O	[16.88]
Total	[100.00]

(1) Bellerberg volcano, Germany; by electron microprobe, H_2O by difference; corresponds to $K_{1.33}Na_{0.72}Ca_{5.32}Sr_{2.36}Ba_{0.26}Al_{17.55}Si_{18.36}O_{72} \cdot 30H_2O$.

Mineral Group: Zeolite group.

Occurrence: In leucite tephrite lava at contacts between lava and calcium-rich xenoliths.

Association: Sanidine, clinopyroxene, pyrrhotite, thompsonite, ettringite, willhendersonite, gismondine, jasmundite, mayenite, srebrodolskite, calcic olivine, portlandite.

Distribution: From the Bellerberg volcano, two km north of Mayen, Eifel district, Germany.

Name: For its occurrence at the Bellerberg (also Bellberg) volcano, Germany.

Type Material: University of Vienna, Vienna, Austria.

References: (1) Rüdinger, B., E. Tillmanns, and G. Hentschel (1993) Bellbergite – a new mineral with the zeolite structure type EAB. Mineral. Petrol., 48, 147–152.

Crystal Data: Orthorhombic. *Point Group:* 222. Radiated or stellate with minute foliated structure; compact, hornlike forms; massive granular; in intricate intergrowths with other minerals.

Physical Properties: *Cleavage:* Perfect on {100}, {010}, {001}. *Fracture:* Conchoidal. Hardness = 4.5–6 D(meas.) = 2.98–3.02 D(calc.) = [3.07]

Optical Properties: Translucent. *Color:* Pale yellow to yellowish brown. *Luster:* Waxy, pearly on perfect cleavage.
Optical Class: Biaxial (–). *Pleochroism:* Weak, in yellows. *Orientation:* X ⊥ {001}. $\alpha = 1.624$ $\beta = 1.650$ $\gamma = 1.650$ 2V(meas.) = ~0°

Cell Data: *Space Group:* $P2_12_12_1$. a = 14.5 b = 17.5 c = 29.1 Z = [16]

X-ray Powder Pattern: Franklin, New Jersey, USA.
3.66 (100), 7.25 (90), 3.58 (90), 2.436 (14), 3.43 (9), 3.30 (6), 3.09 (5)

Chemistry:

	(1)	(2)	(3)
SiO_2	39.00	38.36	37.03
Al_2O_3		0.96	
Fe_2O_3		0.71	
FeO	[3.75]	4.94	
MnO	42.12	39.22	54.64
ZnO	2.86	2.93	
MgO	3.83	3.35	
CaO	trace	0.62	
H_2O^+	8.44	8.01	8.33
H_2O^-		0.60	
Total	[100.00]	99.70	100.00

(1) Franklin, New Jersey, USA; FeO lost, estimated by difference. (2) Do. (3) $Mn_5Si_4O_{10}(OH)_6$.

Occurrence: Layered parallel to the walls of secondary veins of calcite in a metamorphosed stratiform zinc orebody (Franklin, New Jersey, USA); intimately intergrown with other manganese minerals (Olympic Peninsula, Washington, USA).

Association: Calcite, willemite (Franklin, New Jersey, USA); hausmannite, jacobsite, braunite, alleghanyite, barite, grossular, johannsenite, calcite (Olympic Peninsula, Washington, USA); hausmannite, rhodochrosite, inesite (Hale Creek, California, USA).

Distribution: In the USA, at Franklin, Sussex Co., New Jersey; in Washington, on the Olympic Peninsula, an ore mineral at a number of small deposits in Clallan, Jefferson, and Mason Cos.; from Ray, Pinal Co., Arizona; at Hale Creek, Trinity Co., California; and near Golconda, Humboldt Co., Nevada. In the Treburland mine, Altarnun, Cornwall, and near Chillaten, Devon, England. From Langbån, Värmland, Sweden. In Russia, at a number of poorly-defined localities. In the Tokuzawa mine, Fukushima Prefecture; the Ichinomata mine, Kumamoto Prefecture; and at Kunimiyama, Kochi Prefecture, Japan. In the Wessels mine, near Kuruman, Cape Province, South Africa.

Name: For Clarence Sweet Bement (1843–1923), mineral collector of Philadelphia, Pennsylvania, USA.

References: (1) Dana, E.S. (1892) Dana's system of mineralogy, (6th edition), 704. (2) Palache, C. (1935) The minerals of Franklin and Sterling Hill, Sussex County, New Jersey. U.S. Geol. Sur. Prof. Paper 180, 117–118. (3) Kato, T. (1963) New data on the so-called bementite. J. Min. Soc. Japan, 6, 93–103 (in Japanese). (4) (1964) Amer. Mineral., 49, 446–447 (abs. ref. 3).

Crystal Data: Hexagonal. *Point Group:* $\bar{6}2m$. Flat pyramidal crystals, tabular on {0001}, triangular or hexagonal in outline, to 6 cm. *Twinning:* By rotation about [0001].

Physical Properties: *Cleavage:* Poor on {10$\bar{1}$1}. *Fracture:* Conchoidal. Hardness = 6–6.5 D(meas.) = 3.65 D(calc.) = 3.68 Blue fluorescence under SW UV; intense blue cathodoluminescence.

Optical Properties: Transparent to translucent. *Color:* Sapphire-blue, white to colorless, pink; commonly bicolored. *Luster:* Vitreous.
Optical Class: Uniaxial (+). *Pleochroism:* O = colorless; E = purple, indigo, greenish blue.
$\omega = 1.756–1.757$ $\epsilon = 1.802–1.804$

Cell Data: *Space Group:* $P\bar{6}2c$. a = 6.6410(7) c = 9.7597(10) Z = 2

X-ray Powder Pattern: San Benito Co., California, USA. (ICDD 26-1036).
3.71 (100), 2.72 (90), 2.86 (75), 5.73 (65), 2.169 (60), 3.30 (50), 3.13 (50)

Chemistry:

	(1)	(2)	(3)
SiO$_2$	43.68	42.62	43.60
TiO$_2$	20.09	19.44	19.32
BaO	36.33	37.27	37.08
Na$_2$O		0.14	
Total	100.10	99.47	100.00

(1) San Benito Co., California, USA. (2) Do.; by electron microprobe. (3) BaTiSi$_3$O$_9$.

Occurrence: In natrolite veins cutting glaucophane schist in a serpentine body (San Benito Co., California, USA); in a magnesio-riebeckite-quartz-phlogopite-albite dike cutting serpentinite (Ohmi, Japan).

Association: Neptunite, joaquinite, natrolite (San Benito Co., California, USA); ohmilite, bario-orthojoaquinite, leucosphenite (Ohmi, Japan).

Distribution: In the USA, at the Gem mine, Mina Numero Uno, and the Victor claim, San Benito Co., California. From Ohmi, Niigata Prefecture, Japan. At Broken Hill, New South Wales, Australia.

Name: For the occurrence in San Benito Co., California, USA.

Type Material: n.d.

References: (1) Louderback, G.D. (1907) Benitoite, a new California gem mineral. Univ. Calif., Bull. Dept. Geol., 5, 149–153. (2) Louderback, G.D. (1909) Benitoite, its paragenesis and mode of occurrence. Univ. Calif., Bull. Dept. Geol., 5, 331–380. (3) Fischer, K. (1969) Verfeinerung der Kristallstruktur von Benitoit BaTi[Si$_3$O$_9$]. Zeits. Krist., 129, 222–243 (in German with English abs.). (4) Laird, J. and A.L. Albee (1972) Chemical composition and physical, optical and structural properties of benitoite, neptunite and joaquinite. Amer. Mineral., 57, 85–102. (5) Wise, W.S. and R.H. Gill (1977) Minerals of the Benitoite Gem mine. Mineral. Record, 8, 442–452.

$(Fe^{2+}, Fe^{3+}, Al, Mg)_{2-3}(Si, Al)_2O_5(OH)_4$ **Berthierine**

Crystal Data: Monoclinic; in part hexagonal or orthorhombic. *Point Group: m.* Commonly as microcrystalline oolitic aggregates.

Physical Properties: Hardness = Soft. D(meas.) = n.d. D(calc.) = [3.06]

Optical Properties: Transparent. *Color:* Dark olive-green, yellowish green.
Optical Class: [Biaxial.] $n = 1.62–1.65$ 2V(meas.) = n.d.

Cell Data: *Space Group: Cm.* $a = 5.25–5.41$ $b = 9.10–9.33$ $c = 7.06–7.27$ $\beta = 104.5°$ Z = [2]

X-ray Powder Pattern: Stanion Lane pit, Corby, England; 1H.
7.105 (100), 3.546 (100), 2.520 (100), 2.147 (70), 1.560 (70), 1.775 (60), 1.478 (60)

X-ray Powder Pattern: Ayrshire, Scotland; mixed 1M and 1H.
7.04 (100), 3.513 (100), 2.514 (90), 1.552 (70), 2.137 (60), 2.673 (40), 2.399 (40)

Chemistry:

	(1)	(2)
SiO_2	20.84	26.01
TiO_2	0.51	
Al_2O_3	28.02	12.33
Fe_2O_3	5.38	5.32
FeO	32.29	37.07
MnO		0.04
MgO	2.02	6.13
CaO		0.48
H_2O	10.70	10.33
Total	99.76	97.71

(1) Belgorod district, Kursk, Russia; corresponds to $(Fe^{2+}_{1.48}Al_{0.96}Fe^{3+}_{0.22}Mg_{0.16})_{\Sigma=2.82}$ $(Si_{1.15}Al_{0.85})_{\Sigma=2.00}O_{5.04}(OH)_{3.93}$. (2) Kaňk, Czech Republic; corresponds to $(Fe^{2+}_{1.81}Mg_{0.53}$ $Al_{0.36}Fe^{3+}_{0.23})_{\Sigma=2.93}(Si_{1.52}Al_{0.48})_{\Sigma=2.00}O_{5.00}(OH)_{4.00}$.

Polymorphism & Series: 1M and 1H polytypes.

Mineral Group: Kaolinite-serpentine group.

Occurrence: Common in unmetamorphosed marine sediments; in lateritic and polar soils.

Association: Glauconite, chlorite-vermiculite interstratified with mica-smectite, siderite, calcite.

Distribution: Widespread; a few localities for studied material are: at Hayange, Meurthe-et-Moselle, France. In several mines in the Weardale fluorite district, Durham; in the Stanion Lane pit, Corby, Northamptonshire; on the Worth Farm and in the Philpots quarry, West Sussex, England. From the Belgorod district and the Voronezh anticline, Kursk, Russia. In the Chichibu mine, Saitama Prefecture, Japan. On Ellef Ringnes Island, Northwest Territories; from near Picton, Ontario; and at Mont Saint-Hilaire, Quebec, Canada.

Name: Honors the French geologist, Pierre Berthier (1782–1861).

References: (1) Dana, E.S. (1892) Dana's system of mineralogy, (6th edition), 658. (2) Brindley, G.W. (1951) The crystal structure of some chamosite minerals. Mineral. Mag., 29, 502–530. (3) Brindley, G.W. and R.F. Youell (1953) Ferrous chamosite and ferric chamosite. Mineral. Mag., 37, 57–70. (4) Brindley, G.W. (1982) Chemical composition of berthierines – a review. Clays and Clay Minerals, 30, 153–155.

Bertrandite $Be_4Si_2O_7(OH)_2$

Crystal Data: Orthorhombic. *Point Group:* $mm2$. Crystals thin tabular, commonly prismatic to needlelike, to 5 cm; in radial aggregates. *Twinning:* Common on {011} or {021}; twins heart-shaped or V-shaped with axes crossing at angles of about 60° and 120°.

Physical Properties: *Cleavage:* Perfect on {001}; distinct on {100}, {010}, and {110}. Hardness = 6–7 D(meas.) = 2.59–2.60 D(calc.) = [2.61] Pyroelectric.

Optical Properties: Transparent. *Color:* Colorless to slightly yellow. *Luster:* Vitreous, pearly on {001}.
Optical Class: Biaxial (–). *Orientation:* $X = a$; $Y = b$. *Dispersion:* $r < v$, weak.
$\alpha = 1.583–1.591$ $\beta = 1.598–1.605$ $\gamma = 1.608–1.614$ 2V(meas.) = ~75°

Cell Data: *Space Group:* $Cmc2_1$. a = 8.7135(4)   b = 15.268(1)   c = 4.5683(3)   Z = 4

X-ray Powder Pattern: Mica Creek, Mt. Isa, Queensland, Australia.
4.38 (100), 3.19 (90), 2.54 (80), 2.28 (60), 2.22 (50), 3.94 (40), 1.305 (40)

Chemistry:

	(1)	(2)
SiO_2	49.26	50.44
Al_2O_3	trace	
Fe_2O_3	1.40	
BeO	42.0	42.00
H_2O	6.90	7.56
Total	99.56	100.00

(1) Barbin, France. (2) $Be_4Si_2O_7(OH)_2$.

Occurrence: In fissures in granites and associated pegmatites and in miarolitic cavities in greisens; commonly an alteration product of beryl, more rarely as a primary mineral.

Association: Beryl, phenakite, herderite, tourmaline, muscovite, fluorite, quartz.

Distribution: In small amounts at numerous localities. A few affording pure material or larger crystals include: in France, near Nantes, in the Barbin quarries at Petit-Port, Loire-Atlantique. At Val Vigezzo, Piedmont, Italy. From Písek, Czech Republic. In the USA, in the Strickland quarry, Portland, Middlesex Co., Connecticut; in Virginia, at Amelia, Amelia Co.; in Colorado, on Mt. Antero, Chaffee Co. and elsewhere. At Spor Mountain, Juab Co., Utah, a large commercial deposit; in the Pala district, San Diego Co., California. In a large deposit at Sierra de Aguachile, Acuña, Coahuila, Mexico. Large crystals at the Golconda mine, near Governador Valadares, Minas Gerais, Brazil. From Kleine Spitzkopje, Namibia. In the Tae Hwa mine, Chung Cheong Buk Do, South Korea. From Akchatau, Kara-Oba, and Kounrad, Kazakhstan.

Name: For Emile Bertrand (1844–1909), French mineralogist.

Type Material: National School of Mines, Paris, France.

References: (1) Dana, E.S. (1892) Dana's system of mineralogy, (6th edition), 545–546. (2) Vernon, R.H. and K.L. Williams (1960) Bertrandite from Mica Creek, Queensland. Amer. Mineral., 45, 1300–1303. (3) Vlasov, K.A., Ed. (1966) Mineralogy of rare elements, v. II, 89–93. (4) Henderson, W.A. (1975) The bertrandites of Connecticut. Mineral. Record, 6, 114–123. (5) Downs, J.W. and F.K. Ross (1987) Neutron-diffraction study of bertrandite. Amer. Mineral., 72, 979–983.

Crystal Data: Hexagonal. *Point Group:* $6/m\,2/m\,2/m$. Crystals prismatic to tabular, with $\{10\bar{1}0\}$, $\{0001\}$, and $\{11\bar{2}0\}$; may be complexly terminated by pyramids, to 18 m and 180 t. Also radial, trapiche, columnar; granular to compact. *Twinning:* On $\{hk\bar{i}l\}$ forms, rarely.

Physical Properties: *Cleavage:* Imperfect on $\{0001\}$. *Fracture:* Conchoidal to uneven. *Tenacity:* Brittle. Hardness = 7.5–8 VHN = 1190–1450, 1300 average. D(meas.) = 2.63–2.97 increasing with alkali content; 2.7 average. D(calc.) = 2.640

Optical Properties: Transparent to translucent. *Color:* Colorless, white, pale blue to sky-blue, bluish green through green to greenish yellow or yellow, rose to peach, deep pink to raspberry-red; may be zoned; commonly colorless in thin section. *Streak:* White. *Luster:* Vitreous, resinous. *Optical Class:* Uniaxial (–). *Pleochroism:* Weak to distinct; O = colorless, yellowish green, light blue, yellowish red; E = sea-green, blue, purplish red. $\omega = 1.567$–1.610 $\epsilon = 1.565$–1.599; increasing with alkali content.

Cell Data: *Space Group:* $P6/mcc$. a = 9.205–9.274   c = 9.187–9.249   Z = 2

X-ray Powder Pattern: Royalston, Massachusetts, USA.
2.867 (100), 3.254 (95), 7.98 (90), 4.60 (50), 3.99 (45), 3.015 (35), 2.523 (30)

Chemistry:

	(1)	(2)	(3)		(1)	(2)	(3)
SiO$_2$	65.59	63.29	67.07	Na$_2$O	0.26	1.23	
Al$_2$O$_3$	18.25	17.76	18.97	Cs$_2$O		3.79	
FeO	1.16			Rb$_2$O		0.05	
BeO	13.46	10.79	13.96	LOI	0.70	1.10	
Li$_2$O		1.30		Total	99.42	99.31	100.00

(1) Mursinka, Russia; by electron microprobe. (2) Nuristan district, Laghman Province. Afghanistan; by electron microprobe. (3) Be$_3$Al$_2$Si$_6$O$_{18}$.

Occurrence: In granites and granite pegmatites, rarely in nepheline syenites. Also in mafic metamorphic rocks, low- to high-temperature hydrothermal veins and in vugs in rhyolite.

Association: Quartz, feldspar, muscovite, lepidolite, spodumene, amblygonite, tourmaline, topaz, cassiterite, columbite, tantalite.

Distribution: Prominent localities for fine specimens include: in Russia, in the Mursinka-Alabashka area, near Yekaterinburg (Sverdlovsk), Ural Mountains, and from Nerchinsk south to Borzya, Adun-Chilon Mountains, Siberia. At Volhynia, Ukraine. In the Muiâne pegmatite, Alto Ligonha district, Mozambique. From Madagascar, especially around Mt. Bity, south of Antsirabe, with giant crystals from the Malakialina district. In the Rafin-Gabas Hills, near Jos, Nigeria. In the Sandawana-Belingwe area, Mweza Range, Zimbabwe. Exceptional green crystals from the Muzo and Chivor districts, Boyacá Province, Colombia. In Brazil, from Minas Gerais, in large districts around Teófilo Otoni, Governador Valadares, Conselheiro Pena, and Itabira. Around Mingora, Swat district; from Dusso and the Skardu district, Pakistan. In Afghanistan, from Kunar and Nagar, near Karimabad, Hunza district, and around Khenj, Panjshir Valley. In the USA, from the Pala and Rincon districts, San Diego Co., California; at Hiddenite, Alexander Co., North Carolina; fine red crystals in the Wah Wah Mountains, Beaver Co., Utah; on Mt. Antero, Chaffee Co., Colorado; and in the Sawtooth Mountains area, Idaho.

Name: From the Greek *beryllos*, original significance unknown.

References: (1) Dana, E.S. (1892) Dana's system of mineralogy, (6th edition), 405–409. (2) Deer, W.A., R.A. Howie, and J. Zussman (1986) Rock-forming minerals, (2nd edition), v. 1B, disilicates and ring silicates, 372–409. (3) Sinkankas, J. (1981) Emeralds and other beryls. Chilton, Radnor, Pennsylvania, USA, 665 pp. (4) Aurisicchio, C., G. Fioravanti, O. Grubessi, and P.F. Zanazzi (1988) Reappraisal of the crystal chemistry of beryl. Amer. Mineral., 73, 826–837.

Crystal Data: Orthorhombic or monoclinic. *Point Group:* n.d. As fine spherulites, to 3 mm; as drusy segregations and fibrous crusts; powdery masses and coatings on other minerals.

Physical Properties: Hardness = 1 in aggregate. D(meas.) = 2.196 D(calc.) = n.d.

Optical Properties: Transparent. *Color:* White. *Luster:* Silky.
Optical Class: Biaxial (–). $\alpha = 1.50–1.544$ $\beta = 1.553(2)$ $\gamma = 1.52–1.562$ 2V(meas.) = $< 45°$

Cell Data: *Space Group:* n.d. Z = n.d.

X-ray Powder Pattern: Lovozero massif, Russia.
4.01 (100), 2.34 (100), 3.64 (90), 1.351 (80), 3.39 (70), 3.19 (70), 2.90 (70)

Chemistry:

	(1)	(2)
SiO_2	34.10	35.11
TiO_2	trace	
Al_2O_3	1.63	
Fe_2O_3	0.12	
BeO	40.00	43.84
MgO	trace	
CaO	0.50	
Na_2O	2.42	
H_2O^+	18.95	21.05
H_2O^-	3.25	
Total	100.97	100.00

(1) Lovozero massif, Russia. (2) $Be_3SiO_4(OH)_2 \cdot H_2O$.

Occurrence: In the center of a zoned pegmatite cutting arfvedsonite-bearing nepheline syenite (Lovozero massif, Russia); in analcime-rich veins (Ilímaussaq intrusion, Greenland).

Association: Albite, epididymite, bertrandite, natrolite, sodalite (Lovozero massif, Russia); analcime, sørensenite, chkalovite, epididymite, tugtupite (Ilímaussaq intrusion, Greenland).

Distribution: From the Lovozero massif, Kola Peninsula, Russia. At Kvanefjeld and on the Taseq slope, in the Ilímaussaq intrusion, southern Greenland.

Name: In allusion to the BERYLLium in its composition.

Type Material: n.d.

References: (1) Kuz'menko, M.V. (1954) Berillite [beryllite] – a new mineral. Doklady Acad. Nauk SSSR, 99, 51–454 (in Russian). (2) (1955) Amer. Mineral., 40, 787–788 (abs. ref. 1). (3) Vlasov, K.A., Ed. (1966) Mineralogy of rare elements, v. II, 96–98. (4) Andersen, S. (1967) On beryllite and bertrandite from the Ilímaussaq alkaline intrusion, South Greenland. Medd. Grønland, 181(4), 11–26.

Crystal Data: Cubic. *Point Group:* $\overline{4}3m$. As an extremely fine powder.

Physical Properties: Hardness = n.d. D(meas.) = n.d. D(calc.) = 2.813 (synthetic).

Optical Properties: Semitransparent. *Color:* White or gray; colorless in thin section. *Luster:* Powdery, earthy. *Optical Class:* Isotropic. $n = 1.625$

Cell Data: *Space Group:* $I\overline{4}3m$. a = 8.82–8.83 Z = 4

X-ray Powder Pattern: Fuka, Japan.
2.786 (100), 2.753 (95), 3.60 (90), 2.597 (50), 1.559 (50), 3.04 (40), 2.96 (40)

Chemistry:

	(1)	(2)	(3)
SiO_2	28.51	23.77	20.56
TiO_2	0.09	0.92	
Al_2O_3	21.79	23.59	34.89
Fe_2O_3	2.66	6.72	
FeO	0.25	0.20	
MnO	0.03	0.02	
MgO	2.72	2.00	
CaO	35.26	36.89	38.38
Na_2O	0.25	0.14	
K_2O	0.18	0.11	
H_2O^+	8.03	4.79	6.17
H_2O^-	0.43	0.40	
P_2O_5	0.02	0.02	
Total	100.22	99.57	100.00

(1) Fuka, Japan; mixed with vesuvianite. (2) Do.; contaminated by small amounts of gehlenite, vesuvianite, and hydrogrossular. (3) $Ca_2Al_2SiO_6(OH)_2$.

Polymorphism & Series: Dimorphous with kamaishilite.

Occurrence: In skarns in limestones, formed through alteration of gehlenite subjected to later retrograde hydration reactions.

Association: Vesuvianite, hydrogrossular, gehlenite, melilite, calcite.

Distribution: From Fuka, near Bicchu, Okayama Prefecture, and in the Akagané mine, Iwate Prefecture, Japan. At Carneal, Co. Antrim, Ireland.

Name: For Bicchu, the town encompassing the Japanese type locality.

Type Material: Department of Earth Sciences, Okayama University, Okayama, Japan, ONM-01; Institute of Geological Sciences, London, England.

References: (1) Henmi, C., I. Kusachi, K. Henmi, P.A. Sabine, and B.R. Young (1973) A new mineral bicchulite, the natural analogue of gehlenite hydrate, from Fuka, Okayama Prefecture, Japan, and Carneal, County Antrim, Northern Ireland. Mineral. J. (Japan), 7, 243–251. (2) (1974) Amer. Mineral., 59, 1330 (abs. ref. 1). (3) Sahl, K. and N.D. Chatterjee (1977) The crystal structure of bicchulite, $Ca_2[Al_2SiO_6](OH)_2$. Zeits. Krist., 146, 35–41. (4) Gupta, A.K. and N.D. Chatterjee (1978) Synthesis, composition, thermal stability, and thermodynamic properties of bicchulite, $Ca_2[Al_2SiO_6](OH)_2$. Amer. Mineral., 63, 58–65. (5) Sahl, K. (1980) Refinement of the crystal structure of bicchulite, $Ca_2[Al_2SiO_6](OH)_2$. Zeits. Krist., 152, 13–21.

Crystal Data: Monoclinic or triclinic, pseudomonoclinic. *Point Group:* 2 or 1. Crystals prismatic, elongated ∥ [010], striations common ∥ [100], in radiating groups, to 10 cm. Also coarse to fine granular.

Physical Properties: *Cleavage:* Perfect on {100}, good on {001}. *Fracture:* Conchoidal. Hardness = 6   D(meas.) = 2.29–2.30   D(calc.) = 2.29

Optical Properties: Transparent. *Color:* Colorless to white. *Streak:* White. *Luster:* Vitreous.
Optical Class: Biaxial (–). *Orientation:* $Z = b$; $X \wedge c = 28°$–$29°$. *Dispersion:* $r < v$.
$\alpha = 1.509$–1.510 $\beta = 1.520$–1.521 $\gamma = 1.522$–1.523 2V(meas.) = 45°

Cell Data: *Space Group:* $P2_1$. a = 8.613(4)   b = 4.962(2)   c = 7.600(4) $\beta = 114.45(1)°$
Z = 2, or   *Space Group:* $P1$. a = 8.606(1)   b = 4.953(1)   c = 7.599(1) $\alpha = 89.89(2)°$
$\beta = 114.42(2)°$ $\gamma = 89.96(2)°$ Z = 2

X-ray Powder Pattern: Bikita, Zimbabwe.
3.46 (100), 3.37 (100), 4.20 (90), 2.479 (90), 7.87 (80), 6.93 (50), 4.37 (40)

Chemistry:

	(1)	(2)	(3)
SiO_2	55.79	58.7	58.87
Al_2O_3	26.68	25.1	24.98
Fe_2O_3	0.07		
MgO	0.33		
Li_2O	6.51	7.27	7.32
Na_2O	0.10		
K_2O	0.17		
H_2O	9.82	8.98	8.83
Total	99.47	100.05	100.00

(1) Bikita, Zimbabwe. (2) Do.; by electron microprobe, Li by AA, H_2O by TGA.
(3) $LiAlSi_2O_6 \cdot H_2O$.

Mineral Group: Zeolite group, probable.

Occurrence: As a late-formed mineral in fractures in lithium-rich pegmatites.

Association: Eucryptite, quartz, petalite, feldspar, calcite, stilbite, allophane (Bikita, Zimbabwe); albite, quartz, fairfieldite (Foote mine, North Carolina, USA).

Distribution: From Bikita, Zimbabwe. In the Foote mine, Kings Mountain, Cleveland Co., North Carolina, USA.

Name: For the locality, Bikita, Zimbabwe.

Type Material: Harvard University, Cambridge, Massachusetts, USA, 124189.

References: (1) Hurlbut, C.S., Jr. (1957) Bikitaite, $LiAlSi_2O_6 \cdot H_2O$; a new mineral from Southern Rhodesia. Amer. Mineral., 42, 792–797. (2) Hurlbut, C.S., Jr. (1958) Additional data on bikitaite. Amer. Mineral., 43, 768–770. (3) Phinney, W.C. and D.B. Stewart (1961) Some physical properties of bikitaite and its dehydration and decomposition products. U.S. Geol. Sur. Prof. Paper 424, D353–D357. (4) Leavens, P.B., C.S. Hurlbut, Jr., and J.A. Nelen (1968) Eucryptite and bikitaite from King's [sic] Mountain, North Carolina. Amer. Mineral., 53, 1202–1207. (5) Kocman, V., R.I. Gait, and J. Rucklidge (1974) The crystal structure of bikitaite, $Li[AlSi_2O_6] \cdot H_2O$. Amer. Mineral., 59, 71–78. (6) Bissert, G. and F. Liebau (1986) The crystal structure of a triclinic bikitaite, $Li[AlSi_2O_6] \cdot H_2O$, with ordered Al/Si distribution. Neues Jahrb. Mineral., Monatsh., 241–252.

Crystal Data: Monoclinic. *Point Group:* $2/m$. Uncommon in good crystals, tabular or short prismatic, with pseudohexagonal outline, to 3 m. Typically irregular foliated or bent masses; in scaly aggregates or disseminated grains. *Twinning:* On composition plane $\{001\}$, twin axis $[310]$.

Physical Properties: *Cleavage:* $\{001\}$, perfect. *Tenacity:* Brittle to flexible, elastic. Hardness = 2.5–3 D(meas.) = 2.7–3.3 D(calc.) = 3.25

Optical Properties: Semitransparent. *Color:* Dark green, brown, black, reddish brown, light yellow, grayish yellow, brownish green, brown; yellow to reddish brown in thin section. *Streak:* White. *Luster:* Splendent to submetallic, vitreous, pearly on cleavage. *Optical Class:* Biaxial (–). *Pleochroism:* Strong; X = gray-yellow, yellow-brown, orange-brown; $Y = Z$ = dark brown, dark green, dark red-brown. *Orientation:* $Y = b$; $X \wedge c = 0°$–$3°$; $Z \wedge a = 0°$–$9°$. *Dispersion:* $r < v$, Fe-rich; may be $r > v$, Mg-rich; weak. *Absorption:* $Y \simeq Z \gg X$. $\alpha = 1.565$–1.625 $\beta = 1.605$–1.696 $\gamma = 1.605$–1.696 2V(meas.) = $0°$–$25°$

Cell Data: *Space Group:* $C2/m$. a = 5.3 b = 9.2 c = 10.2 $\beta = 100°$ Z = 2

X-ray Powder Pattern: Edenville, Orange Co., New York, USA.
10.1 (100), 3.37 (100), 2.66 (80), 2.45 (80), 2.18 (80), 2.00 (80), 1.67 (80)

Chemistry:

	(1)		(1)		(1)
SiO_2	36.25	MnO	0.49	K_2O	9.57
TiO_2	3.39	MgO	11.80	Cl	0.06
Al_2O_3	13.90	CaO	0.00	H_2O^+	2.80
Fe_2O_3	6.80	Li_2O	0.03	Total	100.00
FeO	14.81	Na_2O	0.10		

(1) Vercelli, Italy; by electron microprobe, average of six analyses on one grain, Fe^{3+} by a semimicrovolumetric method, H_2O by TGA; corresponds to $(K_{0.94}Na_{0.02})_{\Sigma=0.96}$ $(Mg_{1.35}Fe^{2+}_{0.95}Mn_{0.03})_{\Sigma=2.33}(Fe^{3+}_{0.39}Ti_{0.20}Al_{0.05}Li_{0.01})_{\Sigma=0.65}(Si_{2.79}Al_{1.21})_{\Sigma=4.00}$ $[O_{10.55}(OH)_{1.44}Cl_{0.01}]_{\Sigma=12.00}$, in the general structural formula $K(Mg, Fe^{2+})_{3-2}(Al, Fe^{3+}, Ti)_{0-1}$ $(Si_{3-2.5}Al_{1-1.5})_{\Sigma=4}O_{10-11}(OH, F)_{2-1}$.

Polymorphism & Series: $1M$, $2M_1$, $3A$ polytypes; forms a series with phlogopite.

Mineral Group: Mica group.

Occurrence: An important rock-forming mineral under a wide range of conditions. In regionally metamorphosed schists and gneisses, and in contact metamorphosed rocks; in granites to nepheline syenites; less common in extrusive igneous rocks, from rhyolites to basalts. Characteristic of potassic hydrothermal alteration; in detrital sediments.

Association: Quartz, potassic feldspar, plagioclase, nepheline, muscovite, pyroxenes, amphiboles, andalusite, cordierite, garnet, spinel.

Distribution: Good crystals from: in Italy, at Vesuvius and Monte Somma, Campania, and in the Pfitschtal and on Mt. Monzoni, Val di Fassa, Trentino-Alto Adige. At Brevik and Arendal, Norway. From near Miass, Ilmen Mountains, Southern Ural Mountains, Russia. In the USA, from Franklin and Sterling Hill, Ogdensburg, Sussex Co., New Jersey; Monroe, Orange Co., and Russell, St. Lawrence Co., New York; at Easton, Northampton Co., Pennsylvania; from the Pala and Rincon districts, San Diego Co., California. In Canada, in Ontario, at Bancroft, Wakefield, and Otter Lake, Parry Sound; in Quebec, in the Bear Lake mine, Pontiac Co.

Name: For the French mineralogist and physicist, Jean Baptiste Biot (1774–1862).

References: (1) Dana, E.S. (1892) Dana's system of mineralogy, (6th edition), 627–632. (2) Deer, W.A., R.A. Howie, and J. Zussman (1963) Rock-forming minerals, v. 3, sheet silicates, 55–84. (3) Nagelschmidt, G. (1937) X-ray investigations on clays. Part III. Zeits. Krist., 97, 514–521. (4) Brigatti, M.F. and P. Davoli (1990) Crystal-structure refinements of $1M$ plutonic biotites. Amer. Mineral., 75, 305–313.

Crystal Data: Monoclinic. *Point Group: m.* Minutely crystalline, powdery, earthy, massive.

Physical Properties: *Fracture:* Even to flat conchoidal when massive. *Tenacity:* Brittle. Hardness = n.d. D(meas.) = 4.47 D(calc.) = [5.09]

Optical Properties: Semitransparent. *Color:* Yellow, green. *Streak:* Light green. *Optical Class:* Biaxial. $\alpha = 1.93$ $\beta = 1.97$ $\gamma = 2.01$ 2V(meas.) = n.d.

Cell Data: *Space Group: Cm.* $a = 5.21$ $b = 9.02$ $c = 7.74$ $\beta = 100°40'$ Z = 2

X-ray Powder Pattern: Schneeberg, Germany; could be confused with chapmanite. 7.63 (100), 3.87 (100), 2.90 (70), 3.18 (50), 3.58 (35), 2.59 (35), 2.53 (25)

Chemistry:

	(1)	(2)	(3)
SiO_2	23.08	23.9	23.03
Al_2O_3		0.3	
Fe_2O_3	33.33	29.3	30.60
FeO		1.8	
Bi_2O_3	43.26	42.5	44.64
As_2O_3		0.08	
H_2O^+		1.8	1.73
Total	99.67	99.7	100.00

(1) Schneeberg, Germany. (2) Do.; corresponds to $Bi_{0.92}(Fe_{1.86}^{3+}Fe_{0.25}^{2+}Al_{0.03})_{\Sigma=2.14}Si_2O_8(OH)$. (3) $BiFe_2(SiO_4)_2(OH)$.

Occurrence: Probably of hydrothermal origin, in veins cutting shale (Schneeberg, Germany).

Association: Quartz, "chalcedony," bismuth, cobaltite, arsenopyrite, chlorargyrite (Schneeberg, Germany); quartz, bismuth, galena, silver ores (Johanngeorgenstadt, Germany); clinobisvanite (Lodi # 4 claim, California, USA).

Distribution: In Germany, at Schneeberg, Johanngeorgenstadt, and Gersdof, Saxony. In the South Terras mine, St. Stephen-in-Brannel, Cornwall, and at Buckbarrow Beck, Carney Fell, Cumbria, England. From the Lodi # 4 claim, Plumas Co., California, and in the Organ Mountains, Dona Ana Co., New Mexico, USA.

Name: For BISMUTh and iron, FERRum, in the composition.

References: (1) Dana, E.S. (1892) Dana's system of mineralogy, (6th edition), 562. (2) Milton, C., J.M. Alexrod, and B. Ingram (1958) Bismutoferrite, chapmanite, and "hypochlorite". Amer. Mineral., 43, 656–670. (3) Zhukhlistov, A.P. and B.B. Zvyagin (1977) Determination of the crystal structures of chapmanite and bismuthoferrite [bismutoferrite] by high-voltage electron diffraction. Kristallografiya (Sov. Phys. Crystal.), 22, 731–738 (in Russian).

Crystal Data: Monoclinic. *Point Group: 2/m.* Crystals thin and pseudohexagonal, to 1 cm; as lamellae which may form rosettes and as dense, micaceous aggregates.

Physical Properties: *Cleavage:* Perfect micaceous on {001}. *Fracture:* Uneven. Hardness = 4–4.5   VHN = 277–304   D(meas.) = 3.02–3.05   D(calc.) = 3.14

Optical Properties: Transparent to translucent. *Color:* Pearly white, grayish blue, greenish blue, light brown; colorless in thin section, prominently zoned into six sectors. *Luster:* Vitreous, pearly on cleavages.
Optical Class: Biaxial (–). *Orientation:* $X \simeq c$; $Y \simeq a$; $Z = b$. $\alpha = 1.651$ $\beta = 1.659$
$\gamma = 1.661$ 2V(meas.) = 35°–52°

Cell Data: *Space Group: C2/c.* a = 4.98   b = 8.67   c = 18.74 $\beta = \sim 90°$ Z = 4

X-ray Powder Pattern: Mt. Bity, Madagascar.
2.480 (100), 1.45 (100), 2.043 (90), 3.136 (80), 1.878 (70), 4.29 (60), 2.355 (60)

Chemistry:

	(1)	(2)	(3)
SiO$_2$	31.95	33.37	31.04
Al$_2$O$_3$	41.75	36.24	39.51
FeO		0.17	
BeO	2.27	7.30	6.46
MgO	0.13	0.04	
CaO	14.30	14.42	14.48
Li$_2$O	2.73	2.39	3.86
Na$_2$O	0.40	0.29	
K$_2$O	0.16	0.04	
H$_2$O	6.50	5.72	4.65
Total	100.19	99.98	100.00

(1) Mt. Bity, Madagascar; BeO apparently partially determined as Al$_2$O$_3$. (2) Londonderry, Australia. (3) CaLiAl$_2$(AlBeSi$_2$)O$_{10}$(OH)$_2$.

Polymorphism & Series: 2M$_1$ polytype.

Mineral Group: Mica group.

Occurrence: A late-stage mineral in lithium-bearing pegmatites.

Association: Tourmaline, albite, lepidolite (Mt. Bity, Madagascar); bavenite, beryl, albite, cassiterite, columbite, tourmaline (Londonderry, Australia); calcite, albite (Strickland quarry, Connecticut, USA).

Distribution: In Madagascar, from Maharitra, on Mt. Bity. In the Namherere mine, Aukole, Uganda. In the USA, from the Strickland quarry, Portland, Middlesex Co., Connecticut, and the Harding pegmatite, Dixon, Taos Co., New Mexico. On the Huron claims, 16 km north of Point du Bois, Manitoba, Canada. At Pizzo Marcio, Val Vigezzo, Piedmont, Italy. From the Maantienvarsi pegmatite dike, Eräjärvi pegmatite area, Orivesi, Finland. At Tittling, near Passau, Bavaria, Germany. At Londonderry, Western Australia.

Name: For the locality, Mt. Bity, Madagascar.

Type Material: Natural History Museum, Paris, France; Harvard University, Cambridge, Massachusetts, USA, 87680.

References: (1) Lacroix, A. (1908) Les minéraux des filons de pegmatite à tourmaline lithique de Madagascar. Bull. Soc. fr. Minéral., 31, 218–247 (in French). (2) (1950) Amer. Mineral., 35, 1091 (discussion of analysis in ref. 1). (3) Strunz, H. (1956) Bityit, ein berylliumglimmer. Zeits. Krist., 107, 325–330 (in German). (4) Vlasov, K.A., Ed. (1966) Mineralogy of rare elements, v. II, 114–116. (5) Lin, J.-C. and S. Guggenheim (1983) The crystal structure of a Li,Be-rich brittle mica: a dioctahedral-trioctahedral intermediate. Amer. Mineral., 68, 130–142.

Crystal Data: Orthorhombic. *Point Group:* $2/m\ 2/m\ 2/m$. Crystals, chisel-shaped, bladed, to 0.5 mm, elongated along [010], showing {011}, {010}, {102}, {001}, and {100}. As radial hemispheres, to 2 mm, composed of fine-grained polycrystalline aggregates.

Physical Properties: *Fracture:* Conchoidal. *Tenacity:* Brittle. Hardness = ~ 3.5 D(meas.) = 1.998 D(calc.) = 1.994 Altered material fluoresces weak bluish under both LW and SW UV.

Optical Properties: Transparent to translucent. *Color:* Colorless to white, due to alteration. *Streak:* White. *Luster:* Vitreous to dull.
Optical Class: Biaxial (–). *Orientation:* $X = c$; $Y = a$; $Z = b$. *Dispersion:* $r > v$, strong.
$\alpha = 1.480(1)$ $\beta = 1.481(1)$ $\gamma = 1.487(1)$ 2V(meas.) = $25(6)°$

Cell Data: *Space Group: Imma.* $a = 20.236(2)$ $b = 23.798(1)$ $c = 12.798(1)$ $Z = 1$

X-ray Powder Pattern: Near Goble, Oregon, USA.
11.269 (100), 3.855 (86), 3.368 (63), 4.456 (34), 11.884 (33), 4.354 (31), 3.604 (31)

Chemistry:

	(1)
SiO_2	63.01
Al_2O_3	12.60
Fe_2O_3	0.11
MgO	0.07
CaO	5.89
Na_2O	1.23
K_2O	0.10
H_2O	17.0
Total	100.01

(1) Near Goble, Oregon, USA; by electron microprobe, H_2O by C-H-N analysis; corresponds to $(Ca_{7.8}Na_{2.9}K_{0.2}Mg_{0.1})_{\Sigma=11.0}(Si_{77.6}Al_{18.3}Fe_{0.1})_{\Sigma=96.0}O_{192} \cdot 70H_2O$.

Mineral Group: Zeolite group.

Occurrence: In vesicles in one of a series of porphyritic basalt flows intercalated with pyroclastics and minor sediments.

Association: Mordenite, tschernichite, zeolites, apophyllite, calcite, aragonite, "opal."

Distribution: Along Goble Creek, near Goble, Columbia Co., Oregon, USA.

Name: After Robert Maxwell Boggs (1918–), of Seattle, Washington, USA, and his son, Dr. Russell Calvin Boggs (1952–), of Cheney, Washington, USA, mineral collectors.

Type Material: American Museum of Natural History, New York, New York, USA; Royal Ontario Museum, Toronto, Canada.

References: (1) Howard, D.G., R.W. Tschernich, J.V. Smith, and G.L. Klein (1990) Boggsite, a new high-silica zeolite from Goble, Columbia County, Oregon. Amer. Mineral., 75, 1200–1204. (2) Pluth, J.J. and J.V. Smith (1990) Crystal structure of boggsite, a new high-silica zeolite with the first three-dimensional channel system bounded by both 12- and 10-rings. Amer. Mineral., 75, 501–507.

HK(UO₂)(SiO₄)·1.5H₂O $\qquad$ Boltwoodite

HK(UO$_2$)(SiO$_4$)·1.5H$_2$O $\hfill$ **Boltwoodite**

Crystal Data: Monoclinic. *Point Group:* 2. Radiating acicular to fibrous, elongated along [010], to 1 cm.

Physical Properties: *Cleavage:* {010} perfect, {001} imperfect. Hardness = 3.5–4 D(meas.) = ~4.7 D(calc.) = [4.46] Fluoresces dull green in both SW and LW UV; radioactive.

Optical Properties: Transparent to translucent. *Color:* Pale yellow. *Luster:* Pearly; radial aggregates vitreous to silky; dull to earthy in microcrystalline pseudomorphs.
Optical Class: Biaxial (–). *Pleochroism:* Weak; X = colorless; $Y = Z$ = yellow; anomalous blue interference colors. $\alpha = 1.668–1.670$ $\beta = 1.695–1.696$ $\gamma = 1.698–1.703$ 2V(meas.) = Large.

Cell Data: *Space Group:* $P2_1$. $a = 7.073(2)$ $b = 7.064(1)$ $c = 6.638(1)$ $\beta = 105°45(1)'$ $Z = 2$

X-ray Powder Pattern: Delta mine, Utah, USA.
6.81 (100), 3.40 (90), 2.95 (80), 3.54 (70), 2.91 (70), 6.40 (50), 5.45 (50)

Chemistry:

	(1)	(2)	(3)
SiO$_2$	14.8	13.27	14.00
Fe$_2$O$_3$		0.52	
UO$_3$	68.5	67.19	66.64
CaO		1.59	
Na$_2$O	0.4	2.54	
K$_2$O	9.4	7.83	10.97
H$_2$O	6.9		8.39
LOI		6.91	
Total	[100.0]	99.85	100.00

(1) Delta mine, Utah, USA; original total given as 99.88%, after deduction of CuO 9.61% and SO$_3$ 2.12% as brochantite, remainder Al$_2$O$_3$, V$_2$O$_5$, PbO, CaO, MgO 0.34%, insoluble 0.19%, recalculated to 100.0%. (2) Swakopmund, Namibia; loss on ignition taken as H$_2$O; corresponds to (H$_3$O)$_{0.55}$(K$_{0.78}$Na$_{0.39}$Ca$_{0.14}$)$_{\Sigma=1.31}$(UO$_2$)(SiO$_4$)·H$_2$O. (3) HK(UO$_2$)(SiO$_4$)·1.5H$_2$O.

Occurrence: In the outer silicate zone of alteration surrounding hydrated uranyl oxides incrusting primary uraninite; also filling fractures at some distance from primary uraninite. From pegmatite and Colorado-Plateau-type sandstone uranium deposits.

Association: Uraninite, becquerelite, fourmarierite, phosphouranylite, gypsum, fluorite.

Distribution: In the USA, from the Delta mine, Emery Co., Utah; the New Method mine, near Amboy, San Bernardino Co., California; in the Little Indian No. 36 and the Lookout No. 22 mines, Marshall Pass area, Gunnison Co., and elsewhere in Colorado; and in the Williams quarry, near Easton, Northampton Co., Pennsylvania. Exceptionally developed at the Rössing mine, near Swakopmund, Namibia. From near Dalbeattie, Kirkcudbrightshire, Scotland. At Karago, Rwanda. From Quebrada del Tigre, Cordoba Province, and near Guandacol, La Rioja Province, Argentina. From near Myponga, South Australia. In the Togo mine, Tottori Prefecture, Japan. Other minor occurrences are known.

Name: For Bertram Borden Boltwood (1870–1927), radiochemist of Yale University, New Haven, Connecticut, USA, who devised the U-Pb method of measuring geologic time.

Type Material: National Museum of Natural History, Washington, D.C., USA, 112710.

References: (1) Frondel, C. and J. Ito (1956) Boltwoodite, a new uranium silicate. Science, 124, 931. (2) (1957) Amer. Mineral., 42, 307 (abs. ref. 1). (3) Honea, R.M. (1961) New data on boltwoodite, an alkali uranyl silicate. Amer. Mineral., 46, 12–25. (4) Kato, T. and Y. Miura (1974) Cell dimensions of boltwoodite. Mineral. J. (Japan), 7, 400–404. (5) Strunz, H. and C. Tennyson (1981) Symmetry and twinning in boltwoodite. Kristallografiya (Sov. Phys. Crystal.), 26, 1288–1292 (in Russian). (6) Stohl, F.V. and D. K. Smith (1981) The crystal chemistry of the uranyl silicate minerals. Amer. Mineral., 66, 610–625.

Bornemanite $BaNa_4Ti_2NbSi_4O_{17}(F, OH) \cdot Na_3PO_4$

Crystal Data: Orthorhombic. *Point Group:* $2/m \, 2/m \, 2/m$ or $mm2$. As platy aggregates, to 1 cm, of fine leaflets along cleavages and on the surface of lomonosovite and in natrolite.

Physical Properties: *Cleavage:* Perfect on {001}. *Tenacity:* Brittle; fibers are flexible. Hardness = 3.5–4 VHN = 257–283 D(meas.) = 3.47–3.50 D(calc.) = 3.49

Optical Properties: Translucent to transparent. *Color:* Pale yellow. *Luster:* Pearly. *Optical Class:* Biaxial (+). *Pleochroism:* Weak; $X = Y$ = colorless; Z = brownish. *Orientation:* $X = c$; $Y = b$; $Z = a$. $\alpha = 1.682–1.683$ $\beta = 1.687–1.695$ $\gamma = 1.718–1.720$ 2V(meas.) = 40° 2V(calc.) = 66°

Cell Data: *Space Group:* $Ibmm$ or $Ibm2$. a = 5.48(5) b = 7.10(5) c = 48.2(1) Z = 4

X-ray Powder Pattern: Lovozero massif, Russia.
24.1 (100), 8.04 (100), 3.44 (100), 3.02 (100), 2.682 (80), 1.610 (80), 1.781 (70)

Chemistry:

	(1)	(2)		(1)	(2)
SiO_2	23.96	25.00	BaO	12.05	13.00
TiO_2	18.72	18.00	Li_2O	0.10	
ZrO_2	0.20	0.25	Na_2O	19.62	20.00
Al_2O_3	0.55	0.00	K_2O	0.65	0.67
Fe_2O_3		0.30	Rb_2O	0.002	
Nb_2O_5	9.22	8.86	Cs_2O	0.00	
Ta_2O_5		0.66	F	1.52	1.64
FeO	0.17		H_2O^+	2.44	
MnO	2.97	2.48	H_2O^-	0.30	
MgO	0.04	0.06	H_2O		0.70
CaO	0.33	0.77	P_2O_5	6.57	6.80
SrO	0.70	0.68	$-O = F_2$	0.64	0.69
			Total	99.47	99.18

(1–2) Lovozero massif, Russia.

Occurrence: In the natrolite zone of alkalic pegmatites in a differentiated alkalic massif.

Association: Natrolite, lomonosovite.

Distribution: In the Jubilee pegmatite, on Mt. Karnasurt, and on Mt. Alluaiv, Lovozero massif, Kola Peninsula, Russia.

Name: For Irina D. Borneman-Starynkevich, Russian mineralogist specializing in Khibiny and Lovozero minerals.

Type Material: Geology Museum, Kola Branch, Academy of Sciences, Apatity; A.E. Fersman Mineralogical Museum, Academy of Sciences, Moscow, Russia; National School of Mines, Paris, France; National Museum of Natural History, Washington, D.C., USA, 143813.

References: (1) Men'shikov, Y.P., I.V. Bussen, E.A. Goiko, N.I. Zabavnikova, A.N. Mer'kov, and A.P. Khomyakov (1975) Bornemanite, a new silicophosphate of sodium, titanium, niobium and barium. Zap. Vses. Mineral. Obshch., 104, 322–326 (in Russian). (2) (1976) Amer. Mineral., 61, 338 (abs. ref. 1).

Crystal Data: Monoclinic. *Point Group:* 2/m. Pseudohexagonal crystals, to 4 μm, aggregated in coatings on other minerals.

Physical Properties: *Cleavage:* Perfect on {001}; poor {010} parting. *Fracture:* Subconchoidal. Hardness = 2.5–3 in aggregate. D(meas.) = 2.81 (on a mixture). D(calc.) = 2.89 (2M$_1$); 2.90 (1M).

Optical Properties: Nearly opaque. *Color:* White to buff or pale cream. *Streak:* White. *Luster:* Dull, earthy to porcelaneous.
Optical Class: Biaxial (–). *Orientation:* X ∧ c = −1°; Y ∧ a = 2°; Z = b. *Dispersion:* r > v, weak. α = 1.557(2) β = 1.587(2) γ = 1.593(2) 2V(meas.) = 44(2)° 2V(calc.) = 47.5°

Cell Data: *Space Group:* [C2/c] (2M$_1$ polytype by analogy to muscovite). a = 5.075(1) b = 8.794(4) c = 19.815(25) β = 95.59(3)° Z = [2], or *Space Group:* [C2/c.] (1M polytype by analogy to muscovite). a = 5.077(1) b = 8.775(3) c = 10.061(2) β = 101.31(2)° Z = [4]

X-ray Powder Pattern: Little Three mine, California, USA.
3.569 (100), 4.391 (80), 3.008 (80), 2.505 (80), 9.862 (60), 4.239 (40), 4.007 (40)

Chemistry:

	(1)		(1)
SiO$_2$	48.1	Na$_2$O	< 0.05
TiO$_2$	< 0.01	K$_2$O	11.0
B$_2$O$_3$	7.0	Rb$_2$O	0.52
Al$_2$O$_3$	28.1	Cs$_2$O	0.05
Fe$_2$O$_3$	0.1	F	0.76
MnO	0.08	H$_2$O$^+$	4.55
MgO	0.15	H$_2$O$^-$	0.22
CaO	0.1	P$_2$O$_5$	< 0.05
Li$_2$O	0.05	−O = F$_2$	0.32
		Total	100.46

(1) Little Three mine, California, USA; by a combination of electron microprobe, XRF, AA, and ICP-atomic emission spectroscopy; corresponding to (K$_{0.89}$Rb$_{0.02}$Ca$_{0.01}$)$_{\Sigma=0.92}$ (Al$_{1.93}$Li$_{0.01}$Mg$_{0.01}$)$_{\Sigma=1.95}$(Si$_{3.06}$B$_{0.77}$Al$_{0.17}$)$_{\Sigma=4.00}$O$_{9.82}$[(OH)$_{2.02}$F$_{0.16}$]$_{\Sigma=2.18}$.

Polymorphism & Series: 2M$_1$, 1M polytypes.

Mineral Group: Mica group.

Occurrence: A coating on other minerals on the floor of a pegmatite pocket, hydrothermally deposited after rupture of the pocket.

Association: Lepidolite, quartz, microcline, topaz.

Distribution: From the Little Three mine, Ramona district, San Diego Co., California, USA.

Name: Presumably for the BOROn content and relation to *muscovite*.

Type Material: National Museum of Natural History, Washington, D.C., USA, 166821.

References: (1) Foord, E.E., R.F. Martin, J.J. Fitzpatrick, J.E. Taggart, Jr., and J.G. Crock (1991) Boromuscovite, a new member of the mica group, from the Little Three mine pegmatite, Ramona district, San Diego County, California. Amer. Mineral., 76, 1998–2002.

Crystal Data: Orthorhombic (?). *Point Group:* n.d. As divergent sprays of bladed crystals, commonly composed of tiny microcrystals, to 250 μm; as radial aggregates and hemispherules of such crystals, to 3 mm.

Physical Properties: Hardness = $\sim$1 D(meas.) = 2.93(10) D(calc.) = n.d.

Optical Properties: Semitransparent. *Color:* Dark brownish red. *Streak:* Brownish red. *Luster:* Vitreous to submetallic on prism surfaces.
Optical Class: Biaxial (–). *Pleochroism:* Strong; $X = Y$ = red-brown; Z = yellow-brown. *Dispersion:* $r < v$, very strong. *Absorption:* $X = Y > Z$. $\alpha = 1.775(5)$ $\beta = 1.798(3)$ $\gamma = 1.800(3)$ 2V(meas.) = 25°

Cell Data: *Space Group:* n.d. Z = n.d.

X-ray Powder Pattern: Franklin, New Jersey, USA.
11.3 (100), 2.567 (40), 3.548 (30), 2.898 (30), 2.262 (25), 2.238 (25), 1.470 (25)

Chemistry:

	(1)	(2)
SiO_2	20.1	21.56
Al_2O_3	1.0	
Fe_2O_3	0.5	
Mn_2O_3	56.3	56.65
As_2O_5	1.0	
MgO	0.9	
CaO	5.1	6.71
H_2O	[15.1]	15.08
Total	[100.0]	100.00

(1) Franklin, New Jersey, USA; by electron microprobe, H_2O by difference; corresponds to $(Ca_{0.76}Mg_{0.19})_{\Sigma=0.95}(Mn_{5.97}^{3+}Fe_{0.05}^{3+})_{\Sigma=6.02}(Si_{2.80}Al_{0.16}As_{0.07}^{5+})_{\Sigma=3.03}O_{16} \cdot 7.1H_2O$.
(2) $CaMn_6Si_3O_{16} \cdot 7H_2O$.

Occurrence: In a metamorphosed stratiform zinc deposit, on surfaces of ore and on secondary carbonate minerals.

Association: Franklinite, calcite, fluorite, willemite, kittatinnyite.

Distribution: From Franklin, Sussex Co., New Jersey, USA.

Name: For Richard C. Bostwick, collector of Franklin minerals and expert in mineral fluorescence.

Type Material: Harvard University, Cambridge, Massachusetts, 125618; National Museum of Natural History, Washington, D.C., USA, C4222.

References: (1) Dunn, P.J. and P.B. Leavens (1983) Bostwickite, a new calcium manganese silicate hydrate from Franklin, New Jersey. Mineral. Mag., 47, 387–389. (2) (1984) Amer. Mineral., 69, 810 (abs. ref. 1).

Crystal Data: Hexagonal. *Point Group:* $6/m \, 2/m \, 2/m$. In very thin hexagonal plates, with {0001} dominant, to 1 mm. *Twinning:* Common, $\perp$ [0001], observable only by X-ray study.

Physical Properties: *Tenacity:* Brittle. Hardness = n.d. D(meas.) = 2.980   D(calc.) = 3.08 Fluoresces bright bluish white in SW UV.

Optical Properties: Semitransparent. *Color:* Colorless. *Luster:* Vitreous. *Optical Class:* Uniaxial (–). $\omega = 1.567$ $\epsilon = 1.566$

Cell Data: *Space Group:* $P6/mcc$.   a = 10.002(2)   c = 14.263(3)   Z = 2

X-ray Powder Pattern: Kings Mountain, North Carolina, USA.
4.109 (10), 2.905 (9), 7.141 (8), 4.343 (8), 5.714 (7), 8.693 (6), 2.681 (6)

Chemistry:

	(1)	(2)
SiO$_2$	65.8	64.70
SnO$_2$	28.2	27.05
Li$_2$O	3.75	4.02
Na$_2$O	0.74	
K$_2$O	3.72	4.23
Total	102.2	100.00

(1) Kings Mountain, North Carolina, USA; by electron microprobe, Li, Na, and K by flame photometry. (2) KSn$_2$Li$_3$Si$_{12}$O$_{30}$.

Mineral Group: Milarite group.

Occurrence: In the late hydrothermal portions of a Li-Sn-rich pegmatite, in vugs and on flat fracture surfaces.

Association: Bavenite, pyrite, tetrawickmanite, stannian titanite, albite, quartz.

Distribution: From the Foote mine, Kings Mountain, Cleveland Co., North Carolina, USA.

Name: For Dr. Kent Combs Brannock (1923–1973), Kingsport, Tennessee, USA, chemist and mineral collector.

Type Material: National Museum of Natural History, Washington, D.C., USA, 125045.

References: (1) White, J.S., Jr., J.E. Arem, J.A. Nelen, P.B. Leavens, and R.W. Thomssen (1973) Brannockite, a new tin mineral. Mineral. Record, 4, 73–76. (2) (1973) Amer. Mineral., 58, 1111 (abs. ref. 1). (3) Armbruster, T. and R. Oberhänsli (1988) Crystal chemistry of double-ring silicates: structures of sugilite and brannockite. Amer. Mineral., 73, 595–600.

Crystal Data: Tetragonal. *Point Group:* $4/m\ 2/m\ 2/m$. As pyramidal crystals, to 5 cm, striated on {001} and {201} ‖ [010]; also dense granular, massive. *Twinning:* On {112}.

Physical Properties: *Cleavage:* Perfect on {112}. *Fracture:* Uneven to subconchoidal. *Tenacity:* Brittle. Hardness = 6–6.5 VHN = 920–1079, 1025–1196 (100 g load). D(meas.) = 4.72–4.83 D(calc.) = [4.86] Weakly magnetic.

Optical Properties: Opaque. *Color:* Dark, brownish black to steel-gray; in polished section, greyish white with a brownish tinge. *Streak:* Brownish black to steel-gray. *Luster:* Submetallic. *Anisotropism:* Weak in air to distinct in oil, dark to lighter slate gray.
R_1–R_2: (400) 23.4–25.4, (420) 22.7–24.5, (440) 22.0–23.6, (460) 21.5–22.9, (480) 21.0–22.4, (500) 20.5–21.9, (520) 20.0–21.5, (540) 19.6–21.2, (560) 19.3–20.9, (580) 19.0–20.6, (600) 18.8–20.3, (620) 18.6–20.0, (640) 18.5–19.8, (660) 18.3–19.6, (680) 18.2–19.5, (700) 18.1–19.4.

Cell Data: *Space Group:* $I4_1/acd$. a = 9.36–9.45 c = 18.60–18.97 Z = 8

X-ray Powder Pattern: Thuringia, Germany.
2.72 (100), 1.656 (70), 2.14 (50), 2.35 (40), 1.420 (40), 1.074 (40), 3.49 (30)

Chemistry:

	(1)	(2)	(3)		(1)	(2)	(3)
SiO_2	9.89	9.68	9.94	FeO	3.81		
TiO_2		0.09		MnO	78.91	[10.80]	11.73
Al_2O_3		0.35		MgO	0.15	0.13	
Fe_2O_3		2.02		CaO	0.34	0.38	
Mn_2O_3		[75.80]	78.33	O	7.35		
				Total	100.45	[99.25]	100.00

(1) Långban, Sweden; total Mn as MnO. (2) Tirodi, India; by electron microprobe, Mn^{2+} estimated equal to $(Si + Al + Ti) - (Mg + Ca)$. (3) $Mn^{2+}Mn_6^{3+}SiO_{12}$.

Polymorphism & Series: Forms a series with abswurmbachite.

Occurrence: Formed by metamorphism of manganese silicates and oxides; also a product of weathering.

Association: Pyrolusite, jacobsite, hausmannite, bixbyite, rhodonite, spessartine, hematite.

Distribution: Of widespread origin; a few localities for large crystals or pure material are: in Germany, at Öhrenstock and Elgersburg, near Ilmenau, Thuringia, and at Ilfeld, Harz Mountains. From Långban, and Jakobsberg, Värmland, Sweden. In Norway, at Botnedalen. In Italy, at St. Marcel, Val d'Aosta. From Kacharwali, Nagpur district, Maharashtra, and Tirodi, Madhya Pradesh, India. In Brazil, at Miguel Burnier, near Ouro Prêto, Minas Gerais. From mines in the Kalahari manganese field, around Kuruman, Cape Province, South Africa. In the USA, at the Spiller manganese mines, 25 km northeast of Mason, Mason Co., Texas; in the Batesville district, Independence Co., Arkansas; at Cartersville, Bartow Co. Georgia; from Snowmass, Pitkin Co., Colorado; in the Black Diablo mine, Pershing Co., Nevada; and the Fort Seward mine, Humboldt Co., California.

Name: After Kammerrath [Advisor of the Chambers] Braun of Gotha, Germany.

References: (1) Palache, C., H. Berman, and C. Frondel (1944) Dana's system of mineralogy, (7th edition), v. I, 551–554. (2) Moore, P.B. and T. Araki (1976) Braunite: its structure and relationship to bixbyite, and some insights on the genealogy of fluorite derivative structures. Amer. Mineral., 61, 1226–1240. (3) Dasgupta, H.C. and R. Manickavasagam (1981) Chemical and X-ray investigation of braunite from the metamorphosed manganiferous sediments of India. Neues Jahrb. Mineral., Abh., 142, 149–160. (4) Bhattacharyya, P.K., S. Dasgupta, M. Fukuoka, and S. Roy (1984) Geochemistry of braunite and associated phases in metamorphosed non-calcareous manganese ores of India. Contr. Mineral. Petrol., 87, 65–71. (5) Berry, L.G. and R.M. Thompson (1962) X-ray powder data for the ore minerals. Geol. Soc. Amer. Mem. 85, 177.

Crystal Data: Orthorhombic, pseudohexagonal. *Point Group: mm2.* Crystals are squat, barrel- or boat-shaped, with pseudohexagonal cross sections, or as slender prismatic crystals, to 2 mm; also as rounded grains. *Twinning:* Simple contact and cyclical on {110}.

Physical Properties: *Cleavage:* Distinct on {130}. Hardness = n.d. D(meas.) = 3.42 D(calc.) = [3.32]

Optical Properties: Transparent. *Color:* Colorless, gray. *Luster:* Vitreous.
Optical Class: Biaxial (+). *Pleochroism:* In thick sections, X = pale violet; Y = Z = colorless to pale green. *Orientation:* X = b; Y = a; Z = c. α = 1.712–1.725 β = 1.716–1.728 γ = 1.725–1.740 2V(meas.) = 10°–34°

Cell Data: *Space Group: P2nn.* a = 10.909(9) b = 18.34(1) c = 6.739(9) Z = 4

X-ray Powder Pattern: Scawt Hill, Ireland.
2.730 (100), 2.663 (100), 2.259 (80), 1.923 (80), 2.067 (60), 1.574 (60), 1.554 (60)

Chemistry:

	(1)	(2)	(3)
SiO_2	33.08	36.6	35.70
TiO_2	0.34		
Fe_2O_3	0.12		
MnO	3.38		
MgO	6.78	5.1	5.99
CaO	49.23	59.9	58.31
BaO	6.91		
F	0.16		
Total	100.00	[101.6]	100.00

(1) Synthetic, sample separated from slag; recalculated after deducting impurities. (2) Scawt Hill, Ireland; by electron microprobe, original analysis Si 17.1 %, Mg 3.1%, Ca 42.8%, here converted to oxides, traces of Na, Al, P, S. (3) $Ca_7Mg(SiO_4)_4$.

Occurrence: In contact metamorphosed limestones and dolostones intruded by diabase (Scawt Hill, Ireland) or syenite monzonite (Marble Canyon, Texas, USA).

Association: Larnite, spurrite, gehlenite, melilite, perovskite, magnetite (Scawt Hill, Ireland).

Distribution: In the USA, at Marble Canyon, Culberson Co., Texas. In Ireland, at Scawt Hill and Ballycraigy, near Larne, Co. Antrim. At Camas Mòr, Isle of Muck, Inverness-shire, and at Camphouse, Ardnamurchan, Argyllshire, Scotland. In the Hatrurim Formation, Israel.

Name: For Max Albrecht Bredig (1902–), physical chemist, who studied the polymorphism of Ca_2SiO_4.

Type Material: n.d.

References: (1) Tilley, C.E. and H.C.G. Vincent (1948) The occurrence of an orthorhombic high-temperature form of Ca_2SiO_4 (bredigite) in the Scawt Hill contact-zone and as a constituent of slags. Mineral. Mag., 28, 255–271. (2) (1948) Amer. Mineral., 33, 786 (abs. ref. 1). (3) Douglas, A.M.B. (1951) X-ray investigation of bredigite. Mineral. Mag., 29, 875–884. (4) Bridge, T.E. (1966) Bredigite, larnite, and γ dicalcium silicates from Marble Canyon. Amer. Mineral., 51, 1766–1774. (5) Moore, P.B. and T. Araki (1976) The crystal structure of bredigite and the genealogy of some alkaline earth orthosilicates. Amer. Mineral., 61, 74–87. (6) Sarkar, S.L. and J.W. Jeffrey (1978) Electron microprobe analysis of Scawt Hill bredigite-larnite rock. J. Amer. Ceramic Soc., 61, 177–178. (7) Heller, L. and H.F.W. Taylor (1956) Crystallographic data for the calcium silicates. H.M. Stationary Office, London, 14.

Crystal Data: Monoclinic. *Point Group:* $2/m$. Crystals commonly equant or prismatic, striated and elongated along [100], to 1.5 cm. Platy, radial fibrous, and in granular aggregates. *Twinning:* Lamellar || {010}.

Physical Properties: *Cleavage:* {010}, perfect. *Fracture:* Uneven. *Tenacity:* Brittle. Hardness = 5–5.5   D(meas.) = 2.45   D(calc.) = [2.42]

Optical Properties: Transparent. *Color:* White, colorless, yellowish, gray, greenish; colorless in thin section. *Streak:* White. *Luster:* Vitreous to pearly on {010}.
Optical Class: Biaxial (+). *Orientation:* $Z \perp$ (010); $X \wedge c = 19°$–$34°$ in various sectors of the crystal. *Dispersion:* $r > v$, weak, crossed. $\alpha = 1.510$ $\beta = 1.512$ $\gamma = 1.523$
2V(meas.) = 65°

Cell Data: *Space Group:* $P2_1/m$.   a = 6.793(2)   b = 17.573(6)   c = 7.759(2)
$\beta = 94.54(3)°$ Z = 2

X-ray Powder Pattern: Strontian, Scotland.
4.53 (10), 6.15 (9), 2.885 (9), 3.209 (8), 3.867 (7), 4.98 (4), 1.989 (4)

Chemistry:

	(1)	(2)
SiO_2	54.42	54.02
Al_2O_3	15.25	15.86
Fe_2O_3	0.08	0.11
CaO	1.19	0.80
SrO	8.99	11.80
BaO	6.80	3.01
Na_2O		0.21
K_2O		0.14
H_2O	13.22	13.72
Total	99.95	99.67

(1) Strontian, Scotland; corresponds to $(Sr_{0.58}Ba_{0.30}Ca_{0.14})_{\Sigma=1.02}Al_{1.98}Si_{6.00}O_{16} \cdot 4.86H_2O$.
(2) Burpala massif, Russia; corresponds to $(Sr_{0.76}Ba_{0.13}Ca_{0.10}Na_{0.04}K_{0.02}Fe_{0.01})_{\Sigma=1.06}Al_{2.06}$ $Si_{5.94}O_{16} \cdot 4.86H_2O$.

Mineral Group: Zeolite group.

Occurrence: Hydrothermally deposited in druses lining cavities in basalts and schists; more rarely in ore deposits.

Association: Zeolites, calcite, quartz.

Distribution: From Strontian, Argyllshire, Scotland. At St. Christophe, Bourg d'Oisans, Isère, and around Barèges, Hautes-Pyrénées, France. In the Burpala massif, about 120 km north of Lake Baikal, eastern Siberia, Russia. At Yellow Lake, near Olalla, British Columbia, Canada. Other localities are reported but require confirmation.

Name: After Sir David Brewster (1781–1868), Scottish mineralogist.

References: (1) Dana, E.S. (1892) Dana's system of mineralogy, (6th edition), 576–577. (2) Vlasov, K.A., Ed. (1966) Mineralogy of rare elements, v. II, 206–207. (3) Gottardi, G. and E. Galli (1985) Natural zeolites. Springer, 300–305. (4) Strunz, H. and C. Tennyson (1956) "Polymorphie" in der Gruppe der Blätterzeolithe (Heulandit-Stilbit-Epistilbit: Brewsterit). Neues Jahrb. Mineral., Monatsh., 11, 1–9 (in German). (5) Schlenker, J.L., J.J. Pluth, and J.V. Smith (1977) Refinement of the crystal structure of brewsterite, $Ba_{0.5}Sr_{1.5}Al_4Si_{12}O_{32} \cdot 10H_2O$. Acta Cryst., 33, 2907–2910. (6) Akizuki, M. (1987) Crystal symmetry and order-disorder structure of brewsterite. Amer. Mineral., 72, 645–648. (7) Nawaz, R. (1990) Brewsterite: re-investigation of morphology and elongation. Mineral. Mag., 54, 654–656.

Crystal Data: Monoclinic or hexagonal. *Point Group:* n.d. Crystals platy, with poorly developed individuals, less than 1 μm; massive.

Physical Properties: Hardness = 2.5–3   D(meas.) = 3.17(1)   D(calc.) = 3.16

Optical Properties: Semitransparent. *Color:* Dark yellow-green.
Optical Class: Biaxial. $n = 1.635(1)$ 2V(meas.) = n.d.

Cell Data: *Space Group:* n.d. 1M and 3A assumed. $a = 5.286(5)$ $b = 9.133(3)$
$c = 7.31(1)$ $\beta = 104°9(7)'$ Z = 2, or   *Space Group:* n.d. 1A assumed. $a = 5.277(1)$
b = n.d. $c = 7.09(1)$ β = n.d.   Z = 2

X-ray Powder Pattern: Megara, Greece.
7.07 (100), 3.54 (80), 2.62 (18), 2.47 (18), 2.37 (18), 1.524 (17), 4.54 (10)

Chemistry:

	(1)
SiO_2	27.45
TiO_2	0.99
Al_2O_3	24.09
ΣLa_2O_3	0.35
Cr_2O_3	0.17
FeO	1.15
NiO	30.18
MgO	3.18
CaO	0.07
H_2O	[12.37]
Total	[100.00]

(1) Megara, Greece; by electron microprobe, H_2O by difference.

Polymorphism & Series: 1M plus 3A and 1A polytypes assumed.

Mineral Group: Kaolinite-serpentine group.

Occurrence: As coatings on limestone and as veinlets cutting kaolinitic clays at the base of a bauxite deposit developed on karst; also a significant component of weathered ultramafic rock (Megara, Greece).

Association: Bastnäsite, malachite, bayerite (Megara, Greece).

Distribution: In the Marmara bauxite deposit, Megara, Greece. From Victorio, Grant Co., New Mexico, USA.

Name: To honor Dr. George William Brindley (1905–1983), Professor of Mineral Science, Pennsylvania State University, University Park, Pennsylvania, USA.

Type Material: Department of Mineralogy, University of Belgrade, Belgrade, Yugoslavia; National Museum of Natural History, Washington, D.C., USA, 136982.

References: (1) Maksimovic, Z. and D.L. Bish (1978) Brindleyite, a nickel-rich aluminous serpentine mineral analogous to berthierine. Amer. Mineral., 63, 484–489.

Britholite-(Ce) $(Ce, Ca)_5(SiO_4, PO_4)_3(OH, F)$

Crystal Data: Hexagonal. *Point Group:* $6/m$. Crystals are hexagonal prisms, to 1 cm; granular, massive.

Physical Properties: *Fracture:* Conchoidal. Hardness = 5 D(meas.) = 4.20–4.69 D(calc.) = 4.65

Optical Properties: Opaque, transparent only in thin flakes. *Color:* Brown, yellow-brown, blue. *Luster:* Adamantine.
Optical Class: Uniaxial (–) or biaxial (–). *Pleochroism:* May be O = brown; E = colorless. $n = 1.77$–1.81 2V(meas.) = $\leq 44°$

Cell Data: *Space Group:* $P6_3/m$. $a = 9.61(2)$ $c = 7.03(2)$ $Z = 2$

X-ray Powder Pattern: Oka, Canada.
2.836 (vs), 3.480 (s), 2.809 (s), 1.858 (s), 4.116 (m), 3.927 (m), 3.211 (m)

Chemistry:

	(1)	(2)	(3)		(1)	(2)	(3)
SiO_2	16.77	12.28	20.89	Nd_2O_3			5.1
TiO_2		0.09		Fe_2O_3	0.43	0.14	
ThO_2		5.62	20.73	MgO	0.13	0.20	
UO_2			1.47	CaO	11.28	28.84	19.92
Al_2O_3		0.47	0.00	Na_2O	1.85	0.21	
$(Ce, La)_2O_3$	60.54	33.43		F	1.33	2.10	
La_2O_3			7.35	H_2O^+	1.27	0.54	
Ce_2O_3			18.61	P_2O_5	6.48	16.96	4.3
Pr_2O_3			1.64	$-O = F_2$		0.90	
				Total	100.08	99.98	100.01

(1) Naujakasik, Greenland. (2) Oka, Canada. (3) Monte Somma, Italy; by electron microprobe, average of ten analyses; corresponds to $(Ca_{2.77}Ce_{0.88}Th_{0.62}La_{0.35}Nd_{0.24}Pr_{0.08}U_{0.04})_{\Sigma=4.98}$ $(Si_{2.71}P_{0.47})_{\Sigma=3.18}O_{12}(OH)$.

Occurrence: In nepheline syenites, pegmatites, and contact deposits related to them.

Association: Zircon, pyrochlore, titanite, fluorite, diopside, andradite, allanite, vesuvianite.

Distribution: At Naujakasik, in the Ilímaussaq intrusion, southern Greenland. From about three km east of Jamestown, Boulder Co., Colorado, USA. At Oka and Mont Saint-Hilaire, Quebec, Canada. From the Pibnesburg complex, Rustenburg, Transvaal, South Africa. In the San Vito quarry, Monte Somma, Campania, Italy. At Tvedalen, Norway. From Tungpei, not otherwise located in China. In the Ishim complex, northern Kazakhstan. From the Khibiny massif, Kola Peninsula; the Kyshtym district, Ural Mountains; and other less-well-defined localities in Russia.

Name: From the Greek for *weight*, in allusion to its density, and its content of *cerium*.

Type Material: University of Copenhagen, Copenhagen, Denmark; The Natural History Museum, London, England, 85379.

References: (1) Dana, E.S. and W.E. Ford (1909) Dana's system of mineralogy, (6th edition), app. II, 19–20. (2) Gay, P. (1957) An investigation of some rare-earth silicates: cerite, lessingite, beckelite, britholite and stillwellite. Mineral. Mag., 31, 455–468. (3) Hughson, M.R. and J.G. Sen (1964) A thorian intermediate member of the britholite–apatite series. Amer. Mineral., 49, 937–951. (4) Vlasov, K.A., Ed. (1966) Mineralogy of rare elements, v. II, 297–300. (5) Deyu Li, Peiling Wang, and Jiancheng Li (1981) The crystal structure of lessingite [britholite-(Ce)], rich in light rare earth of cerium. Guisuanyan Xuebao, 9(4), 422–432. (6) (1982) Chem. Abs., 226893 (abs. ref. 5). (7) Orlandi, P., N. Perchiazzi, and G. Mannucci (1989) First occurrence of britholite-(Ce) in Italy (Monte Somma, Vesuvius). Eur. J. Mineral., 1, 723–725.

Crystal Data: Hexagonal. *Point Group:* $6/m$. As short prismatic hexagonal crystals, to 1.3 cm; more commonly massive.

Physical Properties: *Cleavage:* Imperfect on $\{0001\}$, $\{10\bar{1}0\}$. *Fracture:* Uneven to splintery. Hardness = 6 D(meas.) = 4.35 D(calc.) = [4.44]

Optical Properties: Translucent to transparent. *Color:* Dark reddish brown; in thin section, yellowish brown. *Streak:* Faint brown. *Luster:* Resinous to dull.
Optical Class: Uniaxial (+). $\omega = 1.728$–1.750 $\epsilon = 1.730$–1.752

Cell Data: *Space Group:* $P6_3/m$. a = 9.43 c = 6.81 Z = 2

X-ray Powder Pattern: Suishoyama pegmatite, Japan.
2.813 (100), 2.753 (90), 2.727 (80), 3.13 (50), 3.09 (50), 3.39 (30), 1.885 (30)

Chemistry:

	(1)	(2)		(1)	(2)
SiO$_2$	22.70	21.80	MgO	0.10	0.07
TiO$_2$		0.04	PbO		0.13
UO$_2$		0.23	CaO	9.58	13.31
ThO$_2$	0.51	1.55	Na$_2$O		0.20
Al$_2$O$_3$	0.75	0.72	K$_2$O		0.06
Y$_2$O$_3$	46.91	37.47	F	0.50	1.48
ΣCe$_2$O$_3$	4.47	6.02	H$_2$O$^+$	0.68	
ΣLa$_2$O$_3$	5.76	10.92	H$_2$O$^-$	0.15	
Fe$_2$O$_3$	1.44	0.69	H$_2$O		1.35
Nb$_2$O$_5$		0.06	CO$_2$	0.10	
FeO	0.79	0.50	P$_2$O$_5$	1.73	2.98
MnO	3.67	0.66	$-$O = F$_2$	0.21	0.62
			Total	[99.63]	99.62

(1) Suishoyama pegmatite, Japan; original total given as 99.58%. (2) "European Russia."

Occurrence: In pegmatites.

Association: Yttrialite, thorogummite, tengerite, allanite.

Distribution: In the Suishoyama pegmatite, near Iisaka, Fukushima Prefecture, and from Shinden, Gifu Prefecture, Japan. From an undefined locality given only as "European Russia."

Name: For its chemical relation to *britholite-(Ce)* and dominant *yttrium* in its composition.

Type Material: n.d.

References: (1) Hata, S. (1938) Abukumalite [britholite-(Y)], a new mineral from pegmatites of Iisaka, Fukushima prefecture. Sci. Pap. Inst. Phys. Chem. Res., Tokyo, 34, 1018–1023.
(2) Omori, K. and S. Hasegawa (1953) Yttrialite and abukumalite [britholite-(Y)] from pegmatite of Suishoyama, Iisaka village, Fukushima, Japan. J. Japan. Assoc. Mineral. Petrol. Econ. Geol., 37, 21–29. (3) Ito, J. (1968) Silicate apatites and oxyapatites. Amer. Mineral., 53, 890–907.
(4) (1970) Introduction to Japanese minerals. Geol. Sur. of Japan, 62–63.

Crystal Data: Monoclinic. *Point Group:* 2 or $2/m$. As crystals similar in habit and interfacial angle to orthoclase, with forms {001}, {010}, {110}, {$\bar{1}$01}, to 0.05 mm. In compact masses pseudomorphous after plagioclase. *Twinning:* Noted.

Physical Properties: *Cleavage:* Good on {001}, distinct on {010}. *Tenacity:* Brittle. Hardness = 5.5 D(meas.) = 2.32(1) D(calc.) = 2.38

Optical Properties: Transparent to translucent. *Color:* Colorless. *Streak:* Light grey, yellow when impure. *Luster:* Vitreous.
Optical Class: Biaxial (+). *Orientation:* $Z = b$; $X \wedge a = 4°$; $Y \wedge c = 19°$. $\alpha = 1.530(2)$
$\beta = 1.531(2)$ $\gamma = 1.534(2)$ 2V(meas.) = n.d.

Cell Data: *Space Group:* $P2_1$ or $P2_1/m$.   a = 8.571   b = 13.032   c = 7.187
$\beta = 112°44(1)'$ Z = 4

X-ray Powder Pattern: Sulfur Bank mine, California, USA.
3.81 (100), 6.52 (95), 3.38 (70), 3.23 (70), 4.33 (65), 3.26 (60), 3.01 (40)

Chemistry:

	(1)	(2)		(1)	(2)
SiO_2	66.89	70.06	Na_2O	0.06	
Al_2O_3	20.09	19.82	K_2O	0.65	
MgO	0.22		$(NH_4)_2O$	8.34	10.12
CaO	0.04		H_2O	3.44	
BaO	0.27		Total	[100.00]	100.00

(1) Sulfur Bank mine, California, USA; probably contains a small amount of reversibly hydrated ammonian montmorillonite accounting for the remaining H_2O content; after deductions for TiO_2, FeS_2, and $H_2O^- = 4\%$, recalculated to 100.00%. (2) $(NH_4)AlSi_3O_8$.

Mineral Group: Feldspar group.

Occurrence: A low-temperature hydrothermal replacement of plagioclase in andesite altered by ammonia-bearing hot springs (Sulfur Bank mine, California, USA); in metasomatized rhyolitic ash-flow tuffs (Cedar Mountains, Nevada, USA); in the sedimentary Phosphoria Formation (Idaho, Wyoming, and Montana, USA).

Association: Plagioclase, sulfur, stibnite, pyrite, marcasite, ammoniojarosite, gypsum, barite, anatase, montmorillonite (Sulfur Bank mine, California, USA); illite, albite, montmorillonite, kaolinite (Phosphoria Formation, Idaho, Wyoming, and Montana, USA).

Distribution: In the USA, at the Sulfur Bank mercury mine, Lake Co., and on Sharon Heights, Menlo Park, Santa Clara Co., California; in the Cedar Mountains, 50 km northwest of Tonopah, Esmeralda Co., Nevada; at a number of exposures of the Meade Peak Member of the Phosphoria Formation in Idaho, Wyoming, and Montana. In the Condor oilshale deposit, near Proserpine, Queensland, Australia. From the Toshichi Spa, Iwate Prefecture, Japan.

Name: To honor Emeritus Professor Arthur Francis Buddington (1890–1980), American geologist, Princeton University, Princeton, New Jersey, USA.

Type Material: National Museum of Natural History, Washington, D.C., USA, 116974, 132920, 147604.

References: (1) Erd, R.C., D.E. White, J.J. Fahey, and D.E. Lee (1964) Buddingtonite, an ammonium feldspar with zeolitic water. Amer. Mineral., 49, 831–850. (2) Gulbrandsen, R.A. (1974) Buddingtonite, ammonium feldspar, in the Phosphoria Formation, southeastern Idaho. J. Res. U.S. Geol. Sur., 2, 693–697. (3) Voncken, J.H.L., H.L.M. van Roermund, A.M.J. van der Eerden, J.B.H. Jansen, and R.C. Erd (1993) Holotype buddingtonite: an ammonium feldspar without zeolitic H_2O. Amer. Mineral., 78, 204–209.

Crystal Data: Hexagonal. *Point Group:* $3m$. Crystals short to long prismatic, in divergent sprays, segmented, up to 4 cm, with forms $\{11\bar{2}0\}$, $\{30\bar{3}0\}$, $\{10\bar{1}1\}$, and $\{02\bar{2}1\}$.

Physical Properties: *Cleavage:* Distinct prismatic. *Tenacity:* Brittle. Hardness = 7 D(meas.) = 3.31(1) D(calc.) = 3.29 Pyroelectric and piezoelectric.

Optical Properties: Translucent. *Color:* Dark brown to almost black with a bronze schiller. *Streak:* Yellow-brown. *Luster:* Vitreous.
Optical Class: Uniaxial (–). *Pleochroism:* O = yellow-brown; E = very pale yellow.
$\omega = 1.735(3)$ $\epsilon = 1.655(3)$

Cell Data: *Space Group:* $R3m$. a = 15.869(2) c = 7.188(2) Z = 3

X-ray Powder Pattern: Mexquitic, Mexico. (ICDD 25-703).
2.563 (100), 2.952 (64), 3.96 (52), 3.47 (48), 6.33 (45), 2.032 (43), 4.20 (40)

Chemistry:

	(1)
SiO_2	33.86
TiO_2	0.55
B_2O_3	10.86
Al_2O_3	30.79
Fe_2O_3	17.62
FeO	1.27
MnO	0.13
MgO	0.13
CaO	0.69
Na_2O	2.46
K_2O	0.07
F	1.86
H_2O^+	0.40
$-O = F_2$	0.78
Total	99.91

(1) Mexquitic, Mexico; corresponds to $(Na_{0.82}Ca_{0.13}K_{0.02})_{\Sigma=0.97}(Fe^{3+}_{2.29}Al_{0.27}Fe^{2+}_{0.18}$ $Ti_{0.07}Mg_{0.03}Mn_{0.02})_{\Sigma=2.86}Al_{6.00}(B_{1.08}O_3)_3Si_{5.85}O_{18}[O_{2.59}F_{1.02}(OH)_{0.46}]_{\Sigma=4.07}$.

Mineral Group: Tourmaline group.

Occurrence: Of probable pneumatolytic origin, in cavities in a rhyolite.

Association: Quartz, potassic feldspar, plagioclase, biotite, muscovite.

Distribution: From near Mexquitic, San Luis Potosi, Mexico.

Name: To honor Professor Martin Julian Buerger (1903–1986), eminent crystallographer, Massachusetts Institute of Technology, Cambridge, Massachusetts, USA.

Type Material: Royal Ontario Museum, Toronto, Canada, M25021; American Museum of Natural History, New York City, New York, 35468; National Museum of Natural History, Washington, D.C., USA, 122200.

References: (1) Donnay, G., C.O. Ingamells, and B. Mason (1966) Buergerite, a new species of tourmaline. Amer. Mineral., 51, 198–199. (2) Tippe, A. and W.C. Hamilton (1971) A neutron diffraction study of the ferric tourmaline, buergerite. Amer. Mineral., 56, 101–113. (3) Dunn, P.J. (1976) Buergerite, uniformity of composition. Amer. Mineral., 61, 1029–1030. (4) Deer, W.A., R.A. Howie, and J. Zussman (1986) Rock-forming minerals, (2nd edition), v. 1B, disilicates and ring silicates, 559–602.

Crystal Data: Triclinic. *Point Group:* $\bar{1}$. As radiating prismatic acicular crystals and radial spherules, to 2 cm. *Twinning:* Interpenetrating on {100} and {010}; polysynthetic.

Physical Properties: *Cleavage:* Good on {100} and {010}. *Fracture:* Conchoidal. Hardness = 4.5 D(meas.) = 2.73 D(calc.) = 2.74

Optical Properties: Transparent. *Color:* Pale pink to colorless. *Luster:* Vitreous. *Optical Class:* Biaxial (+). *Orientation:* $Z' \wedge c = 27°$–$29°$ on {010}; $Z' \wedge c = 46°$–$48°$ on {100}. *Dispersion:* $r > v$, barely perceptible. $\alpha = 1.587(2)$ $\beta = [1.590]$ $\gamma = 1.597(2)$ 2V(meas.) = 70°

Cell Data: *Space Group:* $P\bar{1}$. a = 10.992(4) b = 8.185(2) c = 5.671(1) $\alpha = 93°57'$ $\beta = 90°19'$ $\gamma = 89°51'$ Z = 4

X-ray Powder Pattern: Bultfontein mine, Kimberley, South Africa. 1.93 (100), 8.12 (60), 2.92 (60), 2.88 (60), 2.037 (50), 4.06 (40), 3.50 (40)

Chemistry:

	(1)	(2)
SiO_2	26.50	24.06
B_2O_3		0.02
Al_2O_3	0.72	1.37
CaO	54.20	53.59
Na_2O		0.04
F	8.81	7.90
H_2O^+		10.78
H_2O^-		0.75
H_2O	13.36	
CO_2		4.65
P_2O_5		0.02
$-O = F_2$	3.71	3.33
Total	99.88	99.85

(1) Bultfontein mine, Kimberley, South Africa. (2) Fuka, Japan; corresponds to $(Ca_{2.04}$ $(Si_{0.96}Al_{0.06})_{\Sigma=1.02}O_{2.13}[(OH)_{2.87}F_{1.00}]_{\Sigma=3.87}$.

Occurrence: In a large "horse" of diabase and shale fragments in a kimberlite pipe (Bultfontein mine, Kimberley, South Africa); in a contact zone in thermally metamorphosed limestone (Crestmore, California, USA).

Association: Calcite, apophyllite, natrolite (Bultfontein mine, Kimberley, South Africa); afwillite, scawtite (Crestmore, California, USA); oyelite, scawtite, xonotlite (Fuka, Japan).

Distribution: In South Africa, in Cape Province, in the Bultfontein and Dutoitspan diamond mines, Kimberley; the Jagersfontein diamond mine, Orange River Colony; and the N'Chwaning and Wessels manganese mines, near Kuruman. In the USA, at Crestmore, Riverside Co., California. In the Hatrurim Formation, Israel. In Japan, from Fuka, near Bicchu, and in the Mihara mine, Okayama Prefecture.

Name: For the Bultfontein mine, Kimberley, South Africa, where it was first discovered.

Type Material: Cambridge University, Cambridge; The Natural History Museum, London, England, 1928,78.

References: (1) Parry, J., A.F. Williams, and F.E. Wright (1932) On bultfonteinite, a new fluorine-bearing hydrous calcium silicate from South Africa. Mineral. Mag., 23, 145–162. (2) (1933) Amer. Mineral., 18, 32 (abs. ref. 1). (3) McIver, E.J. (1963) The structure of bultfonteinite, $Ca_4Si_2O_{10}F_2H_6$. Acta Cryst., 16, 551–558. (4) Kusachi, I., C. Henmi, and K. Henmi (1984) An oyelite-bearing vein at Fuka, the town of Bitchu, Okayama Prefecture, Japan. J. Japan. Assoc. Mineral. Petrol. Econ. Geol., 79, 267–275.

$Pb_2(Fe^{3+}, Mn^{3+})Te^{4+}(AlSi_3)O_{12}(OH)_2 \cdot H_2O$ Burckhardtite

Crystal Data: Monoclinic (?), pseudohexagonal. *Point Group:* n.d. Rarely as hexagonal crystals, to 50 μm, typically overgrown on dickite; as crystal rosettes, < 0.2 mm.

Physical Properties: *Cleavage:* Perfect micaceous, basal. Hardness = ~2 D(meas.) = n.d. D(calc.) = 4.96 Paramagnetic.

Optical Properties: Transparent to translucent. *Color:* Carmine to violet-red. *Streak:* Pale red. *Luster:* Adamantine to slightly pearly.
Optical Class: Biaxial (–). *Pleochroism:* X = pale magenta; Y = Z = carmine. $\alpha = 1.82$ $\beta = 1.85$ $\gamma = 1.85$ 2V(meas.) = Small.

Cell Data: *Space Group:* C-centered cell (?). $a = 5.21$ $b = 9.04$ $c = 12.85$ $\beta = 90°$ Z = 2

X-ray Powder Pattern: Moctezuma, Mexico.
3.11 (100), 12.8 (90), 3.70 (90), 2.60 (70), 1.840 (60), 1.590 (50), 2.26 (30b)

Chemistry:

	(1)
SiO_2	20.28
TeO_2	17.03
Al_2O_3	5.44
Fe_2O_3	7.25
Mn_2O_3	1.18
PbO	45.23
H_2O	[3.59]
Total	[100.00]

(1) Moctezuma, Mexico; by electron microprobe, valences of Te, Fe, and Mn inferred from the oxidized state of the deposit, H_2O by difference; corresponds to $Pb_{1.83}(Fe^{3+}_{0.82}Mn^{3+}_{0.14})_{\Sigma=0.96}Te_{0.97}$ $(Si_{3.04}Al_{0.96})_{\Sigma=4.00}O_{12.7} \cdot 1.80H_2O$.

Occurrence: In a breccia, loosely cemented by dickite, in very thin disseminations on quartz fragments.

Association: Moctezumite, zemannite, tellurite, barite, dickite, quartz.

Distribution: In the Moctezuma mine, Moctezuma, Sonora, Mexico.

Name: For Mexican geologist Carlos Burckhardt (1869–1935).

Type Material: University of Delaware, Newark, Delaware; Harvard University, Cambridge, Massachusetts, 125476, 119081, 119085; National Museum of Natural History, Washington, D.C., USA, 136509, 144110, 164347, 164348.

References: (1) Gaines, R.V., P.B. Leavens, and J.A. Nelen (1979) Burckhardtite, a new silicate-tellurite from Mexico. Amer. Mineral., 64, 355–358.

Crystal Data: Monoclinic, pseudo-orthorhombic. *Point Group:* $2/m$. As tablets, elongated along [001] and flattened on {010}, to 5 mm. In fan-shaped aggregates, commonly intimately intergrown with låvenite.

Physical Properties: *Fracture:* Conchoidal. *Tenacity:* Brittle. Hardness = 5–6 D(meas.) = 3.33(15)   D(calc.) = 3.27   Weak yellow-orange fluorescence in X-rays.

Optical Properties: Transparent to translucent. *Color:* Colorless, yellowish. *Streak:* White. Luster: Vitreous.
Optical Class: Biaxial (–). *Orientation:* $X = b$; $Y = c$; $Z = a$.   *Dispersion:* $r < v$, weak.
$\alpha = 1.627(2)$ $\beta = 1.634(2)$ $\gamma = 1.639(2)$ 2V(meas.) = 82.1°   2V(calc.) = 80.1°

Cell Data: *Space Group:* $P2_1/a$. a = 10.1173(8)   b = 10.4446(6)   c = 7.2555(3) $\beta = 90.039(7)°$ Z = 4

X-ray Powder Pattern: Burpala massif, Russia; by Gandolfi camera to exclude låvenite domains.
2.962 (vs), 1.886 (ms), 1.556 (ms), 1.787 (s), 3.035 (m), 3.306 (m), 1.678 (m)

Chemistry:

	(1)
SiO_2	31.82
TiO_2	1.06
ZrO_2	31.11
Y_2O_3	0.32
Nb_2O_5	0.22
FeO	0.43
MnO	0.60
CaO	14.52
Na_2O	13.86
F	8.1
H_2O	1.23
$-O = F_2$	3.41
Total	99.86

(1) Burpala massif, Russia; by electron microprobe, average of four analyses, H_2O by Penfield method; corresponds to $(Na_{1.69}Mn_{0.03}Fe_{0.02}Y_{0.01})_{\Sigma=1.75}Ca_{0.98}(Zr_{0.96}Ti_{0.05}Nb_{0.01})_{\Sigma=1.02}Si_{2.00}O_7$ $[F_{1.61}(OH)_{0.26}]_{\Sigma=1.87} \cdot 0.13H_2O$.

Polymorphism & Series: Dimorphous with låvenite.

Occurrence: In a fenitized hornfelsic sandstone in the contact zone of an alkalic intrusive.

Association: Låvenite, albite, nepheline, aegirine, alkalic amphibole, biotite, catapleiite, astrophyllite, fluorite, loparite.

Distribution: In the Burpala massif, about 120 km north of Lake Baikal, eastern Siberia, Russia.

Name: For its occurrence in the Burpala massif, Russia.

Type Material: A.E. Fersman Mineralogical Museum, Academy of Sciences, Moscow, Russia; Museum of Natural History, University of Pisa, Pisa, Italy; The Natural History Museum, London, England, 1994,5.

References: (1) Khomyakov, A.P., D.Y. Pushcharovskii, I.M. Kulikova, and V.I. Kuz'min (1988) New representative of the hiortdahlite-lavenite mineralogical group. Vestnik Mosk. Univ. Geol., 43(1), 87–92 (English trans. of Russian). (2) (1990) Amer. Mineral., 75, 436–437 (abs. ref. 1). (3) Merlino, S., N. Perchiazzi, A.P. Khomyakhov [Khomyakov], D.Y. Pushcharovskii, I.M. Kulikova, and V.I. Kuzmin (1990) Burpalite, a new mineral from the Burpalinskii massif, North Transbajkal [Transbaikal], USSR: its crystal structure and OD character. Eur. J. Mineral., 2, 177–185.

Crystal Data: Triclinic. *Point Group:* $\bar{1}$. Crystals prismatic, needlelike, to 2.5 cm, or tabular on {001}; commonly fibrous, cleavable massive. *Twinning:* Composition plane {110}, simple twins, uncommon.

Physical Properties: *Cleavage:* Perfect on {100}, good on {110} and {1$\bar{1}$0}, poor on {010}. Hardness = 5.5–6.5   D(meas.) = 3.32–3.43   D(calc.) = 3.421   Pink color fades on exposure to light.

Optical Properties: Transparent to translucent. *Color:* Pale pink to brownish red; in thin section, colorless to yellowish pink. *Luster:* Vitreous.
Optical Class: Biaxial (–). *Pleochroism:* $X = Z$ = orange; Y = rose. *Orientation:* $X \wedge a$ $\simeq 15°$; $Y \wedge b \simeq 35°$; $Z \wedge c \simeq 30°$–$35°$. *Dispersion:* $r < v$, weak, strong crossed dispersion.
$\alpha = 1.640$–1.695 $\beta = 1.651$–1.708 $\gamma = 1.653$–1.710 2V(meas.) = $34°$–$60°$

Cell Data: *Space Group:* $P\bar{1}$. $a = 15.412$ $b = 7.157$ $c = 13.824$ $\alpha = 89°29'$ $\beta = 94°51'$ $\gamma = 102°56'$ $Z = 12$

X-ray Powder Pattern: Broken Hill, Australia.
2.880 (100), 2.989 (60), 3.19 (50), 1.776 (50), 2.227 (40), 1.665 (40), 2.711 (30)

Chemistry:

	(1)	(2)
SiO_2	48.44	48.31
FeO	0.27	1.87
MnO	25.20	33.04
ZnO	0.53	
MgO	0.65	1.90
CaO	25.20	14.93
LOI	0.34	
Total	100.63	100.05

(1) Franklin, New Jersey, USA; corresponding to $(\text{Ca}_{1.66}\text{Mn}^{2+}_{1.31}\text{Mg}_{0.06}\text{Fe}^{2+}_{0.01})_{\Sigma=3.04}\text{Si}_{2.97}\text{O}_9$.
(2) Långban, Sweden; corresponding to $(\text{Mn}^{2+}_{1.74}\text{Ca}_{0.99}\text{Mg}_{0.18}\text{Fe}^{2+}_{0.10})_{\Sigma=3.01}\text{Si}_{3.00}\text{O}_9$.

Occurrence: In manganese ores formed by metamorphism of manganese-bearing sediments with attendant metasomatism; typically associated with skarns.

Association: Rhodonite, tephroite, calcite, glaucochroite, johannsenite, wollastonite, diopside, grossular.

Distribution: "Type" material, from Tetela de Ocampo, Hidalgo, Mexico, is a mixture; well-studied material from other localities include: in the USA, at Franklin, Sussex Co., New Jersey. At Långban and in the Harstig mine, Pajsberg, near Persberg, Värmland, Sweden. At the Treburland mine, Altarnun, Cornwall, and in the Railroad quarry, Meldon, near Okehampton, Devon, England. At Camas Malag, Isle of Skye, Scotland. From Băiţa (Rézbánya), Hungary. At Campiglia, Tuscany, and Schio, Vicenza, Italy. From the N'Chwaning mine, near Kuruman, Cape Province, South Africa. In Australia, at Broken Hill, New South Wales. From the Obori mine, Yamagata Prefecture; the Noda-Tamagawa mine, Iwate Prefecture; the Kanoiri mine, Tochigi Prefecture; the Mikumo mine, Shiga Prefecture; and the Hijikuzu mine, Iwate Prefecture, Japan.

Name: For General Anastasio Bustamente (1780–1853), of Mexico.

References: (1) Larsen, E.S. and E.V. Shannon (1922) Bustamite from Franklin Furnace, New Jersey. Amer. Mineral., 7, 95–100. (2) Deer, W.A., R.A. Howie, and J. Zussman (1978) Rock-forming minerals, (2nd edition), v. 2A, single-chain silicates, 574–585. (3) Peacor, D.R. and M.J. Buerger (1962) Determination and refinement of the crystal structure of bustamite, $\text{CaMnSi}_2\text{O}_6$. Zeits. Krist., 117, 331–343. (4) Harada, K., H. Sekino, K. Nagashima, T. Watanabe, and K. Momoi (1974) High-iron bustamite and fluorapatite from the Broken Hill mine, New South Wales, Australia. Mineral. Mag., 39, 601–604. (5) Ohashi, Y. and L.W. Finger (1978) The role of octahedral cations in pyroxenoid crystal chemistry. I. Bustamite, wollastonite, and the pectolite-schizolite-serandite series. Amer. Mineral., 63, 274–288.

Crystal Data: Orthorhombic. *Point Group:* 222. As crystals, tabular to thin tabular, to 2.5 cm.

Physical Properties: *Cleavage:* Perfect on {001}, imperfect on {100}, poor on {010}. *Fracture:* Brittle. Hardness = 5.5–6 D(meas.) = 3.92 D(calc.) = 4.09

Optical Properties: Semitransparent. *Color:* Pale yellow to brown. *Streak:* White. *Luster:* Vitreous.
Optical Class: Biaxial (+). *Pleochroism:* In pale yellows. *Orientation:* X = a; Y = b; Z = c. *Absorption:* Z > Y ≃ X. $\alpha = 1.743$ $\beta = 1.760(3)$ $\gamma = 1.820(5)$ 2V(meas.) = 58°–62°

Cell Data: *Space Group:* $P2_12_12_1$. a = 10.57(6)   b = 9.69(6)   c = 22.38(10)   Z = [4]

X-ray Powder Pattern: Gomel district, Belarus.
2.783 (100), 3.00 (68), 2.95 (63), 4.42 (59B), 2.91 (52), 2.606 (52), 3.30 (45B)

Chemistry:

	(1)
SiO_2	33.98
TiO_2	11.35
RE_2O_3	23.59
FeO	0.82
MnO	2.58
ZnO	1.58
MgO	0.15
SrO	0.43
BaO	20.58
Na_2O	2.08
K_2O	0.40
F	0.98
H_2O	[1.45]
$-O = F_2$	0.41
Total	[99.56]

(1) Gomel district, Belarus; by electron microprobe, average of four analyses; $RE_2O_3 = La_2O_3$ 8.33%, Ce_2O_3 12.13%, Pr_2O_3 0.58%, Nd_2O_3 2.3%, Sm_2O_3 0.1%, Gd_2O_3 0.15%; corresponds to $(Na_{0.95}K_{0.12})_{\Sigma=1.07}(Mn_{0.52}Zn_{0.27}Fe_{0.16}Mg_{0.05})_{\Sigma=1.00}(Ba_{1.90}Sr_{0.06})_{\Sigma=1.96}RE_{2.03}Ti_{2.01}Si_{8.00}O_{26}$ $[F_{0.73}(OH)_{0.27}]_{\Sigma=1.00} \cdot H_2O$.

Mineral Group: Joaquinite group.

Occurrence: In a salband of a quartz vein in lower Proterozoic metasomatized granosyenites.

Association: Brookite, bastnäsite, montmorillonite, quartz, magnesio-riebeckite, aegirine, albite, leucophanite, titanite.

Distribution: In the Zhitkovitschskii district, west of Gomel, Belarus.

Name: For the occurrence in Belarus (formerly Byelorussia, USSR), and the *cerium* content.

Type Material: A.E. Fersman Mineralogical Museum, Academy of Sciences, Moscow, Russia.

References: (1) Shpanov, E.P., G.N. Netschelyustov, S.V. Baturin, and L.S. Solntseva (1989) Byelorussite-(Ce) – $NaMnBa_2Ti_2Si_8O_{26}(F, OH) \cdot H_2O$ – a new mineral of the joaquinite group. Zap. Vses. Mineral. Obshch., 118(5), 100–107 (in Russian). (2) (1991) Amer. Mineral., 76, 665–666 (abs. ref. 1).

$(Na, K)_7Ca(Si_6Al_6)O_{24}S_{1.5} \cdot H_2O$ **Bystrite**

Crystal Data: Hexagonal. *Point Group:* $3m$. Crystals tabular, to 5 mm; and as irregular grains and aggregates.

Physical Properties: *Cleavage:* $\{10\bar{1}0\}$, good; pinacoidal, weak, perhaps a parting. Hardness = 5   D(meas.) = 2.43(1)   D(calc.) = 2.45

Optical Properties: Semitransparent. *Color:* Deep yellow; in thin section, yellow. *Luster:* Vitreous.
Optical Class: Uniaxial (+). *Pleochroism:* Deep yellow to colorless. *Orientation:* Negative elongation. *Absorption:* $O > E$. $\omega = 1.584$ $\epsilon = 1.660$

Cell Data: *Space Group:* $P31c$. $a = 12.855$ $c = 10.700$ $Z = [2]$

X-ray Powder Pattern: Malaya Bystraya deposit, Russia.
3.720 (100), 3.313 (90), 3.919 (80), 4.824 (70), 2.676 (70)

Chemistry:

	(1)
SiO_2	33.37
Al_2O_3	26.74
CaO	5.10
Na_2O	14.42
K_2O	7.63
Cl	0.25
H_2O	1.62
S	12.03
SO_3	0.50
CO_2	0.32
$-O = (S_3, Cl_2)$	2.06
Total	99.92

(1) Malaya Bystraya deposit, Russia; by electron microprobe, S by wet chemical analysis, H_2O by coulometry; corresponding to $(Na_{5.17}K_{1.80})_{\Sigma=6.97}Ca_{1.01}(Si_{6.17}Al_{5.83})_{\Sigma=12.00}O_{24}[S_{4.17}Cl_{0.08}$ $(CO_3)_{0.08}(SO_4)_{0.07}]_{\Sigma=4.40} \cdot H_2O$.

Mineral Group: Cancrinite group.

Occurrence: In a lazurite deposit.

Association: Lazurite, calcite, diopside.

Distribution: In the Malaya Bystraya lazurite deposit, south of Lake Baikal, Siberia, Russia.

Name: Presumably for the occurrence in the Malaya Bystraya deposit, Russia.

Type Material: A.E. Fersman Mineralogical Museum, Academy of Sciences, Moscow, Russia.

References: (1) Ivanov, V.G. and A.N. Sapozhnikov (1985) Lazurites of the USSR. Nauka, Novosibirsk, 1–172 (in Russian). (2) Sapozhnikov, A.N., V.G. Ivanov, L.F. Piskunova, A.A. Kashaev, L.E. Terentieva, and E.A. Pobedimskaya (1991) Bystrite $Ca(Na, K)_7(Si_6Al_6O_{24})(S_3)_{1.5} \cdot H_2O$ – a new cancrinite-like mineral. Zap. Vses. Mineral. Obshch., 120(3), 97–100 (in Russian). (3) Pobedimskaya, E.A., L.E. Terentieva, A.N. Sapozhnikov, A.A. Kashaev, and G.I. Dorokhova (1991) Crystal structure of bystrite. Doklady Acad. Nauk SSSR, 319, 873–878 (in Russian). (4) (1993) Amer. Mineral., 78, 450 (abs. refs. 1–3).

Bytownite $Na_{0.3-0.1}Ca_{0.7-0.9}Al_{1.7-1.9}Si_{2.3-2.1}O_8$

Crystal Data: Triclinic. *Point Group:* $\bar{1}$. Crystals, flattened along [010], are rare, to 3 cm; commonly as cleavable masses or anhedral grains in massive aggregates. *Twinning:* Common after the Albite, Carlsbad, and Pericline laws, developed in simple to complex combinations.

Physical Properties: *Cleavage:* Perfect on {001}, good on {010}, imperfect on {110}. *Fracture:* Uneven to conchoidal. *Tenacity:* Brittle. Hardness = 6–6.5 D(meas.) = 2.72–2.74 D(calc.) = 2.713–2.726

Optical Properties: Transparent to translucent. *Color:* Colorless, white, gray. *Streak:* White. *Luster:* Vitreous, pearly on cleavages. *Optical Class:* Biaxial (–). *Dispersion:* $r < v$, weak. $\alpha = 1.565–1.573$ $\beta = 1.569–1.580$ $\gamma = 1.578–1.585$ 2V(meas.) = 87° to −78° (low); 81° to -83° (high).

Cell Data: *Space Group:* $P\bar{1}$ or $I\bar{1}$ for certain compositions. $a = 8.178$ $b = 12.870$ $c = 14.187$ $\alpha = 93.5°$ $\beta = 115.9°$ $\gamma = 90.63°$ Z = 8

X-ray Powder Pattern: Rustenburg mine, Transvaal, South Africa.
3.20 (100), 4.03 (80), 3.75 (80), 3.17 (80), 3.62 (70), 2.94 (70b), 2.52 (70)

Chemistry:

	(1)	(2)	(3)		(1)	(2)	(3)
SiO_2	49.06	50.54	45.62	Na_2O	2.57	3.40	1.12
Al_2O_3	32.14	31.70	35.02	K_2O	0.17		
Fe_2O_3	0.27			H_2O^+	0.13		
MgO	0.20			H_2O^-	0.03		
CaO	15.38	14.36	18.24	Total	99.95	100.00	100.00

(1) Rustenburg mine, Transvaal, South Africa. (2) $Na_{0.30}Ca_{0.70}Al_{1.70}Si_{2.30}O_8$.
(3) $Na_{0.10}Ca_{0.90}Al_{1.90}Si_{2.10}O_8$.

Polymorphism & Series: Low- and high-temperature structural modifications are recognized.

Mineral Group: Feldspar group, plagioclase series.

Occurrence: A rock-forming mineral typical of mafic intrusive igneous rocks, as gabbros and anorthosites; as phenocrysts in basalts; rare in metamorphic rocks.

Association: Pyroxenes, olivine.

Distribution: One of the less frequently identified feldspars, although still very common. "Type" material from Ottawa, Ontario, Canada was later shown to be a mixture. Elsewhere in Canada, from the Shawmere anorthosite, Foleyet Township, Ontario, and on Yamaska Mountain, near Abbotsford, Quebec. On Rhum Island, Scotland. At Eycott Hill, near Keswick, Cumberland, England. From Närödal, Norway. In Greenland, from Fiskenæsset and Storo. In the Bushveld complex, Transvaal, South Africa. In the USA, in the Stillwater complex, Montana; from Cornwall, Lebanon Co., and Phoenixville, Chester Co., Pennsylvania; near Lakeview, Lake Co., Oregon; in the Lucky Cuss mine, Tombstone, Cochise Co., Arizona; and from Grants, Valencia Co., New Mexico. From Isa Valley, Western Australia.

Name: For the "type" locality, in Ottawa (formerly Bytown), Ontario, Canada.

References: (1) Dana, E.S. (1892) Dana's system of mineralogy, (6th edition), 325–327, 334–335. (2) Deer, W.A., R.A. Howie, and J. Zussman (1963) Rock-forming minerals, v. 4, framework silicates, 94–165. (3) Phillips, W.R. and D.T. Griffen (1981) Optical mineralogy, 352–360. (4) Goodyear, J. and W.J. Duffin (1954) The identification and determination of plagioclase feldspars by the X-ray powder method. Mineral. Mag., 30, 306–326. (5) Fleet, S.G., S. Chandrasekhar, and H.D. Megaw (1966) The structure of bytownite ('body-centered anorthite'). Acta Cryst., 21, 782–801.

Crystal Data: Hexagonal. *Point Group:* 32. Equant trigonal crystals, to 2 mm, showing two prisms, {11$\overline{2}$0} and {2$\overline{11}$0}, and right and left rhombohedra, {10$\overline{1}$2} and {01$\overline{1}$2}.

Physical Properties: *Fracture:* Conchoidal. Hardness = 4   D(meas.) = 2.68(2) D(calc.) = 2.74

Optical Properties: Semitransparent. *Color:* Pale blue to white when altered. *Luster:* Vitreous. *Optical Class:* Uniaxial (−). $\omega = 1.622(1)$ $\epsilon = 1.619(1)$

Cell Data: *Space Group:* $R32$. a = 20.870(4)   c = 16.002(4)   Z = [24]

X-ray Powder Pattern: Liberty Bell Mountain, Washington, USA. 5.231 (100), 3.013 (30), 3.137 (20), 5.993 (12), 2.607 (9), 2.900 (7), 3.197 (6)

Chemistry:

	(1)	(2)	(3)
SiO$_2$	38.81	41.16	43.58
TiO$_2$	0.09	0.04	
ZrO$_2$	31.64	33.58	29.79
Al$_2$O$_3$	2.61	0.05	
FeO	0.03	0.12	
CuO	0.19	0.42	
CaO	11.25	11.62	13.56
Na$_2$O	0.20	0.13	
H$_2$O	[15.18]	[12.88]	13.07
Total	[100.00]	[100.00]	100.00

(1–2) Liberty Bell Mountain, Washington, USA; by electron microprobe, H$_2$O by difference. (3) CaZrSi$_3$O$_9$·3H$_2$O.

Occurrence: In miarolitic cavities in a peralkalic alaskitic border granite of a batholith.

Association: Microcline, quartz, albite, fluorite, chlorite, bastnäsite, zircon, malachite.

Distribution: From Washington Pass, on Liberty Bell Mountain, Okanogan Co., Washington, USA.

Name: For the *calcium* content and relation to *hilairite*.

Type Material: National Museum of Natural History, Washington, D.C., USA, 161901, 162182.

References: (1) Boggs, R.C. (1988) Calciohilairite, CaZrSi$_3$O$_9$·3H$_2$O, the calcium analogue of hilairite from the Golden Horn batholith, northern Cascades, Washington. Amer. Mineral., 73, 1191–1194.

Calcium catapleiite $CaZrSi_3O_9 \cdot 2H_2O$

Crystal Data: Hexagonal. *Point Group:* [6/m 2/m 2/m.] (by analogy to catapleiite). Massive.

Physical Properties: *Cleavage:* Present. Hardness = 4.5–5 D(meas.) = 2.77 D(calc.) = [2.75]

Optical Properties: Opaque, translucent on thin edges. *Color:* Pale yellow to cream. *Luster:* Vitreous to dull.
Optical Class: Uniaxial (+). $\omega = 1.603$ $\epsilon = 1.639$

Cell Data: *Space Group:* [$P6_3mmc$.] a = 7.40 c = 10.07 Z = [2]

X-ray Powder Pattern: Burpala massif, Russia; very close to catapleiite.
2.96 (100), 3.96 (80), 3.06 (80), 1.975 (80), 1.835 (80), 6.45 (70), 1.740 (70)

Chemistry:

	(1)
SiO_2	44.49
TiO_2	0.06
ZrO_2	31.00
Al_2O_3	0.60
RE_2O_3	0.28
Fe_2O_3	0.36
CaO	13.82
Na_2O	0.32
K_2O	0.10
H_2O^+	9.15
H_2O^-	0.18
Total	100.36

(1) Burpala massif, Russia; leading to $(Ca_{0.98}Na_{0.04})_{\Sigma=1.02}Zr_{1.00}(Si_{2.94}Al_{0.04})_{\Sigma=2.98}O_9 \cdot 2.01H_2O$.

Occurrence: In cavities between crystals of microcline, in syenite pegmatites of a differentiated alkalic massif.

Association: Pyrophanite, pyrochlore, titanian låvenite, loparite, kupletskite, RE-apatite, hiortdahlite, seidozerite.

Distribution: In the Burpala massif, about 120 km north of Lake Baikal, eastern Siberia, Russia.

Name: For its *calcium* content and close relation to *catapleiite*.

Type Material: n.d.

References: (1) Portnov, A.M. (1964) Calcium catapleiite, a new variety of catapleiite. Doklady Acad. Nauk SSSR, 154, 607–609 (in Russian). (2) (1964) Amer. Mineral., 49, 1153 (abs. ref. 1).

$(\mathrm{Mn}^{2+}, \mathrm{Ca})_3(\mathrm{Fe}^{3+}, \mathrm{Al})_2(\mathrm{SiO_4})_3$ **Calderite**

Crystal Data: Cubic. *Point Group:* $4/m\,\bar{3}\,2/m$. Granular, massive.

Physical Properties: Hardness = 7   D(meas.) = 4.05   D(calc.) = 4.07

Optical Properties: Semitransparent. *Color:* Dark reddish brown to dark yellowish; yellow to pale greenish yellow in thin section.
Optical Class: Isotropic; shows slight anisotropism. $n = 1.875$–1.934

Cell Data: *Space Group: Ia3d.* a = 11.81   Z = 8

X-ray Powder Pattern: Otjosondu, Namibia.
2.62 (100), 1.574 (100), 2.92 (90), 2.39 (90), 1.632 (90), 0.968 (90), 1.906 (80)

Chemistry:

	(1)	(2)
SiO_2	35.16	32.4
TiO_2	0.28	
Al_2O_3	9.04	
Fe_2O_3	16.27	31.4
FeO	0.00	
MnO	27.38	29.0
MgO	0.50	
CaO	12.12	8.5
Total	100.75	101.3

(1) Otjosondu, Namibia; by electron microprobe; corresponds to $(\mathrm{Mn}_{1.96}\mathrm{Ca}_{1.10})_{\Sigma=3.06}(\mathrm{Fe}^{3+}_{1.04}\mathrm{Al}_{0.87}$
$\mathrm{Mg}_{0.06}\mathrm{Ti}_{0.02})_{\Sigma=1.99}(\mathrm{Si}_{2.97}\mathrm{Al}_{0.03})_{\Sigma=3.00}\mathrm{O}_{12}$. (2) Wabush district, Canada; by electron microprobe, corresponds to $(\mathrm{Mn}_{2.20}\mathrm{Ca}_{0.82})_{\Sigma=3.02}\mathrm{Fe}^{3+}_{2.02}(\mathrm{Si}_{2.90}\mathrm{Al}_{0.10})_{\Sigma=3.00}\mathrm{O}_{12}$.

Mineral Group: Garnet group.

Occurrence: In a manganese deposit (Otjosondu, Namibia); in an iron formation (Wabush district, Canada).

Association: Pyrolusite, aegirine, rhodonite, hematite, rhodochrosite, quartz, kutnohorite (Wabush district, Canada).

Distribution: From Katkamsandi, Hazaribagh district, Bihar, and at Netra, Balaghat district, Madhya Pradesh, India. From Otjosondu, Namibia. In the Wabush district, near Labrador City, Labrador, Newfoundland, Canada.

Name: For James Calder, an early writer on the geology of India; name first applied to a rock, later transferred to its predominant mineral.

Type Material: n.d.

References: (1) Dana, E.S. (1892) Dana's system of mineralogy, (6th edition), 443.
(2) Vermaas, F.H.S. (1952) Manganese-iron garnet from Otjosondu, South-West Africa.
Mineral. Mag., 29, 946–951. (3) Klein, C., Jr. (1966) Mineralogy and petrology of the
metamorphosed Wabush Iron Formation, southwestern Labrador. J. Petrol., 7, 246–305.
(4) Dunn, P.J. (1979) On the validity of calderite. Can. Mineral., 17, 569–571. (5) Amthauer, G.,
K. Katz-Lehnert, D. Lattard, M. Okrusch, and E. Woermann (1989) Crystal chemistry of natural
Mn^{3+} -bearing calderite-andradite garnets from Otjosondu, SWA/Namibia. Zeits. Krist., 189,
43–56.

Crystal Data: Monoclinic. *Point Group: m.* As crystals, to 10 cm; in platy aggregates, to 20 cm; also granular. *Twinning:* Polysynthetic, the twinning plane at an angle of 8° to the less perfect cleavage.

Physical Properties: *Cleavage:* One, very perfect; another, perfect, at 118° to the first. *Fracture:* Splintery, breaks into long acute-angled or wedge-shaped pieces. *Tenacity:* Brittle. Hardness = n.d. D(meas.) = 2.707 D(calc.) = [2.65]

Optical Properties: Transparent to translucent. *Color:* Greenish yellow to grayish green. *Streak:* White. *Luster:* Vitreous. *Optical Class:* Biaxial (–). *Orientation:* $Y = b$; $Z \wedge$ less perfect cleavage = 2°. $\alpha = 1.534$ $\beta = 1.538$ $\gamma = 1.543$ 2V(meas.) = 58°

Cell Data: *Space Group: Cm.* $a = 18.836$ $b = 7.244$ $c = 12.636$ $\beta = 111.76(2)°$ $Z = 2$

X-ray Powder Pattern: Khibiny massif, Russia.
3.080 (100b), 2.907 (80), 1.641 (80), 4.69 (70), 4.81 (60), 2.359 (60), 4.20 (50)

Chemistry:

	(1)	(2)		(1)	(2)
SiO_2	56.08	55.71	K_2O	8.47	10.63
TiO_2	0.10	0.06	F	2.21	2.17
Al_2O_3	0.55	0.20	Cl	0.22	
Fe_2O_3	1.41	0.72	H_2O^+	1.11	1.25
FeO	0.71	0.36	H_2O^-	0.49	0.60
MnO	0.38	0.41	CO_2	0.20	
MgO	0.05	0.26	P_2O_5	0.04	0.08
CaO	20.95	20.39	$-O = (F, Cl)_2$	0.96	0.91
Na_2O	8.01	7.08	Total	100.02	99.01

(1–2) Khibiny massif, Russia. (3) Murun massif, Russia; analysis not given, stated to correspond to $Na_{2.96}K_{2.94}(Ca_{4.69}Fe_{0.17}Mg_{0.08}Mn_{0.07}Al_{0.05}Ti_{0.01})_{\Sigma=5.07}Si_{12}O_{30}[(OH)_{2.27}F_{1.48}O_{0.25}]_{\Sigma=4.00}$.

Occurrence: In pegmatites in a differentiated alkalic massif (Khibiny massif, Russia); in charoitic rocks (Murun massif, Russia).

Association: Fenaksite, lamprophyllite (Khibiny massif, Russia); tinaksite, miserite, charoite (Murun massif, Russia).

Distribution: On Mts. Yukspor and Rasvumchorr, Khibiny massif, Kola Peninsula, and in the Murun massif, southwest of Olekminsk, Yakutia, Russia.

Name: For CAlcium, sodium, NAtrium, and SIlicon in the chemical composition.

Type Material: National Museum of Natural History, Washington, D.C., USA, 136472.

References: (1) Dorfman, M.D., D.D. Rogachev, Z.I. Goroshchenko, and E.I. Uspenskaya (1959) Canasite, a new mineral. Trudy Mineralog. Muzeya Akad. Nauk SSSR, 9, 158–166 (in Russian). (2) (1960) Amer. Mineral., 45, 253–254 (abs. ref. 1). (3) Rozhdestvenskaya, I.V., L.V. Nikishova, I.I. Bannova, and Y.D. Lasebnik (1987) Canasite: the refinement of crystal structure and comparison with that of miserite. Acta Cryst., A43, C159. (4) Rozhdestvenskaya, I.V., L.V. Nikishova, I.I. Bannova, and Y.D. Lasebnik (1987) [Canasite: crystal structure typomorphism.] Mineral. Zhurnal, 10(4), 31–41 (in Russian with English abs.).

Crystal Data: Hexagonal. *Point Group:* 6. Rarely as prismatic crystals, terminated by a low pyramid, to 2 cm; commonly massive. *Twinning:* Lamellar, rare.

Physical Properties: *Cleavage:* Perfect on $\{10\bar{1}0\}$, poor on $\{0001\}$. *Fracture:* Uneven. *Tenacity:* Brittle. Hardness = 5–6 D(meas.) = 2.42–2.51. D(calc.) = [2.49]

Optical Properties: Transparent to translucent. *Color:* Colorless, white, light blue to light grayish blue, honey-yellow, orange, reddish; colorless in thin section. *Streak:* White. *Luster:* Vitreous, pearly, or greasy.
Optical Class: Uniaxial (–) or (+). $\omega = 1.507$–1.528 $\epsilon = 1.495$–1.503

Cell Data: *Space Group:* $P6_3$. a = 12.58–12.76 c = 5.11–5.20 Z = 1

X-ray Powder Pattern: York River, Bancroft, Canada.
3.21 (100), 4.64 (90), 3.64 (70), 2.099 (70), 1.488 (50), 1.447 (50), 10.92 (40)

Chemistry:

	(1)	(2)	(3)
SiO_2	34.35	33.64	34.25
TiO_2		0.07	
Al_2O_3	29.35	29.82	29.06
Fe_2O_3	0.03		
MgO	0.01	0.39	
CaO	8.11	8.64	10.66
Na_2O	17.66	15.41	17.67
K_2O	0.10	0.50	
Cl	0.21		
H_2O^+	3.02	4.61	
H_2O^-	0.11		
CO_2	6.60	6.79	8.36
SO_3		0.03	
Total	99.55	99.90	100.00

(1) Dungannon Township, Ontario, Canada. (2) Iron Hill, Colorado, USA.
(3) $Na_6Ca_2Al_6Si_6O_{24}(CO_3)_2$.

Mineral Group: Cancrinite group.

Occurrence: A primary mineral in some alkalic igneous rocks, including pegmatites in nepheline syenites; also as an alteration product of nepheline.

Association: Nepheline, sodalite, natrolite, orthoclase, monticellite, titanian andradite.

Distribution: At Miass, Ilmen Mountains, Southern Ural Mountains, and on the Kola Peninsula, Russia. From Tvedalen and the Langesundsfjord, Norway. Found near Loch Borolan, Assynt, Scotland. Around the Laacher See, Eifel district, Germany. From Vesuvius, Campania, Italy. In the USA, from Gardiner and Litchfield, Kennebec Co., Maine; Pennsburg, Chester Co., Pennsylvania; at Iron Hill, Gunnison Co., Colorado; and Point of Rocks, Colfax Co., New Mexico. In Canada, from Bancroft, Egan Chute, and Blue Mountain, Ontario; at Mont Saint-Hilaire, Quebec; and a number of other lesser localities.

Name: For Count Georg Cancrin (1774–1845), German-Russian Minister of Finance of Russia.

References: (1) Dana, E.S. (1892) Dana's system of mineralogy, (6th edition), 427–428.
(2) Deer, W.A., R.A. Howie, and J. Zussman (1963) Rock-forming minerals, v. 4, framework silicates, 310–320. (3) Foit, F.F., Jr. (1973) Cancrinite with a new superstructure from Bancroft, Ontario. Can. Mineral., 11, 940–951. (4) Grundy, H.D. and I. Hassan (1982) The crystal structure of a carbonate-rich cancrinite. Can. Mineral., 20, 239–251.

Crystal Data: Hexagonal. *Point Group: 6mm.* Anhedral, granular, to 3 mm.

Physical Properties: *Fracture:* Conchoidal. *Tenacity:* Brittle. Hardness = 5
D(meas.) = 2.40(2) D(calc.) = 2.39 Fluoresces yellow under UV.

Optical Properties: Transparent to translucent. *Color:* Lilac; colorless in thin section.
Streak: White. *Luster:* Vitreous.
Optical Class: Uniaxial (−). $\omega = 1.509(2)$ $\epsilon = 1.490(2)$

Cell Data: *Space Group:* $P6_3mc.$ a = 12.575(3) c = 5.105(2) Z = 1

X-ray Powder Pattern: Lovozero massif, Russia; resembles cancrinite.
3.22 (100), 3.65 (90), 6.30 (70), 4.61 (50), 2.722 (50), 2.597 (20), 2.402 (20)

Chemistry:

	(1)
SiO_2	43.11
Al_2O_3	24.42
Fe_2O_3	0.33
MnO	0.11
CaO	0.68
Na_2O	21.30
K_2O	0.10
H_2O	5.01
CO_2	4.82
SO_3	0.36
Total	100.24

(1) Lovozero massif, Russia, corresponding to $(Na_{6.89}Ca_{0.12}Fe^{3+}_{0.04}Mg_{0.03}K_{0.02})_{\Sigma=7.10}$
$(Si_{7.20}Al_{4.80})_{\Sigma=12.00}O_{24.10}(CO_3)_{1.10}(SO_4)_{0.04} \cdot 2.79H_2O.$

Mineral Group: Cancrinite group.

Occurrence: In ultra-agpaitic pegmatites in a differentiated alkalic massif.

Association: Potassic feldspar, nepheline, arfvedsonite, aegirine, manaksite, additional minor minerals.

Distribution: From the Lovozero massif, Kola Peninsula, Russia.

Name: For the relatively high ratio of silicon to aluminum, and its relation to *cancrinite*.

Type Material: Vernadskii Geological Museum, Moscow; A.E. Fersman Mineralogical Museum, Academy of Sciences, Moscow, Russia.

References: (1) Khomyakov, A.P., E.I. Semenov, E.A. Pobedimskaya, T.N. Nadezhina, and R.K. Rastsvetaeva (1991) Cancrisilite $Na_7[Al_5Si_7O_{24}]CO_3 \cdot 3H_2O$ – a new mineral of the cancrinite group. Zap. Vses. Mineral. Obshch., 120(6), 80–84 (in Russian). (2) (1993) Amer. Mineral., 78, 1314 (abs. ref. 1).

Crystal Data: Metamict; hexagonal after reconstitution at 725 °C. *Point Group:* 3. As crystals, thick hexagonal prismatic, showing {10$\bar{1}$0}, {0001}, and {10$\bar{1}$3}, to 2 cm.

Physical Properties: *Fracture:* Conchoidal. *Tenacity:* Brittle. Hardness = 6–6.5
D(meas.) = 4.407   D(calc.) = [4.50]

Optical Properties: Translucent. *Color:* Greenish brown, pale yellowish. *Luster:* Vitreous to greasy.
Optical Class: Uniaxial (–). $n = \sim 1.76$; birefringence rather strong.

Cell Data: *Space Group:* P3.   a = 10.67(2)   c = 4.680(2)   Z = 1

X-ray Powder Pattern: Metamict.

Chemistry:

	(1)	(2)
SiO_2	14.21	14.51
ThO_2	0.80	
B_2O_3	[17.16]	16.82
Y_2O_3	52.62	54.55
La_2O_3	2.97	
CeO_2	1.29	
CaO	0.67	
BaO	8.02	12.35
Na_2O	0.25	
K_2O	0.20	
F		3.06
LOI	1.81	
$-O = F_2$		1.29
Total	[100.00]	100.00

(1) Lille Arø Island, Langesundsfjord, Norway; B_2O_3 by difference, corresponds to $(Ba_{0.64}La_{0.22}$ $Na_{0.10}K_{0.05})_{\Sigma=1.01}(Y_{5.74}Ca_{0.15}Ce^{4+}_{0.09}Th_{0.04})_{\Sigma=6.02}Si_{2.91}B_{6.07}O_{24}F_2$. (2) $BaY_6Si_3B_6O_{24}F_2$.

Occurrence: In a small vein in nepheline syenite pegmatite.

Association: Wöhlerite, rosenbuschite, thorite, catapleiite, låvenite, nepheline, sodalite.

Distribution: On Lille Arø Island, in the Langesundsfjord, Norway.

Name: For D. Cappelen of Holden, Norway, a collector who discovered the mineral, and its *yttrium* content.

Type Material: University of Oslo, Oslo, Norway; National Museum of Natural History, Washington, D.C., USA, 160547.

References: (1) Dana, E.S. (1892) Dana's system of mineralogy, (6th edition), 413–414. (2) Vlasov, K.A., Ed. (1966) Mineralogy of rare elements, v. II, 248. (3) Shen, J. and P.B. Moore (1984) Crystal structure of cappelenite, $Ba(Y, RE)_6[Si_3B_6O_{24}]F_2$: a silicoborate sheet structure. Amer. Mineral., 69, 190–195.

Carletonite

$KNa_4Ca_4Si_8O_{18}(CO_3)_4(OH,F) \cdot H_2O$

Crystal Data: Tetragonal. *Point Group:* $4/m\ 2/m\ 2/m$. Crystals, prismatic along [001], to 6 cm; massive.

Physical Properties: *Cleavage:* Perfect on {001}, good on {110}. *Fracture:* Conchoidal. *Tenacity:* Brittle. Hardness = 4–4.5   D(meas.) = 2.45   D(calc.) = 2.426

Optical Properties: Transparent to translucent. *Color:* Pink or pale to dark blue, commonly zoned; colorless in thin flakes. *Streak:* White. *Luster:* Vitreous to pearly, may become slightly waxy after long exposure to air.
Optical Class: Uniaxial (–). *Pleochroism:* Weak; O = very pale blue; E = very pale pinkish brown. $\omega = 1.521(1)$ $\epsilon = 1.517(1)$

Cell Data: *Space Group:* $P4/mbm$. $a = 13.178(3)$ $c = 16.695(4)$ $Z = 4$

X-ray Powder Pattern: Mont Saint-Hilaire, Canada.
8.353 (100), 4.171 (100), 2.903 (90), 2.384 (60), 4.053 (50), 16.705 (40), 4.816(40)

Chemistry:

	(1)	(2)
SiO_2	44.9	44.7
TiO_2	trace	trace
Al_2O_3	0.5	0.6
MgO	0.09	0.13
CaO	19.92	19.97
Na_2O	10.23	10.64
K_2O	3.28	3.31
F	0.70	0.73
H_2O^+		3.51
H_2O^-	0.70	0.63
CO_2		15.2
LOI	19.92	
$-O = F_2$	0.29	0.30
Total	99.95	99.12

(1–2) Mont Saint-Hilaire, Canada; CO_2 by acid evolution-gravimetry, H_2O by direct determination of H; the average corresponds to $K_{0.74}Na_{3.56}(Ca_{3.74}Mg_{0.03})_{\Sigma=3.77}$ $(Si_{7.89}Al_{0.11})_{\Sigma=8.00}O_{18}(CO_3)_{3.65}F_{0.41} \cdot 2.05H_2O$.

Occurrence: In cores of thermally metamorphosed wall-rock xenoliths of shale and interbedded limestone, now hornfels and siliceous marble, in nepheline syenite in an intrusive alkalic gabbro-syenite complex.

Association: Quartz, narsarsukite, calcite, fluorite, ancylite, molybdenite, leucosphenite, lorenzenite, galena, albite, pectolite, arfvedsonite, apophyllite, leifite (hornfels); pectolite, microcline, arfvedsonite, apophyllite (marble).

Distribution: From Mont Saint-Hilaire, Quebec, Canada.

Name: For Carleton University, Toronto, Canada, where it was first studied.

Type Material: Canadian Museum of Nature, Ottawa, Canada, T711.

References: (1) Chao, G.Y. (1971) Carletonite, $KNa_4Ca_4Si_8O_{18}(CO_3)_4(F,OH) \cdot H_2O$, a new mineral from Mount St. Hilaire, Quebec. Amer. Mineral., 56, 1855–1866. (2) Chao, G.Y. (1972) The crystal structure of carletonite, $KNa_4Ca_4Si_8O_{18}(CO_3)_4(F,OH) \cdot H_2O$, a double-sheet silicate. Amer. Mineral., 57, 765–778.

Crystal Data: Monoclinic. *Point Group:* m. Fibers, to 0.2 μm width, elongated along [010], to several cm. Commonly gathered in folded bundles; intergrown in parallel with brucite and chrysotile.

Physical Properties: *Cleavage:* On {001}, very good; parting on {010}. *Tenacity:* Flexible. Hardness = n.d. D(meas.) = 2.63(2) D(calc.) = 2.606

Optical Properties: Transparent. *Color:* Light brown. *Streak:* Whitish. *Luster:* Vitreous, pearly.
Optical Class: Biaxial. *Pleochroism:* Orange-brown $\parallel$ [010]; pale orange-brown $\perp$ [010].
$n = 1.605(5)$ 2V(meas.) = n.d.

Cell Data: *Space Group: Cm.* a = 36.70 b = 9.41 c = 7.291 $\beta = 101.1°$ Z = 2

X-ray Powder Pattern: Near Sampeyre, Italy.
7.17 (100), 3.397 (55), 3.595 (45), 2.562 (40), 2.280 (35), 18.02 (25), 3.637 (20)

Chemistry:

	(1)
SiO_2	35.53
TiO_2	2.24
Al_2O_3	1.07
Cr_2O_3	0.24
FeO	4.03
MnO	0.72
MgO	39.28
H_2O	16.85
Total	99.96

(1) Near Sampeyre, Italy; by electron microprobe, average of 15 analyses, H_2O by TGA; corresponds to $(Mg_{18.88}Fe_{1.08}Ti_{0.54}Mn_{0.20}Cr_{0.06})_{\Sigma=20.76}(Si_{11.46}Al_{0.40})_{\Sigma=11.86}H_{36.26}O_{63.00}$.

Occurrence: In a network of veins crosscutting a metamorphic antigorite serpentinite, in an ophiolite.

Association: Diopside, chrysotile, brucite, magnetite, clinohumite, perovskite, uvarovite (near Sampeyre, Italy).

Distribution: In Italy, in Piedmont, at a number of localities around Sampeyre and Casteldelfino, Varaita Valley; around Crissolo, Po Valley; at Monte Nebin, Maira Valley; and around Viù, Viù Valley. From Taberg, Sweden.

Name: For Professor Carlo Sturani (1938–1976), University of Torino, Torino, Italy.

Type Material: Regional Science Museum, Torino, Italy; National Museum of Natural History, Washington, D.C., USA, 162683.

References: (1) Compagnoni, R., G. Ferraris, and M. Mellini (1985) Carlosturanite, a new asbestiform rock-forming silicate from Val Varaita, Italy. Amer. Mineral., 70, 767–772. (2) Mellini, M., G. Ferraris, and R. Compagnoni (1985) Carlosturanite: HRTEM evidence of a polysomatic series including serpentine. Amer. Mineral., 70, 773–781. (3) Belluso, E. and G. Ferraris (1991) New data on balangeroite and carlosturanite from alpine serpentinites. Eur. J. Mineral., 3, 559–566.

Crystal Data: Orthorhombic. *Point Group:* $2/m\ 2/m\ 2/m$. Crystals acicular, prismatic; fibrous, as radiated tufts, to 2 cm. *Twinning:* On {100}.

Physical Properties: *Cleavage:* Perfect on {010}. *Tenacity:* Very brittle. Hardness = 5.5–6 D(meas.) = 2.935–3.031 D(calc.) = [3.07]

Optical Properties: Semitransparent. *Color:* Yellow. *Luster:* Silky, glistening. *Optical Class:* Biaxial (–). *Pleochroism:* Distinct; $X = Y$ = pale yellow; Z = colorless. $\alpha = 1.624$ $\beta = 1.629$ $\gamma = 1.638$ 2V(meas.) = n.d. 2V(calc.) = 67°

Cell Data: *Space Group:* $Ccca$. $a = 13.714(2)$ $b = 20.079(2)$ $c = 5.105(1)$ $Z = 8$

X-ray Powder Pattern: Fukuzumi mine, Japan. 5.73 (100), 5.08 (70), 2.620 (50), 3.46 (30), 3.04 (30), 3.39 (20), 2.761 (20)

Chemistry:

	(1)	(2)	(3)
SiO_2	36.61	38.38	36.52
TiO_2	trace	trace	
Al_2O_3	29.36	29.45	30.98
Fe_2O_3	1.53	1.47	
V_2O_5		0.06	
FeO	3.05	0.22	
MnO	18.08	17.77	21.55
ZnO		0.12	
MgO		1.92	
CaO	trace		
Na_2O		0.19	
K_2O		0.08	
H_2O^+	11.03		
H_2O^-	0.32		
H_2O		10.19	10.95
Total	99.98	99.85	100.00

(1) Fukuzumi mine, Japan. (2) Near Meuville, Belgium; Loss on ignition taken as H_2O. (3) $MnAl_2Si_2O_6(OH)_4$.

Polymorphism & Series: Forms a series with ferrocarpholite.

Occurrence: In low-grade metamorphosed shales.

Association: Sudoite, manganoan garnet, chlorotoid, fluorite.

Distribution: At Horní Slavkov (Schlaggenwald), Czech Republic. From near Meuville, Ardennes Mountains, Belgium. At Wippra, Harz Mountains, Germany. In England, from the Carrock mine, Caldbeck Fells, Cumbria; at Kit Hill and in the Hingston Down Consols mine, near Callington, and in the Stennagwynn mine, St. Stephen-in-Brannel, Cornwall. In the Fukuzumi mine, Kyoto Prefecture, Japan.

Name: From the Greek for *straw*, in allusion to its color.

References: (1) Dana, E.S. (1892) Dana's system of mineralogy, (6th edition), 549. (2) Yoshimura, T. and Y. Aoki (1966) Carpholite from the Fukuzumi mine, Hyogo Prefecture. J. Mineral. Soc. Japan, 8, 43–48. (3) (1967) Mineral. Abs., 18, 200 (abs. ref. 2). (4) A.-M. Fransolet (1972) Données nouvelles sur la carpholite de Meuville (vallée de la Lienne, Belgique). Bull. Soc. fr. Minéral., 95, 84–97 (in French with English abs.). (5) Lindemann, V.W., R. Wögerbauer, and P. Berger (1979) Die Kristallstruktur von Karpholith. Neues Jahrb. Mineral., Monatsh., 282–287 (in German with English abs.). (6) Ghose, S., P.K. Sen Gupta, R.C. Boggs, and E.O. Schlemper (1989) Crystal chemistry of a nonstoichiometric carpholite, $K_x(Mn_{2-x}Li_x)Al_4Si_4O_{12}(OH)_4F_4$: a chain silicate related to pyroxenes. Amer. Mineral., 74, 1084–1090.

Crystal Data: Monoclinic (?). *Point Group:* n.d. As tabular pseudohexagonal crystals, commonly in rosettes, to 4 mm. Stalactitic, reniform with concentrically radiating fibrous internal structure, or massive.

Physical Properties: *Cleavage:* Perfect on {001}. Hardness = 3–3.5 D(meas.) = 2.83–2.94 D(calc.) = 2.788

Optical Properties: Semitransparent. *Color:* Reddish brown, tan; light brown to yellow in thin section.
Optical Class: Biaxial (–). *Orientation:* $X = c.$ $\alpha = 1.602$–1.624 $\beta = 1.632$–1.650
$\gamma = 1.632$–1.650 2V(meas.) = $\sim 0°$

Cell Data: *Space Group:* n.d. $a = 5.668$ $b = 9.811$ $c = 7.527$ $\beta = 104.52°$ $Z = 2$

X-ray Powder Pattern: Långban, Sweden.
2.49 (100), 7.24 (90), 3.64 (80), 2.78 (70), 2.08 (50), 1.954 (30), 1.617 (20)

Chemistry:

	(1)	(2)		(1)	(2)
SiO_2	32.1	36.35	MgO	1.9	8.01
Al_2O_3		0.21	CaO		0.03
As_2O_3	6.5		F		< 0.05
FeO	1.7	1.09	Cl		0.14
MnO	48.8	42.39	H_2O	[8.1]	[9.22]
ZnO	0.9	< 0.03	$-O = (F, Cl)_2$		[0.08]
			Total	[100.0]	[97.44]

(1) Sterling Hill, New Jersey, USA; by electron microprobe, H_2O by difference. (2) Bald Knob, North Carolina, USA; by electron microprobe, H_2O calculated from stoichiometry.

Occurrence: A product of the metamorphism of manganese-bearing minerals.

Association: Gonyerite (Långban, Sweden); calcite, rhodonite, lead, sarkinite, brandtite (Pajsberg, Sweden); manganoan calcite, tirodite, rhodonite, jacobsite (Bald Knob, North Carolina, USA).

Distribution: From the Harstig mine, Pajsberg, nera Persberg, and at Långban, Värmland, Sweden. In the Molinello and Gambatesa manganese mines, near Chiavari, Val Graveglia, Liguria, Italy. At La Falotta, Graubünden, Switzerland. In the Ködnitz Valley, Tirol, Austria. In the USA, at Franklin and Sterling Hill, Ogdensburg, Sussex Co., New Jersey; at Bald Knob, near Sparta, Alleghany Co., North Carolina; in the Hurricane mine, on the Olympic Penninsula, Washington. From the Ichinomata mine, Kumamoto Prefecture, Japan. At Watson's Beach, southeastern Otago, New Zealand. In the N'Chwaning and Wessels mines, near Kuruman, Cape Province, South Africa.

Name: From the Greek for *walnut* and *felt*, in allusion to its color and habit.

References: (1) Dana, E.S. (1892) Dana's system of mineralogy, (6th edition), 704. (2) Kato, T. (1963) New data on the so-called bementite. J. Japan. Assoc. Mineral. Petrol. Econ. Geol., 49, 93–103 (in Japanese). (3) (1964) Amer. Mineral., 49, 446–447 (abs. ref. 2). (4) Peacor, D.R. and E.J. Essene (1980) Caryopilite–a member of the friedelite rather than the serpentine group. Amer. Mineral., 65, 335–339. (5) Dunn, P.J., D.R. Peacor, J.A. Nelen, and J.A. Norberg (1981) Crystal-chemical data for schallerite, caryopilite and friedelite from Franklin and Sterling Hill, New Jersey. Amer. Mineral., 66, 1054–1062. (6) Bayliss, P. (1981) Unit cell data of serpentine group minerals. Mineral. Mag., 44, 153–156. (7) Phillips, W.R. and D.T. Griffen (1981) Optical mineralogy, 602–603. (8) Guggenheim, S., S.W. Bailey, R.A. Eggleton, and P. Wilkes (1982) Structural aspects of greenalite and related minerals. Can. Mineral., 20, 1–18.

Crystal Data: Triclinic. *Point Group:* 1. As plates, to 0.2 mm, elongated along [100] and flattened ∥ {001}; forms observed are {001}, {100}, {120}, and {5$\bar{3}$0}.

Physical Properties: *Cleavage:* Good on {100} and {001}. Hardness = n.d. D(meas.) = n.d. D(calc.) = [3.03]

Optical Properties: Transparent. *Color:* Pale pink. *Luster:* Vitreous. *Optical Class:* Biaxial. $\alpha = 1.663$ on (001). β = n.d. $\gamma = 1.684$ on (001). 2V(meas.) = n.d.

Cell Data: *Space Group:* $P1$. a = 9.791(8)   b = 10.420(9)   c = 7.076(6) $\alpha = 98.91(8)°$ $\beta = 102.63(8)°$ $\gamma = 84.17(8)°$ Z = 2

X-ray Powder Pattern: Baveno, Italy.
2.821 (s), 3.62 (m), 3.10 (m), 2.968 (m), 1.429 (m), 7.22 (w), 6.75 (w)

Chemistry:

	(1)
SiO_2	51.83
TiO_2	0.08
Al_2O_3	0.11
Sc_2O_3	14.74
FeO	4.50
MnO	2.32
MgO	0.20
CaO	16.83
Na_2O	0.06
H_2O	[2.59]
Total	[93.26]

(1) Baveno, Italy; by electron microprobe, H_2O calculated from stoichiometry.

Occurrence: In a geode in granite.

Association: Quartz, orthoclase, albite, jervisite.

Distribution: From the Diverio quarry, Mt. Mottarone, near Baveno, Piedmont, Italy.

Name: For CAlcium and SCANDium in the composition.

Type Material: Municipal Museum of Natural History, Milan, Italy, 23270.

References: (1) Mellini, M., S. Merlino, P. Orlandi, and R. Rinaldi (1982) Cascandite and jervisite, two new scandium silicates from Baveno, Italy. Amer. Mineral., 67, 599–603. (2) Mellini, M. and S. Merlino (1982) The crystal structure of cascandite, $CaScSi_3O_8(OH)$. Amer. Mineral., 67, 604–609.

Crystal Data: Monoclinic, pseudohexagonal. *Point Group:* $2/m$. Crystals pseudohexagonal, thin tabular on {001}, to 15 cm; as rosettes of crystal plates. *Twinning:* Polysynthetically twinned at 30°, 60°, and 90°.

Physical Properties: *Cleavage:* {100}, perfect; {101} and {102}, imperfect. *Fracture:* Conchoidal. *Tenacity:* Brittle. Hardness = 5.5–6 D(meas.) = 2.65–2.9 D(calc.) = [2.77]

Optical Properties: Transparent, translucent to opaque. *Color:* Colorless, white, light yellow to yellowish brown, pale brown, grayish blue, violet; colorless in thin section. *Streak:* White to pale yellow. *Luster:* Weakly vitreous or dull.
Optical Class: Biaxial (+); may be uniaxial (+). *Dispersion:* $r < v$, moderate.
$\alpha = 1.582 - 1.603$ $\beta = 1.582 - 1.618$ $\gamma = 1.600 - 1.639$ 2V(meas.) = 40° untwinned; 0° twinned.

Cell Data: *Space Group:* $I2/c$. $a = 12.779$ $b = 7.419$ $c = 20.157$ $\beta = 90.41°$ Z = 8

X-ray Powder Pattern: Langesundsfjord, Norway; very close to calcium catapleiite.
3.94 (100), 3.05 (100), 2.96 (100), 2.69 (100), 6.35 (90), 1.969 (60), 5.37 (50)

Chemistry:

	(1)	(2)		(1)	(2)
SiO$_2$	41.56	44.90	MnO	0.00	
TiO$_2$	0.00		CaO	5.21	
ZrO$_2$	32.53	30.69	Na$_2$O	9.74	15.44
FeO	1.02		H$_2$O$^+$	9.35	8.97
			Total	99.41	100.00

(1) Låven Island, Langesundsfjord, Norway. (2) Na$_2$ZrSi$_3$O$_9$·2H$_2$O.

Polymorphism & Series: Dimorphous with gaidonnayite; forms a series with calcium catapleiite.

Occurrence: In syenites and nepheline syenites; in pegmatites, typically the result of metasomatic alteration of eudialyte.

Association: Zircon, leucophanite, rinkite, eudialyte, epididymite, låvenite, astrophyllite, sodalite, natrolite, analcime, aegirine.

Distribution: From Låven Island and other localities around the Langesundsfjord, and at Bratthagen, Norway. In the USA, at Magnet Cove, Hot Springs Co., Arkansas; in the Rocky Boy stock, Bearpaw Mountains, Hill Co., Montana; at Wind Mountain, Otero Co., New Mexico. In Canada, in Quebec, large crystals at Mont Saint-Hilaire, and from the Miron and Varennes quarries, Montreal; along the Ice River valley, 25 km south of Field, British Columbia. In the Ilímaussaq intrusion, at Narssârssuk, and the Kangderdlugssuaq Fjord, Greenland. From the Norra Kärr complex, near Gränna, Sweden. In Russia, in the Khibiny and Lovozero massifs, Kola Peninsula, and the Inagli massif, 30 km west of Aldan, Yakutia. On the Los Islands, Guinea.

Name: From the Greek *kata*, for *with*, and *pleios*, for *many*, because it is always accompanied by other rare minerals.

References: (1) Dana, E.S. (1892) Dana's system of mineralogy, (6th edition), 412–413. (2) Deer, W.A., R.A. Howie, and J. Zussman (1986) Rock-forming minerals, (2nd edition), v. 1B, disilicates and ring silicates, 364–371. (3) Chao, G.Y., J.F. Rowland, and T.T. Chen (1973) The crystal structure of catapleiite. Geol. Soc. Amer. Ann. Meeting, 572 (abs.). (4) Chen, T.T. and G.Y. Chao (1973) Twinning in catapleiite. Geol. Soc. Amer. Ann. Meeting, 573–574 (abs.). (5) Ilyushin, G.D., A.A. Voronkov, V.V. Ilyukhin, N.N. Nevskii, and N.V. Belov (1981) Crystal structure of natural monoclinic catapleiite Na$_2$ZrSi$_3$O$_9$·2H$_2$O. Soviet Physics, Doklady Acad. Nauk SSSR, 260, 623–627 (in Russian). (6) Traill, R.J. and A.P. Sabina (1960) Catalogue of X-ray diffraction patterns and specimen mounts on file at the Geological Survey of Canada. Geol. Survey of Canada, Paper 60-4, 19.

Crystal Data: Orthorhombic. *Point Group:* $2/m$ $2/m$ $2/m$. As prismatic crystals, to 1 mm, elongated $\parallel$ [001]; dominant forms {110} and {101}; as spherulitic rosettes, to 5 mm.

Physical Properties: *Cleavage:* Good on {010}. *Tenacity:* Brittle. Hardness = 3–4 D(meas.) = 2.21–2.31   D(calc.) = 2.33

Optical Properties: Transparent. *Color:* Brilliant sky-blue to greenish blue. *Luster:* Vitreous. *Optical Class:* Biaxial (+). *Pleochroism:* Pronounced; $X = Z$ = colorless; Y = blue. *Orientation:* $X = b$; $Y = a$; $Z = c$. *Dispersion:* $r < v$, extreme. $\alpha = 1.542(2)$ $\beta = 1.544(2)$ $\gamma = 1.551(2)$ 2V(meas.) = 52(2)°

Cell Data: *Space Group:* $Pcmn$.   a = 9.792(2)   b = 13.644(3)   c = 9.629(2)   Z = 4

X-ray Powder Pattern: Owyhee Dam, Oregon, USA. 7.964 (100), 6.854 (50), 6.132 (25), 3.930 (25), 3.420 (25), 2.779 (25), 4.531 (13)

Chemistry:

	(1)	(2)
SiO_2	49.4	53.24
VO_2	17.1	18.38
CaO	11.5	12.42
H_2O	[21.0]	15.96
rem.	0.8	
Total	[99.8]	100.00

(1) Oregon; by XRF, H_2O by estimation; actual H_2O content established by structure analysis. (2) $Ca(VO)Si_4O_{10} \cdot 4H_2O$.

Polymorphism & Series: Dimorphous with pentagonite.

Occurrence: In a brown tuff partly filling a fault fissure (Lake Owyhee State Park, Oregon, USA); in a vesicular basalt and red tuff breccia, as cavity fillings and in calcite veinlets (Chapman quarry, Oregon, USA); in pores of altered basalt breccia and tuffaceous andesite (Poona district, India).

Association: Pentagonite, calcite, heulandite, stilbite, analcime, apophyllite, thomsonite, copper (Oregon, USA); stilbite, calcite, heulandite, mordenite, chalcocite (India).

Distribution: From near Owyhee Dam, Lake Owyhee State Park, Malheur Co., and the Chapman quarry, Columbia Co., Oregon, USA. Remarkable specimens from quarries around Wagholi, Poona district, Maharashtra, India.

Name: For CAlcium, VANadium, and SIlicon in the composition.

Type Material: National Museum of Natural History, Washington, D.C., USA, 120583, 120584, 122769.

References: (1) Staples, L.W., H.T. Evans, Jr., and J.R. Lindsay (1973) Cavansite and pentagonite, new dimorphous calcium vanadium silicate minerals from Oregon. Amer. Mineral., 58, 405–411. (2) Evans, H.T., Jr. (1973) The crystal structures of cavansite and pentagonite. Amer. Mineral., 58, 412–424. (3) Wilke, H.-J., G. Schnorrer-Köhler, and A. Bahle (1989) Cavansit aus Indien. Lapis, 14(1), 39–42 (in German). (4) Kothavala, R. (1991) The Wagholi cavansite locality near Poona, India. Mineral. Record, 22, 415–420.

$Ca_3(Yb, Er)Y_4Si_8O_{20}(CO_3)_6(OH)\cdot 7H_2O$ — Caysichite-(Y)

Crystal Data: Orthorhombic. *Point Group:* $mm2$. Most commonly as a pulverulent coating or stain. As radiating groups of crystals elongated $\parallel$ [001], terminated by {001}; in thin, brittle incrustations with columnar structure.

Physical Properties: *Cleavage:* {010} (?). Hardness = 4.5   VHN = 551   D(meas.) = 3.03   D(calc.) = 3.029   Faint green cathodoluminescence.

Optical Properties: Semitransparent. *Color:* Colorless, white to pale yellow; rarely greenish. *Streak:* White. *Luster:* Vitreous.
Optical Class: Biaxial (–). *Orientation:* $X = b$; $Y = a$; $Z = c$. $\alpha = 1.586(4)$ $\beta = 1.614(1)$ $\gamma = 1.621(1)$ 2V(meas.) = 53°   2V(calc.) = 54°30′

Cell Data: *Space Group:* $Ccm2_1$. $a = 13.282(3)$ $b = 13.925(3)$ $c = 9.724(4)$ Z = 4

X-ray Powder Pattern: Evans-Lou mine, Canada.
6.93 (100), 3.321 (90), 4.38 (60), 4.22 (60), 3.481 (60), 4.87 (40), 2.323 (40)

Chemistry:

	(1)
SiO_2	28.84
Al_2O_3	0.58
Y_2O_3	28.18
RE_2O_3	8.08
CaO	10.04
H_2O	8.6
CO_2	15.7
Total	100.02

(1) Evans-Lou mine, Canada; by electron microprobe, H_2O by gravimetry, CO_2 by titrimetry; $RE_2O_3 = La_2O_3$ 0.08%, CeO_2 0.24%, Pr_2O_3 0.09%, Nd_2O_3 0.30%, Sm_2O_3 0.20%, Eu_2O_3 0.09%, Gd_2O_3 0.51%, Tb_2O_3 0.16%, Dy_2O_3 1.16%, Ho_2O_3 0.66%, Er_2O_3 1.84%, Tm_2O_3 0.21%, Yb_2O_3 2.12%, Lu_2O_3 0.42%; corresponds to $Ca_{2.95}RE_{0.70}Y_{4.11}(Si_{7.90}Al_{0.18})_{\Sigma=8.08}O_{20.36}$ $(CO_3)_{5.87}\cdot 7.86H_2O$.

Occurrence: In perthite surrounding the quartz core of a granite pegmatite, coating fractures and filling cavities in quartz, perthite, or hellandite; apparently of secondary origin.

Association: Quartz, hellandite, fergusonite, kainosite, tengerite, lokkaite, yttrian thorogummite, yttrian spessartine, xenotime, wakefieldite, synchysite-(Y).

Distribution: In the Evans-Lou feldspar mine, Poltimore, 35 km north of Ottawa, Quebec, Canada.

Name: For the chemical composition: *Ca, Y, Si, C, H*.

Type Material: Canadian Museum of Nature, Ottawa; Canadian Geological Survey, Ottawa, Canada, 10402; The Natural History Museum, London, England, 1977,278.

References: (1) Hogarth, D.D., G.Y. Chao, A.G. Plant, and H.R. Steacy (1974) Caysichite, a new silico-carbonate of yttrium and calcium. Can. Mineral., 12, 293–298. (2) (1976) Amer. Mineral., 61, 174–175 (abs. ref. 1). (3) Mellini, M. and S. Merlino (1978) Caysichite: a double crank shaft chain structure. Can. Mineral., 16, 81–88.

Crystal Data: Orthorhombic (?). *Point Group:* n.d. As compact masses of minute radiating fibers.

Physical Properties: Hardness = ~5 D(meas.) = 2.96 D(calc.) = n.d.

Optical Properties: Semitransparent. *Color:* Colorless, white to greenish gray, reddish brown; colorless in thin section. *Luster:* Dull.
Optical Class: Biaxial (+). *Dispersion:* $r > v$, weak. $\alpha = 1.592$–1.595 $\beta = 1.597$–1.60
$\gamma = 1.628$–1.630 2V(meas.) = ~58°

Cell Data: *Space Group:* n.d. Z = n.d.

X-ray Powder Pattern: Cebolla Creek, Colorado, USA. (ICDD 16-695).
2.73 (100), 2.88 (90), 2.59 (70), 1.62 (60), 3.01 (50), 2.90 (40), 2.45 (40)

Chemistry:

	(1)	(2)	(3)		(1)	(2)	(3)
SiO_2	27.06	33.05	31.07	CaO	29.27	35.75	41.06
Al_2O_3	11.49	14.03	18.48	Na_2O	2.10	2.57	0.28
Fe_2O_3	2.81	3.43		K_2O	trace		
FeO	0.17	0.21	1.36	H_2O	5.13	6.27	[7.75]
MgO	3.84	4.69		insol.	18.05		
				Total	99.92	[100.00]	[100.00]

(1) Cebolla Creek, Colorado, USA. (2) Do.; recalculated to 100%, computed free of insolubles, taken as diopside, garnet, vesuvianite, etc., corresponds to $(Ca_{3.79}Mg_{0.69}Na_{0.24}Fe^{2+}_{0.02})_{\Sigma=4.74}$ $(Al_{1.64}Fe^{3+}_{0.36})_{\Sigma=2.00}Si_{3.26}O_{12.08}(OH)_{4.12}$. (3) Tokatoka, New Zealand; by electron microprobe, H_2O by difference; corresponds to $(Ca_{4.77}Fe^{2+}_{0.12}Na_{0.06})_{\Sigma=4.95}Al_{2.36}Si_{3.37}O_{12.40}(OH)_{5.60}$.

Occurrence: An alteration product of melilite in a carbonatite (Iron Hill, Colorado, USA); along a diabase-chalk contact (Scawt Hill, Ireland); in zeolitized plagioclase-rich gneiss, amphibolite, and basalt xenoliths (Letseng-La-Terai kimberlite, Lesotho); a late-stage primary igneous mineral (Kimberley, South Africa).

Association: Melilite, natrolite, calcite, phlogopite, vesuvianite, garnet, diopside.

Distribution: On Iron Hill and in Cebolla Creek, near Powderhorn, Gunnison Co., Colorado, USA. From Scawt Hill, near Larne, Co. Antrim, Ireland. In the Gardiner complex, beyond the head of Kangerdlugssuaq Fjord, Greenland. In the Letseng-La-Terai kimberlite, Lesotho. From the DeBeers diamond mine, Kimberley, Cape Province, South Africa. In the Tokatoka district, about 150 km north of Auckland, New Zealand.

Name: For Cebolla Creek, Colorado, in the drainage of which the mineral was first collected.

Type Material: Harvard University, Cambridge, Massachusetts, 85726, 100857; National Museum of Natural History, Washington, D.C., USA, 87530, C3217, R6451; The Natural History Museum, London, England.

References: (1) Larsen, E.S. and W.T. Schaller (1914) Cebollite, a new mineral. Wash. Acad. Sci. IV, No. 16, 480–482. (2) Larsen, E.S. and E.A. Goranson (1932) The deuteric and later alterations of the uncompahgrite of Iron Hill, Colorado. Amer. Mineral., 17, 343–356. (3) Kruger, F.J. (1980) The occurrence of cebollite in kimberlite and included zeolitized crustal xenoliths. Mineral. Mag., 43, 583–586. (4) Baker, C.K. and P.M. Black (1980) Assimilation and metamorphism at a basalt-limestone contact, Tokatoka, New Zealand. Mineral. Mag., 43, 797–807.

Crystal Data: Monoclinic.   *Point Group:* 2/*m* or 2.   As minute micaceous scales or earthy aggregates.

Physical Properties: *Cleavage:* {001}, perfect.   *Tenacity:* Friable to unctuous.
Hardness = ~2   D(meas.) = 2.95–3.05   D(calc.) = 3.00

Optical Properties: Semitransparent.   *Color:* Blue-green, olive-green, apple-green.
Luster: Dull.
Optical Class: Biaxial (–).   *Pleochroism:* Yellow-green, blue-green.   α = 1.606–1.625.   β = n.d.
γ = 1.579–1.661   2V(meas.) = 5°–8°

Cell Data: *Space Group:* C2/*m* or C2.   a = 5.23(2)   b = 9.06(1)   c = 10.13(2)
β = 100°55(10)′   Z = 2

X-ray Powder Pattern: Wind River area, Washington, USA.
2.580 (100), 4.53 (85), 3.635 (80), 3.087 (80), 2.678 (75), 2.402 (75), 3.318 (70)

Chemistry:

	(1)
SiO$_2$	55.61
Al$_2$O$_3$	0.79
Fe$_2$O$_3$	17.19
FeO	4.02
MnO	0.09
MgO	7.26
CaO	0.21
Na$_2$O	0.19
K$_2$O	10.03
H$_2$O	4.88
Total	100.27

(1) 37 km east of Reno, Storey Co., Nevada, USA; corresponds to $(K_{0.92}Na_{0.03}Ca_{0.02})_{\Sigma=0.97}$ $(Mg_{0.78}Fe^{2+}_{0.24})_{\Sigma=1.02}(Fe^{3+}_{0.93}Al_{0.07})_{\Sigma=1.00}Si_{4.00}O_{10}(OH)_2$.

Polymorphism & Series: 1M polytype.

Mineral Group: Mica group.

Occurrence: Replaces primary ferromagnesian silicate minerals in altered intermediate to mafic volcanic rocks, developed under low-grade zeolite facies metamorphism; as amygdule fillings in basalts or andesites.

Association: Montmorillonite, clinoptilolite, heulandite, laumontite, prehnite, chlorite, quartz, calcite.

Distribution: Many localities; a few for well-characterized material include: on Mt. Baldo, near Verona, Vicenza, and at Val di Fassa, Trentino-Alto Adige, Italy. In the Zillertal, Tirol, Austria. In Scotland, at Scuir Mohr. From Streymoy and Suduroy, Faeroe Islands. In the USA, in the John Day Formation, Grant Co., Oregon; in Mt. Rainier National Park, Pierce Co., Washington; and from Red Rock Canyon, Kern Co., California. On the Pearl Islands, off Nicaragua. In the Hosokura mine, Miyagi Prefecture; at Toyoura, Yamagata Prefecture; Nishikata, Tochigi Prefecture; Kamogawa, Chiba Prefecture; and many other places in Japan.

Name: From the French *celadon*, for *sea green*, its color.

References: (1) Dana, E.S. (1892) Dana's system of mineralogy, (6th edition), 683. (2) Wise, W.S. and H.P. Eugster (1964) Celadonite: synthesis, thermal stability and occurrence. Amer. Mineral., 49, 1031–1083. (3) Buckley, H.A., J.C. Bevan, K.M. Brown, L.R. Johnson, and V.C. Farmer (1978) Glauconite and celadonite: two separate mineral species. Mineral. Mag., 42, 373–382. (4) Cipurskii, S.I. and V.A. Dric (1986) Refining the crystallographic structure of celadonite. Mineral. Zhurnal, 8(3), 32–40 (in Russian with English abs.).

Crystal Data: Monoclinic. *Point Group:* $2/m$. As short prismatic to acicular crystals with large prism faces, to 2 cm; commonly massive. *Twinning:* Simple twins according to the Manebach, Baveno, or Carlsbad laws.

Physical Properties: *Cleavage:* Perfect on {001}, good on {010}, and poor on {110}. *Tenacity:* Brittle. Hardness = 6–6.5 D(meas.) = 3.10–3.39 D(calc.) = 3.26

Optical Properties: Transparent. *Color:* Colorless, white, yellow. *Luster:* Vitreous. *Optical Class:* Biaxial (+). *Orientation:* $Y = b$; $X \wedge c = 3°–5°$; $Z \wedge a = 28°–30°$. $\alpha = 1.579–1.587$ $\beta = 1.583–1.593$ $\gamma = 1.588–1.600$ 2V(meas.) = 83°–92°

Cell Data: *Space Group:* $I2/c$. $a = 8.622(4)$ $b = 13.078(6)$ $c = 14.411(8)$ $\beta = 115.09(2)°$ Z = 8

X-ray Powder Pattern: Jakobsberg, Sweden.
3.47 (100), 3.35 (100), 3.02 (95), 2.582 (75), 3.26 (60), 3.80 (55), 6.52 (50)

Chemistry:

	(1)	(2)	(3)
SiO_2	32.43	31.07	32.00
Al_2O_3	26.55	28.35	27.16
Fe_2O_3	0.12	0.36	
MgO	0.11		
CaO	0.23		
BaO	39.72	39.92	40.84
Na_2O	0.16		
K_2O	0.22	0.24	
F	0.64		
H_2O	0.64		
Total	100.82	[99.94]	100.00

(1) Jakobsberg, Sweden. (2) Aberfeldy, Scotland; by electron microprobe; original total given as 99.99%. (3) $BaAl_2Si_2O_8$.

Polymorphism & Series: Dimorphous with paracelsian; forms a series with orthoclase.

Mineral Group: Feldspar group.

Occurrence: In amphibolite-grade, regional or contact metamorphic, Mn, Ba-rich rocks, some of which probably were submarine exhalatives.

Association: Manganoan aegirine, manganoan biotite, paracelsian, jacobsite, hausmannite, rhodochrosite, rhodonite, rutile, hyalophane, barite, cymrite, taramellite, quartz, zoisite, spessartine, dolomite, muscovite.

Distribution: At Jakobsberg and Långban, Värmland, Sweden. In the Simplon region, Valais, Switzerland. In the USA, from Franklin, Sussex Co., New Jersey; near Big Creek, Fresno Co., and near Incline Village, Mariposa Co., California. At the Benallt mine, Rhiw, Lleyn Peninsula, Wales. From near Aberfeldy, Scotland. In Kazakhstan, from the Burultas deposit, west of Lake Balkhash. In Namibia, at Otjosondu. From Piggery Creek, Broken Hill, New South Wales, Australia. In the Kaso mine, Totiki Prefecture, Japan.

Name: To honor the Swedish astronomer and naturalist, Anders Celsius (1701–1744).

References: (1) Dana, E.S. (1899) Dana's system of mineralogy, (6th edition), app. I, 15. (2) Deer, W.A., R.A. Howie, and J. Zussman (1963) Rock-forming minerals, v. 4, framework silicates, 166–178. (3) Gay, P. (1965) An X-ray powder method for the estimation of (K,Ba) feldspars. Mineral. Mag., 34, 204–213. (4) Griffen, D.T. and P.H. Ribbe (1976) Refinement of the crystal structure of celsian. Amer. Mineral., 61, 414–418. (5) Fortey, N.J. and B. Beddoe-Stephens (1982) Barium silicates in stratabound Ba–Zn mineralization in the Scottish Dalradian. Mineral. Mag., 46, 63–72.

$(Ce, La, Ca)_9(Mg, Fe^{3+})(SiO_4)_6(SiO_3OH)(OH)_3$ Cerite-(Ce)

Crystal Data: Hexagonal. *Point Group:* $3m$. As crudely formed pseudo-octahedral crystals, to 7 mm; massive.

Physical Properties: *Fracture:* Uneven. Hardness = 5–5.5 D(meas.) = 4.75(11) D(calc.) = 4.86

Optical Properties: Translucent to opaque. *Color:* Clove-brown with a reddish tinge; pale lavender-brown to colorless in thin fragments. *Streak:* Grayish white. *Luster:* Resinous. *Optical Class:* Uniaxial (+); anomalously biaxial. $\omega = 1.806–1.817$ $\epsilon = 1.808–1.825$ 2V(meas.) = 0°–25°

Cell Data: *Space Group:* $R3c$. a = 10.779(6) c = 38.061(7) Z = 6

X-ray Powder Pattern: Mountain Pass, California, USA.
2.95 (100), 1.954 (50), 3.47 (42), 2.69 (42), 2.83 (38), 3.31 (33), 3.11 (30)

Chemistry:

	(1)		(1)
SiO_2	22.24	CaO	4.48
TiO_2	0.08	BaO	0.45
Al_2O_3	1.14	Na_2O	0.24
La_2O_3	30.04	K_2O	0.16
Ce_2O_3	35.05	H_2O^+	2.72
Fe_2O_3	0.92	H_2O^-	0.17
MnO	0.20	SO_3	0.14
MgO	1.90	F	0.07
		Total	[100.00]

(1) Mountain Pass, California, USA; recalculated to 100% after deduction of 17.7% bastnäsite; corresponds to $(Ce_{3.90}La_{3.37}Ca_{1.63})_{\Sigma=8.90}(Mg_{0.61}Fe^{3+}_{0.39}Na_{0.14}K_{0.06}Ba_{0.05})_{\Sigma=1.25}$ $Si_{6.83}O_{25.68}(OH)_{3.89}$.

Occurrence: In rare-earth-bearing hydrothermal quartz-barite-carbonatite veins in shonkinite (Mountain Pass, California, USA).

Association: Bastnäsite, allanite, epidote, monazite, törnebohmite, fluorite, uraninite, barite, quartz, galena.

Distribution: In Sweden, at the Bastnäs mine, Riddarhyttan, Västmanland. In the USA, from the Mountain Pass district, San Bernardino Co., California, and in Colorado, at Jamestown, Boulder Co. In Canada, from Papineau Co., and Mont Saint-Hilaire, Quebec. In the Kyshtymsk district, Ural Mountains, Russia.

Name: For its *cerium* content; the element was discovered in this mineral and named for the planetoid *Ceres*, also then just discovered.

Type Material: Harvard University, Cambridge, Massachusetts, USA, 48851.

References: (1) Dana, E.S. (1892) Dana's system of mineralogy, (6th edition), 550. (2) Gay, P. (1957) An X-ray investigation of some rare-earth silicates: cerite, lessingite, beckelite, britholite, and stillwellite. Mineral. Mag., 31, 455–468. (3) Gay, P. (1957) The crystallography of cerite. Amer. Mineral., 42, 429–432. (4) Glass, J.J., H.T. Evans, Jr., M.K. Carron, and F.A. Hillebrand (1958) Cerite from Mountain Pass, San Bernardino County, California. Amer. Mineral., 43, 460–475. (5) Moore, P.B. and J. Shen (1983) Cerite, $RE_9(Fe^{3+}, Mg)(SiO_4)_6(SiO_3OH)(OH)_3$: its crystal structure and relation to whitlockite. Amer. Mineral., 68, 996–1003.

Cervandonite-(Ce) $\qquad$ $(Ce, Nd, La)(Fe^{3+}, Ti, Fe^{2+}, Al)_3(Si, As)_3O_{13}$

Crystal Data: Monoclinic. *Point Group:* m, 2, or $2/m$. As porous, saddle- or rosettelike, rounded, radial crystal aggregates, to 4 mm.

Physical Properties: *Cleavage:* {001}, poor. *Fracture:* Conchoidal. *Tenacity:* Brittle. Hardness = 5 VHN = 450 (50 g load). D(meas.) = n.d. D(calc.) = 4.9

Optical Properties: Opaque, translucent in very thin splinters. *Color:* Black; in transmitted light, yellowish, reddish brown to black. *Streak:* Brownish black. *Luster:* Adamantine. *Optical Class:* Biaxial. *Pleochroism:* Yellowish, reddish brown to black. $n = 1.99–2.00$ 2V(meas.) = n.d.

Cell Data: *Space Group:* Cm, $C2$, or $C2/m$. $a = 11.235(4)$ $b = 19.50(2)$ $c = 7.201(3)$ $\beta = 121.22(3)°$ $Z = 6$

X-ray Powder Pattern: Pizzo Cervandone, Italy.
2.8785 (100), 3.2530 (90), 5.390 (80), 3.0847 (80), 2.7867 (60), 2.6964 (50), 3.5750 (40)

Chemistry:

	(1)	(2)		(1)	(2)
SiO_2	17.65	15.17	Ce_2O_3	10.00	11.63
TiO_2	11.67	11.42	Nd_2O_3	5.16	6.01
SnO_2	1.24	0.00	Fe_2O_3	15.03	15.53
UO_2	1.35	0.33	As_2O_5	20.41	24.18
ThO_2	2.34	1.92	FeO	4.51	4.67
Al_2O_3	3.60	2.95	PbO	0.17	0.16
Y_2O_3	0.94	0.86	CaO	0.25	0.25
La_2O_3	5.14	5.65	Total	[99.46]	[100.73]

(1) Pizzo Cervandone, Italy; by electron microprobe, Fe^{2+}:Fe^{3+} determined by Mössbauer spectroscopy, original total given as 99.47%; corresponds to $(Ce_{0.39}Nd_{0.20}La_{0.20}Y_{0.05}Th_{0.06}U_{0.03}$ $Ca_{0.03}Pb_{0.01})_{\Sigma=0.97}(Fe^{3+}_{1.21}Ti_{0.94}Al_{0.46}Fe^{2+}_{0.40}Sn_{0.05})_{\Sigma=3.06}(Si_{1.89}As^{5+}_{1.14})_{\Sigma=3.03}O_{13}$. (2) Pizzo Cervandone, Switzerland; by electron microprobe, original total given as 100.72%; corresponds to $(Ce_{0.46}Nd_{0.23}La_{0.23}Y_{0.05}Th_{0.05}Ca_{0.03}U_{0.01}Pb_{0.01})_{\Sigma=1.07}(Fe^{3+}_{1.26}Ti_{0.93}Fe^{2+}_{0.42}Al_{0.38})_{\Sigma=2.99}$ $(Si_{1.64}As^{5+}_{1.36})_{\Sigma=3.00}O_{13}$.

Occurrence: In narrow fissure-veins cutting two-mica gneiss.

Association: Senaite, chernovite, rutile, anatase, muscovite, albite, chlorite, tourmaline, magnetite, hematite, quartz, synchysite.

Distribution: On the east flank of Pizzo Cervandone, Alpe Devero, Val d'Aosta, Italy. On the west flank of Cherbadung [Pizzo Cervandone], Binntal, Valais, Switzerland.

Name: After the locality on the border between Italy and Switzerland where the mineral was discovered, Pizzo Cervandone.

Type Material: Natural History Museum, Bern, B3953, B4695, B4880; Natural History Museum, Basel, Switzerland, M30802.

References: (1) Armbruster, T., C. Bühler, S. Graeser, H.A. Stalder, and G. Amthauer (1988) Cervandonite-(Ce), $(Ce, Nd, La)(Fe^{3+}, Fe^{2+}, Ti^{4+}, Al)_3SiAs(Si, As)O_{13}$, a new Alpine fissure mineral. Schweiz. Mineral. Petrog. Mitt., 68, 125–132. (2) (1990) Amer. Mineral., 75, 932 (abs. ref. 1).

Crystal Data: Triclinic. *Point Group:* n.d. As rosettelike intergrowths of curved platy crystals, forming aggregates, to 3 cm.

Physical Properties: *Cleavage:* {001}, perfect. Hardness = ~ 4 VHN = 186 D(meas.) = 3.68 D(calc.) = [3.62]

Optical Properties: Semitransparent. *Color:* Golden brown. *Luster:* Dull.
Optical Class: Biaxial (+). *Pleochroism:* X = yellow-green; Y = yellow to brown; Z = brown.
Orientation: Z = a; $Y \wedge b \simeq 10°$. α = n.d. $\beta = 1.726$ $\gamma = 1.758$ 2V(meas.) = 75°

Cell Data: *Space Group:* n.d. $a = 5.41(1)$ $b = 11.74(2)$ $c = 21.16(4)$ $\alpha = 89°$ $\beta = 90°$ $\gamma = 102°23'$ Z = 2

X-ray Powder Pattern: Alai Range, Tadzhikistan.
10.4 (100), 3.54 (80), 2.79 (80), 2.66 (80), 2.58 (60), 1.772 (40), 4.09 (30)

Chemistry:

	(1)		(1)
SiO_2	33.00	CaO	0.35
TiO_2	8.28	Li_2O	0.46
ZrO_2	1.01	Na_2O	2.46
Al_2O_3	0.52	K_2O	1.15
Fe_2O_3	3.05	Rb_2O	0.18
Nb_2O_5	4.95	Cs_2O	11.60
Ta_2O_5	0.06	F	1.26
FeO	10.00	H_2O^+	1.47
MnO	19.66	$-O = F_2$	0.54
MgO	trace	Total	98.92

(1) Alai Range, Tadzhikistan; corresponds to $(\mathrm{Cs}_{1.22}\mathrm{Na}_{1.14}\mathrm{K}_{0.35}\mathrm{Ca}_{0.09})_{\Sigma=2.80}(\mathrm{Mn}_{3.99}\mathrm{Fe}^{2+}_{2.00}$
$\mathrm{Fe}^{3+}_{0.55}\mathrm{Li}_{0.44})_{\Sigma=6.98}(\mathrm{Ti}_{1.49}\mathrm{Nb}_{0.54}\mathrm{Zr}_{0.12})_{\Sigma=2.15}(\mathrm{Si}_{7.92}\mathrm{Al}_{0.15})_{\Sigma=8.07}\mathrm{O}_{24}[\mathrm{O}_{3.69}(\mathrm{OH})_{2.35}\mathrm{F}_{0.96}]_{\Sigma=7.00}$.

Polymorphism & Series: Forms a series with kupletskite.

Mineral Group: Astrophyllite group.

Occurrence: At the border of aegirine-microcline-quartz pegmatites, and in polylithionite-quartz replacement complexes in pegmatites.

Association: Pyrochlore, stillwellite, tienshanite, sogdianite, thorite.

Distribution: In the [Dara-i-Pioz massif,] Alai Range, Tien Shan, Tadzhikistan.

Name: For the *cesium* in the composition and similarity to *kupletskite*.

Type Material: A.E. Fersman Mineralogical Museum, Academy of Sciences, Moscow, Russia.

References: (1) Efimov, A.F., V.D. Dusmatov, A.A. Ganzeev, and Z.T. Kataeva (1971) Cesium kupletskite, a new mineral. Doklady Acad. Nauk SSSR, 197, 1394–1397 (in Russian). (2) (1972) Amer. Mineral., 57, 328 (abs. ref. 1).

Crystal Data: Triclinic, pseudohexagonal. *Point Group:* $P\bar{1}$; $\bar{3}\,2/m$ pseudohexagonal. In pseudorhombohedral crystals, nearly cubic in aspect, to 4.5 cm; tabular, complex to rounded twinned forms; anhedral, granular, or massive. *Twinning:* About [0001], interpenetrant, simple and repeated, common; contact on $\{10\bar{1}1\}$.

Physical Properties: *Cleavage:* $\{10\bar{1}1\}$, distinct. *Fracture:* Uneven. *Tenacity:* Brittle. Hardness = 4–5   D(meas.) = 2.05–2.20   D(calc.) = 2.035

Optical Properties: Transparent to translucent. *Color:* White, yellow, pink, red, colorless; colorless in thin section. *Streak:* White. *Luster:* Vitreous.
Optical Class: Biaxial (+) or (–) or uniaxial; commonly shows birefringent panelling in six sections. *Orientation:* $X = c$; rarely $Z = c$. $\alpha = 1.478$–1.487 $\beta = $ n.d. $\gamma = 1.480$–1.493
2V(meas.) = $0°$–$32°$

Cell Data: *Space Group:* $P\bar{1}$.   a = 9.40–9.45   b = 9.40–9.44   c = 9.40–9.44
$\alpha = 94.18°-94.33°$ $\beta = 94.08°-94.36°$ $\gamma = 94.07°-94.45°$ Z = [2]

X-ray Powder Pattern: Table Mountain, Colorado, USA; very similar to herschelite.
2.925 (100), 4.32 (75), 9.35 (50), 5.02 (30), 3.87 (30), 2.890 (30), 3.59 (25)

Chemistry:

	(1)	(2)		(1)	(2)
SiO_2	47.56	47.46	K_2O	0.92	
Al_2O_3	20.40	20.13	H_2O^+	16.28	
MgO	0.20		H_2O^-	3.44	
CaO	10.52	11.07	H_2O		21.34
Na_2O	0.32		Total	99.64	100.00

(1) Ritter Hot Spring, Grant Co., Oregon, USA; corresponds to $(Ca_{0.94}K_{0.10}Na_{0.05}Mg_{0.02})_{\Sigma=1.11}$ $Al_{2.01}Si_{3.97}O_{12} \cdot 5.50H_2O$. (2) $CaAl_2Si_4O_{12} \cdot 6H_2O$.

Mineral Group: Zeolite group.

Occurrence: In volcanic rocks as basalts, andesite; rarer in limestones and schists; hydrothermally deposited in cavities and joints in ore veins. In bedded tuff in lake deposits, altered from volcanic glass.

Association: Zeolites, nepheline, melilite, olivine, pyroxenes, amphiboles, axinite, epidote, calcite, tridymite, dolomite.

Distribution: A common zeolite. Fine crystals from Idar-Oberstein, Rhineland-Palatinate, Germany. At Řepčice (Rübendörfel), near Ústí nad Lábem (Aussig), Czech Republic. At a number of localities in Co. Antrim, Ireland. In Scotland, at Kilmalcolm, Renfrewshire. From Haeddin, on Eysturoy; Dalsnipa, on Sandoy; and Skutin, on Nolsoy, Faeroe Islands. Large crystals at Breidhdalsheidhi, Iceland. In the USA, around Paterson, Passaic Co., and Bergen Hill, Hudson Co., New Jersey; on Table Mountain, Jefferson Co., Colorado; at Goble, Columbia Co., and Springfield, Lane Co., Oregon. In the Bay of Fundy district, Nova Scotia, Canada. On Table Mountain, Rosarito Beach, Baja California, Mexico. In the Khandivali quarry, near Bombay, Maharashtra, India. At Richmond and Collingwood, Victoria, and on Fairy Mount, near Kyogle, New South Wales, Australia.

Name: From the Greek *chabazios*, an ancient name of a stone.

References: (1) Dana, E.S. (1892) Dana's system of mineralogy, (6th edition), 589–592. (2) Deer, W.A., R.A. Howie, and J. Zussman (1963) Rock-forming minerals, v. 4, framework silicates, 387–400. (3) Passaglia, E. (1970) The crystal chemistry of chabazites. Amer. Mineral., 55, 1278–1301. (4) Gude, A.J., 3rd and R.A. Sheppard (1966) Silica-rich chabazite from the Barstow Formation, San Bernardino County, Southern California. Amer. Mineral., 51, 909–915. (5) Mazzi, F. and E. Galli (1983) The tetrahedral framework of chabazite. Neues Jahrb. Mineral., Monatsh., 461–480.

Crystal Data: Monoclinic. *Point Group:* $2/m$. In scaly aggregates, foliated or granular; oolitic, made up of very minute radiating crystals; massive. *Twinning:* Twin and composition plane {001}, axis [310].

Physical Properties: *Cleavage:* {001}, perfect. *Tenacity:* Somewhat flexible but inelastic. Hardness = 2–3 D(meas.) = 3.0–3.4 D(calc.) = 3.129 Weakly magnetic.

Optical Properties: Translucent to almost opaque. *Color:* Green, greenish grey, black, brown; in thin section, colorless, yellowish, green. *Streak:* Green to gray. *Luster:* Pearly. *Optical Class:* Biaxial (–). *Pleochroism:* Weak to moderate; X = yellow-green to light brownish green; $Y = Z$ = green to dark green. *Orientation:* $Y = b$; $X \wedge a = 88°–90°$; $Z \wedge a = 0°–2°$: or rarely $Y = b$; $X \wedge a \simeq 0°$; $Z \wedge a \simeq 90°$. *Dispersion:* $r < v$, strong. *Absorption:* $Y \simeq Z > X$; rarely $X \simeq Y > Z$. $\alpha = 1.595–1.671$ $\beta = 1.599–1.684$ $\gamma = 1.599–1.685$ 2V(meas.) = $0°–30°$

Cell Data: *Space Group:* $C2/m$. $a = 5.373$ $b = 9.306$ $c = 14.222$ $\beta = 97°53'$ $Z = 2$

X-ray Powder Pattern: Heřmanovice, Czech Republic. (ICDD 21-1227). 7.05 (100), 3.52 (100), 2.601 (90), 1.551 (90), 2.392 (80), 14.1 (70), 2.554 (70)

Chemistry:

	(1)	(2)		(1)	(2)
SiO_2	26.40	26.65	CaO	0.42	
Al_2O_3	18.23	16.14	Na_2O	0.17	
Fe_2O_3	5.70	6.69	K_2O	0.17	
FeO	25.87	34.43	H_2O^+	10.60	11.42
MnO	0.04		H_2O^-	1.05	0.08
MgO	11.35	4.47	Total	[100.00]	99.88

(1) Wickwar, England; recalculated to 100.00%, total Fe as Fe^{2+} in empirical formula; corresponds to $(Fe^{2+}_{2.82}Mg_{1.84}Ca_{0.05}Na_{0.04}K_{0.02})_{\Sigma=4.77}Al_{1.21}(Si_{2.87}Al_{1.13})_{\Sigma=4.00}O_{10}(OH)_8$.
(2) Schmiedefeld, Germany; total Fe as Fe^{2+} in empirical formula; corresponds to $(Fe^{2+}_{3.87}Mg_{0.76})_{\Sigma=4.63}Al_{1.23}(Si_{3.05}Al_{0.95})_{\Sigma=4.00}O_{10}(OH)_8$.

Polymorphism & Series: Dimorphous with orthochamosite; forms a series with clinochlore.

Mineral Group: Chlorite group.

Occurrence: In sedimentary ironstones; authigenically formed under reducing conditions in the presence of decomposed organic material.

Association: Siderite, kaolinite, quartz, magnetite, pyroxenes, plagioclase, olivine, calcite.

Distribution: Many localities, but careful characterization is necessary. From Chamoson, Valais, and in the Maderantal, Uri, Switzerland. At Heřmanovice and Kladno, Czech Republic. From Schmiedefeld, near Suhl, and Schleiz, Thuringia, Germany. In England, at Frodingham, Lincolnshire; Wickwar, Gloucestershire; at Penzance, Cornwall, and elsewhere. At Knowehead, Co. Antrim, Ireland. In the Arakawa mine, Akita Prefecture, and at Shogase, Tokushima Prefecture, Japan. From Creede, Mineral Co., Colorado; at Hot Springs, Hot Springs Co., Arkansas; and in the Beacon Hill mine, Champion, Marquette Co., Michigan, USA. At Wabana, Newfoundland, Canada.

Name: For the locality at Chamoson, Switzerland.

References: (1) Dana, E.S. (1892) Dana's system of mineralogy, (6th edition), 658. (2) Deer, W.A., R.A. Howie, and J. Zussman (1963) Rock-forming minerals, v. 3, sheet silicates, 131–163. (3) Bayliss, P. (1975) Nomenclature of the trioctahedral chlorites. Can. Mineral., 13, 178–180.

Crystal Data: Tetragonal. *Point Group:* $4/m$. As anhedral grains up to 0.3 mm.

Physical Properties: Hardness = n.d. D(meas.) = 2.8–2.9 D(calc.) = 2.97

Optical Properties: Semitransparent. *Color:* Colorless to white. *Luster:* Vitreous. *Optical Class:* Uniaxial (–). $\omega = 1.653(1)$ $\epsilon = 1.642(1)$

Cell Data: *Space Group:* $I4_1/a$. a = 4.952(1) c = 23.275(6) Z = 4

X-ray Powder Pattern: Taurus Mountains, Turkey.
2.598 (100), 4.171 (70), 3.349 (60), 1.453 (60), 2.235 (50), 4.828 (45), 2.202 (35)

Chemistry:

	(1)	(2)
SiO_2	23.74	23.64
Al_2O_3	38.44	40.12
FeO	0.02	
MgO	0.06	
CaO	21.57	22.07
Na_2O	0.06	
H_2O	[16.11]	14.17
Total	[100.00]	100.00

(1) Taurus Mountains, Turkey; by electron microprobe, average of four analyses, H_2O by difference. (2) $CaAl_2SiO_4(OH)_4$.

Occurrence: In rodingite dikes in an ophiolite.

Association: Vuagnatite, prehnite, hydrogrossular, chlorite, calcite.

Distribution: From Covur Yokusutepe, Taurus Mountains, Burdur Province, Turkey.

Name: For Chantal Sarp, wife of the mineral's discoverer.

Type Material: Natural History Museum, Geneva, Switzerland, 435/1; The Natural History Museum, London, England, 1984,741; National Museum of Natural History, Washington, D.C., USA, 146559.

References: (1) Sarp, H., J. Deferne, and B.W. Liebich (1977) Chantalite, $CaAl_2SiO_4(OH)_4$, a new natural calcium aluminum silicate. Schweiz. Mineral. Petrog. Mitt., 57, 149–156 (in French with English abs.). (2) (1978) Amer. Mineral., 63, 1282. (abs. ref. 1). (3) Liebich, B.W., H. Sarp, and E. Parthé (1979) The crystal structure of chantalite, $CaAl_2(OH)_4SiO_4$. Zeits. Krist., 150, 53–63.

Crystal Data: Monoclinic. *Point Group: m.* Crystals lath-shaped; powdery, compact, massive.

Physical Properties: *Fracture:* Conchoidal to irregular when compact. Hardness = 2.5 when compact. D(meas.) = 3.69–3.75 D(calc.) = [4.29]

Optical Properties: Semitransparent. *Color:* Olive-green to deep yellow. *Streak:* Greenish yellow to deep yellow. *Luster:* Nonmetallic.
Optical Class: Biaxial. $\alpha = 1.85 \perp$ flat face. β = n.d. $\gamma = 1.96 \parallel$ length. 2V(meas.) = n.d.

Cell Data: *Space Group: Cm.* a = 5.19 b = 8.99 c = 7.70 $\beta = 100°40'$ Z = 2

X-ray Powder Pattern: Keeley mine, Canada; could be confused with bismutoferrite.
7.63 (100), 3.58 (100), 3.88 (90), 3.19 (90), 2.90 (70), 2.59 (70), 2.54 (35)

Chemistry:

	(1)	(2)	(3)
SiO_2	27.63	26.34	27.65
Al_2O_3	0.22	0.50	
Fe_2O_3	38.98	38.13	36.74
Sb_2O_3	31.04	30.53	33.54
H_2O^+	1.89	3.62	2.07
H_2O^-	0.31	0.64	
Total	100.07	99.76	100.00

(1) Smilkov, Czech Republic; corresponding to $(Sb_{0.93}^{3+}Fe_{0.07}^{3+})_{\Sigma=1.00}$ $Fe_{2.00}^{2+}(Si_{1.98}Al_{0.02})_{\Sigma=2.00}O_{8.00}(OH)$. (2) Do. (3) $SbFe_2(SiO_4)_2(OH)$.

Occurrence: In veinlets cutting gneisses containing varying amounts of graphite (Smilkov, Czech Republic).

Association: Silver (Keeley mine, Canada).

Distribution: In the Keeley mine, South Lorrain Township, Ontario, Canada. In Mexico, at Velardeña, Durango. At Smilkov, near Votice, Czech Republic. In Germany, at Bräunsdorf, Freiberg, and Schneeberg, Saxony. From near Tafone, Tuscany, Italy. In the Mine de la Bessade, Massiac, Haute-Loire, France. In the Suzuyama mine, Kagoshima Prefecture, Japan.

Name: For Edward John Chapman (1821–1904), Professor of Geology, University of Toronto, Toronto, Canada.

Type Material: Royal Ontario Museum, Toronto, Canada, M14075; Harvard University, Cambridge, Massachusetts, USA, 91501.

References: (1) Walker, T.L. (1924) Chapmanite, a new mineral from South Lorraine, Ontario. Univ. of Toronto, Geol. Studies, 17, 5–8. (2) Walker, T.L. (1924) Chapmanite, a new hydrous ferrous silicoantimonate, from South Lorraine, Ontario. Contr. to Canadian Mineralogy, 5. (3) (1925) Amer. Mineral., 10, 40 (abs. ref. 2) (4) Milton, C., J.M. Axelrod, and B. Ingram (1958) Bismutoferrite, chapmanite, and "hypochlorite". Amer. Mineral., 43, 656–670. (5) Čech, F. and P. Povondra (1963) Study of chapmanite from Smilkov, near Votice (Bohemia, Czechoslovakia). Acta Univ. Carolinae, Geol., 2, 97–114 (in Czech with English abs.). (6) (1964) Amer. Mineral., 49, 1499 (abs. ref. 5). (7) Zhukhlistov, A.P. and B.B. Zvyagin (1977) Determination of the crystal structures of chapmanite and bismuthoferrite [bismutoferrite] by high-voltage electron diffraction. Kristallografiya (Sov. Phys. Crystal.), 22, 731–738 (in Russian).

Crystal Data: Monoclinic. *Point Group:* n.d. Fibrous, massive.

Physical Properties: *Cleavage:* Good in three directions. Hardness = n.d. VHN = 412 (50 g load). D(meas.) = 2.54 D(calc.) = [2.77]

Optical Properties: Semitransparent. *Color:* Shades of deep lilac to violet; colorless in thin section.
Optical Class: Biaxial (+). *Pleochroism:* X = rose; Z = colorless, in thick fragments.
Orientation: $X = b$; $Z \wedge c = 5°$. $\alpha = 1.550(2)$ $\beta = 1.553(2)$ $\gamma = 1.559(2)$
2V(meas.) = 28°–30°

Cell Data: *Space Group:* n.d. $a = 10.7$ $b = 32.0$ $c = 7.25$ $\beta = 113°$ $Z = 18$

X-ray Powder Pattern: Murun massif, Russia.
3.348 (100), 3.134 (85), 12.5 (70), 2.79 (50), 2.71 (35), 3.90 (30), 2.97 (30)

Chemistry:

	(1)	(2)
SiO_2	56.88	58.5
Al_2O_3		0.07
Fe_2O_3	0.12	
FeO		0.01
MnO		0.07
CaO	20.95	20.5
SrO	0.90	0.5
BaO	2.52	2.9
Na_2O	3.77	1.8
K_2O	10.36	8.9
F	0.92	0.7
H_2O^+	4.40	4.7
$-O = F_2$	0.39	[0.15]
Total	100.43	98.5

(1) Murun massif, Russia; corresponds to $(K_{0.93}Ba_{0.07}Sr_{0.03})_{\Sigma=1.03}(Ca_{1.57}Na_{0.51})_{\Sigma=2.08}Si_4O_{10}$ $[(OH)_{0.58}F_{0.28}]_{\Sigma=0.86} \cdot 0.72H_2O$. (2) Do.; by electron microprobe, H_2O by TGA; corresponds to $(K_{0.88}Ba_{0.09}Sr_{0.02})_{\Sigma=0.99}(Ca_{1.71}Na_{0.28})_{\Sigma=1.99}Si_{4.55}O_{10}[(OH)_{0.78}F_{0.18}]_{\Sigma=0.96} \cdot 0.82H_2O$.

Occurrence: In potassic feldspar metasomatites at the contact of nepheline and aegirine syenites with limestones.

Association: Canasite, tinaksite.

Distribution: In the Murun massif, between the Chara and Olekma Rivers, southwest of Olekminsk, Yakutia, Russia.

Name: For the Chara River, Russia, near which it was discovered.

Type Material: University of Rome, Rome, Italy, 24352.

References: (1) Rogova, V.P., Y.G. Rogov, V.A. Drits, and N.N. Kutnetsova (1978) Charoite, a new mineral and a new jewelry stone. Zap. Vses. Mineral. Obshch., 107, 94–100 (in Russian). (2) (1978) Amer. Mineral., 63, 1282 (abs. ref. 1). (3) Kraeff, A., R.P.E. Poorter, and R.D. Schuiling (1980) Additional information on charoite. Neues Jahrb. Mineral., Monatsh., 498–500. (4) Nikishova, L.V., K.A. Lazebnik, and Y.D. Lazebnik (1985) Crystal chemical formula of charoite. Kristallkhim. Strukt. Tipomorfizm Mineral., 100–104. (5) (1988) Amer. Mineral., 73, 198 (abs. ref. 4).

Crystal Data: Hexagonal. *Point Group:* $6/m \ 2/m \ 2/m$. As tabular, euhedral to subhedral crystals, to 0.1 mm; dominant forms are $\{0001\}$, $\{10\bar{1}0\}$, $\{11\bar{2}0\}$, and $\{10\bar{1}2\}$.

Physical Properties: *Tenacity:* Brittle. Hardness = n.d. D(meas.) = n.d. D(calc.) = 2.68

Optical Properties: Transparent. *Color:* Deep blue; in transmitted light, light blue to colorless. *Streak:* White. *Luster:* Vitreous.
Optical Class: Uniaxial (+). *Pleochroism:* O = sky-blue; E = colorless. $\omega = 1.575(1)$
$\epsilon = 1.578(1)$

Cell Data: *Space Group:* $[P6/mcc]$ (by analogy to milarite group). $a = 10.153(4)$
$c = 14.388(6)$ Z = 2

X-ray Powder Pattern: Moon Canyon, Utah, USA; close similarity to other milarite group minerals may require chemical analysis for identification.
5.076 (vs), 3.751 (vs), 3.238 (vs), 7.194 (s), 2.7840 (s), 4.148 (m), 2.9350 (m)

Chemistry:

	(1)	(2)
SiO$_2$	69.95	70.29
TiO$_2$	0.21	0.02
Al$_2$O$_3$	0.24	0.00
Fe$_2$O$_3$	5.28	
FeO	5.40	12.60
MnO	0.23	0.23
MgO	13.64	11.44
Na$_2$O	0.45	0.04
K$_2$O	5.24	4.48
Total	100.64	99.10

(1) Moon Canyon, Utah, USA; by electron microprobe, Fe^{2+}:Fe^{3+} calculated from stoichiometry; corresponds to $(K_{1.14}Na_{0.15})_{\Sigma=1.29}(Mg_{3.48}Fe^{2+}_{0.77}Fe^{3+}_{0.68}Mn_{0.03}Ti_{0.03}Al_{0.01})_{\Sigma=5.00}$ $(Si_{11.96}Al_{0.04})_{\Sigma=12.00}O_{30}$. (2) Cancarix, Spain; by electron microprobe, total Fe as FeO, Fe^{2+}:Fe^{3+} in empirical formula calculated from stoichiometry; corresponds to $(K_{0.99}Na_{0.01})_{\Sigma=1.00}$ $(Mg_{2.96}Fe^{2+}_{1.20}Fe^{3+}_{0.63}Mn_{0.03})_{\Sigma=4.82}Si_{12.18}O_{30}$.

Mineral Group: Milarite group.

Occurrence: A late-crystallizing mineral, nonuniformly distributed in the groundmass of lamproites.

Association: Potassian richterite, diopside, potassic feldspar.

Distribution: From Moon Canyon, east of Francis, Summit Co., Utah, USA. In Spain, at Cancarix, Albacete Province.

Name: In honor of Dr. Felix Chayes of the Geophysical Laboratory, Carnegie Institution, Washington, D.C., USA.

Type Material: Institute for Mineralogy, Ruhr University, Bochum, Germany; National Museum of Natural History, Washington, D.C., USA, 165807.

References: (1) Velde, D., O. Medenbach, C. Wagner, and W. Schreyer (1989) Chayesite, K(Mg, Fe^{2+})$_4$Fe^{3+}[Si$_{12}$O$_{30}$]: a new rock-forming silicate mineral of the osumilite group from the Moon Canyon (Utah) lamproite. Amer. Mineral., 74, 1368–1373.

Crystal Data: Monoclinic. *Point Group:* $2/m$ or m. In small veinlets and leaflets, to 5 mm.

Physical Properties: *Cleavage:* Perfect on {001}. Hardness = 2.5–3.6 on {001}; 2.9–4.5 $\perp$ {001}. D(meas.) = 3.14–3.16 D(calc.) = 3.10

Optical Properties: Translucent. *Color:* Olive-green, dark green. *Streak:* Light green. *Luster:* Pearly on cleavage.
Optical Class: Biaxial (–). $\alpha = 1.640$–1.643 $\beta = 1.686$–1.691 $\gamma = 1.702$–1.704 2V(meas.) = 11°–12°

Cell Data: *Space Group:* $C2/c$ or Cc. $a = 5.29(1)$ $b = 9.182(2)$ $c = 20.023(6)$ $\beta = 95°41(5)'$ $Z = 4$

X-ray Powder Pattern: "Karatau," Kazakhstan.
3.33 (100), 2.607 (70), 1.996 (60), 1.660 (60), 3.01 (50), 1.530 (50), 2.887 (40)

Chemistry:

	(1)	(2)
SiO_2	29.90	30.06
Al_2O_3	25.90	27.24
Fe_2O_3	0.40	0.36
V_2O_4	5.40	5.30
V_2O_3	18.30	18.90
MgO	1.60	trace
BaO	9.35	9.60
Na_2O	1.32	
K_2O	0.70	0.90
H_2O^+	6.00	6.10
H_2O^-	1.10	1.00
Total	99.97	99.46

(1) "Karatau," Kazakhstan; corresponds to $(Ba_{0.28}Na_{0.20}K_{0.07})_{\Sigma=0.55}(V^{3+}_{1.13}Al_{0.65}V^{4+}_{0.30}Mg_{0.18}Fe_{0.02})_{\Sigma=2.28}(Si_{2.30}Al_{1.70})_{\Sigma=4.00}O_{10}(OH)_2$. (2) Do.; corresponds to $(Ba_{0.29}K_{0.09})_{\Sigma=0.38}(V^{3+}_{1.18}Al_{0.76}V^{4+}_{0.29}Fe_{0.02})_{\Sigma=2.25}(Si_{2.30}Al_{1.70})_{\Sigma=4.00}O_{10}(OH)_2$.

Polymorphism & Series: $2M_1$ polytype.

Mineral Group: Mica group.

Occurrence: In small veinlets cutting carbonate rocks interbedded with vanadium-rich lower Paleozoic shales.

Association: n.d.

Distribution: From "northwestern Karatau" [Kara-Tau Mountains], southern Kazakhstan.

Name: For V.V. Chernykh, Professor at the Mining Institute, St. Petersburg, Russia.

Type Material: Alma-Ata, Kazakhstan; A.E. Fersman Mineralogical Museum, Academy of Sciences, Moscow, Russia.

References: (1) Ankinovich, S.G., E.A. Ankinovich, I.V. Rozhdestvenskaya, and V.A. Frank-Kamenetskii (1972) Chernykhite, a new barium-vanadium mica from northwestern Karatau. Zap. Vses. Mineral. Obshch., 101, 451–458 (in Russian). (2) (1973) Amer. Mineral., 58, 966 (abs. ref. 1). (3) Rozhdestvenskaya, I.V. and V.A. Frank-Kamenetskii (1974) Structure of the dioctahedral mica chernykhite. Kristallokhim. Strukt. Miner., 28–33 (in Russian). (4) (1975) Chem. Abs., 69539 (abs. ref. 3). (5) Rozhdestvenskaya, I.V. (1979) Refinement of the structure of chernykhite. Kristallokhim. Strukt. Miner., 66–69 (in Russian). (6) (1979) Chem. Abs., 115676 (abs. ref. 5).

Crystal Data: Orthorhombic. *Point Group:* n.d. As euhedral, thin, square or rectangular plates, to 0.03 mm.

Physical Properties: *Cleavage:* Indistinct || [101]. Hardness = n.d. D(meas.) = n.d. D(calc.) = 2.21

Optical Properties: Semitransparent. *Color:* White. *Luster:* Silky.
Optical Class: Biaxial (+). *Orientation:* X = a; Y = c; Z = b. $\alpha = 1.456$ $\beta = 1.460$
$\gamma = 1.480$ 2V(meas.) = 47°

Cell Data: *Space Group:* n.d. a = 13.70 b = 27.96 c = 9.99 Z = 2

X-ray Powder Pattern: Maine mine, France.
13.91 (100), 3.422 (100), 4.85 (90), 3.982 (60), 3.451 (40), 3.322 (40), 2.908 (30)

Chemistry:

	(1)
SiO_2	3.64
Al_2O_3	15.30
ZnO	3.73
MgO	3.50
CaO	3.92
Na_2O	4.44
K_2O	0.30
H_2O	31.11
SO_3	31.91
Total	97.85

(1) Maine mine, France; Ca, Mg, Zn, Si, and Al by AA, Na and K by flame photometry, S by chromatography, and H_2O by TGA; corresponds to $(Na_{3.75}K_{0.17}Ca_{0.08})_{\Sigma=4.00}(Ca_{1.75}Mg_{0.25})_{\Sigma=2.00}$ $(Mg_{1.87}Zn_{1.20})_{\Sigma=3.07}(Al_{7.85}Mg_{0.15})_{\Sigma=8.00}Si_{1.58}S_{10.42}O_{53.46} \cdot 45.13H_2O$.

Occurrence: As a coating on fluorite.

Association: Fluorite, gypsum.

Distribution: From the Maine fluorite mine, near Autun, Saône-et-Loire, France.

Name: For Professor Ronald Chessex, petrographer, University of Geneva, Geneva, Switzerland.

Type Material: Natural History Museum, Geneva, Switzerland, 435/70.

References: (1) Sarp, H. and J. Deferne (1982) Le chessexite, un nouveau minéral. Schweiz. Mineral. Petrog. Mitt., 62, 337-341 (in French with English abs.). (2) (1984) Amer. Mineral., 69, 406 (abs. ref. 1).

Crystal Data: Orthorhombic. *Point Group:* $mm2$. Radiating sprays of prismatic crystals, to 5 cm; as intergrowths $\|$ {010} in anthophyllite and cummingtonite.

Physical Properties: *Cleavage:* Perfect on {110}, intersecting at $\sim 45°$ and $135°$; breakage along {100} and {010} may be due to parting. Hardness = n.d. D(meas.) = n.d. D(calc.) = [3.08]

Optical Properties: Transparent. *Color:* Colorless to very light pinkish brown; in thin section, colorless.
Optical Class: Biaxial (–). *Orientation:* $X = a$; $Y = b$; $Z = c$. *Dispersion:* $r > v$, weak.
$\alpha = 1.617(5)$ $\beta = 1.632(5)$ $\gamma = 1.640(5)$ 2V(meas.) = $71(2)°$

Cell Data: *Space Group:* $A2_1ma$. $a = 18.6140(3)$ $b = 45.306(1)$ $c = 5.2966(3)$ $Z = 4$

X-ray Powder Pattern: Calculated.
8.609 (100), 3.076 (66), 3.246 (49), 11.327 (41), 2.554 (36), 3.751 (34), 2.779 (31)

Chemistry:

	(1)
SiO_2	57.95
Al_2O_3	0.25
FeO	14.14
MnO	0.99
MgO	24.24
CaO	0.42
Na_2O	0.03
H_2O	[2.60]
Total	[100.62]

(1) Chester, Vermont, USA; by electron microprobe, H_2O assuming (OH) sites filled by $(OH)^{1-}$.

Occurrence: In black wallrock between chlorite and actinolite zones of a metamorphosed ultramafic body.

Association: Jimthompsonite, clinojimthompsonite, anthophyllite, cummingtonite, talc.

Distribution: In the Carleton talc quarry, near Chester, Windsor Co., Vermont, USA.

Name: For the Chester, Vermont, USA, locality.

Type Material: Royal Ontario Museum, Toronto, Canada, M36083; Harvard University, Cambridge, Massachusetts; National Museum of Natural History, Washington, D.C., USA, 145869.

References: (1) Veblen, D.R. and C.W. Burnham (1978) New biopyriboles from Chester, Vermont: I. Descriptive mineralogy. Amer. Mineral., 63, 1000–1009. (2) Veblen, D.R. and C.W. Burnham (1978) New biopyriboles from Chester, Vermont: II. The crystal chemistry of jimthompsonite, clinojimthompsonite, chesterite, and the amphibole-mica reaction. Amer. Mineral., 63, 1053–1073.

Crystal Data: Monoclinic. *Point Group: 2/m.* Small lamellar, lenticular, and prismatic crystals; massive, in lumps to 10 cm. *Twinning:* Twin plane {001}.

Physical Properties: *Fracture:* Irregular to conchoidal. *Tenacity:* Brittle. Hardness = 5.5–6 D(meas.) = 4.53–4.67 D(calc.) = [4.99]

Optical Properties: Opaque except on thin edges. *Color:* Dark reddish brown to black. *Luster:* Resinous to dull.
Optical Class: Biaxial (–). *Pleochroism:* X = almost colorless; Y = reddish brown; Z = dark brown. *Orientation:* $Z \wedge c = 11°–26°$. *Dispersion:* $r < v$, distinct to strong. *Absorption:* Z > Y > X. α = n.d. β = 1.88–2.02 γ = n.d. 2V(meas.) = Medium.

Cell Data: *Space Group:* $P2_1/a$. a = 13.370(3) b = 5.660(2) c = 11.280(3) β = 100°52(2)′ Z = [2]

X-ray Powder Pattern: Meade Co., Kansas, USA; some materials are metamict.
3.17 (100), 3.14 (100), 2.71 (100), 2.16 (50), 1.961 (50), 4.58 (40), 3.46 (40)

Chemistry:

	(1)	(2)		(1)	(2)
SiO_2	19.23	12.04	Nb_2O_5	0.01	
TiO_2	19.61	17.08	FeO	8.85	7.76
ThO_2	0.73	0.82	MnO	1.11	0.50
Al_2O_3	2.17	0.93	MgO	0.05	0.74
Y_2O_3	0.93	1.50	CaO	2.49	3.35
$(La, Dy)_2O_3$	18.77	18.35	$(Na, K)_2O$	0.10	
Ce_2O_3	20.52	25.29	H_2O	0.81	1.50
Fe_2O_3	2.60	9.56	P_2O_5		0.38
Cr_2O_3	0.07		SO_3	0.15	
			Total	98.10	99.80

(1) Ural Mountains, Russia. (2) Mohave Co., Arizona, USA.

Polymorphism & Series: Dimorphous with perrierite.

Occurrence: As an accessory mineral in some granites, and in pegmatites associated with alkalic granite and syenites; in air-fall ash deposits.

Association: Titanite, monazite, apatite, cronstedtite, quartz (Aquarius Mountains, Arizona, USA).

Distribution: From the Ilmen Mountains, Southern Ural Mountains; the Keivy massif, Kola Peninsula; and elsewhere in Russia. At Buer, Björkedalen, Norway. From Glen Bay, Na h-Eagan, and Conachair, in the Isles of St. Kilda, Outer Hebrides, Scotland. In Madagascar, from the Tordendrika-Ifasina region, Androngovato, and Ifasina, Ambatofinandrahana. In the USA, in the Aquarius Mountains, Mohave Co., Arizona; the Jemez Mountains and Pueblo Mesa, Rio Arriba Co., New Mexico; at Onion Creek, Grand Co., Utah; in Yellowstone National Park, Wyoming; from Irish Flats, Meade Co., Kansas. Also in the Spruce Pine district, Mitchell Co., North Carolina; near Martin's Store, Nelson Co., Virginia; and in New Hampshire, from near Stark, Coos Co. A few other localities have been verified.

Name: For General Konstantin Vladimirovich Tschevkin [Chevkin] (1802–1875), Chief of Staff of the Russian Mining Engineers Corps.

References: (1) Dana, E.S. (1892) Dana's system of mineralogy, (6th edition), 718.
(2) Vlasov, K.A., Ed. (1966) Mineralogy of rare elements, v. II, 309–312. (3) Kauffman, A.J., Jr. and H.W. Jaffe (1946) Chevkinite (tscheffkinite) from Arizona. Amer. Mineral., 31, 582–588.
(4) Mitchell, R.S. (1966) Virginia metamict minerals: perrierite and chevkinite. Amer. Mineral., 51, 1394–1405. (5) Izett, G.A. and R.E. Wilcox (1968) Perrierite, chevkinite, and allanite in upper Cenozoic ash beds in the western United States. Amer. Mineral., 53, 1558–1567.

Crystal Data: Orthorhombic, pseudohexagonal. *Point Group: mm2.* Crystals are hemimorphic, spear-shaped, flattened on {010}, and display the pyramid {161}, to 0.4 mm. In aggregates and spherulites of such crystals; as grains and crusts.

Physical Properties: *Cleavage:* Good to perfect on {100}, {010}, and {001}. Hardness = ~3 D(meas.) = 2.56–2.64 D(calc.) = 2.65

Optical Properties: Translucent. *Color:* Pale orange-yellow; in thin section, colorless to yellow or orange-yellow; color sharply zoned near crystal margins. *Streak:* White to pale ocher. *Luster:* Vitreous to pearly.
Optical Class: Biaxial (+). *Pleochroism:* Weak; X = colorless to pale yellow; Z = yellow-orange. *Orientation:* X = a; Y = b; Z = c. *Absorption:* Z > X. $\alpha = 1.594$–1.596 $\beta = 1.600$ $\gamma = 1.613$–1.618 2V(meas.) = n.d. 2V(calc.) = 50°

Cell Data: *Space Group:* $P2_1ab$. a = 8.866(7) b = 31.34(2) c = 4.787(3) Z = 4

X-ray Powder Pattern: Tanno, Italy.
15.7 (100bb), 2.903 (100), 3.28 (75), 4.15 (30), 3.93 (30b), 3.82 (30), 1.944 (30)

Chemistry:

	(1)	(2)		(1)	(2)
SiO_2	52.5	50.0	Na_2O	0.3	0.68
Al_2O_3	3.6	6.76	K_2O		0.04
FeO		2.01	F		0.21
MnO	12.9	10.6	H_2O	11.3	10.4
BeO	9.3	8.72	CO_2		0.83
MgO		0.14	$-O = F_2$		0.09
CaO	10.2	9.69	Total	100.1	99.99

(1) Tanno, Italy; by electron microprobe; Be determined spectrographically, H_2O by TGA; corresponds to $(Ca_{0.97}Na_{0.05})_{\Sigma=1.02}Mn_{0.97}(Be_{1.98}Al_{0.03})_{\Sigma=2.01}(Si_{4.65}Al_{0.35})_{\Sigma=5.00}$ $O_{12.63}(OH)_{2.37} \cdot 2.16H_2O$. (2) Langangen, Norway; by a variety of analytical methods; corresponds to $(Ca_{0.92}Na_{0.12})_{\Sigma=1.04}(Mn_{0.80}Fe_{0.15}Mg_{0.02})_{\Sigma=0.97}(Be_{1.86}Al_{0.14})_{\Sigma=2.00}(Si_{4.43}Al_{0.57})_{\Sigma=5.00}$ $O_{12.45}(OH)_{2.49}F_{0.06} \cdot 1.83H_2O$.

Occurrence: Coating beryl in late-stage pegmatites (Tanno, Italy); in vugs in syenite pegmatites (Norway).

Association: Beryl, bavenite (Tanno, Italy); analcime, aegirine, helvite, eudidymite, natrolite, feldspar (Norway).

Distribution: From Tanno, Chiavenna, Lombardy, Italy. In Norway, in the Heia and Vevja larvikite quarries, near Larvik, Tvedalen district; near Blåfjell, Langangen, Langesundsfjord; and at Bakkane, Brunlanes.

Name: For the Italian locality at Tanno, Chiavenna.

Type Material: Mineralogical Museum, University of Rome, Rome, Italy, 24342; University of Oslo, Norway; National Museum of Natural History, Washington, D.C., USA, 159917.

References: (1) Bondi, M., W.L. Griffin, V. Mattioli, and A. Mottana (1983) Chiavennite, $CaMnBe_2Si_5O_{13}(OH)_2 \cdot 2H_2O$, a new mineral from Chiavenna (Italy). Amer. Mineral., 68, 623–627. (2) Raade, G., R. Åmli, M.H. Mladeck, V.K. Din, A.O. Larsen, and A. Åsheim (1983) Chiavennite from syenite pegmatites in the Oslo Region, Norway. Amer. Mineral., 68, 628–633.

Crystal Data: Orthorhombic. *Point Group:* $mm2$. As complex equant crystals, to 5 mm; massive.

Physical Properties: *Cleavage:* Fair in one direction; indistinct in two other directions. *Fracture:* Uneven to conchoidal. *Tenacity:* Brittle. Hardness = 6 D(meas.) = 2.662 D(calc.) = 2.70

Optical Properties: Transparent to translucent. *Color:* White; colorless in thin section. *Luster:* Vitreous.
Optical Class: Biaxial (+). $\alpha = 1.544$ $\beta = [1.546]$ $\gamma = 1.549$ 2V(meas.) = 62°–82°

Cell Data: *Space Group:* $Fdd2$. a = 21.129(5) b = 6.881(2) c = 21.188(5) Z = 24

X-ray Powder Pattern: Lovozero massif, Russia.
3.97 (vs), 2.46 (vs), 2.41 (vs), 1.79 (s), 1.48 (s), 1.40 (s), 1.11 (s)

Chemistry:

	(1)	(2)
SiO_2	56.81	58.01
Fe_2O_3	0.30	
FeO	0.12	
BeO	12.67	12.07
CaO	0.37	
Na_2O	28.93	29.92
K_2O	0.13	
H_2O^-	0.23	
SO_3	0.22	
Total	99.78	100.00

(1) Lovozero massif, Russia. (2) $Na_2BeSi_2O_6$.

Occurrence: In veins in pegmatite in a differentiated alkalic massif (Lovozero massif, Russia); in sodalite xenoliths in an intrusive alkalic gabbro-syenite complex (Mont Saint-Hilaire, Canada).

Association: Manganoan pectolite, microcline, sodalite, eudialyte, neptunite, ussingite (Lovozero massif, Russia); villiaumite, ussingite, sodalite, aegirine, lovozerite, vuonnemite, sérandite, lueshite, steenstrupine (Mont Saint-Hilaire, Canada).

Distribution: At two locations on Mt. Punkaruaiv, Lovozero massif, and in the Khibiny massif, Kola Peninsula, Russia. On the Kangerdluarssuk Plateau and on the Taseq slope, in the Ilímaussaq intrusion, southern Greenland. From Mont Saint-Hilaire, Quebec, Canada.

Name: For Valery Pavlovich Chkalov (1904–1938), first to fly nonstop from Moscow to the USA across the North Pole.

Type Material: n.d.

References: (1) Gerasimovskii, V.I. (1939) Chkalovite. Doklady Acad. Nauk SSSR, 22, 259–263 (in Russian). (2) (1940) Amer. Mineral., 25, 380 (abs. ref. 1) (3) Vlasov, K.A., M.V. Kuz'menko, and E.M. Es'kova (1966) The Lovozero alkaline massif. Akad. Nauk SSSR, 429–432 (in English). (4) Simonov, M.A., Y.K. Egorov, and N.V. Belov (1975) Refined crystal structure of chkalovite, $Na_2Be(Si_2O_6)$. Doklady Acad. Nauk SSSR, 225, 1319–1322 (in Russian). (5) (1977) Mineral. Abs., 28, 140 (abs. ref. 4). (6) Mandarino, J.A. and V. Anderson (1989) Monteregian Treasures. Cambridge Univ. Press, 57.

Crystal Data: Hexagonal or monoclinic, pseudohexagonal. *Point Group:* $6/m$ or $2/m$. Massive, granular.

Physical Properties: *Cleavage:* Indistinct on $\{0001\}$ and $\{10\bar{1}0\}$. Hardness = ~5 D(meas.) = 3.068   D(calc.) = 3.046

Optical Properties: Translucent. *Color:* Pale rose, rose-pink, orange. *Luster:* Vitreous. *Optical Class:* Uniaxial (−). $\omega = 1.655(2)$ $\epsilon = 1.650(2)$

Cell Data: *Space Group:* $P6_3/m$. a = 9.530(2)   c = 6.914(2)   Z = 2

X-ray Powder Pattern: Crestmore, California, USA.
2.843 (100), 2.751 (70), 3.458 (50), 2.800 (40), 1.964 (40), 1.855 (40), 2.651 (30)

Chemistry:

	(1)		(1)
SiO_2	17.31	Cl	1.64
Al_2O_3	0.13	H_2O^+	0.53
Fe_2O_3	0.22	H_2O^-	0.10
MnO	0.01	CO_2	0.61
MgO	0.47	P_2O_5	3.06
CaO	55.18	SO_3	20.69
F	0.57	$-O = (F, Cl)_2$	0.61
		Total	99.91

(1) Crestmore, California, USA.

Mineral Group: Ellestadite group.

Occurrence: In veinlets cutting contact metamorphosed limestone.

Association: Diopside, wollastonite, vesuvianite, monticellite, okenite, calcite.

Distribution: From Crestmore, Riverside Co., California, USA.

Name: For Dr. Reuben B. Ellestad (1900–), American analytical chemist, University of Minnesota, Minneapolis, Minnesota, USA, and *chlorine* in the chemical composition.

References: (1) Palache, C., H. Berman, and C. Frondel (1951) Dana's system of mineralogy, (7th edition), v. II, 906 [ellestadite]. (2) McConnell, D. (1937) The substitution of SiO_4- and SO_4-groups for PO_4-groups in the apatite structure; ellestadite, the end-member. Amer. Mineral., 22, 977–986. (3) Rouse, R.C. and P.J. Dunn (1982) A contribution to the crystal chemistry of ellestadite and the silicate sulfate apatites. Amer. Mineral., 67, 90–96.

Crystal Data: Monoclinic or triclinic, both pseudohexagonal. *Point Group:* $2/m$ or 1. Rarely as rosettes of pseudohexagonal tabular crystals; commonly coarsely foliated with foliae typically curved or bent; massive. *Twinning:* Common on {001}, may be lamellar; twin axes [100], [110], and [130] observed.

Physical Properties: *Cleavage:* Perfect on {001}, distinct on {110}; parting on {010}. *Tenacity:* Brittle. Hardness = 6.5   D(meas.) = 3.46–3.80   D(calc.) = 3.56

Optical Properties: Translucent. *Color:* Dark gray, greenish gray, greenish black; colorless to green in thin section. *Streak:* White, grayish, or very slightly greenish. *Luster:* Somewhat pearly on cleavage surfaces.
Optical Class: Biaxial (+) or (–). *Pleochroism:* X = olive-green to yellow; Y = grayish blue to blue; Z = colorless to pale greenish yellow. *Orientation:* X or $Y = b$ (monoclinic); $Z \wedge c = 2°$–$30°$; $Y \simeq b$ (triclinic). *Dispersion:* $r > v$, strong. $\alpha = 1.705$–1.730 $\beta = 1.708$–1.734 $\gamma = 1.712$–1.740 2V(meas.) = $\sim 45°$–$70°$

Cell Data: *Space Group:* $C2/c$.   a = 9.50(3)   b = 5.50(2)   c = 18.22(5) $\beta = 101°57(5)'$ Z = 4, or   *Space Group:* $C\bar{1}$.   a = 9.46(1)   b = 5.50(1)   c = 9.15(1) $\alpha = 97.05(2)°$ $\beta = 101.56(2)°$ $\gamma = 90.10(2)°$ Z = 4

X-ray Powder Pattern: Natick, Rhode Island, USA.
4.498 (100), 4.449 (100), 2.963 (90), 1.5813 (80), 2.367 (70), 2.306 (70), 2.639 (50)

Chemistry:

	(1)		(1)
SiO_2	23.91	CaO	0.04
TiO_2	0.20	Na_2O	0.00
Al_2O_3	40.12	K_2O	trace
Fe_2O_3	1.23	F	0.01
FeO	27.06	H_2O^+	7.03
MnO	0.16	H_2O^-	0.01
MgO	0.51	Total	100.28

(1) Natick, Rhode Island, USA; corresponds to $(Fe^{2+}_{1.89}Mg_{0.06}Mn_{0.01})_{\Sigma=1.96}$ $(Al_{3.95}Fe^{3+}_{0.08}Ti_{0.01})_{\Sigma=4.04}Si_{2.00}O_{9.80}(OH)_{3.20}$.

Polymorphism & Series: Forms a series with carboirite; monoclinic and triclinic polytypes are known.

Mineral Group: Chloritoid group.

Occurrence: In regionally metamorphosed pelitic sediments and schists in the biotite, garnet, and lower grade staurolite zones; in quartz-carbonate veins and other hydrothermal environments.

Association: Muscovite, chlorite, staurolite, garnet, kyanite, quartz, mica, rutile.

Distribution: A widely distributed mineral; only a few studied occurrences are noted. In Russia, from Kosoy Brod, south of Yekaterinburg (Sverdlovsk), Ural Mountains. At Salmchâteau, near Ottré, Ardennes Mountains, Belgium. From Nadels and Zermatt, Valais, Switzerland. At Prägraten, Tirol, Austria. In Scotland, along the coast between Stonehaven and Aberdeen; and from Unst, Shetland Islands. Around Tintagel, Cornwall, England. In the USA, at Natick, Kent Co., Rhode Island; Chester, Hampden Co., Massachusetts; in Clove Valley, Duchess Co., New York; from Bull Mountain, Patrick Co., Virginia. In Canada, at Chibougamau and St. Giles, Quebec. From Kalgoorlie, Western Australia.

Name: For its resemblance to members of the *chlorite* group.

References: (1) Dana, E.S. (1892) Dana's system of mineralogy, (6th edition), 640–642. (2) Deer, W.A., R.A. Howie, and J. Zussman (1982) Rock-forming minerals, (2nd edition), v. 1A, orthosilicates, 867–912. (3) Halferdahl, L.B. (1961) Chloritoid: its composition, X-ray and optical properties, stability and occurrence. J. Petrol., 2, 49–135. (4) Hanscomb, R.H. (1980) The structure of triclinic chloritoid and chloritoid polymorphism. Amer. Mineral., 65, 534–539.

Chondrodite $(Mg, Fe^{2+})_5(SiO_4)_2(F, OH)_2$

Crystal Data: Monoclinic. *Point Group:* $2/m$. Crystals rare, varied in habit, typically flattened $\parallel$ [010], to 10 cm. Commonly as rounded grains, massive. *Twinning:* On {001}, common, simple and lamellar; possibly $\parallel$ {015} and {305}.

Physical Properties: *Cleavage:* {100} indistinct; parting on {001}. *Tenacity:* Brittle. Hardness = 6–6.5 D(meas.) = 3.16–3.26 D(calc.) = 3.177

Optical Properties: Transparent to translucent. *Color:* Light yellow, yellow, brown, red; in thin section, pale yellow or brown to colorless. *Luster:* Vitreous to resinous.
Optical Class: Biaxial (+). *Pleochroism:* X = colorless, very pale yellow, brownish yellow; Y = colorless, yellow-green; Z = colorless, pale green. *Orientation:* $Z = b$; $X \wedge c = 22°–31°$.
Dispersion: $r > v$, strong to weak. $\alpha = 1.592–1.643$ $\beta = 1.602–1.655$ $\gamma = 1.621–1.676$
2V(meas.) = 71°–85°

Cell Data: *Space Group:* $P2_1/c$. a = 7.8404(2) b = 4.7284(3) c = 10.2539(3)
$\beta = 109°2'$ Z = 2

X-ray Powder Pattern: Tilly Foster mine, New York, USA. (ICDD 12-527).
2.258 (100), 1.740 (70), 3.02 (45), 2.510 (45), 3.56 (35), 2.758 (35), 2.288 (35)

Chemistry:

	(1)	(2)		(1)	(2)
SiO_2	33.60	32.16	MgO	59.30	53.21
TiO_2	0.06	0.20	CaO	0.00	0.00
Al_2O_3	0.24	0.73	F	6.61	7.11
Fe_2O_3	0.05	0.60	H_2O^+	1.46	1.20
FeO	0.86	6.08	H_2O^-	0.00	0.00
MnO	0.16	1.35	$-O = F_2$	2.78	2.99
			Total	99.56	99.65

(1) Hangelby, Finland; corresponds to $(Mg_{5.13}Fe^{2+}_{0.04}Al_{0.02}Mn_{0.01})_{\Sigma=5.20}Si_{1.95}O_8$ $[F_{1.21}(OH)_{0.56}O_{0.23}]_{\Sigma=2.00}$. (2) Kafveltorp, Sweden; corresponds to $(Mg_{4.74}Fe^{2+}_{0.30}Mn_{0.07}$ $Al_{0.05}Fe^{3+}_{0.03}Ti_{0.01})_{\Sigma=5.20}Si_{1.92}O_8[F_{1.34}(OH)_{0.48}O_{0.18}]_{\Sigma=2.00}$.

Polymorphism & Series: Forms a series with alleghanyite.

Mineral Group: Humite group.

Occurrence: In contact metamorphic zones, in limestones and dolostones, associated with felsic to alkalic plutonic rocks, especially with Fe-B-F metasomatism; in a carbonatite.

Association: Phlogopite, spinel, magnetite, grossular, wollastonite, forsterite, monticellite, cuspidine, diopside, calcite.

Distribution: From Pargas, Hangelby, and Sibbo, Finland. At Kafveltorp, near Kopparberg, Sweden. From Monte Somma and Vesuvius, Campania, Italy. At Le Chipal, Vosges, France. From Bhandara, Maharashtra, India. In the USA, fine crystals from the Tilly Foster mine, Brewster, Putnam Co., and Amity, Orange Co., New York; at Franklin and Sparta, Sussex Co., New Jersey; Johnson Camp, Cochise Co., and the Lakeshore mine, Pinal Co., Arizona; and Crestmore, Riverside Co., California. From Bancroft, Ontario, Canada. In the Loolekop carbonatite, Transvaal, South Africa. A few other localities are known.

Name: From the Greek meaning *grain*, alluding to its occurrence in isolated grains.

References: (1) Dana, E.S. (1892) Dana's system of mineralogy, (6th edition), 536–538. (2) Deer, W.A., R.A. Howie, and J. Zussman (1982) Rock-forming minerals, (2nd edition), v. 1A, orthosilicates, 379–417. (3) Gibbs, G.V., P.H. Ribbe, and C.P. Anderson (1970) The crystal structures of the humite minerals. II. Chondrodite. Amer. Mineral., 55, 1182–1194. (4) Phillips, W.R. and D.T. Griffen (1981) Optical mineralogy, 142–144.

$NaMg_3(Cr^{3+}, Fe^{3+})_6(BO_3)_3Si_6O_{18}(OH)_4$ **Chromdravite**

Crystal Data: Hexagonal. *Point Group: 3m.* In pyramidal acicular crystals, to 0.1 mm.

Physical Properties: Hardness = [~7] (by analogy to the tourmaline group). D(meas.) = 3.40(1)   D(calc.) = [3.39]

Optical Properties: Semitransparent. *Color:* Dark green to nearly black. *Optical Class:* Uniaxial (–). *Pleochroism: O* = dark green; *E* = yellow-green. *Absorption: O > E.* $\omega = 1.778(5)$ $\epsilon = 1.772(5)$

Cell Data: *Space Group: R3m.* a = 16.11   c = 7.27   Z = [3]

X-ray Powder Pattern: Onezhkii basin, Russia.
2.62 (100), 3.58 (75), 3.04 (75), 6.57 (50), 4.05 (50), 2.079 (50), 4.31 (40)

Chemistry:

	(1)
SiO_2	30.75
TiO_2	0.13
B_2O_3	9.00
Al_2O_3	2.92
Fe_2O_3	7.65
V_2O_3	1.46
Cr_2O_3	31.60
MnO	0.19
MgO	9.05
CaO	0.16
Na_2O	2.66
LOI	4.43
Total	[100.00]

(1) Onezhkii basin, Russia; recalculated to 100.00% after correction for 6.5% chromian phengite; corresponds to $(Na_{0.97}Ca_{0.03})_{\Sigma=1.00}(Mg_{2.57}V_{0.22}Al_{0.16}Mn_{0.03}Ti_{0.02})_{\Sigma=3.00}(Cr^{3+}_{4.71}Fe^{3+}_{1.08}Al_{0.21})_{\Sigma=6.00}$ $[(B_{0.97}Al_{0.03})_{\Sigma=1.00}O_3]_3(Si_{5.81}Al_{0.19})_{\Sigma=6.00}O_{18}[(OH)_{3.77}O_{0.23}]_{\Sigma=4.00}$.

Mineral Group: Tourmaline group.

Occurrence: In micaceous metasomatic clay-carbonate rocks.

Association: Chromian phengite, taeniolite, vanadian muscovite, quartz, dolomite.

Distribution: In the Onezhkii basin, central Karelia, Russia.

Name: For CHROMium in the composition and similarity to *dravite*.

Type Material: Mining Institute, St. Petersburg, Russia.

References: (1) Rumantseva, E.V. (1983) Chromdravite, a new mineral. Zap. Vses. Mineral. Obshch., 112, 222–226 (in Russian). (2) (1984) Amer. Mineral., 69, 210 (abs. ref. 1). (3) Deer, W.A., R.A. Howie, and J. Zussman (1986) Rock-forming minerals, (2nd edition), v. 1B, disilicates and ring silicates, 559–602. (4) Dunn, P.J. (1977) Chromium in dravite. Mineral. Mag., 41, 408–410.

Chrysocolla $(Cu, Al)_2H_2Si_2O_5(OH)_4 \cdot nH_2O$

Crystal Data: Orthorhombic (?). *Point Group:* n.d. Crystals acicular, to 5 mm, in radiating clusters; fine fibrous, botryoidal, earthy; commonly cryptocrystalline, opaline, or enamel-like.

Physical Properties: *Fracture:* Conchoidal. *Tenacity:* Brittle to somewhat sectile. Hardness = ∼2–4 D(meas.) = 1.93–2.4 D(calc.) = n.d.

Optical Properties: Translucent to opaque. *Color:* Blue, blue-green, or green; brown to black when impure. *Streak:* White when pure. *Luster:* Vitreous, porcelaneous, earthy. *Optical Class:* Biaxial (–). $\alpha = 1.575$–1.585 $\beta = 1.597$ $\gamma = 1.598$–1.635 2V(meas.) = n.d.

Cell Data: *Space Group:* n.d. $a = 5.72$–5.92 $b = 17.7$–18.0 $c = 8.00$–8.28 $Z =$ n.d.

X-ray Powder Pattern: Locality unknown. (ICDD 27-188).
1.486 (100), 17.9 (80), 2.90 (80), 2.56 (70), 7.9 (60), 4.07 (60), 1.602 (40)

Chemistry:

	(1)	(2)
SiO_2	35.80	39.48
Al_2O_3	2.00	1.91
Fe_2O_3	trace	0.13
MnO		0.88
CuO	42.00	46.93
MgO	0.08	0.47
CaO	1.04	0.52
Na_2O	0.04	
K_2O	0.05	
H_2O^+	10.00	8.29
H_2O^-	9.46	1.31
Total	100.47	99.92

(1) Mednorudyansk, Russia. (2) Kamoya, Zaire.

Occurrence: In the oxidized portions of many copper deposits.

Association: Malachite, tenorite, halloysite, nontronite.

Distribution: A few localities for rich or commercial material include: from Nizhni Tagil, Ural Mountains, Russia. At Libethen (Lubietová), Slovakia. In the Timna (King Solomon's) mine, Israel. From the Star of the Congo mine, Lubumbashi, and at Likasi and Kakanda, Shaba Province, Zaire. In the USA, in Arizona, constitutes parts of large orebodies, as in the Globe-Miami district, Gila Co., Morenci, Greenlee Co., San Manuel and Ray mines, Pinal Co.; in New Mexico, at Santa Rita, Grant Co.; from Utah, in the Tintic district, Juab Co. At Cananea, Sonora, Mexico. In Chile, in the Chuquicamata and Exotica deposits; around Copiapó and Coquimbo. From the Chillagoe district, Queensland, Australia.

Name: From the Greek for *gold* and *glue*, for a substance used in the soldering of gold, including chrysocolla as well as other blue and green minerals.

References: (1) Dana, E.S. (1892) Dana's system of mineralogy, (6th edition), 699–700. (2) Chukhrov, F.V., B.B. Zvyagin, A.I. Gorshkov, L.P. Ermilova, and E.S. Rudnitskaya (1968) Chrysocolla. Izv. Akad. Nauk SSSR, Ser. Geol., no. 6, 29–44 (in Russian). (3) (1969) Amer. Mineral., 54, 993 (abs. ref. 2). (4) Van Oosterwyck-Gastuche, M.-C. (1970) La structure de la chrysocolle. Compt. Rendus Acad. Sci. Paris, 271, 1837–1840 (in French).

Crystal Data: Monoclinic. *Point Group:* $2/m$. In thin to thick pseudohexagonal crystals, with tapering pyramidal faces, to 10 cm. Commonly foliated, fibrous, granular, earthy, massive. *Twinning:* Twin plane {001}; twin axis [310], composition plane {001}.

Physical Properties: *Cleavage:* {001}, perfect. *Tenacity:* Laminae flexible, inelastic. Hardness = 2–2.5   D(meas.) = 2.60–3.02   D(calc.) = 2.628

Optical Properties: Transparent to translucent. *Color:* Grass-green, olive-green, yellowish, white; pink, rose-red; colorless to pale green or yellow in thin section. *Streak:* Greenish white to white. *Luster:* Pearly, greasy, dull.
Optical Class: Biaxial (+) or (–). *Pleochroism:* Distinct; X = light yellow-green to light blue-green; $Y = Z$ = light greenish yellow to light blue-green. *Orientation:* $Y = b$; $Z \wedge c = 2°$–$9°$. *Dispersion:* $r < v$. *Absorption:* $Y \simeq Z > X$ or $Z \simeq Y > Z$. $\alpha = 1.571$–1.588 $\beta = 1.571$–1.588 $\gamma = 1.576$–1.597 2V(meas.) = $0°$–$50°$

Cell Data: *Space Group:* $C2/m$. a = 5.350(3)   b = 9.267(5)   c = 14.27(1) $\beta = 96.35(5)°$ Z = 2

X-ray Powder Pattern: Synthetic (IIb structure); berthierine plus clinochlore easily mistaken for chamosite.
3.57 (100), 2.540 (100), 2.008 (100), 1.539 (100), 14.1 (80), 7.14 (80), 4.76 (80)

Chemistry:

	(1)	(2)		(1)	(2)
SiO_2	33.83	32.12	MgO	34.94	35.36
Al_2O_3	12.95	9.50	CaO		1.24
Fe_2O_3	2.25		H_2O^+	13.11	10.25
Cr_2O_3		7.88	H_2O^-		2.04
FeO	3.02	1.98	Total	100.10	100.37

(1) Zillertal, Austria; corresponds to $(Mg_{4.94}Fe^{2+}_{0.24})_{\Sigma=5.18}(Al_{0.65}Fe^{3+}_{0.16})_{\Sigma=0.81}$ $(Si_{3.21}Al_{0.79})_{\Sigma=4.00}O_{10}(OH)_8$. (2) Deer Park, Wyoming, USA; corresponds to $(Mg_{5.05}Fe^{2+}_{0.16}Ca_{0.13})_{\Sigma=5.34}(Cr_{0.60}Al_{0.15}Fe^{3+}_{0.15})_{\Sigma=0.90}(Si_{3.08}Al_{0.92})_{\Sigma=4.00}O_{10}(OH)_8$.

Polymorphism & Series: Forms a series with chamosite; stacking disorder is common.

Mineral Group: Chlorite group.

Occurrence: A hydrothermal alteration product of amphiboles, pyroxenes, biotite. In chlorite schists, serpentinites, marbles, calc-silicate rocks, amphibolites, less commonly in ultramafic rocks. In ore veins; a detrital component of sediments.

Association: Serpentine, calcite, dolomite, actinolite, biotite, olivine, plagioclase, talc, chromite, uvarovite.

Distribution: Some localities for well-crystallized material are: in the USA, in the Emery mine, Chester, Hampden Co., Massachusetts; at Texas, Lancaster Co., and West Chester, Chester Co., Pennsylvania; in the Tilly Foster mine, Brewster, Putnam Co., New York. In the Zillertal, Tirol, Austria. From the Pfitschtal, Trentino-Alto Adige; Val Malenco, Lombardy; and at Ala, Piedmont, Italy. At Rimpfischwänge, near Zermatt, Valais, Switzerland. From Ojén, Málaga Province, Spain. On Unst, Shetland Islands, Scotland. In the Kop Krom [chrome mine], Kop Mountains, near Aşkale, Turkey. In Russia, in the Ural Mountains, at Akhmatovsk, Berbliouchka, and Hardadinsk; and at Miass, Ilmen Mountains, Southern Ural Mountains.

Name: For its inclined optic axes and the Greek *chloros*, for green, its common color.

References: (1) Dana, E.S. (1892) Dana's system of mineralogy, (6th edition), 644–650, 650–653 [penninite, kämmererite]. (2) Deer, W.A., R.A. Howie, and J. Zussman (1963) Rock-forming minerals, v. 3, sheet silicates, 131–163. (3) Bayliss, P. (1975) Nomenclature of the trioctahedral chlorites. Can. Mineral., 13, 178–180. (4) Rule, A.C. and S.W. Bailey (1987) Refinement of the crystal structure of a monoclinic ferroan clinochlore. Clays and Clay Minerals, 35, 129–138.

Crystal Data: Monoclinic or triclinic. *Point Group:* n.d. Asbestiform, fibrous along [100], curled to cylindrical; also bladed, massive.

Physical Properties: Hardness = 2.5 D(meas.) = 2.53(1) D(calc.) = 2.61

Optical Properties: Semitransparent. *Color:* White, pale green to dark green. *Luster:* Silky in fibrous aggregates.
Optical Class: Biaxial (–). $\alpha = 1.569(2)$ $\beta = [1.569]$ $\gamma = 1.570(2)$ 2V(meas.) = ~42°

Cell Data: *Space Group:* n.d. $a = 5.3129(9)$ $b = 9.120(3)$ $c = 14.637(2)$ $\beta = 93°9.8(6)'$
$Z = 4$

X-ray Powder Pattern: Butler Estate chrome mine, California, USA.
7.31 (100), 3.65 (70), 4.57 (50), 1.535 (50), 2.270 (30), 2.205 (30), 2.092 (30)

Chemistry:

	(1)	(2)		(1)	(2)
SiO_2	42.2	43.37	MnO	0.06	
TiO_2	0.002		NiO	0.04	
Al_2O_3	0.66		MgO	41.7	43.63
Fe_2O_3	1.2		CaO	0.01	
Cr_2O_3	0.02		H_2O^+	13.3	13.00
FeO	0.09		H_2O^-	0.95	
			Total	100.23	100.00

(1) Joe No. 5 pit, California, USA. (2) $Mg_3Si_2O_5(OH)_4$.

Polymorphism & Series: Polymorphous with antigorite, orthochrysotile, lizardite, and parachrysotile; may also be termed chrysotile-$2M_{c1}$.

Mineral Group: Kaolinite-serpentine group.

Occurrence: Intermixed with orthochrysotile in veinlets cutting serpentinite.

Association: Orthochrysotile, lizardite, corundum.

Distribution: Undoubtedly of common occurrence in asbestos deposits, but requires careful characterization for confirmation, which has been accomplished at only a few localities, such as: in the USA, from the Butler Estate chrome mine, Fresno Co., and the Joe No. 5 pit, New Idria, San Benito Co., California; in the Belvidere Mountain quarries, Lowell, Orleans Co., Vermont; and from the Salt River Canyon, near Globe, Gila Co., Arizona. At Thetford Mines, Quebec, Canada. From Quilla, Charsadda Tehsil, Pakistan. In Australia, from Woodsreef, New South Wales.

Name: Chrysotile from the Greek for *golden* and *fiber*; *clino* in reference to the mineral's crystallization in inclined axis crystal systems.

Type Material: n.d.

References: (1) Deer, W.A., R.A. Howie, and J. Zussman (1963) Rock-forming minerals, v. 3, sheet silicates, 170–190. (2) Whittaker, E.J.W. (1956) The structure of chrysotile. II. Clino-chrysotile. Acta Cryst., 9, 855–861. (3) Page, N.J. and R.G. Coleman (1967) Serpentine-mineral analyses and physical properties. U.S. Geol. Sur. Prof. Paper 575-B, B103–B107. (4) Wicks, F.J. and E.J.W. Whittaker (1975) A reappraisal of the structures of the serpentine minerals. Can. Mineral., 13, 227–243. (5) Middleton, A.P. and E.J.W. Whittaker (1976) The structure of Povlen-type chrysotile. Can. Mineral., 14, 301–306. (6) Yada, K. (1979) Microstructures of chrysotile and antigorite by high-resolution electron microscopy. Can. Mineral., 17, 679–691. (7) Bayliss, P. (1981) Unit cell data of serpentine group minerals. Mineral. Mag., 44, 153–156. (8) Wicks, F.J. and D.S. O'Hanley (1988) Serpentine minerals: structures and petrology. In: S.W. Bailey, Ed., Hydrous phyllosilicates. Rev. Mineral. 19, MSA, 91–167.

Crystal Data: Monoclinic. *Point Group:* $2/m$. Crystals prismatic, to 2 mm; lamellar, fibrous, or massive; as stellate clusters. *Twinning:* Polysynthetic on {100}.

Physical Properties: *Cleavage:* Good on {110}, (110) $\wedge$ (1$\bar{1}$0) $\sim$91°. *Tenacity:* [Brittle] (by analogy to enstatite). Hardness = [5–6] D(meas.) = n.d. D(calc.) = 3.210

Optical Properties: Semitransparent. *Color:* [White to brown; colorless in thin section.] *Optical Class:* Biaxial (+). *Orientation:* $X = b$; $Z \wedge c = 20°$–22°. $\alpha = 1.651$ $\beta = 1.654$ $\gamma = 1.660$ 2V(meas.) = 53.5°

Cell Data: *Space Group:* $P2_1/c$ (synthetic). $a = 9.6065$ $b = 8.8146$ $c = 5.1688$ $\beta = 108.335°$ $Z = 4$

X-ray Powder Pattern: Synthetic.
2.873 (100), 2.976 (75), 3.17 (30), 2.452 (30), 2.115 (30), 3.28 (25), 2.518 (18)

Chemistry:

	(1)
SiO$_2$	57.36
TiO$_2$	0.00
Al$_2$O$_3$	0.11
Cr$_2$O$_3$	0.28
FeO	6.32
MnO	0.09
MgO	34.97
CaO	0.26
Na$_2$O	0.00
Total	99.39

(1) Ogasawara Islands, Japan; by electron microprobe, corresponds to $(Mg_{1.81}Fe_{0.18}Ca_{0.01}Cr_{0.01})_{\Sigma=2.01}(Si_{1.99}Al_{0.01})_{\Sigma=2.00}O_6$.

Polymorphism & Series: Dimorphous with enstatite; forms a series with clinoferrosilite.

Mineral Group: Pyroxene group.

Occurrence: As phenocrysts, likely inverted from "protoenstatite," in high-magnesium nonfeldspathic andesites. Exsolved in diopside in ultramafic rocks and in enstatite in high-grade metamorphic hornfelses. An essential constituent of chondrite and achondrite meteorites, likely formed by stress.

Association: Diopside, enstatite, chromite, "hypersthene," glass.

Distribution: From Cape Vogel, Papua New Guinea, and on the inner slope of the Mariana trench. At Chichijima and Mukojima, Ogasawara (Bonin) Islands, Japan. From Népoui, New Caledonia. In the USA, from the Mt. Stuart batholith, central Cascade Mountains, Washington.

Name: For its monoclinic crystal system and chemical identity to *enstatite*.

Type Material: n.d.

References: (1) Dana, E.S. and W.E. Ford (1909) Dana's system of mineralogy, (6th edition), app. II, 30. (2) Deer, W.A., R.A. Howie, and J. Zussman (1978) Rock-forming minerals, (2nd edition), v. 2A, single-chain silicates, 20–161. (3) Stephenson, D.A., C.B. Sclar, and J.V. Smith (1966) Unit cell volumes of synthetic orthoenstatite and low clinoenstatite. Mineral. Mag., 35, 838–846. (4) Komatsu, M. (1980) Clinoenstatite in volcanic rocks from the Bonin Islands. Contr. Mineral. Petrol., 74, 329–338. (5) Ohashi, Y. (1984) Polysynthetically-twinned structures of enstatite and wollastonite. Phys. Chem. Minerals, 10, 217–229.

Clinoferrosilite $(Fe^{2+}, Mg)_2Si_2O_6$

Crystal Data: Monoclinic. *Point Group:* 2/m. As acicular crystals lacking terminal faces. *Twinning:* On {100}.

Physical Properties: *Cleavage:* [Good on {110}, (110) $\wedge$ (1$\bar{1}$0) $\sim$87°] (by analogy to ferrosilite). Hardness = [5–6] D(meas.) = n.d. D(calc.) = 4.068

Optical Properties: Semitransparent. *Color:* Colorless with slight amber tint. *Optical Class:* Biaxial (+). *Orientation:* $Z \wedge c = 31°$. $\alpha = 1.764(2)$ (synthetic $Fe_2Si_2O_6$). $\beta = 1.767(2)$ $\gamma = 1.792(2)$ 2V(meas.) = 25(5)°

Cell Data: *Space Group:* $P2_1/c$ (synthetic $Fe_2Si_2O_6$). a = 9.7085 b = 9.0872 c = 5.2284 $\beta = 108.432°$ Z = 4

X-ray Powder Pattern: Synthetic $Fe_2Si_2O_6$.
3.035 (100), 2.603 (63), 2.909 (55), 2.408 (53), 2.161 (51), 4.605 (49), 3.234 (40)

Chemistry: (1) Material from Lake Naivasha, Kenya, approximates $(Fe_{1.90}Mn_{0.10})_{\Sigma=2.00}Si_2O_6$, with Mg, Al, Ca, Ti, V, Cr, Co, and Ni absent.

Polymorphism & Series: Dimorphous with ferrosilite; forms a series with clinoenstatite.

Mineral Group: Pyroxene group.

Occurrence: As acicular crystals in lithophysae in obsidian (Lake Naivasha, Kenya).

Association: Anorthoclase, magnetite, cristobalite, fayalite, biotite (Lake Naivasha, Kenya).

Distribution: From near Lake Naivasha, Rift Valley, Kenya. In the USA, from the Coso Mountains, Inyo Co., California; and at Obsidian Cliffs, Yellowstone National Park, Wyoming. From Hrafntinnuhryggur, Iceland.

Name: For its monoclinic crystal system and chemical identity with *ferrosilite*.

Type Material: National Museum of Natural History, Washington, D.C., USA, 102793.

References: (1) Deer, W.A., R.A. Howie, and J. Zussman (1978) Rock-forming minerals, (2nd edition), v. 2A, single-chain silicates, 20–161. (2) Bowen, N.L. (1935) Ferrosilite as a natural mineral. Amer. J. Sci., 30, 481–494. (3) (1936) Amer. Mineral., 21, 678 (abs. ref. 2). (4) Lindsley, D.H., B.T.C. Davis, and I.D. MacGregor (1964) Ferrosilite (FeSiO$_3$): synthesis at high pressures and temperatures. Science, 144, 73–74. (5) Bown, M.G. (1965) Re-investigation of clinoferrosilite from Lake Naivasha, Kenya. Mineral. Mag., 34, 66–70.

Crystal Data: Monoclinic. *Point Group: m.* Crystals typically prismatic or tabular, to 4 mm; may be wedge-shaped reflecting the domatic class in which the species crystallizes.

Physical Properties: *Cleavage:* Perfect on {010}. Hardness = 5.5 D(meas.) = 3.28–3.335 D(calc.) = [3.32] Strongly pyroelectric.

Optical Properties: Transparent to translucent. *Color:* Amethystine, colorless to white. *Luster:* Brilliant, glassy; pearly on {010}. *Optical Class:* Biaxial (–). *Orientation:* $Z = b$; $Y \wedge c = 28°$. $\alpha = 1.662$ $\beta = 1.667$ $\gamma = 1.669$ 2V(meas.) = Large.

Cell Data: *Space Group: Cc.* $a = 5.090–5.131$ $b = 15.829–15.928$ $c = 5.386–5.422$ $\beta = 103.39°–103.43°$ Z = 4

X-ray Powder Pattern: Franklin, New Jersey, USA. (ICDD 17-214). 2.76 (100), 3.23 (70), 2.50 (60), 7.81 (50), 3.97 (50), 2.36 (50), 2.47 (40)

Chemistry:

	(1)	(2)	(3)
SiO$_2$	27.22	26.73	27.87
(Fe, Al)$_2$O$_3$	0.28	0.37	
MnO	0.50	1.11	
ZnO	37.44	37.13	37.76
MgO	0.07		
CaO	26.25	26.25	26.01
H$_2$O$^+$	8.56	8.09	8.36
Total	100.32	99.68	100.00

(1) Franklin, New Jersey, USA; average of two analyses. (2) Do. (3) CaZnSiO$_4$·H$_2$O.

Occurrence: In a metamorphosed stratiform zinc orebody (Franklin, New Jersey, USA).

Association: Hancockite, nasonite, glaucochroite, roeblingite, calcite, willemite, axinite, larsenite, hodgkinsonite, franklinite (Franklin, New Jersey, USA); stringhamite, kinoite, apophyllite (Christmas mine, Arizona, USA).

Distribution: From Franklin, Sussex Co., New Jersey; in the Christmas mine, Gila Co., Arizona, USA.

Name: From the Greek *klino*, for *incline*, and *hedra*, for *face*, for the inclined facial character of the crystal morphology.

Type Material: Harvard University, Cambridge, Massachusetts, USA.

References: (1) Dana, E.S. (1899) Dana's system of mineralogy, (6th edition), app. I, 17–18. (2) Palache, C. (1935) The minerals of Franklin and Sterling Hill, Sussex County, New Jersey. U.S. Geol. Sur. Prof. Paper 180, 106–108. (3) Venetopoulos, C.C. and P.J. Rentzeperis (1976) Redetermination of the crystal structure of clinohedrite, CaZnSiO$_4$·H$_2$O. Zeits. Krist., 144, 377–392. (4) Simonov, M.A., E.L. Belokoneva, Y.K. Yegorov-Tismenko, and N.V. Belov (1977) Crystal structure of clinohedrite CaZn[SiO$_4$]·H$_2$O. Doklady Acad. Nauk SSSR, 237, 334–337 (in Russian). (5) (1978) Chem. Abs., 88, 107998 (abs. ref. 4).

Clinoholmquistite $Li_2[(Mg, Fe^{2+})_3 Al_2]Si_8 O_{22}(OH)_2$

Crystal Data: Monoclinic.   *Point Group: 2/m.*   As elongated prismatic crystals.

Physical Properties: *Cleavage:* [Perfect on {110}, intersecting at ~56° and ~124°; partings on {100}, {001}.]   *Tenacity:* [Brittle.]   Hardness = [5–6]   D(meas.) = 3.00   D(calc.) = [3.07]

Optical Properties: [Transparent to translucent.]   *Color:* [Blue.]   *Luster:* [Vitreous.]
Optical Class: Biaxial (–). *Orientation:* $Y = b$; $Z = c$; $X \wedge a = 15°–16°$. $\alpha = 1.610$ $\beta = 1.627$
$\gamma = 1.633$ 2V(meas.) = 55°–61°

Cell Data: *Space Group: P2/m.*   a = 9.80(2)   b = 17.83(3)   c = 5.30(1) $\beta = 109°06'$
Z = 2

X-ray Powder Pattern: "Siberia," Russia.
7.93 (100), 2.985 (100), 2.70 (100), 4.40 (90), 1.371 (90), 1.614 (80), 1.571 (70)

Chemistry:

	(1)		(1)
SiO_2	57.68	CaO	[1.80]
TiO_2	0.00	Li_2O	3.37
Al_2O_3	13.52	Na_2O	1.74
Fe_2O_3	0.44	K_2O	0.28
FeO	5.87	F	1.70
MnO	0.45	H_2O^+	1.67
MgO	9.37	$-O = F_2$	0.71
		Total	[97.18]

(1) "Siberia," Russia; recalculated to remove calcite contamination; corresponds to
$(Li_{1.79}Ca_{0.21})_{\Sigma=2.00}(Mg_{1.93}Fe^{2+}_{0.68}Al_{0.21}Li_{0.08}Fe^{3+}_{0.05}Mn_{0.05})_{\Sigma=3.00}Al_{2.00}Si_8O_{22}(OH)_2$.

Polymorphism & Series: Dimorphous with holmquistite; forms a series with
magnesio-clinoholmquistite and ferro-clinoholmquistite.

Mineral Group: Amphibole (Fe–Mn–Mg) group: $0.1 \leq Mg/(Mg + Fe^{2+}) \leq 0.89$; $(Ca + Na)_B$
< 1.34; Li ≥ 1.0.

Occurrence: Partly replaced by holmquistite.

Association: Holmquistite, calcite.

Distribution: From an undefined locality in Siberia, Russia.

Name: In allusion to its similarity to *holmquistite* and its monoclinic structure.

Type Material: n.d.

References: (1) Ginzburg, I.V. (1965) Holmquistite and its structural variety clinoholmquistite.
Trudy Mineral. Muzeya Akad. Nauk SSSR, 16, 73–89 (in Russian). (2) (1967) Amer. Mineral.,
52, 1585–1586 (abs. ref. 1). (3) Litvin, A.L., I.V. Ginzburg, L.N. Egorova, and A.A. Petrunina
(1975) On the crystal structure of clinoholmquistite. Konst. Svoistva Miner., 9, 3–6 (in Russian).
(4) (1976) Chem. Abs., 85, 65701 (abs. ref. 3).

Crystal Data: Monoclinic. *Point Group:* $2/m$. Crystals complex, highly modified, to 1 cm; massive. *Twinning:* Common on {100}, simple, lamellar.

Physical Properties: *Cleavage:* {100}, poor. *Fracture:* Uneven to subconchoidal. *Tenacity:* Brittle. Hardness = 6   D(meas.) = 3.17–3.35   D(calc.) = 3.279

Optical Properties: Transparent to translucent. *Color:* White, yellow to brown with increasing Ti; in thin section, colorless, pale yellow to golden yellow. *Luster:* Vitreous. *Optical Class:* Biaxial (+). *Pleochroism:* X = golden yellow, yellow-brown, deep reddish yellow; Y = pale yellow, orange-yellow, light yellow; Z = pale yellow, orange-yellow, colorless. *Orientation:* $Z = b$; $X \wedge c = 9°{-}15°$. *Dispersion:* $r > v$, strong. $\alpha = 1.628{-}1.638$ $\beta = 1.641{-}1.654$ $\gamma = 1.662{-}1.674$ 2V(meas.) = 73°–76°

Cell Data: *Space Group:* $P2_1/c$. $a = 13.68$ $b = 4.75$ $c = 10.27$ $\beta = 100°50'$ Z = 2

X-ray Powder Pattern: Hämeenkylä, Finland. (ICDD 14-692).
1.738 (100), 5.02 (70), 3.70 (70), 2.76 (70), 2.54 (70), 2.51 (70), 2.26 (70)

Chemistry:

	(1)	(2)		(1)	(2)
SiO$_2$	36.53	35.90	MgO	54.16	44.16
TiO$_2$	0.26	5.59	CaO		0.01
Al$_2$O$_3$	0.22		F	2.74	0.00
Fe$_2$O$_3$	0.56		H$_2$O$^+$	1.52	2.64
FeO	5.04	11.21	H$_2$O$^-$	0.04	
MnO	0.34	0.50	$-$O = F$_2$	1.15	
			Total	100.26	100.01

(1) Hämeenkylä, Finland; corresponds to $(\text{Mg}_{8.42}\text{Fe}^{2+}_{0.50}\text{Mn}^{2+}_{0.06}\text{Ti}_{0.02})_{\Sigma=9.00}(\text{SiO}_4)_4$ $[\text{F}_{1.04}(\text{OH})_{0.93}\text{O}_{0.03}]_{\Sigma=2.00}$. (2) Franscia, Italy; corresponds to $(\text{Mg}_{7.33}\text{Fe}^{2+}_{1.04}\text{Ti}_{0.47}\text{Mn}^{2+}_{0.05})_{\Sigma=8.89}$ $(\text{SiO}_4)_4[(\text{OH})_{1.04}\text{O}_{0.96}]_{\Sigma=2.00}$.

Mineral Group: Humite group.

Occurrence: In contact metamorphic zones in dolostones and metasomatized limestones adjacent to felsic plutonic rocks; in serpentine and talc schists; in a carbonatite.

Association: Grossular, wollastonite, forsterite, monticellite, cuspidine, fluoborite, ludwigite, dolomite, calcite, talc, biotite, spinel, vesuvianite, sanidine, meionite, nepheline.

Distribution: At Monte Somma and Vesuvius, Campania, and Franscia, Val Malenco, Lombardy, Italy. From Hämeenkylä and Ojamo, Lohja, Finland. At Broadford, Isle of Skye, Scotland. From Llanos de Juanar, Málaga Province, Spain. At the Kangerdlugssuaq Fjord, Greenland. Found near Lake Baikal, eastern Siberia, Russia. From Kuche-Lal, Pamir Mountains, Tadzhikistan. In the USA, at the Tilly Foster iron mine, Brewster, Putnam Co., New York; at Buell Park, near Fort Defiance, Apache Co., Arizona; Crestmore, Riverside Co., and the Twin Lakes region, Fresno Co., California. At Cargill Lake, Ontario, Canada. In the Jacupiranga mine, São Paulo, Brazil.

Name: For its monoclinic crystal system and relation to *humite*.

Type Material: National Museum of Natural History, Washington, D.C., USA, 94997.

References: (1) Dana, E.S. (1892) Dana's system of mineralogy, (6th edition), 538–541. (2) Deer, W.A., R.A. Howie, and J. Zussman (1982) Rock-forming minerals, (2nd edition), v. 1A, orthosilicates, 379–417. (3) Robinson, K., G.V. Gibbs, and P.H. Ribbe (1973) The crystal structures of the humite minerals. IV. Clinohumite and titanclinohumite. Amer. Mineral., 58, 43–49. (4) Kocman, V. and J. Rucklidge (1973) The crystal structure of a titaniferous clinohumite. Can. Mineral., 12, 39–45.

Clinojimthompsonite

$$(Mg, Fe^{2+})_5Si_6O_{16}(OH)_2$$

Crystal Data: Monoclinic. *Point Group:* $2/m$. As very thin lamellae.

Physical Properties: *Cleavage:* [{110} predicted from the structure.] Hardness = n.d. D(meas.) = n.d. D(calc.) = [3.01]

Optical Properties: Semitransparent. *Color:* Colorless to very light pinkish brown; in thin section, colorless.
Optical Class: [Biaxial.] *Orientation:* Extinction angle direction N $\wedge$ c $\simeq 10°$. α = n.d. β = n.d. γ = n.d. 2V(meas.) = n.d.

Cell Data: *Space Group:* $C2/c$. a = 9.874(4)   b = 27.24(3)   c = 5.316(3) $\beta = 109.47(3)°$ Z = 4

X-ray Powder Pattern: Calculated.
8.809 (100), 13.6 (73), 3.083 (60), 2.639 (56), 2.506 (42), 4.704 (33), 4.036 (31)

Chemistry:

	(1)
SiO_2	58.55
Al_2O_3	0.37
FeO	12.13
MnO	0.73
MgO	24.93
CaO	0.50
Na_2O	0.10
H_2O	[2.93]
Total	[100.24]

(1) Chester, Vermont, USA; by electron microprobe, H_2O assuming (OH) sites filled by $(OH)^{1-}$.

Polymorphism & Series: Dimorphous with jimthompsonite.

Occurrence: In the black wallrock between chlorite and actinolite zones of a metamorphosed ultramafic body.

Association: Chesterite, clinojimthompsonite, anthophyllite, cummingtonite, talc.

Distribution: In the Carleton talc quarry, near Chester, Windsor Co., Vermont, USA.

Name: For its monoclinic crystallography and relation to *jimthompsonite*.

Type Material: Royal Ontario Museum, Toronto, Canada, M36083; Harvard University, Cambridge, Massachusetts; National Museum of Natural History, Washington, D.C., USA, 145689.

References: (1) Veblen, D.R. and C.W. Burnham (1978) New biopyriboles from Chester, Vermont: I. Descriptive mineralogy. Amer. Mineral., 63, 1000–1009. (2) Veblen, D.R. (1978) New biopyriboles from Chester, Vermont: II. The crystal chemistry of jimthompsonite, clinojimthompsonite, and chesterite, and the amphibole-mica reaction. Amer. Mineral., 63, 1053–1073.

Crystal Data: Monoclinic. *Point Group:* $2/m$. Habit not described.

Physical Properties: *Fracture:* Conchoidal. Hardness = 4   D(meas.) = 2.85–2.88 D(calc.) = [2.84]

Optical Properties: Semitransparent. *Color:* Pale lilac. *Luster:* Vitreous. *Optical Class:* Biaxial (+). $\alpha = 1.556$–1.557 $\beta = 1.559$–1.561 $\gamma = 1.563$–1.567 2V(meas.) = 75°–80°

Cell Data: *Space Group:* $P2/c$.   a = 7.303(2)   b = 12.201(5)   c = 14.715(4) $\beta = 91°56'$ Z = 8

X-ray Powder Pattern: Mt. Koashva, Russia. 2.724 (100), 2.639 (57), 2.550 (50), 2.031 (50), 1.823 (47), 1.838 (35), 3.97 (12)

Chemistry:

	(1)	(2)	(3)
SiO_2	20.1	20.1	21.45
Ce_2O_3	0.18	0.44	
MnO	0.60	0.54	
CaO	15.65	16.45	20.02
SrO	3.08	1.91	
Na_2O	32.55	32.7	33.19
K_2O	0.09	0.05	
P_2O_5	25.35	25.5	25.34
LOI	1.08	2.55	
Total	98.68	100.24	100.00

(1) Mt. Koashva, Russia; by electron microprobe, contains traces of Ti, Zr, Al, Fe, Mg, Ba, loss on ignition may be H_2O; corresponds to $(Na_{3.04}K_{0.01})_{\Sigma=3.05}(Ca_{0.81}Sr_{0.09}Mn_{0.02})_{\Sigma=0.92}$ $Si_{0.97}P_{1.03}O_{6.95}$. (2) Mt. Yukspor, Russia; by electron microprobe, impurities as in (1), corresponds to $Na_{3.04}(Ca_{0.85}Sr_{0.05}Mn_{0.02})_{\Sigma=0.92}Si_{0.96}P_{1.04}O_{6.97}$. (3) $Na_3CaPSiO_7$.

Occurrence: Apparently as an alteration product of eudialyte in alkalic pegmatites in a differentiated alkalic massif.

Association: Zirsinalite, eudialyte.

Distribution: In Russia, on Mts. Yukspor and Koashva, Khibiny massif, Kola Peninsula.

Name: For its monoclinic crystallography and chemical similarity to *phosinaite*.

Type Material: A.E. Fersman Mineralogical Museum, Academy of Sciences, Moscow, Russia.

References: (1) Khomyakov, A.P., D.Y. Pushcharovskii, and J.G. Ronsbo (1981) Clinophosinaite, $Na_3CaPSiO_7$, a new mineral. Zap. Vses. Mineral. Obshch., 110, 351–355 (in Russian). (2) (1982) Amer. Mineral., 67, 414 (abs. ref. 1). (3) Krutik, V.M, D.Y. Pushcharovskii, A.P. Khomyakov, E.A. Pobedimskaya, and N.V. Belov (1980) Anion radical of mixed type (four $[Si_4O_{12}]$ rings and P orthotetrahedra) in the structure of monoclinic phosinaite. Kristallografiya (Sov. Phys. Crystal.), 25, 240–247 (in Russian).

Clinoptilolite

$$(Na, K, Ca)_{2-3}Al_3(Al, Si)_2Si_{13}O_{36} \cdot 12H_2O$$

Crystal Data: Monoclinic. *Point Group:* $2/m$. Platy crystals, to 1 cm, showing {010}, {001}, {101}, {20$\bar{1}$}; commonly fine-grained, massive.

Physical Properties: *Cleavage:* {010}, perfect. Hardness = 3.5–4   D(meas.) = 2.16(2) D(calc.) = 2.146

Optical Properties: Transparent to translucent. *Color:* Colorless to white; colorless in thin section. *Luster:* Vitreous.
Optical Class: Biaxial (–) or (+). *Orientation:* $Y = b$; $Z \wedge a = 30°–45°$. *Dispersion:* $r < v$, strong. $\alpha = 1.476–1.488$ $\beta = 1.479$ $\gamma = 1.479–1.489$ 2V(meas.) = $32°–48°$

Cell Data: *Space Group:* $C2/m$.   a = 17.633(8)   b = 17.941(8)   c = 7.400(4) $\beta = 116.39°$ Z = 2

X-ray Powder Pattern: Agoura, California, USA.
8.92 (100), 2.974 (80), 3.897 (57), 3.964 (55), 2.728 (33), 3.419 (16), 3.119 (15)

Chemistry:

	(1)	(2)
SiO_2	69.93	68.44
Al_2O_3	11.89	11.86
Fe_2O_3	0.02	
MgO	0.47	0.12
CaO	1.07	3.26
Na_2O	2.96	1.51
K_2O	3.47	2.14
H_2O	[10.19]	[12.67]
Total	[100.00]	[100.00]

(1) Agoura, California, USA; H_2O by difference, corresponds to $(Na_{1.22}K_{0.94}Ca_{0.24}Mg_{0.15})_{\Sigma=2.55}$ $Al_{2.99}Si_{15.01}O_{36} \cdot 11.6H_2O$. (2) Richardson Ranch, Oregon, USA; by electron microprobe, after removal of 0.46% FeO as impurity, H_2O by difference; corresponds to $(Ca_{0.78}Na_{0.65}$ $K_{0.60}Mg_{0.04})_{\Sigma=2.07}Al_{3.10}Si_{14.91}O_{36} \cdot 11.5H_2O$.

Mineral Group: Zeolite group.

Occurrence: Formed by devitrification of volcanic glass, and in cavities in rhyolites and andesites.

Association: Zeolites, montmorillonite, hectorite, thenardite, halite, gaylussite, celadonite, quartz, "opal," calcite.

Distribution: Abundant in the western USA, as in Wyoming, at Mt. Hoodoo, Park Co.; in California, from near Agoura, Los Angeles Co., at Owens Lake, Inyo Co., and 10 km west of Hector, San Bernardino Co.; in Arizona, from northwest of Bowie, Cochise Co., and near Wikieup, Mohave Co.; in the Cuchillo deposit, near Winston, Sierra Co., New Mexico. In Oregon, large crystals from Agate Beach, Lincoln Co., Succor Creek, Malheur Co., and at Richardson Ranch, near Bend, Jefferson Co. At Kamloops Lake, near Kamloops, British Columbia, Canada. From near Kapfenberg, Styria, Austria. On the Vogelsberg, Hesse, Germany. From Val di Fassa and Alpe di Siusi, Trentino-Alto Adige, Italy.

Name: For its inclined optics and relation to *ptilolite* [mordenite].

Type Material: The Natural History Museum, London, England, 1939,314.

References: (1) Schaller, W.T. (1932) The mordenite-ptilolite group; clinoptilolite, a new species. Amer. Mineral., 17, 128–134. (2) Deer, W.A., R.A. Howie, and J. Zussman (1963) Rock-forming minerals, v. 4, framework silicates, 377–385. (3) Wise, W.S., W.J. Nokleberg, and M. Kokinos (1969) Clinoptilolite and ferrierite from Agoura, California. Amer. Mineral., 54, 887–895. (4) Alietti, A. (1972) Polymorphism and crystal chemistry of heulandites and clinoptilolites. Amer. Mineral., 57, 1448–1462. (5) Smyth, J.R., A.T. Spaid, and D.L. Bish (1990) Crystal structures of a natural and Cs-exchanged clinoptilolite. Amer. Mineral., 75, 522–528.

Crystal Data: Monoclinic. *Point Group:* m or $2/m$. Crystals tabular $\perp$ [001], or acicular ∥ [010], up to 5 mm, and as aggregates. *Twinning:* On {001} or along [100], well-developed polysynthetic on a microscopic scale.

Physical Properties: *Cleavage:* Perfect on {001}, poor on {100}. Hardness = 4.5 D(meas.) = 2.58   D(calc.) = 2.69

Optical Properties: Semitransparent. *Color:* Colorless to white; colorless in thin section. *Luster:* Vitreous.
Optical Class: Biaxial. $\alpha = 1.575$ $\beta = 1.580$ $\gamma = 1.585$ 2V(meas.) = n.d.   2V(calc.) = 89.8°

Cell Data: *Space Group:* Cc or $C2/c$. $a = 11.331(9)$ $b = 7.353(7)$ $c = 22.67(2)$ $\beta = 96.59(7)°$ Z = [4]

X-ray Powder Pattern: Fuka, Japan.
11.25 (100), 3.034 (60), 2.794 (60), 3.304 (51), 3.068 (45), 2.811 (41), 3.012 (37)

Chemistry:

	(1)
SiO_2	46.55
TiO_2	0.01
B_2O_3	0.23
Al_2O_3	0.36
Fe_2O_3	0.01
MnO	0.06
MgO	0.11
CaO	39.04
Na_2O	0.02
K_2O	0.10
F	0.18
H_2O	13.75
$-O = F_2$	0.08
Total	100.34

(1) Fuka, Japan; by electron microprobe, wet chemical analysis for B, F, and H_2O; corresponds to $(Ca_{5.29}Mg_{0.02}K_{0.02})_{\Sigma=5.33}(Si_{5.90}Al_{0.05}B_{0.05})_{\Sigma=6.00}[O_{16.54}(OH)_{1.39}F_{0.07}]_{\Sigma=18.00} \cdot 5.1H_2O$.

Polymorphism & Series: Dimorphous with tobermorite.

Occurrence: In gehlenite-spurrite-bearing skarns.

Association: Calcite, tobermorite, plombièrite, apophyllite.

Distribution: At Fuka, near Bicchu, Okayama Prefecture, Japan.

Name: For its monoclinic crystal system and chemical identity to *tobermorite*.

Type Material: National Science Museum, Tokyo, Japan.

References: (1) Henmi, C. and I. Kusachi (1989) Monoclinic tobermorite from Fuka, Bitchu-cho, Okayama Prefecture, Japan. J. Japan. Assoc. Mineral. Petrol. Econ. Geol., 84, 374–379 (in Japanese with English abs.). (2) (1992) Amer. Mineral., 77, 451 (abs. ref. 1). (3) Henmi, C. and I. Kusachi (1992) Clinotobermorite, $Ca_5Si_6(O, OH)_{18} \cdot 5H_2O$, a new mineral from Fuka, Okayama Prefecture, Japan. Mineral. Mag., 56, 353–358. (4) (1993) Amer. Mineral., 78, 672 (abs. ref. 3).

Clinozoisite

$$Ca_2Al_3(SiO_4)(Si_2O_7)O(OH)$$

Crystal Data: Monoclinic. *Point Group:* $2/m$. Crystals prismatic, typically elongated and striated $\parallel$ [010]; commonly coarse to fine granular; also fibrous. *Twinning:* Lamellar on {100}, uncommon.

Physical Properties: *Cleavage:* Perfect on {001}. *Fracture:* Uneven. *Tenacity:* Brittle. Hardness = 6.5 D(meas.) = 3.21–3.38 D(calc.) = 3.364

Optical Properties: Transparent to translucent. *Color:* Colorless, pale yellow, pink, red, gray, green; colorless in thin section.
Optical Class: Biaxial (+). *Orientation:* $Y = b$; $X \wedge a = 0°$–$7°$. *Dispersion:* $r < v$.
$\alpha = 1.670$–1.718 $\beta = 1.670$–1.725 $\gamma = 1.690$–1.734 2V(meas.) = $14°$–$90°$

Cell Data: *Space Group:* $P2_1/m$. $a = 8.879(5)$ $b = 5.583(5)$ $c = 10.155(6)$
$\beta = 115.50(5)°$ Z = 2

X-ray Powder Pattern: Vernirovice, Czech Republic. (ICDD 21-128).
2.89 (100), 2.79 (90), 2.59 (70), 2.68 (60), 2.67 (60), 2.40 (60), 2.29 (60)

Chemistry:

	(1)	(2)		(1)	(2)
SiO_2	40.36	38.01	MgO	0.74	0.01
TiO_2	0.09	trace	CaO	22.62	23.86
Al_2O_3	30.67	31.71	H_2O^+	3.50	1.83
Fe_2O_3	0.61	4.68	H_2O^-	0.00	0.01
FeO	1.77	0.23	Total	100.36	100.34
MnO		trace			

(1) Kälviä, Finland; corresponds to $(Ca_{1.86}Fe^{2+}_{0.11}Mg_{0.08})_{\Sigma=2.05}(Al_{2.78}Fe^{3+}_{0.04})_{\Sigma=2.82}Si_{3.10}O_{12.5}$.
(2) Camaderry Mountain, Ireland; corresponds to $(Ca_{1.96}Fe^{2+}_{0.02})_{\Sigma=1.98}(Al_{2.79}Fe^{3+}_{0.27})_{\Sigma=3.06}$
$(Si_{2.92}Al_{0.08})_{\Sigma=3.00}O_{12.5}(OH)_{0.94}$.

Polymorphism & Series: Dimorphous with zoisite.

Mineral Group: Epidote group.

Occurrence: Typically in low- to medium-grade regionally metamorphosed igneous and sedimentary rocks, as well as in contact metamorphosed calcium-rich sediments; an alteration product of plagioclase feldspars (saussuritization).

Association: Amphiboles, plagioclase, quartz.

Distribution: A widespread mineral. Typical occurrences for good crystals include: at the Goslarwand, near Prägraten, Tirol, Austria. From Vernirovice, Czech Republic. At Kälviä, Finland. From Arendal, Norway. At Camaderry Mountain, Co. Wicklow, Ireland. From Amborompotsy, Madagascar. At Sittampundi, Madras, India. In the USA, from the Belvidere Mountain quarries, Lowell, Orleans Co., Vermont; in northern Spade Spring Canyon, Los Angeles Co., California; in the Nightingale district, Pershing Co., Nevada. In Mexico, from the Juarez district, Baja California, and at Alamos, Sonora. In Canada, from Timmins, Ontario.

Name: For its monoclinic crystallography and relation to *zoisite*.

Type Material: n.d.

References: (1) Dana, E.S. (1899) Dana's system of mineralogy, (6th edition), app. I, 18. (2) Deer, W.A., R.A. Howie, and J. Zussman (1986) Rock-forming minerals, (2nd edition), v. 1B, disilicates and ring silicates, 44–134. (3) Dollase, W. A. (1968) Refinement and comparison of the structures of zoisite and clinozoisite. Amer. Mineral., 53, 1882–1898.

Crystal Data: Monoclinic. *Point Group:* $2/m$.   In tabular pseudohexagonal crystals, complexly twinned; foliated or lamellar radiated; massive. *Twinning:* On {001}, twin axis [310] or [$3\bar{1}0$]; spiral polysynthetic twinning.

Physical Properties: *Cleavage:* {001}, perfect. *Tenacity:* Brittle. Hardness = 3.5 on {001}; 6 ⊥ {001}.   D(meas.) = 3.0–3.1   D(calc.) = 3.096

Optical Properties: Transparent to translucent. *Color:* Colorless, yellow, orange, red-brown, green, brown. *Streak:* White, slightly yellow-gray. *Luster:* Vitreous, pearly, submetallic. *Optical Class:* Biaxial (–). *Pleochroism:* X = colorless, pale orange, red-brown; $Y = Z$ = pale brownish yellow, pale green. *Orientation:* $Y \simeq a$ or $Y = b$; $Z = b$ or $\simeq a$; $X \wedge c = 5°–10°$. *Dispersion:* $r < v$, weak.   *Absorption:* $Y \simeq Z > X$. $\alpha = 1.643–1.648$ $\beta = 1.655–1.662$ $\gamma = 1.655–1.663$ 2V(meas.) = 2°–40°

Cell Data: *Space Group:* $C2/m$. a = 5.204   b = 9.026   c = 9.812 $\beta = 100°20'$ Z = 2

X-ray Powder Pattern: Zlatoust, Russia.
2.56 (100), 3.21 (70), 2.11 (70), 1.505 (60), 9.68 (50), 2.45 (50), 1.485 (50)

Chemistry:

	(1)	(2)		(1)	(2)
SiO_2	18.78	16.74	MgO	20.56	20.03
TiO_2	0.54		CaO	12.90	13.09
Al_2O_3	40.00	42.70	F	2.01	
Fe_2O_3		2.85	H_2O^+	[4.20]	4.49
FeO	1.86	0.41	$-O = F_2$	[0.85]	
MnO	0.00		Total	[100.00]	100.31

(1) Amity, New York, USA; by electron microprobe, H_2O by difference; corresponds to $Ca_{0.97}(Mg_{2.15}Al_{0.70}Fe^{2+}_{0.11}Ti_{0.03})_{\Sigma=2.99}(Al_{2.68}Si_{1.32})_{\Sigma=4.00}O_{9.45}[(OH)_{1.55}F_{0.45}]_{\Sigma=2.00}$.
(2) Crestmore, California, USA; corresponds to $Ca_{0.98}(Mg_{2.09}Al_{0.70}Fe^{3+}_{0.15}Fe^{2+}_{0.02})_{\Sigma=2.96}$ $(Al_{2.83}Si_{1.17})_{\Sigma=4.00}O_{9.92}(OH)_{2.10}$.

Polymorphism & Series: 1M polytype; $2M_1$ and 3A polytypes rare.

Mineral Group: Mica group.

Occurrence: In chlorite schists; in metasomatically altered limestones; in siliceous skarns near contact metamorphic zones.

Association: Talc, spinel, grossular, vesuvianite, clinopyroxene, monticellite, chondrodite, phlogopite, chlorite, quartz, calcite, dolomite.

Distribution: In the USA, around Amity, Edenville, and Warwick, Orange Co., New York; from Crestmore, Riverside Co., California; near Ludwig, Lyon Co., Nevada; at Sulzer, Prince of Wales Island, Alaska. On Mt. Monzoni and at Mts. Castone and Adamello, Trentino-Alto Adige, Italy. In the Pargas district, Finland. From the Akhmatovsk mine, near Zlatoust, Ural Mountains, Russia. In the Chichibu mine, Saitama Prefecture, Japan. On the Ertsberg, Irian Jaya.

Name: For De Witt Clinton (1769–1828), American statesman.

References: (1) Dana, E.S. (1892) Dana's system of mineralogy, (6th edition), 638–640 [seybertite, xanthophyllite]. (2) Deer, W.A., R.A. Howie, and J. Zussman (1963) Rock-forming minerals, v. 3, sheet silicates, 99–102. (3) Forman, S.A., H. Kodama, and S. Abbey (1967) A re-examination of xanthophyllite [clintonite] from the type locality. Can. Mineral., 9, 25–30. (4) MacKinney, J.A., C.I. Mora, and S.W. Bailey (1988) Structure and crystal chemistry of clintonite. Amer. Mineral., 73, 365–375.

Crystal Data: Monoclinic, pseudohexagonal. *Point Group: 2/m.* As irregular to nearly rectangular grains, up to 3 mm.

Physical Properties: *Cleavage:* {010} and {012}, suspected. *Fracture:* Subconchoidal. Hardness = 7.5–8 D(meas.) = n.d. D(calc.) = 2.92

Optical Properties: Transparent. *Color:* Colorless. *Luster:* Vitreous.
Optical Class: Biaxial (+) (synthetic). *Orientation:* $X = b$; $Z \wedge c = 4°–6°$. *Dispersion:* $r < v$.
$\alpha = 1.5940(5)$ $\beta = 1.5955(5)$ $\gamma = 1.5990(5)$ 2V(meas.) = 64° 2V(calc.) = 66°34′

Cell Data: *Space Group: C2/c.* a = 7.143(2) b = 12.383(3) c = 7.143(2)
$\beta = 120.00(3)°$ Z = 16

X-ray Powder Pattern: Synthetic. (ICDD 14-654).
3.09 (100), 3.43 (30), 1.71 (12), 1.345 (12), 2.69 (10), 1.70 (10), 1.545 (10)

Chemistry:

	(1)
SiO$_2$	99.56
TiO$_2$	0.00
Al$_2$O$_3$	0.05
FeO	0.03
MnO	0.00
MgO	0.00
Na$_2$O	0.04
K$_2$O	0.00
Total	99.68

(1) Roberts Victor mine, South Africa; by electron microprobe, average of four analyses.

Polymorphism & Series: Quartz, tridymite, cristobalite, and stishovite are polymorphs.

Occurrence: Formed at pressures in excess of 20 kbar, as fine-grained intergrowths with quartz and silica glass in impact metamorphosed rocks and tektites. A primary phase in grospydite (kyanite-garnet-clinopyroxene) xenoliths in kimberlite, and intergrown with quartz as inclusions in garnet and clinopyroxene within eclogites of high-grade metamorphic terranes.

Association: Quartz, stishovite, silica glass, kyanite, omphacite, garnet, sanidine, muscovite, talc, rutile, dolomite.

Distribution: From Meteor Crater, Coconino Co., Arizona, USA. At the Nördlinger Ries crater, Bavaria, Germany. In the Kara structure, Pay-Khoy, Russia. In Italy, in the Dora-Maira massif, Parigi, near Martiniana Po, Piedmont. From Grytting, Selje district, Norway. In the Wabar craters, Rub' al Khali, Saudi Arabia. At Dabie Mountain, west of Qianshan, Anhwei Province, China. In the Roberts Victor diamond mine, near Kimberley, Cape Province, and near Boshoff, Orange Free State, South Africa.

Name: For Loring Coes, Jr. (1915–), American chemist, who synthesized the material prior to the discovery of a natural occurrence.

Type Material: n.d.

References: (1) Frondel, C. (1962) Dana's system of mineralogy, (7th edition), v. III, silica minerals, 310–316. (2) Sclar, C.B., L.C. Carrison, and C.M. Schwartz (1962) Optical crystallography of coesite. Amer. Mineral., 47, 1292–1302. (3) Gibbs, G.V., C.T. Prewitt, and K.J. Baldwin (1977) A study of the structural chemistry of coesite. Zeits. Krist., 145, 108–123. (4) Smyth, J.R. and C.J. Hatton (1977) A coesite-sanidine grospydite from the Roberts Victor kimberlite. Earth and Planetary Science Letters, 34, 284–290.

Crystal Data: Tetragonal; may be metamict.   *Point Group:* $4/m \ 2/m \ 2/m$.   As crystals, rarely > 20 μm, and in colloform to botryoidal incrustations, radially fibrous. Commonly pulverulent, in aggregates of extremely fine crystallites, or massive.

Physical Properties: *Fracture:* Irregular to subconchoidal, earthy.   *Tenacity:* Brittle or friable.   Hardness = 5–6   D(meas.) = ≤ 5.1 D(calc.) = n.d.   Radioactive.

Optical Properties: Opaque, transparent on very thin edges.   *Color:* Black from included organic matter; pale brown to dark brown in thin section.   *Luster:* Dull to adamantine. *Optical Class:* Isotropic due to fine grain size; uniaxial (+) or (–).   *Pleochroism:* Moderate; pale yellow-brown $\parallel$ length, medium brown $\perp$ length. $n = 1.73$–1.75

Cell Data: *Space Group:* $I4_1/amd$ (synthetic USiO$_4$).   a = 6.979   c = 6.252   Z = 4

X-ray Powder Pattern: Synthetic USiO$_4$; natural material gives broad peaks or may be metamict.
3.48 (100), 4.64 (95), 2.636 (95), 1.803 (70), 2.789 (45), 1.849 (35), 1.738 (35)

Chemistry:

	(1)	(2)	(3)
SiO$_2$	5.20	18.20	16.41
UO$_2$	68.29	81.80	73.75
Al$_2$O$_3$	3.87		
Fe$_2$O$_3$	1.24		
As$_2$O$_5$	4.35		
V$_2$O$_5$	2.85		
Pb	0.12		
H$_2$O	[8.49]		9.84
Total	[94.41]	100.00	100.00

(1) La Sal No. 2 mine, Colorado, USA; H$_2$O from loss on ignition, contains admixed clay, vanadates, organic matter, etc. (2) USiO$_4$. (3) USiO$_4 \cdot$ 2H$_2$O.

Occurrence: In Colorado-Plateau-type black unoxidized U-V deposits, replacing organic material in sandstone; in other sedimentary and hydrothermal vein uranium deposits.

Association: Uraninite, thorite, pyrite, marcasite, roscoelite, clay minerals, amorphous organic matter.

Distribution: Numerous minor occurrences; those mentioned are for well-crystallized or studied material. In the USA, in the La Sal No. 2 mine, Gateway, Mesa Co., and in the Peanut mine, Montrose Co., Colorado; around the Jackpile and Crownpoint mines, Valencia Co., New Mexico; in the Mi Vida, Homestake, and other mines, San Juan Co., Utah. From Wölsendorf, Bavaria; at Niederramstadt, near Schneeberg, Johanngeorgenstadt, and Niederpfannenstiel, Saxony, Germany. At Hüttenberg, Carinthia, Austria. From Jáchymov (Joachimsthal) and Příbram, Czech Republic. In the Geevor, South Terras, and other mines, Cornwall, England.

Name: For Reuben Clare Coffin (1886–1972), Tulsa, Oklahoma, USA, pioneer geologist in the study of uranium deposits of the Colorado Plateau.

Type Material: National Museum of Natural History, Washington, D.C., USA, 112646; The Natural History Museum, London, England, 1956,201.

References: (1) Stieff, L.R., T.W. Stern, and A.M. Sherwood (1956) Coffinite, a uranous silicate with hydroxyl substitution: a new mineral. Amer. Mineral., 41, 675–688. (2) Fuchs, L.H. and E. Gebert (1958) X-ray studies of synthetic coffinite, thorite and uranothorites. Amer. Mineral., 43, 243–248. (3) Dubinchuk, V.T., I.S. Naumova, I.Y. Kravtsova, and G.A. Sidorenko (1981) Determination of the crystal structure of naturally occurring coffinite. Mineral. Zhurnal, 3(4), 81–85 (in Russian with English abs.). (4) (1982) Mineral. Abs., 33, 222 (abs. ref. 3). (5) Hansley, P.A. and J.L. Fitzpatrick (1989) Compositional and crystallographic data on REE-bearing coffinite from the Grants uranium region, northwestern New Mexico. Amer. Mineral., 74, 263–270.

Crystal Data: Hexagonal. *Point Group:* $\bar{3}\ 2/m$. Poorly developed stout hexagonal prisms, to a few tenths mm, lacking terminal faces; most material is strongly altered.

Physical Properties: Hardness = n.d. D(meas.) = 2.844 D(calc.) = 2.79

Optical Properties: Semitransparent. *Color:* Colorless.
Optical Class: Uniaxial (–). $\omega = 1.598(2)$ $\epsilon = 1.598(2)$

Cell Data: *Space Group:* $R\bar{3}m$. $a = 10.429(2)$ $c = 13.149(3)$ $Z = 6$

X-ray Powder Pattern: Mt. Shaheru, Zaire.
2.657 (100), 2.607 (80), 3.304 (70), 3.722 (50), 3.354 (40), 1.861 (40), 4.380 (30)

Chemistry:

	(1)	(2)	(3)		(1)	(2)	(3)
SiO_2	49.78	50.20	50.86	Na_2O	16.14	20.53	17.49
TiO_2	0.32	0.23		K_2O	1.18	0.27	
ZrO_2	0.44			F	1.87		
Al_2O_3	2.45	0.05		Cl	0.30	0.02	
Fe_2O_3	1.86			H_2O^+	1.39		
FeO	0.54	0.35		H_2O^-	0.42		
MnO	0.58	0.17		CO_2	0.00		
MgO	0.41	0.17		P_2O_5	0.02	0.00	
CaO	22.68	27.28	31.65	SO_3	0.19		
SrO	0.00			S		0.02	
BaO	0.09			$-O = (F, Cl)_2$	0.86		
				Total	99.80	99.29	100.00

(1) Mt. Shaheru, Zaire. (2) Oldoinyo Lengai volcano, Tanzania; by electron microprobe.
(3) $Na_2Ca_2Si_3O_9$.

Occurrence: In nephelinite (Mt. Shaheru, Zaire); in nephelinite and ash ejecta (Oldoinyo Lengai volcano, Tanzania).

Association: Götzenite (Mt. Shaheru, Zaire); wollastonite, clinopyroxene, nepheline, melilite, titanian garnet, titanian magnetite (Oldoinyo Lengai, Tanzania).

Distribution: On Mt. Shaheru, the extinct southern cone of Mt. Nyiragongo, Kivu Province, Zaire. From the Oldoinyo Lengai volcano, Tanzania.

Name: For Arthur Delmar Combe, Geological Survey of Uganda.

Type Material: National Museum of Natural History, Washington, D.C., USA, 142981; The Natural History Museum, London, England, 1957, 705.

References: (1) Sahama, T.G. and K. Hytönen (1957) Götzenite and combeite, two new silicates from the Belgian Congo. Mineral. Mag., 31, 503–510. (2) (1958) Amer. Mineral., 43, 791 (abs. ref. 1). (3) Fischer, R.X. and E. Tillmanns (1987) Revised data for combeite, $Na_2Ca_2Si_3O_9$. Acta Cryst., C43, 1852–1854. (4) Dawson, J.B., J.V. Smith, and I.M. Steele (1989) Combeite $(Na_{2.33}Ca_{1.74}\text{others}_{0.12})Si_3O_9$ from Oldoinyo Lengai, Tanzania. J. Geol., 97, 365–372.

Crystal Data: Triclinic. *Point Group:* $\bar{1}$. Pseudohexagonal platy crystals; curved, radial scales, spherulites, barrel-shaped, fibrous. *Twinning:* Around [310], composition plane {001}.

Physical Properties: *Cleavage:* {001}, perfect. *Tenacity:* Flexible but inelastic.
Hardness = 2.5–3.5   D(meas.) = 2.58–2.69   D(calc.) = 2.968

Optical Properties: Transparent to translucent. *Color:* White, yellowish green, pink, brown; in thin section, colorless to pale green or pink. *Luster:* Pearly or silky.
Optical Class: Biaxial (+); rarely with biaxial sectors around a uniaxial core. *Pleochroism:* X = Y = pale green to pink; Z = colorless to pale yellow. *Orientation:* Y = b; X ∧ a = 0°–3°; Z ∧ a = 90°–87°. *Dispersion:* r < v. *Absorption:* X = Y > Z. α = 1.572–1.578 β = 1.579–1.584 γ = 1.589–1.595   2V(meas.) = 0°–80°

Cell Data: *Space Group:* $C\bar{1}$. a = 5.14   b = 8.90   c = 14.15 α = 90°33' β = 90°12' γ = 90°   Z = 2

X-ray Powder Pattern: Londonderry, Western Australia.
2.315 (10), 4.70 (9), 3.52 (9), 14.1 (8), 7.05 (7), 2.505 (7), 1.489 (7)

Chemistry:

	(1)	(2)		(1)	(2)
SiO$_2$	33.40	38.26	CaO	0.45	
Al$_2$O$_3$	47.47	44.28	Li$_2$O	3.12	2.00
Fe$_2$O$_3$	0.00	1.31	(Na, K)$_2$O	0.09	
FeO	0.71	0.48	H$_2$O$^+$	14.98	13.00
MnO	trace		H$_2$O$^-$	0.23	
MgO	0.20		Total	100.65	99.33

(1) Kalbinsky Range, Ural Mountains, Russia; corresponds to $(Li_{1.11}Na_{0.02}K_{0.01})_{\Sigma=1.14}$ $(Al_{3.89}Fe^{2+}_{0.05}Ca_{0.04}Mg_{0.03})_{\Sigma=4.01}(Si_{2.95}Al_{1.05})_{\Sigma=4.00}O_{10}(OH)_8$. (2) Djalair deposit, "Middle Asia," Russia; corresponds to $Li_{0.7}(Al_{3.96}Fe^{3+}_{0.09}Fe^{2+}_{0.04})_{\Sigma=4.09}(Si_{3.38}Al_{0.62})_{\Sigma=4.00}O_{10.35}(OH)_{7.65}$.

Mineral Group: Chlorite group.

Occurrence: A late-stage hydrothermal alteration product of lithium-bearing minerals in pegmatites; a primary hydrothermal vein mineral.

Association: Lepidolite, spodumene, tourmaline, petalite, quartz, albite, microcline.

Distribution: Good examples from: in the USA, at Hebron, Mt. Mica near South Paris, and Buckfield, Oxford Co., Maine; in the Pala and Rincon districts, San Diego Co., California; from the Jeffrey quarry, North Little Rock, Pulaski Co., and Magnet Cove, Hot Springs Co., Arkansas. At Waitabit Creek, northwest of Donald, British Columbia, Canada. From Ogofau, Carmarthenshire, England. At Radkovice and Dobrá Voda, Czech Republic. In the Varuträsk pegmatite, near Skellefteå, Västerbotten, Sweden. From Lipovka, Ural Mountains, Russia. In the Muiâne pegmatite, Alto Ligonha district, Mozambique. At Londonderry, Western Australia. Many additional localities are known.

Name: For Josiah B. Cooke, Jr. (1827–1894), American mineralogist and chemist, Harvard University, Cambridge, Massachusetts, USA.

References: (1) Dana, E.S. (1892) Dana's system of mineralogy, (6th edition), 625. (2) Deer, W.A., R.A. Howie, and J. Zussman (1963) Rock-forming minerals, v. 3, sheet silicates, 131–163. (3) Brown, B.E. and S.W. Bailey (1962) Chlorite polytypism: I. Regular and semi-random one-layer structure. Amer. Mineral., 47, 819–850. (4) Černý, P. (1970) Compositional variations in cookeite. Can. Mineral., 10, 636–647. (5) Vrublevskaja [Vrublevskaya], Z.V., I.S. Delitsin, B.B. Zvyagin, and S.V. Soboleva (1975) Cookeite with a perfect regular structure, formed by bauxite alteration. Amer. Mineral., 60, 1041–1046. (6) Bailey, S.W. and J.S. Lister (1989) Structures, compositions, and X-ray diffraction identification of dioctahedral chlorites. Clays and Clay Minerals, 37, 193–202.

Coombsite $K(Mn^{2+}, Fe^{2+}, Mg)_{13}(Si, Al)_{18}O_{42}(OH)_{14}$

Crystal Data: [Hexagonal] (by analogy to zussmanite). *Point Group:* [3 or $\bar{3}$.] Fibrous crystals, to 20 μm, in patchy spherulitic aggregates.

Physical Properties: Hardness = n.d. D(meas.) = 3.0(1) D(calc.) = 3.063

Optical Properties: Semitransparent. *Color:* Pale yellowish brown.
Optical Class: Uniaxial (–). *Absorption:* Weak; $O > E$. $\omega = 1.619(1)$ $\epsilon = 1.600(1)$

Cell Data: *Space Group:* [$R3$ or $R\bar{3}$.] $a = 11.828(2)$ $c = 29.146(9)$ $Z = 3$

X-ray Powder Pattern: Watson's Beach, New Zealand; closely resembles zussmanite.
9.68 (100), 2.556 (90), 2.793 (70), 2.241 (50), 4.835 (30), 3.241 (25)

Chemistry:

	(1)
SiO_2	45.18
TiO_2	0.01
Al_2O_3	3.59
FeO	3.28
MnO	38.19
MgO	1.08
CaO	0.03
Na_2O	0.08
K_2O	2.19
H_2O	[6.37]
Total	[100.00]

(1) Watson's Beach, New Zealand; by electron microprobe, H_2O by difference; corresponds to $(K_{1.01}Na_{0.06})_{\Sigma=1.07}(Mn_{11.75}Fe_{1.00}Mg_{0.58})_{\Sigma=13.33}(Si_{16.41}Al_{1.54})_{\Sigma=17.95}O_{42}(OH)_{14}$.

Occurrence: In a manganese-rich rock in metagraywacke-argillite in the pumpellyite-prehnite facies.

Association: Rhodonite, quartz, rhodochrosite, kutnohorite, manganoan calcite, spessartine, apatite, parsettensite, caryopilite.

Distribution: At Watson's Beach, southeastern Otago, New Zealand.

Name: For Professor Douglas Saxon Coombs (1924–), mineralogist and petrologist, University of Otago, Dunedin, New Zealand.

Type Material: University of Otago, Dunedin, New Zealand; National Museum of Natural History, Washington, D.C., USA.

References: (1) Sameshima, T. and Y. Kawachi (1991) Coombsite, Mn analogue of zussmanite, and associated Mn-silicates, parsettensite and caryopilite, from southeast Otago, New Zealand. New Zealand J. Geol. Geophys., 34, 329–335. (2) (1992) Amer. Mineral., 77, 671 (abs. ref. 1).

Crystal Data: Orthorhombic, pseudohexagonal. *Point Group:* $2/m\ 2/m\ 2/m$. Crystals short prismatic, striated $\parallel$ [001], to 18 cm; typically granular to compact, massive. *Twinning:* Common on {110}, {130}, simple, lamellar, cyclical.

Physical Properties: *Cleavage:* Fair on {100}, poor on {001} and {010}. *Fracture:* Subconchoidal. *Tenacity:* Brittle. Hardness = 7–7.5   D(meas.) = 2.60–2.66 D(calc.) = 2.505

Optical Properties: Transparent to translucent. *Color:* Blue, smoky blue, bluish violet; greenish, yellowish brown, gray; colorless to very pale blue in thin section. *Luster:* Vitreous. *Optical Class:* Biaxial (+) or (–). *Pleochroism:* X = pale yellow, green; Y = violet, blue-violet; Z = pale blue. *Orientation:* $X = c$; $Y = a$; $Z = b$. *Dispersion:* $r < v$, weak to marked. *Absorption:* $Z > Y > X$. $\alpha = 1.527$–1.560 $\beta = 1.532$–1.574 $\gamma = 1.537$–1.578 2V(meas.) = 35°–106°

Cell Data: *Space Group:* $Cccm$.   a = 17.079(3)   b = 9.730(2)   c = 9.356(2)   Z = 4

X-ray Powder Pattern: Synthetic.
8.45 (100), 8.52 (95), 3.039 (65), 3.035 (65), 3.132 (55), 3.012 (55), 4.09 (50)

Chemistry:

	(1)	(2)		(1)	(2)
SiO_2	50.2	49.46	MgO	12.8	12.06
TiO_2	< 0.01	0.01	CaO	0.23	0.03
Al_2O_3	33.5	33.58	Na_2O	0.26	0.14
Fe_2O_3	0.14	0.14	K_2O	0.14	0.30
FeO	0.84	2.12	H_2O^+	1.69	1.71
MnO	0.06	0.08	H_2O^-	0.12	0.10
			Total	99.98	99.73

(1) White Well, Western Australia; corresponds to $(Mg_{1.91}Fe^{2+}_{0.08}Na_{0.05}Ca_{0.02}K_{0.02}Mn_{0.01})_{\Sigma=2.09}$ $Al_{3.95}Si_{5.02}O_{18}$. (2) Smith Ridge, Boehls Butte quadrangle, Idaho, USA; corresponds to $(Mg_{1.81}Fe^{2+}_{0.18}K_{0.04}Ca_{0.03}Mn_{0.01})_{\Sigma=2.07}Al_{3.99}Si_{4.98}O_{18}$.

Polymorphism & Series: Dimorphous with indialite; forms a series with sekaninaite.

Occurrence: In thermally metamorphosed argillaceous sediments and high-grade regionally metamorphosed schists, gneisses, and granulites; in mafic igneous rocks and granites; detrital.

Association: Sillimanite, potassic feldspar, muscovite, biotite, corundum, spinel, garnet, andalusite.

Distribution: Widely distributed; some localities for abundant material are: at Bodenmais, Bavaria, Germany. From Orijärvi and Leppävirta, Finland. In Norway, at Kragerö, near Tvedestrand, and Akland, Söndeled, near Risör. In the Mt. Bity district, south of Antsirabe, Madagascar. On the Treasure Casket claims, Fungure Reserve, Zimbabwe. Found around Namib, Namibia. Crystals from the gem gravels of Sri Lanka. Around Tiruchchirappalli and Coimbatore, Tamil Nadu, India. In the Harts Range, Northern Territory, Australia. In the USA, at Richmond, Cheshire Co., New Hampshire; and at Haddam, Middlesex Co., Connecticut. Fine crystals from Thompson, Manitoba, Canada.

Name: For Pierre Louis A. Cordier (1777–1861), French geologist who first studied the species.

References: (1) Dana, E.S. (1892) Dana's system of mineralogy, (6th edition), 419–421. (2) Deer, W.A., R.A. Howie, and J. Zussman (1986) Rock-forming minerals, (2nd edition), v. 1B, disilicates and ring silicates, 410–540. (3) (1961) NBS Mono. 25, 28. (4) Pryce, M.W. (1973) Low-iron cordierite in phlogopite schist from White Well, Western Australia. Mineral. Mag., 39, 241–243. (5) Cohen, J.P., F.K. Ross, and G.V. Gibbs (1977) An X-ray and neutron diffraction study of low hydrous cordierite. Amer. Mineral., 62, 67–78. (6) Armbruster, T. (1986) Role of Na in the structure of low-cordierite: a single-crystal X-ray study. Amer. Mineral., 71, 746–757.

Corrensite $(Ca, Na, K)(Mg, Fe, Al)_9(Si, Al)_8O_{20}(OH)_{10} \cdot nH_2O$

Crystal Data: n.d. *Point Group:* n.d. As aggregates of clay-sized particles.

Physical Properties: *Cleavage:* Perfect basal. *Tenacity:* Unctuous. Hardness = n.d. D(meas.) = n.d. D(calc.) = n.d.

Optical Properties: Semitransparent. *Color:* Yellowish green. *Luster:* Dull, earthy. *Optical Class:* Biaxial (−). *Pleochroism:* X = yellowish ocher; Y = Z = brownish green. $\alpha = 1.560–1.585$ β = n.d. $\gamma = 1.582–1.612$ 2V(meas.) = Small.

Cell Data: *Space Group:* n.d. Z = n.d.

X-ray Powder Pattern: Wellington Formation, Lyon Co., Kansas, USA; a 30 Å d-spacing changes to: air dried, 28–29; ethylene glycol, 32–33; on heating to ~500 °C, 23–29. 14.5 (100), 1.54 (90), 2.57 (80), 7.26 (70), 4.60 (60), 3.61 (50), 2.44 (50)

Chemistry:

	(1)	(2)		(1)	(2)
SiO_2	37.2	34.	MgO	18.9	24.
TiO_2	0.4		CaO	1.0	
Al_2O_3	15.5	13.	Na_2O	0.2	
Fe_2O_3	6.7	7.	K_2O	1.4	
MnO	0.01		H_2O	18.4	22.
			Total	99.7	[100.]

(1) Sparta, Tennessee, USA; total Fe as Fe_2O_3, traces of Zr, V, and Cu determined spectrographically. (2) Do.; recalculated to 100% after removal of impurities.

Polymorphism & Series: A 1:1 mixed-layer interstratification of trioctahedral chlorite with either a trioctahedral vermiculite or a trioctahedral smectite.

Occurrence: From diverse sedimentary environments, which produced clastic, carbonate, volcanoclastic, or evaporite rocks. Also a hydrothermal alteration product, may be formed under retrograde diagenetic conditions; in soils.

Association: Illite, chlorite, laumontite, gypsum, anhydrite, quartz, dolomite.

Distribution: Probably fairly widely distributed; a few localities with material known to meet the definition are: from Zaiserweiher, three km from Maulbronn, Baden-Württemberg, and at the Hünstollen, near Holserode, Lower Saxony, Germany. At Wilkesley, near Audlem, Cheshire, and elsewhere in England. From Hillhouse quarry, Ayrshire, Scotland. In the USA, from Juniper Canyon, Moffat Co., Colorado; in the Wellington Formation, in Lyon Co., Kansas; in the County rock quarry, Sparta, White Co., Tennessee; and from a number of localities in west-central Montana, as between Bowman's Corner and Wolf Creek, Lewis and Clark Co. From Packwood, Western Australia. From the Yamanaka area, Shimane Prefecture, and in the Yoshino mine, Yamagata Prefecture, Japan.

Name: For Professor Carl Wilhelm Correns (1893–1980), Director of the Sedimentary Petrography Institute, Göttingen University, Göttingen, Germany.

Type Material: National Museum of Natural History, Washington, D.C., USA, 107373, 162265.

References: (1) Lippmann, F. (1954) Über einen Keuperton von Zaisersweiher bei Maulbronn. Heidelberg. Beitr. Mineral. Petrog., 4, 130–134 (in German). (2) (1955) Amer. Mineral., 40, 137 (abs. ref. 1). (3) Peterson, M.N.A. (1961) Expandable chloritic clay minerals from Upper Mississippian carbonate rocks of the Cumberland Plateau in Tennessee. Amer. Mineral., 46, 1245–1269. (4) Kopp, O.C. and S.M. Fallis (1974) Corrensite in the Wellington Formation, Lyons, Kansas. Amer. Mineral., 59, 623–624. (5) Yoshimura, T. and S. Okubo (1988) Refractive indices and birefringence of corrensite and the related minerals. Mineral. J. (Japan), 14, 145–149. (6) Reynolds, R.C., Jr. (1988) Mixed layer chlorite minerals. In: S.W. Bailey, Ed., Hydrous phyllosilicates. Rev. Mineral. 19, MSA, 601–629. (7) Shau, Y.-H., D.R. Peacor, and E.J. Essene (1990) Corrensite and mixed-layer chlorite/corrensite in metabasalt from northern Taiwan: TEM/AEM, EMPA, XRD, and optical studies. Contr. Mineral. Petrol., 105, 123–142.

Crystal Data: Orthorhombic. *Point Group:* n.d. Very thin, lath-shaped crystals showing few forms, with sharp terminations, to 2 mm; radiating sprays of crystals and compact balls.

Physical Properties: *Cleavage:* {010}, perfect. *Tenacity:* Brittle. Hardness = Soft. D(meas.) = 2.14(2) D(calc.) = 2.05

Optical Properties: Transparent. *Color:* Colorless, white, gray, blue-gray. *Luster:* Pearly. *Optical Class:* Biaxial (–). *Orientation:* X = a; Y = b; Z = c. $\alpha = 1.505–1.513$ $\beta = 1.509–1.516$ $\gamma = 1.509–1.518$ 2V(meas.) = 44°–53°

Cell Data: *Space Group:* n.d. $a = 11.27(1)$ $b = 15.25(1)$ $c = 12.61(3)$ Z = 6

X-ray Powder Pattern: Goble, Oregon, USA.
15.2 (100), 3.81 (35), 2.964 (35), 2.934 (25), 3.052 (20), 5.08 (17), 7.62 (15)

Chemistry:

	(1)	(2)
SiO_2	42.73	46.37
Al_2O_3	24.32	25.87
Fe_2O_3		0.08
CaO	12.86	15.12
Na_2O	0.70	0.80
K_2O		0.10
H_2O	22.8	21.0
Total	103.41	109.34

(1) Goble, Oregon, USA; by electron microprobe, H_2O by weight loss to 600 °C; corresponds to $(Ca_{0.96}Na_{0.10})_{\Sigma=1.06}Al_{2.00}Si_{2.99}O_{10} \cdot 5.6H_2O$. (2) Superior, Arizona, USA; by electron microprobe; H_2O by weight loss to 600 °C; corresponds to $(Ca_{1.04}Na_{0.10}K_{0.01})_{\Sigma=1.15}Al_{1.96}Si_{2.98}O_{10} \cdot 5.0H_2O$.

Mineral Group: Zeolite group.

Occurrence: In cavities in basalts and scoria, typically the only zeolite within a single vesicle.

Association: Zeolites, apophyllite.

Distribution: In the USA, at Goble, Columbia Co., at Beech Creek, Grant Co., and at Spray, Wheeler Co., Oregon; on Capitol Peak, Thurston Co., Washington; from eight km south of Superior, Pinal Co., Arizona; and at Table Mountain, Jefferson Co., Colorado. In Canada, at Monte Lake, British Columbia. From Dunseverick, Ballyclare, and elsewhere in Co. Antrim, Ireland. At Kingsburgh, Isle of Skye, Scotland. In Mjoädalsá Canyon, near Hvammur, Iceland. From Dalsnipa, on Sandoy and Satan, on Streymoy, Faeroe Islands. At Kuniga, Oki Islands, Japan. From Flinders and Jindivick, Victoria, Australia. A number of other localities are known.

Name: For John Cowles (1907–1985), amateur mineralogist of Rainier, Oregon, USA.

Type Material: University of California, Santa Barbara, California, 6720, 6721; National Museum of Natural History, Washington, D.C., USA, 135026.

References: (1) Wise, W.S. and R.W. Tschernich (1975) Cowlesite, a new Ca-zeolite. Amer. Mineral., 60, 951–956. (2) Vezzalini, G., G. Artioli, S. Quartieri, and H. Foy (1992) The crystal chemistry of cowlesite. Mineral. Mag., 56, 575–579.

Crystal Data: Orthorhombic. *Point Group:* [2/m 2/m 2/m or mm2.] Fibrous crystals, elongated along [001] and flattened on {010}, to 0.5 mm. As densely packed spherules of crystals; in masses of matted and tangled fibers.

Physical Properties: *Cleavage:* Poor on {010}. Hardness = 2.5 for spherules, greater for crystals. D(meas.) = 4.1(1) D(calc.) = 4.01

Optical Properties: Transparent to translucent. *Color:* Pale green to yellow-green; yellow in thin section. *Streak:* Light green.
Optical Class: Biaxial (+). *Pleochroism:* X = yellow-green; Y = green; Z = yellow-green.
Orientation: $X = a$; $Y = b$; $Z = c$. *Absorption:* $Z = X > Y$. $\alpha = 1.737(2)$ $\beta = 1.747(2)$
$\gamma = 1.768(2)$ 2V(meas.) = 69° 2V(calc.) = 70°

Cell Data: *Space Group:* $mmmCb$ [diffraction symbol]. $a = 12.483–12.497$
$b = 21.375–21.395$ $c = 7.283$ $Z = 4$

X-ray Powder Pattern: Tiger, Arizona, USA.
10.726 (100), 6.024 (50), 4.067 (50), 3.555 (50b), 3.013 (50), 2.982 (50), 2.696 (50)

Chemistry:

	(1)	(2)
SiO_2	25.5	25.59
Fe_2O_3	12.3	13.61
Al_2O_3	2.1	
CuO	13.5	13.56
ZnO	1.2	
PbO	37.0	38.03
H_2O	8.8	9.21
Total	100.4	100.00

(1) Tiger, Arizona, USA; average of several microchemical analyses, corresponding to $Pb_{1.98}$ $(Cu_{2.02}Zn_{0.18})_{\Sigma=2.20}(Fe^{3+}_{1.85}Al_{0.05})_{\Sigma=1.90}Si_{5.08}O_{17.18} \cdot 5.85H_2O$. (2) $Pb_2Cu_2Fe_2Si_5O_{17} \cdot 6H_2O$.

Occurrence: In the oxidized zone of a base-metal deposit, in andesite breccia loosely cemented with iron oxides and wulfenite (Tiger, Arizona, USA).

Association: Mimetite, dioptase, fluorite, willemite, wulfenite, descloizite, murdochite (Tiger, Arizona, USA); ajoite, fluorite (Potter-Cramer property, Arizona, USA).

Distribution: In the USA, at the Mammoth-St. Anthony mine, Tiger, Pinal Co., the Potter-Cramer property, Wickenburg, Maricopa Co., and a prospect in the Tucson Mountains, Pima Co., Arizona; from near Gold Point, Esmeralda Co., Nevada. In Mexico, at Caborca and Munihuaza, near Alamos, Sonora. In the Cruz del Sur mine, Rio Negro Province, Argentina.

Name: For Dr. Saville Cyrus Creasey (1917–), economic geologist, U.S. Geological Survey, expert on Arizona mineral deposits.

Type Material: The Natural History Museum, London, England, 1976,412; University of Arizona, Tucson, Arizona; Harvard University, Cambridge, Massachusetts, 117000; National Museum of Natural History, Washington, D.C., USA, 147661.

References: (1) Williams, S.A. and R.A. Bideaux (1975) Creaseyite, $Cu_2Pb_2(Fe, Al)_2$ $Si_5O_{17} \cdot 6H_2O$. Mineral. Mag., 40, 227–231. (2) (1976) Amer. Mineral., 61, 503 (abs. ref. 1).

Crystal Data: Tetragonal, pseudocubic. *Point Group:* 422. As pseudo-octahedral crystals, to 4 mm, with {110} and {331}, rarely pseudocubic. Commonly dendritic to skeletal; as spherulites to several cm; fibrous or microcrystalline ("opal"), massive. *Twinning:* On {111}, common, interpenetrant, polysynthetic, repeated.

Physical Properties: *Tenacity:* Brittle. Hardness = 6–7   D(meas.) = 2.32–2.36 D(calc.) = 2.33

Optical Properties: Transparent. *Color:* Colorless, white, milky white to yellowish; in transmitted light, colorless. *Luster:* Vitreous.
Optical Class: Uniaxial (–). $\omega = 1.487$ $\epsilon = 1.484$

Cell Data: *Space Group:* $P4_12_12$. $a = 4.9709(1)$ $c = 6.9278(2)$ Z = 4

X-ray Powder Pattern: Synthetic.
4.05 (100), 2.485 (20), 2.841 (13), 3.135 (11), 1.870 (7), 2.465 (5), 2.118 (5)

Chemistry:

	(1)
SiO_2	[99.13]
TiO_2	0.38
Al_2O_3	0.18
FeO	0.09
MnO	< 0.02
MgO	< 0.03
CaO	< 0.02
Na_2O	0.05
K_2O	0.17
P_2O_5	< 0.03
Total	[100.00]

(1) Mare Imbrium, Moon; by electron microprobe, SiO_2 by difference.

Polymorphism & Series: Quartz, tridymite, coesite, and stishovite are polymorphs; inverts from high- or β-cristobalite at 268 °C or below.

Occurrence: In vesicles and lithophysae; a late-crystallizing phase in basaltic to rhyolitic volcanic rocks; from acid-sulfate-type hydrothermal alteration of volcanic rocks; precipitated by hot springs. By contact metamorphism of sandstone; developed during diagenesis, recrystallized from siliceous sedimentary rocks.

Association: Tridymite, quartz, sanidine, anorthoclase, fayalite, magnetite, kaolinite, alunite, "opal."

Distribution: On Cerro San Cristóbal, near Pachuca, Hidalgo, and in the Santín mine, Santa Caterina, Guanajuato, Mexico. In the USA, at Glass Mountain, Little Lake, and Sugarloaf Mountain, near Coso Hot Springs, Inyo Co., California; near Crater Lake, Klamath Co., Oregon; widespread in the San Juan Mountains, San Juan Co., Colorado. At Mayen and Mendig, Eifel district, Germany. From Sárospatak, Hungary. In the Ellora Caves, Maharashtra, India. From the Tokatoka district, about 150 km north of Auckland, New Zealand. In Japan, at Goroyama, Nagano Prefecture; Futo, Shizuoka Prefecture; and many other places.

Name: For the first noted occurrence at Cerro San Cristóbal, Mexico.

References: (1) Frondel, C. (1962) Dana's system of mineralogy, (7th edition), v. III, silica minerals, 273–286. (2) Deer, W.A., R.A. Howie, and J. Zussman (1963) Rock-forming minerals, v. 4, framework silicates, 179–230. (3) (1960) NBS Circ. 539, 10, 48. (4) Mason, B.F. (1972) Lunar tridymite and cristobalite. Amer. Mineral., 57, 1530–1535. (5) Pluth, J.J., J.V. Smith, and J. Faber, Jr. (1985) Crystal structure of low cristobalite at 10, 293, and 473 K: variation of framework geometry with temperature. J. Appl. Physics, 57, 1045–1049.

Crystal Data: Triclinic, monoclinic, or hexagonal. *Point Group:* See polytypes.
As hexagonal prisms, tapering toward one end and vertically striated. In divergent fibrous groups; cylindroidal, reniform, massive. *Twinning:* Twin axis [001]; interpenetrating trigonal pyramids produce six-rayed composite crystals.

Physical Properties: *Cleavage:* Perfect on {001}. *Tenacity:* Thin laminae are elastic.
Hardness = 3.5   D(meas.) = 3.34–3.35   D(calc.) = n.d.

Optical Properties: Nearly opaque. *Color:* Coal-black to brownish black; thin laminae emerald-green by transmitted light, brown to yellow-brown on edges. *Streak:* Dark olive-green.
Luster: Brilliant.
Optical Class: Biaxial; pseudo-uniaxial (–). *Pleochroism:* In browns and blacks. $\omega = 1.72$–1.80
2V(meas.) = n.d.

Cell Data: *Space Group:* $P1$. $a = 5.472(8)$ $b = 9.467(19)$ $c = 14.241(39)$
$\alpha = 90.015(20)°$ $\beta = 90.042(18)°$ $\gamma = 89.952(15)°$ Z = 4

X-ray Powder Pattern: Wheal Jane, Cornwall, England. (ICDD 17-470).
7.09 (100), 3.54 (85), 2.722 (50), 2.439 (40), 1.586 (40), 2.305 (16), 1.680 (16)

Chemistry:

	(1)	(2)
SiO_2	16.42	15.04
Al_2O_3	0.90	
Fe_2O_3	29.72	39.97
FeO	41.86	35.97
CaO	1.32	
H_2O	10.17	9.02
Total	100.39	100.00

(1) Herja, Romania. (2) $Fe_2^{2+}Fe^{3+}(SiFe^{3+})O_5(OH)_4$.

Polymorphism & Series: Stacking polytypes have space groups:
$P1$, Cc, Cm, $P31m$, $P6_3cm$, $P6_3$, $P3_1$, $P31c$, and $R3$.

Mineral Group: Kaolinite-serpentine group.

Occurrence: A low-temperature hydrothermal product in ore veins.

Association: Siderite, pyrite, sphalerite, clinochlore, quartz.

Distribution: At Příbram, Kutná Hora, Chvaletice, Chyňava, and other places in the Czech Republic. Exceptional groups from Herja (Kisbánya), Romania. From Hagental, Harz Mountains, Germany. In the Salsigne gold mine, north of Carcassonne, Aude, France. In England, from Wheal Jane and Wheal Maudlin, and around St. Just, Cornwall. In the USA, at Long Hill, Fairfield Co., Connecticut; in the Cornucopia mine, Nye Co., Nevada. From Santa Eulalia, Chihuahua, Mexico. At Conghonas do Campo, Minas Gerais, Brazil. From Llallagua, Bolivia. Additional minor occurrences are known.

Name: To honor the Swedish mineralogist and chemist, Axel Fredrik Cronstedt (1722–1765).

References: (1) Dana, E.S. (1892) Dana's system of mineralogy, (6th edition), 656–657.
(2) Deer, W.A., R.A. Howie, and J. Zussman (1963) Rock-forming minerals, v. 3, sheet silicates, 164–169. (3) Frondel, C. (1962) Polytypism in cronstedtite. Amer. Mineral., 47, 781–783.
(4) Steadman, R. and P.M. Nuttall (1963) Polymorphism in cronstedtite. Acta Cryst., 16, 1–8.
(5) Steadman, R. and P.M. Nuttall (1964) Further polymorphism in cronstedtite. Acta Cryst., 17, 404–406. (6) Geiger, C.A., D.L. Henry, S.W. Bailey, and J.J. Maj (1983) Crystal structure of cronstedtite-2H_2. Clays and Clay Minerals, 31, 97–108. (7) Phillips, W.R. and D.T. Griffen (1981) Optical mineralogy, 458–459.

Crystal Data: Monoclinic. *Point Group: 2/m.* [As elongated prismatic crystals, columnar to acicular,] observed as rims on other amphiboles *Twinning:* Rarely, simple or multiple || {100}.

Physical Properties: *Cleavage:* Perfect on {110}, producing intersections of 58° and 122°. *Fracture:* [Conchoidal to uneven.] *Tenacity:* [Brittle.] Hardness = 6 D(meas.) = 3.15–3.30 D(calc.) = 3.223

Optical Properties: Semitransparent. *Color:* Gray, blue; intense blue and violet in thin section. *Luster:* [Vitreous.]
Optical Class: Biaxial (–). *Pleochroism:* Strong; X = yellow; Y = blue; Z = violet. *Orientation:* $Z = b$; $Y \wedge c = -10°$ to $-8°$; $X \wedge a = 4°$–$6°$. *Dispersion:* $r > v$, extreme. $\alpha = 1.632$–1.655 $\beta = 1.642$–1.664 $\gamma = 1.645$–1.666 2V(meas.) = 0°–90°

Cell Data: *Space Group: C2/m.* $a = 9.647$ $b = 17.905$ $c = 5.316$ $\beta = 103.60°$ Z = 2

X-ray Powder Pattern: The Monument, Anglesey, Wales. (ICDD 31-1312).
8.44 (100), 3.15 (90), 2.728 (25), 2.824 (20), 3.29 (14), 2.604 (14), 4.52 (12)

Chemistry:

	(1)		(1)
SiO_2	55.10	Na_2O	6.38
TiO_2	0.68	K_2O	0.09
Al_2O_3	4.27	F	0.02
Fe_2O_3	10.61	Cl	0.03
FeO	9.78	H_2O^+	1.96
MnO	0.48	H_2O^-	0.06
MgO	8.86	$-O = (F, Cl)_2$	0.02
CaO	1.43	Total	99.73

(1) Berkeley Hills, Alameda Co., California, USA; corresponding to $(Na_{1.78}Ca_{0.22}K_{0.02})_{\Sigma=2.02}$ $(Mg_{1.90}Fe^{2+}_{1.18}Mn_{0.06})_{\Sigma=3.14}(Fe^{3+}_{1.15}Al_{0.65}Ti_{0.07}Ca_{0.03})_{\Sigma=1.90}(Si_{7.92}Al_{0.08})_{\Sigma=8.00}O_{22}$ $[(OH)_{1.88}Cl_{0.01}F_{0.01}]_{\Sigma=1.90}$.

Polymorphism & Series: The intermediate member of the glaucophane-riebeckite series.

Mineral Group: Amphibole (alkali) group: $0.3 \leq Fe^{3+}/(Fe^{3+} + Mg) \leq 0.69$; $(Na + K)_A < 0.5$; $Na_B \geq 1.34$.

Occurrence: Common in metamorphic rocks of the blueschist and greenschist facies; formed by retrograde alteration of eclogite.

Association: Glaucophane, lawsonite, pumpellyite, chlorite, muscovite, almandine, epidote, barroisite, ferro-barroisite, omphacite, quartz, albite, aragonite.

Distribution: A few localities for well studied material include: in the USA, from many places in the Franciscan Formation, California, and on Kodiak Island, Alaska. From Heiko Kogenda, Korea. In Pakistan, from near Topsin and Shangla. From the Kotu district, Tokusim Prefecture; in the Nukabira mine and at Horokanai, Hokkaido; and at Takahama, Kumamoto Prefecture, Japan. In the Gantertal, Valais, Switzerland. On Monument Hill, Anglesey, Wales. From Knockormal, Ayrshire, Scotland. At Volno, Serbia.

Name: For Charles Whitman Cross (1854–1949), American geologist, U.S. Geological Survey.

Type Material: University of California, Berkeley, California; Harvard University, Cambridge, Massachusetts, 85057, 85058, 12873, 112666; National Museum of Natural History, Washington, D.C., USA, 94336, 166511, 166512.

References: (1) Dana, E.S. (1899) Dana's system of mineralogy, (6th edition), app. I, 20. (2) Deer, W.A., R.A. Howie, and J. Zussman (1963) Rock-forming minerals, v. 2, chain silicates, 333–351. (3) Borg, I.Y. (1967) Optical properties and cell parameters in the glaucophane-riebeckite series. Contr. Mineral. Petrol., 15, 67–92. (4) Phillips, W.R. and D.T. Griffen (1981) Optical mineralogy, 237–241.

Cummingtonite $(Mg, Fe^{2+})_2(Mg, Fe^{2+})_5Si_8O_{22}(OH)_2$

Crystal Data: Monoclinic. *Point Group:* $2/m$. As bladed, columnar, or fibrous crystals and fibrous aggregates, to 20 cm. *Twinning:* Simple or multiple twinning $\parallel \{100\}$.

Physical Properties: *Cleavage:* Good on $\{110\}$, intersecting at 54° and 126°. *Tenacity:* Brittle. Hardness = 5–6 D(meas.) = 3.1–3.6 D(calc.) = 3.3

Optical Properties: Transparent to translucent. *Color:* Dark green, brown, gray, beige; colorless to pale green in thin section. *Luster:* Vitreous. *Optical Class:* Biaxial (+). *Pleochroism:* With increasing Fe content, weak; $X = Y =$ colorless; Z = pale green. *Orientation:* $Y = b$; $Z \wedge c = -21°$ to $-16°$; $X \wedge a = -9°$ to $-3°$. *Dispersion:* $r < v$, weak. $\alpha = 1.632–1.663$ $\beta = 1.638–1.677$ $\gamma = 1.655–1.697$ 2V(meas.) = 70°–90°

Cell Data: *Space Group:* $C2/m$. $a = 9.516(5)$ $b = 18.139(10)$ $c = 5.311$ $\beta = 102.1°$ $Z = 2$

X-ray Powder Pattern: Labrador City, Canada.
8.30 (100b), 3.06 (90), 3.26 (80), 2.754 (70), 1.403 (60), 9.12 (50), 2.623 (50)

Chemistry:

	(1)	(2)		(1)	(2)		(1)	(2)
SiO_2	51.53	53.84	FeO	16.91	25.63	Na_2O	0.65	0.15
TiO_2	0.31	0.01	MnO	0.22	0.19	K_2O	0.00	
Al_2O_3	5.02	1.61	MgO	20.84	17.44	H_2O^+	2.15	
Fe_2O_3	0.82		CaO	1.34	0.41	H_2O^-	0.64	
						Total	100.43	99.28

(1) Geodh Dubh, Scotland; corresponds to $(Mg_{4.44}Fe^{2+}_{2.02}Al_{0.21}Ca_{0.20}Na_{0.18}Fe^{3+}_{0.09}$ $Ti_{0.03}Mn_{0.03})_{\Sigma=7.20}(Si_{7.36}Al_{0.64})_{\Sigma=8.00}O_{22}(OH)_{2.05}$. (2) Orfordville Formation, Mt. Cube quadrangle, Vermont, USA; by electron microprobe, corresponds to $(Mg_{3.77}Fe^{2+}_{3.11}Al_{0.08}Ca_{0.06}$ $Na_{0.04}Mn_{0.02})_{\Sigma=7.08}(Si_{7.80}Al_{0.20})_{\Sigma=8.00}O_{22}(OH)_2$.

Polymorphism & Series: Forms a series with magnesio-cummingtonite and grunerite.

Mineral Group: Amphibole (Fe–Mn–Mg) group: $0.3 \leq Mg/(Mg + Fe^{2+}) \leq 0.69$; $(Ca + Na)_B$ < 1.34; Li < 1.0; Mn < 0.5.

Occurrence: Commonly in medium-grade regionally metamorphosed rocks; characteristic of metamorphosed iron formations; a late-stage mineral in some gabbros and norites; rarely in silicic volcanic rocks.

Association: "Hornblende," anthophyllite, actinolite, ferro-actinolite, tremolite, arfvedsonite, magnesio-arfvedsonite, gedrite, glaucophane, quartz, garnet.

Distribution: Well-characterized material from: in the USA, at Cummington, Hampshire Co., Massachusetts; the Homestake mine, Lead, Pennington Co., South Dakota; in the Jackson County Iron Formation, Wisconsin. In the Wabush Iron Formation, Labrador City, Labrador, Newfoundland, Canada. From the Miyamori district, Iwate Prefecture, and the Hitachi mine, Ibaragi Prefecture, Japan. In the Garabal Hill-Glen Fyne complex, Scotland. At Teisko, Finland. From Nordmard, Värmland, Sweden.

Name: For the occurrence at Cummington, Massachusetts, USA.

References: (1) Dana, E.S. (1892) Dana's system of mineralogy, (6th edition), 386, 390. (2) Deer, W.A., R.A. Howie, and J. Zussman (1963) Rock-forming minerals, v. 2, chain silicates, 235–248. (3) Ghose, S. (1961) The crystal structure of a cummingtonite. Acta Cryst., 14, 622–627. (4) Klein, C. (1964) Cummingtonite-grunerite series: a chemical, optical and X-ray study. Amer. Mineral., 49, 963–982. (5) Spear, F.S. (1980) The gedrite-anthophyllite solvus and the composition limits of orthoamphibole from the Post Pond Volcanics, Vermont. Amer. Mineral., 65, 1103–1118. (6) Ghose, S. and Y. Hexiong (1989) Mn-Mg distribution in a C2/m manganoan cummingtonite: crystal-chemical considerations. Amer. Mineral., 74, 1091–1096. (7) Phillips, W.R. and D.T. Griffen (1981) Optical mineralogy, 225–258.

Crystal Data: Tetragonal. *Point Group:* $4/m\ 2/m\ 2/m$. Crystals tabular on {001}, to 0.1 mm.

Physical Properties: *Cleavage:* Perfect on {001}. *Tenacity:* Brittle. Hardness = $\sim$5 D(meas.) = 3.08(6)   D(calc.) = 3.09

Optical Properties: Transparent to translucent. *Color:* Azure. *Luster:* Vitreous. *Optical Class:* Uniaxial (–). *Pleochroism:* O = blue; E = nearly colorless, pale rose. $\omega = 1.633(3)$ $\epsilon = 1.590(3)$

Cell Data: *Space Group:* $P4/ncc$.   a = 7.30(1)   c = 15.12(2)   Z = 4

X-ray Powder Pattern: Synthetic.
3.29 (100), 3.78 (90), 3.00 (90), 3.36 (80), 3.19 (50), 2.270 (50), 7.63 (40)

Chemistry:

	(1)	(2)
SiO$_2$	64.44	63.92
Al$_2$O$_3$	2.12	
Fe$_2$O$_3$	0.39	
CuO	12.09	21.16
CaO	12.19	14.92
Na$_2$O	2.52	
K$_2$O	1.06	
H$_2$O$^+$	2.59	
CO$_2$	1.18	
SO$_3$	1.08	
Total	99.66	100.00

(1) Vesuvius, Italy; sample contaminated with 13% quartz. (2) CaCuSi$_4$O$_{10}$.

Occurrence: Intimately mixed with quartz (Vesuvius, Italy); in a mudstone xenolith ejected from a scoria cone (Sattelberg volcanic cone, Germany).

Association: Quartz.

Distribution: In Italy, on Vesuvius, Campania. From Wheal Edward, Cornwall, England. At Summit Rock, near Diamond Lake, Douglas Co., Oregon, USA. From the Sattelberg volcanic cone, near Brenk, Eifel district, Germany. At Messina, Transvaal, South Africa.

Name: For a high copper content and presumed similarity to *rivaite*.

Type Material: National Museum of Natural History, Washington, D.C., USA, 135505.

References: (1) Minguzzi, C. (1938) Cuprorivaite: Un nuovo minerale. Period. Mineral., 9(3), 333–345 (in Italian). (2) (1939) Amer. Mineral., 24, 350 (abs. ref. 1). (3) Pabst, A. (1959) Structures of some tetragonal silicates. Acta Cryst., 12, 733–739. (4) Mazzi, F. and A. Pabst (1962) Reexamination of cuprorivaite. Amer. Mineral., 47, 409–411.

Cuprosklodowskite $(H_3O)_2Cu(UO_2)_2(SiO_4)_2 \cdot 2H_2O$

Crystal Data: Triclinic. *Point Group:* $\bar{1}$. As needles, to 3 cm, commonly somewhat flattened on {010} and elongated along [100]. In radiating groups of crystals, matted fibers, and thin crusts.

Physical Properties: *Cleavage:* {100}. Hardness = n.d. D(meas.) = 3.85(2) D(calc.) = 3.83 Radioactive.

Optical Properties: Transparent to translucent. *Color:* Yellowish green to grass-green or greenish yellow. *Luster:* Dull to silky in aggregates.
Optical Class: Biaxial (–). *Pleochroism:* X = nearly colorless; Y = Z = yellowish green. *Orientation:* X = a; Y = c; Z = b. *Dispersion:* r > v. $\alpha = 1.654–1.655$ $\beta = 1.664–1.667$ $\gamma = 1.664–1.667$ 2V(meas.) = Very small to 0°.

Cell Data: *Space Group:* $P\bar{1}$. $a = 7.052(5)$ $b = 9.267(8)$ $c = 6.655(5)$ $\alpha = 109°14(3)'$ $\beta = 89°50(3)'$ $\gamma = 110°1(4)'$ $Z = 1$

X-ray Powder Pattern: Jáchymov, Czech Republic.
8.18 (100), 4.09 (90), 2.97 (80), 4.82 (70), 3.52 (60), 2.21 (60), 2.72 (50)

Chemistry:

	(1)	(2)	(3)
SiO_2	14.4	13.40	13.94
UO_3	64.7	64.96	66.38
CuO	9.0	8.99	9.23
H_2O	12.1	11.72	10.45
Total	100.2	99.07	100.00

(1) Shaba Province, Zaire; Fe_2O_3 trace. (2) Jáchymov, Czech Republic; Fe_2O_3, PbO, and CaO in traces. (3) $(H_3O)_2Cu(UO_2)_2(SiO_4)_2 \cdot 2H_2O$.

Occurrence: A secondary mineral formed by alteration of earlier uranium minerals, both *in situ* and after transport of solutions.

Association: Becquerelite, brochantite, uranophane, kasolite, vandenbrandeite, liebigite, uranophane-beta.

Distribution: In Zaire, from Shaba Province, at Kalongwe and Kambove, with exceptional specimens from the Shinkolobwe mine, Kasolo, and the Musonoi mine, Kolwezi. From Jáchymov (Joachimsthal), Czech Republic. At Johanngeorgenstadt, Saxony, Germany. From West Wheal Owles, St. Just, Cornwall, England. In the USA, in New Mexico, from the Sanostee area, San Juan Co., and in the Woodrow mine, Laguna district, Valencia Co.; in Utah, in the Frey No. 4 mine, Frey Canyon, the Posey mine, Red Canyon, and the Happy Jack mine, White Canyon, San Juan Co. In Canada, from the Nicholson mine, Lake Athabasca, Saskatchewan. At Kenzan, Okayama Prefecture, Japan.

Name: A name supplied by Buttgenbach while introducing Vaes' paper, in the mistaken belief that it was the copper analog of *sklodowskite*.

Type Material: n.d.

References: (1) Vaes, J.P. (1933) Sur un minéral de Kalongwe (Katanga). Ann. Soc. Geol. Belg., Bull. 10, 56, 331–332 (in French). (2) (1934) Amer. Mineral., 19, 235 (abs. ref. 1). (3) Frondel, C. (1958) Systematic mineralogy of uranium and thorium. U.S. Geol. Sur. Bull. 1064, 304–307. (4) Bariand, P., M.T. Le Bihan, and Y. Gillet (1967) Étude cristallographique de la cuprosklodowskite. Bull. Soc. fr. Minéral., 90, 259 (in French). (5) Rosenzweig, A. and R.S. Ryan (1975) Refinement of the crystal structure of cuprosklodowskite, $Cu[(UO_2)_2 (SiO_3OH)_2] \cdot 6H_2O$. Amer. Mineral., 60, 448–453. (6) Stohl, F.V. and D.K. Smith (1981) The crystal chemistry of the uranyl silicate minerals. Amer. Mineral., 66, 610–625.

Crystal Data: Monoclinic. *Point Group:* 2/m. Crystals spear-shaped, minute.
Twinning: Simple, lamellar, polysynthetic on {100}.

Physical Properties: *Cleavage:* Very good on {001}; distinct on {110}. *Fracture:* Uneven.
Tenacity: Brittle. Hardness = 5–6 D(meas.) = 2.85–2.96 D(calc.) = 2.978

Optical Properties: Transparent to translucent. *Color:* Pale rose-red; in thin section, rose-red
to colorless. *Luster:* Vitreous.
Optical Class: Biaxial (+). *Orientation:* $Y = b$; $Z \wedge c = 6°$. *Dispersion:* $r > v$,
marked, inclined. *Absorption:* Moderate. $\alpha = 1.586–1.592$ $\beta = 1.595$ $\gamma = 1.598–1.606$
2V(meas.) = Large.

Cell Data: *Space Group:* $P2_1/a$. $a = 10.906(5)$ $b = 10.521(6)$ $c = 7.518(3)$
$\beta = 109.30(3)°$ Z = 4

X-ray Powder Pattern: Monte Somma, Italy.
3.062 (100), 2.943 (35), 3.259 (30), 2.900 (30), 2.873 (30), 3.034 (25), 2.018 (18)

Chemistry:

	(1)	(2)	(3)
SiO$_2$	31.79	32.36	32.79
MnO		0.71	
MgO	0.00		
CaO	59.16	61.37	61.21
Na$_2$O	trace	0.48	
K$_2$O	trace	0.27	
F	10.05	9.05	10.37
Cl	trace		
H$_2$O$^+$	0.28		
H$_2$O$^-$	0.16		
CO$_2$	0.62		
$-$O = (F, Cl)$_2$	4.23	3.81	4.37
rem.	1.53		
Total	99.36	100.43	100.00

(1) Monte Somma, Italy; remainder is trivalent oxides. (2) Franklin, New Jersey, USA.
(3) Ca$_4$Si$_2$O$_7$F$_2$.

Occurrence: In tuff ejecta (Monte Somma, Italy); in contact metamorphosed limestone
(Franklin, New Jersey, USA); in melilite skarn (Dupezeh Mountain, Iraq).

Association: Augite, "hornblende," diopside, grossular, biotite, phlogopite, monticellite,
wollastonite, calcite, spinel, magnetite, perovskite.

Distribution: At Monte Somma and Vesuvius, Campania, and in the Alban Hills, as at Ariccia,
Lazio, Italy. From Carlingford, Co. Louth, Ireland. At Broadford and Camas Malag, Isle of Skye,
and Camas Mór, Isle of Muck, Scotland. From the Bellerberg volcano, two km north of Mayen,
Eifel district, Germany. In the USA, at Franklin, Sussex Co., New Jersey; in the Alder district,
Custer Co., Idaho; at Crestmore, Riverside Co., California; on Cascade Mountain, Adirondack
Mountains, Essex Co., New York. On Dupezeh Mountain, near Hero Town, Qala-Diza region,
Iraq. In the Hatrurim Formation, Israel. From Fuka, near Bicchu, and in the Sampo and Mihara
mines, Kawakami, Okayama Prefecture, Japan.

Name: From the Greek *cuspis*, for a *spear*, the characteristic shape of the twinned crystals.

References: (1) Dana, E.S. (1892) Dana's system of mineralogy, (6th edition), 533. (2) Tilley,
C.E. (1947) Cuspidine from dolomite contact skarns, Broadford, Skye. Mineral. Mag., 28,
90–95. (3) Van Valkenberg, A. and G.F. Rynders (1958) Synthetic cuspidine. Amer. Mineral.,
43, 1195–1202. (4) Saburi, S., A. Kawahara, C. Henmi, I. Kusachi, and K. Kihara (1977) The
refinement of the crystal structure of cuspidine. Mineral. J. (Japan), 8, 286–298.

Cymrite

$BaAl_2Si_2(O, OH)_8 \cdot H_2O$

Crystal Data: Monoclinic, pseudohexagonal. *Point Group:* 2. As thin pseudohexagonal platelets and prisms, to 7 mm; also fibrous.

Physical Properties: *Cleavage:* Perfect on {001}, imperfect on {110}. *Fracture:* Irregular. *Tenacity:* Brittle. Hardness = 2–3 D(meas.) = 3.413(5) D(calc.) = 3.447 Greenish white fluorescence under UV.

Optical Properties: Translucent. *Color:* Colorless when pure; commonly dark green or brown due to inclusions of alteration products. *Luster:* Vitreous to satiny.
Optical Class: Biaxial (–); pseudouniaxial. $\alpha = 1.611$ $\beta = 1.619$ $\gamma = 1.621$
2V(meas.) = 0°–5°

Cell Data: *Space Group:* $P2_1$. $a = 5.324$–5.334 $b = 36.6$ $c = 7.662$–7.705 $\beta = 90°$
Z = [8]

X-ray Powder Pattern: Brooks Range, Alaska, USA.
2.96 (100), 3.96 (90), 2.67 (70), 7.71 (50), 2.24 (40), 2.21 (40), 1.850 (40)

Chemistry:

	(1)	(2)	(3)
SiO_2	37.65	30.3	30.54
TiO_2		0.0	
Al_2O_3	14.94	25.9	25.91
Fe_2O_3	9.26		
MnO	0.86		
BaO	31.50	38.7	38.97
K_2O		0.03	
H_2O^+	5.31	[4.58]	4.58
Total	99.52	[99.5]	100.00

(1) Benallt mine, Wales. (2) Pacheco Pass, California, USA; by electron microprobe, theoretical H_2O added. (3) $BaAl_2Si_2O_8 \cdot H_2O$.

Occurrence: In veinlets cutting hydrothermal manganese silicate ore (Benallt mine, Wales); in a copper-bearing deposit in dolostones and siderites (Ruby Creek, Alaska, USA); in jadeite graywacke near the contact of an ultramafic rock and the Franciscan Formation (San Benito Co., California, USA); a product of high-pressure metamorphism of manganese-rich rocks (Andros Island, Greece).

Association: Ganophyllite (Benallt mine, Wales); calcite, albite, lawsonite (San Benito Co., California, USA); hyalotekite, banalsite, hyalophane, hedyphane, manganoan biotite (Långban, Sweden).

Distribution: In the Benallt mine, Rhiw, Lleyn Peninsula, Wales. From Långban, Värmland, Sweden. In the USA, from Ruby Creek, Brooks Range, Alaska, and near Pacheco Pass, San Benito Co., California. On Andros Island, Cyclades Islands, Greece. From the Lianyuan-Shaoyang area, Hunan Province, China. In the Shiramaru mine, Tokyo Prefecture, Japan. At Northwest Nelson, South Island, New Zealand. In the Saureisk deposit, Polar Ural Mountains, and near Lake Baikal, eastern Siberia, Russia. From the Black Rock and Hotazel mines, near Kuruman, Cape Province, South Africa. A number of other occurrences are known.

Name: From *Cymru*, the Welsh name for *Wales*, the locality of discovery.

Type Material: The Natural History Museum, London, England, 1944,36 and 1944,48.

References: (1) Campbell Smith, W., F.A. Bannister, and M.H. Hey (1949) Cymrite, a new barium mineral from the Benallt manganese mine, Rhiw, Carnarvonshire. Mineral. Mag., 28, 676–681. (2) (1950) Amer. Mineral., 35, 135 (abs. ref. 1). (3) Runnells, D.D. (1964) Cymrite in a copper deposit, Brooks Range, Alaska. Amer. Mineral., 49, 158–165. (4) Essene, E.J. (1967) An occurrence of cymrite in the Franciscan Formation, California. Amer. Mineral., 52, 1885–1890. (5) Drits, V.A., A.A. Kashaev, and G.V. Sokolova (1975) Crystal structure of cymrite. Kristallografiya (Sov. Phys. Crystal.), 20, 280–286 (in Russian).

Crystal Data: Monoclinic. *Point Group:* $2/m$. As bladed prismatic crystals, commonly fibrous, to 5 mm, and radiating aggregates. *Twinning:* On {110}, rare cyclical eightlings.

Physical Properties: *Cleavage:* Perfect on {001}, {100}. *Tenacity:* Brittle. Hardness = 4–4.5   D(meas.) = 2.165–2.206   D(calc.) = 2.138–2.141

Optical Properties: Transparent. *Color:* Colorless. *Luster:* Vitreous. *Optical Class:* Biaxial (+) or (−). *Orientation:* $X = b$; $Z \wedge c = 35°$. $\alpha = 1.484$–1.492 $\beta = 1.482$–1.496 $\gamma = 1.489$–1.500 2V(meas.) = 65°–73°

Cell Data: *Space Group:* $C2/m$.   a = 18.625   b = 7.508   c = 10.247 $\beta = 108.056°$ Z = 1

X-ray Powder Pattern: Elba, Italy.
3.452 (100), 3.204 (100), 1.873 (75), 8.90 (50), 6.91 (50), 4.97 (50), 4.88 (50)

Chemistry:

	(1)	(2)		(1)	(2)
SiO_2	63.20	64.53	Na_2O	1.20	2.16
Al_2O_3	14.31	13.77	K_2O	1.92	1.31
Fe_2O_3		0.01	Cs_2O	0.96	
MgO	0.00	0.00	H_2O^+		11.23
CaO	5.49	5.60	H_2O^-		1.04
SrO	0.13		H_2O	[12.79]	
			Total	[100.00]	99.65

(1) Elba, Italy; by electron microprobe, H_2O by difference; corresponds to $(Ca_{3.52}K_{0.73}Na_{0.70}$ $Cs_{0.24}Sr_{0.04})_{\Sigma=5.23}Al_{10.10}Si_{37.86}O_{96} \cdot nH_2O$. (2) Hatsuneura, Ogasawara Islands, Japan.

Mineral Group: Zeolite group.

Occurrence: Of hydrothermal origin in basalt, quartz veins, and pegmatites; in hydrothermally altered pumiceous tuffaceous sediments.

Association: Mordenite, heulandite, ferrierite.

Distribution: From San Piero in Campo, Elba, and on the Alpe di Siusi, Trentino–Alto Adige, Italy. From near Zvezdel, Rhodope Mountains, Bulgaria. In Japan, at Hatsuneura, Ogasawara Islands; Hokiya-dake, Nagano Prefecture; and in the Onoyama mine, Kagoshima Prefecture. In the USA, from Altoona, Wahkiakum Co., Washington; at Cape Lookout, Tillamook Co., and Agate Beach, Lincoln Co., Oregon; and in the Lower Geyser basin, Yellowstone National Park, Wyoming. A number of other localities are known.

Name: For Antonio D'Achiardi (1839–1902), Italian mineralogist, whose son described the mineral.

Type Material: National Museum of Natural History, Washington, D.C., USA, 123998.

References: (1) D'Achiardi, G. (1906) Zeoliti del filone della Speranza presso S. Piero in Campo (Elba). Mem. Soc. Tosc. Sci. Nat., 22, 150–165 (in Italian). (2) Berman, H. (1925) Notes on dachiardite. Amer. Mineral., 10, 421–428. (3) Galli, E. (1965) Lo spettro di polvere della dachiardite. Period. Mineral., 34, 129–135 (in Italian). (4) Bonardi, M. (1979) Composition of type dachiardite from Elba: a re-examination. Mineral. Mag., 43, 548–549. (5) Nishido, H. and R. Otsuka (1981) Chemical composition and physical properties of dachiardite group zeolites. Mineral. J. (Japan), 10, 371–384. (6) Quartieri, S., G. Vezzalini, and A. Alberti (1990) Dachiardite from Hokiya-dake: evidence of a new topology. Eur. J. Mineral., 2, 187–193.

Crystal Data: Triclinic. *Point Group:* $\bar{1}$. Crystals short prismatic, to 0.5 mm, dominated by $\{\bar{1}01\}$ and pinacoids $\parallel$ to $[001]$. *Twinning:* Not uncommon, with $\{100\}$ as the composition plane.

Physical Properties: *Cleavage:* Distinct on $\{\bar{1}01\}$ and $\{010\}$, indistinct on $\{100\}$. *Tenacity:* Brittle. Hardness = 7.5 D(meas.) = 2.84(2) D(calc.) = 2.81 Strongly cathodoluminescent.

Optical Properties: Transparent. *Color:* Colorless, white, pink, brown. *Luster:* Vitreous. *Optical Class:* Biaxial (–). *Orientation:* $X \wedge c = 7°$. *Dispersion:* $r > v$, very weak. $\alpha = 1.575(2)$ $\beta = 1.590(2)$ $\gamma = 1.601(2)$ 2V(meas.) = $72(2)°$

Cell Data: *Space Group:* $P\bar{1}$. a = 7.371(3) b = 7.730(3) c = 6.912(2) $\alpha = 106°14(2)'$ $\beta = 111°27(2)'$ $\gamma = 100°0(2)'$ Z = 1

X-ray Powder Pattern: Ascension Island.
4.20 (vs), 3.58 (vs), 3.08 (vs), 5.90 (s), 4.31 (s), 6.54 (m), 3.36 (m)

Chemistry:

	(1)	(2)		(1)	(2)
SiO_2	61.85	63.07	MgO		0.10
TiO_2		1.23	CaO		0.02
ZrO_2	21.70	19.37	BaO		0.11
Al_2O_3		0.06	Na_2O	1.75	0.04
Fe_2O_3	0.37		K_2O	14.60	15.60
FeO		0.28	H_2O	0.64	
MnO		0.02	P_2O_5		0.04
			Total	100.91	99.94

(1) Ascension Island; corresponds to $(K_{1.79}Na_{0.32})_{\Sigma=2.11}Zr_{1.01}Si_{5.95}O_{15}$. (2) Dalsfjorden, Norway; by electron microprobe, average of ten analyses; corresponds to $(K_{1.90}Fe_{0.02}Mg_{0.02}Na_{0.01})_{\Sigma=1.95}(Zr_{0.90}Ti_{0.09}Al_{0.01})_{\Sigma=1.00}Si_{6.01}O_{15}$.

Occurrence: A rare accessory in ejected blocks of alkalic granite in trachytic and basaltic tuffs (Ascension Island); in syenite (Serra de Agua de Pau, Azores); in lamproite (Cancarix, Spain).

Association: Quartz, aegirine, arfvedsonite, aenigmatite (Ascension Island); sanidine, arfvedsonite, quartz, aegirine, fayalite, astrophyllite, pyrrhotite (Serra de Agua de Pau, Azores); pyrophanite, elpidite, monazite, janhaugite, kupletskite (Gjerdingen, Norway).

Distribution: On Green Mountain and Middleton Peak, Ascension Island, southern Atlantic Ocean. From Serra de Agua de Pau, São Miguel, Azores. Near Cancarix, Albacete Province, Spain. At Dalsfjorden, Sunnfjord, and Gjerdingen, near Oslo, Norway. In the Straumsvola complex, Dronning Maud Land, Antarctica. From the Murun massif, southwest of Olekminsk, and other less-well-defined localities in the Aldan Shield, Yakutia, Russia.

Name: For Reginald Aldworth Daly (1871–1957), Professor of Geology, Harvard University, Cambridge, Massachusetts, USA.

Type Material: The Natural History Museum, London, England, 64685; National Museum of Natural History, Washington, D.C., USA, 113106.

References: (1) Van Tassel, R. and M.H. Hey (1952) Dalyite, a new potassium zirconium silicate from Ascension Island, Atlantic. Mineral. Mag., 29, 850–857. (2) (1952) Amer. Mineral., 37, 1071 (abs. ref. 1). (3) Fleet, S.G. (1965) The crystal structure of dalyite. Zeits. Krist., 121, 349–368. (4) Robins, B., H. Furnes, and P. Ryan (1983) A new occurrence of dalyite. Mineral. Mag., 47, 93–94.

Crystal Data: Cubic. *Point Group:* $\bar{4}3m$. As octahedral and dodecahedral crystals, to 10 cm; as irregular segregations.

Physical Properties: *Cleavage:* Poor on {111}, {1$\bar{1}$1}. *Fracture:* Subconchoidal to uneven. *Tenacity:* Brittle. Hardness = 5.5–6   D(meas.) = 3.28–3.46   D(calc.) = 3.36

Optical Properties: Semitransparent. *Color:* Yellow, pink, reddish brown, red; in thin section, colorless to pink. *Luster:* Vitreous or greasy. *Optical Class:* Isotropic. $n = 1.747–1.771$

Cell Data: *Space Group:* $P\bar{4}3n$.   a = 8.20–8.23   Z = 2

X-ray Powder Pattern: Needlepoint Mountain, Canada.
3.35 (100), 1.932 (70), 1.012 (60), 2.193 (50), 1.118 (50), 1.043 (50), 3.68 (40)

Chemistry:

	(1)	(2)	(3)	(4)
SiO$_2$	29.48	34.08	30.76	32.26
FeO	37.53	30.66	44.27	51.44
MnO	11.53	13.28	7.37	
ZnO	4.87	2.97	0.89	
BeO	14.17	16.32	[13.42]	13.42
CaO			0.07	
S	5.04	5.61	5.41	5.74
$-$O = S	2.52	2.80	[2.70]	2.86
Total	100.10	100.12	[99.49]	100.00

(1) Redruth, England; corresponds to $(Fe_{2.96}^{2+}Mn_{0.92}^{2+}Zn_{0.34})_{\Sigma=4.22}Be_{3.32}Si_{2.78}O_{12}S$. (2) Imalka, Russia; corresponds to $(Fe_{2.28}^{2+}Mn_{1.00}^{2+}Zn_{0.19})_{\Sigma=3.47}Be_{3.48}Si_{3.02}O_{12}S$. (3) Bartlett, New Hampshire, USA; by electron microprobe, BeO calculated from stoichiometry. (4) $Fe_4Be_3(SiO_4)_3S$.

Polymorphism & Series: Forms two series, with genthelvite, and with helvite.

Occurrence: In granites, granite pegmatites, zones of contact metamorphism, skarns, and gneisses.

Association: Magnetite, garnet, fluorite (skarn); albite, cassiterite, pyrite, muscovite (tin-bearing pegmatites); arsenopyrite, quartz, chlorite (hydrothermal deposits).

Distribution: In the USA, at Rockport and Gloucester, Essex Co., Massachusetts; on Moat Mountain, Conway, Carroll Co., New Hampshire; at Iron Mountain, Sierra Co., New Mexico; and from the Black Hills, south of Jerome, Yavapai Co., Arizona. From Needlepoint Mountain, McDame area, British Columbia, and on Walrus Island, James Bay, Quebec, Canada. In Sweden, from Yxsjö, Örebro. At Redruth, St. Just, Lanlivery, Falmouth, and Lanivet, Cornwall, England. In Russia, from Imalka, Transbaikal, and other less-well-defined localities. On Mt. Francisco, Ribawa area, Western Australia. In the Mihara mine, Hiroshima Prefecture, Japan.

Name: To honor James Dwight Dana (1813–1895), American geologist and mineralogist, Yale University, New Haven, Connecticut, USA.

Type Material: Harvard University, Cambridge, Massachusetts, 85384; National Museum of Natural History, Washington, D.C., USA, 124353; The Natural History Museum, London, England, 1976,422.

References: (1) Dana, E.S. (1892) Dana's system of mineralogy, (6th edition), 435. (2) Deer, W.A., R.A. Howie, and J. Zussman (1963) Rock-forming minerals, v. 4, framework silicates, 303–309. (3) Vlasov, K.A., Ed. (1966) Mineralogy of rare elements, v. II, 119–126. (4) Glass, J.J., R.H. Jahns, and R.E. Stevens (1944) Helvite and danalite from New Mexico and the helvite group. Amer. Mineral., 29, 163–191. (5) Thompson, R.M. (1957) Danalite from British Columbia. Can. Mineral., 6, 68–71. (6) Dunn, P.J. (1976) Genthelvite and the helvine group. Mineral. Mag., 40, 627–636. (7) Hassan, I. and H.D. Grundy (1985) The crystal structures of helvite group minerals, $(Mn, Fe, Zn)_8(Be_6Si_6O_{24})S_2$. Amer. Mineral., 70, 186–192.

Crystal Data: Orthorhombic. *Point Group:* $2/m \; 2/m \; 2/m$. Crystals prismatic, rhomboidal to square in cross section, to 50 cm; also as disseminated masses.

Physical Properties: *Cleavage:* Indistinct on $\{001\}$. *Fracture:* Uneven to subconchoidal. *Tenacity:* Brittle. Hardness = 7–7.25   D(meas.) = 2.93–3.02   D(calc.) = 2.99

Optical Properties: Transparent to translucent. *Color:* Colorless, white, wine-yellow, yellowish brown, greenish; colorless in thin section. *Streak:* White. *Luster:* Vitreous to greasy. *Optical Class:* Biaxial (–). *Dispersion:* $r < v$, strong. $\alpha = 1.627–1.633$ $\beta = 1.630–1.636$ $\gamma = 1.633–1.639$ 2V(meas.) = $\sim 90°$

Cell Data: *Space Group:* $Pnam$. $a = 8.038(3)$ $b = 8.752(5)$ $c = 7.730(3)$ $Z = 4$

X-ray Powder Pattern: Maglovec Hill, Prešov, Slovakia.
3.57 (100), 2.961 (80), 2.743 (65), 2.655 (55), 2.729 (40), 1.435 (40), 3.65 (35)

Chemistry:

	(1)
SiO_2	49.18
B_2O_3	27.65
Al_2O_3	0.42
FeO	0.26
MnO	0.02
MgO	0.11
CaO	22.10
SrO	0.03
Na_2O	0.07
H_2O	0.24
Total	100.08

(1) Maglovec Hill, Prešov, Slovakia; corresponds to $(Ca_{0.97}Na_{0.01}Mg_{0.01}Fe_{0.01})_{\Sigma=1.00}$
$(B_{1.95}Al_{0.02})_{\Sigma=1.97}Si_{2.01}O_{8.00}$.

Occurrence: In granite and metamorphosed carbonate rocks associated with hydrothermal activity; in evaporites.

Association: Titanite, axinite, tourmaline, mica, quartz, grossular, albite, fluorite, apophyllite, stilbite, datolite, bakerite, dolomite, calcite, anhydrite, gypsum.

Distribution: In the USA, at Danbury, Fairfield Co., Connecticut; and at Russell, St. Lawrence Co., New York. In Mexico, large crystals from Charcas, San Luis Potosí; and in the La Verde mine, near La Huerta, Baja California. From Chapare, Cochabamba, Bolivia. In Switzerland, on the Piz Valatscha, Mt. Scopi, south of Disentis, Graubünden. In Madagascar, very large crystals from Anjanabonoina; at Maharitra, Mt. Bity; at Imalo, near Mania, south of the Betafo district; and at Sahasonjo, northeast of Andina. In Russia, from Alabashka, near Mursinka, Ural Mountains; large crystals from Vostochnaïa, Siberia, and at Dal'negorsk, Primorski Krai; numerous other minor occurrences. From the Mogok district, Myanmar (Burma). In Japan, from the Obira mine, Bungo, Oita Prefecture, and at Toroku, Miyazaki Prefecture.

Name: For the locality at Danbury, Connecticut, USA.

References: (1) Dana, E.S. (1892) Dana's system of mineralogy, (6th edition), 490–492.
(2) Černý, P. (1973) The occurrence of danburite in xenoliths in andesite from Maglovec, near Prešov, Czechoslovakia. Acta Univ. Carol., Geol., Rost vol., 111–118 (in Czech with English abs.).
(3) Phillips, M.W., G.V. Gibbs, and P.H. Ribbe (1974) The crystal structure of danburite: a comparison with anorthite, albite, and reedmergnerite. Amer. Mineral., 59, 79–85.

Mn$_2$(Fe^{2+}, Mg)$_5$Si$_8$O$_{22}$(OH)$_2$ **Dannemorite**

Crystal Data: Monoclinic. *Point Group:* 2/m. Crystals prismatic, fibrous, to 1 cm; columnar or asbestiform.

Physical Properties: *Cleavage:* Perfect on {110}, with intersections at ~55° and ~125°. *Tenacity:* [Brittle.] Hardness = [5–6] D(meas.) = 3.50 D(calc.) = [3.35]

Optical Properties: Transparent to translucent. *Color:* Yellowish brown, greenish gray, colorless. *Luster:* Silky to vitreous.
Optical Class: Biaxial (–). *Pleochroism:* X = red-violet; Y = pale violet; Z = blue. *Orientation:* $Y = b$; $Z \wedge c = 15°$. *Dispersion:* $r < v$. $\alpha = 1.666$ $\beta = 1.682$ $\gamma = 1.698$ 2V(meas.) = 88°

Cell Data: *Space Group:* $C2/m$. a = 9.561 (ICDD 23-302). b = 18.28 c = 5.348 $\beta = 102.09°$ Z = 2

X-ray Powder Pattern: Väster Silfberg, Sweden. (ICDD 23-302).
8.40 (100), 3.08 (60), 3.28 (20), 2.764 (18), 4.70 (12), 2.637 (10), 2.201 (10)

Chemistry:

	(1)	(2)
SiO$_2$	50.74	51.6
TiO$_2$	0.06	0.13
Al$_2$O$_3$	0.88	0.65
Fe$_2$O$_3$	1.80	
FeO	24.13	23.9
MnO	7.38	8.1
MgO	10.57	11.1
CaO	2.00	0.83
Na$_2$O	0.22	1.1
K$_2$O	0.08	0.09
F	0.07	
H$_2$O$^+$	1.94	
Total	99.87	97.5

(1) Uttersvik, Sweden; corresponds to $(Fe^{2+}_{3.10}Mg_{2.42}Mn_{0.96}Ca_{0.33}Fe^{3+}_{0.21}Na_{0.07}Ti_{0.01}K_{0.01})_{\Sigma=7.11}$ $(Si_{7.80}Al_{0.16})_{\Sigma=7.96}O_{22}[(OH)_{1.99}F_{0.03}]_{\Sigma=2.02}$. (2) Haute-Maurienne, France; by electron microprobe.

Polymorphism & Series: Forms a series with tirodite.

Mineral Group: Amphibole (Fe–Mn–Mg) group: Mg/(Mg + Fe^{2+}) < 0.5; (Ca + Na)$_B$ < 1.34; Li < 1.0; Mn ≥ 0.5.

Occurrence: An uncommon mineral in metamorphosed iron-poor manganiferous rocks.

Association: Calcite, quartz, garnet.

Distribution: From Dannemora, Uppland; Uttersvik and Nävekvarn, Södermanland; in the Brunsjö mine, near Grythyttan, Örebro; and at Väster Silfberg, Värmland, Sweden. At Haute-Maurienne, Isère, France. From Guarulhos, São Paulo, Brazil. At Paddy's River mine, Australian Capital Territory, and from Broken Hill, New South Wales, Australia.

Name: For the original locality, Dannemora, Sweden.

References: (1) Dana, E.S. (1892) Dana's system of mineralogy, (6th edition), 386, 391, 395. (2) Deer, W.A., R.A. Howie, and J. Zussman (1963) Rock-forming minerals, v. 2, chain silicates, 239. (3) Mottana, A. (1986) Blueschist-facies metamorphism of manganiferous cherts: a review of the alpine occurrences. In: B.W. Evans and E.H. Brown, Eds., Blueschists and eclogites, Geol. Soc. Amer. Memoir 164, 267–299.

Crystal Data: Hexagonal. *Point Group:* $6/m \, 2/m \, 2/m$ (probable). As masses to 5 cm.

Physical Properties: Hardness = 5 D(meas.) = 2.92 D(calc.) = 2.80

Optical Properties: Semitransparent. *Color:* Colorless, white, rarely brownish or pale to deep blue.
Optical Class: Unixial (−). *Pleochroism:* O = violet; E = blue. $\omega = 1.580(2)$ $\epsilon = 1.575(2)$

Cell Data: *Space Group:* $P6/mcc$ (probable). $a = 10.32$ $c = 14.39$ $Z = 2$

X-ray Powder Pattern: Dara-i-Pioz massif, Tadzhikistan.
3.26 (100), 2.93 (65), 7.09 (60), 2.56 (55), 4.13 (50), 2.76 (45), 4.43 (40)

Chemistry:

	(1)
SiO_2	63.65
ZrO_2	5.00
RE_2O_3	0.96
Fe_2O_3	1.85
Nb_2O_5	0.90
MnO	8.25
ZnO	7.85
CaO	0.57
Li_2O	1.74
Na_2O	2.96
K_2O	5.14
LOI	0.58
Total	99.45

(1) Dara-i-Pioz massif, Tadzhikistan; corresponds to $K_{1.23}(Na_{1.08}Li_{0.58}Ca_{0.11})_{\Sigma=1.77}$ $Li_{0.73}(Mn_{1.31}Zn_{1.10})_{\Sigma=2.41}(Zr_{0.46}Fe_{0.26}Nb_{0.07})_{\Sigma=0.79}Si_{12}[O, (OH)]_{30}$.

Mineral Group: Milarite group.

Occurrence: In an alkalic massif.

Association: Aegirine, quartz, sogdianite, eudialyte, manganoan pectolite, polylithionite.

Distribution: In the Dara-i-Pioz massif, Alai Range, Tien Shan, Tadzhikistan.

Name: For the occurrence in the Dara-i-Pioz massif, Tadzhikistan.

Type Material: Institute of Mineralogy and Geochemistry of Rare Elements, Moscow; A.E. Fersman Mineralogical Museum, Academy of Sciences, Moscow, Russia.

References: (1) Semenov, E.I., V.D. Dusmatov, A.P. Khomyakov, A.A. Voronkov, and M.E. Kazakova (1975) Darapiosite [darapiozite], a new mineral of the milarite group. Zap. Vses. Mineral. Obshch., 104, 583–585 (in Russian). (2) (1976) Amer. Mineral., 61, 1053–1054 (abs. ref. 1).

Crystal Data: Monoclinic. *Point Group:* $2/m$. Crystals typically short prismatic, with prominent {110}, {011}, or tabular on {010}, to 12 cm. Botryoidal or globular, having columnar structure; granular to compact; cryptocrystalline.

Physical Properties: *Fracture:* Conchoidal to uneven. *Tenacity:* Brittle. Hardness = 5–5.5 D(meas.) = 2.96–3.00 D(calc.) = [3.00] Commonly fluoresces blue under SW UV.

Optical Properties: Transparent to translucent, rarely opaque. *Color:* Colorless or white, commonly with a greenish tinge; may be grayish, yellow, green, red, pink; in thin section, colorless. *Luster:* Vitreous, rarely subresinous on fracture surface.
Optical Class: Biaxial (–). *Orientation:* $Y = b$; $Z \wedge c = -1°$ to $-4°$. *Dispersion:* $r > v$, weak.
$\alpha = 1.622–1.626$ $\beta = 1.649–1.654$ $\gamma = 1.666–1.670$ 2V(meas.) = $72°–75°$

Cell Data: *Space Group:* $P2_1/c$. a = 4.832(4) b = 7.608(4) c = 9.636(8) $\beta = 90.40(7)°$ Z = 4

X-ray Powder Pattern: St. Andreasberg, Germany. (ICDD 11-70).
3.11 (100), 2.855 (65), 2.189 (60), 3.76 (45), 1.875 (40), 1.644 (40), 2.986 (35)

Chemistry:

	(1)
SiO$_2$	38.00
B$_2$O$_3$	19.38
Al$_2$O$_3$	0.65
Fe$_2$O$_3$	0.28
MnO	0.39
MgO	0.09
CaO	35.39
H$_2$O$^+$	5.71
H$_2$O$^-$	0.26
Total	100.15

(1) Silver Harbour, Lake Superior, Ontario, Canada; corresponds to $(Ca_{1.02}Mg_{0.02}Mn_{0.01})_{\Sigma=1.05}$ $B_{0.90}(Si_{1.03}Al_{0.02})_{\Sigma=1.05}O_{4.05}(OH)_{0.95}$.

Mineral Group: Gadolinite group.

Occurrence: A secondary mineral in mafic igneous rocks; in geodes in tuffs; in skarns in limestones; in serpentinites and "hornblende" schists; from some ore veins.

Association: Calcite, prehnite, zeolites, danburite, axinite, garnet.

Distribution: Widespread; some localities for fine specimens are: in Norway, from Arendal, and north of Hardangerfjiord. At St. Andreasberg, Harz Mountains, Germany. In Italy, from Serra dei Zanchetti, near Bologna, Emilia-Romagna, and from Alpe di Siusi, Trentino-Alto Adige. From the Kratzenberg, Habachtal, Austria. In Japan, from the Iwato copper mine, Miyazaki Prefecture. As exceptional crystals from Dal'negorsk, Primorski Krai, Russia. In the USA, large crystals from the Lane quarry, Westfield, Hampden Co., Massachusetts; in the Roncari quarry, East Granby, Hartford Co., Connecticut; from Bergen Hill, Hudson Co., Great Notch, Essex Co., and at Paterson and Prospect Park, Passaic Co., New Jersey; in the Goose Creek quarry, near Leesburg, Loudoun Co., Virginia. Very large crystals from Charcas, San Luis Potosí, Mexico.

Name: From the Greek *to divide*, in reference to the granular structure of a massive variety.

References: (1) Dana, E.S. (1892) Dana's system of mineralogy, (6th edition), 502–505. (2) Deer, W.A., R.A. Howie, and J. Zussman (1963) Rock-forming minerals, v. 3, sheet silicates, 171–175. (3) Foit, F.F., Jr., M.W. Phillips, and G.V. Gibbs (1973) A refinement of the crystal structure of datolite, CaBSiO$_4$(OH). Amer. Mineral., 58, 909–914.

Crystal Data: Triclinic. *Point Group:* $\bar{1}$ or 1. As pseudohexagonal grains up to 5 mm. *Twinning:* Twins parallel or complex, with the composition plane parallel to the twin axis.

Physical Properties: *Fracture:* Conchoidal. Hardness = 5   D(meas.) = 2.76   D(calc.) = 2.754

Optical Properties: Semitransparent. *Color:* Colorless. *Luster:* Vitreous. *Optical Class:* Biaxial (+). $\alpha = 1.623(2)$ $\beta = $ n.d. $\gamma = 1.668(2)$ 2V(meas.) = n.d.

Cell Data: *Space Group:* $P\bar{1}$ or $P1$. a = 7.26–7.272   b = 7.480–7.51   c = 6.910–6.92 $\alpha = 105.55°-105.6°$ $\beta = 112.82°-112.9°$ $\gamma = 99.4°-99.42°$ Z = [1]

X-ray Powder Pattern: Murun massif, Russia.
3.51 (100), 3.002 (100), 2.615 (90), 2.789 (70), 5.88 (50), 4.09 (50), 3.34 (50)

Chemistry:

	(1)	(2)	(3)
SiO_2	67.47	66.70	67.44
TiO_2	15.07	14.48	14.94
ZrO_2		0.09	
FeO	trace	0.15	
MgO	trace		
Na_2O	trace	0.02	
K_2O	17.40	17.67	17.62
Total	99.94	99.11	100.00

(1) Murun massif, Russia; by electron microprobe, average of four analyses; corresponds to $K_{1.98}Ti_{1.01}Si_{6.01}O_{15.03}$. (2) Smoky Butte, Montana, USA; by electron microprobe. (3) $K_2TiSi_6O_{15}$.

Occurrence: In contact metamorphic rocks in an alkalic massif (Murun massif, Russia); of magmatic origin, lining vugs in lamproites (Smoky Butte, Montana, USA).

Association: Aegirine, pectolite, titanite, quartz, potassic feldspar, calcite (Murun massif, Russia); phlogopite, diopside, leucite, analcime, olivine, pseudobrookite, potassian richterite, Ca-Sr carbonates, priderite, barite (Smoky Butte, Montana, USA).

Distribution: From near the Davan Spring, in the Murun massif, southwest of Olekminsk, Yakutia, Russia. At Smoky Butte, Garfield Co., Montana, USA.

Name: For the Davan Spring, in the Murun massif, Russia.

Type Material: A.E. Fersman Mineralogical Museum, Academy of Sciences, Moscow, Russia.

References: (1) Lazebnik, K.A., Y.D. Lazebnik, and V.J. Mokhotko (1984) Davanite, $K_2TiSi_6O_{15}$, a new alkali titanosilicate. Zap. Vses. Mineral. Obshch., 113, 95–97 (in Russian). (2) (1985) Amer. Mineral., 70, 214–215 (abs. ref. 1). (3) Wagner, C. and D. Velde (1986) Davanite, $K_2TiSi_6O_{15}$, in the Smoky Butte (Montana) lamproites. Amer. Mineral., 71, 1473–1475. (4) Raade, G. (1987) Davanite, $K_2TiSi_6O_{15}$, in the Smoky Butte (Montana) lamproites: discussion of X-ray powder data. Amer. Mineral., 72, 1014–1015.

Crystal Data: Monoclinic. *Point Group:* $2/m$. Asbestiform; as small bundles of extremely fine fibers, to several cm; these may be folded or imbricated.

Physical Properties: *Cleavage:* Good on {100}. *Fracture:* Cross fractures on {010}. *Tenacity:* Brittle. Hardness = 2–3   D(meas.) = 3.30–3.38   D(calc.) = 3.34

Optical Properties: Transparent to translucent. *Color:* Creamy white to very pale rose; colorless to faint yellow in thin section.
Optical Class: Biaxial (–). *Orientation:* $Z = b$; $X \simeq \perp$ {100}. $\alpha = 1.660(5)$ $\beta = 1.684(2)$ $\gamma = 1.690(2)$ 2V(meas.) = 48°–70°

Cell Data: *Space Group:* $P2_1/m$. $a = 9.518(6)$ $b = 5.753(2)$ $c = 12.04(1)$ $\beta = 108.00(5)°$ $Z = 2$

X-ray Powder Pattern: Ottré, Belgium.
3.511 (100), 2.870 (60), 3.103 (45), 4.290 (40), 5.719 (35), 2.840 (35), 8.51 (30)

Chemistry:

	(1)	(2)	(3)
SiO_2	37.82	37.45	37.84
TiO_2		trace	
Al_2O_3	46.88	48.09	48.15
Fe_2O_3	1.10		
Cr_2O_3		trace	
FeO		1.29	
MnO	9.08	9.14	11.17
CuO	0.79		
ZnO	0.49		
MgO	0.44	0.45	
H_2O	[2.83]	[2.82]	2.84
P_2O_5	0.35		
Total	[99.78]	[99.24]	100.00

(1) Ottré, Belgium; by electron microprobe, total Fe as Fe_2O_3, H_2O calculated from stoichiometry. (2) Recht, Belgium; by electron microprobe, H_2O calculated from stoichiometry. (3) $MnAl_6Si_4O_{17}(OH)_2$.

Occurrence: In quartz veins cutting Mn, Al-rich metapelites, derived from shales subjected to low-grade metamorphism.

Association: Quartz, pyrophyllite, ottrélite, andalusite, sudoite, kaolinite, rutile, dickite (Ottré, Belgium); chloritoid, hematite, chlorite (Sart-Close, Belgium).

Distribution: In Belgium, in the Stavelot massif, at Ottré, at Sart-Close, near Salmchâteau, at Regne, and at Recht.

Name: For Charles Joseph Davreux (1800–1863), Belgian pharmacist and natural scientist, Professor of Mineralogy at the University of Liège, Belgium.

Type Material: Royal Institute of Natural Sciences of Belgium, Brussels, Belgium.

References: (1) Dana, E.S. (1892) Dana's system of mineralogy, (6th edition), 706. (2) Fransolet, A.-M. and P. Bourguignon (1976) Précisions minéralogiques sur la davreuxite. Compt. Rendus Acad. Sci. Paris, 283, 295–296 (in French). (3) (1978) Amer. Mineral., 63, 795 (abs. ref. 2). (4) Fransolet, A.-M., K. Abraham, and K. Sahl (1984) Davreuxite: a reinvestigation. Amer. Mineral., 69, 777–782. (5) Sahl, K., P.G. Jones, and G.M. Sheldrick (1984) The crystal structure of davreuxite, $MnAl_6Si_4O_{17}(OH)_2$. Amer. Mineral., 69, 783–787.

Crystal Data: Hexagonal. *Point Group:* $6/m$. As prismatic hexagonal crystals, striated lengthwise, up to 4 cm.

Physical Properties: *Cleavage:* Perfect on $\{10\overline{1}0\}$, perfect to indistinct on $\{0001\}$. *Fracture:* Conchoidal to uneven. *Tenacity:* Brittle. Hardness = 5.5–6 D(meas.) = 2.42–2.53 D(calc.) = [2.50]

Optical Properties: Transparent to translucent. *Color:* Colorless to white; colorless in thin section. *Luster:* Vitreous.
Optical Class: Uniaxial (+). $\omega = 1.515$–1.519 $\epsilon = 1.519$–1.522

Cell Data: *Space Group:* $P6_3/m$. a = 12.705(4) c = 5.368(3) Z = 1

X-ray Powder Pattern: Vesuvius, Italy; can be distinguished from quadridavyne only by single-crystal diffraction.
4.80 (100), 3.67 (100), 3.28 (100), 2.670 (60), 2.121 (60), 2.756 (50), 2.652 (50)

Chemistry:

	(1)	(2)
SiO_2	32.21	32.23
Al_2O_3	29.22	28.98
CaO	12.60	10.36
Na_2O	10.14	11.01
K_2O	6.79	7.11
Cl	6.71	6.25
CO_2		1.26
SO_3	4.43	4.11
$-O = Cl_2$	1.51	1.56
Total	100.59	99.75

(1) Monte Somma, Italy; corresponds to $(Na_{3.54}Ca_{2.43}K_{1.56})_{\Sigma=7.53}(Al_{6.20}Si_{5.80})_{\Sigma=12.00}O_{24}$ $[Cl_{2.05}(SO_4)_{0.60}]_{\Sigma=2.65}$. (2) Do.; corresponds to $(Na_{3.86}Ca_{2.01}K_{1.64})_{\Sigma=7.51}(Al_{6.18}Si_{5.82})_{\Sigma=12.00}O_{24}$ $[Cl_{2.13}(SO_4)_{0.56}(CO_3)_{0.31}]_{\Sigma=3.00}$.

Mineral Group: Cancrinite group.

Occurrence: In volcanic ejecta and in leucite-rich lavas.

Association: Nepheline, leucite.

Distribution: From Monte Somma and Vesuvius, Campania, and Pitigliano, Tuscany, Italy. On the Island of Zabargad (Zabirget or St. Johns), in the Red Sea, Egypt.

Name: For the great English chemist, Sir Humphrey Davy (1778–1829).

References: (1) Dana, E.S. (1892) Dana's system of mineralogy, (6th edition), 428–429. (2) Deer, W.A., R.A. Howie, and J. Zussman (1963) Rock-forming minerals, v. 4, framework silicates, 310–320. (3) Bariand, P., F. Cesbron, and R. Giraud (1968) Une nouvelle espèce minérale: l'afghanite de Sar-e-Sang, Badakhshan, Afghanistan. Comparison avec les minéraux du groupe de la cancrinite. Bull. Soc. fr. Minéral., 91, 34–42 (in French with English abs.). (4) Bonaccorsi, E., S. Merlino, and M. Pasero (1990) Davyne: its structural relationships with cancrinite and vishnevite. Neues Jahrb. Mineral., Monatsh., 97–112.

$(Fe^{2+}, Mn^{2+})_6(Fe^{3+}, Al)_3Si_6O_{20}(OH)_5$ **Deerite**

Crystal Data: Monoclinic. *Point Group:* $2/m$. Crystals acicular, lozenge-shaped in cross section, to 1 mm. *Twinning:* Submicroscopic with twin axis [001], pervasive.

Physical Properties: *Cleavage:* Good on {110}. Hardness = n.d. D(meas.) = 3.837 D(calc.) = [3.86]

Optical Properties: Nearly opaque, transparent on thin edges. *Color:* Black.
Optical Class: Biaxial. *Pleochroism:* Slight; X = dark brown; $Y = Z$ = dark brown-black.
Orientation: $Z = c$. $\alpha = 1.840(1)$ β = n.d. $\gamma = 1.870(1)$ 2V(meas.) = n.d.

Cell Data: *Space Group:* $P2_1/a$. $a = 10.786(8)$ $b = 18.88(2)$ $c = 9.564(9)$
$\beta = 107.45(5)°$ $Z = 4$

X-ray Powder Pattern: Laytonville district, California, USA. (ICDD 19-421).
9.03 (100), 3.01 (70), 2.64 (55), 2.54 (25), 3.22 (20), 2.37 (20), 9.45 (16)

Chemistry:

	(1)	(2)	(3)	(4)
SiO_2	33.44	33.40	33.27	33.50
Al_2O_3	0.92	0.16	0.41	
Fe_2O_3	22.09	20.89	21.41	22.26
FeO	36.77	34.79	35.64	40.06
MnO	2.86	5.72	0.74	
MgO	0.15	0.25	0.64	
H_2O	4.25	[4.25]	[4.25]	4.19
Total	100.48	[99.46]	[96.36]	100.00

(1) Laytonville district, California, USA. (2) Do.; by electron microprobe. (3) Ambin massif, Italy; by electron microprobe; H_2O and Fe^{2+}:Fe^{3+} in (2–3) assumed same as (1). (4) $Fe_6^{2+}Fe_3^{3+}Si_6O_{20}(OH)_5$.

Occurrence: In meta-ironstones and Cu-Fe sulfide deposits, in blueschist facies metasediments formed at high pressure and low temperature.

Association: Howieite, zussmanite, stilpnomelane, spessartine, riebeckite, quartz, aegirine, grunerite, aragonite, manganoan siderite, ferroan kutnohorite (Laytonville district, California, USA).

Distribution: In the USA, in California, in the Laytonville quarry, at Covelo, and at Burn's Flats, Mendocino Co.; at Ward Creek, Cazadero, Sonoma Co.; and at Panoche Pass, San Benito Co.; in Oregon, at Wild Horse Lookout, Curry Co., and in the Powers quarry, Coos Co. From Salbertrand, Viafiorcia, and Beth-Ghinivert, Piedmont, Italy. At Termignon, Haute-Savoie, France. From Sifnos, Greece. At Süpüren-Karaalan, Eskişehir Province, Turkey. From the Diahöt Valley, and at Ouegoa Koumec, New Caledonia. A few other occurrences are known.

Name: To honor Professor William Alexander Deer (1910–), mineralogist-petrologist, Cambridge University, Cambridge, England.

Type Material: National Museum of Natural History, Washington, D.C., USA, 109455; The Natural History Museum, London, England, 1964,543.

References: (1) Agrell, S.O., M.G. Bown, and D. McKie (1965) Deerite, howieite and zussmanite, three new minerals from the Franciscan of the Laytonville district, Mendocino County, California. MSA meeting, Bozeman, Montana, July 26–31, 1964. Amer. Mineral., 50, 278 (abs.). (2) Agrell, S.O. and M. Gay (1970) De la deerite dans les Alpes franco-italiennes. Bull. Soc. fr. Minéral., 93, 263–264 (in French). (3) Fleet, M.E. (1977) The crystal structure of deerite. Amer. Mineral., 62, 990–998. (4) Muir Wood, R. (1979) The iron-rich blueschist facies minerals: I. Deerite. Mineral. Mag., 43, 251–259.

Defernite

$Ca_6(CO_3)_{2-x}(SiO_4)_x(OH)_7(Cl, OH)_{1-2x}$ (x ≤ 0.5)

Crystal Data: Orthorhombic. *Point Group:* $2/m \; 2/m \; 2/m$. As anhedral platy crystals, to 2 cm, flattened on {010}; in fan-shaped forms.

Physical Properties: *Cleavage:* Perfect on {010}; distinct on {100}. *Fracture:* Conchoidal to splintery. Hardness = n.d. D(meas.) = 2.34–2.5 D(calc.) = 2.31–2.42

Optical Properties: Transparent. *Color:* Colorless to deep red or rose-brown. *Streak:* Pink. *Luster:* Vitreous.
Optical Class: Biaxial (–). *Pleochroism:* Strong; X = ocher-brown; Y = brownish orange to ocher-brown; Z = colorless to yellow. *Orientation:* $X = c$; $Y = b$; $Z = a$. *Dispersion:* $r > v$, weak. $\alpha = 1.544(2)$ $\beta = 1.580(2)$ $\gamma = 1.584(2)$ 2V(meas.) = 38(2)°

Cell Data: *Space Group:* $Pnam$. a = 17.82–17.86 b = 22.62–22.78 c = 3.63–3.66 Z = 8

X-ray Powder Pattern: Güneyce-İkizdere, Turkey.
11.37 (100), 2.899 (50), 3.045 (40), 8.29 (35), 2.418 (35), 1.962 (35), 5.68 (30)

Chemistry:

	(1)	(2)
SiO_2	1.2	3.6
FeO		0.1
MnO		1.2
MgO		0.1
CaO	62.9	65.0
Cl	3.3	3.5
H_2O^+	[19.94]	11.47
CO_2	13.4	16.3
$-O = Cl_2$	0.74	0.8
Total	[100.00]	100.5

(1) Güneyce-İkizdere, Turkey; by electron microprobe, average of 18 analyses on three samples, H_2O by difference; corresponds to $Ca_{6.00}(CO_3)_{1.62}(SiO_4)_{0.1}(OH)_{7.82}Cl_{0.50} \cdot 2.00H_2O$ (with zeolitic H_2O). (2) Kombat mine, Namibia; preferred values obtained from analysis by AA, electron microprobe, Heraeus CHN analyzer, and the Penfield method; corresponds to $(Ca_{5.90}Mn_{0.09}$ $Mg_{0.01})_{\Sigma=6.00}(CO_3)_{1.88}(SiO_4)_{0.30}(OH)_{6.48}Cl_{0.50}$.

Occurrence: In skarn at a granite-limestone contact (Güneyce-İkizdere, Turkey); throughout a body of unbanded, granular massive hausmannite (Kombat mine, Namibia).

Association: Vesuvianite, wollastonite, andradite, diopside, calcite, rustumite, spurrite, hillebrandite (Güneyce-İkizdere, Turkey); hausmannite, hillebrandite, brucite, crednerite, barite, calcite, vesuvianite, jacobsite, hematite, copper (Kombat mine, Namibia).

Distribution: From the Güneyce-İkizdere area, Trabzon Province, Turkey. In the Kombat mine, 49 km south of Tsumeb, Namibia.

Name: For Jacques Deferne, Curator of Mineralogy, Museum of Natural History, Geneva, Switzerland.

Type Material: Museum of Natural History, Geneva, Switzerland, 435/30; National Museum of Natural History, Washington, D.C., USA, 163241.

References: (1) Sarp, H., M.F. Taner, J. Deferne, H. Bizouard, and B.W. Liebich (1980) La defernite, $Ca_6(CO_3)_2(OH, Cl)_8 \cdot nH_2O$, un nouveau carbonate de calcium chloro-hydroxylé. Bull. Minéral., 103, 185–189 (in French with English abs.). (2) (1980) Amer. Mineral., 65, 1066 (abs. ref. 1). (3) Liebich, B.W. and H.S. Sarp (1985) La structure cristalline de la defernite. Schweiz. Mineral. Petrog. Mitt., 65, 153–158 (in French with English abs.). (4) Peacor, D.R., H. Sarp, P.J. Dunn, J. Innes, and J.A. Nelen (1988) Defernite from the Kombat mine, Namibia: a second occurrence, structure refinement, and crystal chemistry. Amer. Mineral., 73, 888–893.

$(K, Na)_{10}Ca_5Al_6Si_{32}O_{80}(Cl_2, F_2, SO_4)_3 \cdot 18H_2O$ Delhayelite

Crystal Data: Orthorhombic. *Point Group:* $2/m \ 2/m \ 2/m$ or $mm2$. Crystals platy.

Physical Properties: *Cleavage:* Distinct on {010}. Hardness = n.d. D(meas.) = 2.60(3) D(calc.) = [2.50]

Optical Properties: Semitransparent. *Color:* Colorless; in thin section, abnormally bluish gray in sections cut normal to [010].
Optical Class: Biaxial (–). *Orientation:* $X = a$; $Y = c$; $Z = b$. $\alpha = 1.532(2)$ $\beta = 1.532(2)$ $\gamma = 1.532(2)$ 2V(meas.) = 83(3)°

Cell Data: *Space Group:* $Pmmn$ or $Pmn2_1$.   a = 24.86(1)   b = 7.07(2)   c = 13.06(2) Z = 1

X-ray Powder Pattern: Mt. Shaheru, Zaire.
3.078 (100), 12.3 (35), 6.158 (25), 3.482 (10), 2.961 (5), 2.788 (5), 1.760 (5)

Chemistry:

	(1)		(1)
SiO_2	52.60	K_2O	9.27
TiO_2	0.09	F	0.33
Al_2O_3	9.22	Cl	3.91
Fe_2O_3	2.72	H_2O^+	5.93
MnO	0.07	H_2O^-	3.35
MgO	1.03	SO_3	1.31
CaO	7.99	$-O = (F, Cl)_2$	1.01
Na_2O	3.20	Total	100.01

(1) Mt. Shaheru, Zaire; total Fe as Fe_2O_3; corrected for nepheline 2%.

Occurrence: In a kalsilite-bearing melilite-nepheline lava (Mt. Shaheru, Zaire).

Association: Kalsilite, nepheline, götzenite, combeite, kirschsteinite (Mt. Shaheru, Zaire); fenaksite, eudialyte, lomonosovite (Khibiny massif, Russia).

Distribution: On Mt. Shaheru, the extinct southern cone of Mt. Nyiragongo, Kivu Province, Zaire. On Mts. Rasvumchorr and Yukspor, Khibiny massif, Kola Penninsula, Russia.

Name: For Fernard Delhaye (1880–1946), Belgian geologist, a pioneer in the geological exploration of the northern Kivu region of Zaire.

Type Material: n.d.

References: (1) Sahama, T.G. and K. Hytönen (1959) Delhayelite, a new silicate from the Belgian Congo. Mineral. Mag., 32, 6–9. (2) (1959) Amer. Mineral., 44, 1321–1322 (abs. ref. 1). (3) Cannillo, E., G. Rossi, and L. Ungaretti (1969) The crystal structure of delhayelite. Rend. Soc. Ital. Mineral. Petrol., 26, 63–75 (in English).

Delindeite

$(Na, K)_3(Ba, Ca)_4(Ti, Fe, Al)_6Si_8O_{26}(OH)_{14}$

Crystal Data: Monoclinic. *Point Group:* $2/m$. As lath-shaped crystals or flakes, forming compact spherulitic aggregates, to 1 mm. *Twinning:* Submicroscopic on {100}, common.

Physical Properties: *Cleavage:* {001}, good. *Fracture:* Irregular. *Tenacity:* Brittle. Hardness = n.d. D(meas.) = 3.3(1) D(calc.) = 3.70(1)

Optical Properties: Translucent. *Color:* Pale pinkish gray. *Luster:* Resinous, pearly. *Optical Class:* Biaxial (+). $\alpha = 1.790(5)$ $\beta = 1.825(5)$ $\gamma = [1.982]$ 2V(meas.) = Moderate. 2V(calc.) = 54°

Cell Data: *Space Group:* $C2/m$ or subgroup. a = 21.617(13) b = 6.816(5) c = 5.383(3) $\beta = 94.03(5)°$ Z = 1

X-ray Powder Pattern: Diamond Jo quarry, Arkansas, USA.
10.80 (100), 2.888 (31), 3.083 (28), 3.54 (24), 2.806 (20), 2.262 (18), 2.753 (16)

Chemistry:

	(1)
SiO_2	27.10
TiO_2	23.02
Al_2O_3	1.10
Fe_2O_3	2.54
MgO	0.00
CaO	0.61
BaO	33.05
Na_2O	3.79
K_2O	1.47
H_2O	[7.32]
Total	[100.00]

(1) Diamond Jo quarry, Arkansas, USA; by electron microprobe, total Fe as Fe_2O_3, H_2O by difference; corresponds to $(Na_{2.16}K_{0.55})_{\Sigma=2.71}(Ba_{3.80}Ca_{0.19})_{\Sigma=3.99}(Ti_{5.08}Fe_{0.56}Al_{0.34})_{\Sigma=5.98}$ $(Si_{7.96}Al_{0.04})_{\Sigma=8.00}[O_{32.83}(H_2O)_{7.17}]_{\Sigma=40.00}$.

Occurrence: In vugs and miarolitic cavities, as a weathering product of titaniferous nepheline syenite.

Association: Pectolite, barite, labuntsovite, lourenswalsite, pyroxene, titanite, sphalerite, potassic feldspar.

Distribution: Found in the Diamond Jo quarry, Magnet Cove, Hot Spring Co., Arkansas, USA.

Name: In honor of amateur mineralogist Henry deLinde of Mabelvale, Arkansas, USA, owner of the Diamond Jo quarry.

Type Material: n.d.

References: (1) Appleman, D.E., H.T. Evans, Jr., G.L. Nord, E.J. Dwornik, and C. Milton (1987) Delindeite and lourenswalsite, two new titanosilicates from the Magnet Cove region, Arkansas. Mineral. Mag., 51, 417–425. (2) (1988) Amer. Mineral., 73, 1493–1494 (abs. ref. 1).

Crystal Data: Triclinic. *Point Group:* $\overline{1}$. Crystals are bladed, in thin veinlets and patches, to 2 mm.

Physical Properties: Hardness = n.d. D(meas.) = n.d. D(calc.) = 2.975

Optical Properties: Semitransparent (?). *Color:* [White (?).]
Optical Class: Biaxial (–) (synthetic). $\alpha = 1.650$ $\beta = 1.661$ $\gamma = 1.664$ 2V(meas.) = $< 30°$

Cell Data: *Space Group:* $P\overline{1}$ (synthetic). $a = 6.815$ $b = 6.934$ $c = 12.882$ $\alpha = 90.67°$
$\beta = 97.70°$ $\gamma = 98.14°$ $Z = 2$

X-ray Powder Pattern: Synthetic. (ICDD 29-376).
2.821 (100), 2.551 (80), 2.806 (70), 3.059 (50), 2.856 (50), 2.982 (45), 2.730 (45)

Chemistry: No chemical analysis appears to have been published.

Occurrence: In late-stage, fracture-controlled replacement veins cutting thermally metamorphosed limestones.

Association: Calcite, spurrite, kilchoanite, foshagite, tilleyite.

Distribution: From near Kilchoan, Ardnamurchan, Argyllshire, Scotland.

Name: For Della Martin Roy (1926–), Pennsylvania State University, University Park, Pennsylvania, USA.

Type Material: National Museum of Natural History, Washington, D.C., USA, 119426.

References: (1) Agrell, S.O. (1965) Polythermal metamorphism of limestones at Kilchoan, Ardnamurchan. Mineral. Mag., 34, 1–15. (2) (1965) Amer. Mineral., 50, 2104–2105 (abs. ref. 1).

Crystal Data: Monoclinic. *Point Group:* n.d. As acicular crystals, in parallel columnar aggregates, to 15 cm.

Physical Properties: *Cleavage:* Cleavage or parting observed ⊥ elongation of crystals. *Fracture:* Splintery. *Tenacity:* Brittle. Hardness = 4–5 D(meas.) = 2.76 D(calc.) = 2.81

Optical Properties: Semitransparent. *Color:* Light gray to greenish gray; colorless to faint yellow in thin section. *Luster:* Pearly to silky.
Optical Class: Biaxial (+). α = 1.565–1.569 β = 1.566–1.570 γ = 1.574–1.578
2V(meas.) = 20°–30°

Cell Data: *Space Group:* n.d. a = 30.92(7) b = 7.20(3) c = 18.27(5) β = 95° Z = 20

X-ray Powder Pattern: Khibiny massif, Russia.
3.32 (100), 2.75 (100), 3.24 (90), 3.03 (90), 3.65 (80), 3.08 (80), 2.79 (80)

Chemistry:

	(1)	(2)		(1)	(2)
SiO_2	49.83	50.30	BaO		0.23
TiO_2	0.13	0.12	Li_2O	0.01	
Al_2O_3	0.65	0.83	Na_2O	2.66	2.26
Fe_2O_3	0.55		K_2O	8.94	9.30
Nb_2O_5	0.14		Rb_2O	0.09	
FeO	0.06	0.31	F	3.95	3.00
MnO	0.83	0.53	H_2O^+	1.16	n.d.
MgO	0.35	0.08	H_2O^-	0.16	n.d.
CaO	31.16	31.50	P_2O_5	0.04	
SrO	0.35	0.65	$-O = F_2$	1.62	1.23
			Total	99.44	[97.88]

(1) Khibiny massif, Russia; after deduction of small amounts of contaminant minerals, corresponds to $(K_{0.68}Na_{0.32})_{\Sigma=1.00}(Ca_{1.95}Mn_{0.04}Sr_{0.01})_{\Sigma=2.00}Si_{3.00}O_8[F_{0.54}(OH)_{0.46}]_{\Sigma=1.00}$.
(2) Murun massif, Russia; by electron microprobe, average of 11 grains, original total given as 98.88%; corresponds to $(K_{0.69}Na_{0.26}Mn_{0.05}Fe_{0.02}Mg_{0.01})_{\Sigma=1.03}(Ca_{1.97}Sr_{0.02}Mn_{0.01})_{\Sigma=2.00}$ $(Si_{2.94}Al_{0.05})_{\Sigma=2.99}O_8F_{0.56}$.

Occurrence: In nepheline syenite in a differentiated alkalic massif (Khibiny massif, Russia).

Association: Nepheline, potassic feldspar, aegirine, fluorite, apatite, biotite, yuksporite (Khibiny massif, Russia): aegirine, kalsilite, feldspar (Murun massif, Russia).

Distribution: On Mts. Eveslogchorr and Yukspor, Khibiny massif, Kola Peninsula, and in the Murun massif, southwest of Olekminsk, Yakutia, Russia.

Name: Honors Alexander Petrovich Denisov, Russian geologist.

Type Material: Geology Museum, Kola Branch, Academy of Sciences, Apatity; A.E. Fersman Mineralogical Museum, Academy of Sciences, Moscow, Russia; The Natural History Museum, London, England, 1994,7.

References: (1) Men'shikov, Y.P. (1984) Denisovite, $Ca_4(K_{1.4}Na_{0.6})_2Si_6O_{16}(F, OH)_2$, a new mineral from the Khibina massif. Zap. Vses. Mineral. Obshch., 113, 718–723 (in Russian).
(2) (1985) Amer. Mineral., 70, 1329 (abs. ref. 1). (3) Konev, A.A., Y.I. Vorob'ev, L.F. Paradina, and A.N. Sapozhnikov (1987) Denisovite from the Murun pluton, its second find in the world. Doklady Acad. Nauk SSSR, 293, 196–198 (in Russian).

Crystal Data: Monoclinic. *Point Group: m.* As pseudohexagonal crystals, to 2 mm, which may be elongated along [100]. As aggregates of platelets resembling books; commonly as compact masses of microscopic crystals.

Physical Properties: *Cleavage:* Perfect on {001}. *Tenacity:* Flexible but inelastic. Hardness = 2–2.5 D(meas.) = 2.60 D(calc.) = [2.62]

Optical Properties: Transparent. *Color:* White, but will assume coloration from included impurities. *Luster:* Satiny.
Optical Class: Biaxial (+). *Orientation:* $X \wedge c = 8°–14°$; $Y \wedge a = 14°–20°$. *Dispersion:* $r > v$.
$\alpha = 1.560–1.564$ $\beta = 1.561–1.566$ $\gamma = 1.566–1.570$ 2V(meas.) = 50°–80°

Cell Data: *Space Group: Cc.* a = 5.150(1) b = 8.940(1) c = 14.424(2) $\beta = 96°44(1)'$
Z = 4

X-ray Powder Pattern: Schuylkill, Pennsylvania, USA. (ICDD 10-446).
7.15 (100), 3.580 (100), 2.326 (90), 4.124 (70), 3.799 (60), 2.510 (50), 1.975 (50)

Chemistry:

	(1)	(2)		(1)	(2)
SiO$_2$	46.86	46.14	MgO	0.09	
TiO$_2$	0.51		CaO	0.22	
Al$_2$O$_3$	37.12	39.61	Na$_2$O	0.07	
Fe$_2$O$_3$	1.43		K$_2$O	0.60	
FeO	0.06		H$_2$O	13.06	13.91
			Total	100.02	99.66

(1) Anglesey, Wales. (2) Pine Knot colliery, Pennsylvania, USA; corresponds to Al$_{2.02}$Si$_{1.99}$O$_5$(OH)$_{4.00}$.

Polymorphism & Series: Halloysite, kaolinite, and nacrite are polymorphs.

Mineral Group: Kaolinite-serpentine group.

Occurrence: Commonly of hydrothermal origin along veins derived in part from the alteration of aluminosilicate minerals; also as an authigenic sedimentary mineral.

Association: Quartz, "chalcedony."

Distribution: Widespread; some localities for studied materials follow: at Amlwch and the Pant-y-Gaseg quarry, Trwynbychan, Anglesey, Wales. From Mád, Hungary. In the Iza Cave, Mt. Rodna, Romania. From Mas D'Alary, Lodève, Hérault, France. At Postmasburg and Barkly East, Cape Province, and in the Middelburg district, Transvaal, South Africa. In the USA, on Red Mountain, near Ouray, San Juan Co., Colorado; in the Mineral Mountain area, near St. George, Washington Co., Utah; the Pine Knot colliery, Schuylkill, Schuylkill Co., Pennsylvania; and on Bruin Creek, Elliott Co., Kentucky. At San Juanito and Cusihuiriáchic, Chihuahua, Mexico.

Name: For Allan Brugh Dick (1833–1926), Scottish metallurgical chemist.

Type Material: n.d.

References: (1) Ross, C.S. and P.F. Kerr (1930) Dickite, a kaolin mineral. Amer. Mineral., 15, 34–39. (2) Newnham, R.E. (1961) A refinement of the dickite structure and some remarks on polymorphism in kaolin minerals. Mineral. Mag., 32, 683–704. (3) Grim, R.E. (1953) Clay Mineralogy. McGraw Hill, 370. (4) Deer, W.A., R.A. Howie, and J. Zussman (1963) Rock-forming minerals, v. 3, sheet silicates, 194–212. (5) Sen Gupta, P.K., E.O. Schlemper, W.D. Johns, and F. Ross (1984) Hydrogen positions in dickite. Clays and Clay Minerals, 32, 483–485.

Crystal Data: Monoclinic. *Point Group:* $2/m$. As prismatic crystals with nearly square cross sections, to 50 cm; granular, columnar, lamellar massive. *Twinning:* Simple or multiple twins on {100} or {010}, common.

Physical Properties: *Cleavage:* Distinct on {110}, (110) $\wedge$ ($1\bar{1}0$) $\sim87°$; partings on {100} and probably {010}. *Fracture:* Uneven to conchoidal. *Tenacity:* Brittle. Hardness = 5.5–6.5 D(meas.) = 3.22–3.38   D(calc.) = 3.278

Optical Properties: Transparent to opaque. *Color:* Colorless, white, yellow, pale to dark green, black; colorless in thin section. *Streak:* White, gray, gray-green. *Luster:* Vitreous or dull.
Optical Class: Biaxial (+). *Orientation:* $Y = b$; $Z \wedge c = -38°$ on (010); $X \wedge a = -22°$.
Dispersion: $r > v$, weak to moderate. $\alpha = 1.664$ $\beta = 1.672$ $\gamma = 1.694$ 2V(meas.) = 59°

Cell Data: *Space Group:* $C2/c$. $a = 9.746$ $b = 8.899$ $c = 5.251$ $\beta = 105.63°$ Z = 4

X-ray Powder Pattern: Schwartzenstein, Austria. (ICDD 11-654).
2.991 (100), 2.528 (40), 2.893 (30), 2.518 (30), 3.23 (25), 2.952 (25), 1.625 (25)

Chemistry:

	(1)	(2)		(1)	(2)		(1)	(2)
SiO_2	54.66	54.09	FeO	0.07	1.47	K_2O		0.15
TiO_2		0.28	MnO	0.02	0.09	H_2O^+	0.22	0.22
Al_2O_3	0.07	1.57	MgO	18.78	16.96	H_2O^-		0.08
Fe_2O_3	0.68	0.74	CaO	25.85	21.10	rem.		0.49
Cr_2O_3		2.03	Na_2O		1.37	Total	100.35	100.64

(1) Juva, Finland; corresponds to $Ca_{1.00}(Mg_{1.01}Fe^{3+}_{0.02})_{\Sigma=1.03}Si_{1.98}O_6$. (2) Dutoitspan mine, Kimberley, Cape Province, South Africa; corresponds to $(Ca_{0.82}Na_{0.05}Fe^{2+}_{0.04}Mg_{0.04}K_{0.01})_{\Sigma=0.96}$ $(Mg_{0.88}Cr_{0.06}Al_{0.03}Fe^{3+}_{0.02}Ti_{0.01})_{\Sigma=1.00}(Si_{1.96}Al_{0.04})_{\Sigma=2.00}O_6$.

Polymorphism & Series: Forms two series, with hedenbergite, and with johannsenite.

Mineral Group: Pyroxene group.

Occurrence: Typical of metamorphosed siliceous Ca, Mg-rich rocks of the pyroxene-hornfels or epidote-amphibolite facies; common in skarns, Ca, Mg-rich gneisses and schists, and some kimberlites and peridotites. Less common in alkalic olivine basalts and andesites.

Association: Calcite, forsterite, chondrodite, monticellite, clinohumite, scapolite, wollastonite, grossular, vesuvianite, tremolite, quartz.

Distribution: Selected localities for fine crystals follow: at Schwarzenstein, Zillertal, and near Prägraten, Tirol, Austria. From Ala, Piedmont, and St. Marcel, Val d'Aosta, Italy. At Otokumpu, Finland. In Russia, at the Akhmatovsk deposit, near Zlatoust, Ural Mountains; large crystals in the Inagli massif, 30 km west of Aldan, Yakutia; and along the Slyudyanka River, near Lake Baikal, Siberia. In Canada, many localities; in Ontario, at Bird's Creek, Eganville, Dog's Lake, Littlefield, and Burgess; in Quebec, at Wakefield, Brompton Lake, near Magog, and in the Jeffrey mine, Asbestos. In the USA, at DeKalb, St. Lawrence Co., Natural Bridge, Jefferson Co., Sing Sing, near Ossining, Westchester Co., New York; and at Ducktown, Polk Co., Tennessee. At Ampandrandava and Andranodambo, Taolañaro (Fort Dauphin), Madagascar. Large gemmy crystals from the Kunlun Mountains, Sinkiang Uighur Autonomous Region, China. From Tange-Achin, Kandahar Province, Afghanistan. Found near Jaipur, Rajasthan, India.

Name: From the Greek for *double* and *appearance*, apparently for two possible orientations of the prism zone.

References: (1) Dana, E.S. (1892) Dana's system of mineralogy, (6th edition), 351–359. (2) Deer, W.A., R.A. Howie, and J. Zussman (1978) Rock-forming minerals, (2nd edition), v. 2A, single-chain silicates, 198–293.

Crystal Data: Hexagonal. *Point Group:* $\bar{3}$. Commonly in prismatic to rhombohedral crystals, terminated by {02$\bar{2}$1}, to 4 cm. As indistinct crystalline aggregates and massive.

Physical Properties: *Cleavage:* Perfect on {10$\bar{1}$1}. *Fracture:* Conchoidal to uneven. *Tenacity:* Brittle. Hardness = 5   D(meas.) = 3.28–3.35   D(calc.) = [3.30]   Pyroelectric.

Optical Properties: Transparent to translucent. *Color:* Emerald-green, blue-green. *Streak:* Green. *Luster:* Vitreous.
Optical Class: Uniaxial (+), with three or six sectors in basal section. $\omega = 1.652$–1.658 $\epsilon = 1.704$–1.710

Cell Data: *Space Group:* $R\bar{3}$.   a = 14.566   c = 7.778   Z = 18

X-ray Powder Pattern: Tsumeb, Namibia. (ICDD 33-487).
2.600 (100), 2.443 (60), 7.29 (50), 4.07 (30), 4.90 (25), 2.714 (20), 2.117 (20)

Chemistry:

	(1)	(2)
SiO$_2$	38.93	38.11
CuO	49.51	50.46
H$_2$O	11.27	11.43
Total	99.71	100.00

(1) Unknown locality. (2) CuSiO$_2$(OH)$_2$.

Occurrence: In the oxidized zone of some copper deposits.

Association: Chrysocolla, malachite, mimetite, wulfenite, cerussite, hemimorphite, fluorite, quartz.

Distribution: From Altyn-Tyube, 50 km east of Karaganda, Kirghiz Steppe, Kazakhstan. At Băiţa (Rézbánya), Romania. In the USA, in Arizona, at the Mammoth-St. Anthony mine, Tiger, at Ray, and in the Table Mountain mine, Galiuro Mountains, Pinal Co., and from the Harquehala mine, Harquehala Mountains, Yuma Co.; in California, from the Blue Bell claims, near Baker, San Bernardino Co. At Chiviquin, Cordoba Province, Argentina. In Chile, from the Mina La Verde, Copiapó. Fine crystals from around Renéville, Mindouli, and Pimbi, Congo Republic. At Tantara and the Mashamba West mine, Shaba Province, Zaire. As splendid large crystals from Tsumeb, Guchab, and the Omaue deposit, Kaokoveld, Namibia. From Mavoyo, Angola. In Zimbabwe, at several mines in the Lomagundi district, the Inez mine, Hartley district, and the Midway mine, Fort Victoria. Many other minor localities are known.

Name: From the Greek for *through* and *to see*, as the cleavage directions may be seen on looking through the crystal.

References: (1) Dana, E.S. (1892) Dana's system of mineralogy, (6th edition), 463–464. (2) Ribbe, P.H., G.V. Gibbs, and M.M. Hamil (1977) A refinement of the structure of dioptase, Cu$_6$[Si$_6$O$_{18}$]•6H$_2$O. Amer. Mineral., 62, 807–811.

Dissakisite-(Ce)　　　　　　　　$Ca(Ce, La)MgAl_2(SiO_4)(Si_2O_7)O(OH)$

Crystal Data: Monoclinic. *Point Group:* $2/m$. Rarely crystallized; as anhedral grains, up to 0.6 mm.

Physical Properties: Hardness = n.d. D(meas.) = 3.75(15)　D(calc.) = 3.97–4.02

Optical Properties: Transparent. *Color:* Pale yellow-brown in thin section.
Luster: Vitreous.
Optical Class: Biaxial (+). *Pleochroism:* Weak; X = pale brown; $Y = Z$ = light yellow-brown.
Orientation: $Y = b$; $Z \wedge a = 23.7°$. *Dispersion:* $r < v$, medium. *Absorption:* $Y = Z > X$.
$\alpha = 1.735(3)$　$\beta = 1.741(3)$　$\gamma = 1.758(3)$　2V(meas.) = 64.2(3)°　2V(calc.) = 62°

Cell Data: *Space Group:* $P2_1/m$. a = 8.905(1)　b = 5.684(1)　c = 10.113(1)
$\beta = 114.62(2)°$　Z = 2

X-ray Powder Pattern: Balchen Mountain, Antarctica.
2.698 (100), 2.910 (90), 2.622 (60), 3.50 (50), 2.842 (50), 9.1 (40), 2.177 (40)

Chemistry:

	(1)	(2)		(1)	(2)
SiO_2	32.09	31.85	MnO	0.00	0.27
TiO_2	0.89	0.02	MgO	6.84	7.42
ThO_2	0.08		CaO	10.75	9.57
Al_2O_3	17.67	10.09	CdO	0.10	
Y_2O_3	0.04		F	0.21	0.87
RE_2O_3	12.87	15.50	H_2O	[1.54]	1.79
Ce_2O_3	16.94	11.66	P_2O_5	0.02	
Fe_2O_3		5.83	$-O = F_2$	0.09	0.37
FeO	1.80	5.34	Total	[101.75]	99.84

(1) Balchen Mountain, Antarctica; by electron microprobe, H_2O calculated, $RE_2O_3 = La_2O_3$ 9.80%, Nd_2O_3 2.08%, Pr_2O_3 0.86%, Sm_2O_3 0.13%, $Eu_2O_3 < 0.10\%$, $Gd_2O_3 < 0.10\%$, $Dy_2O_3 < 0.05\%$, $Er_2O_3 < 0.05\%$; corresponds to $Ca_{1.08}(Ce_{0.57}RE_{0.45})_{\Sigma=1.02}(Mg_{0.93}Fe_{0.14})_{\Sigma=1.07}$ $(Al_{1.91}Ti_{0.06})_{\Sigma=1.97}Si_3O_{12.25}[(OH)_{0.69}F_{0.06}]_{\Sigma=0.75}$. (2) Östanmossa mine, Sweden; corresponds to $Ca_{1.00}(Ce_{0.42}RE_{0.55})_{\Sigma=0.97}(Al_{1.16}Mg_{1.08}Fe^{3+}_{0.43}Fe^{2+}_{0.43}Mn_{0.02})_{\Sigma=3.12}Si_{1.03}O_{12}[(OH)_{1.16}F_{0.27}]_{\Sigma=1.43}$.

Mineral Group: Epidote group.

Occurrence: Formed in marble, under amphibolite-facies metamorphism, at about 600 °C and 7 kbar (Balchen Mountain, Antarctica).

Association: Calcite, dolomite, forsterite, clinohumite, phlogopite, chlorite, ilmenite-geikelite, spinel, zircon, pyrrhotite (Balchen Mountain, Antarctica).

Distribution: On Balchen Mountain, Sør Rondane Mountains, east Antarctica. From the Östanmossa mine, Norberg, Sweden. At the Fedorovskoye deposit, Yakutia, Russia. From Outokumpu, Finland. Well crystallized from Luzenac, Ariège, France. In the Donghai district, Kiangsu Province, China.

Name: From the Greek for *twice over*, for a magnesium analog of allanite being described twice.

Type Material: National Museum of Natural History, Washington, D.C., USA, 168421.

References: (1) Grew, E.S., E.J. Essene, D.R. Peacor, S.-C. Su, and M. Asami (1991) Dissakisite-(Ce), a new member of the epidote group and the Mg analogue of allanite-(Ce), from Antarctica. Amer. Mineral., 76, 1990–1997. (2) Rouse, R.C. and D.R. Peacor (1993) The crystal structure of dissakisite-(Ce), the Mg analogue of allanite-(Ce). Can. Mineral., 31, 153–157.

$Cu^{1+}Mn_{14}^{2+}Fe^{3+}(As^{3+}O_3)_5(As^{5+}O_4)(SiO_4)_2(OH)_6$ **Dixenite**

Crystal Data: Hexagonal. *Point Group:* 3. In aggregates of thin flakes; massive.

Physical Properties: *Cleavage:* Perfect, basal micaceous. Hardness = 3–4 D(meas.) = 4.36 D(calc.) = 4.375

Optical Properties: Translucent. *Color:* Deep red-brown to nearly black; intense red in thin sheets. *Luster:* Metallic to resinous.
Optical Class: Uniaxial (+). $n = 1.96(2)$

Cell Data: *Space Group:* $R3$. a = 8.233(4) c = 37.499(1) Z = 3

X-ray Powder Pattern: Långban, Sweden. (ICDD 19-426).
2.92 (100), 4.10 (90), 2.37 (80), 2.40 (55), 3.90 (50), 2.83 (50), 3.31 (45)

Chemistry:

	(1)
SiO_2	5.31
Fe_2O_3	3.75
Mn_2O_3	8.05
As_2O_3	32.16
MnO	43.35
CuO	3.49
MgO	0.32
CaO	0.39
Na_2O	0.13
K_2O	0.14
H_2O	2.80
P_2O_5	0.02
Total	99.91

(1) Långban, Sweden; cationic charges from structural study; corresponds to $Cu_{0.85}^{1+}(Mn_{13.58}^{2+}$ $Mn_{0.22}^{3+}Mg_{0.15})_{\Sigma=13.95}Fe_{0.91}^{3+}As_{4.99}^{3+}As_{1.30}^{5+}Si_{1.71}O_{27}(OH)_6$.

Occurrence: In a manganese orebody in serpentine.

Association: Hematite, adelite, lead, domeykite, magnussonite.

Distribution: At Långban, Värmland, Sweden.

Name: From the Greek *di*, for *two*, and *xenos*, for *stranger*, in reference to the then unique association of silica and arsenious oxide in the mineral.

Type Material: n.d.

References: (1) Flink, G. (1920) Trigonit och dixenit, två nya mineral från Långbanshytte gruvor. Geol. Fören. Förhandl. Stockholm, 42, 436–439 (in Swedish). (2) (1921) Amer. Mineral., 6, 93 (abs. ref. 1). (3) Wickman, F.E. (1951) From the notes of the late K. Johansson. VII. A revised chemical analysis of dixenite from Långban. Geol. Fören. Förhandl. Stockholm, 73, 637–638. (4) Moore, P.B. and T. Araki (1981) Dixenite, $Cu^{1+}Mn_{14}^{2+}Fe^{3+}(OH)_6(As^{3+}O_3)_5$ $(Si^{4+}O_4)_2(As^{5+}O_4)$: metallic $[As_4^{3+}Cu^{1+}]$ clusters in an oxide matrix. Amer. Mineral., 66, 1263–1273.

Crystal Data: Hexagonal. *Point Group:* $6/m \; 2/m \; 2/m$. Crystals tabular, crudely hexagonal, to 0.7 mm.

Physical Properties: *Cleavage:* {0001}, perfect. Hardness = 6 D(meas.) = n.d. D(calc.) = 2.747

Optical Properties: Semitransparent. *Color:* Colorless. *Optical Class:* Uniaxial (+). $\omega = 1.575$ $\epsilon = 1.580$

Cell Data: *Space Group:* $P6/mmm$. a = 5.122(5) c = 14.781(5) Z = 2

X-ray Powder Pattern: Chelyabinsk coal basin, Russia.
3.73 (100), 2.57 (80), 2.85 (70), 1.847 (70), 4.48 (60), 3.83 (60)

Chemistry:

	(1)
SiO_2	43.89
Al_2O_3	35.39
MgO	0.01
CaO	19.29
Na_2O	0.32
K_2O	0.03
Total	98.93

(1) Chelyabinsk coal basin, Russia; by electron microprobe, average of four analyses; corresponds to $(Ca_{0.96}Na_{0.03})_{\Sigma=0.99}Al_{1.95}Si_{2.05}O_8$.

Polymorphism & Series: Trimorphous with anorthite and svyatoslavite.

Mineral Group: Feldspar group.

Occurrence: On fracture surfaces in blocks from burned coal dumps.

Association: Cordierite, mullite, anorthite, wollastonite, tridymite, fayalite, fassaite, norbergite-chondrodite, graphite, iron sulfides.

Distribution: In the Chelyabinsk coal basin, Southern Ural Mountains, Russia.

Name: For Dmitrii Steinberg, noted petrologist.

Type Material: A.E. Fersman Mineralogical Museum, Academy of Sciences, Moscow, Russia.

References: (1) Chesnokov, B.V., E.V. Lotova, E.N. Nigmatulina, V.S. Pavlyuchenko, and A.F. Bushmakin (1990) Dmisteinbergite $CaAl_2Si_2O_8$ (hexagonal) – a new mineral. Zap. Vses. Mineral. Obshch., 119(5), 43–45 (in Russian). (2) (1992) Amer. Mineral., 77, 446–447 (abs. ref. 1). (3) Takéuchi, Y. and G. Donnay (1959) The crystal structure of hexagonal $CaAl_2Si_2O_8$. Acta Cryst., 12, 465–470.

Crystal Data: Monoclinic. *Point Group:* $2/m$. As subhedral crystals, to 0.3 mm; as fibrous and felted radiating aggregates of thin laths; as irregularly rounded grains, massive. *Twinning:* Simple twins common, "similar to a type exhibited by allanite."

Physical Properties: Hardness = n.d. D(meas.) = 3.9 D(calc.) = [3.86]

Optical Properties: Translucent. *Color:* Brown; very light brown with a pinkish tinge in thin section. *Luster:* Vitreous.
Optical Class: Biaxial (+). $\alpha = 1.715$ $\beta = 1.718$ $\gamma = 1.733$ 2V(meas.) = n.d.

Cell Data: *Space Group:* $P2_1/m$. $a = 8.934(18)$ $b = 5.721(7)$ $c = 10.176(22)$ $\beta = 114.31(12)°$ $Z = 2$

X-ray Powder Pattern: Östanmossa mine, Sweden.
2.915 (100), 2.709 (70), 2.852 (30), 9.29 (20), 3.52 (20), 2.150 (20), 3.26 (15)

Chemistry:

	(1)
SiO$_2$	32.4
Al$_2$O$_3$	8.9
La$_2$O$_3$	6.0
Ce$_2$O$_3$	13.2
Pr$_2$O$_3$	2.6
Nd$_2$O$_3$	6.0
Sm$_2$O$_3$	2.4
Gd$_2$O$_3$	2.1
FeO	3.3
MgO	13.1
CaO	9.2
F	3.0
H$_2$O	2.02
$-$O = F$_2$	1.3
Total	102.9

(1) Östanmossa mine, Sweden; by electron microprobe, H$_2$O from Geijer (1927); corresponds to $(Ca_{0.91}Ce_{0.45}La_{0.20}Nd_{0.20}Pr_{0.09}Sm_{0.08}Gd_{0.06})_{\Sigma=1.99}(Mg_{1.81}Fe_{0.25})_{\Sigma=2.06}Al_{0.97}Si_3O_{10.99}$ $[(OH)_{1.25}F_{0.88}]_{\Sigma=1.13}$.

Mineral Group: Epidote group.

Occurrence: In tactite replacement deposits developed during metamorphism of dolomitic limestone.

Association: Tremolite, norbergite, magnetite, dolomite, calcite.

Distribution: In the Östanmossa mine, Norberg, Västmanland, Sweden.

Name: Honors Professor Wayne A. Dollase, University of California, Los Angeles, California, USA, for his crystal chemical research on minerals of the epidote group, and its *cerium* content.

Type Material: National Museum of Natural History, Washington, D.C., USA, R6505.

References: (1) Peacor, D.R. and P.J. Dunn (1988) Dollaseite-(Ce) (magnesium orthite redefined): structure refinement and implications for F + M^{2+} substitutions in epidote-group minerals. Amer. Mineral., 73, 838–842. (2) Geijer, P. (1927) Some mineral associations from the Norberg district. Sveriges Geologiska Undersökning, 20, 1–32.

Crystal Data: Monoclinic. *Point Group:* 2. As flaky aggregates.

Physical Properties: *Cleavage:* One perfect. *Tenacity:* Flexible. Hardness = 2.5
D(meas.) = 2.63   D(calc.) = [2.66]

Optical Properties: Semitransparent. *Color:* White. *Luster:* Pearly.
Optical Class: Biaxial (+). *Orientation:* Z = c. *Dispersion:* $r > v$, marked. $\alpha = 1.728$
$\beta = 1.729$ $\gamma = 1.735$ 2V(meas.) = 52°

Cell Data: *Space Group:* C2. a = 5.174   b = 8.956   c = 14.26 $\beta = 97.83°$ Z = [2]

X-ray Powder Pattern: n.d.

Chemistry:

	(1)	(2)	(3)
SiO_2	33.64	34.73	34.40
Al_2O_3	45.02	48.08	51.85
Fe_2O_3	1.38	1.04	
MgO	1.75	0.80	
CaO	2.03		
Li_2O	trace	trace	
Na_2O	1.30		
H_2O	15.20	14.58	13.75
Total	100.32	99.23	100.00

(1) Utrennyaia shaft, Donets basin, Ukraine. (2) Uralskaya vein, Donets basin, Ukraine.
(3) $Al_{4.33}(Si_3Al)O_{10}(OH)_8$.

Polymorphism & Series: 1a-2 polytype.

Mineral Group: Chlorite group.

Occurrence: In the lode walls and crushed zones in ore veins and as coatings on slickensides of coals. In soils and sediments in very aluminum-rich environments.

Association: n.d.

Distribution: From a number of places in the Donets basin, as at Nagol'no Tarasovka, Ukraine. On Novaya Zemlya, Arctic Ocean. From Kesselberg, Black Forest, Germany. At Saint-Paul-de-Fenouillet, Pyrénées-Orientales, France. From Namivu, Alto Ligonha district, Mozambique. At Szabo Bluff, Scott Glacier, Antarctica.

Name: For occurrences in the DONets BASin, Ukraine.

Type Material: n.d.

References: (1) Lazerenko, E.K. (1940) Donbassites, a new group of minerals from the Donets basin. Doklady Acad. Nauk SSSR, 28, 509–521 (in Russian). (2) (1941) Amer. Mineral., 26, 349 (abs. ref. 1). (3) Aleksandrova, V.A., V.A. Drits, and G.V. Sokolova (1972) Structural features of dioctahedral one-packet chlorite. Kristallografiya (Sov. Phys. Crystal.), 17, 525–532 (in Russian). (4) Bailey, S.W. and J.S. Lister (1989) Structures, compositions, and X-ray diffraction identification of dioctahedral chlorites. Clays and Clay Minerals, 37, 193–202.

Crystal Data: Orthorhombic. *Point Group:* $2/m\ 2/m\ 2/m$. Interlocking grains up to 3 mm.

Physical Properties: *Cleavage:* Perfect on {110}. Hardness = 5–6 D(meas.) = 3.36(1) D(calc.) = 3.403(2)

Optical Properties: Semitransparent. *Color:* Pale buff, yellow-orange; faint pink in thin section. *Luster:* Vitreous.
Optical Class: Biaxial (–). *Orientation:* $Z = c$. $\alpha = 1.677(2)$ $\beta = 1.684(2)$ $\gamma = 1.692(2)$
2V(meas.) = 88(5)°

Cell Data: *Space Group: Pbca.* a = 18.384(11)   b = 8.879(7)   c = 5.226(3)   Z = 8

X-ray Powder Pattern: Balmat, New York, USA.
3.18 (100), 2.896 (60), 1.495 (11), 1.479 (11), 4.03 (10), 3.09 (10), 2.961 (10)

Chemistry:

	(1)
SiO$_2$	55.12
Al$_2$O$_3$	0.23
FeO	0.14
MnO	18.48
MgO	26.31
CaO	0.69
Na$_2$O	0.03
Total	[101.00]

(1) Balmat, New York, USA; by electron microprobe, original total given as 100.00%;
corresponding to $(Mg_{1.41}Mn_{0.56}Ca_{0.03})_{\Sigma=2.00}(Si_{1.98}Al_{0.01})_{\Sigma=1.99}O_{5.99}$.

Polymorphism & Series: Dimorphous with kanoite.

Mineral Group: Pyroxene group.

Occurrence: In manganese-rich siliceous marbles metamorphosed to the upper amphibolite facies.

Association: Tirodite, tourmaline, ferrian braunite, manganoan dolomite, hedyphane, anhydrite.

Distribution: In the Balmat No. 4 mine, Balmat, St. Lawrence Co., New York, USA.

Name: To honor Dr. Donald R. Peacor, University of Michigan, Ann Arbor, Michigan, USA.

Type Material: Harvard University, Cambridge, Massachusetts, 124237; National Museum of Natural History, Washington, D.C., USA, 159862.

References: (1) Petersen, E.U., L.M. Anovitz, and E.J. Essene (1984) Donpeacorite, (Mn, Mg)MgSi$_2$O$_6$, a new orthopyroxene and its proposed phase relations in the system MnSiO$_3$ – MgSiO$_3$ – FeSiO$_3$. Amer. Mineral., 69, 472–480.

Crystal Data: Triclinic, pseudomonoclinic by twinning. *Point Group:* $\bar{1}$ or 1. As anhedral to prismatic grains, to 0.1 mm; some grains exhibit hopper and skeletal crystal habit. *Twinning:* Twinned by two-fold rotation about the pseudomonoclinic [010] axis.

Physical Properties: *Cleavage:* Good on {010} and {001}. *Fracture:* Irregular. *Tenacity:* Brittle. Hardness = $\sim$5 D(meas.) = n.d. D(calc.) = 3.959

Optical Properties: Nearly opaque. *Color:* Dark red-brown to dark brown. *Streak:* Gray. *Luster:* Submetallic.
Optical Class: Biaxial. *Pleochroism:* In ultrathinned sections, very strong; X = red-orange-brown; Y = yellowish brown; Z = greenish brown. *Absorption:* Extreme. $\alpha = 1.82(1)$ $\beta = 1.84(1)$ $\gamma = 1.86(1)$ 2V(meas.) = 90°

Cell Data: *Space Group:* $P\bar{1}$ or $P1$. $a = 10.505(3)$ $b = 10.897(3)$ $c = 9.019(1)$ $\alpha = 106.26(2)°$ $\beta = 95.16(2)°$ $\gamma = 124.75(2)°$ Z = 2

X-ray Powder Pattern: Durham ranch, Wyoming, USA.
2.971 (100), 2.558 (80), 2.515 (80), 2.125 (60), 1.511 (30), 1.482 (30), 8.1 (20)

Chemistry:

	(1)	(2)
SiO_2	11.16	14.37
TiO_2	0.56	
Al_2O_3	24.85	24.39
Fe_2O_3	41.65	38.19
Cr_2O_3	0.05	
FeO	2.77	
MnO	0.19	
MgO	5.57	9.64
CaO	13.63	13.41
Na_2O	0.02	
K_2O	0.02	
Total	100.47	100.00

(1) Durham ranch, Wyoming, USA; by electron microprobe, Fe^{2+}:Fe^{3+} calculated from stoichiometry; corresponding to $(Ca_{1.99}Na_{0.01})_{\Sigma=2.00}(Mg_{1.19}Fe^{2+}_{0.33}Fe^{3+}_{0.29}Ca_{0.10}$ $Ti_{0.06}Mn_{0.02}Cr_{0.01})_{\Sigma=2.00}Fe^{3+}_{4.00}Al_{4.00}(Si_{1.60}Al_{0.20}Fe^{3+}_{0.20})_{\Sigma=2.00}O_{20}$. (2) $Ca_2Mg_2Fe_4Al_4Si_2O_{20}$.

Mineral Group: Aenigmatite group.

Occurrence: A product of oxidizing, high-temperature, low-pressure metamorphism of alkalic rocks, in a pyrometamorphic zone in sediments.

Association: Esseneite, titanian andradite, magnetite-magnesioferrite-spinel, plagioclase, gehlenite-åkermanite, wollastonite, ulvöspinel, nepheline, apatite, ferroan sahamalite.

Distribution: From the Durham ranch, Powder River basin, 13 km northeast of Reno Junction and 25 km south of Gillette, Campbell Co., Wyoming, USA.

Name: To honor Dr. John A. Dorr, Jr., Professor of Geology, University of Michigan, Ann Arbor, Michigan, USA, in recognition of his regional geologic research in Wyoming.

Type Material: University of Michigan, Ann Arbor, Michigan; National Museum of Natural History, Washington, D.C., USA, 163357.

References: (1) Cosca, M.A., R.R. Rouse, and E.J. Essene (1988) Dorrite [$Ca_2(Mg_2Fe_4^{3+})$ $(Al_4Si_2)O_{20}$], a new member of the aenigmatite group from a pyrometamorphic melt-rock. Amer. Mineral., 73, 1440–1448.

Crystal Data: Hexagonal. *Point Group: 3m.* Crystals equant or short to long prismatic, with dominant pyramid faces, to 20 cm. Striated $\parallel$ [0001], cross sections typically triangular with curved convex sides. Commonly radiating, granular, and massive. *Twinning:* Rare, on $\{10\bar{1}1\}$ and $\{40\bar{4}1\}$.

Physical Properties: *Cleavage:* $\{11\bar{2}0\}$, $\{10\bar{1}1\}$, very poor. *Fracture:* Uneven to conchoidal. *Tenacity:* Brittle. Hardness = 7   D(meas.) = 3.03–3.18   D(calc.) = 3.038   Pyroelectric and piezoelectric; may weakly fluoresce under SW UV.

Optical Properties: Transparent to translucent. *Color:* Brown to black, red, yellow, blue, green, colorless, white; colorless to yellow in thin section. *Streak:* Light brown, white. *Luster:* Vitreous to resinous.
Optical Class: Uniaxial (–). *Pleochroism:* Very strong; O = pale yellow; E = colorless, yellowish, greenish, brownish. *Absorption:* $O > E$. ω = 1.634–1.661 ϵ = 1.612–1.632

Cell Data: *Space Group: R3m.* a = 15.94–15.98   c = 7.19–7.23   Z = 3

X-ray Powder Pattern: Dobrawa, Slovenia.
2.576 (100), 3.99 (85), 2.961 (85), 4.22 (65), 3.48 (60), 2.040 (45), 1.920 (35)

Chemistry:

	(1)		(1)
SiO_2	36.52	MgO	11.25
TiO_2	0.17	CaO	0.42
B_2O_3	10.32	Na_2O	2.34
Al_2O_3	33.41	K_2O	0.57
FeO	0.30	F	0.12
MnO	0.57	H_2O^+	3.76
		Total	99.75

(1) Dobrawa, Slovenia; corresponds to $(Na_{0.73}K_{0.12}Ca_{0.07})_{\Sigma=0.92}(Mg_{2.70}Al_{0.35}Mn_{0.08}Fe^{2+}_{0.04}Ti_{0.02})_{\Sigma=3.19}Al_{6.00}(B_{0.96}O_3)_3Si_{5.88}O_{18}[(OH)_{3.70}O_{0.20}F_{0.10}]_{\Sigma=4.00}$.

Polymorphism & Series: Forms two series, with schorl, and with elbaite.

Mineral Group: Tourmaline group.

Occurrence: Commonly in metamorphosed limestones or mafic igneous rocks with metasomatically introduced boron; rarely in pegmatites; as authigenic overgrowths in sedimentary rocks.

Association: Quartz, calcite, dolomite, epidote, microcline, albite, muscovite, fluorite, titanite.

Distribution: Found around Dravograd (Unterdrauberg), Slovenia. From Beura, Val d'Ossola, Piedmont, Italy. At Arendal and Snarum, Norway. From Outokumpu, and Kaavi, Kuopio, Finland. In the USA, fine crystals from Gouverneur, Pierrepont, and Macomb, St. Lawrence Co., New York; at Franklin and Hamburg, Sussex Co., New Jersey; and Newry, Oxford Co., Maine. In Canada, in Hull Township, Quebec; at Wilberforce, Ontario; and many other places. From Brumado, Bahia, Brazil. Large crystals from Yinnietharra, 800 km north of Perth, Western Australia. At Osarara, Narok district, Kenya. From Gujarkot, Nepal. Increasingly distinguished from other tourmaline group members.

Name: For the district along the Drava (Drave) River in Austria and Slovenia.

Type Material: National Museum of Natural History, Washington, D.C., USA, R18133, R17274.

References: (1) Dana, E.S. (1892) Dana's system of mineralogy, (6th edition), 551–558 [tourmaline]. (2) Deer, W.A., R.A. Howie, and J. Zussman (1986) Rock-forming minerals, (2nd edition), v. 1B, disilicates and ring silicates, 559–602. (3) (1964) NBS Mono. 25, 47.

Dumortierite

$Al_7(BO_3)(SiO_4)_3O_3$

Crystal Data: Orthorhombic. *Point Group:* $2/m \, 2/m \, 2/m$. As fibrous or columnar crystals, having generally poor face development, to 6 cm; as coarsely crystalline to intimate parallel aggregates of needles; massive. *Twinning:* Common on {110}, which may produce trillings.

Physical Properties: *Cleavage:* Distinct on {100}, poor on {110}; parting on {001}. Hardness = 7–8.5   D(meas.) = 3.21–3.41   D(calc.) = 3.45

Optical Properties: Transparent to translucent. *Color:* Bright smalt-blue to greenish blue, violet, red-violet, purple, brown. *Luster:* Vitreous to dull.
Optical Class: Biaxial (–). *Pleochroism:* Strong; X = deep blue or violet; Y = yellow to red-violet or nearly colorless; Z = colorless or very pale blue. *Orientation:* $X = c$; $Y = b$; $Z = a$. *Dispersion:* $r > v$, strong. $\alpha = 1.659$–1.686 $\beta = 1.684$–1.722 $\gamma = 1.686$–1.723
2V(meas.) = 13°–55°

Cell Data: *Space Group: Pmcn.* $a = 11.828(1)$ $b = 20.243(3)$ $c = 4.7001(5)$ $Z = 4$

X-ray Powder Pattern: Sri Lanka.
5.85 (s), 2.09 (ms), 5.06 (m), 3.43 (m), 3.22 (m), 2.91 (mb), 4.26 (mw)

Chemistry:

	(1)	(2)
SiO_2	29.99	31.52
TiO_2	1.56	
B_2O_3	[6.09]	6.09
Al_2O_3	61.49	62.39
Fe_2O_3	0.06	
MgO	0.10	
CaO	0.02	
H_2O	[0.60]	
Total	[99.91]	100.00

(1) Dehesa, California, USA; by electron microprobe, B_2O_3 and H_2O inferred from structure determination. (2) $Al_7(BO_3)(SiO_4)_3O_3$.

Occurrence: In aluminum-rich regionally metamorphosed rocks, disseminated and in veinlets cutting schists; also in pegmatitic veins.

Association: Quartz, cordierite, kyanite, andalusite, sillimanite, muscovite, rutile.

Distribution: In France, from near Beaunan and at Chaponost, near Lyon, Rhône. At Wolfshau, near Schmiedeberg, Silesia, Poland. From Kaňk, near Kutná Hora, Czech Republic. In the USA, from near Oreana, Humboldt Co., Nevada; at Clip, near Yuma, and Quartzsite, La Paz Co., Arizona; Petaca, Rio Arriba Co., New Mexico; in California, from near Dehesa, San Diego Co., and from near Ogilby, Imperial Co. In Canada, from Ashby Township, Addington Co., Ontario. From the Erongo Mountains, Namibia. On Madagascar, many places for fine examples, as at Soavina, north of Atofinandrahana; Saharina; Ambatolahinanahary; Ambositra; and Riamfotsy. In India, at Mogra, Bhandara district, Maharashtra. Many other minor localities are known.

Name: For Eugène Dumortier (1802–1873?), French paleontologist, of Lyons, France.

References: (1) Dana, E.S. (1892) Dana's system of mineralogy, (6th edition), 558. (2) Claringbull, G.F. and M.H. Hey (1958) New data for dumortierite. Mineral. Mag., 31, 901–907. (3) Moore, P.B. and T. Araki (1978) Dumortierite, $Si_3B[Al_{6.75}\square_{0.25}O_{17.25}(OH)_{0.75}]$: a detailed structure analysis. Neues Jahrb. Mineral., Abh., 132, 231–241. (4) Alexander, V.D., D.T. Griffen, and T.J. Martin (1986) Crystal chemistry of some Fe- and Ti-poor dumortierites. Amer. Mineral., 71, 786–794.

Crystal Data: Monoclinic. *Point Group:* $2/m$. Crystals prismatic, up to 5 mm, prism zone striated $\parallel$ [001]. Forms are $\{111\}$, $\{210\}$, $\{410\}$, $\{201\}$, $\{\overline{2}01\}$, $\{001\}$, and $\{100\}$. As minute lathlike inclusions along $\{110\}$ in andradite.

Physical Properties: *Fracture:* Conchoidal. Hardness = 5.5   D(meas.) = 2.93(1) D(calc.) = 2.931

Optical Properties: Transparent to translucent. *Color:* Colorless to white. *Luster:* Vitreous. *Optical Class:* Biaxial (+). *Orientation:* $X = b$; $Y \wedge c = 23.5°$ *Dispersion:* $r > v$, marked. $\alpha = 1.584$ $\beta = 1.586$ $\gamma = 1.600$ 2V(meas.) = $\sim 35°$

Cell Data: *Space Group:* $P2_1/a$.   a = 15.829(7)   b = 7.721(3)   c = 7.438(3) $\beta = 101°34(3)'$ Z = 2

X-ray Powder Pattern: Kings Mountain, North Carolina, USA.
4.812 (100), 5.257 (90), 7.31 (80), 3.021 (80), 3.353 (60), 6.905 (50), 5.944 (50)

Chemistry:

	(1)
SiO_2	46.75
SnO_2	18.59
Al_2O_3	14.07
CaO	14.2
H_2O	6.7
Total	100.3

(1) Kings Mountain, North Carolina, USA; by microchemical analysis, corresponds to $Ca_{1.96}Sn^{4+}_{0.95}Al_{2.12}Si_{6.03}H_{5.77}O_{22}$.

Occurrence: Probably of hydrothermal origin, in a seam in spodumene-bearing pegmatite (Foote mine, North Carolina, USA); as exsolved inclusions in stanniferous andradite (Kitel'skoye deposit, Russia).

Association: Tetrawickmanite, bavenite, quartz, albite (Foote mine, North Carolina, USA); andradite (Kitel'skoye deposit, Russia).

Distribution: In the Foote mine, Kings Mountain, Cleveland Co., North Carolina, USA. From the Kitel'skoye tin skarn deposit, north of Lake Lagoda, Russia.

Name: For Jack Eaker, mineral collector of Kings Mountain, North Carolina, USA, who discovered the mineral.

Type Material: The Natural History Museum, London, England, 1968,204; National Museum of Natural History, Washington, D.C., USA, 120301, 121143.

References: (1) Leavens, P.B., J.S. White, Jr., and M.H. Hey (1970) Eakerite, a new tin silicate. Mineral. Record, 1, 92–96. (2) (1971) Amer. Mineral., 56, 637–638 (abs. ref. 1). (3) Kossiakoff, A.A. and P.B. Leavens (1976) The crystal structure of eakerite, a calcium-tin silicate. Amer. Mineral., 61, 956–962. (4) Gaydukova, V.S., I.M. Grigor'yev, V.T. Dubinchuk, A.M. Il'yasov, V.I. Kuz'min, and G.A. Sidorenko. (1982) Exsolution of stanniferous garnets. Doklady Acad. Nauk SSSR, 250, 151–153 (in Russian). (5) (1983) Mineral. Abs., 34, 164 (abs. ref. 4).

Eckermannite $NaNa_2[(Mg, Fe^{2+})_4Al]Si_8O_{22}(OH)_2$

Crystal Data: Monoclinic. *Point Group:* $2/m$. As elongated prismatic crystals. Also as fibrous aggregates. *Twinning:* Simple or multiple twinning $\parallel \{100\}$.

Physical Properties: *Cleavage:* Perfect on $\{110\}$, intersecting at $56°$ and $124°$; parting on $\{010\}$. *Fracture:* [Uneven.] *Tenacity:* [Brittle.] Hardness = 5–6 D(meas.) = 3.0–3.3 D(calc.) = [3.14]

Optical Properties: Translucent to opaque. *Color:* Black, dark green, bluish green; pale bluish green in thin section. *Luster:* [Vitreous.]
Optical Class: Biaxial (–). *Pleochroism:* Strong; X = bluish green, yellow, indigo; Y = light bluish green, yellow-brown, gray-violet; Z = pale green, deep green, pale brownish green, colorless. *Orientation:* $Y = b$; $X \wedge c = 30°$–$50°$; $Z \wedge a = 45°$–$70°$. *Dispersion:* $r > v$, strong. *Absorption:* $X > Y > Z$. $\alpha = 1.610$–1.640 $\beta = 1.625$–1.650 $\gamma = 1.630$–1.655 2V(meas.) = $15°$–$80°$

Cell Data: *Space Group:* $C2/m$. a = 9.799 b = 17.833 c = 5.273 $\beta = 104.18°$ Z = 2

X-ray Powder Pattern: Synthetic $Na_3Mg_4AlSi_8O_{22}(OH)_2$ (ICDD 20-386).
3.10 (100), 2.708 (80), 3.40 (70), 3.25 (70), 2.500 (60), 2.965 (50), 2.164 (50)

Chemistry:

	(1)	(2)		(1)	(2)
SiO_2	58.65	57.10	CaO	1.40	0.31
TiO_2		0.35	Li_2O		1.15
Al_2O_3	5.98	6.19	Na_2O	9.30	9.77
Fe_2O_3	2.37	8.01	K_2O	1.10	2.38
FeO	1.34	2.69	F		2.69
MnO		0.34	H_2O^+	2.20	0.50
ZnO		0.59	H_2O^-		0.08
MgO	18.56	9.13	$-O = F_2$		1.13
			Total	100.90	100.15

(1) Tawmaw, Myanmar; corresponds to $(Na_{2.43}Ca_{0.20}K_{0.19})_{\Sigma=2.82}(Mg_{3.73}Al_{0.86}Fe^{3+}_{0.24}Fe^{2+}_{0.15})_{\Sigma=4.98}$ $(Si_{7.91}Al_{0.09})_{\Sigma=8.00}O_{22}(OH)_{1.98}$. (2) Norra Kärr complex, Sweden; corresponds to $(Na_{2.66}K_{0.42}Ca_{0.05})_{\Sigma=3.13}(Mg_{1.91}Fe^{3+}_{0.85}Li_{0.65}Fe^{2+}_{0.32}Zn_{0.06}Mn_{0.04}Ti_{0.04})_{\Sigma=3.87}Al_{1.02}Si_{8.02}O_{22}$ $[F_{1.19}(OH)_{0.47}]_{\Sigma=1.66}$.

Polymorphism & Series: Forms a series with ferro-eckermannite.

Mineral Group: Amphibole (alkali) group: $Fe^{2+}/(Fe^{2+} + Mg) < 0.5$; $Fe^{3+}/(Fe^{3+} + Al^{vi}) < 0.5$; $(Na + K)_A \geq 0.5$; $Na_B \geq 1.34$.

Occurrence: In alkalic plutonic igneous rocks; rarely in alkalic volcanics and nepheline-bearing pegmatites.

Association: Nepheline, albite, aegirine, katophorite.

Distribution: From the Norra Kärr complex, near Gränna, Sweden. At Tawmaw, Myitkyina-Mogaung district, Kachin State, northern Myanmar (Burma). In the Noda-Tamagawa mine, Iwate Prefecture, Japan. At Camp Albion, Boulder Co., Colorado, USA.

Name: To honor Professor Claes Walther Harry von Eckermann (1886–1969), Swedish petrologist, Stockholm, Sweden.

Type Material: n.d.

References: (1) Adamson, O.J. (1942) Eckermannite, a new alkali amphibole. Preliminary note. Geol. Fören. Förhandl. Stockholm, 64, 329–334. (2) (1944) Amer. Mineral., 29, 455 (abs. ref. 1). (3) Deer, W.A., R.A. Howie, and J. Zussman (1963) Rock-forming minerals, v. 2, chain silicates, 364–374. (4) Phillips, W.R. and D.T. Griffen (1981) Optical mineralogy, 243–245.

Crystal Data: Monoclinic. *Point Group:* $2/m$. Commonly as well-formed prismatic crystals, to 2.5 cm; fibrous; as reaction rims on pyroxenes. *Twinning:* Simple or multiple twinning ∥ {100}.

Physical Properties: *Cleavage:* Good on {110}, intersecting at 56° and 124°; partings on {100}, {001}. *Tenacity:* [Brittle.] Hardness = [5–6] D(meas.) = 3.05–3.37 D(calc.) = 3.06

Optical Properties: Semitransparent. *Color:* White, gray, pale green; color zoning is commonly seen in thin section. *Luster:* Vitreous.
Optical Class: Biaxial (–). *Pleochroism:* [Distinct, in greens, blue-greens, and yellow-browns.]
Orientation: $Y = b$; $Z \wedge c = 18°–34°$. *Dispersion:* $r > v$, weak. *Absorption:* $Z > Y > X$.
$\alpha = 1.622–1.665$ $\beta = 1.632–1.678$ $\gamma = 1.641–1.684$ 2V(meas.) = 50°–82°

Cell Data: *Space Group:* $C2/m$. a = 9.837(4) b = 17.954(6) c = 5.307(2)
$\beta = 105.18(2)°$ Z = 2

X-ray Powder Pattern: Franklin, New Jersey, USA. (ICDD 23-1405).
3.120 (100), 8.43 (80), 3.267 (40), 2.699 (20), 2.800 (18), 3.377 (12), 9.01 (10)

Chemistry:

	(1)	(2)		(1)	(2)
SiO_2	50.91	50.55	Na_2O	3.40	2.35
TiO_2	0.08	0.51	K_2O	0.11	0.54
Al_2O_3	8.68	6.90	F		1.86
Fe_2O_3	1.61		H_2O^+	1.74	1.27
FeO	3.61	1.30	H_2O^-	0.17	
MnO	0.12	0.07	P_2O_5	0.05	
MgO	19.38	22.06	$-O = F_2$		0.78
CaO	10.25	13.30	Total	100.11	99.93

(1) Kotaki, Japan; corresponds to $(Na_{0.70}K_{0.02})_{\Sigma=0.72}(Ca_{1.54}Fe^{2+}_{0.22}Na_{0.22}Mn_{0.02})_{\Sigma=2.00}$ $(Mg_{4.05}Al_{0.57}Fe^{2+}_{0.20}Fe^{3+}_{0.17}Ti_{0.01})_{\Sigma=5.00}(Si_{7.13}Al_{0.87})_{\Sigma=8.00}O_{22}(OH)_{1.63}$. (2) Franklin Marble, Orange Co., New York, USA; by electron microprobe, F by wet chemical analysis; corresponds to $(Na_{0.63}K_{0.10})_{\Sigma=0.73}(Ca_{1.99}Mn_{0.01})_{\Sigma=2.00}(Mg_{4.58}Al_{0.18}Fe^{2+}_{0.15}Ti_{0.05})_{\Sigma=4.96}(Si_{7.05}Al_{0.95})_{\Sigma=8.00}$ $O_{22}[(OH)_{1.18}F_{0.82}]_{\Sigma=2.00}$.

Polymorphism & Series: Forms a series with ferro-edenite.

Mineral Group: Amphibole (calcic) group: $Mg/(Mg + Fe^{2+}) \geq 0.5$; $(Na + K)_A \geq 0.5$; $Na_B < 0.67$; $(Ca + Na)_B \geq 1.34$; $6.75 \leq Si \leq 7.25$.

Occurrence: In intermediate plutonic igneous and medium-grade metamorphic rocks, as amphibolites and marbles.

Association: Titanite, mica, chondrodite (Edenville, New York, USA).

Distribution: In the USA, perhaps from Edenville, Orange Co., New York, and Franklin and Sterling Hill, Ogdensburg, Sussex Co., New Jersey. At Tory Hill, near Bancroft, and at Wilberforce, Ontario, Canada. From Kotaki, Niigata Prefecture, Japan.

Name: For the locality at Edenville, New York, USA, although it is not certain to occur there.

References: (1) Dana, E.S. (1892) Dana's system of mineralogy, (6th edition), 386, 391. (2) Deer, W.A., R.A. Howie, and J. Zussman (1963) Rock-forming minerals, v. 2, chain silicates, 263–314. (3) Kohn, J.A. and J.E. Comeforo (1955) Synthetic asbestos investigations, II: X-ray and other data on synthetic fluor-richterite, -edenite, and -boron edenite. Amer. Mineral., 40, 410–421. (4) Kearns, L.E., L.E. Kite, P.B. Leavens, and J.A. Nelen (1980) Fluorine distribution in the hydrous silicate minerals of the Franklin Marble, Orange County, New York. Amer. Mineral., 65, 557–562.

Crystal Data: Monoclinic. *Point Group:* $2/m$, m, or 2. As sheaves of ill-formed platy crystals, with dominant $\{100\}$, to 200 μm. Also as cryptocrystalline nodular masses, which may be hollow. *Twinning:* Polysynthetic on $\{100\}$.

Physical Properties: *Fracture:* Irregular to subconchoidal. *Tenacity:* Brittle. Hardness = 4 VHN = 153–217, 192 average (100 g load). D(meas.) = 9.4(3)   D(calc.) = 9.11 Photosensitive, darkening on exposure to UV, IR, X-rays, and visible light.

Optical Properties: Translucent. *Color:* Lemon-yellow to orange-yellow when fresh; dark olive-green, through lighter yellowish green, to dark green-brown on exposed surface; in reflected light, gray to slightly lighter gray, with pale lemon-yellow internal reflections; lemon-yellow in transmitted light. *Streak:* Pale yellowish green. *Luster:* Vitreous, resinous when nodular. *Optical Class:* Biaxial. *Pleochroism:* Weak. *Absorption:* Strong. $\alpha = [2.10]$ β = n.d. $\gamma = [2.58]$ 2V(meas.) = n.d.
R: (400) 16.7, (420) 17.1, (440) 17.0, (460) 16.4, (480) 15.8, (500) 15.3, (520) 14.9, (540) 14.6, (560) 14.3, (580) 14.2, (600) 14.0, (620) 14.0, (640) 13.8, (660) 13.8, (680) 13.7, (700) 13.6

Cell Data: *Space Group:* $C2/m$, Cm, or $C2$.   a = 11.755(3)   b = 7.678(2)   c = 5.991(2) $\beta = 111.73(3)°$ Z = 2

X-ray Powder Pattern: Terlingua, Texas, USA.
3.160 (100), 2.715 (63), 1.872 (36), 2.952 (34), 3.027 (27), 2.321 (24), 6.28 (20)

Chemistry:

	(1)	(2)	(3)
SiO_2	8.6	7.9	8.76
Hg_2O	89.6	91.0	91.24
Total	98.2	[98.9]	100.00

(1) Socrates mine, California, USA; by electron microprobe, average of five analyses.
(2) Terlingua, Texas, USA; by electron microprobe, average of three analyses, originally given as Si 3.7%, Hg 87.5%, here recalculated to oxides. (3) $Hg_6Si_2O_7$.

Occurrence: A secondary mineral, probably resulting from reaction between mercury and quartz under unknown conditions.

Association: Mercury, cinnabar, montroydite, terlinguaite, eglestonite, calcite, quartz, barite.

Distribution: In the Socrates mercury mine, Sonoma Co., and the Clear Creek mercury mine, New Idria district, San Benito Co., California; at Terlingua, Brewster Co., Texas, USA. From the San Luis mine, Huahuaxtla, Guerrero, Mexico.

Name: For Dr. Edgar Herbert Bailey (1914–1983), distinguished geologist and mercury specialist with the U. S. Geological Survey.

Type Material: Canadian Museum of Nature, Ottawa, Canada, 65531; The Natural History Museum, London, England, 1906,190.

References: (1) Roberts, A.C., M. Bonardi, R.C. Erd, A.J. Criddle, C.J. Stanley, G. Cressey, R.J. Angel, and J.H.G. Laflamme (1990) Edgarbaileyite, the first known silicate of mercury, from California and Texas. Mineral. Record, 21, 215–220. (2) (1990) Amer. Mineral., 75, 1431–1432 (abs. ref. 1). (3) Angel, R.J., G. Cressey, and A. Criddle (1990) Edgarbaileyite, $Hg_6Si_2O_7$: the crystal structure of the first mercury silicate. Amer. Mineral., 75, 1192–1196.

Crystal Data: Orthorhombic, pseudotetragonal. *Point Group:* 222, $\bar{4}2m$ pseudotetragonal. Prismatic crystals exhibit pseudotetragonal pyramid or sphenoid forms, to 10 cm; also massive. *Twinning:* On {110} and about [001].

Physical Properties: *Cleavage:* {110}, perfect. Hardness = 4–4.5 D(meas.) = 2.73–2.78 D(calc.) = 2.75–2.80 Pyroelectric and piezoelectric.

Optical Properties: Transparent to translucent. *Color:* White, grayish, pink; colorless in thin section. *Luster:* Vitreous.
Optical Class: Biaxial (–). *Orientation:* X = c; Y = b; Z = a. *Dispersion:* r < v, strong.
α = 1.535–1.541 β = 1.542–1.553 γ = 1.545–1.557 2V(meas.) = 54°–62°

Cell Data: *Space Group:* $P2_12_12$; $P\bar{4}2_1m$ pseudotetragonal. a = 9.550(10) b = 9.665(10) c = 6.523(5) Z = 2

X-ray Powder Pattern: Kilpatrick, Scotland. (ICDD 25-61).
3.576 (100), 6.51 (80), 2.741 (75), 5.38 (60), 4.79 (50), 4.69 (50), 2.589 (45)

Chemistry:

	(1)	(2)
SiO$_2$	35.14	36.3
Al$_2$O$_3$	20.12	20.6
BaO	31.18	28.6
K$_2$O		0.32
H$_2$O	13.16	12.8
Total	99.60	98.62

(1) Bölet mine, Sweden; corresponds to Ba$_{1.03}$Al$_{2.01}$Si$_{2.98}$O$_{10.00}$·3.72H$_2$O. (2) Ice River, Canada; by electron microprobe, H$_2$O by TGA; corresponds to (Ba$_{0.96}$K$_{0.04}$)$_{\Sigma=1.00}$Al$_{2.08}$Si$_{3.12}$O$_{10.00}$·4.00H$_2$O.

Mineral Group: Zeolite group.

Occurrence: In cavities in mafic igneous rocks and nepheline syenites; in carbonatites; in hydrothermal veins.

Association: Thomsonite, analcime, natrolite, harmotome, brewsterite, prehnite, calcite.

Distribution: At several quarries in the Kilpatrick Hills, Dumbartonshire, Scotland. From Disgwylfa Hill, Shropshire, England. Large crystals from the Bölet mine, Västergötland, Sweden. At Staré Ransko, Czech Republic. In the Podol'skoye and other deposits, Southern Ural Mountains, and in the Khibiny massif, Kola Peninsula, Russia. Along Ash Creek, in Mendocino Co. near the Sonoma Co. line, California, USA. In the Brunswick No. 12 mine, near Bathurst, New Brunswick; along the Ice River valley, 25 km south of Field, British Columbia; and at Mont Saint-Hilaire, Quebec, Canada. In the Jacupiranga mine, São Paulo, Brazil.

Name: For a Mr. Edington, Glasgow, Scotland, the discoverer of the mineral.

References: (1) Dana, E.S. (1892) Dana's system of mineralogy, (6th edition), 599. (2) Deer, W.A., R.A. Howie, and J. Zussman (1963) Rock-forming minerals, v. 4, framework silicates, 358–376. (3) Van Reeuwijk, L.P. (1972) High-temperature phases of zeolites of the natrolite group. Amer. Mineral., 57, 499–510. (4) Galli, E. (1976) Crystal structure refinement of edingtonite, Acta Cryst., 32, 1623–1627. (5) Mazzi, F., E. Galli, and G. Gottardi (1984) Crystal structure refinement of two tetragonal edingtonites. Neues Jahrb. Mineral., Monatsh., 373–382. (6) (1985) Amer. Mineral., 70, 1333–1334 (abs. ref. 5). (7) Grice, J.D., R.A. Gault, and H.G. Ansell (1984) Edingtonite: the first two Canadian occurrences. Can. Mineral., 22, 253–258. (8) Belitsky, I.A., S.P. Gabuda, W. Joswig, and H. Fuess (1986) Study of the structure and dynamics of water in the zeolite edingtonite at low temperature by neutron diffraction and NMR-spectroscopy. Neues Jahrb. Mineral., Monatsh., 541–551.

Eggletonite

$$(Na, K, Ca)_2(Mn, Fe)_8(Si, Al)_{12}O_{29}(OH)_7 \cdot 11H_2O$$

Crystal Data: Monoclinic. *Point Group:* $2/m$ or m. Crystals display a pseudohexagonal cross section, elongated along [100], with {011} and {001} common. As clusters of divergent sprays of radiating acicular prisms, to 1.5 mm. *Twinning:* On {001}.

Physical Properties: *Cleavage:* Perfect on {001}. *Tenacity:* Very brittle. Hardness = 3–4 D(meas.) = 2.76   D(calc.) = 2.76

Optical Properties: Semitransparent. *Color:* Dark to golden brown, red-brown. *Streak:* Light brown. *Luster:* Vitreous.
Optical Class: Biaxial (–). *Pleochroism:* Very weak; X = pale brown to colorless; $Y = Z$ = pale yellow-brown. *Orientation:* $Z = b$; $X \simeq c'$; $Y \simeq a$. *Dispersion:* $r < v$, weak. *Absorption:* $Z \simeq Y > X$. $\alpha = 1.566(2)$ $\beta = 1.606(2)$ $\gamma = 1.606(2)$ 2V(meas.) = 9(3)°

Cell Data: *Space Group:* $I2/a$ or Ia. $a = 5.554$ $b = 13.72$ $c = 25.00$ $\beta = 93.95°$ Z = 2

X-ray Powder Pattern: Big Rock quarry, Arkansas, USA.
12.4 (100), 3.13 (30), 2.691 (25), 2.600 (20), 2.462 (20), 3.45 (15), 2.854 (15)

Chemistry:

	(1)
SiO_2	41.5
Al_2O_3	7.6
FeO	3.0
MnO	31.4
ZnO	0.2
MgO	0.4
CaO	1.5
Na_2O	1.7
K_2O	1.3
H_2O	[11.4]
Total	[100.0]

(1) Big Rock quarry, Arkansas, USA; by electron microprobe, H_2O by difference; corresponds to $(Na_{0.82}K_{0.40}Ca_{0.39})_{\Sigma=1.61}(Mn_{6.61}Fe_{0.61}Al_{0.56}Mg_{0.16}Zn_{0.08})_{\Sigma=8.02}$ $(Si_{10.33}Al_{1.67})_{\Sigma=12.00}[O_{28.92}(OH)_{3.08}]_{\Sigma=32.00}(OH)_{4.00} \cdot 10.66H_2O$.

Occurrence: In miarolitic pockets in nepheline syenite pegmatite.

Association: Albite, natrolite, apophyllite (Big Rock quarry, Arkansas, USA); pyrophanite, kupletskite (3M quarry, Arkansas, USA).

Distribution: In the Big Rock and 3M quarries, Little Rock, Pulaski Co., Arkansas, USA.

Name: To honor Dr. Richard A. Eggleton, Australian National University, Canberra, Australia.

Type Material: National Museum of Natural History, Washington, D.C., USA, 137143.

References: (1) Peacor, D.R., P.J. Dunn, and W.B. Simmons (1984) Eggletonite, the Na analogue of ganophyllite. Mineral. Mag., 48, 93–96. (2) (1985) Amer. Mineral., 70, 436 (abs. ref. 1).

Crystal Data: Hexagonal. *Point Group:* 6/m 2/m 2/m. Crystals platy to prismatic, to 1 mm, with dominant {10$\overline{1}$0} and {0001}, and {11$\overline{2}$0} and {10$\overline{1}$2}; in druses.

Physical Properties: Hardness = n.d. D(meas.) = n.d. D(calc.) = 2.67

Optical Properties: Transparent. *Color:* Colorless to faintly yellow or green. *Streak:* White. *Luster:* Vitreous.
Optical Class: Uniaxial (+). ω = 1.5430–1.5445 ϵ = 1.5443–1.5458

Cell Data: *Space Group:* P6/mcc. a = 10.155 c = 14.223(6) Z = 2

X-ray Powder Pattern: Bellerberg volcano, Germany; essentially identical with roedderite.
3.26 (10), 3.75 (9), 4.43 (6), 7.07 (5), 5.11 (5), 4.14 (5), 2.91 (5)

Chemistry:

	(1)
SiO$_2$	71.06
TiO$_2$	0.06
Al$_2$O$_3$	0.79
Cr$_2$O$_3$	0.06
FeO	0.48
MnO	0.46
CuO	0.08
ZnO	0.34
MgO	16.25
Na$_2$O	6.48
K$_2$O	4.18
Total	100.24

(1) Bellerberg volcano, Germany; by electron microprobe, microchemical analysis showed Li$_2$O 0.1%; corresponding to (Na$_{2.12}$K$_{0.09}$)$_{\Sigma=2.21}$(Mg$_{4.08}$Al$_{0.14}$Fe$_{0.07}$Mn$_{0.07}$Zn$_{0.04}$ Cr$_{0.01}$Ti$_{0.01}$Cu$_{0.01}$)$_{\Sigma=4.43}$(Si$_{11.98}$Al$_{0.02}$)$_{\Sigma=12.00}$O$_{30}$.

Polymorphism & Series: Forms a series with roedderite.

Mineral Group: Milarite group.

Occurrence: In vesicles in contact metamorphosed basement gneiss xenoliths in leucite tephrite.

Association: Tridymite, hematite, pseudobrookite, pyroxene, amphibole, quartz, sanidine.

Distribution: From the Bellerberg volcano, two km north of Mayen, Eifel district, Germany.

Name: For the locality in the Eifel district, Rhineland-Palatinate, Germany.

Type Material: Institute for Mineralogy, Ruhr University, Bochum, Germany; National Museum of Natural History, Washington, D.C., USA, 162496.

References: (1) Abraham, K., W. Gebert, O. Medenbach, W. Schreyer, and G. Hentschel (1980) KNa$_2$Mg$_{4.5}$[Si$_{12}$O$_{30}$], ein neues Mineral der Milaritgruppe aus der Eifel, mit Natrium in Oktaederposition. Fortschr. Mineral., 58, Beiheft 1, 3–4 (abs., in German). (2) (1981) Amer. Mineral., 66, 218 (abs. ref. 1). (3) Abraham, K., W. Gebert, O. Medenbach, W. Schreyer, and G. Hentschel (1983) Eifelite, KNa$_3$Mg$_4$Si$_{12}$O$_{30}$, a new mineral of the osumilite group with octahedral sodium. Contr. Mineral. Petrol., 82, 252–258.

Crystal Data: Tetragonal; may be metamict. *Point Group:* $4/m\,2/m\,2/m$. As poorly-formed pyramidal crystals, to 1 cm, striated $\parallel$ [100], showing {101}, {110}, {001}, and {100}; as clusters of grains; rounded massive.

Physical Properties: *Cleavage:* Distinct on {101}, indistinct on {001}. *Fracture:* Irregular. *Tenacity:* Brittle. Hardness = 4.5–5 D(meas.) = 2.95–3.28 D(calc.) = 3.36 Emits α, β, and γ radiation.

Optical Properties: Transparent to turbidly translucent. *Color:* Colorless, light to dark green, yellow-green to yellowish brown, straw-yellow, red from inclusions; in thin section, dark brown. *Streak:* White. *Luster:* Vitreous.
Optical Class: Uniaxial (–), may be biaxial (–). $n = 1.595$–1.597 (metamict). $\omega = 1.580(3)$ $\epsilon = 1.568(3)$ 2V(meas.) = $10°$–$15°$

Cell Data: *Space Group:* $I422$. a = 7.483(3) c = 14.893(6) Z = 2

X-ray Powder Pattern: Tombstone Mountains, Canada.
4.14 (100), 3.343 (96), 3.265 (65), 6.70 (61), 7.45 (58), 2.642 (54), 1.796 (26)

Chemistry:

	(1)	(2)	(3)
SiO_2	55.6	47.6	56.10
ThO_2	27.6	37.8	30.81
UO_2	2.1	1.1	
Al_2O_3	trace	0.8	
Fe_2O_3	0.5		
FeO		0.4	
MnO	trace	0.2	
PbO	0.8		
MgO	trace		
CaO	13.7	9.6	13.09
Total	100.3	97.5	100.00

(1) Eheliyagoda district, Sri Lanka. (2) Tombstone Mountains, Canada; by electron microprobe, heated specimen; separate wet chemical analysis gives F 3.4% and H_2O 7.8%; corresponds to $(Ca_{1.91}Fe_{0.06}Mn_{0.03})_{\Sigma=2.00}(Th_{0.90}U_{0.05})_{\Sigma=0.95}Si_8O_{20}$. (3) $Ca_2ThSi_8O_{20}$.

Occurrence: Detrital (Sri Lanka); in a glacial erratic syenitic boulder (Tombstone Mountains, Canada); in volcanic ejecta (Case Collina, Italy).

Association: Fluorite, garnet, quartz, microcline, clinopyroxene, apatite, sodic plagioclase, hematite, thorogummite, zircon, titanite (Tombstone Mountains, Canada); quartz, feldspar, pyroxene (Case Collina, Italy).

Distribution: From the Eheliyagoda and Okkampitiya districts, near Ratnapura, Sri Lanka. In the Tombstone Mountains, Yukon Territory, Canada. At Case Collina, Pitigliano, near Grosetto, Tuscany, and on the Vico volcano, near Vetralla, Lazio, Italy. From the Murun massif, southwest of Olekminsk, Yakutia, Russia.

Name: For the mineral's discoverer, F.L.D. Ekanayake, of Colombo, Sri Lanka.

Type Material: Canadian Geological Survey, Ottawa, Canada, 62722; The Natural History Museum, London, England, 1961,42; National Museum of Natural History, Washington, D.C., USA, 148771.

References: (1) Anderson, B.W., G.F. Claringbull, R.J. Davis, and D.K. Hill (1961) Ekanite, a new metamict mineral from Ceylon. Nature, 190, 997. (2) (1961) Amer. Mineral., 46, 1516 (abs. ref. 1). (3) Gübelin, E.J. (1962) Ekanite. The Gemmologist, 31, 142–152, 165–169. (4) Szymański, J.T., D.R. Owens, A.C. Roberts, H.G. Ansell, and G.Y. Chao (1982) A mineralogical study and crystal-structure determination of nonmetamict ekanite, $ThCa_2Si_8O_{20}$. Can. Mineral., 20, 65–75.

Crystal Data: Hexagonal. *Point Group:* $3m$. Crystals prismatic to acicular, with prominent trigonal prism and pyramid, to 1.6 m, commonly hemimorphic, striated $\parallel$ [0001]. Also radial, fibrous, and massive. *Twinning:* Rare, on $\{10\bar{1}1\}$ or $\{40\bar{4}1\}$.

Physical Properties: *Cleavage:* $\{11\bar{2}0\}$, $\{10\bar{1}1\}$, very poor. *Fracture:* Uneven to conchoidal. *Tenacity:* Brittle. Hardness = 7   D(meas.) = 2.90–3.10   D(calc.) = 3.069   Pyroelectric and piezoelectric.

Optical Properties: Transparent to translucent. *Color:* Green, blue, red, orange, yellow, colorless, zoning common parallel to trigonal outline; colorless in thin section. *Luster:* Vitreous to resinous.
Optical Class: Uniaxial (–); under strain may show slight biaxiality. *Pleochroism:* O = pink, pale green, pale to deep blue; E = colorless, yellow, olive-green, purplish. *Absorption:* $O > E$.
$\omega = 1.633$–1.651 $\epsilon = 1.615$–1.630

Cell Data: *Space Group:* $R3m$.   a = 15.80–15.93   c = 7.09–7.13   Z = 3

X-ray Powder Pattern: Mt. Apatite, Auburn, Androscoggin Co., Maine, USA. (ICDD 26-964).
2.560 (100), 2.931 (90), 3.96 (80), 3.45 (70), 4.20 (60), 2.029 (50), 4.96 (35)

Chemistry:

	(1)		(1)
SiO_2	37.89	CaO	0.07
TiO_2	0.04	Li_2O	1.66
B_2O_3	10.28	Na_2O	2.43
Al_2O_3	43.85	F	0.10
FeO	0.11	H_2O^+	3.47
MnO	0.11	$-O = F_2$	0.04
		Total	99.97

(1) Elba, Italy; corresponds to $(Na_{0.74}Ca_{0.01})_{\Sigma=0.75}(Al_{2.11}Li_{1.05}Fe^{2+}_{0.01}Mn_{0.01})_{\Sigma=3.18}$ $Al_{6.00}(B_{0.93}O_3)_3Si_{5.94}O_{18}[(OH)_{3.63}O_{0.32}F_{0.05}]_{\Sigma=4.00}$.

Polymorphism & Series: Forms a series with dravite.

Mineral Group: Tourmaline group.

Occurrence: In granites, granite pegmatites, and some metamorphic rocks; in high-temperature hydrothermal veins; detrital in sediments.

Association: Quartz, albite, lepidolite, microcline, garnet, muscovite, beryl, apatite, spodumene.

Distribution: Many localities; some for exceptional specimens follow. In Italy, around San Piero in Campo, Elba. In the Yekaterinburg (Sverdlovsk) district, Ural Mountains, Russia. From Korgal and Mawi, Laghman Province, Afghanistan. In Pakistan, from around Gilgit and Stak Nala. At Ampantsikahitra, Anjanabonoina, Antandrokomby, Maharitra on Mt. Bity, and elsewhere on Madagascar. In the Alto Ligonha district, Mozambique. At Karibib and Usakos, Namibia. In the USA, from Mt. Mica near South Paris, Newry, and elsewhere in Oxford Co., Maine; at Haddam, Middlesex Co., Connecticut; and from the Pala and Mesa Grande districts, San Diego Co., California. In Brazil, from a large district around Araçuaí-Itinga-Salinas, and the Itatiaia district, Governador Valadares, Minas Gerais; at São José da Batalha, Paraíba.

Name: From the occurrence on the Island of Elba, Italy.

References: (1) Dana, E.S. (1892) Dana's system of mineralogy, (6th edition), 551–558 [tourmaline]. (2) Vernadsky [Vernadskii], V.I. (1913) Über die chemische Formel der Turmaline. Zeits. Krist., 53, 273–288 (in German). (3) Deer, W.A., R.A. Howie, and J. Zussman (1986) Rock-forming minerals, (2nd edition), v. 1B, disilicates and ring silicates, 559–602. (4) Donnay, G. and R. Barton, Jr. (1972) Refinement of the crystal structure of elbaite and the mechanism of tourmaline solid solution. Tschermaks Mineral. Petrog. Mitt., 18, 273–286.

Crystal Data: Hexagonal. *Point Group:* 6. Rarely as elongated prisms, to 1 cm, with hexagonal cross section. Typically as anhedral inclusions in pyrope crystals.

Physical Properties: *Tenacity:* Brittle. Hardness = 6.5 D(meas.) = 3.15–3.22 D(calc.) = 3.10–3.17 Faint blue cathodoluminescence.

Optical Properties: Transparent. *Color:* Purple to lilac; some individuals exhibit smoky or pink cores. *Luster:* Vitreous.
Optical Class: Uniaxial (–). *Pleochroism:* Vivid; O = colorless; E = colorless to deep lilac.
$\omega = 1.655$–1.679 $\epsilon = 1.654$–1.670

Cell Data: *Space Group:* $P6_3$. $a = 12.255(8)$ $c = 4.932(4)$ $Z = 1$

X-ray Powder Pattern: Parigi, Italy.
3.11 (100), 3.54 (75), 2.653 (70), 3.06 (55), 2.183 (55), 3.61 (50), 2.337 (30)

Chemistry:

	(1)	(2)	(3)
SiO_2	39.13	32.61	40.11
TiO_2	3.96	0.58	6.67
ZrO_2	0.00	2.13	
Al_2O_3	24.91	20.64	25.52
FeO	0.19	0.41	
MgO	22.05	25.80	20.18
H_2O			7.52
P_2O_5	0.42	8.26	
Total	90.66	90.43	100.00

(1) Parigi, Italy; by electron microprobe, analysis of a most intensely colored crystal; colorometric analysis yielded H_2O 8.0(4)%. (2) Do.; a less intensely colored crystal. (3) $Mg_6TiAl_6Si_8O_{28}(OH)_{10}$.

Occurrence: As inclusions in pyrope porphyroblasts developed during high-pressure (25–30 kbar), medium-temperature (700 °C–800 °C) metamorphism of continental crustal rocks.

Association: Pyrope, kyanite, talc, clinochlore, rutile, zircon, sodic amphibole.

Distribution: In the Dora-Maira massif, Parigi, near Martiniana Po, Piedmont, Italy.

Name: Honors Professor François Ellenberger, Paris, France, for his geological work in the western Alps.

Type Material: University of P. and M. Curie, Paris; National School of Mines, Paris, France; Institute for Mineralogy, Ruhr University, Bochum, Germany; National Museum of Natural History, Washington, D.C., USA, 163497.

References: (1) Chopin, C., R. Klaska, O. Medenbach, and D. Dron (1986) Ellenbergerite, a new high-pressure Mg-Al-(Ti,Zr)-silicate with a novel structure based on face-sharing octahedra. Contr. Mineral. Petrol., 92, 316–321. (2) (1988) Amer. Mineral., 73, 190–191 (abs. ref. 1).

Crystal Data: Orthorhombic. *Point Group:* $2/m \, 2/m \, 2/m$. Crystals prismatic, elongated ‖ [001], to 30 cm, striated ‖ [010]. In fan-shaped aggregates, columnar, fine fibrous, massive. *Twinning:* Noted.

Physical Properties: *Cleavage:* Perfect on {110}. *Fracture:* Splintery. Hardness = 5 D(meas.) = 2.52–2.62 D(calc.) = 2.59

Optical Properties: Opaque to translucent and transparent. *Color:* White, colorless, yellowish, brownish, green, brick-red from inclusions. *Luster:* Silky or vitreous. *Optical Class:* Biaxial (+). *Orientation:* $X = c$; $Y = b$; $Z = a$. *Dispersion:* $r < v$. $\alpha = 1.556$–1.563 $\beta = 1.565$–1.569 $\gamma = 1.574$–1.577 2V(meas.) = $76°$–$89°$

Cell Data: *Space Group:* $Pbcm$. a = 7.14(2) b = 14.68(1) c = 14.65(1) Z = 4

X-ray Powder Pattern: Narssârssuk, Greenland.
3.26 (100), 3.12 (70), 1.945 (70), 1.759 (70), 1.590 (70), 1.352 (70), 1.492 (60)

Chemistry:

	(1)	(2)	(3)
SiO$_2$	59.44	57.13	60.11
TiO$_2$	trace	0.05	
ZrO$_2$	20.48	20.33	20.55
Nb$_2$O$_5$		1.43	
FeO	0.14	0.14	
CaO	0.17	0.43	
Na$_2$O	10.41	9.89	10.33
K$_2$O	0.13	0.19	
F		0.12	
Cl	0.15	0.18	
H$_2$O$^+$	5.72	9.94	9.01
H$_2$O$^-$	3.89		
Total	100.53	99.83	100.00

(1) Narssârssuk, Greenland. (2) Lovozero massif, Russia. (3) Na$_2$ZrSi$_6$O$_{15}$·3H$_2$O

Occurrence: In albitized nepheline syenite and associated pegmatites, aegirine-rich granite, and fenites, in a differentiated alkalic massif (Lovozero massif, Russia).

Association: Albite, quartz, aegirine, epididymite, labuntsovite (Lovozero massif, Russia).

Distribution: At Narssârssuk and in the Ilímaussaq intrusion, southern Greenland. In Russia, from the Lovozero and Khibiny massifs, Kola Peninsula; in the Ulan-Erge massif, southeastern Tuva; and in the Burpala massif, about 120 km north of Lake Baikal, eastern Siberia. In large crystals from Tarbagatai, eastern Kazakhstan. In the Khan-Bogdinskii granite massif, Gobi, Mongolia. At Gjerdingen, near Oslo, Norway. On Rockall Island, North Atlantic. In Canada, at Mont Saint-Hilaire and near Varennes, Quebec, and in the Strange Lake complex, southeast of Lac Brisson, Quebec and Labrador, Newfoundland. From Washington Pass, Okanogan Co., Washington, USA.

Name: From the Greek *hope*, named in anticipation of discovery of other interesting minerals.

Type Material: University of Copenhagen, Copenhagen, Denmark.

References: (1) Dana, E.S. (1899) Dana's system of mineralogy, (6th edition), app. I, 24. (2) Vlasov, K.A., Ed. (1966) Mineralogy of rare elements, v. II, 365–370. (3) Cannillo, E., G. Rossi, and L. Ungaretti (1973) The crystal structure of elpidite. Amer. Mineral., 58, 106–109. (4) Sapozhnikov, A.N. and A.A. Kashaev (1978) Features of the crystal structure of calcium-containing elpidite. Kristallografiya (Sov. Phys. Crystal.), 23, 52–56 (in Russian).

Crystal Data: Orthorhombic, pseudohexagonal. *Point Group:* $2/m\ 2/m\ 2/m$.   Euhedral crystals, tabular ∥ {010} and elongated along [100], to 2 mm; in pseudohexagonal twins; as equidimensional grains. *Twinning:* Interpenetration triplets with {110} as twin plane.

Physical Properties: Hardness = 5–6   D(meas.) = 2.76(7)   D(calc.) = 2.81

Optical Properties: Transparent. *Color:* Colorless to creamy pinkish. *Luster:* Vitreous. *Optical Class:* Biaxial (–). *Orientation:* X = b; Y = a; Z = c.   *Dispersion:* $r > v$, very strong. $\alpha = 1.596(1)$ $\beta = 1.597(1)$ $\gamma = 1.597(1)$ 2V(meas.) = 0°–30°

Cell Data: *Space Group: Acam.*   a = 10.072(3)   b = 17.337(6)   c = 14.004(3)   Z = 8

X-ray Powder Pattern: Island of Igdlutalik, Greenland.
4.352 (100), 3.209 (80), 3.501 (70), 3.090 (70), 4.087 (60), 3.192 (60), 2.875 (60)

Chemistry:

	(1)
SiO_2	70.75
TiO_2	0.55
ZrO_2	0.10
Al_2O_3	1.34
Fe_2O_3	12.13
MnO	0.03
MgO	0.10
CaO	0.00
Li_2O	2.78
Na_2O	11.98
K_2O	0.00
Total	99.76

(1) Island of Igdlutalik, Greenland; by electron microprobe, Li by flame photometry; corresponds to $Na_{3.96}Li_{1.91}(Fe_{1.56}^{3+}Al_{0.27}Ti_{0.07}Mg_{0.03}Zr_{0.01})_{\Sigma=1.94}Si_{12.05}O_{30}$.

Mineral Group: Milarite group.

Occurrence: A minor constituent of a peralkalic trachyte dike, in flow-banded aegirine-albite-rich layers.

Association: Albite, aegirine, quartz, riebeckite, micas, zircon, pectolite, apatite, calcite, titanian narsarsukite, zincian nordite, thorite, opaque oxides.

Distribution: On the Island of Igdlutalik, Julianehåb district, Greenland.

Name: For Dr. Charles Henry Emeleus, University of Durham, Durham, England.

Type Material: University of Copenhagen, Copenhagen, Denmark; Department of Geology, Edinburgh University, Edinburgh, Scotland; The Natural History Museum, London, England, 1987,118; National Museum of Natural History, Washington, D.C., USA, 146464.

References: (1) Upton, B.G.J., P.G. Hill, O. Johnsen, and O.V. Petersen (1978) Emeleusite: a new LiNaFe[III] silicate from south Greenland. Mineral. Mag., 42, 31–34. (2) (1979) Amer. Mineral., 64, 242 (abs. ref. 1). (3) Johnsen, O., K. Nielsen, and I. Sotofte (1978) The crystal structure of emeleusite, a novel example of sechser-doppelkette. Zeits. Krist., 147, 297–306.

Crystal Data: Monoclinic. *Point Group: m.* Microscopic tubes, to 15 μm, forming massive aggregates; porcelaneous massive.

Physical Properties: *Cleavage:* {001}, probable. *Fracture:* Conchoidal when massive. Hardness = 2–2.5 (?) D(meas.) = 2.0–2.2 D(calc.) = [2.13]

Optical Properties: Semitransparent. *Color:* White; gray, green, yellow, red, blue from included impurities. *Luster:* Pearly, waxy, or dull. *Optical Class:* Biaxial. $n = 1.53–1.55$ 2V(meas.) = n.d.

Cell Data: *Space Group: Cc.* $a = 5.14(4)$ $b = 8.90(4)$ $c = 20.7(1)$ $\beta = 99.7°$ Z = [2]

X-ray Powder Pattern: Locality unknown. (ICDD 9-451).
4.42 (100), 10.1 (90), 3.34 (90), 1.481 (90), 2.56 (80), 1.678 (80), 1.283 (70)

Chemistry:

	(1)	(2)
SiO$_2$	44.46	40.85
TiO$_2$	0.15	
Al$_2$O$_3$	36.58	34.66
Fe$_2$O$_3$	0.36	
FeO	0.07	
MgO	0.18	
CaO	0.19	
Na$_2$O	0.01	
K$_2$O	0.51	
H$_2$O$^+$	13.38	24.49
H$_2$O$^-$	4.05	
P$_2$O$_5$	0.18	
Total	100.12	100.00

(1) Bedford, Indiana, USA. (2) Al$_2$Si$_2$O$_5$(OH)$_4$·2H$_2$O.

Mineral Group: Kaolinite-serpentine group; called halloysite in European literature.

Occurrence: Formed by weathering or hydrothermal alteration of aluminosilicate minerals, such as feldspars. A precipitate from acid sulfate groundwater. Dehydrates to halloysite irreversibly above 400 °C.

Association: Kaolinite, allophane, alunite, halloysite, montmorillonite.

Distribution: Material considered as standards from: in the USA, in the North Gardiner mine, Bedford, Lawrence Co., Indiana; from near Anamosa, Jones Co., Iowa; in the Guadalupe Mountains caves, Otero Co., New Mexico; in the East Tintic district, Utah Co., Utah. Many localities for halloysite would likely provide endellite as well.

Name: For Kurt Endell (1887–?), German geologist, one of the discoverers of the mineral.

Type Material: n.d.

References: (1) Alexander, L.T., G.T. Faust, S.B. Hendricks, H. Insley, and H.F. McMurdie (1943) Relationship of the clay minerals halloysite and endellite. Amer. Mineral., 28, 1–18. (2) Bates, T.F., F.A. Hildebrand, and A. Swineford (1950) Morphology and structure of endellite and halloysite. Amer. Mineral., 35, 463–484. (3) Faust, G.T. (1955) The endellite–halloysite nomenclature. Amer. Mineral., 40, 1110–1118. (4) Deer, W.A., R.A. Howie, and J. Zussman (1963) Rock-forming minerals, v. 3, sheet silicates, 194–212. (5) Kohyama, N., K. Fukushima, and A. Fukami (1978) Observation of the hydrated form of tubular halloysite by an electron microscope equipped with an environmental cell. Clays and Clay Minerals, 26, 25–40.

Crystal Data: Orthorhombic. *Point Group:* $2/m\ 2/m\ 2/m$. Crystals prismatic, to 40 cm; commonly lamellar, fibrous, or massive. *Twinning:* Simple and lamellar twinning on {100}.

Physical Properties: *Cleavage:* Good, {210}, (210) $\wedge$ ($2\bar{1}0$) ~88°; partings on {100} and {010}. *Fracture:* Uneven. *Tenacity:* Brittle. Hardness = 5–6 D(meas.) = 3.2–3.9 D(calc.) = 3.189

Optical Properties: Translucent to opaque. *Color:* White, grayish, yellowish, greenish, olive-green, brown; colorless in thin section. *Streak:* White to grayish. *Luster:* Vitreous to pearly on cleavages.
Optical Class: Biaxial (+). *Orientation:* $X = b$; $Y = a$; $Z = c$. *Dispersion:* $r < v$, weak to moderate. $\alpha = 1.649$–1.667 $\beta = 1.653$–1.671 $\gamma = 1.657$–1.680 2V(meas.) = $55°$–$90°$

Cell Data: *Space Group: Pbca.* $a = 18.23$ $b = 8.84$ $c = 5.19$ $Z = 8$

X-ray Powder Pattern: Bamle, Norway.
3.175 (100), 2.878 (55), 2.540 (25), 1.488 (25), 2.497 (18), 2.477 (18), 1.473 (18)

Chemistry:

	(1)		(1)
SiO_2	58.48	MnO	0.02
TiO_2	trace	MgO	34.71
Al_2O_3	0.88	CaO	0.50
Fe_2O_3	0.72	Na_2O	0.23
Cr_2O_3	0.25	K_2O	0.08
FeO	3.93	H_2O^-	0.21
		Total	100.01

(1) Maliba Matso, Lesotho; corresponds to $(Mg_{1.77}Fe^{2+}_{0.11}Al_{0.04}Ca_{0.02}Na_{0.02}Fe^{3+}_{0.02})_{\Sigma=1.98}Si_{2.00}O_6$.

Polymorphism & Series: Dimorphous with clinoenstatite; forms a series with ferrosilite.

Mineral Group: Pyroxene group.

Occurrence: In pyroxenites, peridotites, and dunites; in ultramafic inclusions in alkalic olivine basalts and kimberlite; in mafic volcanics, rarely in felsic volcanics. Characteristic of charnockites and the granulite metamorphic facies, in regionally metamorphosed rocks and metagabbros. Common in chondrite, achondrite, and stony-iron meteorites.

Association: Olivine, phlogopite, clinopyroxene, diopside, spinel, pyrope.

Distribution: Prominent localities for well-studied material include: from Ždár (Zdjarberg), Czech Republic. At Ødegaard and Kjörrestad, Bamle, Norway. From The Lizard, Cornwall, England. At Dawros, Co. Connaught, Ireland. From Kupferberg, Bavaria, and the Bellerberg volcano, two km north of Mayen, Eifel district, Germany. At Kraubath, Styria, Austria. Fine crystals from Mbeya, Tanzania. At Liahobong, Lesotho. In the USA, at Webster, Jackson Co., and on Corundum Hill, Macon Co., North Carolina; at Wood's chrome mine and Texas, Lancaster Co., and elsewhere in Pennsylvania. Gem material from Embilipituya, Sri Lanka.

Name: From the Greek for *opponent*, in allusion to its refractory nature under the blowpipe.

References: (1) Dana, E.S. (1892) Dana's system of mineralogy, (6th edition), 346–348. (2) Deer, W.A., R.A. Howie, and J. Zussman (1978) Rock-forming minerals, (2nd edition), v. 2A, single-chain silicates, 20–162. (3) Pollack, S.S. and W.D. Ruble (1964) X-ray identification of ordered and disordered orthoenstatite. Amer. Mineral., 49, 983–992.

NaLiAl$_2$(Al$_2$Si$_2$)O$_{10}$(OH)$_2$ — Ephesite

Crystal Data: Triclinic, pseudomonoclinic. *Point Group:* 1. As books and flakes, to 13 mm. *Twinning:* Commonly twinned about [310] or [$\overline{3}10$].

Physical Properties: *Cleavage:* {001}, perfect. *Tenacity:* Brittle. Hardness = 3.5–4 D(meas.) = 2.984   D(calc.) = 2.965

Optical Properties: Translucent. *Color:* Pink. *Luster:* Vitreous, pearly on cleavage. *Optical Class:* Biaxial (–). *Dispersion:* $r < v$. $\alpha = 1.592$ $\beta = 1.624$ $\gamma = 1.625$ 2V(meas.) = 28°

Cell Data: *Space Group:* C1. $a = 5.123(2)$ $b = 8.872(3)$ $c = 19.307(3)$ $\alpha = 89.97(2)°$ $\beta = 95.15(2)°$ $\gamma = 89.96(2)°$ Z = 4

X-ray Powder Pattern: Postmasburg district, South Africa.
3.20 (100), 9.59 (65), 1.922 (25), 2.52 (20), 4.81 (16), 1.478 (16), 4.41 (10)

Chemistry:

	(1)	(2)	(3)
SiO$_2$	30.70	30.86	30.97
TiO$_2$		0.03	
Al$_2$O$_3$	55.67	51.68	52.55
Fe$_2$O$_3$		0.47	
FeO		0.03	
MnO		0.12	
MgO		0.09	
CaO	2.55	0.02	
BaO		0.17	
Li$_2$O		3.52	3.85
Na$_2$O	5.52	7.44	7.99
K$_2$O	1.10	0.15	
H$_2$O$^+$	4.91	4.92	4.64
H$_2$O$^-$		0.06	
Total	100.45	99.57	100.00

(1) Gumuch-dagh, Turkey; Li perhaps analyzed as Na and K. (2) Postmasburg district, South Africa. (3) NaLiAl$_2$(Al$_2$Si$_2$)O$_{10}$(OH)$_2$.

Polymorphism & Series: 2M$_1$, 1M polytypes.

Mineral Group: Mica group.

Occurrence: In an emery deposit (Gumuch-dagh, Turkey); a minor constituent in manganese oxide ore (Postmasburg district, South Africa).

Association: Corundum, magnetite (Gumuch-dagh, Turkey); diaspore, braunite, bixbyite (Postmasburg district, South Africa).

Distribution: Some material from the original locality at Gumuch-dagh, near Ephesus, Turkey was later determined to be a mixture, but has not been reexamined by modern methods. From the Postmasburg district, and in the Wessels mine, near Kuruman, Cape Province, South Africa. In the Ilímaussaq intrusion, southern Greenland.

Name: For Ephesus, Turkey, near where it may occur.

References: (1) Dana, E.S. (1892) Dana's system of mineralogy, (6th edition), 707. (2) Schaller, W.T., M.K. Carron, and M. Fleischer (1967) Ephesite, Na(Li, Al$_2$)(Al$_2$Si$_2$)O$_{10}$(OH)$_2$, a trioctahedral member of the margarite group, and related brittle micas. Amer. Mineral., 52, 1689–1696. (3) Gallagher, M.J. and J.R. Hawkes (1966) Beryllium minerals from Rhodesia and Uganda. Bull. Geol. Sur. Great Britain, 25, 59–75. (4) Slade, P.G., P.K. Schultz, and C. Dean (1987) Refinement of the ephesite structure in C1 symmetry. Neues Jahrb. Mineral., Monatsh., 275–287.

Crystal Data: Orthorhombic. *Point Group:* $2/m\ 2/m\ 2/m$. As pseudohexagonal crystals, tabular on {001}, also elongated along [010], to 2 cm. Micaceous, spherulitic, fine to coarse granular, porcelaneous massive. *Twinning:* On {001}, common, simple and polysynthetic; also as trillings.

Physical Properties: *Cleavage:* Perfect on {001}, distinct on {100}. *Fracture:* Uneven to conchoidal when massive. Hardness = 5.5 D(meas.) = 2.55–2.61 D(calc.) = 2.56

Optical Properties: Transparent to translucent. *Color:* Colorless, white, also tinted violet, blue, or yellow when finely crystalline. *Luster:* Pearly on cleavages, vitreous on fractures. *Optical Class:* Biaxial (+) or (–). *Orientation:* X = a; Y = c; Z = b. *Dispersion:* $r > v$, distinct. $\alpha = 1.536$–1.544 $\beta = 1.51$–1.544 $\gamma = 1.542$–1.546 2V(meas.) = $16°$–$32°$ 2V(calc.) = $26°$

Cell Data: *Space Group:* $Pnma$. a = 12.74(1) b = 13.63(1) c = 7.33(1) Z = 8

X-ray Powder Pattern: Narssârssuk, Greenland.
3.37 (10), 3.08 (10), 2.96 (10), 1.790 (6), 1.634 (6), 2.48 (5), 1.537 (5)

Chemistry:

	(1)	(2)	(3)
SiO_2	73.74	71.79	73.49
Al_2O_3		1.94	
Fe_2O_3		0.12	
BeO	10.56	10.45	10.20
MgO		trace	
CaO		0.16	
Na_2O	12.88	11.43	12.64
K_2O		0.30	
H_2O^+	3.73	4.14	3.67
H_2O^-		0.03	
Total	100.91	100.36	100.00

(1) Narssârssuk, Greenland. (2) Mt. Karnasurt, Russia. (3) $NaBeSi_3O_7(OH)$.

Polymorphism & Series: Dimorphous with eudidymite.

Occurrence: A late-stage mineral in nepheline syenite pegmatites.

Association: Albite, aegirine, eudialyte, elpidite, neptunite, tugtupite, leucophanite, natrolite, analcime.

Distribution: At Narssârssuk and in the Ilímaussaq intrusion, southern Greenland. On Arø Island, in the Langesundsfjord, and in the Vevja quarry, Tvedalen, Norway. On the Kola Peninsula, in the Lovozero massif, on Mts. Karnasurt and Alluaiv, and at Kuivchorr, and in the Khibiny massif; from the Murun massif, southwest of Olekminsk, Yakutia, Russia. At Věžná, Czech Republic. From Mont Saint-Hilaire, Quebec, Canada. In the USA, near Quincy, Norfolk Co., Massachusetts. On Mt. Malosa, Zomba Plateau, Malawi.

Name: The prefix *epi* from the Greek for *near*, referring to the dimorphous relation with *eudidymite*.

Type Material: University of Copenhagen, Copenhagen, Denmark.

References: (1) Dana, E.S. (1899) Dana's system of mineralogy, (6th edition), app. I, 24–25. (2) Vlasov, K.A., M.V. Kuz'menko, and E.M. Es'kova (1966) The Lovozero alkaline massif. Akad. Nauk SSSR, 433–438 (in English). (3) Petersen, O.V. (1966) Crossed axial plane dispersion in epididymite. Amer. Mineral., 51, 916–919. (4) Mandarino, J.A. and D.C. Harris (1969) Epididymite from Mont St. Hilaire, Quebec. Can. Mineral., 9, 706–709. (5) Robinson, P.D. and J.H. Fang (1970) The crystal structure of epididymite. Amer. Mineral., 55, 1541–1549.

Crystal Data: Monoclinic. *Point Group:* $2/m$. Crystals prismatic, to 35 cm, commonly elongated and striated $\parallel$ [010]. Fibrous, coarse to fine granular, massive. *Twinning:* On {100}, contact, lamellar, common.

Physical Properties: *Cleavage:* Perfect on {001}, imperfect on {100}. *Fracture:* Uneven. *Tenacity:* Brittle. Hardness = 6–7   D(meas.) = 3.38–3.49   D(calc.) = 3.391–3.464

Optical Properties: Transparent to nearly opaque. *Color:* Pistachio-green to pale green, yellow, yellowish green, greenish yellow, greenish black; yellow-green in thin section. *Luster:* Vitreous, pearly, somewhat resinous.
Optical Class: Biaxial (–). *Pleochroism:* Strong; X = colorless, pale yellow, pale green; Y = greenish yellow; Z = yellowish green. *Orientation:* $Y = b$.   *Dispersion:* $r > v$. $\alpha = 1.715$–1.751
$\beta = 1.725$–1.784 $\gamma = 1.734$–1.797 2V(meas.) = $90°$–$116°$

Cell Data: *Space Group:* $P2_1/m$.   a = 8.8877(14)   b = 5.6275(8)   c = 10.1517(12)
$\beta = 115.383(14)°$ Z = 2

X-ray Powder Pattern: Bourg d'Oisans, [France].
2.900 (100), 2.679 (100), 2.688 (70), 4.02 (50), 2.599 (50), 2.460 (50), 3.40 (40)

Chemistry:

	(1)		(1)
SiO_2	36.52	MnO	0.00
TiO_2	0.00	MgO	0.00
Al_2O_3	20.97	CaO	23.05
Fe_2O_3	17.22	H_2O^+	1.98
FeO	0.45	Total	100.19

(1) Westfield, Hampden Co., Massachusetts, USA; corresponds to $(Ca_{1.99}Fe^{2+}_{0.03})_{\Sigma=2.02}Al_{1.94}Fe^{3+}_{1.04}$ $(Si_{2.94}Al_{0.06})_{\Sigma=3.00}O_{12}(OH)_{1.06}$.

Polymorphism & Series: Forms a series with clinozoisite.

Mineral Group: Epidote group.

Occurrence: Characteristic of several facies of regionally metamorphosed rocks and some contaminated felsic igneous rocks; in contact zones between igneous and calcareous sedimentary rocks; from alteration of plagioclase (saussuritization).

Association: Zeolites, amphiboles, plagioclase feldspars, quartz, actinolite, calcite (greenschist facies); amphiboles, vesuvianite, scapolite, talc, wollastonite, pyroxenes, garnet (epidote-hornfels facies); pumpellyite, glaucophane, lawsonite, riebeckite, garnet, omphacite (blueschist facies).

Distribution: A few notable occurrences follow for this widespread mineral. From Bourg d'Oisans, Isère, France. At Arendal, Norway. From Traversella, Piedmont, Italy. Exceptional crystals from the Knappenwand, Untersulzbachtal, Salzburg, Austria. In the USA, fine crystals from around Sulzer, Prince of Wales Island, Alaska; on Garnet Hill, Calaveras Co., California; at the Calumet mine, Chaffee Co., Colorado; in the Seven Devils district, Adams Co., Idaho; and from near Hawthorne, Mineral Co., Nevada. In Mexico, at San Quentin, Baja California. From Naukluft Farm, Rehoboth, Namibia. In Pakistan, exceptional crystals from Baghicha, near Skardu, and at Turmik, Gilgit district.

Name: From the Greek for *increase*, the base of the prism with one longer side.

References: (1) Dana, E.S. (1892) Dana's system of mineralogy, (6th edition), 516–521. (2) Deer, W.A., R.A. Howie, and J. Zussman (1986) Rock-forming minerals, (2nd edition), v. 1B, disilicates and ring silicates, 44–134. (3) Seki, Y. (1959) Relation between chemical composition and lattice constants of epidote. Amer. Mineral., 44, 720–730. (4) Gabe, E.J., J.C. Portheine, and S.H. Whitlow (1973) A reinvestigation of the epidote structure: confirmation of the iron location. Amer. Mineral., 58, 218–223. (5) Kvick, Å, J.J. Pluth, J.W. Richardson, Jr., and J.V. Smith (1988) The ferric iron distribution and hydrogen bonding in epidote: a neutron diffraction study at 15 K. Acta Cryst., 44, 351–355.

Crystal Data: Monoclinic. *Point Group:* $2/m$, m, or 2. Crystals prismatic, to 3 cm; in spherulitic or sheaflike aggregates. *Twinning:* Always on {100}, pseudo-orthorhombic; on {110} to form penetration crosses.

Physical Properties: *Cleavage:* {010}, perfect. *Fracture:* Uneven. *Tenacity:* Brittle. Hardness = 4 D(meas.) = 2.22–2.28 D(calc.) = 2.266 Piezoelectric.

Optical Properties: Transparent to translucent. *Color:* Colorless, white, pinkish, yellowish; colorless in thin section. *Luster:* Vitreous.
Optical Class: Biaxial (–). *Orientation:* $Y = b$; $Z \wedge c \simeq -10°$; $X \wedge a = 11°$.
Dispersion: $r < v$. $\alpha = 1.485–1.505$ $\beta = 1.497–1.515$ $\gamma = 1.497–1.519$ 2V(meas.) = ~44°

Cell Data: *Space Group:* $C2/m$, Cm, or $C2$. $a = 9.08(1)$ $b = 17.74(1)$ $c = 10.25(1)$ $\beta = 124.54(5)°$ Z = 3

X-ray Powder Pattern: Locality unknown. (ICDD 19-213).
3.45 (100), 8.89 (90), 3.21 (90), 3.87 (70), 4.91 (65), 6.89 (60), 2.917 (60)

Chemistry:

	(1)
SiO_2	57.79
Al_2O_3	17.62
Fe_2O_3	0.02
MgO	0.02
CaO	8.21
Na_2O	1.39
K_2O	0.06
H_2O^+	12.21
H_2O^-	3.10
Total	100.42

(1) Fossarfell, Iceland; corresponds to $(Ca_{0.90}Na_{0.27}K_{0.01})_{\Sigma=1.18}Al_{2.12}Si_{5.89}O_{16} \cdot 5.20H_2O$.

Polymorphism & Series: Dimorphous with goosecreekite.

Mineral Group: Zeolite group.

Occurrence: In cavities in basalts and gneisses.

Association: Zeolites, quartz.

Distribution: While relatively rare, many localities are known. Exceptional crystals from around the Berufjord, Iceland. Found near San Piero in Campo, Elba, Italy. At Giebelsbach, near Fiesch, Valais, Switzerland. In Japan, at Kuroiwa, Niigata Prefecture; Yugawara, Kanagawa Prefecture; and elsewhere. Large crystals from Nasik and Khandivali quarry, Bombay, Maharashtra, India. In the USA, large crystals from Kosmos, near Morton, Lewis Co., Washington; at Goble, Columbia Co., Oregon.

Name: From the Greek, *epi* for *near*, and the quite similar mineral *stilbite*.

References: (1) Dana, E.S. (1892) Dana's system of mineralogy, (6th edition), 577–578. (2) Deer, W.A., R.A. Howie, and J. Zussman (1963) Rock-forming minerals, v. 4, framework silicates, 377–385. (3) Perrotta, A.J. (1967) The crystal structure of epistilbite. Mineral. Mag., 36, 480–490. (4) Slaughter, M. and W.T. Kane (1969) The crystal structure of a disordered epistilbite. Zeits. Krist., 130, 68–87. (5) Galli, E. and R. Rinaldi (1974) The crystal chemistry of epistilbites. Amer. Mineral., 59, 1055–1061. (6) Akizuki, M. and H. Nishido (1988) Epistilbite: symmetry and twinning. Amer. Mineral., 73, 1434–1439.

Crystal Data: Triclinic. *Point Group:* 1. As crystals, rectangular and tabular on {001}, to 5 cm. As irregular plates and curved lamellar masses.

Physical Properties: *Cleavage:* Perfect on {001}, distinct on {110}. *Tenacity:* Very brittle to friable. Hardness = 1.5–3   D(meas.) = 2.65–2.89   D(calc.) = [3.18]

Optical Properties: Opaque to translucent, transparent in thin sheets. *Color:* White, gray, yellow-gray, tan, light brown; in thin section, colorless. *Luster:* Pearly to silky. *Optical Class:* Biaxial (–). *Orientation:* $Z \wedge c = 70°$. *Dispersion:* $r > v$, inclined. $\alpha = 1.610$ $\beta = 1.650–1.720$ $\gamma = 1.682–1.770$ 2V(meas.) = $\sim 60°–80°$

Cell Data: *Space Group:* $P1$. $a = 5.41$ $b = 7.08$ $c = 12.07$ $\alpha = 103°03'$ $\beta = 96°03'$ $\gamma = 88°36'$ Z = 2

X-ray Powder Pattern: Ilímaussaq intrusion, Greenland.
4.322 (10), 2.99 (9), 2.869 (9), 12.00 (8), 5.902 (7), 1.790 (7), 2.155 (6)

Chemistry:

	(1)	(2)		(1)	(2)
SiO_2	27.59	25.60	CaO	0.77	0.00
TiO_2	7.22	14.55	Na_2O	17.59	4.97
ZrO_2		0.00	K_2O	trace	0.70
Al_2O_3		2.05	F	1.98	0.00
Nb_2O_5	33.56	31.40	H_2O^+	9.26	7.35
Ta_2O_5		0.90	H_2O^-	1.75	8.20
FeO	0.20	0.90	P_2O_5		2.10
MnO	0.30	1.20	$-O = F_2$	0.83	
MgO	0.13	0.00	Total	99.52	99.92

(1) Julianehåb district, Greenland; material partly altered. (2) Lovozero massif, Russia.

Occurrence: In alkalic pegmatites, albitites, sodalite xenoliths, and hydrothermal veins.

Association: Aegirine, albite, murmanite, sphalerite, manganoan pectolite, neptunite, steenstrupine, sodalite, eudialyte, nenadkevichite.

Distribution: At a number of localities in the Ilímaussaq intrusion, southern Greenland. From the Lovozero massif, Kola Peninsula, Russia. At Mont Saint-Hilaire, Quebec, Canada.

Name: From the Greek for *letter*, in allusion to the flat rectangular crystal habit and white color.

Type Material: University of Copenhagen, Copenhagen, Denmark, 319, 320.

References: (1) Dana, E.S. and W.E. Ford (1909) Dana's system of mineralogy, (6th edition), app. II, 39. (2) Vlasov, K.A., Ed. (1966) Mineralogy of rare elements, v. II, 562–564. (3) Khalilov, A.P., Y.S. Makarov, K.S. Mamedov, and L.A. P'yanzina (1965) Crystal structure of minerals of the murmanite-lomonosovite group. Doklady Acad. Nauk SSSR, 162, 179–182 (in Russian). (4) Karup-Møller, S. (1986) Epistolite from the Ilímaussaq alkaline complex in South Greenland. Neues Jahrb. Mineral., Abh., 155, 289–304. (5) Mandarino, J.A. and V. Anderson (1989) Monteregian Treasures. Cambridge Univ. Press, 76.

Crystal Data: Monoclinic. *Point Group:* $2/m$. As thick plates, tabular $\parallel \{100\}$, intimately intergrown with orthoericssonite, to 2 cm.

Physical Properties: *Cleavage:* Perfect on $\{100\}$, fair on $\{011\}$. *Tenacity:* Very brittle. Hardness = 4.5 D(meas.) = 4.21 D(calc.) = 4.38 Weakly magnetic.

Optical Properties: Translucent. *Color:* Deep reddish black; in thin section, brown to yellowish brown. *Streak:* Rich brown.
Optical Class: Biaxial (+); properties indistinguishable from orthoericssonite. *Pleochroism:* $X =$ pale greenish tan; $Y =$ red-brown; $Z =$ deep brown. *Orientation:* $X = b$; $Y = c$; $Z = a$.
Dispersion: $r > v$, perceptible. *Absorption:* $Z > Y > X$. $\alpha = 1.807(5)$ $\beta = 1.833(5)$ $\gamma = 1.89(1)$ 2V(meas.) = 43°

Cell Data: *Space Group:* $C2/m$. a = 20.46 b = 7.03 c = 5.34 $\beta = 95°30'$ Z = 4

X-ray Powder Pattern: Långban, Sweden.
3.510 (100), 10.12 (60), 2.780 (60), 1.752 (60), 1.597 (60), 5.08 (50), 3.398 (50)

Chemistry:

	(1)	(2)
SiO_2	20.83	23.83
Fe_2O_3	14.47	15.83
As_2O_5	1.23	
MnO	24.38	28.14
PbO	1.40	
BaO	29.81	30.41
H_2O	1.52	1.79
Total	93.64	100.00

(1) Långban, Sweden; by electron microprobe, intergrown with orthoericssonite, oxidation states and H_2O separately determined. (2) $BaMn_2FeOSi_2O_7(OH)$.

Polymorphism & Series: Dimorphous with orthoericssonite.

Occurrence: A rare mineral, embedded in a fine-grained manganoan aegirine zone in well-banded tephroite-rhodonite-manganoan aegirine skarn, in a metamorphosed manganese orebody.

Association: Orthoericssonite, tephroite, rhodonite, manganoan aegirine, hedyphane, andradite, hausmannite, långbanite, richterite.

Distribution: From Långban, Värmland, Sweden.

Name: For John E. Ericsson (1803–1889), Swedish-American inventor and designer of the iron-clad ship "Monitor," who was born at Långban, Sweden.

Type Material: National Museum of Natural History, Washington, D.C., USA, 120061; The Natural History Museum, London, England, 1987,118.

References: (1) Moore, P.B. (1971) Ericssonite and orthoericssonite, two new members of the lamprophyllite group from Långban, Sweden. Lithos, 4, 137–145. (2) (1971) Amer. Mineral., 56, 2157 (abs. ref. 1).

Crystal Data: Hexagonal. *Point Group:* $6/m\ 2/m\ 2/m$. Crystals prismatic, to 15 mm; in radiating groups or overgrowing offretite or levyne; typically fibrous, wool-like.

Physical Properties: *Cleavage:* Poor prismatic. *Fracture:* Splintery. Hardness = 3.5–4
D(meas.) = 2.08   D(calc.) = 2.13

Optical Properties: Transparent to translucent. *Color:* White; colorless in thin section.
Streak: White. *Luster:* Vitreous to silky.
Optical Class: Uniaxial (+). $\omega = 1.458$–1.480 $\epsilon = 1.458$–1.468

Cell Data: *Space Group:* $P6_3mmc$.   a = 13.21–13.26   c = 15.04–15.12   Z = 2

X-ray Powder Pattern: Tecopa, California, USA; close to offretite.
11.41 (100), 6.61 (73), 4.322 (67), 3.746 (65), 2.860 (60), 2.812 (52), 2.839 (50)

Chemistry:

	(1)	(2)		(1)	(2)
SiO_2	57.40	60.67	Na_2O	1.45	4.39
Al_2O_3	15.60	12.90	K_2O	3.40	4.09
Fe_2O_3		1.35	H_2O^+		7.69
FeO		0.09	H_2O^-		6.94
MgO	1.11	1.09	H_2O	17.58	
CaO	2.92	0.65	rem.		0.14
			Total	99.46	100.00

(1) Durkee, Oregon, USA; corresponds to $(Ca_{0.75}K_{0.52}Mg_{0.40}Na_{0.34})_{\Sigma=2.01}Al_{4.40}Si_{14.10}O_{36} \cdot$
$14.00H_2O$. (2) Tecopa, California, USA; remainder is TiO_2 0.09%, MnO 0.03%, P_2O_5 0.02%;
corresponds to $(Na_{0.99}K_{0.60}Mg_{0.38}Ca_{0.16})_{\Sigma=2.13}Fe_{0.25}Al_{3.54}Si_{13.71}O_{36} \cdot 11.34H_2O$.

Mineral Group: Zeolite group.

Occurrence: In basalts; in altered rhyolitic tuffs and lake-bed sediments formed by alteration of vitric ash fallen into the lakes.

Association: Zeolites, "opal," quartz, celadonite, montmorillonite, dolomite, calcite.

Distribution: Some localities for relatively pure material are: in the USA, at Tecopa, Inyo Co., California; in Nevada, at Pine Valley, Eureka Co., Eastgate, Churchill Co., Jersey Valley, Pershing Co., and Reese River, Lander Co.; in Oregon, at Cape Lookout, Tillamook Co., Agate Beach, Lincoln Co., and Durkee, Baker Co. From the Parkgate quarry, Templepatrick, Co. Antrim, Ireland. At Hvalstod and in the Berufjord area, Iceland. At Moeraki, Otago, South Island, New Zealand. From Mazé, Niigata Prefecture, Japan. Many other localities are known.

Name: For the Greek word for *wool*, recalling its common appearance.

Type Material: Harvard University, Cambridge, Massachusetts, 86532; National Museum of Natural History, Washington, D.C., USA, R4066, 133333.

References: (1) Dana, E.S. (1899) Dana's system of mineralogy, (6th edition), app. I, 25.
(2) Staples, L.W. and J.A. Gard (1959) The fibrous zeolite erionite; its occurrence, unit cell, and structure. Mineral. Mag., 32, 261–281. (3) Sheppard, R.A. and A.J. Gude, 3d (1969) Chemical composition and physical properties of the related zeolites offretite and erionite. Amer. Mineral., 54, 875–886. (4) Gard, J.A. and J.M. Tait (1971) Refinement of the crystal structure of erionite. In: Molecular sieves, Leuven Univ. Press, 94. (5) Wise, W.S. and R.W. Tschernich (1976) The chemical composition and origin of the zeolites offretite, erionite, and levyne. Amer. Mineral., 61, 853–863.

Erlianite

$$(Fe^{2+}, Mg)_4(Fe^{3+}, V^{3+})_2Si_6O_{15}(OH, O)_8$$

Crystal Data: Orthorhombic. *Point Group:* $2/m\ 2/m\ 2/m$ or $mm2$. As fibers, flakes, and lathlike aggregates, to 2 cm.

Physical Properties: *Cleavage:* Perfect on {001} and {100}. Hardness = 3.7 D(meas.) = 3.11 D(calc.) = [3.17]

Optical Properties: Opaque. *Color:* Black; brown in thin section. *Streak:* Brownish black. *Luster:* Silky.
Optical Class: Biaxial (–). *Pleochroism:* Slight; X = light brown; $Y = Z$ = dark brown. *Orientation:* $X = b$; $Y = c$; $Z = a$. *Dispersion:* $r < v$, weak. *Absorption:* $Z = Y > X$. $\alpha = 1.667$ $\beta = 1.674$ $\gamma = 1.679$ 2V(meas.) = 56°–59°

Cell Data: *Space Group:* $Pmmn$ or $Pm2_1n$. a = 23.20(1) b = 9.20(1) c = 13.18(1) Z = 6

X-ray Powder Pattern: Harhada mine, China.
11.5 (100), 2.89 (60), 2.61 (60), 3.05 (50), 2.52 (50), 1.560 (50), 2.42 (30)

Chemistry:

	(1)		(1)
SiO_2	38.80	MgO	1.00
TiO_2	0.38	CaO	0.83
Al_2O_3	0.19	K_2O	0.08
Fe_2O_3	21.26	Na_2O	0.09
V_2O_5	1.15	H_2O^+	7.65
FeO	26.67	H_2O^-	0.90
MnO	0.55	P_2O_5	0.05
		Total	99.60

(1) Harhada mine, China; by colorimetric microanalysis, corresponds to $(Fe^{2+}_{3.33}Fe^{3+}_{0.36}$ $Mg_{0.22}Mn_{0.07})_{\Sigma=3.98}(Fe^{3+}_{1.89}V_{0.11})_{\Sigma=2.00}(Si_{5.79}Fe^{3+}_{0.14}Ti_{0.04}Al_{0.03})_{\Sigma=6.00}O_{15}[O,(OH)]_8$.

Occurrence: As coatings along fault surfaces cutting a low-grade metamorphosed volcanic-sedimentary iron formation; presumably formed in a high-pressure, low-temperature environment.

Association: Minnesotaite, stilpnomelane, quartz, magnetite, siderite, albite, deerite.

Distribution: In the Harhada iron mine, along the Jining-Erlian railway, Inner Mongolia, China.

Name: Presumably for the town of Erlian, China.

Type Material: n.d.

References: (1) Feng, X. and R. Yang (1986) Erlianite, a new vanadium- and iron-bearing silicate mineral. Mineral. Mag., 50, 285–289. (2) (1987) Amer. Mineral., 72, 1023–1024 (abs. ref. 1).

Crystal Data: Triclinic. *Point Group:* $\bar{1}$. Forms parallel fibrous aggregates, fibers elongated along [001], to 3 cm; in disseminated elongated grains, to 1 cm.

Physical Properties: *Cleavage:* {100} and {010}, perfect. *Fracture:* Fibrous. Hardness = 2–3   D(meas.) = 2.75(2)   D(calc.) = 2.73

Optical Properties: Translucent; transparent in thin section. *Color:* Olive-green with brown and yellow tinges. *Streak:* White. *Luster:* Vitreous.
Optical Class: Biaxial (+). *Pleochroism:* Strong; $X = Y$ = light green or yellow; Z = dark olive-green. *Orientation:* $X \wedge c = 86°$; $Y \wedge c = 73°$; $Z \wedge c = 17°$. *Dispersion:* $r > v$.
$\alpha = 1.569(2)$ $\beta = 1.574(2)$ $\gamma = 1.590(2)$ 2V(meas.) = 58(2)°   2V(calc.) = 59°

Cell Data: *Space Group:* $P\bar{1}$.   a = 10.244(2)   b = 11.924(3)   c = 5.276(3)
$\alpha = 103.491(2)°$ $\beta = 96.960(3)°$ $\gamma = 91.945(3)°$ Z = 1

X-ray Powder Pattern: Khibiny massif, Russia.
11.58 (10), 2.990 (10), 2.709 (8), 1.652 (8), 2.608 (7), 2.459 (6), 2.160 (6)

Chemistry:

	(1)
SiO_2	47.1
TiO_2	3.0
FeO	6.1
MnO	4.7
MgO	0.5
CaO	0.1
Na_2O	12.4
K_2O	13.6
H_2O	[12.5]
Total	[100.0]

(1) Khibiny massif, Russia; by electron microprobe, H_2O by difference; corresponds to $Na_{4.08}K_{2.95}$ $(Fe_{0.87}Mn_{0.68}Ti_{0.38}Mg_{0.13}Ca_{0.02})_{\Sigma=2.08}Si_8O_{19.98}(OH)_{4.04} \cdot 5.06H_2O$.

Occurrence: In unweathered ultra-agpaitic pegmatites in a differentiated alkalic massif.

Association: Orthoclase, nepheline, sodalite, aegirine, alkalic amphibole, villiaumite, natrite, rasvumite, vuonnemite, astrophyllite, shcherbakovite, kazakovite, koashvite, thermonatrite, nacaphite.

Distribution: From Mts. Rasvumchorr and Koashva, Khibiny massif, Kola Peninsula, Russia.

Name: For Vadim Victorovich Ershov (1939–1989), of the Moscow Mining Institute, Moscow, Russia.

Type Material: Geology Museum, Kola Branch, Academy of Sciences, Apatity; A.E. Fersman Mineralogical Museum, Academy of Sciences, Moscow, Russia.

References: (1) Khomyakov, A.P., Y.P. Men'shikov, R.K. Rastsvetaeva, and G.N. Nechelyustov (1993) Ershovite $Na_4K_3(Fe, Mn, Ti)_2Si_8O_{20}(OH)_4 \cdot 4H_2O$ – a new mineral. Zap. Vses. Mineral. Obshch., 122(1), 116–120 (in Russian). (2) (1994) Amer. Mineral., 79, 1010 (abs. ref. 1). (3) Rastsvetaeva, R.K., O.Y. Rekhlova, and A.P. Khomyakov (1991) Crystal structure of a new natural Na,K,Fe-silicate. Kristallografiya (Sov. Phys. Crystal.), 36, 892–897 (in Russian).

Crystal Data: Cubic. *Point Group:* n.d. In granular crystals, to 0.5 mm.

Physical Properties: *Fracture:* Subconchoidal. Hardness = 5.8–6.5 D(meas.) = 2.35
D(calc.) = 2.34

Optical Properties: Transparent. *Color:* Colorless. *Streak:* White. *Luster:* Vitreous.
Optical Class: Isotropic. $n = 1.502(1)$

Cell Data: *Space Group:* n.d. $a = 5.975$ $Z = 1$

X-ray Powder Pattern: Altay mine, China.
1.798 (100), 1.996(80), 3.443 (20), 2.988 (20), 2.674 (20)

Chemistry:

	(1)	(2)
SiO$_2$	77.86	79.50
Al$_2$O$_3$	1.45	
FeO	0.04	
CaO	2.82	
Na$_2$O	17.98	20.50
Total	100.15	100.00

(1) Altay mine, China; by electron microprobe, average of six analyses. (2) Na$_2$Si$_4$O$_9$.

Occurrence: As linings of miarolitic cavities in a Ta-Nb-Be-bearing granite pegmatite.

Association: Topaz, albite, muscovite, quartz, apatite, garnet.

Distribution: From the Altay pegmatite mine, Fuyun Co., 600 km northeast of Urumchi,
Sinkiang Uighur Autonomous Region, China.

Name: For the Ertixi River, near the occurrence in China.

Type Material: The Geological Museum, Chengdu Geological College, [Ch'engdu], China.

References: (1) Zhang Rubo, Han Fengming, and Du Chonliang (1985) Ertixiite – a new
mineral from the Altay pegmatite mine, Xinjiang, China. Geochemistry (China), 4, 192–195.
(2) (1986) Amer. Mineral., 71, 1544 (abs. ref. 1). (3) (1988) Mineral. Mag., 52, 724 (abs. ref. 1).

Crystal Data: Monoclinic. *Point Group:* $2/m$. Granular aggregates, massive, to 9 cm.

Physical Properties: *Cleavage:* Distinct on {010} and {100}; poor on {101}.
Fracture: Conchoidal. Hardness = 5–5.5 D(meas.) = 4.28–4.42 D(calc.) = 4.25 Brilliant
yellow fluorescence under SW UV; kelly green cathodoluminescence.

Optical Properties: Opaque, transparent in small grains. *Color:* White, darkening on
exposure to light. *Luster:* Dull to slightly greasy.
Optical Class: Biaxial (–). $\alpha = 1.760$–1.762 $\beta = 1.769$–1.770 $\gamma = 1.769$–1.774
2V(meas.) = 5°–40°

Cell Data: *Space Group:* $P2_1/n$. a = 17.628(4) b = 8.270(3) c = 30.52(2) $\beta = 90°$
Z = 12

X-ray Powder Pattern: Franklin, New Jersey, USA.
3.017 (100), 2.534 (75), 7.62 (45), 1.944 (45), 2.367 (40), 2.884 (33), 3.363 (23)

Chemistry:

	(1)	(2)
SiO_2	24.10	25.3
FeO	0.48	
MnO	0.57	0.6
ZnO	30.61	32.3
PbO	27.63	26.8
MgO	0.23	0.4
CaO	16.36	16.4
H_2O^-	0.12	
Total	100.10	101.8

(1) Franklin, New Jersey, USA; corresponding to $Pb_{1.23}(Ca_{2.89}Mn_{0.08}Mg_{0.06})_{\Sigma=3.03}Zn_{3.73}$
$Si_{3.98}O_{16}$. (2) Do.; by electron microprobe, average of 10 analyses, corresponding to $Pb_{1.15}$
$(Ca_{2.80}Mg_{0.09}Mn_{0.08})_{\Sigma=2.97}Zn_{3.80}Si_{4.04}O_{16}$.

Occurrence: In a metamorphosed stratiform zinc orebody.

Association: Willemite, zincite, hardystonite, glaucochroite, andradite, franklinite, clinohedrite,
leucophoenicite, larsenite, copper.

Distribution: From Franklin, Sussex Co., New Jersey, USA.

Name: For Professor Esper Signius Larsen, Jr. (1879–1961), American petrologist and
mineralogist, Harvard University, Cambridge, Massachusetts, USA.

Type Material: n.d.

References: (1) Moore, P.B. and P.H. Ribbe (1965) A study of "calcium-larsenite" renamed
esperite. Amer. Mineral., 50, 1170–1178. (2) Palache, C., L.H. Bauer, and H. Berman (1928)
Larsenite, calcium-larsenite and associated minerals at Franklin, N.J. Amer. Mineral., 13, 334–340.
(3) Palache, C. (1935) The minerals of Franklin and Sterling Hill, Sussex County, New Jersey.
U.S. Geol. Sur. Prof. Paper 180, 81–82. (4) Dunn, P.J. (1985) The lead silicates from Franklin,
New Jersey: occurrence and composition. Mineral. Mag., 49, 721–727.

Crystal Data: Monoclinic. *Point Group:* $2/m$. As prismatic crystals, up to 8 mm.

Physical Properties: *Cleavage:* Perfect on {110}. Hardness = 6 D(meas.) = n.d.
D(calc.) = 3.54

Optical Properties: Transparent in thin crystals. *Color:* Reddish brown, becoming darker
with increasing Fe content. *Streak:* White. *Luster:* Vitreous.
Optical Class: Biaxial (–). *Pleochroism:* X = lemon-yellow; Y = greenish yellow; Z =
apple-green. *Orientation:* $Y = b$; $Z \wedge c = 9°$. *Dispersion:* $r < v$, strong. $\alpha = 1.795(5)$
$\beta = 1.815(5)$ $\gamma = 1.825(5)$ 2V(meas.) = 77(5)°

Cell Data: *Space Group:* $C2/c$. a = 9.79(1) b = 8.822(9) c = 5.37(1) $\beta = 105.81(9)°$
Z = 4

X-ray Powder Pattern: Durham ranch, Wyoming, USA.
3.000 (100), 2.526 (70), 2.960 (60), 2.554 (40), 2.576 (30), 1.545 (30), 1.430 (25)

Chemistry:

	(1)
SiO_2	29.51
TiO_2	0.99
Al_2O_3	17.95
Fe_2O_3	23.89
FeO	0.69
MnO	0.11
MgO	2.68
CaO	23.40
Na_2O	0.14
Total	99.36

(1) Durham ranch, Wyoming, USA; by electron microprobe, average of 43 analyses on several
grains, $Fe^{2+}:Fe^{3+}$ calculated from normalized formula; corresponds to $(Ca_{1.01}Na_{0.01})_{\Sigma=1.02}$
$(Fe^{3+}_{0.72}Mg_{0.16}Al_{0.04}Ti_{0.03}Fe^{2+}_{0.02})_{\Sigma=0.97}(Si_{1.19}Al_{0.81})_{\Sigma=2.00}O_6$.

Mineral Group: Pyroxene group.

Occurrence: A high-temperature, low-pressure, oxidized and quenched crystallization product
derived from fused sediments contiguous to naturally combusted coal seams.

Association: Anorthite, melilite, magnetite-hercynite, glass.

Distribution: At Durham ranch, in the Powder River basin, 13 km northeast of Reno Junction
and 25 km south of Gillette, Campbell Co., Wyoming, USA.

Name: For Dr. Eric J. Essene, Professor at the University of Michigan, Ann Arbor, Michigan,
USA, and discoverer of the first specimens.

Type Material: National Museum of Natural History, Washington, D.C., USA, 163357.

References: (1) Cosca, M.A. and D.R. Peacor (1987) Chemistry and structure of esseneite
$(CaFe^{3+}AlSiO_6)$, a new pyroxene produced by pyrometamorphism. Amer. Mineral., 72, 148–156.

Crystal Data: Monoclinic. *Point Group: 2/m.* As prismatic crystals, flattened on {100}, commonly morphologically complex, to 10 cm.

Physical Properties: *Cleavage:* Perfect on {010}, imperfect on {110} and {001}. *Fracture:* Conchoidal. *Tenacity:* Brittle. Hardness = 7.5   VHN = 1310   D(meas.) = 2.99–3.10 D(calc.) = 3.115

Optical Properties: Transparent to translucent. *Color:* Colorless, white, pale green to deep yellowish green, greenish blue, pale blue to deep blue, rarely mottled; colorless to pale blue or pale green in thin section. *Streak:* White. *Luster:* Vitreous, somewhat pearly on cleavages. *Optical Class:* Biaxial (+). *Pleochroism:* May be marked in shades of deep blue. *Orientation:* $Z \wedge c = 41°$. *Dispersion:* $r > v$, distinct. $\alpha = 1.651–1.653$ $\beta = 1.655–1.657$ $\gamma = 1.669–1.675$ 2V(meas.) = ~50°   2V(calc.) = 46°

Cell Data: *Space Group:* $P2_1/a$.   a = 4.763(5)   b = 14.29(2)   c = 4.618(5) $\beta = 100°15(5)'$ Z = 4

X-ray Powder Pattern: Villa Rica, Brazil.
7.146 (100), 3.219 (50), 3.836 (35), 2.773 (35), 2.444 (35), 2.543 (25), 1.991 (18)

Chemistry:

	(1)	(2)
SiO$_2$	42.05	41.60
Al$_2$O$_3$	33.97	34.76
Fe$_2$O$_3$	0.12	
FeO		0.28
BeO	17.51	16.95
Na$_2$O		0.13
K$_2$O		0.04
H$_2$O	6.35	5.95
Total	[100.00]	99.71

(1) Ottré, Belgium; recalculated to 100.00% after deduction of 3.5% quartz; corresponds to $Be_{1.00}Al_{0.95}Si_{1.00}O_4(OH)_{1.01}$. (2) Santo do Encoberto, Brazil; by electron microprobe; corresponds to $(Be_{0.99}Fe_{0.01}Na_{0.01})_{\Sigma=1.01}Al_{1.00}Si_{1.01}O_4(OH)_{0.96}$.

Occurrence: The product of decomposition of beryl in pegmatites, and in low-temperature alpine veins.

Association: Feldspar, quartz, topaz, beryl, mica, calcite, ankerite, chlorite.

Distribution: Fine crystals occur from: the Sanarka River, Orenburg district, and the Kamenka River, Southern Ural Mountains, Russia. At Epprechtstein, Bavaria, Germany. In the Muiâne pegmatite, Alto Ligonha district, Mozambique. Found near Mikese Station, Morogoro district, Tanzania. From Karoi, Miami district, Zimbabwe. In Brazil, in Minas Gerais, around Villa Rica, Boa Vista, Santa do Encoberto, Dom Bosco, Ouro Prêto; Mãe dos Homens, in Capelinha; and many other localities. From the Gachalá mine, Las Cruces, near Chivor, Colombia. A number of other minor localities are known.

Name: From the Greek for *easily* and *fracture* regarding its perfect cleavage.

References: (1) Dana, E.S. (1892) Dana's system of mineralogy, (6th edition), 508–509. (2) Biscoe, J. and B.E. Warren (1930) The structure of euclase HBeAlSiO$_5$. Zeits. Krist., 86, 292–297. (3) Mrose, M.E. and O. von Knorring (1959) The mineralogy of väyrynenite, (Mn, Fe)Be(PO$_4$)(OH). Zeits. Krist., 112, 275–288. (4) Graziani, G. and G. Guidi (1980) Euclase from Santa do Encoberto, Minas Gerais, Brazil. Amer. Mineral., 65, 183–187. (5) Anderson, S. (1980) Euclase. J. Gemmology, 17, 18–29. (6) Hanson, A. (1985) Découverte d'euclase dans un filon de quartz à Ottré, massif de Stavelot, Belgique. Bull. Minéral., 108, 139–143 (in French with English abs.).

Crystal Data: Hexagonal. *Point Group:* $\bar{3}$. Rare euhedral crystals, to 3 cm; common forms are $\{10\bar{1}0\}$, $\{11\bar{2}0\}$ and $\{0001\}$. In coarse crystalline aggregates and massive.

Physical Properties: *Cleavage:* $\{0001\}$ and $\{10\bar{1}0\}$. *Fracture:* Conchoidal. Hardness = 6.5 D(meas.) = 2.657–2.666 D(calc.) = 2.654–2.661 Fluoresces pink to red or orange under SW UV.

Optical Properties: Transparent to translucent. *Color:* Colorless. *Optical Class:* Uniaxial (+). $\omega = 1.570$–1.573 $\epsilon = 1.583$–1.587

Cell Data: *Space Group:* $R\bar{3}$. $a = 13.47$–13.54 $c = 9.00$–9.01 $Z = 18$

X-ray Powder Pattern: Bikita, Zimbabwe.
3.362 (100), 3.965 (90), 2.733 (70), 2.545 (60), 2.373 (50), 1.797 (50), 6.736 (30)

Chemistry:

	(1)	(2)	(3)
SiO$_2$	49.32	47.97	47.68
Al$_2$O$_3$	40.43	40.28	40.46
Li$_2$O	9.72	11.45	11.86
Na$_2$O	0.24	0.26	
K$_2$O	0.01		
Total	99.72	99.96	100.00

(1) Bikita, Zimbabwe. (2) Kings Mountain, North Carolina, USA. (3) LiAlSiO$_4$.

Occurrence: In lithium-rich pegmatites.

Association: Albite, spodumene, petalite, amblygonite, lepidolite, quartz.

Distribution: In the USA, at Branchville, Fairfield Co., Connecticut; Center Strafford, Strafford Co., New Hampshire; in the Harding pegmatite, Dixon, Taos Co., New Mexico; in the Midnight Owl pegmatite, Maricopa Co., Arizona; and in the Foote mine, Kings Mountain, Cleveland Co., North Carolina. From the Tanco pegmatite, Bernic Lake, Manitoba, and near Nakina, Ontario, Canada. In Zimbabwe, large amounts at Bikita, with many other minor localities. From the Haapaluoma quarry, near Peräseinäjoki, Finland.

Name: From the Greek for *well* and *concealed*, in reference to its occurrence as intimate intergrowths with albite.

References: (1) Dana, E.S. (1892) Dana's system of mineralogy, (6th edition), 426. (2) Hurlbut, C.S., Jr. (1962) Eucryptite from Bikita, Southern Rhodesia. Amer. Mineral., 47, 557–559. (3) Leavens, P.B., C.S. Hurlbut, Jr., and J.A. Nelen (1968) Eucryptite and bikitaite from King's [sic] Mountain, North Carolina. Amer. Mineral., 53, 1202–1207. (4) Černý, P. (1972) The Tanco pegmatite at Bernic Lake, Manitoba. VII. Eucryptite. Can. Mineral., 11, 708–713. (5) Hesse, K.-F. (1985) Crystal structures of natural and synthetic α-eucryptite, LiAlSiO$_4$. Zeits. Krist., 172, 147–151.

$Na_4(Ca, Ce)_2(Fe^{2+}, Mn^{2+})ZrSi_8O_{22}(OH, Cl)_2(?)$ Eudialyte

Crystal Data: Hexagonal. *Point Group:* $\bar{3}\ 2/m$. Crystals short rhombohedral with {0001} dominant, to long prismatic, up to 10 cm. More commonly as irregular masses and vein fillings.

Physical Properties: *Cleavage:* Perfect to indistinct on {0001}, imperfect on {11$\bar{2}$0}. *Fracture:* Uneven. *Tenacity:* Brittle. Hardness = 5–6   D(meas.) = 2.74–3.10   D(calc.) = n.d.

Optical Properties: Translucent. *Color:* Brown, yellow-brown, yellow, pink, rose-red, cherry-red, red; colorless in thin section. *Luster:* Vitreous to dull. *Optical Class:* Uniaxial (+) or (–). *Pleochroism:* Weak; O = colorless, pink, pale yellow; E = pink to colorless. ω = 1.588–1.636 ϵ = 1.588–1.658

Cell Data: *Space Group:* $R\bar{3}m$.   a = 13.95–14.29   c = 29.89–30.49   Z = 12

X-ray Powder Pattern: Kipawa Lake, Canada; could be mistaken for alluaivite.
2.82 (100), 2.94 (90), 3.10 (80), 4.25 (50), 4.05 (40), 3.35 (40), 3.19 (40)

Chemistry:

	(1)	(2)		(1)	(2)
SiO_2	50.35	50.14	CaO	9.74	11.18
TiO_2	0.38	0.46	SrO	0.11	0.47
ZrO_2	11.80	11.83	Na_2O	12.53	14.06
Al_2O_3	0.44	0.07	K_2O	0.43	1.39
RE_2O_3	6.40	0.37	F	0.23	
Fe_2O_3	0.19	0.50	Cl	1.47	1.82
Nb_2O_5	0.69	0.11	H_2O^+	1.64	1.07
FeO	2.41	5.32	H_2O^-		0.12
MnO	1.34	0.60	P_2O_5	0.03	
MgO	0.13	0.24	S		0.04
			$-O = (F, Cl)_2$	0.43	0.41
			Total	99.88	99.38

(1) Kipawa Lake, Canada; corresponds to $Na_{3.85}(Ca_{1.65}RE_{0.19}K_{0.17})_{\Sigma=2.01}(Fe^{2+}_{0.32}Mn^{2+}_{0.18}$ $Nb_{0.06}Mg_{0.03})_{\Sigma=0.59}(Zr_{0.91}Al_{0.08}Ti_{0.04}Fe^{3+}_{0.01})_{\Sigma=1.04}Si_{8.02}O_{22}[(OH)_{1.73}Cl_{0.39}]_{\Sigma=2.12}$. (2) Khibiny massif, Russia; corresponds to $Na_{4.29}(Ca_{1.88}K_{0.28}RE_{0.02})_{\Sigma=2.18}(Fe^{2+}_{0.70}Mn^{2+}_{0.08}Mg_{0.06}Nb_{0.01})_{\Sigma=0.85}$ $(Zr_{0.98}Fe^{3+}_{0.13}Ti_{0.05}Al_{0.01})_{\Sigma=1.17}Si_{7.90}O_{22}[(OH)_{1.12}Cl_{0.43}]_{\Sigma=1.55}$.

Occurrence: In nepheline syenites, alkalic granites, and associated pegmatites; may be a major constituent, of both magmatic and late-stage pneumatolytic origin.

Association: Microcline, nepheline, aegirine, lamprophyllite, lorenzenite, murmanite, arfvedsonite, sodalite, aenigmatite, rinkite, låvenite, titanite, titanian magnetite.

Distribution: Numerous localities, a number affording fine crystals. In Greenland, on the Kangerdluarssuk Plateau, in the Ilímaussaq intrusion, at Narssârssuk, and elsewhere. Around the Langesundsfjord, Norway. In Russia, from the Lovozero and Khibiny massifs, Kola Peninsula; at Barynga Ridge, Taimyr; in the Burpala massif, about 120 km north of Lake Baikal, eastern Siberia. In the USA, from Magnet Cove, Hot Spring Co., Arkansas; at Point of Rocks, Colfax Co., and in a large deposit at Pajarito Mountain, Otero Co., New Mexico. In Canada, from Seal Lake and the Red Wine complex, Labrador, Newfoundland; in the Sheffield Lake complex, Kipawa River, Villedieu Township, and from Mont Saint-Hilaire, Quebec. From Tenerife, Canary Islands.

Name: From the Greek for *easily* and *dissolved*, in allusion to its ready dissolution in acids.

References: (1) Dana, E.S. (1892) Dana's system of mineralogy, (6th edition), 409–412. (2) Deer, W.A., R.A. Howie, and J. Zussman (1986) Rock-forming minerals, (2nd edition), v. 1B, disilicates and ring silicates, 348–363. (3) Vlasov, K.A., Ed. (1966) Mineralogy of rare elements, v. II, 355–364. (4) Rastsvetaeva, R.K. and B.E. Borutzky (1988) Crystallochemical traits of eudialyte in the light of new structural data. Mineral. Zhurnal, 10(1), 48–57 (in Russian). (5) Pol'shin, E.V., A.N. Platonov, B.E. Borutzky, M.N. Taran, and R.K. Rastsvetaeva (1991) Optical and Mössbauer study of minerals of the eudialyte group. Phys. Chem. Minerals, 18, 117–125.

Crystal Data: Monoclinic. *Point Group:* $2/m$. Crystals tabular, to 3 cm, commonly twinned; spherulitic, micaceous, coarse and fine granular, massive. *Twinning:* On {001}, simple twins, common, may be polysynthetic.

Physical Properties: *Cleavage:* Perfect on {001}, imperfect on {$\bar{5}$51}. *Fracture:* Uneven to conchoidal. Hardness = 6 D(meas.) = 2.55 D(calc.) = [2.57]

Optical Properties: Transparent to translucent. *Color:* White or colorless, also grayish blue, blue, violet, yellow; colorless in thin section. *Luster:* Vitreous, pearly on cleavage.
Optical Class: Biaxial (+). *Orientation:* $Y = b$; $Z \wedge c = -58.5°$. *Dispersion:* $r > v$, distinct.
$\alpha = 1.545$ $\beta = 1.546$ $\gamma = 1.549–1.551$ 2V(meas.) = 25°–30°

Cell Data: *Space Group:* $C2/c$. a = 12.63(1) b = 7.38(1) c = 14.02(1) $\beta = 103°43(5)'$
Z = 8

X-ray Powder Pattern: Langesundsfjord, Norway.
3.163 (10), 3.398 (8), 3.074 (8), 6.35 (6), 2.999 (6), 2.848 (6), 3.687 (5)

Chemistry:

	(1)	(2)	(3)
SiO_2	72.94	73.11	73.49
Al_2O_3	0.57		
FeO	0.29		
BeO	11.20	10.12	10.20
CaO	0.44		
Na_2O	10.56	12.24	12.64
H_2O^+	0.35		
H_2O^-	4.07	3.79	3.67
Total	100.42	99.26	100.00

(1) Langesundsfjord, Norway. (2) Mt. Alluaiv, Russia. (3) $NaBeSi_3O_7(OH)$.

Polymorphism & Series: Dimorphous with epididymite.

Occurrence: A late-stage mineral in alkalic nepheline syenite pegmatites.

Association: Albite, neptunite, elpidite, natrolite, analcime, fluorite.

Distribution: On Arø Island, in the Langesundsfjord, Norway. In the Ilímaussaq intrusion, southern Greenland. At Meldon, Okehampton, Devon, England. On Mt. Alluaiv, Lovozero massif, and Mt. Yukspor, Khibiny massif, Kola Peninsula; in the Burpala massif, 120 km north of Lake Baikal, eastern Siberia; and other less-well-defined localities in Russia. From Seal Lake, Labrador, Newfoundland, and at Mont Saint-Hilaire, Quebec, Canada. On Mt. Malosa, Zomba Plateau, Malawi.

Name: From the Greek for *well* and *twinned*, in allusion to its common occurrence in twins.

References: (1) Dana, E.S. (1892) Dana's system of mineralogy, (6th edition), 313–314.
(2) Vlasov, K.A., Ed. (1966) Mineralogy of rare elements, v. II, 135-140. (3) Nickel, E.H. (1963) Eudidymite from Seal Lake, Labrador, Newfoundland. Can. Mineral., 7, 643–649. (4) Fang, J.H., P.D. Robinson, and Y. Ohya (1972) Determination of the crystal structure of eudidymite and its dimorphic relationship to epididymite. Amer. Mineral., 57, 1345–1354.

Crystal Data: Cubic. *Point Group:* $\overline{4}3m$. As tetrahedral crystals, to 1 mm, showing {100}, {110}, {111}, and {211}. *Twinning:* Interpenetrant, may be multiple.

Physical Properties: *Cleavage:* Very imperfect on {110}. *Fracture:* Uneven. *Tenacity:* Rather brittle. Hardness = 4.5   D(meas.) = 6.62–6.76   D(calc.) = [6.76]

Optical Properties: Transparent to opaque. *Color:* Dark brown, yellowish gray, grayish white, straw-yellow, colorless; colorless to pale brown in thin section. *Streak:* White or yellowish gray. *Luster:* Resinous or adamantine.
Optical Class: Isotropic; anomalously biaxial (–). $n = 2.05$

Cell Data: *Space Group:* $I\overline{4}3d$.   a = 10.29   Z = [4]

X-ray Powder Pattern: Synthetic.
3.20 (100), 2.70 (100), 4.13 (80), 2.070 (70), 1.990 (70), 1.664 (70), 1.584 (70)

Chemistry:

	(1)	(2)
SiO$_2$	16.52	16.21
Bi$_2$O$_3$	82.23	83.79
rem.	1.15	
Total	99.90	100.00

(1) Schneeberg, Germany; remainder is Fe$_2$O$_3$ and P$_2$O$_5$. (2) Bi$_4$(SiO$_4$)$_3$.

Occurrence: In bismuth-rich hydrothermal veins.

Association: Bismuth, quartz.

Distribution: In Germany, at Schneeberg and Johanngeorgenstadt, Saxony. From Dognecea (Dognaczka), Romania. At Lanlivery, Cornwall, England. From the Southwick cliffs, near Dalbeattie, Kirkcudbrightshire, Scotland. In Canada, from the Evans-Lou mine, near St.-Pierre-de-Wakefield, Quebec.

Name: From the Greek for *good* and *dissolved*, in reference to its low point of fusion.

Type Material: Mining Academy, Freiberg, Germany, 23972.

References: (1) Dana, E.S. (1892) Dana's system of mineralogy, (6th edition), 436.
(2) Menzer, G. (1931) Die Kristallstruktur von Eulytin. Zeits. Krist., 78, 136–163 (in German).
(3) Durif-Varambon, A. (1959) Étude de la substitution du silicium dans quelques types d'orthosilicates. Bull. Soc. fr. Minéral., 82, 285–314 (in French). (4) Segal, D.J., R.P. Santoro, and R.E. Newnham (1966) Neutron diffraction study of BiSi$_3$O$_{12}$. Zeits. Krist., 123, 73–76.
(5) Phillips, W.R. and D.T. Griffen (1981) Optical mineralogy, 412.

Falcondoite $(Ni, Mg)_4Si_6O_{15}(OH)_2 \cdot 6H_2O$

Crystal Data: Orthorhombic. *Point Group:* n.d.   As slightly schistose, friable masses of minute fibrous particles, less than 10 μm in length.

Physical Properties: *Tenacity:* Friable.   Hardness = 2–3   D(meas.) = 1.9(1)   D(calc.) = 2.54

Optical Properties: Transparent to translucent. *Color:* Whitish green. *Optical Class:* Biaxial. $n = <1.55$; birefringence ~0.01–0.02.   2V(meas.) = n.d.

Cell Data: *Space Group:* n.d.   a = 13.5   b = 26.9   c = 5.24   Z = 4

X-ray Powder Pattern: Near Bonao, Dominican Republic.
12.2 (100), 2.58 (35), 3.33 (30), 2.62 (30), 2.44 (30), 3.19 (25), 2.39 (20)

Chemistry:

	(1)	(2)
SiO_2	45.93	49.2
Al_2O_3	0.39	
Cr_2O_3	0.45	
FeO	1.99	< 0.03
NiO	26.7	23.6
MgO	8.46	9.26
LOI	15.0	[18.0]
Total	98.92	[100.0]

(1) Near Bonao, Dominican Republic; corresponds to $(Ni_{2.66}Mg_{1.55}Fe_{0.22}Al_{0.06}Cr_{0.04})_{\Sigma=4.53}$ $Si_{5.71}O_{16} \cdot 6.22H_2O$. (2) Do.; average of five analyses by electron microprobe and other methods; loss on ignition by difference, taken as H_2O; corresponds to $(Ni_{2.32}Mg_{1.68})_{\Sigma=4.00}Si_6O_{16} \cdot 7.30H_2O$.

Occurrence: In garnierite veins which cut an extensive laterite, associated with a serpentinized harzburgite massif (near Bonao, Dominican Republic).

Association: Sepiolite, talc, quartz (near Bonao, Dominican Republic).

Distribution: From near Bonao, Dominican Republic. In the USA, from near Glamis, Imperial Co., California.

Name: For the familiar name "Falcondo" of the company operating the mine in the Dominican Republic in which the mineral was first found.

Type Material: Royal Ontario Museum, Toronto, Canada, M34324.

References: (1) Springer, G. (1976) Falcondoite, nickel analogue of sepiolite. Can. Mineral., 14, 407–409.

Crystal Data: Cubic. *Point Group:* $4/m\,\bar{3}\,2/m$. Crystals octahedra or rarely trisoctahedra, to 4 mm. *Twinning:* On {111}, contact and penetration twins.

Physical Properties: *Cleavage:* {111}, perfect. *Fracture:* Uneven to conchoidal. *Tenacity:* Brittle. Hardness = 4.5–5   D(meas.) = 1.92–1.93   D(calc.) = 2.09

Optical Properties: Transparent to translucent. *Color:* White, colorless, brown; colorless in thin section. *Streak:* White. *Luster:* Vitreous to adamantine. *Optical Class:* Isotropic. $n = 1.466$–1.480

Cell Data: *Space Group:* $Fd3m$. a = 24.638–24.65   Z = 32

X-ray Powder Pattern: San Bernardino Co., California, USA.
3.757 (100), 14.2 (95), 5.64 (83), 3.292 (64), 4.355 (52), 2.904 (36), 2.373 (36)

Chemistry:

	(1)	(2)
SiO_2	46.12	48.59
Al_2O_3	16.81	15.97
MgO		0.79
CaO	4.79	4.93
Na_2O	5.09	3.02
K_2O		0.16
H_2O	27.02	[26.54]
Total	99.83	[100.00]

(1) Kaiserstuhl, Germany. (2) San Bernardino Co., California, USA; by electron microprobe, average of 15 analyses, H_2O by difference, corresponding to $(Na_{0.52}Ca_{0.47}Mg_{0.10}K_{0.02})_{\Sigma=1.11}$ $Al_{1.78}Si_{4.24}O_{12} \cdot 16H_2O$.

Mineral Group: Zeolite group.

Occurrence: Rare, in vesicles of basalts, phonolites, and tuffs; formed by palagonization or authigenically.

Association: Zeolites, olivine, augite, nepheline.

Distribution: In Germany, from Sasbach, Kaiserstuhl, and elsewhere in Baden-Württemberg; from the Pflasterkaute, near Eisenach, Thuringia; at Annerod, near Giessen, and at a number of localities around the Vogelsberg volcano, Hesse. From Aci Reale and Aci Castello, Sicily, Italy. On the Cima Dome, near Valley Wells, San Bernardino Co., California, and at several localities on Oahu, Hawaii, USA. From Davis Hill and Khartoum, Ontario; at Laurel, Hincks Bridge, and Notre-Dame-de-la-Salette, and at the Daisy Mica mine, Ottawa Co., Quebec, Canada. A few other localities are known.

Name: For Barthélemy Faujas de Saint Fond (1741–1819), French mineralogist and vulcanologist.

References: (1) Dana, E.S. (1892) Dana's system of mineralogy, (6th edition), 598–599.
(2) Deer, W.A., R.A. Howie, and J. Zussman (1963) Rock-forming minerals, v. 4, framework silicates, 392–393. (3) Bergerhoff, G., W.H. Baur, and W. Nowacki (1958) Über die Kristallstruktur des Faujasites. Neues Jahrb. Mineral., Monatsh., 193–200 (in German). (4) Baur, W.H. (1964) On the cation and water positions in faujasite. Amer. Mineral., 49, 697–704.
(5) Wise, W.S. (1982) New occurrence of faujasite in southeastern California. Amer. Mineral., 67, 794–798.

Crystal Data: Orthorhombic. *Point Group:* $2/m\ 2/m\ 2/m$. Crystals are thick to thin, tabular $\parallel$ {100}, typically with wedge-shaped terminations, producing eight-sided cross sections. Commonly granular, compact, or massive. *Twinning:* On {100}; also on {031}, as trillings.

Physical Properties: *Cleavage:* {010}, {100}, imperfect. *Fracture:* Conchoidal. *Tenacity:* Brittle. Hardness = 6.5–7 D(meas.) = 4.392 D(calc.) = [4.40]

Optical Properties: Transparent. *Color:* Greenish yellow, yellow-brown, brown; pale yellow to amber in thin section. *Streak:* White. *Luster:* Vitreous to resinous on fractures. *Optical Class:* Biaxial (–). *Pleochroism:* Faint; $X = Z$ = pale yellow; Y = yellow-orange, reddish brown. *Orientation:* $X = b$; $Y = c$; $Z = a$. *Dispersion:* $r > v$, weak. $\alpha = 1.827$ $\beta = 1.869$ $\gamma = 1.879$ 2V(meas.) = 48°

Cell Data: *Space Group:* *Pbnm* (synthetic). $a = 4.8211$ $b = 10.4779$ $c = 6.0889$ $Z = 4$

X-ray Powder Pattern: Rockport, Massachusetts, USA. (ICDD 20-1139). 2.501 (100), 2.828 (90), 1.777 (90), 3.55 (80), 2.565 (70), 1.523 (70), 1.516 (70)

Chemistry:

	(1)
SiO_2	29.83
FeO	69.48
MnO	0.28
H_2O^+	0.34
H_2O^-	0.04
Total	99.97

(1) Brocken massif, Germany; corresponds to $(Fe_{1.97}^{2+}Mn_{0.01}^{2+})_{\Sigma=1.98}Si_{1.01}O_4$.

Polymorphism & Series: Forms two series, with forsterite, and with tephroite.

Mineral Group: Olivine group.

Occurrence: In ultramafic volcanic and plutonic rocks, less commonly in felsic plutonic rocks; rarely in granite pegmatite; in lithophysae in obsidian. In metamorphosed iron-rich sediments and impure carbonate rocks.

Association: Augite, plagioclase, microcline, quartz, apatite, magnetite, ilmenite, spinel, hedenbergite, arfvedsonite, amphiboles, almandine, tridymite, grunerite.

Distribution: Among the occurrences of relatively pure material are: in the USA, in the St. Peter's Dome area, El Paso Co., Colorado; at Obsidian Cliff, Yellowstone National Park, Wyoming; from Coso Hot Springs, Inyo Co., California; at Rockport, Essex Co., Massachusetts; from Monroe, Orange Co., New York; in the Iron Hill mine, Cumberland, Providence Co., Rhode Island. In the Blue Bell mine, Riondel, British Columbia, Canada. From the Brocken massif, Harz Mountains, Germany. At Raftsund, Norway. In the Skaergaard and Ilímaussaq intrusions, Greenland. On Red Rock Ridge, Nevy Fjord, Palmer Peninsula, Antarctica.

Name: After Faial (Fayal) Island in the Azores, thought to occur in a local volcanic rock, but probably obtained from slag carried as ship's ballast.

References: (1) Dana, E.S. (1892) Dana's system of mineralogy, (6th edition), 456–457, 451–456 [chrysolite]. (2) Deer, W.A., R.A. Howie, and J. Zussman (1982) Rock-forming minerals, (2nd edition), v. 1A, orthosilicates, 3–336. (3) Smyth, J.R. (1975) High temperature crystal chemistry of fayalite. Amer. Mineral., 60, 1092–1097.

Crystal Data: Triclinic. *Point Group:* $\bar{1}$. As pseudohexagonal tabular crystals resembling muscovite crystals, to 5 cm.

Physical Properties: *Cleavage:* Perfect micaceous on {001}. Hardness = n.d. D(meas.) = 2.43–2.58   D(calc.) = [2.43]

Optical Properties: Transparent to translucent. *Color:* Colorless, whitish to pale raspberry-red. *Luster:* Vitreous, satiny, pearly.
Optical Class: Biaxial (–). *Orientation:* X = c; Y = b; Z = a. *Dispersion:* $r < v$, distinct.
$\alpha = 1.522$ $\beta = 1.530$ $\gamma = 1.531$ 2V(meas.) = 32°

Cell Data: *Space Group:* $C\bar{1}$.   a = 9.676(2)   b = 16.706(1)   c = 13.233(2) $\alpha = 93.35°$ $\beta = 114.96°$ $\gamma = 90.03°$ Z = 2

X-ray Powder Pattern: Tur'yii Peninsula, Russia.
2.93 (100b), 2.97 (90), 1.826 (90b), 11.7 (80), 6.0 (80), 3.13 (80b), 4.21 (70)

Chemistry:

	(1)
SiO_2	62.99
Al_2O_3	3.30
CaO	15.80
Na_2O	8.00
K_2O	3.80
LOI	5.33
Total	99.22

(1) Tur'yii Peninsula, Russia; corresponding to $K_{1.08}Ca_{4.08}Na_{3.72}$ $(Si_{15.20}Al_{0.80})_{\Sigma=16.00}O_{36.20}(OH)_{3.80} \cdot 6H_2O$.

Occurrence: In fine veinlets in fenitized sandstone (Tur'yii Peninsula, Russia).

Association: Narsarsukite, quartz, apophyllite (Tur'yii Peninsula, Russia).

Distribution: On the Tur'yii Peninsula, Kola Peninsula, and in the Murun massif, southwest of Olekminsk, Yakutia, Russia.

Name: For Evgraf Stepanovich Fedorov (1853–1919), eminent Russian crystallographer.

Type Material: n.d.

References: (1) Kukharenko, A.A., M.P. Orlova, and A.G. Bulakh (1965) The Caledonian ultrabasic alkalic rocks and carbonatites of the Kola Peninsula and northern Karelia. Izd. Nedra, Moscow, 479–481 (in Russian). (2) (1967) Amer. Mineral., 52, 561–562 (abs. ref. 1). (3) Sokolova, G.V., A.A. Kashaev, V.A. Drits, and V.V. Ilyukhin (1983) The crystal structure of fedorite. Kristallografiya (Sov. Phys. Crystal.), 28, 170–172 (in Russian).

Fenaksite

$(K, Na, Ca)_4(Fe^{2+}, Fe^{3+}, Mn^{2+})_2Si_8O_{20}(OH, F)$

Crystal Data: Triclinic. *Point Group:* $\bar{1}$. As grains up to 4 cm.

Physical Properties: *Cleavage:* Two intersecting at an angle of 122°. Hardness = 5–5.5 D(meas.) = 2.744 D(calc.) = 2.74

Optical Properties: Transparent to translucent. *Color:* Light rose. *Luster:* Pearly on cleavages.
Optical Class: Biaxial (+) [*sic*]. *Orientation:* $Z = b$; $Z \wedge$ cleavage 1 = 20°; $Y \wedge$ cleavage 2 = 49°; $X \wedge$ cleavage 2 = 8°. $\alpha = 1.541$ $\beta = 1.560$ $\gamma = 1.567$ 2V(meas.) = 84° 2V(calc.) = [61°]

Cell Data: *Space Group:* $P\bar{1}$. a = 6.98 b = 8.24 c = 9.98 $\alpha = 114.20°$ $\beta = 80.22°$ $\gamma = 115.60°$ Z = 2

X-ray Powder Pattern: Khibiny massif, Russia.
3.03 (100), 3.55 (70), 3.44 (70), 2.46 (70), 2.88 (60), 2.71 (60), 1.875 (60)

Chemistry:

	(1)		(1)
SiO_2	60.54	CaO	0.74
TiO_2	0.04	Na_2O	7.51
Al_2O_3	0.66	K_2O	11.71
Fe_2O_3	1.54	F	0.47
FeO	12.49	H_2O^+	0.67
MnO	2.49	H_2O^-	0.78
MgO	0.70	$-O = F_2$	0.20
		Total	100.14

(1) Khibiny massif, Russia.

Occurrence: In pegmatites associated with an ijolite-urtite intrusion in a differentiated alkalic massif.

Association: Albite, nepheline, "aegirine-augite," delhayelite, eudialyte.

Distribution: On Mts. Rasvumchorr and Yukspor, Khibiny massif, Kola Peninsula, Russia.

Name: Presumably for *Fe, Na, K, Si* in the composition.

Type Material: n.d.

References: (1) Dorfman, M.D., D.D. Rogachev, Z.I. Goroshchenko, and A.V. Mokretsova (1959) Fenaksite–a new mineral. Trudy Mineral. Muzeya Akad. Nauk SSSR, 152–157 (in Russian). (2) (1960) Amer. Mineral., 45, 252–253 (abs. ref. 1). (3) (1960) Mineral. Abs., 14, 414 (abs. ref. 1). (4) Golovachev, V.P., Y.N. Drozdov, E.A. Kuz'min, and N.V. Belov (1970) The crystal structure of phenaxite [fenaksite] $FeNaK(Si_4O_{10}) - KFeNa(Si_4O_{10})$. Doklady Acad. Nauk SSSR, 194, 818–820 (in Russian).

Crystal Data: Monoclinic. *Point Group:* n.d. As flaky to tabular micaceous grains, to 40 μm, and as fibrous to acicular grains forming massive planar aggregates. *Twinning:* Common.

Physical Properties: *Cleavage:* Perfect on {001}. Hardness = n.d. D(meas.) = n.d. D(calc.) = 3.46 (synthetic).

Optical Properties: Semitransparent. *Color:* Light reddish brown or brownish red. *Optical Class:* Biaxial. *Pleochroism:* X = light reddish brown; Y = Z = pale yellow-green; other grains display X = brownish red; Y = Z = pale greenish brown; varies with Al content. $\alpha = 1.653$–1.677 $\beta = $ n.d. $\gamma = 1.691$–1.721 2V(meas.) = $0°$–$10°$

Cell Data: *Space Group:* n.d. $a = 5.43(1)$ $b = 9.40(1)$ $c = 10.33(1)$ $\beta = 100°4(10)'$ Z = [2]

X-ray Powder Pattern: Synthetic $KFe_3^{2+}Fe^{3+}Si_3O_{10}(OH)_2$.
10.163 (100), 3.388 (25), 2.674 (13), 3.721 (9), 2.959 (8), 2.028 (6), 2.026 (3)

Chemistry:

	(1)	(2)
SiO_2	38.50	35.49
TiO_2	0.02	0.01
Al_2O_3	5.43	1.42
Fe_2O_3	6.84	13.25
Cr_2O_3	0.02	0.02
FeO	23.62	29.77
MnO	0.03	0.03
NiO	0.02	0.04
MgO	12.05	7.24
CaO	0.04	0.05
Na_2O	0.06	0.02
K_2O	8.46	8.33
Total	95.09	95.67

(1) Hamersley Range, Western Australia; by electron microprobe, average of 18 points, Fe_2O_3 estimated from tetrahedral Fe^{3+}; corresponds to $(K_{0.86}Na_{0.01})_{\Sigma=0.87}(Fe_{1.58}^{2+}Mg_{1.44})_{\Sigma=3.02}$ $(Al_{0.51}Fe_{0.41}^{3+})_{\Sigma=0.92}Si_{3.08}O_{10}(OH)_2$. (2) Do.; average of 16 points, corresponds to $K_{0.90}(Fe_{2.11}^{2+}Mg_{0.92})_{\Sigma=3.03}(Fe_{0.85}^{3+}Al_{0.14})_{\Sigma=0.99}Si_{3.01}O_{10}(OH)_2$.

Polymorphism & Series: 1M polytype.

Mineral Group: Mica group.

Occurrence: A product of very low-grade metamorphism of banded iron formation, perhaps by potassium metasomatism of stilpnomelane.

Association: Quartz, hematite, magnetite, stilpnomelane, riebeckite, ankerite.

Distribution: In the Colonial mine, Wittenoom, Hamersley Range, Western Australia.

Name: Originally as the synthetic FERRIc iron analog of *annite*.

Type Material: n.d.

References: (1) Miyano, T. and S. Miyano (1982) Ferri-annite from the Dales Gorge member iron-formations, Wittenoom area, Western Australia. Amer. Mineral., 67, 1179–1194. (2) Wones, D.R. (1963) Phase equilibria of "ferrianite", $KFe_3^{+2}Fe^{3+}Si_3O_{10}(OH)_2$. Amer. J. Sci., 261, 581–596. (3) Donnay, G., N. Morimoto, H. Takeda, and J.D.H. Donnay (1964) Trioctahedral one-layer micas. I. Crystal structure of a synthetic iron mica. Acta Cryst., 17, 1369.

Crystal Data: Orthorhombic; rarely monoclinic. *Point Group:* $2/m\ 2/m\ 2/m$ or $2/m$. Radiating groups of thin lath-shaped crystals, tabular on $\{100\}$, elongated along $[001]$, to 2.5 cm.

Physical Properties: *Cleavage:* $\{100\}$, perfect. Hardness = 3–3.5 D(meas.) = 2.136 D(calc.) = 2.11

Optical Properties: Transparent to translucent. *Color:* White, colorless, pinkish, orange to red. *Streak:* White. *Luster:* Vitreous to pearly.
Optical Class: Biaxial (+). *Orientation:* $X = a$; $Y = b$; $Z = c$. $\alpha = 1.473$–1.479
$\beta = 1.474$–1.483 $\gamma = 1.477$–1.486 2V(meas.) = $50°$

Cell Data: *Space Group: Immm.* $a = 18.90$–19.22 $b = 14.14$–14.16 $c = 7.48$–7.54
$Z = 2$, or *Space Group: $P2_1/n$.* $a = 18.886(9)$ $b = 14.182(6)$ $c = 7.470(5)$ $Z = 2$

X-ray Powder Pattern: Agoura, California, USA.
3.537 (100), 3.778 (65), 9.47 (50), 7.07 (38), 3.977 (35), 3.943 (35), 3.313 (35)

Chemistry:

	(1)	(2)		(1)	(2)
SiO_2	70.14	71.21	BaO	0.38	
Al_2O_3	11.45	9.84	Na_2O	1.86	1.59
Fe_2O_3	0.13	0.05	K_2O	0.63	2.85
MgO	2.56	1.70	H_2O^+		4.25
CaO	0.15		H_2O^-		8.63
SrO	0.18		H_2O	[12.52]	
			Total	[100.00]	100.12

(1) Kamloops Lake, Canada; by electron microprobe, H_2O by difference; corresponding to $(Na_{0.78}K_{0.18})_{\Sigma=0.96}(Mg_{0.82}Ca_{0.04}Ba_{0.04}Sr_{0.02})_{\Sigma=0.92}(Si_{15.11}Al_{2.90})_{\Sigma=18.01}O_{36} \cdot 9H_2O$.
(2) Itomuka mine, Japan; by electron microprobe, H_2O by Penfield method, corresponding to $(K_{0.78}Na_{0.66})_{\Sigma=1.44}Mg_{0.54}(Si_{15.48}Al_{2.52})_{\Sigma=18.00}O_{36} \cdot 9.41H_2O$.

Mineral Group: Zeolite group.

Occurrence: In seams and vesicles in deeply weathered basalts, flow breccias, and andesites; in tuffaceous sediments.

Association: Zeolites, plagioclase, "hypersthene," augite, montmorillonite, "chalcedony," calcite.

Distribution: In Canada, in British Columbia, at Kamloops Lake, near Kamloops, and at Monte Lake and Pinaus Lake, near Westwold. In the USA, from Altoona, Wahkiakum Co., Washington; near Agoura, Los Angeles Co., and on Silver Mountain, southwest of Markleeville, Alpine Co., California. From Albero Basso and Val Timonchio, Vicenza, Italy. At Weitendorf, near Wildon, Styria, Austria. From Reichweiler, Rhineland-Palatinate, Germany. On Phillip Island, Victoria, and at Unanderra, New South Wales, Australia. In the Itomuka mine, Hokkaido, and at Tadumi, Fukushima Prefecture, Japan. A number of other occurrences are known.

Name: For Dr. Walter Frederick Ferrier (1865–1950), geologist and mining engineer with the Canadian Geological Survey.

Type Material: McGill University, Montreal, Canada, RMF3099.

References: (1) Graham, R.P.D. (1918) Ferrierite, a new zeolitic mineral from British Columbia, with notes on some other Canadian minerals. Trans. Royal Soc. Canada, 12(3), 185–201. (2) (1919) Amer. Mineral., 4, 90 (abs. ref. 1). (3) Wise, W.S., W.J. Nokleberg, and M. Kokinos (1969) Clinoptilolite and ferrierite from Agoura, California. Amer. Mineral., 54, 887–895. (4) Yajima, S. and T. Nakamura (1971) New occurrence of ferrierite. Mineral. J. (Japan), 6, 343–364. (5) Wise, W.S. and Tschernich, R.W. (1976) Chemical composition of ferrierite. Amer. Mineral., 61, 60–66. (6) Gramlich-Meier, R., V. Gramlich, and W.H. Meier (1985) The crystal structure of the monoclinic variety of ferrierite. Amer. Mineral., 70, 619–623. (7) Alberti, A. and C. Sabelli (1987) Statistical and true symmetry of ferrierite (1987) Zeits. Krist., 178, 249–256.

Crystal Data: [Monoclinic] (by analogy to pyrophyllite). *Point Group:* [2/m.] Fine-scaly, granular, compact.

Physical Properties: *Tenacity:* Waxy. Hardness = 1.5–2 D(meas.) = 2.97–3.01 D(calc.) = [3.05]

Optical Properties: Semitransparent. *Color:* Brownish yellow.
Optical Class: Biaxial. *Pleochroism:* X = Y = greenish; Z = bright yellow. $\alpha = 1.650$–1.660 $\beta = 1.676$–1.678 $\gamma = 1.686$–1.688 2V(meas.) = Small.

Cell Data: *Space Group:* n.d. $a = 5.26$ $b = 9.10$ $c = 19.1$ $\beta = 95°30'$ $Z = [4]$

X-ray Powder Pattern: The Strassenschacht, Germany.
4.54 (10), 9.6 (8), 1.518 (8), 3.17 (7), 2.62 (4), 2.47 (4), 1.725 (3)

Chemistry: (1) The Strassenschacht, Germany; by electron microprobe, analysis not given, stated to correspond to $Fe_{1.96}^{3+}Mg_{0.11}Ca_{0.05}(Si_{3.80}Al_{0.13}Fe_{0.07}^{3+})_{\Sigma=4.00}O_{10}(OH)_2 \cdot H_2O$. (2) Mt. Tologay, Kazakhstan; by electron microprobe, analysis not given, stated to correspond to $Fe_{1.97}^{3+}Ca_{0.18}(Na,K)_{0.03}Mg_{0.02}(Si_{3.74}Al_{0.23})_{\Sigma=3.97}O_{10}(OH)_2 \cdot 1.5H_2O$.

Polymorphism & Series: [Isostructural with the 2M modification of pyrophyllite.]

Occurrence: Found on museum specimens.

Association: n.d.

Distribution: From the Strassenschacht, near Eibenstock, Saxony, Germany. On Mt. Tologay, not otherwise located in Kazakhstan.

Name: Presumably for its FERRIc iron content and relation to *pyrophyllite*.

Type Material: Moscow University, Moscow, Russia.

References: (1) Chukhrov, F.V., B.B. Svyagin, V.A. Drits, A.I. Gorshkov, L.P. Ermilova, E.A. Goilo, and E.S. Rudnitskaya (1979) On ferripyrophyllite and related phases. Izv. Akad. Nauk SSSR, Ser. Geol. 2, 5–20 (in Russian). (2) (1979) Chem. Abs., 91, 24056 (abs. ref. 1). (3) Tschuchrow [Chukhrov], F.W., B.B. Swjagin [Svyagin], W.A. Driz [Drits], A.I. Gorschkow [Gorshkov], L.P. Jermilowa [Ermilova], E.A. Goilo, and E.S. Rudnizkaja [Rudnitskaya] (1979) Über Ferripyrophyllit. Chem. Erde, 38, 324–330 (in German). (4) (1980) Mineral. Abs., 31, 355–356 (abs. ref. 3).

Ferrisurite

$$(Pb, Ca)_{2-3}(CO_3)_{1.5-2}(OH, F)_{0.5-1} \cdot (Fe^{3+}, Al)_2Si_4O_{10}(OH)_2 \cdot nH_2O$$

Crystal Data: Monoclinic, pseudo-orthorhombic. *Point Group:* 2 or $2/m$. As crystals, flattened on {010}, tapering to fibrous, elongated ∥ [100]; in compact radial aggregates, to 2 mm.

Physical Properties: *Cleavage:* Perfect on {010}. *Tenacity:* Very flexible. Hardness = 2–2.5 D(meas.) = 4.0(1) D(calc.) = 3.89

Optical Properties: Transparent to translucent. *Color:* Forest-green to light yellow-green. *Streak:* Greenish yellow to olive-green. *Luster:* Silky. *Optical Class:* Biaxial (+). *Pleochroism:* X = yellow; Y = brown; Z = light green. *Orientation:* X = c; Y = b; Z = a. $\alpha = 1.757(3)$ $\beta = 1.763(3)$ $\gamma = 1.773(3)$ 2V(meas.) = n.d. 2V(calc.) = 76°

Cell Data: *Space Group:* $P2_1$ or $P2_1/m$. a = 5.241(1) b = 9.076(5) c = 16.23(1) $\beta = 90.03(7)°$ Z = 2

X-ray Powder Pattern: Shirley Ann claim, California, USA.
4.53 (100), 3.240 (90), 2.612 (80), 2.272 (50), 16.1 (40), 3.727 (35), 5.40 (25), 1.715 (25)

Chemistry:

	(1)		(1)
SiO_2	26.6	BaO	0.1
Al_2O_3	3.2	Na_2O	0.3
Fe_2O_3	10.5	F	0.8
FeO	0.8	H_2O	3.5
PbO	42.7	CO_2	8.2
MgO	0.2	$-O = F_2$	0.3
CaO	3.4	Total	100.0

(1) Shirley Ann claim, California, USA; by electron microprobe, FeO by potentiometric titration, $Fe^{2+}:Fe^{3+}$ by wet chemical analysis, CO_2 by coulometric titration, H_2O by Karl Fischer titration; corresponding to $(Pb_{1.73}Ca_{0.55}Na_{0.09}Ba_{0.01})_{\Sigma=2.38}(CO_3)_{1.68}[(OH)_{0.51}F_{0.38}]_{\Sigma=0.89} \cdot (Fe^{3+}_{1.19}Al_{0.57}Fe^{2+}_{0.10}Mg_{0.05})_{\Sigma=1.91}Si_4O_{10}(OH)_2 \cdot 0.50H_2O$.

Occurrence: In an oxidized lead deposit in contact metamorphosed impure limestone, as an oxidation product of silicates proximate to galena.

Association: Galena, pyrite, chalcopyrite, covellite, chalcocite, quartz, calcite, hematite, cerussite, mimetite, wulfenite, malachite.

Distribution: At the Shirley Ann claim, Ubehebe district, west of Death Valley, Inyo Co., California, USA.

Name: For a predominance of Fe^{3+} over Al and its relation to *surite*.

Type Material: Natural History Museum of Los Angeles Co., Los Angeles, California; National Museum of Natural History, Washington, D.C., USA, 170288.

References: (1) Kampf, A.R., L.L. Jackson, G.B. Sidder, E.E. Foord, and P.M. Adams (1992) Ferrisurite, the Fe^{3+} analogue of surite, from Inyo County, California. Amer. Mineral., 77, 1107–1111.

Crystal Data: Monoclinic. *Point Group:* $2/m$. [As bladed or columnar crystals; also radiating fibrous; granular, massive.] *Twinning:* Simple or lamellar twinning ‖ {100}.

Physical Properties: *Cleavage:* Perfect on {110}, intersecting at 56° and 124°; parting on {100}. *Tenacity:* Brittle. Hardness = 5–6 D(meas.) = 3.24–3.48 D(calc.) = [3.34]

Optical Properties: Transparent to translucent. *Color:* Dark green; in thin section, yellow-green or blue-green. *Luster:* Vitreous.
Optical Class: Biaxial (–). *Pleochroism:* Moderate to strong, in yellows and greens.
Orientation: $Y = b$; $Z \wedge c = 12.5°–17°$. *Dispersion:* $r > v$, weak. *Absorption:* $Z > Y \geq X$.
$\alpha = 1.646–1.686$ $\beta = 1.656–1.696$ $\gamma = 1.666–1.704$ 2V(meas.) = 71°–81°

Cell Data: *Space Group:* $C2/m$. $a = 9.891(1)$ $b = 18.200(1)$ $c = 5.305(1)$
$\beta = 104.64(1)°$ Z = 2

X-ray Powder Pattern: Prieska, South Africa. (ICDD 23-118).
8.58 (100), 2.728 (40), 3.157 (35), 2.033 (30), 2.546 (25), 4.57 (20), 2.613 (18)

Chemistry:

	(1)	(2)		(1)	(2)
SiO_2	51.0	49.89	Na_2O	0.35	0.11
TiO_2	0.04	1.04	K_2O	0.16	0.08
Al_2O_3	2.0	0.75	F	0.14	
Fe_2O_3	2.8		H_2O^+	1.6	
FeO	19.9	30.62	H_2O^-	0.11	
MnO	2.4	0.07	CO_2	0.19	
MgO	8.6	4.59	P	0.09	
CaO	10.7	10.99	$-O = F_2$	0.06	
			Total	100.02	98.14

(1) Cumberland, Rhode Island, USA; corresponds to $(Ca_{1.74}Mn_{0.17}Na_{0.10}K_{0.03})_{\Sigma=2.04}$ $(Fe^{2+}_{2.53}Mg_{1.94}Fe^{3+}_{0.31}Mn_{0.14}Al_{0.08})_{\Sigma=5.00}(Si_{7.72}Al_{0.28})_{\Sigma=8.00}O_{22}(OH)_2$. (2) Jackson County Iron Formation, Wisconsin, USA; by electron microprobe, corresponds to $(Ca_{1.86}Na_{0.03}K_{0.02})_{\Sigma=1.91}$ $(Fe^{2+}_{3.98}Mg_{1.08}Fe^{3+}_{0.06}Ti_{0.01}Mn_{0.01}Al_{0.01})_{\Sigma=5.15}(Si_{7.87}Al_{0.13})_{\Sigma=8.00}O_{22}(OH)_2$.

Polymorphism & Series: Forms a series with tremolite and actinolite.

Mineral Group: Amphibole (calcic) group: $Mg/(Mg + Fe^{2+}) < 0.50$; $(Na + K)_A < 0.5$; $Na_B < 0.67$; $(Ca + Na)_B \geq 1.34$; $Si \geq 7.5$.

Occurrence: In iron-rich greenschist and blueschist facies metamorphic rocks; in metamorphosed iron formations; as part of a vein assemblage cutting Ca-Fe-Si tactites.

Association: Hedenbergite (replacement), andradite, ilvaite (contact metamorphic); cummingtonite, quartz, magnetite, riebeckite, biotite, hematite (iron formations).

Distribution: Probably widespread, but many locality references lack qualifying chemical analyses. In the USA, in the Jackson County Iron Formation, Wisconsin, and at Cumberland, Providence Co., Rhode Island. From Black Perry Mountain, New South Wales, Australia. At Prieska, Cape Province, South Africa.

Name: For *ferrous* iron in its composition and similarity to *actinolite*.

Type Material: n.d.

References: (1) Mitchell, J.T., F.D. Bloss, and G.V. Gibbs (1971) Examination of the actinolite structure and four other C2/m amphiboles in terms of double bonding. Zeits. Krist., 133, 273–300. (2) Kimball, K.L. and F.S. Spear (1984) Metamorphic petrology of the Jackson County Iron Formation, Wisconsin. Can. Mineral., 22, 605–619. (3) Deer, W.A., R.A. Howie, and J. Zussman (1963) Rock-forming minerals, v. 2, chain silicates, 249–262. (4) Phillips, W.R. and D.T. Griffen (1981) Optical mineralogy, 229–231.

Ferroan pargasite $NaCa_2[(Fe^{2+}, Mg)_4Al](Si_6Al_2)O_{22}(OH)_2$

Crystal Data: Monoclinic. *Point Group:* $2/m$. [As stout prismatic crystals; granular.] *Twinning:* [Simple or multiple twinning $\parallel$ {100}.]

Physical Properties: *Cleavage:* [Perfect on {110}, intersecting at $\sim 56°$ and $\sim 124°$; partings on {100}, {001}.] *Tenacity:* [Brittle.] Hardness = 5–6 D(meas.) = [3.2–3.44] D(calc.) = n.d.

Optical Properties: Semitransparent. *Color:* Dark green to black; light brown to bluish green in thin section. *Luster:* [Vitreous.]
Optical Class: Biaxial (−). *Pleochroism:* Distinct from colorless to greens and light browns. *Orientation:* $Y = b$; $Z \wedge c = 25°$. $\alpha = 1.674$ $\beta = 1.688$ $\gamma = 1.698$ 2V(meas.) = $77°$–$89°$

Cell Data: *Space Group:* $C2/m$.   a = 9.91–9.93 (synthetic ferroan pargasite). b = 18.00–18.08   c = 5.29–5.32 $\beta = 105.35°$ Z = 2

X-ray Powder Pattern: n.d.

Chemistry:

	(1)		(1)
SiO_2	42.06	MgO	11.19
TiO_2	1.54	CaO	11.48
Al_2O_3	14.54	Na_2O	2.35
Fe_2O_3	4.10	K_2O	0.18
FeO	10.14	H_2O^+	2.01
MnO	0.23	H_2O^-	0.16
		Total	99.98

(1) Tamasato, Japan; corresponds to $(Na_{0.67}K_{0.03})_{\Sigma=0.70}Ca_{1.81}(Mg_{2.45}Fe^{2+}_{1.24}Fe^{3+}_{0.31}Mn_{0.03})_{\Sigma=4.03}$ $(Al_{0.69}Ti_{0.17}Fe^{3+}_{0.14})_{\Sigma=1.00}(Si_{6.18}Al_{1.82})_{\Sigma=8.00}O_{22.04}(OH)_{1.96}$.

Polymorphism & Series: Forms a series with pargasite and ferro-pargasite.

Mineral Group: Amphibole (calcic) group: $0.3 \leq Mg/(Mg + Fe^{2+}) \leq 0.69$; $Fe^{3+} \leq Al^{vi}$; $(Na + K)_A \geq 0.5$; $Na_B < 0.67$; $(Ca + Na)_B \geq 1.34$; Si < 6.25; Ti < 0.5.

Occurrence: In metamorphosed limestones and mafic igneous rocks, altered ultramafic rocks, and in fresh mafic igneous rocks.

Association: n.d.

Distribution: From Tamasato, Tagozu, Kitakami Mountainland, Iwate Prefecture, Japan. While undoubtedly widespread, no other material appears to have been analyzed.

Name: For its FERROus iron content and position intermediate between *pargasite* and *ferro-pargasite*.

Type Material: n.d.

References: (1) Gilbert, M.C. (1966) Synthesis and stability relationships of ferropargasite. Amer. J. Sci., 264, 698–742. (2) Charles, R.W. (1980) Amphiboles on the join pargasite-ferropargasite. Amer. Mineral., 65, 996–1001. (3) Deer, W.A., R.A. Howie, and J. Zussman (1963) Rock-forming minerals, v. 2, chain silicates, 263–314, esp. 298. (4) Leake, B.E. (1968) A catalog of analyzed calciferous and subcalciferous amphiboles together with their nomenclature and associated minerals. Geol. Soc. Amer. Special Paper 98, 210 p. [analysis 1181].

Crystal Data: Triclinic. *Point Group:* $\bar{1}$. Crystals typically flattened, axe-head-shaped, to 20 cm; granular, massive.

Physical Properties: *Cleavage:* Good on {100}, poor on {001}, {110}, and {011}. *Fracture:* Uneven to conchoidal. *Tenacity:* Brittle. Hardness = 6.5–7 D(meas.) = 3.25–3.28 D(calc.) = [3.33]

Optical Properties: Transparent to translucent. *Color:* Clove-brown, brown, plum-blue, pearl-gray; colorless to pale brown or blue in thin section. *Luster:* Vitreous. *Optical Class:* Biaxial (–). $\alpha = 1.674$–1.682 $\beta = 1.682$–1.690 $\gamma = 1.685$–1.693 2V(meas.) = 65°–75°

Cell Data: *Space Group:* $P\bar{1}$. $a = 7.1437(4)$ $b = 9.1898(6)$ $c = 8.9529(4)$ $\alpha = 91.857(6)°$ $\beta = 98.188(5)°$ $\gamma = 77.359(4)°$ $Z = 2$

X-ray Powder Pattern: Isère, France.
2.812 (100), 3.16 (90), 3.46 (80), 6.30 (70), 3.68 (60), 3.28 (60), 2.998 (60)

Chemistry:

	(1)	(2)	(3)		(1)	(2)	(3)
SiO₂	43.14	41.97	42.16	ZnO	0.04		
TiO₂	0.10			MgO	1.34	0.66	
B₂O₃	6.12	[6.14]	6.11	CaO	19.76	19.18	19.67
Al₂O₃	16.70	17.24	17.88	Na₂O	0.36		
Fe₂O₃	1.28			K₂O	0.23		
FeO	7.12	10.41	12.60	H₂O⁺	1.56	[1.57]	1.58
MnO	1.66	2.61		Total	99.41	[99.78]	100.00

(1) Durango, Mexico. (2) Rosebery district, Tasmania, Australia; by electron microprobe, B₂O₃ interpolated from end members, H₂O calculated from stoichiometry. (3) Ca₂FeAl₂BSi₄O₁₅(OH).

Polymorphism & Series: Forms a series with manganaxinite.

Mineral Group: Axinite group; Ca > 1.5 per formula unit, Fe > Mn.

Occurrence: In low- to high-grade regionally metamorphosed rocks, in contact metamorphosed rocks, and in pegmatites.

Association: Prehnite, andradite, hedenbergite, zoisite, actinolite, datolite, tourmaline, vesuvianite, calcite, albite, quartz.

Distribution: Many localities, even for fine crystals. On Roscommon Cliff and in the Botallack mine, St. Just, Cornwall, England. From Bourg d'Oisans, Isère, France. In the Maderantal, Uri, Switzerland. From the Puiva deposit, near Saranpaul, Polar Ural Mountains; at Zlatoust, Ural Mountains; and elsewhere in Russia. At Skardu, Pakistan. In the Obira mine, Bungo, Oita Prefecture; the Toroku mine, Miyazaki Prefecture; and elsewhere in Japan. In the USA, from Coarse Gold, Madera Co., near Yreka, Siskiyou Co., and New Melones Lake, Calaveras Co., California; and from Elkhorn, Jefferson Co., Montana. In Canada, found near Hope, British Columbia, and in the Moneta mine, Timmins, Ontario. From Trinidad, Baja California, Mexico. At Vitoria da Conquista, Bahia, Brazil.

Name: From the Greek for axe, referring to the common crystal shape, and for dominant FERROan iron, in a mineral of the *axinite* group.

References: (1) Dana, E.S. (1892) Dana's system of mineralogy, (6th edition), 527–529. (2) Deer, W.A., R.A. Howie, and J. Zussman (1986) Rock-forming minerals, (2nd edition), v. 1B, disilicates and ring silicates, 603–623. (3) Milton, C., F.A. Hildebrand, and A.M. Sherwood (1953) The identity of tinzenite with manganoan axinite. Amer. Mineral., 38, 1148–1158. (4) Swinnea, J.S., H. Steinfink, L.E. Rendon-DiazMiron, and S. Enciso de la Vega (1981) The crystal structure of a Mexican axinite. Amer. Mineral., 66, 428–431.

Ferro-barroisite $[NaCa][(Fe^{2+}, Mg)_3Al_2](Si_7Al)O_{22}(OH)_2$

Crystal Data: Monoclinic. *Point Group:* $2/m$. Prismatic; also as rims and cores of mixed amphibole crystals.

Physical Properties: *Cleavage:* [Perfect on {110}, intersecting at $\sim56°$ and $\sim124°$; partings on {100}, {001}.] *Tenacity:* [Brittle.] Hardness = [5–6] D(meas.) = n.d. D(calc.) = n.d.

Optical Properties: Semitransparent. *Color:* Deep bluish green to blue in thin section. *Luster:* [Vitreous.]
Optical Class: [Biaxial.] *Pleochroism:* Strong in blues and bluish green. $\alpha = $ n.d. $\beta = $ n.d. $\gamma = $ n.d. 2V(meas.) = n.d.

Cell Data: *Space Group:* [$C2/m$.] a = n.d. b = n.d. c = n.d. $\beta = $ n.d. Z = n.d.

X-ray Powder Pattern: n.d.

Chemistry:

	(1)	(2)	(3)
SiO_2	49.2	42.14	46.71
TiO_2	0.14	1.15	1.81
Al_2O_3	8.86	3.75	2.05
FeO	21.9	35.14	34.08
MnO	0.27	1.28	0.94
MgO	7.03	0.13	0.93
CaO	5.55	6.86	5.56
Na_2O	4.20	2.71	4.12
K_2O	0.55	0.87	1.21
Total	97.70	94.03	97.41

(1) Klamath Mountains, California, USA; by electron microprobe, $Fe^{2+}:Fe^{3+}$ from stoichiometry and charge balance; corresponding to $(Na_{1.19}Ca_{0.87}K_{0.10})_{\Sigma=2.16}(Fe^{2+}_{1.69}Mg_{1.53}Mn_{0.03})_{\Sigma=3.25}$ $(Fe^{3+}_{1.00}Al_{0.73}Ti_{0.02})_{\Sigma=1.75}(Si_{7.20}Al_{0.80})_{\Sigma=8.00}O_{22}(OH)_2$. (2) Cauro-Bastelica, Corsica; by electron microprobe. (3) Iskou ring complex, Niger; by electron microprobe.

Polymorphism & Series: Forms a series with barroisite.

Mineral Group: Amphibole (sodic-calcic) group: $Mg/(Mg + Fe^{2+}) < 0.5$; $(Na + K)_A < 0.5$; $0.67 \leq Na_B \leq 1.33$; $(Ca + Na)_B \geq 1.34$; $Si < 7.5$.

Occurrence: A primary mineral in some alkalic granites and syenite ring complexes; from blueschist facies metavolcanic rocks.

Association: Crossite (metavolcanic).

Distribution: On Condrey Mountain, Klamath Mountains, Del Norte Co., California, USA. From Cauro-Bastelica, Corsica. In the Iskou ring complex, Aïr, Niger.

Name: For *ferroan* iron in its composition and similarity to *barroisite*.

Type Material: n.d.

References: (1) Giret, A., B. Bonin, and J.-M. Leger (1980) Amphibole compositional trends in oversaturated alkaline plutonic ring-complexes. Can. Mineral., 18, 481–495. (2) Helper, M.A. (1986) Deformation and high P/T metamorphism in the central portion of the Condrey Mountain window, north-central Klamath Mountains, California and Oregon. In: B.W. Evans and E.H. Brown, Eds., Blueschists and eclogites. Geol. Soc. Amer. Mem. 164, 125–141.

Crystal Data: Triclinic. *Point Group:* $\bar{1}$. Crystalline, massive. *Twinning:* Simple twins on {100}.

Physical Properties: *Cleavage:* {100}, perfect; {110}, {1$\bar{1}$0}, good. Hardness = 6
D(meas.) = 3.09 D(calc.) = [3.09]

Optical Properties: Semitransparent. *Color:* Colorless, pink to brown.
Optical Class: Biaxial. *Orientation:* $X' \wedge c = 44°$ in section perpendicular to the zone of
cleavages. $\alpha = 1.640$ β = n.d. $\gamma = 1.653$ 2V(meas.) = 60(3)°

Cell Data: *Space Group:* $A\bar{1}$. a = 7.862 b = 7.253 c = 13.967 $\alpha = 89°44'$ $\beta = 95°28'$
$\gamma = 103°29'$ Z = 6

X-ray Powder Pattern: Kagata or Ofuku mine, Japan. (ICDD 29-336).
3.270 (100), 3.049 (80), 2.278 (65), 3.470 (60), 3.84 (55), 2.696 (30), 7.67 (25)

Chemistry:

	(1)	(2)	(3)
SiO_2	50.00	53.30	49.8
TiO_2	trace		
Al_2O_3			0.0
Fe_2O_3	0.00		
FeO	9.29	10.44	9.0
MnO	1.22	1.58	1.6
MgO	0.00	0.06	0.1
CaO	38.86	34.63	38.9
H_2O	0.00		
insol.	0.45		
Total	99.82	100.01	99.4

(1) Isle of Skye, Scotland; corresponding to $(Ca_{1.65}Fe_{0.31}Mn_{0.04})_{\Sigma=2.00}Si_2O_6$. (2) Do.; by electron
microprobe, corresponding to $(Ca_{1.58}Fe_{0.38}Mn_{0.04})_{\Sigma=2.00}Si_2O_6$. (3) Ofuku mine, Japan; by
electron microprobe, corresponds to $(Ca_{1.64}Fe_{0.30}Mn_{0.06})_{\Sigma=2.00}Si_2O_6$.

Occurrence: Surrounding chert nodules in a skarn in dolostone (Isle of Skye, Scotland).

Association: Hedenbergite, grossular-andradite (Isle of Skye, Scotland).

Distribution: At Camas Malag, Isle of Skye, Scotland. From Scawt Hill, near Larne,
Co. Antrim, Ireland. In the Kagata, Ofuku, and Ohta mines, Yamaguchi Prefecture; in the
Kasugayama and Tsuchiarashi deposits, near Iida, Nagano Prefecture; on Kurodaké Peak, near
Toyama, Yoyama Prefecture; in the Tsumo mine, Shimane Prefecture; in the Horado mine, Gifu
Prefecture; and at a number of other localities in Japan. From the Wessels mine, near Kuruman,
Cape Province, South Africa.

Name: For FERROan iron in its composition and its relation to *bustamite*.

Type Material: n.d.

References: (1) Deer, W.A., R.A. Howie, and J. Zussman (1978) Rock-forming minerals, (2nd
edition), v. 2A, single-chain silicates, 575–585. (2) Tilley, C.E. (1948) On iron-wollastonites
in contact-skarns: an example from Skye. Amer. Mineral., 33, 736–738. (3) Rapoport, P.A.
and C.W. Burnham (1973) Ferrobustamite: the crystal structures of two Ca,Fe bustamite-type
pyroxenoids. Zeits. Krist., 138, 419–438. (4) Shimazaki, H. and T. Yamanaka (1973) Iron-
wollastonite from skarns and its stability relation in the $CaSiO_3 - CaFeSi_2O_6$ join. Geochem. J.,
7, 67–79. (5) Yamanaka, T., R. Sadanaga, and Y. Takéuchi (1977) Structural variation in the
ferrobustamite solid solution. Amer. Mineral., 62, 1216–1224. (6) Shimazaki, H. and M. Bunno
(1978) Subsolidus skarn equilibria in the system $CaSiO_3 - CaMgSi_2O_6 - CaFeSi_2O_6 - CaMnSi_2O_6$.
Can. Mineral., 16, 539–545.

Ferrocarpholite $\qquad\qquad\qquad\qquad\qquad\qquad$ $(\text{Fe}^{2+}, \text{Mg})\text{Al}_2\text{Si}_2\text{O}_6(\text{OH})_4$

Crystal Data: Orthorhombic. *Point Group:* $2/m\ 2/m\ 2/m$. Crystals prismatic, elongated along [001], or as aggregates of parallel fibers, to 5 cm.

Physical Properties: *Cleavage:* Perfect on {010}, indistinct on {110}. Hardness = 5.5 D(meas.) = 3.04 D(calc.) = 3.05

Optical Properties: Transparent to translucent. *Color:* Dark green, bright green, grayish green; colorless in thin section.
Optical Class: Biaxial (–). *Pleochroism:* X = yellowish to greenish; Y = colorless to yellowish; Z = pale bluish to bluish green. *Orientation:* $X = b$; $Y = a$; $Z = c$. *Dispersion:* $r > v$, weak to strong. *Absorption:* $X > Y > Z$. $\alpha = 1.614$–1.628 $\beta = 1.627$–1.644 $\gamma = 1.632$–1.647 2V(meas.) = $49°$–$70°$

Cell Data: *Space Group:* $Ccca$. a = 13.797(9) b = 20.20(2) c = 5.116(5) Z = 8

X-ray Powder Pattern: Haute-Ubaye district, France.
5.04 (100), 5.69 (70), 3.355 (30), 2.603 (25), 3.019 (20), 1.680 (11), 3.44 (10)

Chemistry:

	(1)	(2)
SiO_2	37.59	38.56
TiO_2	0.22	0.30
Al_2O_3	29.39	31.68
Fe_2O_3	2.07	1.65
FeO	17.98	12.21
MnO	0.14	0.32
MgO	2.52	4.65
H_2O	10.08	11.02
Total	[99.99]	[100.39]

(1) West of Tomata, Celebes, Indonesia; after deduction of CaO 0.13%, Na_2O 0.14%, K_2O 0.09%, SO_3 0.12%; corresponds to $(\text{Fe}^{2+}_{0.81}\text{Mg}_{0.20}\text{Mn}_{0.01})_{\Sigma=1.02}(\text{Al}_{1.87}\text{Fe}^{3+}_{0.08}\text{Ti}_{0.01})_{\Sigma=1.96}\text{Si}_{2.03}\text{O}_{9.87}\text{H}_{3.64}$.
(2) Haute-Ubaye district, France; after deduction of Na_2O 0.23% and K_2O 0.45% as mica; corresponds to $(\text{Fe}^{2+}_{0.53}\text{Mg}_{0.36}\text{Mn}_{0.01})_{\Sigma=0.90}(\text{Al}_{1.95}\text{Fe}^{3+}_{0.06})_{\Sigma=2.01}\text{Si}_{2.01}\text{O}_6(\text{OH})_4$.

Polymorphism & Series: Forms two series, with carpholite, and with magnesiocarpholite.

Occurrence: In quartz veins in blueschists of low metamorphic grade, formed from felsic tuffs at about 250 °C and 3 kbar.

Association: Quartz, glaucophane, lawsonite, pumpellyite, jadeite, stilpnomelane.

Distribution: At several localities west of Tomata, Celebes, Indonesia. In the Diahot region, New Caledonia. From Italy, at Colle Ciarbonet, Piedmont. From the Vanoise massif, Savoie, and Pic du Pelvat, Haute-Ubaye district, Alpes-de-Haute Provence, France. At Ruwi, Oman.

Name: The FERROus iron analogue of *carpholite*.

Type Material: National Museum of Natural History, Washington, D.C., USA, 106754.

References: (1) De Roever, W.P. (1951) Ferrocarpholite, the hitherto unknown ferrous iron analogue of carpholite proper. Amer. Mineral., 36, 736–745. (2) De Roever, W.P. and C. Kieft (1971) Additional data on ferrocarpholite from Sulawesi (Celebes), Indonesia. Amer. Mineral., 56, 1976–1982. (3) Steen, D. and J. Bertrand (1977) Sur la préscence de ferrocarpholite associée aux schistes à glaucophane de Haute-Ubaye (Basses-Alpes, France). Schweiz. Mineral. Petrog. Mitt., 57, 157–168 (in French with English abs.). (4) Ferraris, G., G. Ivaldi, and B. Goffé (1992) Structural study of a magnesian ferrocarpholite: are carpholites monoclinic? Neues Jahrb. Mineral., Monatsh., 337–347.

Crystal Data: Monoclinic. *Point Group:* $2/m$. Commonly as well-formed prismatic crystals; fibrous; as reaction rims on pyroxenes. *Twinning:* [Simple or multiple twinning ‖ {100}.]

Physical Properties: *Cleavage:* [Good on {110}, intersecting at ~56° and ~124°; partings on {100}, {001}.] *Tenacity:* [Brittle.] Hardness = [5–6] D(meas.) = ~3.4 D(calc.) = n.d.

Optical Properties: Semitransparent. *Color:* Dark green, black. *Luster:* [Vitreous.] *Optical Class:* Biaxial (–). *Pleochroism:* Distinct; X = yellow; Y = green; Z = dark green. *Orientation:* $Y = b$; $Z \wedge c \simeq 15°$. *Dispersion:* Weak. $\alpha = 1.65$–1.70 $\beta = 1.66$–1.71 $\gamma = 1.67$–1.73 2V(meas.) = 20°–90°

Cell Data: *Space Group:* [$C2/m$.] a = n.d. b = n.d. c = n.d. β = n.d. Z = n.d.

X-ray Powder Pattern: n.d.

Chemistry:

	(1)	(2)
SiO_2	45.74	43.25
TiO_2	1.29	0.27
Al_2O_3	4.98	6.07
Fe_2O_3		1.19
FeO	27.86	31.64
MnO	1.14	1.05
MgO	5.96	1.49
CaO	10.02	10.65
Na_2O	1.99	1.34
K_2O	0.54	0.87
Total	99.52	97.82

(1) La Tabatière, Canada; by electron microprobe, $Fe^{2+}:Fe^{3+}$ from stoichiometry and charge balance; corresponds to $(Na_{0.55}K_{0.11})_{\Sigma=0.66}(Ca_{1.66}Mn_{0.15}Fe^{2+}_{0.14}Na_{0.05})_{\Sigma=2.00}(Fe^{2+}_{3.47}Mg_{1.38}Ti_{0.15})_{\Sigma=5.00}(Si_{7.09}Al_{0.91})_{\Sigma=8.00}O_{22}(OH)_2$. (2) Tibchi ring complex, Nigeria; by electron microprobe, $Fe^{2+}:Fe^{3+}$ calculated from charge balance; corresponds to $(Na_{0.42}K_{0.18})_{\Sigma=0.60}(Ca_{1.85}Fe^{2+}_{0.15})_{\Sigma=2.00}(Fe^{2+}_{4.15}Mg_{0.36}Al_{0.18}Fe^{3+}_{0.14}Mn_{0.14}Ti_{0.03})_{\Sigma=5.00}(Si_{7.01}Al_{0.99})_{\Sigma=8.00}O_{22}(OH)_2$.

Polymorphism & Series: Forms a series with edenite.

Mineral Group: Amphibole (calcic) group: $Mg/(Mg + Fe^{2+}) < 0.5$; $(Na + K)_A \geq 0.5$; $Na_B < 0.67$; $(Ca + Na)_B \geq 1.34$; $6.75 \leq Si \leq 7.25$.

Occurrence: An early or deuteric phase replacing pyroxene in syenitic ring complexes; from medium-grade metamorphic amphibolites and gneisses; from plutonic igneous rocks.

Association: Pyroxene, ferro-actinolite, monazite, magnetite, fluorite (Tibchi ring complex, Nigeria).

Distribution: From the Baie-des-Moutons complex, La Tabatière, Quebec, Canada. At the Tibchi ring complex, Nigeria. Undoubtedly occurs at other localities but qualifying analyses appear lacking.

Name: For its high *ferrous* iron content and relation to *edenite*.

Type Material: n.d.

References: (1) Lalonde, A.E. and R.F. Martin (1983) The Baie-des-Moutons syenitic complex, La Tabatière, Québec, II. The ferromagnesian minerals. Can. Mineral., 21, 81–91. (2) Ike, E.C., P. Bowden, and R.F. Martin (1985) Amphibole in the porphyries of the Tibchi anorogenic ring-complex, Nigeria: product of deuteric adjustments. Can. Mineral., 23, 447–456. (3) Phillips, W.R. and D.T. Griffen (1981) Optical mineralogy, 221.

Ferro-gedrite

$$(Fe^{2+}, Mg)_2[(Fe^{2+}, Mg)_3Al_2](Si_6Al_2)O_{22}(OH)_2$$

Crystal Data: Orthorhombic. *Point Group: 2/m 2/m 2/m.* As prismatic to bladed crystals; also fibrous.

Physical Properties: *Cleavage:* Perfect on {210}, intersecting at 54° and 126°. *Tenacity:* Brittle. Hardness = 5.5–6 D(meas.) = 3.566 D(calc.) = 3.562

Optical Properties: Transparent to translucent. *Color:* Gray, brown. *Luster:* [Vitreous.] *Optical Class:* Biaxial (–). *Pleochroism:* X = pale green; Y = brownish green; Z = greenish blue. *Orientation:* X = a; Y = b; Z = c. *Dispersion:* $r < v$. $\alpha = 1.690-1.695$ $\beta = 1.705-1.710$ $\gamma = 1.718-1.725$ 2V(meas.) = ~82°

Cell Data: *Space Group: Pnma.* a = 18.514 b = 17.945 c = 5.315 Z = 4

X-ray Powder Pattern: Mt. Yakushi, Japan.
8.23 (100), 3.043 (75), 3.221 (20), 8.95 (12), 4.632 (10), 2.679 (10), 2.581 (10)

Chemistry:

	(1)	(2)		(1)	(2)
SiO_2	38.41	36.53	CaO	0.03	0.11
TiO_2	0.48	0.07	Na_2O	1.16	2.06
Al_2O_3	19.72	18.80	K_2O	0.04	0.00
Fe_2O_3	2.21		F	0.13	
FeO	33.54	39.18	H_2O^+	2.06	
MnO	2.30	0.42	$-O = F_2$	0.06	
MgO	0.03	0.30	Total	[100.05]	97.47

(1) Mt. Yakushi, Japan; adjusted for removal of estimated 6% chamosite contamination, original total given as 99.92%; corresponds to $(Fe^{2+}_{4.40}Na_{0.33}Mn_{0.31}Mg_{0.01}K_{0.01})_{\Sigma=5.06}(Al_{1.68}Fe^{3+}_{0.35}$ $Ti_{0.06})_{\Sigma=2.09}(Si_{6.03}Al_{1.97})_{\Sigma=8.00}O_{22}[(OH)_{2.15}F_{0.06}]_{\Sigma=2.21}$. (2) Kawai mine, Japan; by electron microprobe.

Polymorphism & Series: Forms a series with magnesio-gedrite and gedrite.

Mineral Group: Amphibole (Fe–Mn–Mg) group; $Mg/(Mg + Fe^{2+}) < 0.10$; $(Ca + Na)_B < 1.34$; Li < 1.0; Si < 7.0.

Occurrence: In contact metamorphosed pelitic rocks.

Association: Chamosite, chlorite, andalusite, cordierite, garnet, muscovite, biotite, spinel, quartz, labradorite, magnetite, graphite.

Distribution: On Mt. Yakushi, Miyamori district, Kitakami Mountainland, Iwate Prefecture, and in the Kawai mine, Ena, Gifu Prefecture, Japan. From Fiskenæsset, Greenland. At Turbeoshov, Shuyeverskoya, Karelia, Russia. From Spitzenberg, Harz Mountains, Germany.

Name: For *ferrous* iron in its composition and similarity to *gedrite*.

Type Material: n.d.

References: (1) Seki, Y. and M. Yamasaki (1957) Aluminian ferroanthophyllite from the Kitakami Mountainland, northeastern Japan. Amer. Mineral., 42, 506–520. (2) Matsubara, S., A. Kato, and M. Nomura (1979) The occurrence of ferrogedrite from the Kawai mine, Ena, Gifu Prefecture, Japan. Bull. National Science Museum, Tokyo, Ser. C, Geol., 6, 107–113. (3) (1982) Mineral. Abs., 33, 298 (abs. ref. 2). (4) Phillips, W.R. and D.T. Griffen (1981) Optical mineralogy, 223–225.

Crystal Data: Monoclinic. *Point Group:* $2/m$. [Elongated prismatic crystals.] *Twinning:* [Simple or multiple twinning ∥ {100}.]

Physical Properties: *Cleavage:* [Perfect on {110}, intersecting at ∼58° and ∼122°.] *Fracture:* [Conchoidal to uneven.] *Tenacity:* [Brittle.] Hardness = [6] D(meas.) = 3.18–3.34 D(calc.) = 3.224

Optical Properties: Semitransparent. *Color:* Blue to bluish gray. *Luster:* [Vitreous.] *Optical Class:* Biaxial (–). *Pleochroism:* Vivid; stronger with increasing iron content; X = colorless; Y = violet; Z = blue. *Orientation:* Y = b; Z ∧ c = 6°. *Dispersion:* r < v, weak. $\alpha = 1.62$–1.64 $\beta = 1.64$–1.67 $\gamma = 1.64$–1.68 2V(meas.) = 42°–45°

Cell Data: *Space Group:* $C2/m$. a = 9.587(4) b = 17.832(7) c = 5.315(2) $\beta = 103.47(3)°$ Z = 2

X-ray Powder Pattern: Bouehndep, New Caledonia. (ICDD 27-714). 8.31 (100), 3.06 (80), 2.756 (30), 2.701 (25), 4.47 (20), 3.23 (20), 3.35 (15)

Chemistry:

	(1)	(2)		(1)	(2)
SiO_2	54.63	53.8	MnO	0.08	0.1
TiO_2	0.06		MgO	4.75	3.9
Al_2O_3	11.02	10.1	CaO	0.98	0.9
Fe_2O_3	2.76	4.0	Na_2O	6.25	6.8
FeO	16.02	17.88	K_2O	0.01	
			Total	[96.56]	97.48

(1) Calabria, Italy; by electron microprobe, Fe^{2+}:Fe^{3+} from Mössbauer spectroscopy, original total given as 97.56%; corresponding to $(Na_{1.75}Ca_{0.15})_{\Sigma=1.90}(Fe^{2+}_{1.94}Al_{1.83}Mg_{1.03}Fe^{3+}_{0.31}$ $Ti_{0.01}Mn_{0.01})_{\Sigma=5.13}(Si_{7.94}Al_{0.06})_{\Sigma=8.00}O_{22}(OH)_2$. (2) Bouehndep, New Caledonia; by electron microprobe, FeO determined by wet chemical analysis, corresponding to $(Na_{1.91}Ca_{0.14})_{\Sigma=2.05}$ $(Fe^{2+}_{2.17}Mg_{0.84})_{\Sigma=3.01}(Al_{1.51}Fe^{3+}_{0.44}Mn_{0.01})_{\Sigma=1.96}(Si_{7.79}Al_{0.21})_{\Sigma=8.00}O_{22}(OH)_2$.

Polymorphism & Series: Forms a series with glaucophane.

Mineral Group: Amphibole (alkali) group: $Fe^{2+}/(Fe^{2+} + Mg) \geq 0.5$; $Fe^{3+}/(Fe^{3+} + Al^{vi}) < 0.3$; $(Na + K)_A < 0.5$; $Na_B \geq 1.34$.

Occurrence: In metamorphic rocks of the high-pressure, low-temperature blueschist facies, commonly derived from either siliceous sedimentary, alkalic basaltic, or felsic volcanic predecessors.

Association: Paragonite, lawsonite, albite, spessartine, almandine, epidote, omphacite.

Distribution: Widespread. Representative occurrences include: from Bouehndep, and near Ouégoa, New Caledonia. At Champ de Praz, Val d'Aosta, Italy. From near Menai Bridge, Anglesey, Wales. In Scotland, from near Heriot. On the Ile de Groix, Bay of Biscay, off Morbihan, France.

Name: For its high *ferrous* iron content and similarity to *glaucophane*.

Type Material: n.d.

References: (1) Hoffmann, C. (1972) Natural and synthetic ferroglaucophane. Contr. Mineral. Petrol., 34, 135–149. (2) Hawthorne, F.C. (1979) The crystal chemistry of the amphiboles. X. Refinement of the crystal structure of ferroglaucophane and an ideal polyhedral model for clinoamphiboles. Can. Mineral., 17, 1–10. (3) Hoffmann, C. and K. Katz (1982) Trend surface analysis of some physical properties of alkali (sodic) amphiboles. Lithos, 15, 17–25. (4) Phillips, W.R. and D.T. Griffen (1981) Optical mineralogy, 237–241.

Ferro-hornblende $\qquad$ $Ca_2[(Fe^{2+},Mg)_4Al](Si_7Al)O_{22}(OH)_2$

Crystal Data: Monoclinic. *Point Group:* $2/m$. As prismatic crystals, to 10 cm, and as rims on pyroxenes. *Twinning:* [Simple or lamellar twinning ‖ {100}.]

Physical Properties: *Cleavage:* Perfect on {110}, with intersections at 56° and 124°; may show partings on {100}, {010}. *Tenacity:* [Brittle.] Hardness = [5–6] D(meas.) = 3.12–3.30 D(calc.) = n.d.

Optical Properties: Semitransparent. *Color:* Green, greenish brown. *Luster:* [Vitreous.] *Optical Class:* Biaxial (–). *Pleochroism:* X = light brown, greenish yellow; Y = light green, olive-green; Z = dark green, bluish green. *Orientation:* $Y = b$, $Z \wedge c \simeq 18°$. $\alpha = 1.641$–1.704 $\beta = 1.657$–1.718 $\gamma = 1.662$–1.727 2V(meas.) = 57°–79°

Cell Data: *Space Group:* [$C2/m$.] a = n.d. b = n.d. c = n.d. β = n.d. Z = n.d.

X-ray Powder Pattern: n.d.

Chemistry:

	(1)	(2)		(1)	(2)
SiO_2	45.24	42.73	MgO	8.90	0.40
TiO_2	1.26	3.32	CaO	10.21	9.20
Al_2O_3	10.77	7.35	Na_2O	1.34	1.09
Fe_2O_3	0.41	7.05	K_2O	0.42	0.90
FeO	19.33	25.86	H_2O^+	1.50	0.75
MnO	0.75	0.48	H_2O^-	0.00	
			Total	100.13	99.13

(1) Centennial mine, Broken Hill, New South Wales, Australia; corresponds to $(Ca_{1.65}Na_{0.39}K_{0.08})_{\Sigma=2.12}(Fe^{2+}_{2.43}Mg_{2.00}Al_{0.72}Ti_{0.14}Mn_{0.09}Fe^{3+}_{0.05})_{\Sigma=5.43}(Si_{6.81}Al_{1.19})_{\Sigma=8.00}$ $O_{22}[(OH)_{1.50}O_{0.50}]_{\Sigma=2.00}$. (2) Mayurbhanj district, Orissa, India; corresponds to $(Ca_{1.60}Na_{0.34}$ $K_{0.19})_{\Sigma=2.13}(Fe^{2+}_{3.51}Fe^{3+}_{0.86}Ti_{0.40}Al_{0.34}Mg_{0.10}Mn_{0.06})_{\Sigma=5.27}(Si_{6.93}Al_{1.07})_{\Sigma=8.00}O_{22}(OH)_{0.81}$.

Polymorphism & Series: Forms a series with magnesio-hornblende.

Mineral Group: Amphibole (calcic) group: $Mg/(Mg + Fe^{2+}) < 0.50$; $(Na + K)_A < 0.5$; $Na_B < 0.67$; $(Ca + Na)_B \geq 1.34$; $6.50 \leq Si \leq 7.24$.

Occurrence: From granites, granodiorites, and metabasalts; common in amphibolites and schists. As reaction rims on ferroan hedenbergite.

Association: Hedenbergite (granite); biotite, epidote, albite, quartz (amphibolite).

Distribution: Very widespread, but many locality references lack qualifying chemical analyses. A few historic localities for well-crystallized material include: at Monte Somma and Vesuvius, Campania, Italy. From Pargas, Finland. At Kragerö, Arendal, and around the Langesundsfjord, Norway. From Bílina and Schima, Czech Republic. In the USA, from Franklin and Sterling Hill, Ogdensburg, Sussex Co., New Jersey; from Edwards, Pierrepont, and Gouverneur, St. Lawrence Co., New York. From Bancroft, Pakenham, and Eganville, Ontario, Canada. From Broken Hill, New South Wales, Australia.

Name: For its *ferrous* iron content and from the German for *horn* and *to deceive*, in allusion to its similarity to valuable minerals in ores.

References: (1) Dana, E.S. (1892) Dana's system of mineralogy, (6th edition), 385–398 [hornblende, in part]. (2) Leake, B.E. (1968) A catalog of analyzed calciferous and subcalciferous amphiboles together with their nomenclature and associated minerals. Geol. Soc. Amer. Special Paper 98, 210 p. [analysis 754]. (3) Iyengar, S.V.P. and K.K. Basu (1980) Ferro-hornblende, an iron-rich end member from Simlipal Granite, Mayurbhanj District, Orissa. J. Geol. Soc. India, 21, 558–561.

Crystal Data: Monoclinic. *Point Group:* $2/m$. [Prismatic.] *Twinning:* [Simple or multiple twinning $\parallel$ {100}.]

Physical Properties: *Cleavage:* [Perfect on {110}, with intersections at 56° and 124°; partings on {100}, {001}.] *Tenacity:* [Brittle.] Hardness = [5–6] D(meas.) = ~3.2–3.42 D(calc.) = n.d.

Optical Properties: Semitransparent. *Color:* [Dark brown to black.] *Luster:* [Vitreous.] *Optical Class:* [Biaxial (–).] *Pleochroism:* X = yellow; Y = impure green; Z = greenish brown. *Orientation:* [Y = b.] *Dispersion:* [r > v.] $\alpha = 1.677$ $\beta = $ n.d. $\gamma = 1.709$ 2V(meas.) = 72°

Cell Data: *Space Group:* [$C2/m$.] a = n.d. b = n.d. c = n.d. $\beta = $ n.d. Z = n.d.

X-ray Powder Pattern: n.d.

Chemistry:

	(1)
SiO_2	36.95
TiO_2	6.25
Al_2O_3	15.02
Fe_2O_3	3.62
FeO	19.40
MgO	4.34
CaO	10.25
Na_2O	1.34
K_2O	1.33
H_2O^+	1.26
H_2O^-	0.19
Total	[99.95]

(1) Koraput, India; original total given as 99.98%; corresponds to $(Na_{0.40}K_{0.26})_{\Sigma=0.66}Ca_{1.84}$ $(Fe^{2+}_{2.51}Mg_{1.00}Ti_{0.73}Al_{0.48}Fe^{3+}_{0.42})_{\Sigma=5.14}(Si_{5.74}Al_{2.26})_{\Sigma=8.00}O_{22}[(OH)_{1.30}O_{0.70}]_{\Sigma=2.00}$.

Polymorphism & Series: Forms a series with kaersutite.

Mineral Group: Amphibole (calcic) group: $Mg/(Mg + Fe^{2+}) < 0.50$; $Na_B < 0.67$; $(Ca + Na)_B \geq 1.34$; $Si < 6.5$; $Ti \geq 0.5$.

Occurrence: From syenites, camptonites, and essexites.

Association: Alkalic feldspar, oligoclase, biotite, ilmenite-magnetite.

Distribution: From Koraput, Orissa, India.

Name: For its high *ferrous* iron composition and similarity to *kaersutite*.

Type Material: n.d.

References: (1) Bose, M.K. (1964) Amphiboles in alkaline rocks of Koraput, Orissa. Mineral. Mag., 33, 912–917.

Crystal Data: Monoclinic. *Point Group:* $2/m$. As homoaxial intergrowths with grunerite. *Twinning:* [Simple or multiple twinning $\parallel \{100\}$.]

Physical Properties: *Cleavage:* [Perfect on $\{110\}$, with intersections at $\sim 56°$ and $\sim 124°$; partings on $\{100\}$, $\{001\}$.] *Tenacity:* [Brittle.] Hardness = 5–6 $\quad$ D(meas.) = n.d. D(calc.) = 3.44 (synthetic ferro-pargasite).

Optical Properties: Semitransparent. *Color:* Bluish green in thin section. *Luster:* [Vitreous.] *Optical Class:* Biaxial (–). *Pleochroism:* Moderate. $\alpha = 1.700$ $\quad \beta = 1.713$ $\quad \gamma = 1.718$ 2V(meas.) = n.d.

Cell Data: *Space Group:* $C2/m$. $a = 9.953(5)$ (synthetic ferro-pargasite). $\quad b = 18.152(3)$ $c = 5.330(2)$ $\quad \beta = 105.3(1)°$ $\quad Z = 2$

X-ray Powder Pattern: Synthetic ferro-pargasite. 8.50 (100), 3.15 (80), 2.718 (60), 2.607 (40), 2.57 (35), 3.40 (25), 2.36 (25)

Chemistry:

	(1)
SiO_2	38.65
Al_2O_3	16.50
FeO	27.75
MnO	0.39
MgO	2.03
CaO	10.30
Na_2O	1.80
K_2O	0.50
Total	97.92

(1) Flowerdale, Scotland; by electron microprobe, corresponds to $(Na_{0.55}K_{0.10})_{\Sigma=0.65}Ca_{1.73}$ $(Fe^{2+}_{3.65}Mg_{0.48}Mn_{0.05})_{\Sigma=4.18}Al_{1.13}(Si_{6.07}Al_{1.93})_{\Sigma=8.00}O_{22}(OH)_2$.

Polymorphism & Series: Forms a series with pargasite.

Mineral Group: Amphibole (calcic) group: $Mg/(Mg + Fe^{2+}) < 0.3$; $Fe^{3+} \leq Al^{vi}$; $(Na + K)_A \geq$ 0.5; $Na_B < 0.67$; $(Ca + Na)_B \geq 1.34$; $Si < 6.25$; $Ti < 0.5$.

Occurrence: In an amphibolite facies metamorphosed banded iron formation.

Association: Grunerite, almandine, biotite, ferroan clinochlore, magnetite, quartz.

Distribution: From Flowerdale, near Gairloch, Scotland.

Name: For its high *ferrous* iron content and relation to *pargasite*.

Type Material: n.d.

References: (1) Williams, P.J. (1986) Petrology and origin of iron-rich silicate-magnetite-quartz rocks from Flowerdale near Gairloch, Wester Ross. Scottish J. Geol., 22, 1–12. (2) Gilbert, M.C. (1966) Synthesis and stability relationships of ferropargasite. Amer. J. Sci., 264, 698–742. (3) Charles, R.W. (1980) Amphiboles on the join pargasite-ferropargasite. Amer. Mineral., 65, 996–1001.

Crystal Data: Hexagonal. *Point Group:* n.d. As subhedral to euhedral platy crystals, to 1 mm; in radiating crystal aggregates.

Physical Properties: *Cleavage:* {0001}, good. Hardness = n.d. D(meas.) = n.d. D(calc.) = [3.14]

Optical Properties: Semitransparent. *Color:* In transmitted light, colorless. *Optical Class:* Uniaxial (–). $\omega = 1.677(2)$ $\epsilon = 1.652(2)$

Cell Data: *Space Group:* n.d. $a = 13.33(3)$ $c = 7.11(2)$ $Z = [2]$

X-ray Powder Pattern: Pegmont deposit, Australia.
2.675 (10), 7.13 (8), 3.564 (6), 2.243 (6), 1.833 (4), 1.667 (4), 1.513 (4)

Chemistry:

	(1)
SiO_2	34.17
Al_2O_3	0.00
FeO	49.54
MnO	4.36
MgO	0.64
Cl	4.00
H_2O	[8.19]
$-O = Cl_2$	0.90
Total	[100.00]

(1) Pegmont deposit, Australia; by electron microprobe, H_2O by difference; corresponds to $(Fe_{7.14}Mn_{0.64}Mg_{0.16})_{\Sigma=7.94}Si_{5.89}O_{14.42}[(OH)_{9.41}Cl_{1.17}]_{\Sigma=10.58}$.

Polymorphism & Series: Forms a series with manganpyrosmalite.

Occurrence: Intergrown with sulfides, formed during prograde metamorphism of a stratiform Pb-Zn deposit.

Association: Fayalite, greenalite, galena, sphalerite, clinopyroxene, "hornblende," grunerite, garnet, biotite, magnetite, apatite.

Distribution: From the Pegmont lead-zinc deposit, 175 km southeast of Mt. Isa, Queensland, Australia.

Name: For its high FERROus iron content and relation to *manganpyrosmalite*; *pyrosmalite* from the Greek for *fire* and *odor*, for the odor when heated.

Type Material: n.d.

References: (1) Vaughan, J.P. (1986) The iron end-member of the pyrosmalite series from the Pegmont lead-zinc deposit, Queensland. Mineral. Mag., 50, 527–531. (2) Vaughan, J.P. (1987) Ferropyrosmalite and nomenclature in the pyrosmalite series. Mineral. Mag., 51, 174. (3) (1988) Amer. Mineral., 73, 933–934 (abs. refs. 1 and 2).

Ferro-richterite　　　　　　　　　　$Na[NaCa](Fe^{2+}, Mg)_5Si_8O_{22}(OH)_2$

Crystal Data: Monoclinic. *Point Group:* $2/m$. Crystals prismatic; in reticulated masses; as overgrowths on clinopyroxene. *Twinning:* [Simple or multiple twinning ∥ {100}.]

Physical Properties: *Cleavage:* [Perfect on {110}, intersecting at ~56° and ~124°; partings on {100}, {001}.] *Fracture:* [Uneven.] *Tenacity:* [Brittle.] Hardness = 5–6　D(meas.) = n.d. D(calc.) = 3.46 (synthetic ferro-richterite).

Optical Properties: Semitransparent. *Color:* Brown to green. *Streak:* [White.] *Luster:* [Vitreous.]
Optical Class: Biaxial (–). *Pleochroism:* Strong; X = yellow; Y = blue-black; Z = dark blue-green. *Orientation:* $Y = b$. *Dispersion:* $r < v$. $\alpha = 1.688$　$\beta = 1.699$　$\gamma = 1.704$
2V(meas.) = 35°

Cell Data: *Space Group:* [$C2/m$.]　$a = 9.982(7)$ (synthetic ferro-richterite).　　$b = 18.223(6)$
$c = 5.298(5)$　$\beta = 103°44(7)'$　$Z = 2$

X-ray Powder Pattern: Synthetic ferro-richterite.
8.58 (100), 2.739 (70), 3.18 (65), 2.540 (50), 3.43 (35), 2.615 (35), 3.32 (25)

Chemistry:

	(1)	(2)		(1)	(2)
SiO_2	48.85	51.15	MnO	1.31	1.84
TiO_2	1.02	1.33	MgO	4.20	8.08
Al_2O_3	2.39	1.35	CaO	6.10	5.04
Fe_2O_3		0.82	Na_2O	4.37	6.62
Cr_2O_3	0.04		K_2O	1.24	1.10
FeO	30.19	20.98	Total	[99.71]	98.31

(1) Baie-des-Moutons, Quebec, Canada; by electron microprobe, $Fe^{2+}:Fe^{3+}$ calculated from stoichiometry, original total given as 99.68%, corresponding to $(Na_{0.59}K_{0.25})_{\Sigma=0.84}$ $(Ca_{1.02}Na_{0.73}Mn_{0.17}Fe^{2+}_{0.08})_{\Sigma=2.00}(Fe^{2+}_{3.85}Mg_{0.98}Ti_{0.12}Al_{0.05})_{\Sigma=5.00}(Si_{7.61}Al_{0.39})_{\Sigma=8.00}$ $O_{22}(OH)_2$. (2) Kangerdlugssuaq Fjord, Greenland; by electron microprobe, $Fe^{2+}:Fe3+$ calculated; corresponding to $(Na_{0.77}K_{0.21})_{\Sigma=0.98}(Ca_{1.82}Na_{0.18})_{\Sigma=2.00}(Fe^{2+}_{2.67}Mg_{1.83}Mn_{0.24}Ti_{0.15}$ $Fe^{3+}_{0.09}Al_{0.02})_{\Sigma=5.00}(Si_{7.78}Al_{0.22})_{\Sigma=8.00}O_{22}(OH)_2$.

Polymorphism & Series: Forms a series with richterite.

Mineral Group: Amphibole (sodic-calcic) group: $Mg/(Mg + Fe^{2+}) < 0.5$; $(Na + K)_A \geq 0.5$; $0.67 \leq Na_B \leq 1.33$; $(Ca + Na)_B \geq 1.34$; Si ≥ 7.5.

Occurrence: A rare accessory mineral in sodium-rich syenites and their volcanic equivalents; replacing ferromagnesian minerals in granite.

Association: Feldspar, pyroxenes, zircon, fayalite, ferro-actinolite.

Distribution: From the Baie-des-Moutons complex, La Tabatière, Quebec, Canada. At the Kangerdlugssuaq Fjord, Greenland. From the Tibchi ring complex, Nigeria.

Name: For the *ferrous* iron in its composition and similarity to *richterite*.

Type Material: n.d.

References: (1) Deer, W.A., R.A. Howie, and J. Zussman (1963) Rock-forming minerals, v. 2, chain silicates, 352–358. (2) Charles, R.W. (1974) The physical properties of the Mg–Fe richterites. Amer. Mineral., 59, 518–528. (3) Lalonde, A.E. and R.F. Martin (1983) The Baie-des-Moutons syenitic complex, La Tabatière, Québec: II. The ferromagnesian minerals. Can. Mineral., 21, 81–91. (4) Brooks, C.K. and R.C.O. Gill (1982) Compositional variation in the pyroxenes and amphiboles of the Kangerdlugssuaq intrusion, East Greenland: further evidence for the crustal contamination of syenite magma. Mineral. Mag., 45, 1–9. (5) Ike, E.C., P. Bowden, and R.F. Martin (1985) Amphibole in the porphyries of the Tibchi anorogenic ring-complex, Nigeria: product of deuteric adjustments. Can. Mineral., 23, 447–456.

Crystal Data: Orthorhombic. *Point Group:* $2/m\ 2/m\ 2/m$. As coarse anhedral to subhedral grains, up to 5 mm. *Twinning:* On {100}.

Physical Properties: *Cleavage:* Good on {210}, (210) ∧ (2$\bar{1}$0) ∼88°; partings on {100}, {010}. Hardness = 5–6   D(meas.) = 3.96   D(calc.) = [3.87]

Optical Properties: Semitransparent. *Color:* Green, dark brown; greenish or reddish in thin section.
Optical Class: Biaxial (–). *Orientation:* X = b. *Dispersion:* $r < v$, strong. $\alpha = 1.755$–1.768 $\beta = 1.763$–1.770 $\gamma = 1.772$–1.788   2V(meas.) = 55°–90°

Cell Data: *Space Group:* Pbca.   a = 18.431   b = 9.080   c = 5.238   Z = 8

X-ray Powder Pattern: Oribi Gorge, South Africa. (ICDD 19-607).
3.21 (100), 2.892 (75), 2.504 (55), 2.577 (50), 2.126 (50), 2.979 (45), 4.58 (35)

Chemistry:

	(1)
SiO_2	44.43
TiO_2	0.12
Al_2O_3	2.96
Fe_2O_3	0.70
FeO	44.91
MnO	1.20
MgO	3.38
CaO	1.69
Na_2O	0.07
K_2O	0.05
Total	99.51

(1) Bauchi, Nigeria; corresponds to $(\text{Fe}^{2+}_{1.60}\text{Mg}_{0.21}\text{Ca}_{0.08}\text{Mn}_{0.04}\text{Fe}^{3+}_{0.02}\text{Na}_{0.01})_{\Sigma=1.96}$ $(\text{Si}_{1.90}\text{Al}_{0.10})_{\Sigma=2.00}\text{O}_6$.

Polymorphism & Series: Dimorphous with clinoferrosilite; forms a series with enstatite.

Mineral Group: Pyroxene group.

Occurrence: In medium- to high-grade metamorphosed iron formations.

Association: Magnetite, hematite, ferroan diopside, quartz, almandine.

Distribution: From Bauchi, Nigeria. At Tiris, Mauritania. From Oribi Gorge, Marble Delta, Natal, South Africa. In the USA, in Montana, around Copper Mountain and Carmichael Creek, in the Tobacco Root Mountains; in the Ruby Creek area, in the Gravelly Range, Madison Co.; also the Carter Creek area, in the Ruby Mountains, Beaverhead and Madison Cos.; and in Utah, at Hanksville, Wayne Co. On Arcedeckne Island, District of Franklin, Arctic Canada.

Name: For FERROus iron and SILicon in the composition; originally a "normative" mineral in rocks.

References: (1) Deer, W.A., R.A. Howie, and J. Zussman (1978) Rock-forming minerals, (2nd edition), v. 2A, single-chain silicates, 20–161. (2) Turnock, A.C., D.H. Lindsley, and J.E. Grover (1973) Synthesis and unit cell parameters of Ca–Mg–Fe pyroxenes. Amer. Mineral., 58, 50–59. (3) Sueno, S., M. Cameron, and C.T. Prewitt (1976) Orthoferrosilite: high temperature crystal chemistry. Amer. Mineral., 61, 38–53.

Ferro-tschermakite

$$Ca_2[(Fe^{2+}, Mg)_3Al_2](Si_6Al_2)O_{22}(OH)_2$$

Crystal Data: Monoclinic. *Point Group: 2/m.* [Prismatic.] *Twinning:* [Simple or multiple twinning ∥ {100}.]

Physical Properties: *Cleavage:* [Perfect on {110}, with intersections at ∼56° and ∼124°; partings on {100}, {001}.] *Tenacity:* [Brittle.] Hardness = [5–6] D(meas.) = 3.323–3.353 D(calc.) = [3.25]

Optical Properties: Transparent to translucent. *Color:* Green; green to tan in thin section. *Luster:* [Vitreous.]
Optical Class: Biaxial (–). *Pleochroism:* Distinct; X = light tan; Y = green; Z = blue-green. *Orientation:* $Y = b$; $Z \wedge c \simeq 7°$–14°. *Absorption:* $Z = Y > X$. $\alpha = 1.660$–1.694 $\beta = 1.680$–1.732 $\gamma = 1.688$–1.736 2V(meas.) = 45°–75°

Cell Data: *Space Group: C2/m.* a = 9.864 b = 18.130 c = 5.331 $\beta = 104.95°$ Z = 2

X-ray Powder Pattern: n.d.

Chemistry:

	(1)	(2)		(1)	(2)
SiO_2	40.51	40.43	Na_2O	1.05	0.89
TiO_2	1.42	1.25	K_2O	1.00	1.01
Al_2O_3	13.39	12.62	F	0.13	0.90
Fe_2O_3	5.32	4.24	Cl	0.04	
FeO	17.16	22.15	H_2O^+	1.78	1.76
MnO	0.35	0.28	H_2O^-	0.02	0.05
MgO	6.47	3.80	$-O = (F, Cl)_2$	0.06	0.38
CaO	11.43	11.60	Total	100.01	100.60

(1) Emeryville, New York, USA; corresponds to $(Ca_{1.87}Na_{0.31}K_{0.20})_{\Sigma=2.38}(Fe^{2+}_{2.20}Mg_{1.48}Fe^{3+}_{0.61}Al_{0.60}Ti_{0.16}Mn_{0.04})_{\Sigma=5.09}(Si_{6.20}Al_{1.80})_{\Sigma=8.00}O_{22}[(OH)_{1.81}O_{0.12}F_{0.06}Cl_{0.01}]_{\Sigma=2.00}$. (2) Lake's Grave, Australia; corresponds to $(Ca_{1.92}Na_{0.27}K_{0.20})_{\Sigma=2.39}(Fe^{2+}_{2.86}Mg_{0.87}Fe^{3+}_{0.49}Al_{0.53}Ti_{0.15}Mn_{0.04})_{\Sigma=4.94}(Si_{6.24}Al_{1.76})_{\Sigma=8.00}O_{21.75}[(OH)_{1.81}F_{0.44}]_{\Sigma=2.25}$.

Polymorphism & Series: Forms a series with tschermakite.

Mineral Group: Amphibole (calcic) group: $Mg/(Mg + Fe^{2+}) < 0.5$; $(Na + K)_A < 0.5$; $Na_B < 0.67$; $(Ca + Na)_B \geq 1.34$; $Si < 6.25$; $Ti < 0.5$.

Occurrence: A product of medium- to high-pressure metamorphism of highly aluminous rocks, producing schists, gneisses, or amphibolites; from differentiated mafic plutonic igneous rocks.

Association: Almandine, chlorite, biotite, muscovite, quartz, anorthite, apatite (metamorphic).

Distribution: From Emeryville, St. Lawrence Co., New York, USA. At Lake's Grave, Broken Hill, New South Wales, Australia.

Name: For *ferrous* iron in its composition and relation to *tschermakite*.

Type Material: n.d.

References: (1) Deer, W.A., R.A. Howie, and J. Zussman (1963) Rock-forming minerals, v. 2, chain silicates, 263–314. (2) Leake, B.E. (1968) A catalog of analyzed calciferous and subcalciferous amphiboles together with their nomenclature and associated minerals. Geol. Soc. Amer. Special Paper 98, 210 p. [analyses 882 and 885]. (3) Phillips, W.R. and D.T. Griffen (1981) Optical mineralogy, 232–234.

Crystal Data: Triclinic, pseudotetragonal. *Point Group:* $\bar{1}$ or 1. Crystals commonly distorted, flattened, pseudotetragonal, to 2 cm. *Twinning:* Possibly on {001}.

Physical Properties: *Fracture:* Uneven. Hardness = 5–5.5 D(meas.) = 3.44–3.46 D(calc.) = [3.43]

Optical Properties: Transparent to translucent. *Color:* Dark brown to golden yellow. *Streak:* White with pale brownish tint. *Luster:* Vitreous. *Optical Class:* Biaxial (–). $\alpha = 1.873–1.886$ $\beta = 1.930$ $\gamma = 1.914–1.939$ 2V(meas.) = $0°–7°$

Cell Data: *Space Group:* $P\bar{1}$ or $P1$. a = 7.210(1) b = 7.213(2) c = 20.451(3) $\alpha = 95.15(3)°$ $\beta = 95.60(2)°$ $\gamma = 89.04(5)°$ Z = [8]

X-ray Powder Pattern: Khibiny massif, Russia.
3.058 (100), 2.815 (60), 1.518 (55), 1.801 (50), 1.687 (45), 1.552 (40), 2.530 (35)

Chemistry:

	(1)
SiO_2	22.46
TiO_2	17.08
Nb_2O_5	21.79
CaO	28.80
Na_2O	6.92
F	4.23
$-O = F_2$	1.78
Total	99.50

(1) Khibiny massif, Russia; by electron microprobe, average of several analyses, H_2O in empirical analysis to sum to 13 cations; corresponds to $(Ca_{2.82}Na_{1.21})_{\Sigma=4.03}(Ti_{1.18}Nb_{0.86})_{\Sigma=2.04}Si_2O_{10.89}$ $[F_{1.15}(OH)_{0.96}]_{\Sigma=2.11}$.

Occurrence: In aegirine-rich nepheline pegmatites in a differentiated alkalic massif (Khibiny massif, Russia).

Association: Feldspar, pectolite, apatite, aegirine, lamprophyllite, rinkite, sulfides.

Distribution: In the Khibiny massif, Kola Peninsula, Russia. From Üdersdorf, Eifel district, Germany.

Name: For Academician Aleksandr Evgenevich Fersman (1883–1945), eminent Russian mineralogist, geochemist, and gemologist.

Type Material: n.d.

References: (1) Labuntsov, A.N. (1929) La fersmanite – un nouveau minéral des Monts Chibines [Khibiny massif]. Doklady Acad. Nauk SSSR, 12, 297–301 (in French). (2) (1931) Amer. Mineral., 16, 92 (abs. ref. 1). (3) Machin, M.P. (1977) Fersmanite, $(Ca, Na)_4(Ti, Nb)_2Si_2O(F, OH)_2$: a restudy. Can. Mineral., 15, 87–91. (4) Saf'yanov, Y.N., R.I. Bochkova, and V.V. Ilyukhin (1984) The crystal structure of fersmanite. Kristallografiya (Sov. Phys. Crystal.), 29, 56–59 (in Russian). (5) Vlasov, K.A., Ed. (1966) Mineralogy of rare elements, v. II, 564–566.

Feruvite

$$(Ca, Na)(Fe^{2+}, Mg)_3(Al, Mg)_6(BO_3)_3Si_6O_{18}(OH)_4$$

Crystal Data: Hexagonal. *Point Group:* $3m$. Subhedral to anhedral grains, to 2 mm.

Physical Properties: *Fracture:* Conchoidal. *Tenacity:* Brittle. Hardness = $\sim$7
D(meas.) = 3.207(9)   D(calc.) = 3.21

Optical Properties: Semitransparent. *Color:* Dark brown-black. *Streak:* Gray.
Luster: Vitreous to dull.
Optical Class: Uniaxial (–). *Pleochroism:* Strong; O = light brown; E = very dark brown.
$\omega = 1.687(1)$ $\epsilon = 1.669(1)$

Cell Data: *Space Group:* $R3m$. a = 16.012(2)   c = 7.245(2)   Z = 3

X-ray Powder Pattern: Cuvier Island, New Zealand.
2.586 (100), 2.979 (80), 4.24 (60), 4.00 (60), 3.50 (60), 2.051 (50), 6.43 (40)

Chemistry:

	(1)	(2)
SiO_2	33.33	33.33
TiO_2	2.19	2.19
B_2O_3	11.25	[9.93]
Al_2O_3	23.38	23.38
Fe_2O_3		[2.56]
FeO	13.56	[11.26]
MnO	0.07	0.07
MgO	7.80	7.80
CaO	3.30	3.30
Na_2O	1.16	1.16
K_2O	0.05	0.05
H_2O	[3.48]	[3.43]
Total	[99.57]	[98.46]

(1) Cuvier Island, New Zealand; by electron microprobe, average of five analyses,
total Fe as FeO, H_2O calculated from stoichiometry. (2) Do.; B_2O_3, Fe_2O_3, FeO, and
H_2O calculated from stoichiometry, original total given as 98.45%; corresponding to
$(Ca_{0.62}Na_{0.39}K_{0.01})_{\Sigma=1.02}(Fe^{2+}_{1.53}Mg_{1.21}Ti_{0.29}Mn_{0.01})_{\Sigma=3.04}(Al_{4.72}Mg_{0.82}Fe^{3+}_{0.34}Fe^{2+}_{0.12})_{\Sigma=6.00}$
$(BO_3)_3(Si_{5.83}Al_{0.10})_{\Sigma=5.93}O_{18}(OH)_4$.

Mineral Group: Tourmaline group.

Occurrence: By hydrothermal replacement of silicates in a pegmatitic rock.

Association: Dravite, quartz, microcline, chlorapatite, pyrite.

Distribution: On Cuvier Island, New Zealand.

Name: For iron, FERrum, in the formula and the similarity to *uvite*.

Type Material: Canadian Museum of Nature, Ottawa, Canada.

References: (1) Grice, J.D. and G.W. Robinson (1989) Feruvite, a new member of
the tourmaline group, and its crystal structure. Can. Mineral., 27, 199–203. (2) (1990)
Amer. Mineral., 75, 706–707 (abs. ref. 1).

Crystal Data: Tetragonal. *Point Group: 4/m 2/m 2/m.* Crystals tabular to prismatic, to 20 cm, commonly pseudocubic with {100}, {001}, and modified by {111}; prism zone deeply striated ∥ [001]; granular. *Twinning:* Rare on {111}.

Physical Properties: *Cleavage:* Perfect on {001}, imperfect on {110}. *Fracture:* Uneven. *Tenacity:* Brittle. Hardness = 4.5–5   D(meas.) = 2.33–2.37   D(calc.) = 2.37

Optical Properties: Transparent to translucent. *Color:* Colorless, white, pink, pale, yellow, green; in thin section, colorless. *Luster:* Vitreous, pearly on {001}. *Optical Class:* Uniaxial (+). *Dispersion:* High, may be anomalous. $\omega = 1.530$–1.536 $\epsilon = 1.532$–1.538

Cell Data: *Space Group: P4/mnc.* a = 8.963(2)   c = 15.804(2)   Z = 2

X-ray Powder Pattern: St. Andreasberg, Germany. (ICDD 19-82).
7.81 (100), 2.49 (90), 3.57 (60), 3.90 (50), 2.95 (50b), 4.51 (45), 3.17 (40)

Chemistry:

	(1)	(2)
SiO$_2$	51.98	52.98
Al$_2$O$_3$	0.36	
CaO	25.10	24.73
K$_2$O	4.45	5.19
Na$_2$O	0.61	
F	2.20	2.09
H$_2$O	16.25	15.89
$-$O = F$_2$	0.92	0.88
Total	100.03	100.00

(1) Phoenix mine, Keweenaw Co., Michigan, USA; corresponding to $(K_{0.85}Na_{0.16})_{\Sigma=1.01}$ $Ca_{4.05}(Si_{7.83}Al_{0.07})_{\Sigma=7.90}O_{20}F_{1.05} \cdot 8.1H_2O$. (2) $KCa_4Si_8O_{20}F \cdot 8H_2O$.

Polymorphism & Series: Forms a series with hydroxyapophyllite.

Occurrence: A secondary mineral in amygdules or druses in basalts; in cavities in granite; in tactite and other metamorphic rocks; a late-stage hydrothermal mineral in some mineral deposits.

Association: Zeolites, datolite, pectolite, calcite, quartz.

Distribution: Many localities worldwide, but distinction from hydroxyapophyllite requires chemical analysis. Fine specimens with large crystals from the Teigarhorn, Berufjord, Iceland. On the Alpe di Siusi, Trentino-Alto Adige, Italy. At St. Andreasberg, Harz Mountains, Germany. From Korsnäs, Finland. Exceptional groups from India, in the Poona, Nasik, and Bombay districts, Maharashtra. In the USA, from French Creek, Chester Co., and Cornwall, Lebanon Co., Pennsylvania; at Paterson, Passaic Co., and at Bergen Hill, Hudson Co., New Jersey; from the Fairfax quarry, Centreville, Fairfax Co., Virginia. Very large crystals from near Bento Gonçalves, Rio Grande do Sul, Brazil.

Name: For the preponderance of *fluorine* in the composition, and *apophyllite*, from the Greek for *away from* and *leaf*, in allusion to its manner of exfoliating on heating.

References: (1) Dana, E.S. (1892) Dana's system of mineralogy, (6th edition), 566–569. (2) Deer, W.A., R.A. Howie, and J. Zussman (1963) Rock-forming minerals, v. 3, sheet silicates, 258–266. (3) Colville, A.A., C.P.A. Anderson, and P.M. Black (1971) Refinement of the crystal structure of apophyllite. I. X-ray diffraction and physical properties. Amer. Mineral., 56, 1222–1233. (4) Bartl, H. and G. Pfeifer (1976) Neutronenbeugungsanalyse des Apophyllit $KCa_4(Si_4O_{10})_2(F/OH) \cdot 8H_2O$. Neues Jahrb. Mineral., Monatsh., 58–65 (in German with English abs.). (5) Dunn, P.J. and W.E. Wilson (1978) Nomenclature revisions in the apophyllite group: hydroxyapophyllite, apophyllite, fluorapophyllite. Mineral. Record, 3, 95–98.

Crystal Data: Hexagonal. *Point Group:* $6/m$. As acicular or hexagonal prismatic, poorly-terminated crystals, to 3 mm, and as fine-grained aggregates.

Physical Properties: *Fracture:* Conchoidal. Hardness = 4.5 D(meas.) = 3.03(1) D(calc.) = 3.090

Optical Properties: Transparent to translucent. *Color:* Blue to pale bluish green; colorless in thin section. *Streak:* White with weak bluish tint. *Luster:* Vitreous to greasy. *Optical Class:* Uniaxial (–). $\omega = 1.638(2)$ $\epsilon = 1.632(2)$

Cell Data: *Space Group:* $P6_3/m$. a = 9.485(2) c = 6.916(2) Z = 4

X-ray Powder Pattern: Kopeysk, Russia.
2.84 (100), 2.74 (90), 1.852 (80), 1.954 (70), 1.729 (70), 2.80 (60), 2.28 (60)

Chemistry:

	(1)
SiO_2	15.30
Al_2O_3	1.84
Fe_2O_3	0.11
MnO	0.18
MgO	1.38
CaO	55.00
Na_2O	0.33
K_2O	0.1
F	3.60
H_2O^+	0.30
CO_2	0.66
P_2O_5	1.31
SO_3	20.75
$-O = F_2$	1.52
Total	[99.34]

(1) Kopeysk, Russia; after correction for impurities, MgO, Al_2O_3, 4% CaO, Fe_2O_3, Na_2O, and H_2O^+, original total given as 99.24%; corresponds to $(Ca_{9.96}Mn_{0.04})_{\Sigma=10.00}[(SO_4)_{2.84}(SiO_4)_{2.80}(PO_4)_{0.20}(CO_3)_{0.16}]_{\Sigma=6.00}F_{2.08}$.

Polymorphism & Series: Forms a series with hydroxylellestadite.

Occurrence: Formed in burned fragments of petrified wood in coal dumps (Kopeysk, Russia).

Association: Lime, periclase, magnesioferrite, hematite, srebrodolskite, anhydrite.

Distribution: From mines around Kopeysk, Chelyabinsk coal basin, Southern Ural Mountains, Russia. From the Bellerberg volcano, two km north of Mayen, Eifel district, Germany. At Crestmore, Riverside Co., California, and Franklin, Sussex Co., New Jersey, USA.

Name: For *fluorine* in the chemical composition and analogy to *ellestadite*.

Type Material: A.E. Fersman Mineralogical Museum, Academy of Sciences, Moscow, Russia.

References: (1) Chesnokov, B.V., L.F. Bazhenova, and A.F. Bushmakin (1987) Fluorellestadite $Ca_{10}[(SO_4), (SiO_4)]_6F_2$ – a new mineral. Zap. Vses. Mineral. Obshch., 116, 743–746 (in Russian). (2) (1989) Amer. Mineral., 74, 502–503 (abs. ref. 1). (3) Rouse, R.C. and P.J. Dunn (1982) A contribution to the crystal chemistry of ellestadite and the silicate sulfate apatites. Amer. Mineral., 67, 90–96.

Crystal Data: Hexagonal. *Point Group: 3m.* Crystals prismatic, elongated and striated ∥ [0001], with triangular cross section.

Physical Properties: *Fracture:* Irregular. *Tenacity:* Brittle. Hardness = ~7 D(meas.) = 3.17   D(calc.) = 3.14

Optical Properties: Translucent in thin splinters. *Color:* Bluish black. *Streak:* Grayish white. *Luster:* Vitreous.
Optical Class: Uniaxial (–). *Pleochroism:* Strong; O = pale lavender; E = dark blue. $\omega = 1.664(1)$ $\epsilon = 1.642(1)$

Cell Data: *Space Group: R3m.* $a = 15.967(2)$ $c = 7.126(1)$ Z = 3

X-ray Powder Pattern: "Southern California," USA.
2.573 (100), 3.452 (91), 6.338 (84), 2.944 (71), 4.212 (48), 3.989 (38), 2.038 (29)

Chemistry:

	(1)
SiO_2	35.90
B_2O_3	[10.37]
Al_2O_3	34.90
FeO	11.45
MnO	1.71
MgO	0.21
CaO	0.03
Li_2O	[0.31]
Na_2O	0.75
H_2O	[3.56]
Total	[99.19]

(1) "Southern California," USA; by electron microprobe, average of 10 analyses; Ti, Cu, K, F not detected, B_2O_3, Li_2O, and H_2O from stoichiometry to fill their respective sites; corresponds to $Na_{0.25}(Fe_{1.60}Al_{0.89}Mn_{0.24}Li_{0.22}Mg_{0.05})_{\Sigma=3.00}Al_{6.00}(BO_3)_3Si_{6.01}O_{18}(OH)_4$.

Mineral Group: Tourmaline group.

Occurrence: Probably in granite pegmatites.

Association: The original specimens are loose crystals without matrix.

Distribution: Found as museum specimens designated only as from "southern California," USA. [White Queen mine, Pala district, San Diego Co., California, USA.]

Name: To honor Franklin F. Foit, Jr. (1942–), of Washington State University, Pullman, Washington, USA, for his work on tourmaline group minerals.

Type Material: Canadian Museum of Nature, Ottawa, Canada, 81512.

References: (1) MacDonald, D.J., F.C. Hawthorne, and J.D. Grice (1993) Foitite, $\square[Fe_2^{2+}(Al, Fe^{3+})]Al_6Si_6O_{18}(BO_3)_3(OH)_4$, a new alkali-deficient tourmaline: description and crystal structure. Amer. Mineral., 78, 1299–1303.

Crystal Data: Orthorhombic. *Point Group:* $2/m \; 2/m \; 2/m$. In euhedral or subhedral crystals, typically thick, with striations $\parallel$ elongation, with wedge-shaped terminations, to 17 cm. Commonly granular or compact massive. *Twinning:* On $\{100\}$, $\{011\}$, $\{012\}$.

Physical Properties: *Cleavage:* $\{010\}$, $\{100\}$, imperfect. *Fracture:* Conchoidal. *Tenacity:* Brittle. Hardness = 7 D(meas.) = 3.275 D(calc.) = 3.271

Optical Properties: Transparent to translucent. *Color:* Green, yellowish, lemon-yellow, white, grayish, blue-gray; colorless in thin section. *Streak:* White. *Luster:* Vitreous. *Optical Class:* Biaxial (+). *Orientation:* $X = b$; $Y = c$; $Z = a$. *Dispersion:* $r > v$. $\alpha = 1.635$ $\beta = 1.651$ $\gamma = 1.670$ 2V(meas.) = 82°

Cell Data: *Space Group:* *Pbnm* (synthetic). $a = 4.7540$ $b = 10.1971$ $c = 5.9806$ $Z = 4$

X-ray Powder Pattern: Ojamo, Finland. (ICDD 7-74). 2.458 (100), 3.883 (70), 2.512 (70), 2.768 (60), 5.10 (50), 2.269 (40), 2.250 (30)

Chemistry:

	(1)
SiO_2	41.72
FeO	1.11
MgO	57.83
Total	100.66

(1) Mogok district, Myanmar; corresponds to $(Mg_{2.02}Fe^{2+}_{0.02})_{\Sigma=2.04}Si_{0.98}O_4$.

Polymorphism & Series: Trimorphous with ringwoodite and wadsleyite; forms two series, with fayalite, and with tephroite.

Mineral Group: Olivine group.

Occurrence: In mafic and ultramafic igneous rocks and thermally metamorphosed impure dolomitic limestones.

Association: Enstatite, plagioclase, phlogopite, magnetite, chromite, antigorite, dolomite, brucite, diopside, corundum, amphiboles, calcite, spinel, augite.

Distribution: Localities for near end member material or good crystals include: at Monte Somma and Vesuvius, Campania, Italy. From Ojamo, Lohja, Finland. At Møre og Ramsdel, Norway. Large crystals from the Kovdor massif, Kola Peninsula, and at Zlatoust and Mt. Itkul, Ural Mountains, Russia. In the Eifel district, Germany, at Forstberg. In the USA, in Arizona, on Peridot Mesa, four km southwest of San Carlos, Gila Co., and at Buell Park, Apache Co. In the Mogok district and at Pyaung Gaung, Myanmar (Burma). Exceptional crystals from Suppatt, near Basham, North-West Frontier Province, Pakistan. At Kingiti, Mpapwa, Tanzania. Large gem crystals from the Island of Zabargad (Zabirget or St. Johns), in the Red Sea, Egypt.

Name: After Adolarius Jacob Forster (1739–1806), English mineral collector and dealer.

References: (1) Dana, E.S. (1892) Dana's system of mineralogy, (6th edition), 450–451, 451–456 [chrysolite]. (2) Deer, W.A., R.A. Howie, and J. Zussman (1982) Rock-forming minerals, (2nd edition), v. 1A, orthosilicates, 3–336. (3) Hazen, R.M. (1976) Effects of temperature and pressure on the crystal structure of forsterite. Amer. Mineral., 61, 1280–1293.

Crystal Data: Monoclinic. *Point Group:* $2/m$, 2, or m. Crystals elongated along [010], to 5 cm; fibrous and in compact masses.

Physical Properties: *Cleavage:* Distinct on {001}. *Tenacity:* Brittle. Hardness = 3 when massive. D(meas.) = 2.73 D(calc.) = 2.74

Optical Properties: Translucent. *Color:* White; colorless in thin section. *Luster:* Silky. *Optical Class:* Biaxial (+). *Orientation:* Positive elongation, parallel extinction. $\alpha = 1.594$ $\beta = 1.594$ $\gamma = 1.598$ 2V(meas.) = $\sim$60°

Cell Data: *Space Group:* A-centered. $a = 10.32$ $b = 7.36$ $c = 14.07$ $\beta = 106.4°$ $Z = 4$

X-ray Powder Pattern: Crestmore, California, USA.
2.92 (vvs), 1.74 (vs,b), 6.8 (ms), 4.95 (ms), 3.37 (ms), 2.30 (ms), 2.16 (ms)

Chemistry:

	(1)	(2)
SiO$_2$	36.60	42.66
(Al, Fe)$_2$O$_3$	0.50	
MgO	2.61	
CaO	48.91	53.08
H$_2$O	8.88	4.26
CO$_2$	2.67	
Total	100.17	100.00

(1) Crestmore, California, USA. (2) Ca$_4$Si$_3$O$_9$(OH)$_2$.

Occurrence: In thin veins in thermally altered limestone (Crestmore, California, USA; Kilchoan, Scotland) or melilite skarn (Dupezeh Mountain, Iraq).

Association: Hillebrandite, calcite, vesuvianite, garnet, thaumasite (Crestmore, California, USA); merwinite, larnite, kilchoanite (Kilchoan, Scotland); perovskite, grossular, schorlomite, monticellite, wollastonite, phlogopite, spinel, cuspidine, baddeleyite, baghdadite, pyrrhotite, djerfisherite, valleriite (Dupezeh Mountain, Iraq).

Distribution: In the USA, at Crestmore, Riverside Co., California. From near Kilchoan, Ardnamurchan, Argyllshire, Scotland. On Dupezeh Mountain, near Hero Town, Qala-Diza region, Iraq. From the Hatrurim Formation, Israel. At Kushiro, Hiroshima Prefecture, and in the Akagané mine, Iwate Prefecture, Japan. In the Wessels mine, near Kuruman, Cape Province, South Africa.

Name: For William Frederick Foshag (1894–1956), Curator of the Smithsonian mineral collections, Washington, D.C., USA, who studied Crestmore minerals.

Type Material: National Museum of Natural History, Washington, D.C., USA, 95229.

References: (1) Eakle, A.S. (1925) Foshagite, a new silicate from Crestmore, California. Amer. Mineral., 10, 97–99. (2) Heller, L. and H.F.W. Taylor (1956) Crystallographic data for the calcium silicates. H.M. Stationary Office, London, 53–56. (3) Gard, J.A. and H.F.W. Taylor (1958) Foshagite: Composition, unit cell and dehydration. Amer. Mineral., 43, 1–15. (4) Gard, J.A. and H.F.W. Taylor (1960) The crystal structure of foshagite. Acta Cryst., 13, 785–793.

Crystal Data: Monoclinic or orthorhombic.   *Point Group:* n.d.   Crystals, scaly and thin plates, tabular on {100} and elongated and striated parallel [001]; as radial and spheroidal aggregates.

Physical Properties: *Cleavage:* Perfect on {100}.   Hardness = 2.5–3   D(meas.) = 2.5   D(calc.) = n.d.

Optical Properties: Semitransparent.   *Color:* White.
Optical Class: Biaxial (–).   *Dispersion:* $r < v$. $\alpha = 1.535$ $\beta = 1.542$ $\gamma = 1.549$
2V(meas.) = 12°–18°

Cell Data: *Space Group:* n.d.   Z = n.d.

X-ray Powder Pattern: n.d.

Chemistry:

	(1)	(2)
SiO_2	32.65	35.09
RE_2O_3	1.89	
CaO	45.45	49.13
Na_2O	0.40	
H_2O^-	0.16	
H_2O		15.78
LOI	16.66	
Total	97.21	100.00

(1) Mt. Yukspor, Russia. (2) $Ca_3Si_2O_7 \cdot 3H_2O$.

Occurrence: In veins.

Association: Calcite, mesolite.

Distribution: On Mt. Yukspor, Khibiny massif, Kola Peninsula, Russia.

Name: For its relation to *foshagite* and "centrallassite" [gyrolite].

Type Material: n.d.

References: (1) Chirvinsky, P. (1936) Foshallasite [*sic*] from the Chibina-tundra [Khibiny massif]. Acad. Sci. USSR, Vernadsky [Vernadskii] Jubilee Volume, 757 (in Russian). (2) (1938) Amer. Mineral., 23, 667 (abs. ref. 1). (3) (1938) Mineral. Abs., 7, 10 (abs. ref. 1). (4) Heller, L. and H.F.W. Taylor (1956) Crystallographic data for the calcium silicates. H.M. Stationary Office, London, 50.

Crystal Data: Monoclinic. *Point Group:* n.d. Fibrous to porcelaneous massive.

Physical Properties: Hardness = 3.5–4   D(meas.) = 3.08–3.10   D(calc.) = 3.44–3.54

Optical Properties: Opaque to translucent. *Color:* Yellowish white to bluish. *Streak:* White to pale green. *Luster:* Silky.
Optical Class: Biaxial (–). $n = 1.61$–1.62   2V(meas.) = 15°–20°

Cell Data: *Space Group:* n.d. $a = 5.372$ $b = 9.246$ $c = 7.273$ $\beta = 103°33'$ Z = 2

X-ray Powder Pattern: Vieille-Montagne, Belgium.
7.03 (100), 3.53 (70), 2.650 (40), 2.490 (30), 2.385 (30), 1.540 (30), 4.60 (20)

Chemistry:

	(1)	(2)	(3)
SiO_2	20.86	24.8	32.78
Al_2O_3	16.62	17.3	21.71
Fe_2O_3	0.52		
CuO	0.50	5.2	16.80
ZnO	47.04	40.95	14.50
MgO	0.07		0.24
CaO	0.35	0.34	0.37
K_2O			0.08
H_2O^+	13.13	[12.8]	[13.52]
H_2O^-	1.13		
Total	100.22	[101.39]	[100.00]

(1) Vieille-Montagne, Belgium; corresponds to $(Zn_{2.29}Al_{0.66}Cu_{0.03}Mg_{0.01})_{\Sigma=2.99}$ $(Si_{1.37}Al_{0.63})_{\Sigma=2.00}O_5(OH)_{4.01}$. (2) Silver Bill mine, Arizona, USA; by electron microprobe, H_2O by loss on ignition; corresponds to $(Zn_{1.84}Al_{0.77}Cu_{0.24})_{\Sigma=2.85}(Si_{1.51}Al_{0.49})_{\Sigma=2.00}O_5(OH)_4$. (3) Mohawk mine, California, USA; H_2O by difference, corresponds to $(Al_{1.18}Cu_{0.69}Zn_{0.58})_{\Sigma=2.45}$ $(Si_{1.79}Al_{0.21})_{\Sigma=2.00}O_5(OH)_4$.

Mineral Group: Kaolinite-serpentine group.

Occurrence: In oxidized zinc-bearing mineral deposits.

Association: Smithsonite (Vieille-Montagne, Belgium); gebhardite, willemite, cerussite (Tsumeb, Namibia); sauconite (Silver Bill mine, Arizona, USA).

Distribution: Original locality not known with certainty, but thought to be one of the mines of Vieille-Montagne (Altenberg), near Moresnet, Belgium. From Tsumeb, Namibia. At Laurium, Greece. From the Copperthwaite vein, Swaledale, North Yorkshire, England. In the USA, in the Silver Bill mine, Gleeson, Cochise Co., Arizona; the Blanchard mine, Socorro Co., New Mexico; and the Mohawk mine, San Bernardino Co., California. From Mina Ojuela, Mapimi, Durango, Mexico.

Name: To honor Julien Jean Joseph Fraipont (1857–1910), and Charles Fraipont, of Liège, Belgium, both geologists and paleontologists.

Type Material: University of Liège, Liège, Belgium, 13727.

References: (1) Cesàro, G. (1927) Sur la fraipontite, silicate basique hydraté de zinc et d'aluminum. Ann. Soc. Geol. Belgique, 50, 106–110 (in French). (2) (1928) Amer. Mineral., 13, 492 (abs. ref. 1). (3) Fransolet, A.-M. and P. Bourguignon (1975) Données nouvelles sur la fraipontite de Moresnet (Belgique). Bull. Soc. fr. Minéral., 98, 235–244 (in French with English abs.). (4) (1977) Amer. Mineral., 62, 175 (abs. ref. 3). (5) Foord, E.E., J.E. Taggart, and N.M. Conklin (1983) Cuprian fraipontite and sauconite from the Defiance-Silver Bill mines, Gleeson, Arizona. Mineral. Record, 14, 131–132. (6) Wise, W.S. (1990) The mineralogy of the Mohawk mine, San Bernardino Co., California. San Bernardino Co. Mus. Assoc. Quarterly, 37(1), 31 pp.

Franciscanite

$$Mn_3^{2+}V_{1-x}^{5+}(SiO_4)(O, OH)_3 \quad (x = 0.5)$$

Crystal Data: Hexagonal. *Point Group:* 3. As irregular grains to 1.0 mm.

Physical Properties: *Fracture:* Uneven. Hardness = ~ 4 D(meas.) = 4.1(3) D(calc.) = 3.97 May be weakly magnetic, possibly resulting from tiny opaque inclusions.

Optical Properties: Transparent. *Color:* Cherry-red when fresh, darkening to brownish red with exposure. *Streak:* Brownish red. *Luster:* Vitreous. *Optical Class:* Uniaxial (+). *Pleochroism:* Strong; O = wine-red; E = dark red to nearly black. *Absorption:* Intense; $E > O$. $\omega = 1.856(3)$ $\epsilon = 1.882(3)$

Cell Data: *Space Group:* $P3$. a = 8.1518(3) c = 4.8091(2) Z = 2

X-ray Powder Pattern: Pennsylvania mine, California, USA.
2.331 (100), 3.105 (90), 2.844 (90), 1.785 (70), 1.538 (50), 2.668 (40), 3.97 (30)

Chemistry:

	(1)
SiO_2	18.4
Fe_2O_3	0.0
Al_2O_3	0.0
WO_3	0.0
Sb_2O_5	0.0
As_2O_5	0.0
V_2O_5	13.5
MnO	64.9
MgO	0.0
H_2O	[3.4]
Total	[100.2]

(1) Pennsylvania mine, California, USA; by electron microprobe, H_2O by analogy to welinite; corresponds to $Mn_{2.98}V_{0.49}(SiO_4)(O, OH)_{\sim 3}$.

Occurrence: As sparse, irregular segregations within a sheared sonolite-bearing assemblage in chert.

Association: Sonolite, hausmannite, braunite, gageite.

Distribution: From the Pennsylvania mine, San Antonio Valley, Santa Clara Co., California, USA.

Name: For the Franciscan complex of California, USA, in which it was found.

Type Material: National Museum of Natural History, Washington, D.C., USA, 163041.

References: (1) Dunn, P.J., D.R. Peacor, R.C. Erd, and R.A. Ramik (1986) Franciscanite and örebroite, two new minerals from California and Sweden, related to redefined welinite. Amer. Mineral., 71, 1522–1526. (2) Pertlik, F. (1986) The crystal structure of franciscanite, $Mn_3(V_x\square_{1-x})(SiO_4)(O, OH)_3$, [x $\cong$ 0.5]. Neues Jahrb. Mineral., Monatsh., 493–499.

$Ca_2(Fe^{3+}, Al)Mn^{3+}Mn_3^{2+}Zn_2Si_2O_{10}(OH)_8$ **Franklinfurnaceite**

Crystal Data: Monoclinic. *Point Group:* 2. As very thin, platy crystals, tabular on {001}, to 0.3 mm, with dominant {001} showing serrated edges; as polycrystalline aggregates of subparallel individuals.

Physical Properties: *Cleavage:* Perfect on {001}. *Tenacity:* Extremely brittle. Hardness = 3   D(meas.) = 3.66   D(calc.) = 3.737

Optical Properties: Translucent. *Color:* Dark brown; in thin section, brown to very dark brown. *Streak:* Brown. *Luster:* Vitreous.
Optical Class: Biaxial (–). *Pleochroism:* Intense; X = very dark brown; Y = brown; Z = deep brown. *Orientation:* Z = b; Y ∧ c = 29°. *Dispersion:* r < v, moderate. *Absorption:* Strong; X ≫ Z > Y. α = 1.792(4) β = 1.798(4) γ = 1.802(4) 2V(meas.) = 79° 2V(calc.) = 78.5°

Cell Data: *Space Group:* C2. a = 5.483(7) b = 9.39(3) c = 14.51(1) β = 97.04(8)°
Z = 2

X-ray Powder Pattern: Franklin, New Jersey, USA.
2.305 (100), 2.707 (80b), 2.602 (70b), 14.4 (50), 3.35 (50), 3.60 (40), 3.20 (40)

Chemistry:

	(1)	(2)
SiO_2	14.5	15.0
Al_2O_3	0.8	0.4
Fe_2O_3	9.4	9.3
Mn_2O_3	9.5	10.6
MnO	17.0	19.0
ZnO	22.7	24.1
MgO	3.6	0.6
CaO	14.3	12.6
H_2O	[8.2]	8.4
Total	[100.0]	100.0

(1) Franklin, New Jersey, USA; by electron microprobe, Fe_2O_3 shown present by microchemical tests, $Mn^{2+}:Mn^{3+}$ from crystal structure, H_2O by difference; corresponds to $Ca_2(Fe_{0.76}^{3+}Al_{0.24})_{\Sigma=1.00}(Mn_{0.97}^{3+}Mg_{0.03})_{\Sigma=1.00}(Mn_{0.54}^{2+}Mg_{0.28}Zn_{0.18})_{\Sigma=2.00}Zn_2Si_2O_{10}(OH)_8$.
(2) Do; H_2O by TGA, corresponds to $Ca_{1.88}(Fe_{0.97}^{3+}Al_{0.06})_{\Sigma=1.03}Mn_{1.12}^{3+}Mn_{2.24}^{2+}Mg_{0.12}Zn_{2.47}Si_{2.08}O_{10.22}(OH)_{7.78}$.

Occurrence: A late-stage mineral in vugs in a metamorphosed stratiform zinc deposit.

Association: Willemite, clinohedrite, hodgkinsonite, hetaerolite, franklinite, barite, rhodonite.

Distribution: From Franklin, Sussex Co., New Jersey, USA.

Name: After Franklin Furnace, the former name for Franklin, New Jersey, USA.

Type Material: National Museum of Natural History, Washington, D.C., USA, C6309, R19144.

References: (1) Dunn, P.J., D.R. Peacor, R.A. Ramik, S.-C. Su, and R.C. Rouse (1987) Franklinfurnaceite, a $Ca-Fe^{3+}-Mn^{3+}-Mn^{2+}$ zincosilicate isotypic with chlorite, from Franklin, New Jersey. Amer. Mineral., 72, 812–815. (2) Peacor, D.R., R.C. Rouse, and S.W. Bailey (1988) Crystal structure of franklinfurnaceite: a tri-dioctahedral zincosilicate intermediate between chlorite and mica. Amer. Mineral., 73, 876–887.

Franklinphilite

$$(K, Na)_4(Mn^{2+}, Mg, Zn)_{48}$$
$$(Si, Al)_{72}(O, OH)_{216} \cdot 6H_2O$$

Crystal Data: Triclinic, pseudohexagonal. *Point Group:* 1 or $\bar{1}$. Platy crystals, to 0.5 mm, in very fine-grained tangled radial aggregates.

Physical Properties: *Cleavage:* Imperfect on {001}. *Tenacity:* Brittle. Hardness = 4 D(meas.) = 2.6–2.8 D(calc.) = 2.66

Optical Properties: Translucent to nearly opaque; transparent to translucent in thin section. *Color:* Very dark brown to black. *Streak:* Light brown. *Luster:* Vitreous to slightly resinous. *Optical Class:* Biaxial (–). *Pleochroism:* Distinct; X = pale yellow; $Y = Z$ = deep brown. *Orientation:* $X \wedge (001) \simeq 6°$. $\alpha = 1.545(5)$ $\beta = 1.583(3)$ $\gamma = 1.583(3)$ 2V(meas.) = 10(3)° 2V(calc.) = 0°

Cell Data: *Space Group:* $P1$ or $P\bar{1}$; orthohexagonal cell is $a = 5.521(4)$ $b = 9.560(6)$ $c = 36.57(5)$ $Z = 3/8$

X-ray Powder Pattern: Franklin, New Jersey, USA.
12.3 (100), 2.583 (40), 2.737 (30), 2.362 (30), 1.594 (30), 1.580 (30), 4.79 (20)

Chemistry:

	(1)
SiO_2	44.0
Al_2O_3	3.6
Fe_2O_3	7.8
MnO	22.3
MgO	6.4
ZnO	5.9
Na_2O	0.4
K_2O	1.5
H_2O	[8.1]
Total	[100.0]

(1) Franklin, New Jersey, USA.; by electron microprobe, H_2O by difference, corresponding to $(K_{2.6}Na_{1.1})_{\Sigma=3.7}(Mn_{26.1}Mg_{13.2}Zn_{6.0}Fe^{3+}_{2.7})_{\Sigma=48.0}(Si_{60.8}Al_{5.8}Fe^{3+}_{5.4})_{\Sigma=72.0}$ $[O_{163.2}(OH)_{52.8}]_{\Sigma=216.0} \cdot nH_2O$.

Polymorphism & Series: Forms a series with lennilenapeite.

Occurrence: In a veinlet crosscutting a breccia of calcite, aegirine, and serpentine in a metamorphosed stratiform zinc deposit.

Association: Friedelite, nelenite, rhodonite, tirodite.

Distribution: At Franklin, Sussex Co., New Jersey, USA.

Name: From the locality Franklin, New Jersey, USA, and the Greek for *friend*, for the association of elements characteristic of Franklin.

Type Material: National Museum of Natural History, Washington, D.C., USA, 167390.

References: (1) Dunn, P.J., D.R. Peacor, and S.-C. Su (1992) Franklinphilite, the manganese analog of stilpnomelane, from Franklin, New Jersey. Mineral. Record, 23, 465–468. (2) (1993) Amer. Mineral., 78, 672–673 (abs. ref. 1).

Crystal Data: Hexagonal. *Point Group:* $\bar{3}\, 2/m$, $3m$, or 32. As squat prisms, to 1 cm, roughly hexagonal in shape, not of measurable quality.

Physical Properties: *Cleavage:* {0001}, distinct. Hardness = 5   D(meas.) = 2.46–2.52 D(calc.) = 2.52–2.57

Optical Properties: Transparent to opaque. *Color:* White to colorless. *Luster:* Pearly. *Optical Class:* Uniaxial (+). $\omega = 1.504$–1.510 $\epsilon = 1.506$–1.512

Cell Data: *Space Group:* $P\bar{3}m1$, $P3m1$, or $P321$. $a = 12.884(9)$ $c = 26.580(21)$ $Z = 1$

X-ray Powder Pattern: Pitigliano, Italy.
3.72 (100), 3.59 (43), 3.81 (42), 3.56 (39), 2.148 (29), 3.302 (17), 3.054 (16)

Chemistry:

	(1)	(2)		(1)	(2)
SiO_2	32.44	31.85	K_2O	4.24	7.23
Al_2O_3	25.21	25.13	Cl	0.36	0.13
Fe_2O_3	0.04	0.10	H_2O^+	1.88	1.70
MgO	0.14	0.26	CO_2	1.54	1.98
CaO	12.08	10.44	SO_3	10.65	10.22
Na_2O	11.50	10.99	$-O = Cl_2$	0.08	0.03
			Total	[100.00]	[100.00]

(1) Pitigliano, Italy; by a variety of techniques including AA and XRF, recalculated to 100%; corresponding to $(Na_{4.31}Ca_{2.50}Mg_{0.04}Fe_{0.01})_{\Sigma=6.86}(Si_{6.26}Al_{5.74})_{\Sigma=12.00}O_{24}$ $[(SO_4)_{1.54}(OH)_{0.70}(CO_3)_{0.41}Cl_{0.12}]_{\Sigma=2.77} \cdot 0.86 H_2O$. (2) Ariccia, Italy; methods as for (1), corresponding to $(Na_{4.16}Ca_{2.18}Mg_{0.08}Fe_{0.01})_{\Sigma=6.43}(Si_{6.22}Al_{5.78})_{\Sigma=12.00}O_{24}[(SO_4)_{1.50}(OH)_{0.64}$ $(CO_3)_{0.53}Cl_{0.04}]_{\Sigma=2.71} \cdot 0.79 H_2O$.

Mineral Group: Cancrinite group.

Occurrence: In ejected metasomatized pumice blocks, thought to be the product of a syntectic process between a trachytic magma and carbonate rocks at the volcanic vent.

Association: Diopside, vesuvianite, afghanite, liottite (Pitigliano, Italy); calcite, leucite (Ariccia, Italy).

Distribution: In Italy, in the Pitigliano quarry, near Grosseto, Tuscany; and at Sacrofano and Ariccia, near Rome, Lazio.

Name: For Marco Franzini, Professor of Mineralogy, University of Pisa, Pisa, Italy.

Type Material: University of Pisa, Pisa, 3208; University of Modena, Modena, Italy.

References: (1) Merlino, S. and P. Orlandi (1977) Franzinite, a new mineral phase from Pitigliano (Italy). Neues Jahrb. Mineral., Monatsh., 163–167. (2) (1977) Amer. Mineral., 62, 1259 (abs. ref. 1). (3) Leoni, L., M. Mellini, S. Merlino, and P. Orlandi (1979) Cancrinite-like minerals: new data and crystal chemical considerations. Rend. Soc. Ital. Mineral. Petrol., 35, 713–719.

Crystal Data: Tetragonal. *Point Group:* 4mm. As subhedral to euhedral tetragonal crystals, slightly elongated along [001], bounded principally by {110} and {001}, to 3 mm.

Physical Properties: *Cleavage:* Fair on {001}. Hardness = 3–4 D(meas.) = 4.43(2) D(calc.) = 4.45 Fluoresces pale yellow under SW UV.

Optical Properties: Transparent to translucent. *Color:* Lemon-yellow to canary-yellow; yellow to colorless in thin section. *Streak:* White. *Luster:* Vitreous.
Optical Class: Uniaxial (–). *Pleochroism:* O = colorless; E = yellow, with anomalous blue interference colors. $\omega = 1.775(3)$ $\epsilon = 1.765$

Cell Data: *Space Group:* P4bm. a = 8.518(2) c = 5.211(1) Z = 2

X-ray Powder Pattern: Synthetic.
3.077 (100), 3.301 (45), 2.697 (25), 3.816 (20), 2.607 (20), 2.151 (20), 1.874 (20)

Chemistry:

	(1)	(2)	(3)
SiO_2	21.8	23.8	23.72
TiO_2	17.2	15.2	15.76
Al_2O_3	0.12		
FeO	1.0	0.77	
MnO	0.027	0.074	
MgO	< 0.05	0.19	
CaO	0.14	0.55	
SrO	0.28		
BaO	59.4	59.4	60.52
K_2O	0.00	0.00	
H_2O	0.00	0.00	
Total	[100.0]	[100.0]	100.00

(1–2) Rush Creek area, California, USA; by D-C arc spectrography, recalculated to 100%.
(3) $\text{Ba}_2\text{TiOSi}_2\text{O}_7$.

Occurrence: Disseminated in gneissic metamorphic rocks composed mainly of sanbornite and quartz (Rush Creek, California, USA).

Association: Quartz, sanbornite, celsian, taramellite, diopside, pyrrhotite (Rush Creek, California, USA); bario-orthojoaquinite, benitoite, baotite, natrolite (Esquire mine #1, California, USA); gillespite, sanbornite, taramellite, pellyite, muirite, barite (Gunn claim, Canada).

Distribution: In the USA, in California, from the Rush Creek area and at the Esquire mine #1, Big Creek, Fresno Co.; at the Victor mine, Clear Creek, and the Gem mine, San Benito Co. From the Gunn claim, Itsy Mountains, near Macmillan Pass, Yukon Territory, Canada. At Graulai and Üdersdorf, Eifel district, Germany.

Name: For the original localities in Fresno Co., California, USA.

Type Material: California Division of Mines & Geology, San Francisco, California, USA.

References: (1) Alfors, J.T., M.C. Stinson, R.A. Matthews, and A. Pabst. (1965) Seven new barium minerals from eastern Fresno County, California. Amer. Mineral., 50, 314–340. (2) Moore, P.B. and S.J. Louisnathan (1969) The crystal structure of fresnoite, $\text{Ba}_2(\text{TiO})\text{Si}_2\text{O}_7$. Zeits. Krist., 130, 438–448. (3) (1971) NBS Mono. 25, 19.

Crystal Data: Monoclinic, pseudohexagonal. *Point Group: 2/m.* As pseudohexagonal tablets, may be hemimorphic, to 1 cm; more rarely as slender needles or as fibrous or lamellar aggregates; stalactitic, massive, cryptocrystalline. *Twinning:* Universal, repeated about [110] and [$\bar{1}$10], composition plane {001}, giving pseudorhombohedral symmetry.

Physical Properties: *Cleavage:* Perfect on {0001}. Hardness = 4–5 D(meas.) = 3.04–3.06 D(calc.) = 3.066

Optical Properties: Translucent to transparent. *Color:* Pale pink, brownish red, dark red, brown. *Streak:* Pale rose. *Luster:* Vitreous, pearly on basal plane. *Optical Class:* Biaxial; sensibly uniaxial (–). $\omega = 1.654$–1.656 $\epsilon = 1.620$–1.625 2V(meas.) = n.d.

Cell Data: *Space Group: C2/m.* $a = 23.33(5)$ $b = 13.396(8)$ $c = 7.447(4)$ $\beta = 105.08(8)°$ $Z = [4]$

X-ray Powder Pattern: Franklin, New Jersey, USA; nearly identical with mcgillite. 2.56 (100), 7.17 (90), 3.60 (70), 2.88 (60), 1.676 (60), 2.115 (40), 2.408 (30)

Chemistry:

	(1)	(2)	(3)
SiO_2	36.35	34.45	34.6
Al_2O_3	0.03		
FeO	5.27	0.68	3.4
MnO	46.67	51.48	48.4
ZnO	0.32	1.43	1.0
MgO	0.05	1.54	1.4
CaO	trace		0.1
Cl	3.08	0.29	2.9
H_2O	8.86	9.74	[8.9]
$-O = Cl_2$	0.69	0.06	0.7
Total	99.94	99.55	[100.0]

(1–2) Franklin, New Jersey, USA. (3) Sterling Hill, New Jersey, USA; by electron microprobe, H_2O by difference; corresponding to $(Mn_{7.01}^{2+}Fe_{0.49}Mg_{0.36}Zn_{0.13}Ca_{0.02})_{\Sigma=8.01}Si_{5.92}O_{15}$ $[(OH)_{10.15}Cl_{0.84}]_{\Sigma=10.99}$.

Occurrence: In metamorphosed Mn-Fe deposits.

Association: Rhodochrosite, alabandite (Adervielle, France); franklinite, willemite, calcite, rhodochrosite, barite, chlorite, sphalerite (Franklin, New Jersey, USA); calcite, kutnohorite, andradite, hausmannite (Kuruman, South Africa).

Distribution: In France, at Adervielle, Louron Valley, Hautes-Pyrénées, and Costabonne, Pyrénées-Orientales. From the Harstig mine, Pajsberg, near Persberg, Värmland, and in the Sjö mine, near Grythyttan, Örebro, Sweden. In the Ködnitz Valley, Tirol, Austria. At Dshumart and Kamya, Atasui region, Kazakhstan. In the USA, at Franklin and Sterling Hill, Ogdensburg, Sussex Co., New Jersey, and the Sunnyside mine, Silverton district, San Juan Co., Colorado. From the N'Chwaning mine, near Kuruman, Cape Province, South Africa. At Broken Hill, New South Wales, Australia.

Name: For the noted French chemist and mineralogist, Charles Friedel (1832–1899).

References: (1) Dana, E.S. (1892) Dana's system of mineralogy, (6th edition), 465, 1035. (2) Frondel, C. and L.H. Bauer (1953) Manganpyrosmalite and its polymorphic relation to friedclite and schallerite. Amer. Mineral., 38, 755–760. (3) Dunn, P.J., D.R. Peacor, J.A. Nelen, and J.A. Norberg (1981) Crystal-chemical data for schallerite, caryopilite and friedelite from Franklin and Sterling Hill, New Jersey. Amer. Mineral., 66, 1054–1062. (4) Ozawa, T., Y. Takéuchi, T. Takahata, G. Donnay, and J.D.H. Donnay (1983) The pyrosmalite group of minerals. II. The layer structure of mcgillite and friedelite. Can. Mineral., 21, 7–17.

Crystal Data: Orthorhombic. *Point Group:* $mm2$ or $2/m\ 2/m\ 2/m$. As flaky crystals, to 0.2 mm.

Physical Properties: Hardness = ~4 D(meas.) = 2.770(5) D(calc.) = 2.77

Optical Properties: Translucent. *Color:* White to pale brown; colorless in transmitted light. *Optical Class:* Biaxial. $\alpha = 1.59(5)$ $\beta = 1.60(6)$ $\gamma = 1.62(9)$ 2V(meas.) = $\sim90°$

Cell Data: *Space Group:* $Bm2_1b$, $B2mb$, or $Bmmb$. $a = 5.48(1)$ $b = 3.78(1)$ $c = 23.42(3)$ $Z = 2$

X-ray Powder Pattern: Fuka, Japan.
2.854 (100), 3.084 (90), 2.926 (65), 2.338 (30), 1.756 (30b), 5.86 (25), 3.904 (20)

Chemistry:

	(1)	(2)		(1)	(2)
SiO_2	29.09	28.98	K_2O	0.01	0.02
TiO_2	0.00	0.00	F	0.32	0.43
Al_2O_3	0.55	0.27	H_2O^+	4.45	4.26
Fe_2O_3	0.10	0.14	H_2O^-	0.23	0.39
MnO	0.00	0.00	CO_2	10.32	10.22
MgO	0.14	0.02	P_2O_5	0.01	0.07
CaO	54.40	54.81	$-O = F_2$	0.13	0.18
Na_2O	0.17	0.05	Total	99.66	[99.48]

(1) Fuka, Japan; corresponding to $(Ca_{3.97}Na_{0.02}Mg_{0.01}Fe_{0.01})_{\Sigma=4.01}(Si_{1.98}Al_{0.04})_{\Sigma=2.02}O_{6.03}$ $(CO_3)_{0.96}[(OH)_{2.02}F_{0.07}]_{\Sigma=2.09}$. (2) Mihara, Japan; original total given as 99.38%, corresponding to $(Ca_{4.03}Fe_{0.01}Na_{0.01})_{\Sigma=4.05}(Si_{1.99}Al_{0.02})_{\Sigma=2.01}O_{6.08}(CO_3)_{0.96}[(OH)_{1.95}F_{0.09}]_{\Sigma=2.04}$.

Occurrence: A retrograde mineral and alteration product in skarns formed from metasomatism of limestone.

Association: Cuspidine, xonotlite, calcite, spurrite, hillebrandite, scawtite, foshagite, wollastonite, fluorite, gehlenite, perovskite, grossular, hydrogrossular, vesuvianite, monticellite.

Distribution: In Japan, at Fuka, near Bicchu, and Mihara, Okayama Prefecture, and at Kushiro, Hiroshima Prefecture.

Name: For the locality where it was first discovered, Fuka, Japan.

Type Material: Department of Earth Sciences, Okayama University, Okayama, ONM-02; National Science Museum, Tokyo, Japan; National Museum of Natural History, Washington, D.C., USA, 136583.

References: (1) Henmi, C., I. Kusachi, A. Kawahara, and K. Henmi (1977) Fukalite, a new calcium carbonate silicate hydrate mineral. Mineral. J. (Japan), 8, 374–381. (2) (1978) Amer. Mineral., 63, 793 (abs. ref. 1).

Crystal Data: Metamict; monoclinic after recrystallization. *Point Group:* $2/m$. As irregular highly-fractured masses, to 2 cm.

Physical Properties: *Fracture:* Conchoidal. Hardness = [6.5–7] (by analogy to gadolinite group). D(meas.) = 4.20   D(calc.) = 4.90

Optical Properties: Translucent. *Color:* Black; in thin section, olive-green.
Luster: Vitreous.
Optical Class: Isotropic. $n = 1.78$

Cell Data: *Space Group:* $P2/c$ after recrystallization by heating at 700 °C. $a = 4.82(2)$
$b = 7.58(2)$ $c = 10.01(3)$ $\beta = 90°28(16)'$ Z = [2]

X-ray Powder Pattern: Buer, Norway; hydrothermally recrystallized at 700 °C and 2 kb for 48 hours. (ICDD 29-1409).
2.88 (100), 4.81 (90), 2.60 (80), 2.59 (80), 3.18 (70), 3.00 (70), 3.59 (60)

Chemistry:

	(1)		(1)
SiO_2	23.17	FeO	10.03
TiO_2	0.14	MnO	1.33
B_2O_3	0.55	BeO	8.83
Al_2O_3	0.05	MgO	0.33
Y_2O_3	6.78	CaO	2.67
La_2O_3	14.00	Na_2O	0.10
Ce_2O_3	21.25	K_2O	0.53
RE_2O_3	12.05	Total	[101.81]

(1) Buer, Norway; by a combination of wet chemical analysis, electron microprobe, and emission spectroscopy; RE_2O_3 = Pr_2O_3 2.95%, Nd_2O_3 4.36%, Sm_2O_3 0.39%, Eu_2O_3 0.07%, Gd_2O_3 0.92%, Tb_2O_3 0.29%, Dy_2O_3 1.01%, Ho_2O_3 0.39%, Er_2O_3 0.76%, Tm_2O_3 0.13%, Yb_2O_3 0.65%, Lu_2O_3 0.13%; corresponds to $(Ce_{0.68}La_{0.45}Y_{0.32}RE_{0.71})_{\Sigma=2.16}(Fe^{2+}_{0.74}Mn_{0.10}Mg_{0.04}Ti_{0.01})_{\Sigma=0.89}$ $(Be_{1.86}B_{0.08}Al_{0.01})_{\Sigma=1.95}Si_{2.04}O_{10}$.

Mineral Group: Gadolinite group.

Occurrence: In syenite pegmatite veins along a contact between basalt and monzonite.

Association: Aegirine, pyrochlore, zircon, apatite, titanite, pyrophanite, magnetite, loparite, chevkinite, biotite, microcline, helvite, molybdenite, albite, apophyllite, quartz, calcite.

Distribution: From a dump at Buer, nine km southeast of Skien, Bjørkedalen region, and in the Bakken quarry, Tvedalen, Norway.

Name: For its *cerium* content and relation to *gadolinite*-(Y).

Type Material: n.d.

References: (1) Segalstad, T.V. and A.O. Larsen (1978) Gadolinite-(Ce) from Skien, southwestern Oslo region, Norway. Amer. Mineral., 63, 188–195.

Gadolinite-(Y)

$Y_2Fe^{2+}Be_2Si_2O_{10}$

Crystal Data: Monoclinic; commonly metamict. *Point Group:* $2/m$. Crystals rough; typically prismatic, terminated by {001}, {111}, {221}, and many other forms, to 25 cm; commonly massive.

Physical Properties: *Fracture:* Conchoidal or splintery. *Tenacity:* Brittle. Hardness = 6.5–7 D(meas.) = 4.36–4.77 non-metamict. D(calc.) = 4.41

Optical Properties: Opaque, nearly transparent in thin splinters. *Color:* Black, greenish black, brown; grass-green to olive-green in thin splinters. *Streak:* Greenish gray. *Luster:* Vitreous to greasy.
Optical Class: Biaxial (+); isotropic when metamict. *Orientation:* $Z \wedge c = 10°$.
Dispersion: $r < v$, strong. $\alpha = 1.77$–1.78 β = n.d. $\gamma = 1.78$–1.82 2V(meas.) = 85°

Cell Data: *Space Group:* $P2_1/a$. $a = 10.000(2)$ $b = 7.565(2)$ $c = 4.768(1)$ $\beta = 90.31(2)°$ $Z = 2$

X-ray Powder Pattern: Synthetic.
2.818 (100), 2.816 (100), 4.735 (80), 2.932 (70), 2.558 (60), 2.539 (55), 3.118 (50)

Chemistry:

	(1)	(2)	(3)		(1)	(2)	(3)
SiO_2	24.35	22.13	23.36	Fe_2O_3	2.03	3.47	
ThO_2	0.30	0.89	0.00	FeO	11.39	10.43	12.08
Y_2O_3	45.96	22.24	28.55	BeO	10.17	7.19	[9.71]
$(Dy, La)_2O_3$	3.06	21.23	0.21	CaO	0.30	0.48	0.03
Ce_2O_3	1.65	11.10	2.54	Na_2O	0.17	0.46	
RE_2O_3			[21.56]	H_2O	0.52	0.86	[0.46]
				Total	99.90	100.48	[98.50]

(1) Ytterby, Sweden. (2) Douglas Co., Colorado, USA. (3) Near Yokkaichi, Mie Prefecture, Japan; by electron microprobe, $RE_2O_3 = Pr_2O_3$ [1.2%], Nd_2O_3 4.55%, Sm_2O_3 2.82%, Eu_2O_3 [0.6%], Gd_2O_3 2.93%, Tb_2O_3 [1.1%], Dy_2O_3 4.06%, Ho_2O_3 1.98%, Er_2O_3 1.54%, Tm_2O_3 [0.2%], Yb_2O_3 0.58%, Lu_2O_3 trace, BeO and H_2O from stoichiometry; corresponding to $(Y_{1.30}RE_{0.70})_{\Sigma=2.00}Fe^{2+}_{0.86}Be_{2.00}Si_{2.00}[O_{9.72}(OH)_{0.28}]_{\Sigma=10.00}$.

Mineral Group: Gadolinite group.

Occurrence: In granite and alkalic granite pegmatites.

Association: Allanite, fluorite, fergusonite, yttrialite, chevkinite, zircon.

Distribution: Numerous localities worldwide: in Sweden, from Ytterby, on Resarö Island, near Waxholm, and at Kolsva, Västmanland. On Hitterö Island, Flekkefjord; at Slobrekka and Dauren, Iveland; and elsewhere in Norway. From Piz Blas, Val Nalps, also Piz Rondadura and Pizza Rotondo, Val Bedretto, Ticino, Switzerland. In Austria, at Böckstein, near Badgastein, Salzburg. In the USA, in the Clear Creek pegmatite, Burnet Co., and the Baringer Hill pegmatite and at Rode Ranch, near Bluffton, Llano Co., Texas; in Arizona, near Hackberry, Mohave Co.; on Devil's Head Mountain, Douglas Co., and near Lake George, Park Co., Colorado. In Canada, from Loughborough Township, Frontenac Co., Ontario.

Name: For the Finnish chemist, Johan Gadolin (1760–1852), who discovered yttrium.

References: (1) Dana, E.S. (1892) Dana's system of mineralogy, (6th edition), 509–512. (2) Vlasov, K.A., Ed. (1966) Mineralogy of rare elements, v. II, 235–239. (3) Ito, J. and S.S. Hafner (1974) Synthesis and study of gadolinites. Amer. Mineral., 59, 700–708. (4) Miyawaki, R., I. Nakai, and K. Nagashima (1984) A refinement of the crystal structure of gadolinite. Amer. Mineral., 69, 948–953.

Crystal Data: Monoclinic, pseudotetragonal, or triclinic. *Point Group: 2/m* or $\bar{1}$. Minute laths or saddlelike crystals grouped radially, in bundles, or as matted fibers.

Physical Properties: *Cleavage:* {110}, pronounced. Hardness = 3 D(meas.) = 3.46–3.584 D(calc.) = 3.599

Optical Properties: Transparent to translucent. *Color:* Colorless, pale pink, pale brown. *Luster:* Highly vitreous.
Optical Class: Biaxial (–). *Orientation:* Z ∥ fiber length. *Dispersion:* $r < v$, extreme.
$\alpha = 1.723(3)$ $\beta = 1.734(3)$ $\gamma = 1.736(3)$ 2V(meas.) = n.d.

Cell Data: *Space Group: P2/n.* $a = 19.42$ $b = 19.42$ $c = 9.84$ $\beta = 89.5°$ Z = [1], or *Space Group: P$\bar{1}$.* $a = 14.17$ $b = 14.07$ $c = 9.84$ $\alpha = 76.5°$ $\beta = 76.6°$ $\gamma = 86.9°$ Z = [1]

X-ray Powder Pattern: Franklin, New Jersey, USA.
6.87 (100), 2.758 (80), 2.707 (80), 3.44 (60), 3.25 (60), 2.556 (60), 1.6742 (60)

Chemistry:

	(1)	(2)	(3)
SiO_2	24.71	23.58	23.90
Al_2O_3		0.15	
FeO		0.03	0.20
MnO	50.19	53.74	51.06
ZnO	8.76	3.96	4.30
MgO	11.91	9.95	11.34
CaO			0.19
H_2O	[4.43]	8.24	[9.01]
Total	[100.00]	99.65	[100.00]

(1) Franklin, New Jersey, USA; H_2O by difference. (2) Do. (3) Do.; by electron microprobe, average of seven analyses, H_2O by difference; corresponds to $(Mn_{28.95}Mg_{11.32}Zn_{2.13}Ca_{0.14}Fe_{0.11})_{\Sigma=42.65}Si_{16}O_{54.53}(OH)_{40.23}$.

Polymorphism & Series: 2M, 1A polytypes.

Occurrence: A late-stage low- to medium-temperature mineral, implanted on other species in fissures and solution cavities, in a metamorphosed stratiform zinc orebody (Franklin, New Jersey, USA).

Association: Zincite, willemite, pyrochroite, leucophoenicite, calcite, chlorophoenicite (Franklin, New Jersey, USA).

Distribution: In the USA, at Franklin, Sussex Co., New Jersey, and in the Pennsylvania mine, San Antonio Valley, Santa Clara Co., California. In the Wessels mine, near Kuruman, Cape Province, South Africa.

Name: For Robert B. Gage, of Trenton, New Jersey, USA, who analyzed the first specimens.

Type Material: National Museum of Natural History, Washington, D.C., USA, R6444, 86845.

References: (1) Phillips, A.H. (1910) Gageite, a new mineral from Franklin, New Jersey. Amer. J. Sci., 30, 283–284. (2) Palache, C. (1935) The minerals of Franklin and Sterling Hill, Sussex County, New Jersey. U.S. Geol. Sur. Prof. Paper 180, 111. (3) Moore, P.B. (1968) Relations of the manganese-calcium silicates, gageite and harstigite. Amer. Mineral., 53, 309–315. (4) Moore, P.B. (1969) A novel octahedral framework structure: gageite. Amer. Mineral., 54, 1005–1017. (5) Dunn, P.J. (1979) The chemical composition of gageite: an empirical formula. Amer. Mineral., 64, 1056–1058. (6) Ferraris, G., M. Mellini, and S. Merlino (1987) Electron-diffraction and electron microscopy study of balangeroite and gageite: crystal structures, polytypism, and fiber texture. Amer. Mineral., 72, 382–391.

Gaidonnayite

$Na_2ZrSi_3O_9 \cdot 2H_2O$

Crystal Data: Orthorhombic. *Point Group: mm2.* Crystals well-formed, slightly bladed, flattened on {010}, and elongated and heavily striated along [100], to 4 mm. *Twinning:* Common about [012] with irregular composition plane.

Physical Properties: *Fracture:* Conchoidal. *Tenacity:* Brittle. Hardness = ~ 5 D(meas.) = 2.67(1)   D(calc.) = 2.70   Commonly fluoresces bright green fluorescence under SW and LW UV.

Optical Properties: Transparent to opaque. *Color:* Colorless, white, beige, light brown, pale yellowish green. *Streak:* White. *Luster:* Vitreous.
Optical Class: Biaxial (+). *Orientation:* X = a; Y = b; Z = c. $\alpha = 1.573$–1.575
$\beta = 1.590$–1.592 $\gamma = 1.599$–1.605 2V(meas.) = $53°$–$59°$ 2V(calc.) = $55°$–$62°$

Cell Data: *Space Group:* $P2_1nb$. a = 11.740(3)   b = 12.820(3)   c = 6.691(1)   Z = 4

X-ray Powder Pattern: Mont Saint-Hilaire, Canada; very similar to georgechaoite.
3.124 (100), 5.93 (80), 5.84 (80), 3.094 (80), 5.63 (50), 2.931 (40), 1.637 (40)

Chemistry:

	(1)	(2)
SiO_2	42.51	42.5
TiO_2	0.42	0.5
ZrO_2	30.21	27.7
Nb_2O_5	3.00	1.4
CaO		0.6
Na_2O	13.11	8.7
K_2O	2.20	6.4
H_2O	9.25	[12.2]
Total	100.70	[100.0]

(1) Mont Saint-Hilaire, Canada, by electron microprobe, H_2O by TGA; corresponding to $(Na_{1.72}K_{0.19})_{\Sigma=1.91}Zr_{1.00}Nb_{0.09}Ti_{0.02}Si_{2.88}O_9 \cdot 2.10H_2O$. (2) Narssârssuk, Greenland; by electron microprobe, H_2O by difference; corresponding to $(Na_{1.19}K_{0.58}Ca_{0.05})_{\Sigma=1.82}Zr_{0.95}Nb_{0.04}Ti_{0.03}Si_{3.00}O_9 \cdot nH_2O$.

Polymorphism & Series: Dimorphous with catapleiite.

Occurrence: In miarolitic cavities in nepheline syenite and altered pegmatite dikes in an intrusive alkalic gabbro-syenite complex (Mont Saint-Hilaire, Canada); in augite syenite (Narssârssuk, Greenland).

Association: Siderite, analcime, hilairite, aegirine, calcite, albite, catapleiite, zircon, pyrochlore, ancylite, burbankite (Mont Saint-Hilaire, Canada); vlasovite, gittinsite, apophyllite (Kipawa River, Canada); natrolite, aegirine, albite, siderite (Narssârssuk, Greenland).

Distribution: In Canada, from Mont Saint-Hilaire, and in the Sheffield Lake complex, Kipawa River, Villedieu Township, Quebec. In Greenland, at Narssârssuk. From the Lovozero and Khibiny massifs, Kola Peninsula, Russia.

Name: For Professor Gabrielle Donnay (1920–1987), mineralogist of McGill University, Montreal, Quebec, Canada.

Type Material: Canadian Museum of Nature, Ottawa, T73/2-1, T73/2-2; Royal Ontario Museum, Toronto, Canada, 34803.

References: (1) Chao, G.Y. and D.H. Watkinson (1974) Gaidonnayite, $Na_2ZrSi_3O_9 \cdot 2H_2O$, a new mineral from Mont St. Hilaire, Quebec. Can. Mineral., 12, 316–319. (2) Mandarino, J.A. and B.D. Sturman (1978) The identity of α-catapleiite and gaidonnayite. Can. Mineral., 16, 195–198. (3) Chao, G.Y. (1985) The crystal structure of gaidonnayite $Na_2ZrSi_3O_9 \cdot 2H_2O$. Can. Mineral., 23, 11–15.

Crystal Data: Hexagonal. *Point Group:* 3. As simple hexagonal prisms, tabular on {0001}, typically in subparallel groupings.

Physical Properties: *Cleavage:* {10$\overline{1}$0} and {0001}, distinct. *Fracture:* Uneven. *Tenacity:* Very brittle. Hardness = 3   D(meas.) = 5.7   D(calc.) = [5.69]

Optical Properties: Semitransparent. *Color:* Colorless to gray. *Luster:* Resinous to vitreous.
Optical Class: Uniaxial (+). $\omega = 1.910$ $\epsilon = 1.945$

Cell Data: *Space Group:* $P3$.   a = 9.82   c = 10.13   Z = 1

X-ray Powder Pattern: Jakobsberg, Sweden.
3.06 (10), 3.53 (9), 3.38 (8), 2.71 (8), 4.92 (7), 4.43 (7), 1.985 (7)

Chemistry:

	(1)	(2)	(3)
SiO$_2$	19.8	19.7	18.64
MnO	2.1	2.3	2.44
PbO	66.5	67.4	69.25
CaO	11.5	10.9	9.67
Total	99.9	100.3	100.00

(1) Franklin, New Jersey, USA; by electron microprobe. (2) Jakobsberg, Sweden; by electron microprobe. (3) Pb$_9$Ca$_5$MnSi$_9$O$_{33}$.

Occurrence: In skarn assemblages (Jakobsberg, Sweden); in manganese ores in a metamorphosed stratiform zinc orebody (Franklin, New Jersey, USA).

Association: Tephroite, lead, jacobsite, calcite, phlogopite, macedonite (Jakobsberg, Sweden); clinohedrite, willemite, andradite, franklinite (Franklin, New Jersey, USA).

Distribution: At Långban, in the Harstig mine, Pajsberg, near Persberg, and at Jakobsberg, Värmland, Sweden. From Franklin, Sussex Co., New Jersey, USA.

Name: From the Greek for *luster*, in allusion to its appearance.

References: (1) Dana, E.S. (1892) Dana's system of mineralogy, (6th edition), 422. (2) Dunn, P.J. (1979) Ganomalite from Franklin, New Jersey. Mineral. Record, 10, 47–48. (3) Dunn, P.J., D.R. Peacor, J.W. Valley, and C.A. Randall (1985) Ganomalite from Franklin, New Jersey, and Jakobsberg, Sweden: new chemical and crystallographic data. Mineral. Mag., 49, 579–592. (4) (1987) Amer. Mineral., 72, 1028 (abs. ref. 3). (5) Dunn, P.J. (1985) The lead silicates from Franklin, New Jersey: occurrence and composition. Mineral. Mag., 49, 721–727. (6) Welin, E. (1968) X-ray powder data for minerals from Långban and the related mineral deposits of Central Sweden. Arkiv Mineral. Geol., 4, 499–541.

Ganophyllite $(K, Na)_6(Mn, Al, Mg)_{24}(Si, Al)_{40}O_{96}(OH)_{16} \cdot 21H_2O$

Crystal Data: Monoclinic. *Point Group:* $2/m$. Short to long prismatic crystals, to 2.5 cm, forming rosettes; as six-sided flakes.

Physical Properties: *Cleavage:* Perfect micaceous on {001}; distinct on {100}, {010}. Hardness = 4–4.5   D(meas.) = 2.77–2.84   D(calc.) = 2.875

Optical Properties: Transparent to translucent. *Luster:* Vitreous, brilliant. *Color:* Light brown, pale pink.
Optical Class: Biaxial (–). *Pleochroism:* Dark to pale yellow-brown. *Orientation:* $X \simeq c$; $Y \simeq a$; $Z = b$. *Dispersion:* $r < v$. $\alpha = 1.571$ $\beta = 1.610$ $\gamma = 1.611$ 2V(meas.) = Small.

Cell Data: *Space Group:* $A2/a$. $a = 16.60$ $b = 27.13$ $c = 50.18$ $\beta = 93.96°$ Z = 8

X-ray Powder Pattern: Harstig mine, Sweden.
12.5 (100), 3.14 (25), 2.696 (14), 3.46 (10), 2.980 (10), 2.598 (10), 2.465 (10)

Chemistry:

	(1)	(2)	(3)		(1)	(2)	(3)
SiO_2	39.67	40.0	40.4	MgO	0.20	0.5	0.2
Al_2O_3	7.95	7.9	7.9	CaO	1.11	1.0	1.5
Fe_2O_3	0.90			BaO		0.6	0.3
FeO		0.4	0.3	Na_2O	2.18	1.3	1.1
MnO	35.15	34.0	34.1	K_2O	2.70	3.2	2.6
ZnO		0.2	0.6	H_2O^+	9.79	[10.9]	[11.0]
PbO	0.20			Total	99.85	[100.0]	[100.0]

(1) Pajsberg, Sweden. (2) Do.; by electron microprobe, H_2O by difference. (3) Franklin, New Jersey, USA; by electron microprobe, average of 10 analyses, H_2O by difference.

Occurrence: In manganese-rich portions of metamorphosed Zn-Mn mineral deposits.

Association: Calcite, rhodonite, caryopilite, barite, lead, garnet, manganoan biotite, pyrophanite (Harstig mine, Sweden); rhodonite, willemite, bustamite, axinite, clinohedrite, datolite, roeblingite, charlesite (Franklin, New Jersey, USA); parsettensite, caryopilite (Molinello mine, Italy).

Distribution: In Sweden, from the Harstig mine, Pajsberg, near Persberg, and at Långban, Värmland; and in the Sjö mine, near Grythyttan, Örebro. At the Benallt and Nant mines, Rhiw, Lleyn Peninsula, Wales. From the Molinello and Gambatesa manganese mines, near Chiavari, Val Graveglia, Liguria, Italy. In the USA, in the Maple-Hovey deposits, west of Bridgewater, Aroostook Co., Maine, and at Franklin, Sussex Co., New Jersey. From Mont Saint-Hilaire, Quebec, Canada. In the Ananai mine, Nagaoka, Kochi Prefecture; the Noda-Tamagawa mine, Iwate Prefecture; the Kumahata mine, Shiga Prefecture; and the Osu and Yonoyama mines, Kita, Ehime Prefecture, Japan. From Broken Hill, New South Wales, Australia.

Name: From the Greek for *luster* and *leaf*, for the high luster on cleavages.

References: (1) Dana, E.S. (1892) Dana's system of mineralogy, (6th edition), 564–565. (2) Smith, M.L. and C. Frondel (1968) The related layered minerals ganophyllite, bannisterite and stilpnomelane. Mineral. Mag., 36, 893–913. (3) Kato, T. (1980) The crystal structure of ganophyllite; monoclinic subcell. Mineral. J. (Japan), 10, 1–13. (4) Dunn, P.J., D.R. Peacor, J.E. Nelen, and R.A. Ramik (1983) Ganophyllite from Franklin, New Jersey; Pajsberg, Sweden; and Wales: new chemical data. Mineral. Mag., 47, 563–566. (5) Eggleton, R.A. and S. Guggenheim (1986) A re-examination of the structure of ganophyllite. Mineral. Mag., 50, 307–315.

Crystal Data: Monoclinic. *Point Group:* $2/m$. Crystals steep bipyramidal, prismatic, to 3 mm, bounded by {110}, {1$\bar{1}$0}, {21$\bar{1}$}, and {2$\bar{1}\bar{1}$}; commonly striated.

Physical Properties: *Cleavage:* Perfect on {001}. Hardness = ~6 D(meas.) = 3.68 D(calc.) = 3.89

Optical Properties: Transparent. *Color:* Colorless. *Streak:* White. *Luster:* Vitreous. *Optical Class:* Biaxial. *Orientation:* Y = b; Z $\wedge$ [101] (the edge of the {$\bar{1}$11} form) = 33°; one optic axis is nearly normal to this edge. $\alpha = 1.620(3)$ $\beta = 1.633(3)$ $\gamma = 1.640(3)$ 2V(meas.) = 53°–57°

Cell Data: *Space Group:* $C2/c$. a = 14.639(3) b = 8.466(2) c = 14.438(3) $\beta = 114.21(2)°$ Z = 4

X-ray Powder Pattern: Uintah Co., Utah, USA.
3.05 (100), 3.64 (80), 2.026 (60), 2.873 (45), 2.757 (45), 6.13 (30), 4.23 (30)

Chemistry:

	(1)	(2)	(3)
SiO$_2$	14.6	13.49	13.49
B$_2$O$_3$	24.0		27.35
MgO	1.2		
CaO	7.7		
BaO	46.1	51.41	51.64
Na$_2$O		3.50	3.48
H$_2$O	[5.9]		4.04
rem.	0.06		
Total	[99.56]		100.00

(1) Uintah Co., Utah, USA; contains visible impurities, original total given as 100.1%, H$_2$O by loss on ignition, remainder is trivalent oxides. (2) Do.; by electron microprobe, partial analyses on pure material, Na by AA. (3) Ba$_3$NaSi$_2$B$_7$O$_{16}$(OH)$_4$.

Occurrence: Of authigenic origin, in dolomitic shale in drill core from the Green River Formation (Uintah Co., Utah, USA).

Association: Nahcolite, shortite, searlesite, wurtzite (Uintah Co., Utah, USA).

Distribution: In the USA, from several drill holes in Uintah Co., Utah; in the Kramer borate deposit, Kern Co., and at Searles Lake, San Bernardino Co., California.

Name: For Professor Robert Minard Garrels (1916–1988), American geochemist and educator.

Type Material: n.d.

References: (1) Milton, C., J.M. Axelrod, and F.S. Grimaldi (1955) New mineral, garrelsite (Ba$_{.65}$Ca$_{.29}$Mg$_{.06}$)$_4$H$_6$Si$_2$B$_6$O$_{20}$, from the Green River Formation, Utah. Bull. Geol. Soc. Amer., 66, 1597 (abs.) (2) Milton, C. and A. Pabst (1974) Garrelsite, NaBa$_3$Si$_2$B$_7$O$_{16}$(OH)$_4$ from the Green River Formation of Utah. J. Res. U.S. Geol. Sur., 2, 213–218. (3) Ghose, S., C. Wan, and H.H. Ulbrich (1976) Structural chemistry of borosilicates. I. Garrelsite NaBa$_3$Si$_2$B$_7$O$_{16}$(OH)$_4$: a silicoborate with the pentaborate [B$_5$O$_{12}$]$^{9-}$ polyanion. Acta Cryst., 32, 824–832.

Crystal Data: Tetragonal; possibly orthorhombic, pseudotetragonal. *Point Group:* $\bar{4}2m$. As radiating crystal aggregates completely filling amygdules, characteristically radially jointed; rare single crystals, to 4 mm, are dipyramidal. *Twinning:* Observed optically at a very fine scale; may be a domain structure.

Physical Properties: *Cleavage:* Two at 90°, prismatic. Hardness = 4.5–5 D(meas.) = 2.13–2.17 D(calc.) = 2.201

Optical Properties: Transparent. *Color:* [Colorless to white, yellowish, grayish.] *Luster:* Vitreous to greasy. *Optical Class:* Uniaxial (+) or (–) to biaxial; may be zoned, changing optical properties from the core to the extremities. $\omega = 1.500–1.515$ $\epsilon = 1.502–1.512$

Cell Data: *Space Group:* $I\bar{4}m2$. $a = 9.9266(2)$ $c = 10.3031(3)$ $Z = [0.5]$

X-ray Powder Pattern: Co. Down, Ireland.
4.12 (100), 3.14 (100), 2.66 (100), 7.15 (80), 4.95 (80), 4.07 (60), 3.22 (60)

Chemistry:

	(1)	(2)
SiO_2	43.21	45.15
Al_2O_3	24.20	23.35
Fe_2O_3	0.02	0.02
CaO	10.64	10.32
BaO	trace	0.09
Na_2O	2.94	2.96
K_2O	0.54	0.16
H_2O	18.62	18.74
Total	100.17	100.79

(1) Glenariff, Ireland; corresponds to $(Na_{2.52}K_{0.30})_{\Sigma=2.82}Ca_{5.08}Al_{12.70}Si_{19.24}O_{64} \cdot 27.64H_2O$.
(2) Skessa, Reydarfjord, Iceland; corresponds to $(Na_{2.52}K_{0.08})_{\Sigma=2.60}(Ca_{4.86}Ba_{0.02})_{\Sigma=4.88}Al_{12.10}Si_{19.84}O_{64} \cdot 27.46H_2O$.

Mineral Group: Zeolite group.

Occurrence: As amygdule fillings in silica-undersaturated olivine basalt flows (Co. Antrim, Ireland); in lavas rich in phenocrysts of bytownite and in non-porphyritic olivine basalt flows (Iceland); in diabase porphyry (Angara River, Russia).

Association: Zeolites.

Distribution: In Ireland, from the Glenariff Valley, Co. Antrim, and at a number of other localities towards Belfast. At The Storr, Isle of Skye, Scotland. In Iceland, at numerous localities between Berufjord and Seydisfjord. In Russia, near Nevon, on the Angara River, 50 km below the Ilim River, Siberia. At Gignat, Puy-de-Dôme, France. In the Höwenegg quarry, Hegau, Baden-Württemberg, Germany. From San Giorgio di Perlena, Vicenza, Italy. On Table Mountain, Jefferson Co., Colorado, and from Goble, Columbia Co., Oregon, USA. At Davis Hill, Ontario, and around Laurel and at Mont Saint-Hilaire, Quebec, Canada. A few other localities are known.

Name: For the Garron Plateau area, Co. Antrim, Ireland.

Type Material: n.d.

References: (1) Walker, G.P.L. (1962) Garronite, a new zeolite, from Ireland and Iceland. Mineral. Mag., 33, 173–186. (2) Barrer, R.M., F.W. Bultitude, and I.S. Kerr (1959) Some properties of, and a structural scheme for, the harmotome zeolites. J. Chem. Soc., 294, 1521–1528. (3) (1963) Amer. Mineral., 48, 711–712 (abs. refs. 1 and 2). (4) Artioli, G. (1992) The crystal structure of garronite. Amer. Mineral., 77, 189–196.

Crystal Data: Orthorhombic. *Point Group:* $2/m\ 2/m\ 2/m$. As bladed and prismatic crystals; fibrous and semi-sheaflike aggregates.

Physical Properties: *Cleavage:* Perfect on {210}, intersecting at 54° and 126°, imperfect on {010}, {100}. *Tenacity:* Brittle. Hardness = 5.5–6 D(meas.) = 3.15–3.259 D(calc.) = 3.18–3.334

Optical Properties: Transparent to translucent. *Color:* White, gray, brown, green; colorless, gray, or brown in thin section. *Luster:* Vitreous.
Optical Class: Biaxial (+). *Pleochroism:* Weak to moderate, in browns, yellows, or gray.
Orientation: X = a; Y = b; Z = c. $\alpha = 1.625–1.690$ $\beta = 1.635–1.705$ $\gamma = 1.643–1.718$
2V(meas.) = 70°–90°

Cell Data: *Space Group:* $Pnma$. a = 18.531–18.601 b = 17.741–17.839 c = 5.249–5.284
Z = 4

X-ray Powder Pattern: Grafton, Maine, USA.
3.06 (100), 8.27 (80), 3.23 (70), 8.97 (50), 4.48 (40), 3.65 (40), 3.35 (40)

Chemistry:

	(1)	(2)		(1)	(2)
SiO_2	40.71	43.46	MgO	10.66	15.17
TiO_2	0.29	0.58	CaO	0.05	0.78
Al_2O_3	18.73	14.18	Na_2O	1.59	2.02
Fe_2O_3	0.90		K_2O	0.06	0.05
FeO	24.39	20.35	H_2O^+	2.75	
MnO	0.14	0.49	H_2O^-	0.15	
			Total	100.42	[97.08]

(1) Grafton, Maine, USA; corresponds to $(Fe^{2+}_{3.00}Mg_{2.34}Al_{1.23}Na_{0.46}Fe^{3+}_{0.10}Ti_{0.03}Mn_{0.02}$ $Ca_{0.01}K_{0.01})_{\Sigma=7.20}(Si_{5.99}Al_{2.01})_{\Sigma=8.00}O_{22}(OH)_{2.70}$. (2) Tallan Lake sill, Peterborough Co., Ontario, Canada; by electron microprobe, original total given as 97.07%; corresponding to $(Mg_{3.31}Fe^{2+}_{2.12}Al_{0.82}Na_{0.57}Fe^{3+}_{0.37}Mn_{0.06}Ti_{0.06})_{\Sigma=7.31}(Si_{6.37}Al_{1.63})_{\Sigma=8.00}O_{22}(OH)_2$.

Polymorphism & Series: Forms a series with magnesio-gedrite and ferro-gedrite.

Mineral Group: Amphibole (Fe–Mn–Mg) group: $0.1 \leq Mg/(Mg + Fe^{2+}) \leq 0.89$; $(Ca + Na)_B$ < 1.34; Li < 1.0; Si < 7.0; Al^{iv} > 0.99.

Occurrence: Widespread in medium- to high-grade metamorphic rocks; in metasomatized contact metamorphic rocks.

Association: Garnet, cordierite, "hornblende," anthophyllite, cummingtonite, sapphirine, sillimanite, kyanite, quartz, staurolite, biotite.

Distribution: From Gèdre, Héas Valley, Haut Pyrénées, France. At Strathy, Sutherlandshire, and Glen Urquhart, Inverness-shire, Scotland. From Snarum, Bjordammen, and Bamle, Norway. At Schisshyttan, near Väster Silfberg, Värmland, Sweden. From Shuyeverskoya, Karelia, Russia. In the USA, at Haddam, Middlesex Co., Connecticut; Grafton, Oxford Co., Maine; Masons' Mountain, Macon Co., North Carolina; and Powder Mill Point, Berkeley, Alameda Co., California. At Budsbrook, Western Australia. In the Wakamatsu mine, Tottori Prefecture, and at Iratsuyama, Ehime Prefecture, Japan.

Name: For the occurrence near Gèdre, France.

References: (1) Dana, E.S. (1892) Dana's system of mineralogy, (6th edition), 384–385. (2) Milton, D.J. and J. Ito (1961) Gedrite from Oxford County, Maine. Amer. Mineral., 46, 734–740. (3) Papike, J.J. and M. Ross (1970) Gedrites: crystal structures and intra-crystalline cation distributions. Amer. Mineral., 55, 1945–1972. (4) Hawthorne, F.C., J.L. Griep, and L. Curtis (1980) A three-amphibole assemblage from the Tallan Lake sill, Peterborough County, Ontario. Can. Mineral., 18, 275–284. (5) Phillips, W.R. and D.T. Griffen (1981) Optical mineralogy, 223–225.

Crystal Data: Tetragonal. *Point Group:* $\overline{4}2m$. Crystals are commonly short prismatic, resembling octahedrally modified cubes; granular, massive. *Twinning:* On {100}, {001}; lamellar on {001}.

Physical Properties: *Cleavage:* Distinct on {001}, poor on {110}. *Fracture:* Uneven, splintery to conchoidal. *Tenacity:* Brittle. Hardness = 5–6 D(meas.) = 3.038 D(calc.) = 3.03

Optical Properties: Transparent to translucent and opaque. *Color:* Colorless, brown, yellowish, greyish green; colorless to pale yellow in thin section. *Streak:* White to grayish white. *Luster:* Vitreous to resinous.
Optical Class: Uniaxial (–). *Pleochroism:* May show anomalous Berlin blue interference colors. *Absorption:* Weak; $O > E$. $\omega = 1.669$ (synthetic). $\epsilon = 1.658$

Cell Data: *Space Group:* $P\overline{4}2_1m$ (synthetic). a = 7.6850(4) c = 5.0636(3) Z = 2

X-ray Powder Pattern: Crestmore, California, USA.
2.848 (100), 1.818 (75), 1.921 (64), 3.066 (43), 2.437 (38), 1.768 (36), 2.738 (32)

Chemistry:

	(1)	(2)	(3)
SiO_2	30.09	22.86	21.91
TiO_2	trace		
Al_2O_3	21.67	32.09	37.19
Fe_2O_3	1.36	3.03	
FeO	2.14		
MnO	0.04		
MgO	3.87	0.85	
CaO	38.36	41.25	40.90
Na_2O	0.75		
K_2O	0.16		
H_2O^+	1.64		
Total	100.08	100.08	100.00

(1) Val di Fassa, Italy. (2) Carneal, Ireland; by electron microprobe. (3) $Ca_2Al(AlSi)O_7$.

Polymorphism & Series: Forms a series with åkermanite.

Mineral Group: Melilite group.

Occurrence: In contact metamorphosed impure limestones; in calcium-rich ultramafic volcanic rocks.

Association: Dolomite, calcite, augite, olivine, nepheline, leucite, merwinite, spurrite, diopside, ferroan spinel, phlogopite, pyrope, grossular, antigorite, pyroxenes, apatite, vesuvianite.

Distribution: Occurrences are typically of intermediate series members. Relatively pure examples have been found at: Mt. Monzoni, Val di Fassa, and Canzòcoli, Val di Fiemme, Trentino-Alto Adige, Italy. On Scawt Hill, near Larne, and at Carneal, Co. Antrim, Ireland. At Camas Mòr, Isle of Muck, and Camphouse, Ardnamurchan, Argyllshire, Scotland. From the Bellerberg volcano, two km north of Mayen, Eifel district, Germany. In the USA, at Crestmore, Riverside Co., California; from the Tres Hermanas district, Luna Co., New Mexico; and at Iron Hill, Gunnison Co., Colorado. From the Terneras mine, Velardeña, Durango, Mexico.

Name: After the German chemist, Adolf Ferdinand Gehlen (1775–1815).

References: (1) Dana, E.S. (1892) Dana's system of mineralogy, (6th edition), 476–477. (2) Deer, W.A., R.A. Howie, and J. Zussman (1986) Rock-forming minerals, (2nd edition), v. 1B, disilicates and ring silicates, 285–334. (3) Sabine, P.A., M.T. Styles, and B.R. Young (1982) Gehlenite, an exomorphic mineral from Carneal, Co. Antrim, Northern Ireland. U.K. Inst. of Geol. Sci., Report 82-1, 61–63. (4) (1982) Mineral. Abs., 33, 297 (abs. ref. 3). (5) Swainson, I.P., M.T. Dove, W.S. Schmahl, and A. Putnis (1992) Neutron powder diffraction study of the åkermanite–gehlenite solid solution series. Phys. Chem. Minerals, 19, 185–195.

Crystal Data: Cubic. *Point Group:* $\bar{4}3m$. As tetrahedra and tristetrahedra, to 5 cm. Also as irregular segregations, to 25 cm.

Physical Properties: *Cleavage:* Poor on {111}, {1$\bar{1}$1}. *Fracture:* Uneven. *Tenacity:* Brittle. Hardness = 6.5 D(meas.) = 3.44–3.70 D(calc.) = 3.70 May fluoresce intense green under LW and SW UV, then phosphorescent.

Optical Properties: Transparent. *Color:* Red, pink, yellow, pale green, emerald-green; colorless or pink in thin section. *Luster:* Vitreous.
Optical Class: Isotropic. $n = 1.738$–1.752

Cell Data: *Space Group:* $P\bar{4}3n$. $a = 8.10$–8.15 $Z = 2$

X-ray Powder Pattern: Mont Saint-Hilaire, Canada.
3.320 (100), 1.916 (80), 2.168 (70), 2.567 (65), 1.657 (65), 1.483 (50), 1.435 (50)

Chemistry:

	(1)	(2)		(1)	(2)		(1)	(2)
SiO$_2$	30.26	30.70	CuO	0.30		LOI	0.21	
Al$_2$O$_3$		0.18	ZnO	46.20	40.56	$-$O = S	2.78	2.74
FeO	6.81	11.73	BeO	12.70	12.39	Total	100.41	100.04
MnO	1.22	1.72	S	5.49	5.50			

(1) St. Peter's Dome, Colorado, USA; average of two analyses, corresponds to $(Zn_{3.31}Fe^{2+}_{0.56}$ $Mn_{0.10}Cu_{0.02})_{\Sigma=3.99}Be_{2.91}Si_{2.97}O_{11.72}S_{1.00}$. (2) Jos, Nigeria; corresponds to $(Zn_{2.93}Fe^{2+}_{0.96}$ $Mn_{0.14})_{\Sigma=4.03}Be_{2.92}(Si_{3.02}Al_{0.04})_{\Sigma=3.06}O_{11.99}S_{1.01}$.

Polymorphism & Series: Forms two series, with danalite, and with helvite.

Occurrence: In miarolitic cavities in granite pegmatites; in alkaline granites and syenites, greisens, and skarns.

Association: Phenakite, bertrandite, hambergite, siderite, sphalerite, willemite, gahnite, topaz, zircon, quartz, microcline, zeolites.

Distribution: From West Cheyenne Cañon, near St. Peter's Dome, El Paso Co., Colorado, and Cumberland, Providence Co., Rhode Island, USA. Large crystals from Mont Saint-Hilaire, Quebec, Canada. At Lågendalen, Tvedalen, Brevik, and on Låven Island, in the Langesundsfjord area, Norway. At Wheal Treburland, Altarnun, Cornwall, England. From Coire an Lochain, Cairngorm Mountains, Banffshire, Scotland. In the Jos-Bukuru complex, Nigeria. From Cerro Blancom Tanti, Cordoba, Argentina. In the Ilímaussaq intrusion and the Motzfeldt centre, Greenland. From Russia, in the Lovozero and Khibiny massifs, Kola Peninsula, and at Titkäranta, Karelia.

Name: The prefix for Frederick August Ludwig Karl Wilhelm Genth (1820–1893), German-American mineralogist, who described a zinc-rich *helvite* (danalite), later renamed genthelvite.

Type Material: Pennsylvania State University, University Park, Pennsylvania; National Museum of Natural History, Washington, D.C., USA, 127180; The Natural History Museum, London, England, 1976,427.

References: (1) Dana, E.S. (1899) Dana's system of mineralogy, (6th edition), app. I, 1032 [danalite]. (2) Glass, J.J., R.H. Jahns, and R.E. Stevens (1944) Helvite and danalite from New Mexico and the helvite group. Amer. Mineral., 29, 163–191. (3) Deer, W.A., R.A. Howie, and J. Zussman (1963) Rock-forming minerals, v. 4, framework silicates, 303–309. (4) Vlasov, K.A., Ed. (1966) Mineralogy of rare elements, v. II, 119–126. (5) Dunn, P.J. (1976) Genthelvite and the helvine group. Mineral. Mag., 40, 627–636. (6) Hassan, I. and H.D. Grundy (1985) The crystal structures of helvite group minerals, $(Mn, Fe, Zn)_8(Be_6Si_6O_{24})S_2$. Amer. Mineral., 70, 186–192. (7) Burt, D.M. (1988) Stability of genthelvite, $Zn_4(BeSiO_4)_3S$: an exercise in chalcophilicity using exchange operators. Amer. Mineral., 73, 1384–1394.

Crystal Data: Orthorhombic. *Point Group:* mm2. As well-formed single and twinned hemimorphic crystals, to 1 mm; forms include {010}, {011}, {100}, {$\overline{1}$00}, {$\overline{1}$01}, {101}, and {120}. *Twinning:* Common, by interpenetration of individuals related by a half-turn about [023]; the [100] axes of twinned crystals are parallel, with [010] of one 14° from [001] of the other.

Physical Properties: *Fracture:* Conchoidal. Hardness = 5 D(meas.) = 2.70(2) D(calc.) = 2.689

Optical Properties: Semitransparent. *Color:* Colorless to white. *Optical Class:* Biaxial (–). *Orientation:* X = a; Y = b; Z = c. α = 1.578(1) β = 1.597(1) γ = 1.606(1) 2V(meas.) = 67° 2V(calc.) = 68°

Cell Data: *Space Group:* P2$_1$nb. a = 11.836(4) b = 12.940(6) c = 6.735(4) Z = 4

X-ray Powder Pattern: Wind Mountain, New Mexico, USA; very similar to gaidonnayite. 3.12 (100), 6.46 (73), 5.95 (70), 5.67 (52), 5.83 (32), 2.829 (22), 2.201 (21)

Chemistry:

	(1)	(2)
SiO$_2$	43.18	43.16
TiO$_2$	0.11	
ZrO$_2$	29.03	29.51
FeO	0.15	
Na$_2$O	7.54	7.42
K$_2$O	10.75	11.28
H$_2$O	[9.21]	8.63
Total	[99.97]	100.00

(1) Wind Mountain, New Mexico, USA; by electron microprobe, average of several analyses, H$_2$O estimated by analogy to gaidonnayite. (2) NaKZrSi$_3$O$_9$·2H$_2$O.

Occurrence: In miarolitic cavities in analcime-bearing nepheline syenite (Wind Mountain, New Mexico, USA); in a carbonatite, in dolomitic veinlets cutting metasomatically altered pyroxenites, apparently altering from catapleiite (Vuoriyärvi complex, Russia).

Association: Microcline, nepheline, aegirine, catapleiite, monazite, chlorite (Wind Mountain, New Mexico, USA); dolomite, strontianite, phlogopite, barite, komkovite, pyrite (Vuoriyärvi complex, Russia).

Distribution: On Wind Mountain, Otero Co., New Mexico, USA. In the Vuoriyärvi carbonatite complex, Kola Peninsula, Russia. From Pocos dos Caldas, Minas Gerais, Brazil.

Name: Honors Professor George Y. Chao, of Carleton University, Toronto, Canada, for his studies of zirconium silicates.

Type Material: National Museum of Natural History, Washington, D.C., USA, 161902.

References: (1) Boggs, R.C. and S. Ghose (1985) Georgechaoite NaKZrSi$_3$O$_9$·2H$_2$O, a new mineral species from Wind Mountain, New Mexico. Can. Mineral., 23, 1–4. (2) Ghose, S. and P. Thakur (1985) The crystal structure of georgechaoite NaKZrSi$_3$O$_9$·2H$_2$O. Can. Mineral., 23, 5–10. (3) (1986) Amer. Mineral., 71, 227 (abs. refs. 1 and 2).

Crystal Data: Orthorhombic. *Point Group:* $2/m \; 2/m \; 2/m$. As mats and sprays of bunched prismatic crystals, to 1 cm; as single crystals typically bounded by {100}, {010}, {110}, lacking terminations.

Physical Properties: *Cleavage:* Good on {010}. Hardness = 4.5 D(meas.) = 3.68(2) D(calc.) = 3.66

Optical Properties: Translucent to opaque. *Color:* White to very pale pink. *Streak:* White. *Luster:* Vitreous to subadamantine, silky when matted.
Optical Class: Biaxial (–). *Orientation:* X = b; Y = c; Z = a. $\alpha = 1.665(2)$ $\beta = 1.675(2)$ $\gamma = 1.678(2)$ 2V(meas.) = 50°–60°

Cell Data: *Space Group: Bbcm.* a = 8.185(7) b = 18.65(2) c = 6.256(6) Z = 8

X-ray Powder Pattern: Sterling Hill, New Jersey, USA.
2.598 (100), 9.326 (85), 3.418 (80), 2.758 (75), 2.332 (75), 2.983 (60), 4.806 (50)

Chemistry:

	(1)	(2)	(3)	(4)
SiO_2	23.9	23.2		22.84
FeO	0.06			
MnO	22.1	21.1		20.22
ZnO	31.7	29.6		30.94
MgO	16.6	19.5		19.15
Na_2O	0.04			
H_2O			4.12	6.85
Total	94.4	93.4		100.00

(1) Sterling Hill, New Jersey, USA; after deducting 3.93% of calcite as $CaCO_3$. (2) Do.; by electron microprobe. (3) Do.; H_2O by Penfield method. (4) $(Mg_{1.25}Mn_{0.75})_{\Sigma=2.00}ZnSiO_4(OH)_2$.

Occurrence: In a hydrothermal vein cutting franklinite-willemite ore in a metamorphosed stratiform zinc orebody.

Association: Calcite, manganpyrosmalite, sphalerite.

Distribution: From Sterling Hill, Ogdensburg, Sussex Co., New Jersey, USA.

Name: For Ewald Gerstmann, mineral collector of Franklin, New Jersey, USA.

Type Material: Harvard University, Cambridge, Massachusetts; National Museum of Natural History, Washington, D.C., USA, 135926, 137021.

References: (1) Moore, P.B. and T. Araki (1977) Gerstmannite, a new zinc silicate mineral and a novel cubic close-packed oxide structure. Amer. Mineral., 62, 51–59.

Crystal Data: Monoclinic, probable. *Point Group:* n.d. As spherules of radial fibers, to 0.3 mm.

Physical Properties: *Tenacity:* Waxy or gummy. Hardness = 2 D(meas.) = 2.72(5) D(calc.) = [2.54]

Optical Properties: Transparent or translucent. *Color:* Chrysocolla-green, inclining to pale blue-green. *Luster:* Nonmetallic.
Optical Class: Biaxial (−). *Pleochroism:* Weak in drab grayish green. *Orientation:* Blades extinguish up to 8° from Z ∥ length. *Absorption:* $Z > X = Y$. $\alpha = 1.560$ $\beta = 1.635$ $\gamma = 1.635$ 2V(meas.) = Very small.

Cell Data: *Space Group:* n.d. $a = 13.38$ $b = 19.16$ $c = 9.026$ $\beta = \sim 90°$ $Z = [4]$

X-ray Powder Pattern: Christmas, Arizona, USA.
13.4 (100), 7.786 (50), 4.790 (40), 3.897 (40), 10.97 (30), 6.684 (30), 3.315 (30)

Chemistry:

	(1)	(2)
SiO_2	41.5	40.77
MnO	0.5	
CuO	36.2	44.97
MgO	2.3	
CaO	3.8	
H_2O	14.6	14.26
Total	98.9	100.00

(1) Christmas, Arizona, USA; average of two closely agreeing analyses. (2) $Cu_5Si_6O_{17} \cdot 7H_2O$.

Occurrence: A retrograde metamorphic or mesogene mineral formed at the expense of a prograde calc-silicate and sulfide assemblage; in tactites, commonly incrusting fractures; also filling cracks or interstices in diopside grains.

Association: Kinoite, apachite, stringhamite, junitoite, clinohedrite, xonotlite, diopside, apophyllite, calcite, tobermorite.

Distribution: From the Christmas copper mine, Gila Co., Arizona, USA.

Name: For Gila Co., Arizona, USA, in which it was found.

Type Material: The Natural History Museum, London, England, 1980,533; University of Arizona, Tucson, Arizona; National Museum of Natural History, Washington, D.C., USA, 150201.

References: (1) Cesbron, F.P. and S.A. Williams (1980) Apachite and gilalite, two new copper silicates from Christmas, Arizona. Mineral. Mag., 43, 639–641. (2) (1980) Amer. Mineral., 65, 1065 (abs. ref. 1).

Crystal Data: Tetragonal. *Point Group:* $4/m\,2/m\,2/m$. As crystalline masses, to 2 cm, and embedded grains.

Physical Properties: *Cleavage:* Very good on {001}, poor on {100}; very poor on {110}. *Tenacity:* Brittle. Hardness = 3–4 D(meas.) = 3.390–3.402 D(calc.) = 3.404

Optical Properties: Translucent. *Color:* Deep red. *Streak:* Pink. *Luster:* Vitreous. *Optical Class:* Uniaxial (–). *Pleochroism:* Strong; O = colorless; E = red. ω = 1.619–1.621 ϵ = 1.618–1.620

Cell Data: *Space Group:* $P4/ncc$. a = 7.5164(6) c = 16.0768(10) Z = 4

X-ray Powder Pattern: Dry Delta, Alaska, and near Incline Village, California, USA. (ICDD 3-402).
3.39 (100), 4.41 (70), 3.22 (70), 3.09 (50), 2.65 (5), 2.39 (50), 8.02 (40)

Chemistry:

	(1)	(2)
SiO$_2$	50.08	51.63
TiO$_2$	0.00	
Al$_2$O$_3$	0.34	
Fe$_2$O$_3$	0.56	
Mn$_2$O$_3$	0.14	
FeO	14.60	15.43
BaO	31.02	32.94
LOI	0.82	
insol.	2.20	
Total	99.76	100.00

(1) Dry Delta, Alaska, USA; nearly pure material, corrected for assumed oxidation of FeO to Fe$_2$O$_3$, loss on ignition taken as H$_2$O. (2) BaFeSi$_4$O$_{10}$.

Occurrence: In a glacial moraine (Dry Delta, Alaska, USA); probably from a contact metamorphic zone (near Incline Village, California, USA).

Association: Celsian, hedenbergite, quartz (Dry Delta, Alaska, USA); sanbornite, celsian, diopside, tourmaline, quartz, pyrrhotite (Incline Village, California, USA); pellyite, sanbornite, taramellite, fresnoite, muirite, barite (Itsy Mountains, Canada).

Distribution: In the USA, near the head of Dry Delta, Alaska Range, Alaska; from near Incline Village, Mariposa Co., and on the Esquire No. 7 claim, Big Creek, Fresno Co., California. On the Gunn claim, in the Itsy Mountains, near Macmillan Pass, Yukon Territory, Canada. In the La Madrelena mine, Tres Pozos, Baja California, Mexico.

Name: For Frank Gillespie, of Richardson, Alaska, USA, who discovered the first specimen.

Type Material: Harvard University, Cambridge, Massachusetts, 90622; National Museum of Natural History, Washington, D.C., USA, 94347.

References: (1) Schaller, W.T. (1922) Gillespite, a new mineral. J. Wash. Acad. Sci., 123, 7–8. (2) Schaller, W. (1929) The properties and associated minerals of gillespite. Amer. Mineral., 14, 319–322. (3) Pabst A. (1943) Crystal structure of gillespite, BaFeSi$_4$O$_{10}$. Amer. Mineral., 28, 372–390. (4) Hazen, R.M. and C.W. Burnham (1974) The crystal structure of gillespite I and II: a structure determination at high pressure. Amer. Mineral., 59, 1166–1176.

Crystal Data: Monoclinic. *Point Group:* $2/m$. Crystals bipyramidal, composite, to 2 cm. In stellate or radiating spherulitic aggregates; powdery. *Twinning:* On the normal to {100}, giving pseudotetragonal or pseudo-orthorhombic individuals.

Physical Properties: *Cleavage:* {$\overline{2}$32}, distinct. *Fracture:* Subconchoidal. Hardness = 4.5 D(meas.) = 2.20–2.26 D(calc.) = 2.28 Piezoelectric.

Optical Properties: Transparent to translucent. *Color:* White, grayish, bluish white, reddish; colorless in thin section. *Luster:* Vitreous.
Optical Class: Biaxial (–); sectored. *Orientation:* $Y = b$; $Z \wedge c = 42.5°$. $\alpha = 1.525$–1.540 $\beta = 1.531$–1.544 $\gamma = 1.541$–1.550 2V(meas.) = 82°–86°

Cell Data: *Space Group:* $P2_1/c$. $a = 10.02$ $b = 10.62$ $c = 9.84$ $\beta = 92°25'$ $Z = 2$

X-ray Powder Pattern: Round Top volcano, Hawaii, USA.
2.706 (100), 4.25 (70), 3.19 (70), 4.93 (60), 3.13 (60), 7.26 (55), 2.738 (55)

Chemistry:

	(1)	(2)	(3)
SiO_2	33.89	38.03	33.45
Al_2O_3	28.14	26.61	28.38
Fe_2O_3	0.00	trace	
CaO	13.96	12.81	15.61
SrO	0.025		
BaO	0.27	0.00	
Na_2O		1.32	
K_2O	2.86	0.07	
H_2O	20.76	21.58	22.56
Total	99.91	100.42	100.00

(1) Capo di Bove, Italy; corresponds to $(Ca_{1.78}K_{0.44})_{\Sigma=2.22}Al_{3.95}Si_{4.03}O_{16} \cdot 8.24H_2O$.
(2) Eyrarfjäll, Reydarfjord, Iceland; corresponds to $(Ca_{1.59}Na_{0.30}K_{0.02})_{\Sigma=1.91}$ $Al_{3.63}Si_{4.40}O_{16} \cdot 8.34H_2O$. (3) $Ca_2Al_4Si_4O_{16} \cdot 9H_2O$.

Mineral Group: Zeolite group.

Occurrence: In cavities in nepheline and olivine basalt and leucite tephrite.

Association: Zeolites, calcite, chlorite, quartz.

Distribution: At Capo di Bove and elsewhere around Rome, Lazio, Italy. In Germany, from the Hohenberg, near Bühne, Westphalia; on the Frauenberg, near Fulda, Hesse; at the Schlauroth, Görlitz, Saxony; on the Arensberg, near Zilsdorf; and from the Schellkopf, near Brenk, and elsewhere in the Eifel district. From near Zálezly (Salesel), Czech Republic. Found near the Gorner glacier, near Zermatt, Valais, Switzerland. In Ireland, at many localities in Co. Antrim, as in the Bruslee quarry, Ballyclare. On Iceland, many localities around Reydarfjord, Fáskrúdsfjord, and Fagridalur. At Round Top volcano, Oahu, and Alexander Dam, Kauai, Hawaii, USA. From Concepción del Oro, Zacatecas, Mexico. A number of other localities are known.

Name: For Professor Carlo Giuseppe Gismondi (1762–1824), Italian mineralogist, Rome, Italy, who first examined the mineral.

References: (1) Dana, E.S. (1892) Dana's system of mineralogy, (6th edition), 586–587. (2) Deer, W.A., R.A. Howie, and J. Zussman (1963) Rock-forming minerals, v. 4, framework silicates, 401–407. (3) Walker, G.P.L. (1962) Low-potash gismondine from Ireland and Iceland. Mineral. Mag., 33, 187–201. (4) Fischer, K. (1963) The crystal structure determination of the zeolite gismondite. $CaAl_2Si_2O_8 \cdot 4H_2O$. Amer. Mineral., 48, 664–672. (5) Iijima, A. and K. Harada (1969) Authigenic zeolites in zeolitic palagonite tuffs on Oahu, Hawaii. Amer. Mineral., 54, 182–197. (6) Nawaz, R. (1980) Morphology, twinning, and optical orientation of gismondine. Mineral. Mag., 43, 841–844. (7) Vezzalini, G. and R. Oberti (1984) The crystal chemistry of gismondines: the non-existence of K-rich gismondines. Bull. Minéral., 107, 805–812.

Crystal Data: Monoclinic. *Point Group:* 2. As radiating sheaves of crystals and fibrous masses, with individual fibers to about 0.3 mm. In fine intergrowths with vlasovite and apophyllite.

Physical Properties: Hardness = 3.5–4   D(meas.) = n.d.   D(calc.) = 3.624

Optical Properties: Translucent. *Color:* Chalky white; colorless in thin section.
Luster: Vitreous.
Optical Class: Biaxial (–). *Orientation:* $X \wedge c = 5°–10°$; Y and $Z \simeq \perp$ to elongation.
$\alpha = 1.720(2)$ $\beta = 1.736(2)$ $\gamma = 1.738(2)$ 2V(meas.) = 20°–40°

Cell Data: *Space Group:* $C2$. $a = 6.852(2)$ $b = 8.659(1)$ $c = 4.686(2)$ $\beta = 101.69(2)°$
Z = 2

X-ray Powder Pattern: Kipawa River, Canada.
3.155 (100), 3.232 (80), 3.026 (80), 2.660 (80), 1.677 (70), 5.32 (60), 2.226 (50)

Chemistry:

	(1)
SiO$_2$	40.8
ZrO$_2$	40.3
CaO	18.4
Total	99.5

(1) Kipawa River, Canada; by electron microprobe, corresponding to $Ca_{0.98}Zr_{0.98}Si_{2.03}O_7$.

Occurrence: In eudialyte-rich pegmatitic lenses in a regionally metamorphosed agpaitic syenite complex (Kipawa River, Canada); an alteration product of eudialyte in syenite, quartz syenite, and alkalic granite (Pajarito Mountain, New Mexico, USA).

Association: Apophyllite, vlasovite, eudialyte, fluorite, graphite, calcite, apatite, "opal" (Kipawa River, Canada); eudialyte, zircon, chlorite (Pajarito Mountain, New Mexico, USA).

Distribution: From the [Sheffield Lake complex,] Kipawa River, Villedieu Township, Quebec, and in the Strange Lake complex, southeast of Lac Brisson, Quebec and Labrador, Newfoundland, Canada. On Pajarito Mountain, Mescalero Indian Reservation, Otero Co., New Mexico, USA.

Name: For Professor John Gittins, Canadian petrologist, University of Toronto, Toronto, Ontario, Canada.

Type Material: Canadian Geological Survey, Ottawa, 19558; University of Toronto, Toronto; Royal Ontario Museum, Toronto, Canada, M37321; Harvard University, Cambridge, Massachusetts, 117030, 117031; National Museum of Natural History, Washington, D.C., USA, 147002.

References: (1) Ansell, H.G., A.C. Roberts, A.G. Plant, and B.D. Sturman (1980) Gittinsite, a new calcium zirconium silicate from the Kipawa agpaitic syenite complex, Quebec. Can. Mineral., 18, 201–203. (2) (1981) Amer. Mineral., 66, 1274–1275 (abs. ref. 1). (3) Roelofsen-Ahl, J.N. and R.C. Peterson (1989) Gittinsite: a modification of the thortveitite structure. Can. Mineral., 27, 703–708.

Giuseppettite

$(Na, K, Ca)_{7-8}(Si, Al)_{12}O_{24}(SO_4, Cl)_{1-2}$

Crystal Data: Hexagonal. *Point Group:* $6mm$, $\bar{6}m2$, or $6/m\,2/m\,2/m$. As anhedral grains, in veinlets a few mm thick.

Physical Properties: Hardness = 6–7 D(meas.) = 2.35 D(calc.) = 2.365

Optical Properties: Transparent. *Color:* Pale blue-violet.
Optical Class: Uniaxial (+). $\omega = 1.491$ $\epsilon = 1.507$

Cell Data: *Space Group:* $P6_3mc$, $P\bar{6}2c$, or $P6_3/mmc$. a = 12.850(1) c = 42.22(3) Z = 8

X-ray Powder Pattern: Sacrofano, Italy.
3.712 (100), 3.446 (80), 3.126 (70), 2.141 (66), 6.42 (62), 2.640 (62), 4.318 (53)

Chemistry:

	(1)
SiO_2	33.25
Al_2O_3	28.56
Fe_2O_3	0.03
CaO	4.85
Na_2O	14.37
K_2O	8.00
Cl	0.78
SO_3	9.92
$-O = Cl_2$	0.18
Total	99.58

(1) Sacrofano, Italy; by electron microprobe, SO_3 confirmed by IR; corresponds to $(Na_{5.0}K_{1.8}Ca_{1.0})_{\Sigma=7.8}(Al_{6.05}Si_{5.95})_{\Sigma=12.00}O_{24}(SO_4)_{1.8}Cl_{0.25}$.

Mineral Group: Cancrinite group.

Occurrence: As veinlets in a block of sanidinite volcanic ejecta.

Association: Potassic feldspar, nepheline, haüyne, biotite, kalsilite.

Distribution: At Sacrofano, in the Biachella Valley, Lazio, Italy.

Name: Honors Giuseppe Giuseppetti, Professor of Mineralogy, University of Pavia, Pavia, Italy.

Type Material: Mineralogy Museum, Institute of Mineralogy and Petrology, University of Pavia, Pavia, Italy.

References: (1) Mazzi, F. and C. Tadini (1981) Giuseppettite, a new mineral from Sacrofano (Italy), related to the cancrinite group. Neues Jahrb. Mineral., Monatsh., 103–110. (2) (1982) Amer. Mineral., 67, 415 (abs. ref. 1).

Crystal Data: Orthorhombic. *Point Group:* $2/m \ 2/m \ 2/m$. Crystals commonly long prismatic, to 1 cm, typically unterminated, alone or in columnar aggregates; forms include {100}, {010}, {110}, {120}, {103}, {021}, {111}, {121}. Coarse-grained cleavable, fine-grained massive. *Twinning:* Penetration and contact twins, rarely triplets at about 60°, with {011} as twin plane.

Physical Properties: *Cleavage:* Very poor on {001}. *Fracture:* Conchoidal. Hardness = ~6 D(meas.) = 3.407 D(calc.) = 3.465

Optical Properties: Semitransparent. *Color:* Bluish green; small crystals white or pinkish; may be brown when massive, perhaps due to manganese oxides. *Luster:* Vitreous. *Optical Class:* Biaxial (–). *Dispersion:* $r > v$, marked. $\alpha = 1.682–1.686$ $\beta = 1.716–1.722$ $\gamma = 1.729–[1.735]$ 2V(meas.) = 60°51′

Cell Data: *Space Group:* $Pbnm$. a = 4.92(3) b = 11.19(2) c = 6.51(2) Z = 4

X-ray Powder Pattern: Franklin, New Jersey, USA. (ICDD 14-376). 1.85 (100), 2.69 (80), 2.63 (80), 3.69 (60), 2.96 (60), 1.13 (50), 1.11 (50)

Chemistry:

	(1)	(2)
SiO$_2$	31.48	32.2
FeO	trace	
MnO	38.00	37.5
ZnO		0.8
PbO	1.74	
MgO		0.2
CaO	28.95	29.5
Total	100.17	100.2

(1) Franklin, New Jersey, USA; neglecting PbO, corresponds to Ca$_{0.99}$Mn$_{1.02}$Si$_{1.00}$O$_4$. (2) Do.; by electron microprobe, corresponds to (Ca$_{0.98}$Zn$_{0.02}$Mg$_{0.01}$)$_{\Sigma=1.01}$Mn$_{0.99}$Si$_{1.00}$O$_4$.

Occurrence: In a metamorphosed stratiform zinc deposit (Franklin, New Jersey, USA); in skarn at the contact between diabase and marble (Anakit Creek, Russia); in calc-silicate rocks with manganese ores (Kuruman, South Africa).

Association: Nasonite, willemite, andradite, hardystonite, tephroite, clinohedrite, esperite, leucophoenicite, hodgkinsonite, diopside, cuspidine, calcite, franklinite, zincite (Franklin, New Jersey, USA).

Distribution: At Franklin, Sussex Co., New Jersey, USA. From near Anakit Creek, at the mouth of the Lower Tunguska River, central Siberia, Russia. In the Wessels mine, near Kuruman, Cape Province, South Africa. From the Kombat mine, 49 km south of Tsumeb, Namibia.

Name: From the Greek for *blue green* and *color*, in allusion to its color.

Type Material: n.d.

References: (1) Dana, E.S. (1899) Dana's system of mineralogy, (6th edition), app. I, 29; Dana, E.S. and W.E. Ford (1909) Dana's system of mineralogy, (6th edition), app. II, 46. (2) Palache, C. (1935) The minerals of Franklin and Sterling Hill, Sussex County, New Jersey. U.S. Geol. Sur. Prof. Paper 180, 79–80. (3) O'Mara, J.H. (1951) Unit cell and space group of glaucochroite. Amer. Mineral., 36, 918. (4) Pertsev, N.N. and I.P. Laputina (1974) Glaucochroite in the Anakit skarns, Lower Tunguska. Doklady Acad. Nauk SSSR, 216, 1379–1382 (in Russian). (5) Lager, G.A. and E.P. Meagher (1978) High-temperature structural study of six olivines. Amer. Mineral., 63, 365–377. (6) Leavens, P.B., P.J. Dunn, and D.M. Burt (1987) Glaucochroite (olivine, CaMnSiO$_4$) from Franklin, New Jersey: its composition, occurrence, and formation. Amer. Mineral., 72, 423–428.

Glauconite

$$(K, Na)(Fe^{3+}, Al, Mg)_2(Si, Al)_4O_{10}(OH)_2$$

Crystal Data: Monoclinic. *Point Group:* m or $2/m$. In grains or pellets composed partly of aggregates of small micaceous crystals.

Physical Properties: *Cleavage:* {001}, perfect. Hardness = 2 D(meas.) = 2.4–2.95 D(calc.) = 2.903

Optical Properties: Translucent to nearly opaque. *Color:* Grass-green, yellow-green, blue-green; in thin section, green, yellow, olive-green. *Luster:* Dull, glistening. *Optical Class:* Biaxial (–). *Pleochroism:* X = yellow-green, green; $Y = Z$ = deeper yellow, bluish green. *Orientation:* $Y = b$; $Z \simeq a$; $X \wedge a \simeq 10°$. *Dispersion:* $r > v$. *Absorption:* $Y \simeq Z > X$. $\alpha = 1.592$–1.610 $\beta = 1.614$–1.641 $\gamma = 1.614$–1.641 2V(meas.) = $0°$–$20°$

Cell Data: *Space Group:* Cm or $C2/m$. $a = 5.25$–5.31 $b = 9.09$–9.19 $c = 10.03$–10.15 $\beta = \sim 100°$ $Z = 2$

X-ray Powder Pattern: Anzoategni, Venezuela. (ICDD 9-439).
10.1 (100), 2.587 (100), 4.53 (80), 3.33 (60), 2.396 (60), 1.511 (60), 3.63 (40)

Chemistry:

	(1)	(2)		(1)	(2)
SiO_2	49.29	46.52	CaO	0.74	0.51
TiO_2	0.12		Na_2O	0.12	0.19
Al_2O_3	3.17	4.61	K_2O	6.02	7.65
Fe_2O_3	21.72	24.76	H_2O^+	7.21	5.83
FeO	3.19	2.02	H_2O^-	4.60	3.20
MgO	3.85	4.65	P_2O_5	0.32	0.08
			Total	100.35	100.02

(1) Whare Flat, Otago, New Zealand; corresponds to $(K_{0.60}Ca_{0.06}Na_{0.02})_{\Sigma=0.68}(Fe^{3+}_{1.27}Mg_{0.44}Fe^{2+}_{0.21}Al_{0.11})_{\Sigma=2.03}(Si_{3.82}Al_{0.18})_{\Sigma=4.00}O_{10}(OH)_2$. (2) Makhtesh Ramon, Israel; corresponds to $(K_{0.76}Ca_{0.04}Na_{0.03})_{\Sigma=0.83}(Fe^{3+}_{1.33}Mg_{0.54}Fe^{2+}_{0.13}Al_{0.06})_{\Sigma=2.06}(Si_{3.64}Al_{0.36})_{\Sigma=4.00}O_{10}(OH)_2$.

Polymorphism & Series: 1M polytype.

Mineral Group: Mica group.

Occurrence: Altered from detrital biotite or other parent materials, by marine diagenesis in shallow water under reducing conditions; especially in loosely consolidated sandstone (greensand), and impure limestone and siltstone.

Association: Quartz, feldspars, glaucophane, dolomite, siderite, calcite, ankerite, pyrite, limonite.

Distribution: Worldwide in marine sediments. A few localities for studied material are: on the Otago Peninsula, New Zealand. From the Flinders Ranges and Yorke Peninsula, South Australia. Around Sandwich and Folkestone, Kent, and Hollybush Hill, Worcester, England. At Villers-sur-Mer, Calvados, France. From Verona, Vicenza, Italy. At a number of localities in Israel. From the N'Chwaning mine, near Kuruman, Cape Province, South Africa. In the USA, particularly widespread in Alabama; from Burnet Co., Texas; both on and offshore in several of the coastal counties of California; from Baker Co., Oregon; in the Black Hills of South Dakota.

Name: From the Greek *glaucos*, for *bluish green*.

References: (1) Dana, E.S. (1892) Dana's system of mineralogy, (6th edition), 683–684. (2) Deer, W.A., R.A. Howie, and J. Zussman (1963) Rock-forming minerals, v. 3, sheet silicates, 35–41. (3) Bentor, Y.K. and M. Kastner (1965) Notes on the mineralogy and origin of glauconite. J. Sed. Petrol., 35, 155–166. (4) Buckley, H.A., J.C. Bevan, K.M. Brown, L.R. Johnson, and V.C. Farmer (1978) Glauconite and celadonite: two separate mineral species. Mineral. Mag., 42, 373–382.

Crystal Data: Monoclinic. *Point Group:* $2/m$. As prismatic crystals; columnar, fibrous, or granular aggregates; massive. *Twinning:* Simple or multiple twinning $\parallel$ {100}.

Physical Properties: *Cleavage:* Perfect on {110}, intersecting at $\sim56°$ and $\sim124°$; partings on {010}, {001}. *Fracture:* Conchoidal to uneven. *Tenacity:* Brittle. Hardness = 6 D(meas.) = 3.08–3.22   D(calc.) = 3.132

Optical Properties: Translucent. *Color:* Gray, lavender-blue, commonly zoned; lavender-blue to colorless in thin section. *Streak:* Blue-gray. *Luster:* Vitreous to pearly. *Optical Class:* Biaxial (–). *Pleochroism:* Vivid; X = yellow to colorless; Y = violet to lavender; Z = blue. *Orientation:* Y = b; $Z \wedge c = -7°$ to $-6°$, $X \wedge a \simeq 8°$. *Dispersion:* $r < v$, weak. $\alpha = 1.594–1.630$ $\beta = 1.612–1.648$ $\gamma = 1.619–1.652$ 2V(meas.) = 0°–50°

Cell Data: *Space Group:* $C2/m$.   a = 9.595   b = 17.798   c = 5.307 $\beta = 103.66°$ Z = 2

X-ray Powder Pattern: Sebastopol quadrangle, California, USA. (ICDD 20–453). 8.26 (100), 3.06 (65), 2.693 (60), 4.45 (25), 3.38 (25), 2.937 (25), 2.523 (25)

Chemistry:

	(1)	(2)		(1)	(2)		(1)	(2)
SiO_2	58.04	56.28	FeO	6.12	10.34	K_2O	0.02	0.11
TiO_2	0.66	0.17	MnO	0.07	0.25	F	0.02	
Al_2O_3	10.31	12.16	MgO	11.71	8.41	Cl	0.01	
Fe_2O_3	2.89	1.72	CaO	1.37	0.62	H_2O^+	1.98	
Cr_2O_3		0.11	Na_2O	6.97	7.04	H_2O^-	0.00	
						Total	100.17	97.21

(1) Tiburon Peninsula, California, USA; corresponds to $(Na_{1.96}Ca_{0.04})_{\Sigma=2.00}(Mg_{2.39}Al_{1.82}Fe^{2+}_{0.61}Fe^{3+}_{0.18})_{\Sigma=5.00}Si_8O_{22}(OH)_2$. (2) Kodiak Islands, Alaska, USA; by electron microprobe, $Fe^{2+}:Fe^{3+}$ calculated; corresponds to $(Na_{1.90}Ca_{0.09}K_{0.02})_{\Sigma=2.01}(Al_{1.82}Mg_{1.74}Fe^{2+}_{1.20}Fe^{3+}_{0.18}Mn_{0.03}Ti_{0.02}Cr_{0.01})_{\Sigma=5.00}(Si_{7.83}Al_{0.17})_{\Sigma=8.00}O_{22}(OH)_2$.

Polymorphism & Series: Forms a series with ferro-glaucophane.

Mineral Group: Amphibole (alkali) group: $Fe^{2+}/(Fe^{2+} + Mg) < 0.5$; $Fe^{3+}/(Fe^{3+} + Al^{vi}) < 0.3$; $(Na + K)_A < 0.5$; $Na_B \geq 1.34$.

Occurrence: Characteristic of the blueschist facies, in former subduction zones in mountain belts; in the greenschist facies and in eclogites that have undergone retrograde metamorphism.

Association: Crossite, chlorite, epidote, pumpellyite, lawsonite, omphacite, jadeite, actinolite, barroisite, cummingtonite, aragonite.

Distribution: Widespread in some mountain belts. On Syra Island, Cyclades Islands, Greece. At numerous sites in the California Coast Ranges, as on the Tiburon Peninsula and at Vonsen Ranch, Marin Co., at Glaucophane Ridge, Panoche Valley, San Benito Co., and near Valley Ford, Sonoma Co.; in the Kodiak Islands, Alaska, USA. At St. Marcel, Val d'Aosta, and Piollore (Biella), Piedmont, Italy. On Anglesey, Wales. In Japan, at Ubuzan, Aichi Prefecture, and Otakiyama, Tokushima Prefecture.

Name: From the Greek for *bluish green* and *to appear*.

References: (1) Dana, E.S. (1892) Dana's system of mineralogy, (6th edition), 399. (2) Deer, W.A., R.A. Howie, and J. Zussman (1963) Rock-forming minerals, v. 2, chain silicates, 333–351. (3) Borg, I.Y. (1967) Optical properties and cell parameters in the glaucophane-riebeckite series. Contr. Mineral. Petrol., 15, 67–92. (4) Papike, J.J. and J.R. Clark (1968) The crystal structure and cation distribution of glaucophane. Amer. Mineral., 53, 1156–1173. (5) Hoffmann, C. and K. Katz (1982) Trend surface analysis of some physical properties of alkali (sodic) amphiboles. Lithos, 15, 17–25. (6) Roeske, S.M. (1986) Field relations and metamorphism of the Raspberry Schist, Kodiak Islands, Alaska. In: B.W. Evans and E.H. Brown, Eds., Blueschists and eclogites, Geol. Soc. Amer. Memoir 164, 169–184. (7) Phillips, W.R. and D.T. Griffen (1981) Optical mineralogy, 237–241.

Crystal Data: Hexagonal. *Point Group:* $6/m \, 2/m \, 2/m$. As euhedral crystals, pyramidal, tabular, or rhombohedral, striated $\parallel$ (0001) or less commonly, $\parallel$ [0001], to 4 cm. Rarely in radiating aggregates or granular. *Twinning:* Penetration twins on $\{10\bar{1}1\}$, common.

Physical Properties: *Cleavage:* $\{10\bar{1}0\}$, distinct; parting on $\{0001\}$. *Fracture:* Uneven. *Tenacity:* Brittle. Hardness = 4.5 D(meas.) = 2.02–2.17 D(calc.) = 2.098 Piezoelectric.

Optical Properties: Transparent to translucent, opaque. *Color:* Colorless, white, reddish white, salmon-red, yellowish, greenish white; colorless in thin section. *Luster:* Vitreous. *Optical Class:* Uniaxial (–) or (+); may be anomalously biaxial. $\omega = 1.476$–1.494 $\epsilon = 1.474$–1.480

Cell Data: *Space Group:* $P6_3/mmc$. a = 13.75–13.80 c = 9.97–10.08 Z = 4

X-ray Powder Pattern: Montecchio Maggiore, Italy.
4.106 (100), 11.908 (63), 2.978 (55), 2.690 (44), 3.227 (41), 7.68 (29), 5.026 (28)

Chemistry:

	(1)	(2)
SiO_2	50.00	46.57
Al_2O_3	19.17	21.12
CaO	2.95	7.04
SrO	0.58	0.25
Na_2O	7.20	4.35
K_2O	0.10	0.27
H_2O	20.00	20.40
Total	100.00	100.00

(1) Montecchio Maggiore, Italy; by electron microprobe, corresponds to $(Na_{1.16}Ca_{0.26}$ $Sr_{0.03}K_{0.01})_{\Sigma=1.46}Al_{1.88}Si_{4.16}O_{12} \cdot 5.54H_2O$. (2) South quarry, White Head, Co. Antrim, Ireland; by electron microprobe, corresponds to $(Na_{0.71}Ca_{0.64}Sr_{0.03}K_{0.01})_{\Sigma=1.39}Al_{2.10}Si_{3.92}O_{12} \cdot 5.73H_2O$.

Mineral Group: Zeolite group.

Occurrence: Formed from sodium-rich fluids, in basalts and related igneous rocks, also pegmatites.

Association: Zeolites, calcite, aragonite, quartz.

Distribution: Widespread in small amounts. A few localities for well studied material follow. From Montecchio Maggiore, Vicenza, Italy. At Glenarm and elsewhere in Co. Antrim, Ireland. On the Isle of Skye, Scotland. At Pyrgos, Cyprus. In the USA, from Bergen Hill, Hudson Co., and Great Notch, Paterson, and Prospect Park, Passaic Co., New Jersey; at Springfield, Lane Co., Oregon. On Pinnacle Rock, Five Islands, and Two Islands, Nova Scotia, Canada. In the Ilímaussaq intrusion, southern Greenland. At Sarbay-Sokolov, Kazakhstan. Large crystals from Bekiady, Madagascar. From around Flinders, Victoria, Australia.

Name: For Christian Gottlob Gmelin (1792–1860), German mineralogist and chemist of Tübingen, Germany.

References: (1) Dana, E.S. (1892) Dana's system of mineralogy, (6th edition), 593–594. (2) Deer, W.A., R.A. Howie, and J. Zussman (1963) Rock-forming minerals, v. 4, framework silicates, 387–400. (3) Passaglia, E., D. Pongiluppi, and G. Vezzalini (1978) The crystal chemistry of gmelinites. Neues Jahrb. Mineral., Monatsh., 310–322. (4) Galli, E., E. Passaglia, and P.F. Zanazzi (1982) Gmelinite: structural refinements of sodium-rich and calcium-rich natural crystals. Neues Jahrb. Mineral., Monatsh., 145–155. (5) Vezzalini, G., S. Quartieri, and E. Passaglia (1990) Crystal structure of a K-rich natural gmelinite and comparison with the other refined gmelinite samples. Neues Jahrb. Mineral., Monatsh., 504–516.

Crystal Data: Orthorhombic, pseudotetragonal, or tetragonal. *Point Group: mm2* or 422. Individual crystals very rare, spikelike pyramidal, to 2 mm; more commonly as clusters of lath-shaped or fibrous crystals, elongated along [001]; massive. *Twinning:* On {110}, inferred from X-ray studies.

Physical Properties: *Fracture:* Brittle. Hardness = ~4 D(meas.) = 2.194 on a mixture with impurities. D(calc.) = 2.147

Optical Properties: Transparent to translucent. *Color:* Chalky white to light brown. *Streak:* White. *Optical Class:* Uniaxial (–) or biaxial (–). *Orientation:* X = c. $\omega = 1.494(3)$ $\epsilon = 1.489(3)$ 2V(meas.) = n.d.

Cell Data: *Space Group: Pmn2$_1$.* a = 10.108(1) b = 9.766(1) c = 10.171(1) Z = 1, or *Space Group: P4$_2$2$_1$2.* a = 10.115(3) c = 9.766(3) Z = 1

X-ray Powder Pattern: Gobbins area, Ireland.
7.11 (100b), 4.116 (100), 3.201 (100), 3.106 (80), 2.699 (80b), 5.056 (50), 2.651 (40)

Chemistry:

	(1)	(2)
SiO_2	49.21	44.82
Al_2O_3	23.64	22.81
Fe_2O_3	0.04	
MgO	1.00	
CaO	1.58	2.51
SrO	0.36	
BaO	0.12	
Na_2O	9.85	6.01
K_2O	0.66	7.73
H_2O	[13.54]	16.12
Total	[100.00]	100.00

(1) Gobbins area, Ireland; by electron microprobe, average of six analyses; H_2O by difference, originally given as 13.55%; corresponds to $(Na_{3.98}Ca_{0.35}Mg_{0.31}Sr_{0.04}Ba_{0.01}Fe_{0.01})_{\Sigma=4.70}K_{0.18}Al_{5.80}Si_{10.25}O_{32} \cdot 9.41H_2O$. (2) Two-Mouth Cave, Ireland; by electron microprobe, corresponds to $(Na_{2.50}Ca_{0.59})_{\Sigma=3.09}K_{2.11}Al_{6.17}Si_{9.93}O_{32} \cdot 12H_2O$.

Mineral Group: Zeolite group.

Occurrence: A secondary mineral in amygdules in basalts (Co. Antrim, Ireland); in cavities in sodalite syenite in an intrusive alkalic gabbro-syenite complex (Mont Saint-Hilaire, Canada).

Association: Gmelinite, calcite (Co. Antrim, Ireland); tetranatrolite, sérandite (Mont Saint-Hilaire, Canada).

Distribution: On Island Magee, from Two Mouth Cave, at the coastal escarpment near Hills Port, south of The Gobbins, and at Dunseverick, near Giant's Causeway, Co. Antrim, Ireland. From Mont Saint-Hilaire, Quebec, Canada.

Name: For the type locality at The Gobbins, Ireland.

Type Material: Ulster Museum, Belfast, Ireland, I7881; The Natural History Museum, London, England, 1982,203; National Museum of Natural History, Washington, D.C., USA, 149432.

References: (1) Nawaz, R. and J.F. Malone (1982) Gobbinsite, a new zeolite mineral from Co. Antrim, N. Ireland. Mineral. Mag., 46, 365–369. (2) (1983) Amer. Mineral., 68, 642–643 (abs. ref. 1). (3) Nawaz, R. (1983) New data on gobbinsite and garronite. Mineral. Mag., 47, 567–568. (4) McCusker, L.B., C. Baerlocher, and R. Nawaz (1985) Rietveld refinement of the crystal structure of the new zeolite mineral gobbinsite. Zeits. Krist., 171, 281–289. (5) Mandarino, J.A. and V. Anderson (1989) Monteregian Treasures. Cambridge Univ. Press, 97.

Goldmanite

$$Ca_3(V^{3+}, Al, Fe^{3+})_2(SiO_4)_3$$

Crystal Data: Cubic. *Point Group:* $4/m \, \bar{3} \, 2/m$. As dodecahedra, to 0.1 mm; in anhedral grains.

Physical Properties: Hardness = 7   D(meas.) = 3.74–3.77   D(calc.) = 3.765

Optical Properties: Transparent. *Color:* Dark green to brownish green, commonly sector zoned; green to brown in thin section.
Optical Class: Isotropic to weakly anisotropic. $n = 1.821(1)$

Cell Data: *Space Group:* $Ia3d$. $a = 12.06$ $Z = 8$

X-ray Powder Pattern: Sandy mine, New Mexico, USA.
2.688 (100), 3.005 (65), 1.607 (49), 2.453 (38), 1.951 (22), 1.667 (18), 2.357 (16)

Chemistry:

	(1)	(2)
SiO_2	36.6	35.58
Al_2O_3	4.9	
Fe_2O_3	5.4	0.24
V_2O_3	18.3	28.66
Cr_2O_3		0.88
MnO	0.3	
MgO	0.7	
CaO	33.3	33.06
Total	99.5	98.42

(1) Sandy mine, New Mexico, USA; corresponds to $(Ca_{2.91}Mg_{0.08}Mn_{0.02})_{\Sigma=3.01}$ $(V_{1.20}Al_{0.47}Fe_{0.33})_{\Sigma=2.00}Si_{2.99}O_{12}$. (2) North Sea; by electron microprobe, corresponds to $Ca_{2.98}(V_{1.94}Cr_{0.06}Fe^{3+}_{0.01})_{\Sigma=2.01}Si_{3.00}O_{12}$.

Mineral Group: Garnet group.

Occurrence: In a small metamorphosed U-V deposit in sandstone, embedded in vanadium-rich clays and calcite (Sandy mine, New Mexico, USA); in calcareous metapelites and tungsten-bearing skarns (Coat-an-Noz, France); detrital.

Association: Quartz, calcite, diopside, vesuvianite, wollastonite, epidote, grossular-andradite, biotite, spinel, plagioclase, pyrrhotite, roscoelite, montmorillonite (Sandy mine, New Mexico, USA); plagioclase, titanite, clinopyroxene, actinolite, chlorite, pyrite, arsenopyrite (Coat-an-Noz, France); rhodonite, braunite (Yamoto mine, Japan).

Distribution: In the Sandy mine, near Laguna, Valencia Co., New Mexico, USA. From Coat-an-Noz, Côtes-du-Nord, France. In a deep well in the United Kingdom sector of the North Sea. At Ishimskaya Luka, northern Kazakhstan. In the Yamoto mine, Kagoshima Prefecture, Japan.

Name: For Dr. Marcus Isaac Goldman (1881–1965), sedimentary petrologist, U.S. Geological Survey.

Type Material: National Museum of Natural History, Washington, D.C., USA, 121717, 121953.

References: (1) Moench, R.H. and R. Meyrowitz (1964) Goldmanite, a vanadium garnet from Laguna, New Mexico. Amer. Mineral., 49, 644–655. (2) Novak, G.A. and G.V Gibbs (1971) The crystal chemistry of the silicate garnets. Amer. Mineral., 56, 791–825. (3) Benkerrou, C. and M. Fonteilles (1989) Vanadian garnets in calcareous metapelites and skarns at Coat-an-Noz, Belle-Isle-en-Terre (Côtes du Nord), France. Amer. Mineral., 74, 852–858. (4) Hallsworth, C.R., A. Livingstone, and A.C. Morton (1992) Detrital goldmanite from the Palaeocene of the North Sea. Mineral. Mag., 56, 117–120.

Crystal Data: Tetragonal. *Point Group:* $\overline{4}2m$. Fibrous crystals, at the centers of radiating spherulites, to 3 cm; massive.

Physical Properties: Hardness = 5   D(meas.) = 2.25–2.36   D(calc.) = 2.33

Optical Properties: Translucent. *Color:* White, yellowish to salmon-red. *Luster:* Silky. *Optical Class:* Biaxial (+) or (–); commonly zoned. *Orientation:* X = c. α = 1.497–1.508 β = 1.498–1.510 γ = 1.499–1.513   2V(meas.) = 50°

Cell Data: *Space Group:* $I\overline{4}2d$.   a = 13.21(1)   c = 6.622(4)   Z = 2

X-ray Powder Pattern: Chaux de Bergonne, France; may be confused with natrolite and tetranatrolite.
2.92 (100), 5.93 (80), 6.70 (60), 4.44 (60), 4.74 (50), 3.23 (50), 3.12 (40)

Chemistry:

	(1)	(2)	(3)
SiO$_2$	43.45	43.20	44.58
Al$_2$O$_3$	27.91	27.90	25.22
CaO	6.95	3.61	6.94
Na$_2$O	8.69	13.16	7.66
H$_2$O	[13.00]	11.74	15.60
Total	[100.00]	99.61	100.00

(1) Chaux de Bergonne, France; by electron microprobe, H$_2$O by difference; corresponding to Na$_{2.22}$Ca$_{0.98}$Al$_{4.32}$Si$_{5.71}$O$_{20}$•5.70H$_2$O. (2) Aci Trezza, Sicily, Italy; corresponds to Na$_{3.5}$Ca$_{0.5}$Al$_{4.5}$ Si$_{5.9}$O$_{20.8}$•5.35H$_2$O. (3) Na$_2$CaAl$_4$Si$_6$O$_{20}$•7H$_2$O.

Mineral Group: Zeolite group.

Occurrence: In cavities in basalt, leucite tephrite, and altered skarn.

Association: Zeolites, calcite.

Distribution: Well characterized material from: in France, at Chaux de Bergonne, Gignat, Puy de Dôme. In Italy, from Capo di Bove, near Rome, Lazio; and at Aci Castello, Aci Trezza, Osilo, and other places on Sardinia. From Arendal, Tvedalen, at Brevik and elsewhere around the Langesundsfjord, Norway. At Klöch, Styria, Austria. From the Schellkopf, near Brenk, Eifel district, and in the Höwenegg quarry, Hegau, Baden-Württemberg, Germany. From Allt Ribhein, Fiskavaig Bay, Isle of Skye, Scotland. In the USA, at Crestmore, Riverside Co., California, and around Honolulu, Oahu, Hawaii. At Mazé, Niigata Prefecture, Japan. A number of other localities are known.

Name: For Ferdinand Gonnard, French mineralogist, Lyons, France, who first noted the mineral.

Type Material: The Natural History Museum, London, England, 1930,166.

References: (1) Dana, E.S. (1899) Dana's system of mineralogy, (6th edition), app. I, 30. (2) Hey, M.H. and F.A. Bannister (1932) Studies on the zeolites. Part II. Thomsonite (including faroelite) and gonnardite. Mineral. Mag., 23, 51–125. (3) Meixner, H., M.H. Hey, and A.A. Moss. (1956) Some new occurrences of gonnardite. Mineral. Mag., 31, 265–271. (4) Mazzi, F., A.O. Larsen, G. Gottardi, and E. Galli (1986) Gonnardite has the tetrahedral framework of natrolite: experimental proof with a sample from Norway. Neues Jahrb. Mineral., Monatsh., 219–228. (5) Nawaz, R. (1988) Gonnardite and disordered natrolite-group minerals: their distinction and relations with mesolite, natrolite and thomsonite. Mineral. Mag., 52, 207–219.

Crystal Data: Orthorhombic, pseudohexagonal. *Point Group:* n.d. In rounded, radial aggregates of lath- and plate-shaped crystals, to several mm; as tiny rosettes of intergrown pseudohexagonal plates.

Physical Properties: *Cleavage:* {001}, perfect. *Tenacity:* Cleavage foliae are flexible but inelastic. Hardness = 2.5 D(meas.) = 3.01(4) D(calc.) = 3.03

Optical Properties: Translucent. *Color:* Deep brown with reddish internal reflections, light chocolate-brown; in transmitted light, pale to dark brown. *Optical Class:* Biaxial (–). *Pleochroism:* X = dark brown; Z = light brown. $\alpha = 1.646$ $\beta = 1.664$ $\gamma = 1.664$ 2V(meas.) = 0°

Cell Data: *Space Group:* n.d. $a = 5.47$ $b = 9.46$ $c = 28.8$ $Z = [4]$

X-ray Powder Pattern: Långban, Sweden.
7.23 (10), 3.61 (8), 4.79 (5), 1.574 (5), 14.6 (3), 2.697 (3), 1.634 (3)

Chemistry:

	(1)	(2)
SiO_2	33.06	32.35
Al_2O_3	0.58	0.57
Fe_2O_3	9.42	
FeO		7.30
MnO	33.83	35.50
ZnO	0.42	0.47
PbO	0.56	
MgO	11.55	10.26
CaO	0.07	0.08
F		< 0.05
Cl		0.05
H_2O^+	10.31	[9.89]
H_2O^-	0.02	
$-O = (F, Cl)_2$		[0.05]
Total	99.82	[96.47]

(1) Långban, Sweden; corresponds to $(Mn^{2+}_{3.25}Mg_{1.95}Fe^{3+}_{0.64}Zn_{0.04}Pb_{0.02}Ca_{0.01})_{\Sigma=5.91}$ $(Si_{3.75}Fe^{3+}_{0.17}Al_{0.08})_{\Sigma=4.00}[O_{10.20}(OH)_{7.80}]_{\Sigma=18.00}$. (2) Do.; by electron microprobe, H_2O calculated from stoichiometry; corresponds to $(Mn^{2+}_{3.64}Mg_{1.85}Fe_{0.74}Zn_{0.04}Ca_{0.01})_{\Sigma=6.28}$ $(Si_{3.92}Al_{0.08})_{\Sigma=4.00}O_{10}[(OH)_{7.99}Cl_{0.01}]_{\Sigma=8.00}$.

Mineral Group: Chlorite group.

Occurrence: In hydrothermal veinlets cutting skarn (Långban, Sweden).

Association: Caryopilite, bementite, berzeliite, barite, garnet (Långban, Sweden).

Distribution: From Långban, and at the Harstig mine, Pajsberg, near Persberg, Värmland, Sweden. In the Wessels mine, near Kuruman, Cape Province, South Africa.

Name: In honor of Forest A. Gonyer, analytical chemist in the Department of Mineralogy and Petrography, Harvard University, Cambridge, Massachusetts, USA.

Type Material: National Museum of Natural History, Washington, D.C., USA, 106913; The Natural History Museum, London, England, 1956,190–191.

References: (1) Frondel, C. (1955) Two chlorites: gonyerite and melanolite. Amer. Mineral., 40, 1090–1094. (2) Peacor, D.R. and E.J. Essene (1980) Caryopilite – a member of the friedelite rather than the serpentine group. Amer. Mineral., 65, 335–339.

Crystal Data: Monoclinic. *Point Group:* 2. As equant euhedral crystals, highly curved, to 4 cm; in polycrystalline aggregates.

Physical Properties: *Cleavage:* {010}, perfect. Hardness = ~4.5   D(meas.) = 2.21 D(calc.) = 2.23

Optical Properties: Transparent. *Color:* Colorless to white. *Streak:* White. *Luster:* Vitreous to pearly on crystal faces. *Optical Class:* Biaxial (–). *Orientation:* $Y = b$; $Z \wedge c = 46°$. $\alpha = 1.495(2)$ $\beta = 1.498(2)$ $\gamma = 1.502(2)$ 2V(meas.) = 82(5)°

Cell Data: *Space Group:* $P2_1$. $a = 7.401(3)$ $b = 17.439(36)$ $c = 7.293(3)$ $\beta = 105.44(4)°$ $Z = 2$

X-ray Powder Pattern: Goose Creek quarry, Virginia, USA. 4.53 (100), 7.19 (50), 5.59 (50), 4.91 (50), 3.350 (40), 3.526 (25), 3.277 (25)

Chemistry:

	(1)
SiO$_2$	59.3
Al$_2$O$_3$	17.2
CaO	9.3
H$_2$O	15.0
Total	100.8

(1) Goose Creek quarry, Virginia, USA; by electron microprobe, H$_2$O by DTA-TGA analysis; corresponding to Ca$_{1.01}$Al$_{2.05}$Si$_6$O$_{16.09}$·5.06H$_2$O.

Polymorphism & Series: Dimorphous with epistilbite.

Mineral Group: Zeolite group.

Occurrence: A late-stage mineral in vugs and fractures in a Triassic diabase (Goose Creek quarry, Virginia, USA); in cavities in basalt (Nasik, India).

Association: Prehnite, actinolite, chlorite, epidote, babingtonite, quartz, titanite, stilbite, albite, apophyllite (Goose Creek quarry, Virginia, USA); quartz (Nasik, India).

Distribution: In the Goose Creek quarry, Leesburg, Loudoun Co., Virginia, USA. Exceptional crystals from the Pandulena quarry, Nasik, Mararashtra, India. In the Oberbaumühle quarry, Windischeschenbach, Bavaria, Germany.

Name: For the initially described occurrence in the Goose Creek quarry, Virginia, USA.

Type Material: American Museum of Natural History, New York City, New York; Harvard University, Cambridge, Massachusetts, 117087; National Museum of Natural History, Washington, D.C., USA, 145880; Canadian Geological Survey, Ottawa; Royal Ontario Museum, Toronto, Canada; The Natural History Museum, London, England, 1980,579.

References: (1) Dunn, P.J., D.R. Peacor, N. Newberry, and R.A. Ramik (1980) Goosecreekite, a new calcium aluminum silicate hydrate, possibly related to brewsterite and epistilbite. Can. Mineral., 18, 323–327. (2) (1981) Amer. Mineral., 66, 1275 (abs. ref. 1). (3) Rouse, R.C. and D.R. Peacor (1986) Crystal structure of the zeolite mineral goosecreekite CaAl$_2$Si$_6$O$_{16}$·5H$_2$O. Amer. Mineral., 71, 1494–1501.

Götzenite $Na_2Ca_5Ti(Si_2O_7)_2F_4$

Crystal Data: Triclinic. *Point Group:* $\bar{1}$. Crystals prismatic, acicular, to 1 cm, may be skeletal. *Twinning:* Lamellar twinning on {001}, twin axis [010], ubiquitous.

Physical Properties: *Cleavage:* Perfect on {100}, good on {001}. Hardness = n.d. D(meas.) = 3.03–3.14   D(calc.) = [2.84]

Optical Properties: Transparent. *Color:* Colorless, white to honey-yellow. *Luster:* Greasy. *Optical Class:* Biaxial (+). *Orientation:* $Z' \wedge c \simeq 58°$. *Dispersion:* $r > v$, strong. $\alpha = 1.651–1.660$ $\beta = 1.653–1.662$ $\gamma = 1.659–1.670$ 2V(meas.) = 52°–84°   2V(calc.) = 53.5°

Cell Data: *Space Group:* $P\bar{1}$. $a = 9.667$ $b = 5.731$ $c = 7.334$ $\alpha = 90°$ $\beta = 101.05°$ $\gamma = 101.31°$ Z = 1

X-ray Powder Pattern: Mt. Shaheru, Zaire.
3.100 (100), 2.986 (100), 1.911 (50), 2.648 (40), 2.511 (25), 1.696 (25), 3.994 (15)

Chemistry:

	(1)	(2)	(3)		(1)	(2)	(3)
SiO_2	32.50	32.34	34.01	BaO	0.09	0.00	
TiO_2	9.72	8.74	11.31	Na_2O	4.85	6.32	8.77
ZrO_2		0.19		K_2O	0.14	0.09	
Al_2O_3	4.26	0.45		F	8.33	9.15	10.76
RE_2O_3		1.84		Cl	0.15	0.00	
Fe_2O_3	0.35	0.02		H_2O^+	0.26	0.57	
Nb_2O_5		3.36		H_2O^-	0.14	0.04	
FeO	0.45	0.14		CO_2	0.00		
MnO	0.07	0.62		P_2O_5	0.01	0.00	
MgO	0.29	0.04		SO_3	0.19	0.00	
CaO	41.80	38.95	39.68	$-O = (F, Cl)_2$	3.54	3.85	4.53
SrO	0.00	0.87		Total	100.06	99.88	100.00

(1) Mt. Shaheru, Zaire. (2) Lovozero massif, Russia. (3) $Na_2Ca_5Ti(Si_2O_7)_2F_4$.

Occurrence: In a nephelinite from the wall of the crater of an extinct volcano (Mt. Shaheru, Zaire); in hornfels and marble xenoliths in an intrusive alkalic gabbro-syenite complex (Mont Saint-Hilaire, Canada).

Association: Combeite (Mt. Shaheru, Zaire); aegirine, apatite, fersmanite, pectolite (Kola Peninsula, Russia); cancrinite, fluorite, titanite (Azov region, Ukraine).

Distribution: On Mt. Shaheru, the extinct southern cone of Mt. Nyiragongo, Kivu Province, Zaire. In Russia, from Mt. Yukspor, Khibiny massif, and the Lovozero massif, Kola Peninsula. At Barkevik, Langesundsfjord, Norway. From the Azov region, Ukraine. Near Kirchberg, Saxony; in the Kaiserstuhl, Baden-Württemberg; and around the Laacher See, Eifel district, Germany. At Mont Saint-Hilaire, Quebec, Canada.

Name: For Count Gustav Adolph von Götzen, a German who climbed Mt. Shaheru, Zaire, in 1894.

Type Material: The Natural History Museum, London, England, 1957,702; National Museum of Natural History, Washington, D.C., USA, 142981.

References: (1) Sahama, T.G. and K. Hytönen (1957) Götzenite and combeite, two new silicates from the Belgian Congo. Mineral. Mag., 31, 503–510. (2) (1958) Amer. Mineral., 43, 790 (abs. ref. 1). (3) Sahama, G. (1960) Identity of calcium rinkite and götzenite. Amer. Mineral., 45, 221–224. (4) Cannillo, E., F. Mazzi, and G. Rossi (1972) Crystal structure of götzenite. Kristallografiya (Sov. Phys. Crystal.), 16, 1167–1172 (in Russian). (5) Bulakh, A.G. and Y.L. Kapustin (1973) Götzenite from alkaline rocks of the Tur'ev Peninsula (Kola Peninsula). Zap. Vses. Mineral. Obshch., 102, 464–466 (in Russian). (6) (1974) Mineral. Abs., 25, 130 (abs. ref. 5). (7) Deer, W.A., R.A. Howie, and J. Zussman (1986) Rock-forming minerals, (2nd edition), v. 1B, disilicates and ring silicates, 343–347.

Crystal Data: Orthorhombic. *Point Group:* $2/m\ 2/m\ 2/m$. As anhedral elongated individuals, to 8 cm, typically strongly corroded.

Physical Properties: *Cleavage:* {100}, more perfect, and {010}. Hardness = 7.5 D(meas.) = 2.98–2.99 D(calc.) = [3.00]

Optical Properties: Transparent to translucent. *Color:* Bluish green, greenish blue with increasing Fe; in thin section, colorless to blue or green. *Luster:* Vitreous, somewhat pearly on the cleavage.
Optical Class: Biaxial (–). *Pleochroism:* X = deep blue to blue-green; Y = colorless; Z = light greenish blue to deep green. *Orientation:* X = a; Y = c; Z = b. *Dispersion:* $r < v$, very strong. *Absorption:* Z > X > Y. $\alpha = 1.590$–1.602 $\beta = 1.618$–1.636 $\gamma = 1.623$–1.639 2V(meas.) = 27°–32°

Cell Data: *Space Group:* Pbnm. $a = 10.335(2)$ $b = 10.978(2)$ $c = 5.760(2)$ Z = 4

X-ray Powder Pattern: Sakatelo, Madagascar.
5.17 (vvs), 5.04 (vvs), 5.482 (vs), 2.744 (vs), 2.166 (s), 3.708 (ms), 2.584 (ms)

Chemistry:

	(1)	(2)	(3)		(1)	(2)	(3)
SiO_2	20.39	19.9	19.77	MnO	0.04	0.12	
TiO_2	0.13	0.02		MgO	12.04	9.2	6.63
B_2O_3	11.57	n.d.	11.45	CaO	0.00	0.00	
Al_2O_3	52.12	50.35	50.33	Na_2O	0.04	0.01	
Fe_2O_3	0.80			K_2O	0.09	0.04	
FeO	2.87	7.75	11.82	Total	100.09	87.39	100.00

(1) Sakatelo, Madagascar; corresponds to $(\mathrm{Mg}_{0.87}\mathrm{Fe}^{2+}_{0.12}\mathrm{Fe}^{3+}_{0.03})_{\Sigma=1.02}\mathrm{Al}_{3.00}(\mathrm{B}_{0.97}\mathrm{O}_3)(\mathrm{Si}_{0.99}\mathrm{O}_4)\mathrm{O}_2$.
(2) Mt. Amiata, Italy; by electron microprobe, corresponds to $(\mathrm{Mg}_{0.69}\mathrm{Fe}_{0.32})_{\Sigma=1.01}$ $\mathrm{Al}_{2.98}(\mathrm{BO}_3)(\mathrm{Si}_{1.00}\mathrm{O}_4)\mathrm{O}_2$. (3) $(\mathrm{Mg}, \mathrm{Fe})\mathrm{Al}_3(\mathrm{BO}_3)(\mathrm{SiO}_4)\mathrm{O}_2$ with Mg:Fe = 1:1.

Occurrence: A rare accessory mineral in aluminous boron-rich rocks, thermally and high-grade regionally metamorphosed under low pressure; in pegmatites, aplites, gneisses; in xenoliths.

Association: Quartz, potassic feldspar, plagioclase, biotite, garnet, "hypersthene," spinel, corundum, sillimanite, andalusite, cordierite, tourmaline, kornerupine, sapphirine, serendibite, sinhalite.

Distribution: In large crystals from Andrahomana, southwest of Taolañaro (Fort Dauphin); at Marotrana, Amboasara; from Vohiboly, near Behara; Ampamatoa; and elsewhere on Madagascar. In the Kachebere Hills, Mchinji district, Malawi. From Tizi-Ouchen, Béjaia, Algeria. From near Gananoque, Ontario, Canada. At Johnsburg, Warren Co., New York, USA. On Mt. Amiata and Mt. Cimino, Tuscany, Italy. Around Vestpolltind, and on Almjotheia, 12 km north of Moi, Norway. At Maratakka, Surinam. On Cuvier Island, New Zealand. Known from a number of other minor localities.

Name: For Alfred Grandidier (1836–1912), French naturalist and explorer, an authority on Madagascar.

Type Material: Harvard University, Cambridge, Massachusetts, USA, 86382.

References: (1) Dana, E.S. and W.E. Ford (1909) Dana's system of mineralogy, (6th edition), app. II, 47. (2) McKie, D. (1965) The magnesium aluminum borosilicates: kornerupine and grandidierite. Mineral. Mag., 34, 346–357. (3) Stephenson, D.A. and P.B. Moore (1968) The crystal structure of grandidierite, $(\mathrm{Mg}, \mathrm{Fe})\mathrm{Al}_3\mathrm{SiBO}_9$. Acta Cryst., 24, 1518–1522. (4) Van Bergen, M.J. (1980) Grandidierite from aluminous metasedimentary xenoliths within acid volcanics, a first record in Italy. Mineral. Mag., 43, 651–658.

Crystal Data: Monoclinic. *Point Group:* n.d. Rarely as minute crystals, more commonly as rounded grains, to 1 mm; as porphyroblasts, oolites; cryptocrystalline.

Physical Properties: Hardness = n.d. D(meas.) = 2.85–3.15 D(calc.) = n.d. Moderately magnetic.

Optical Properties: Semitransparent. *Color:* Green, light yellow-green; blue-green in transmitted light.
Optical Class: Isotropic when fine grained. *Pleochroism:* Pale yellow to green. $n = 1.670$

Cell Data: *Space Group:* n.d. $a = 5.54$ $b = 9.55$ $c = 7.44$ $\beta = 104°20'$ $Z = 2$

X-ray Powder Pattern: Mesabi Range, Minnesota, USA.
2.571 (100), 7.12 (80), 3.559 (80), 1.593 (60), 2.184 (40), 1.553 (40), 2.849 (20)

Chemistry:

	(1)	(2)	(3)
SiO_2	33.58	34.7	36.5
Al_2O_3		0.90	0.25
Fe_2O_3	11.16		
FeO	45.19	47.3	40.2
MnO		0.15	8.71
MgO		4.98	3.75
H_2O	10.07	n.d.	n.d.
Total	[100.00]	88.03	89.41

(1) Mesabi Range, Minnesota, USA; recalculated to 100% after deduction of SiO_2 and other impurities; corresponds to $(Fe^{2+}_{2.34}Fe^{3+}_{0.45})_{\Sigma=2.79}Si_{2.18}O_5(OH)_{3.28}$. (2) Do.; by electron microprobe. (3) Bluebell mine, Riondel, Canada; by electron microprobe.

Polymorphism & Series: Group A (1M, 2M$_1$, 3A) and Group C (1A, 2A, 3R) polytypes may be intergrown.

Mineral Group: Kaolinite-serpentine group.

Occurrence: A primary phase in some banded iron formations.

Association: Minnesotaite, stilpnomelane, siderite, chamosite, quartz, pyrite.

Distribution: Some localities for studied material follow. Widespread in the Biwabik Iron Formation, Mesabi Range, St. Louis Co., Minnesota, USA. In Canada, in the Sokoman Iron Formation, Howells River area, Labrador, Newfoundland; in the Gunflint Iron Formation, northeast of Gunflint Lake, Ontario; and in the Bluebell mine, Riondel, British Columbia. In the Matsumo mine, Kochi prefecture, Japan. From the Weld Range, northwest of Cue, Western Australia. In the Salsigne mine, north of Carcassonne, Aude, France. At Glenluce, Wigtownshire, Scotland. From the San Valentin mine, Cartagena district, La Union, Murcia Province, Spain.

Name: In allusion to its *green* color.

Type Material: National Museum of Natural History, Washington, D.C., USA, 91160, 124953.

References: (1) Dana, E.S. and W.E. Ford (1909) Dana's system of mineralogy, (6th edition), app. II, 47. (2) Deer, W.A., R.A. Howie, and J. Zussman (1963) Rock-forming minerals, v. 3, sheet silicates, 164–169. (3) Jolliffe, F. (1935) A study of greenalite. Amer. Mineral., 20, 405–425. (4) Gruner, J.W. (1936) The structure and chemical composition of greenalite. Amer. Mineral., 21, 449–455. (5) Bayliss, P. (1981) Unit cell data of serpentine group minerals. Mineral. Mag., 44, 153–156. (6) Guggenheim, S., S.W. Bailey, R.A. Eggleton, and P. Wilkes (1982) Structural aspects of greenalite and related minerals. Can. Mineral., 20, 1–18.

Crystal Data: Cubic. *Point Group:* $4/m\ \overline{3}\ 2/m$. Commonly in dodecahedra or trapezohedra, up to 15 cm, with striated faces. Also granular, compact, and massive.

Physical Properties: *Cleavage:* Parting on {110} rarely observed. *Fracture:* Uneven to conchoidal. *Tenacity:* Brittle. Hardness = 6.5–7 D(meas.) = 3.594 D(calc.) = 3.594 May fluoresce weak golden yellow under LW or SW UV.

Optical Properties: Transparent to opaque. *Color:* Yellow-green, pale to dark green; golden yellow, pink, red, orange, brownish red, yellowish brown; colorless, white, gray, black; may be sectored: colorless in thin section. *Streak:* White. *Luster:* Vitreous to resinous. *Optical Class:* Isotropic; weak strain birefringence. $n = 1.734$

Cell Data: *Space Group:* $Ia3d$. a = 11.851 Z = 8

X-ray Powder Pattern: Georgetown, Placer Co., California, USA. (ICDD 26-292). 2.647 (100), 1.581 (50), 2.959 (25), 1.921 (25), 1.643 (25), 2.417 (20), 2.321 (18)

Chemistry:

	(1)	(2)
SiO_2	39.30	40.02
TiO_2	0.00	
Al_2O_3	21.93	22.63
Fe_2O_3	0.80	
Cr_2O_3	0.13	
FeO	0.28	
MgO	trace	
CaO	37.10	37.35
H_2O	0.30	
Total	99.84	100.00

(1) Georgetown, Placer Co., California, USA; corresponds to $(Ca_{3.00}Fe^{2+}_{0.02})_{\Sigma=3.02}$ $(Al_{1.92}Fe^{3+}_{0.05})_{\Sigma=1.97}(Si_{2.97}Al_{0.03})_{\Sigma=3.00}O_{12}$. (2) $Ca_3Al_2(SiO_4)_3$.

Polymorphism & Series: Forms three series: with andradite; with hibschite and katoite; and with uvarovite.

Mineral Group: Garnet group.

Occurrence: In contact and regionally metamorphosed calcareous rocks, or rocks which have undergone calcium metasomatism; in some schists and serpentinites.

Association: Calcite, dolomite, epidote, clinozoisite, wollastonite, scapolite, vesuvianite, diopside, tremolite, quartz.

Distribution: Many localities, even for fine crystals and gem material. In Russia, near Chernyshevsk, at the confluence of the Vilyui and Akhtaragdy Rivers, Yakutia. In the Ala Valley, Piedmont, Italy. From the Zillertal, Tirol, Austria. At Ocna de Fier, Romania. In the Jeffrey mine, Asbestos, Quebec, Canada. In the USA, at Standish, Cumberland Co., Maine, and at the Belvidere Mountain quarries, Lowell, Orleans Co., Vermont. In California, from Crestmore, Riverside Co., the Old Cosumnes copper mine, El Dorado Co., San Carlos mine, Mazourka Canyon, Inyo Range, Inyo Co., and Santa Rosa, Sonoma Co.; on Vesper Peak, Sultan basin, Snohomish Co., Washington. In Mexico, large crystals from Xalostoc and Morelos, Chihuahua, and from the Sierra de la Cruces, near Lake Jaco, Coahuila. Gems from the Merelani Hills, southeast of Arusha, Tanzania. In the Taita Hills, near Voi, Kenya.

Name: From *Ribes Grossularium* or gooseberry; the botanical name, in reference to a typical greenish color.

References: (1) Dana, E.S. (1892) Dana's system of mineralogy, (6th edition), 437–447. (2) Deer, W.A., R.A. Howie, and J. Zussman (1982) Rock-forming minerals, (2nd edition), v. 1A, orthosilicates, 468–698, esp. 603–616. (3) Novak, G.A. and G.V Gibbs (1971) The crystal chemistry of the silicate garnets. Amer. Mineral., 56, 791–825.

Grumantite $NaHSi_2O_5 \cdot H_2O$

Crystal Data: Orthorhombic. *Point Group: mm2.* As irregular crystals; compact massive.

Physical Properties: *Cleavage:* Perfect on {110}; another unspecified cleavage was observed. *Fracture:* Steplike. Hardness = 4–5 D(meas.) = 2.21 D(calc.) = 2.26 Weak bluish white fluorescence in UV.

Optical Properties: Transparent to slightly turbid in crystals. *Color:* White in aggregates. *Streak:* White. *Luster:* Vitreous, silky to dull in aggregates. *Optical Class:* Biaxial (+). *Orientation:* X = a; Y = c; Z = b. *Dispersion:* $r > v$, weak. $\alpha = 1.494(2)$ $\beta = 1.507(2)$ $\gamma = 1.523(2)$ 2V(meas.) = 85(2)° 2V(calc.) = 84.9°

Cell Data: *Space Group: Fdd2.* a = 15.979 b = 18.25 c = 7.169 Z = 16

X-ray Powder Pattern: Mt. Alluaiv, Russia.
3.505 (100), 3.006 (100), 6.20 (50), 6.05 (50), 4.46 (50), 3.087 (50), 3.346 (25)

Chemistry:

	(1)	(2)
SiO_2	68.17	67.44
Na_2O	17.20	17.39
H_2O	14.65	15.17
Total	100.02	100.00

(1) Mt. Alluaiv, Russia; corresponds to $Na_{0.98}H_{1.02}Si_2O_5 \cdot 0.93H_2O$. (2) $NaHSi_2O_5 \cdot H_2O$.

Occurrence: In veins cutting alkalic feldspathoid-bearing pegmatites in a differentiated alkalic massif.

Association: Ussingite, makatite, kazakovite, tisinalite, nordite, sodalite, nepheline, potassic feldspar, arfvedsonite, aegirine, eudialyte, loparite.

Distribution: On Mt. Alluaiv, Lovozero massif, Kola Peninsula, Russia.

Name: From an old Russian name for the Spitsbergen Archipelago.

Type Material: A.E. Fersman Mineralogical Museum, Academy of Sciences, Moscow, Russia.

References: (1) Khomyakov, A.P., M.F. Korobitsyn, T.A. Kurova, and G.E. Cherepivskaya (1987) Grumantite $NaHSi_2O_5 \cdot H_2O$ – a new mineral. Zap. Vses. Mineral. Obshch., 116, 244–248 (in Russian). (2) (1988) Amer. Mineral., 73, 440 (abs. ref. 1). (3) Pushcharovsky [Pushcharovskii], D.Y., T.N. Nadezhina, N.A. Yamnova, and R.L. Rastsvetaieva [Rastsvetaeva] (1988) New types of tetrahedral radicals in mineral structures: $Be(OH)_2$ and grumantite. Zeits. Krist., 185, 612.

$Fe_2^{2+}(Fe^{2+}, Mg)_5Si_8O_{22}(OH)_2$ **Grunerite**

Crystal Data: Monoclinic. *Point Group: 2/m.* Commonly as acicular or fibrous radiating crystals; may be asbestiform. *Twinning:* Simple or multiple twinning common ∥ {100}, producing narrow twin lamellae.

Physical Properties: *Cleavage:* Perfect on {110}, with intersections of ~56° and ~124°. *Tenacity:* Brittle. Hardness = 5–6 D(meas.) = 3.40–3.60 D(calc.) = 3.531

Optical Properties: Translucent. *Color:* Dark green to brown, gray to greenish gray; colorless to pale green or brown in thin section. *Luster:* Vitreous, silky when fibrous.
Optical Class: Biaxial (–). *Pleochroism:* With increasing Fe content, X = pale yellow; Y = pale yellow-brown; Z = pale brown. *Orientation:* $Y = b$; $Z \wedge c = -16°$ to $-12°$; $X \wedge a = -3°$ to $2°$. *Dispersion:* $r > v$, weak. $\alpha = 1.663$–1.688 $\beta = 1.677$–1.709 $\gamma = 1.697$–1.729
2V(meas.) = 80°–90°

Cell Data: *Space Group: C2/m.* a = 9.5642(7) b = 18.393(2) c = 5.3388(3)
$\beta = 101.892(3)°$ Z = 2

X-ray Powder Pattern: Wabush Iron Formation, White Lake, Newfoundland, Canada.
8.33 (100), 3.06 (70), 2.756 (70), 2.189 (50), 4.13 (40), 3.26 (40), 2.628 (40)

Chemistry:

	(1)	(2)		(1)	(2)
SiO_2	49.33	47.54	MgO	6.65	0.04
TiO_2	0.02		CaO	0.18	0.00
Al_2O_3	0.39	0.20	Na_2O	0.12	0.29
Fe_2O_3		0.71	K_2O	0.20	0.11
FeO	40.94	47.25	F		0.01
MnO	0.54	2.14	H_2O^+	1.54	1.55
			Total	99.91	99.84

(1) Wabush Iron Formation, Newfoundland, Canada; corresponds to $(Fe_{5.50}^{2+}Mg_{1.59}Mn_{0.08}Na_{0.04}$ $K_{0.04}Ca_{0.03})_{\Sigma=7.28}(Si_{7.92}Al_{0.08})_{\Sigma=8.00}O_{22}(OH)_{1.64}$. (2) Rockport, Essex Co., Massachusetts, USA; corresponds to $(Fe_{6.63}^{2+}Mn_{0.30}Fe_{0.09}^{3+}Na_{0.09}K_{0.02}Mg_{0.01}Al_{0.01})_{\Sigma=7.15}(Si_{7.97}Al_{0.03})_{\Sigma=8.00}O_{22}(OH)_{1.74}$.

Polymorphism & Series: Forms a series with magnesio-cummingtonite and cummingtonite.

Mineral Group: Amphibole (Fe–Mn–Mg) group: $Mg/(Mg + Fe^{2+}) < 0.3$; $(Ca + Na)_B < 1.34$; $Li < 1.0$; $Mn < 0.5$.

Occurrence: Common in medium- to high-grade metamorphosed iron formations. A product of contact metamorphism, and in some blueschist facies metaquartzites.

Association: Magnetite, hematite, hedenbergite, riebeckite, fayalite, garnet, quartz.

Distribution: Many localities. From Collobrières, Var, France. At Loch Duich, Ross-shire, and Glen Beag, Glenelg, Inverness-shire, Scotland. From the Botallack mine, St. Just, Cornwall, England. In Portugal, at Vila Real. At Strömshult, near Tunaberg, Sweden. Commercial production from the Penge area, Transvaal, South Africa. From around Michigamme, Marquette Co., Michigan, USA. In Canada, in the Labrador City area, Labrador, Newfoundland. At Gualilan, Argentina.

Name: To honor the French chemist, Louis Emmanuel Gruner (1809–1883), who provided the first analysis.

References: (1) Dana, E.S. (1892) Dana's system of mineralogy, (6th edition), 386, 391. (2) Deer, W.A., R.A. Howie, and J. Zussman (1963) Rock-forming minerals, v. 2, chain silicates, 235–248. (3) Klein, C. (1964) Cummingtonite-grunerite series: a chemical, optical and X-ray study. Amer. Mineral., 49, 963–982. (4) Finger, L.W. (1969) The crystal structure and cation distribution of a grunerite. MSA Special Paper 2, 95–100. (5) Chakraborty, K.L. (1963) Relationship of anthophyllite, cummingtonite and mangano-cummingtonite in the metamorphosed Wabush Iron-formation, Labrador. Can. Mineral., 7, 738–750. (6) Phillips, W.R. and D.T. Griffen (1981) Optical mineralogy, 225–228.

Crystal Data: Tetragonal. *Point Group:* $\bar{4}2m$. As thin square tablets, to 3 mm, with prominent {001}, also {011}, {111}, and {110}.

Physical Properties: *Cleavage:* Perfect on {010}, distinct on {001}, poor on {110}. *Fracture:* Irregular. Hardness = ~5 D(meas.) = 3.034 D(calc.) = 3.03 Strongly piezoelectric.

Optical Properties: Transparent. *Color:* Colorless. *Luster:* Vitreous. *Optical Class:* Uniaxial (+). $\omega = 1.664(1)$ $\epsilon = 1.672(1)$

Cell Data: *Space Group:* $P\bar{4}2m$. a = 7.48(2) c = 5.044(3) Z = [2]

X-ray Powder Pattern: Gugia, China.
2.765 (10), 1.709 (7), 1.485 (7), 5.25 (4), 2.97 (4), 2.359 (4), 2.315 (4)

Chemistry:

	(1)	(2)	(3)		(1)	(2)	(3)
SiO_2	45.26	44.90	46.70	Na_2O		0.72	
TiO_2	0.02	trace		K_2O		0.20	
$(Zr, Hf)O_2$	0.45			F		0.25	
Al_2O_3	1.08	2.17		Cl	0.73	0.18	
Fe_2O_3	0.03	0.11		H_2O^+	0.40	0.90	
MnO	0.11	0.07		H_2O^-	0.02	0.36	
BeO	8.89	9.49	9.72	P_2O_5		0.08	
MgO	0.39	0.38		$-O = (F, Cl)_2$	0.17	0.15	
CaO	42.94	40.09	43.58	Total	100.15	[99.75]	100.00

(1) Gugia, China. (2) Do.; original total given as 99.79%. (3) $Ca_2BeSi_2O_7$.

Polymorphism & Series: Dimorphous with jeffreyite.

Mineral Group: Melilite group.

Occurrence: In cavities in skarns and melanite adjacent to an alkalic syenite.

Association: Orthoclase, vesuvianite, aegirine, titanite, apatite, prehnite.

Distribution: Near the village of Gugia, otherwise unlocated in China.

Name: For the locality near Gugia, China.

Type Material: n.d.

References: (1) Chi-Jui Peng, Rung-Lung Tsao, and Zu-Rung Zou (1962) Gugiaite, $Ca_2BeSi_2O_7$, a new beryllium mineral and its relation to the melilite group. Scientia Sinica, 11, 977–988 (in English). (2) (1963) Amer. Mineral., 48, 211–212 (abs. ref. 1). (3) Kimata, M. and H. Ohashi (1982) The crystal structure of synthetic gugiaite, $Ca_2BeSi_2O_7$. Neues Jahrb. Mineral., Abh., 143, 210–222.

Crystal Data: Triclinic, pseudohexagonal. *Point Group:* $\bar{1}$. As spherical to radial masses, to 5 cm; massive, platy or plumose aggregates; as fibrous layers. *Twinning:* At a microscopical scale by rotation of 120° and 240° ⊥ (001).

Physical Properties: *Cleavage:* Perfect basal; poor prismatic. *Tenacity:* Flakes flexible. Hardness = 3–4   D(meas.) = 2.388–2.390   D(calc.) = 2.40

Optical Properties: Transparent to translucent, opaque. *Color:* Colorless to white, may be grayish, tan, yellowish, brownish, or pale green from inclusions; colorless in thin section. *Luster:* Vitreous to pearly.
Optical Class: Uniaxial (–). $\omega = 1.540–1.550$ $\epsilon = 1.535–1.536$

Cell Data: *Space Group:* $P\bar{1}$. $a = 9.74(1)$ $b = 9.74(1)$ $c = 22.40(2)$ $\alpha = 95.7(1)°$ $\beta = 91.5(1)°$ $\gamma = 120.0(1)°$ $Z = 4$

X-ray Powder Pattern: Bombay, India; {00*l*} reflections only are sharp, may resemble truscottite or reyerite.
22. (vs), 11.0 (s), 4.20 (s), 1.82 (s), 3.65 (ms), 2.80 (ms), 8.4–7.4 (md)

Chemistry:

	(1)	(2)
SiO_2	50.70	50.88
Al_2O_3	1.48	1.75
Fe_2O_3		0.25
MgO	0.18	0.74
CaO	33.24	31.34
Na_2O		1.34
H_2O	14.18	[13.68]
Total	99.78	[99.98]

(1) Isle of Skye, Scotland; corresponds to $Ca_{15.80}Mg_{0.12}(Si_{23.20}Al_{0.80})_{\Sigma=24.00}O_{60}(OH)_{7.04} \cdot$ $18.12H_2O$. (2) Qarusait, Greenland; by XRF and AA, H_2O by weight loss; corresponds to $Na_{1.18}Ca_{15.22}Mg_{0.50}Fe_{0.08}(Si_{23.07}Al_{0.93})_{\Sigma=24.00}O_{60}(OH)_{7.92} \cdot 16.72H_2O$.

Occurrence: As replacements of wall rocks and in vugs, amygdules, and veinlets in basalts; in hydrothermally altered rhyolites and sediments; in some ore deposits.

Association: Apophyllite, stilbite, laumontite, thomsonite, okenite, tobermorite, xonotlite, calcite.

Distribution: At Portree, Isle of Skye and 'S Airde Beinn, Isle of Mull, Scotland. From Ballhenry, Co. Antrim, Ireland. At Ortano, Elba, and at Monte Biaena, Pannone, Trentino-Alto Adige, Italy. At Ljósá, Eysturoy, and Gjónoyri, Streymoy, Faeroe Islands. From Qarusait and Tapaursak, Greenland. In India, from Nasik, Poona, on Bombay Island and around Bombay, Maharashtra. In Japan, at Sawayama Lake, Nagano Prefecture; Irakawa, Yamagata Prefecture; and Otemo, Niigata Prefecture. In the USA, from Fort Point, San Francisco Co., and in the New Almaden mine, Santa Clara Co., California; from Yellowstone National Park, Wyoming; in the Goose Creek quarry, Leesburg, Loudoun Co., Virginia. From 40 km southwest of Cape Blomiden, between Margaretville and Port George, Nova Scotia, Canada. A few other localities are known.

Name: From the Greek for *round*, in reference to its common habit.

References: (1) Dana, E.S. (1892) Dana's system of mineralogy, (6th edition), 566. (2) Mackay, A.L. and H.F.W. Taylor (1953) Gyrolite. Mineral. Mag., 30, 80–91. (3) Merlino, S. (1988) Gyrolite: its crystal structure and crystal chemistry. Mineral. Mag., 52, 377–387.

Crystal Data: Tetragonal. *Point Group:* $4/m\ 2/m\ 2/m$. Euhedral to irregular crystals and fragments, to 1 cm, heavily zoned with zircon, the outermost portions of which represent this species.

Physical Properties: Hardness = n.d. D(meas.) = n.d. D(calc.) = 6.97 (synthetic).

Optical Properties: Transparent to translucent. *Color:* Orange-red, brownish yellow, rarely colorless.
Optical Class: [Uniaxial.] ω = n.d. ϵ = n.d.

Cell Data: *Space Group:* $I4_1/amd$ (synthetic). $a = 6.5725(7)$ $c = 5.9632(4)$ $Z = 4$

X-ray Powder Pattern: Synthetic; cannot be distinguished from zircon.
3.29 (100), 2.512 (70), 4.43 (60), 1.705 (55), 2.638 (25), 2.057 (20), 2.324 (18)

Chemistry:

	(1)	(2)
SiO$_2$	28.32	27.20
ZrO$_2$	3.28	1.21
HfO$_2$	69.78	72.52
Total	101.38	100.93

(1) Muiâne mine, Mozambique; by electron microprobe, corresponding to $(Hf_{0.80}Zr_{0.06})_{\Sigma=0.86}$ $Si_{1.14}O_4$. (2) Do.; by electron microprobe, corresponding to $(Hf_{0.86}Zr_{0.02})_{\Sigma=0.88}Si_{1.12}O_4$.

Occurrence: In tantalum-bearing granite pegmatites.

Association: Cookeite, albite.

Distribution: At the Morro Conco, Moneia, and Muiâne mines, Morrua area, Zambézia district, Mozambique.

Name: For *hafnium* in the composition.

Type Material: n.d.

References: (1) Correia Neves, J.M., J.E. Lopes Nunes, and T.G. Sahama (1974) High hafnium members of the zircon-hafnon series from the granite pegmatites of Zambézia, Mozambique. Contr. Mineral. Petrol., 48, 73–80. (2) (1976) Amer. Mineral., 61, 175 (abs. ref. 1). (3) Salt, D.J. and G. Hornung (1967) Synthesis and X-ray study of hafnium silicate. J. Amer. Ceram. Soc., 50, 549–550. (4) Speer, J.A. and B.J. Cooper (1982) Crystal structure of synthetic hafnon, HfSiO$_4$, comparison with zircon and the actinide orthosilicates. Amer. Mineral., 67, 804–808.

Crystal Data: Triclinic. *Point Group:* $\bar{1}$. Crystals, < 1 mm, slender needles and plates, elongated along [001] with {010}, {100}, and {$\bar{1}$10}. *Twinning:* On {100}.

Physical Properties: *Cleavage:* {010}, rather perfect; {100}, faint. *Fracture:* Brittle. Hardness = 5 D(meas.) = 3.148 D(calc.) = 3.157

Optical Properties: Semitransparent. *Color:* Wine-yellow, honey-yellow, colorless; light wine-yellow in thin section. *Luster:* Vitreous to adamantine.
Optical Class: Biaxial (+). *Pleochroism:* X = colorless; Z = wine-yellow. *Dispersion:* $r > v$, strong. *Absorption:* $Z > Y > X$. $n = \sim 1.7$, birefringence ~ 0.012. 2V(meas.) = Large.

Cell Data: *Space Group:* $P\bar{1}$. a = 5.676(2) b = 7.259(3) c = 9.586(3) $\alpha = 101.08(4)°$ $\beta = 101.14(4)°$ $\gamma = 90.27(4)°$ Z = 1

X-ray Powder Pattern: Hradiště Mountain, Czech Republic.
2.961 (100), 3.073 (90), 1.896 (70), 3.961 (50), 2.628 (50), 2.493 (50), 1.822 (50)

Chemistry:

	(1)		(1)
SiO_2	32.07	Nb_2O_5	1.38
TiO_2	8.06	Ta_2O_5	0.22
ZrO_2	6.58	MnO	2.06
Al_2O_3	0.03	CaO	32.05
La_2O_3	0.60	Na_2O	7.46
Ce_2O_3	1.01	F	12.09
Fe_2O_3	1.25	$-O = F_2$	5.09
		Total	99.77

(1) Hradiště Mountain, Czech Republic; by electron microprobe, average of fourteen analyses; corresponds to $(Na_{3.52}Ca_{0.48})_{\Sigma=4.00}Ca_{8.00}RE_{0.15}(Ti_{1.52}Zr_{0.80}Mn_{0.45}Fe_{0.25}Nb_{0.09}Ta_{0.02}Al_{0.01})_{\Sigma=3.14}$ $Si_{7.80}[O_{27.68}F_{8.32}]_{\Sigma=36.00}$.

Occurrence: In cavities and embedded in the groundmass of a phonolite.

Association: Aegirine.

Distribution: From Hradiště Mountain, near Mildenau, Czech Republic.

Name: For the locality, formerly named Hoher Hain, in the Czech Republic.

Type Material: Charles University, Prague, Czech Republic, 20530.

References: (1) Dana, E.S. (1899) Dana's system of mineralogy, (6th edition), app. I, 31–32. (2) Vlasov, K.A., Ed. (1966) Mineralogy of rare elements, v. II, 385. (3) Johan, Z. and F. Čech (1989) New data on hainite, $Na_2Ca_4[(Ti, Zr, Mn, Fe, Nb, Ta)_{1.50}\square_{0.50}](Si_2O_7)_2F_4$ and its crystallochemical relationship with götzenite, $Na_2Ca_5Ti(Si_2O_7)_2F_4$. Compt. Rendus Acad. Sci. Paris, 308, 1237–1242 (in French with English abs.). (4) (1990) Amer. Mineral., 75, 936 (abs. ref. 3).

Crystal Data: Monoclinic. *Point Group:* $2/m$ (probable). As spherulites consisting of minute, bladelike grains; also as single flakelike grains, flattened on {100}.

Physical Properties: *Cleavage:* Good on {100}. Hardness = 3.5 D(meas.) = 3.35 D(calc.) = [4.93] Fluoresces weakly dull green under UV; radioactive.

Optical Properties: Semitransparent. *Color:* Pale yellow to greenish yellow. *Luster:* Pearly on the cleavage.
Optical Class: Biaxial (−). *Pleochroism:* Y = pale yellow; Z = colorless. *Orientation:* $Y = b$.
Dispersion: $r > v$, very strong. $\alpha = 1.533–1.571$ $\beta = 1.572–1.580$ $\gamma = 1.573–1.582$
2V(meas.) = 15°–20°

Cell Data: *Space Group:* $P2/c$ (probable). a = 15.4 b = 7.05 c = 7.10 $\beta = 107°52'$
Z = [2]

X-ray Powder Pattern: Coso Mountains, California, USA.
9.14 (100), 4.56 (60), 4.42 (60), 3.19 (50), 3.11 (50), 7.05 (40), 3.54 (40)

Chemistry:

	(1)	(2)
SiO_2	33.1	33.42
UO_3	52.8	53.03
CaO	5.4	5.20
H_2O	8.7	8.35
Total	[100.0]	100.00

(1) Coso Mountains, California, USA; inseparably combined with metahaiweeite, average of four analyses, recalculated to 100.0% after deduction of insolubles, mainly quartz.
(2) $Ca(UO_2)_2Si_6O_{15} \cdot 5H_2O$.

Occurrence: On fracture surfaces in granite and in voids in neighboring loosely consolidated lake bed sediments (Coso Mountains, California, USA); along fractures in tourmaline-bearing granite (Perus, Brazil).

Association: Autunite, uranophane, uranophane-beta, phosphuranylite, torbernite (Perus, Brazil).

Distribution: From the Haiwee Reservoir, Coso Mountains, Inyo Co., California, USA. At Perus, 25 km north of São Paulo, Brazil. On Portezuelo Hill, Ranquil district, Mendoza Province, Argentina. From Badgastein, Salzburg, Austria.

Name: For the Haiwee Reservoir, Coso Mountains, California, USA, above which the mineral was first found to occur.

Type Material: n.d.

References: (1) McBurney, T.C. and J. Murdoch (1959) Haiweeite, a new uranium mineral from California. Amer. Mineral., 44, 839–843. (2) de Camargo, W.G.R. and D.P. Svisero (1969) Haiweeite, a new occurrence in Brazil. Amer. Mineral., 54, 966–969.

Crystal Data: Monoclinic. *Point Group: m.* Massive, claylike. On a microscopic scale, tubular, average 0.04 μm; tubes split open and unroll to form laths or spatula shapes; spherical.

Physical Properties: *Cleavage:* {001}, probable. *Fracture:* Conchoidal. Hardness = 2–2.5 D(meas.) = 2.55–2.565   D(calc.) = [2.57]

Optical Properties: Semitransparent. *Color:* White; gray, green, blue, yellow, red from included impurities. *Luster:* Pearly, waxy, or dull.
Optical Class: Biaxial. $n = 1.55$–1.56   2V(meas.) = n.d.

Cell Data: *Space Group: Cc.* $a = 5.14(4)$ $b = 8.90(4)$ $c = 14.9(1)$ $\beta = 101.9°$ $Z = [4]$

X-ray Powder Pattern: Indiana, USA. (ICDD 29-1489).
10.0 (100), 4.36 (70), 3.35 (40), 2.54 (35), 1.480 (30), 1.672 (14), 1.281 (8)

Chemistry:

	(1)	(2)		(1)	(2)
SiO$_2$	43.98	46.55	CaO	0.32	
TiO$_2$	0.01		Na$_2$O	0.14	
Al$_2$O$_3$	38.46	39.49	K$_2$O	0.48	
FeO	0.03		H$_2$O$^+$	14.59	13.96
MgO	trace		H$_2$O$^-$	2.58	
			Total	100.59	100.00

(1) Eureka, Utah, USA. (2) Al$_2$Si$_2$O$_5$(OH)$_4$.

Polymorphism & Series: Dickite, kaolinite, and nacrite are polymorphs.

Mineral Group: Kaolinite-serpentine group; called endellite in European literature.

Occurrence: A product of hydrothermal alteration or surface weathering of aluminosilicate minerals, as feldspars. Formed by dehydration of endellite above 110 °C.

Association: Kaolinite, endellite, allophane, alunite.

Distribution: Undoubtedly widespread. Probably at Angleur, Liège, Belgium, the "type" locality, now inaccessible. From Valença, Portugal. At Bergnersreuth, near Wunsiedel, Bavaria, Germany. In the USA, from Wagon Wheel Gap, Mineral Co., Colorado; at Gonzales, Gonzales Co., Texas; near Bedford, Lawrence Co., Indiana; at Horse Cove, Hart Co., Kentucky; in the Corundum Hill mine, Macon Co., North Carolina. Also from the Dragon Consolidated iron mine, Eureka, Tintic district, Juab Co., Utah; in a number of porphyry copper deposits in Arizona, as at Ray, Pinal Co., Globe-Miami, Gila Co., and others.

Name: To honor Baron Omalius d'Halloy (1707–1789), who first observed the mineral.

References: (1) Dana, E.S. (1892) Dana's system of mineralogy, (6th edition), 688–689. (2) Alexander, L.T., G.T. Faust, S.B. Hendricks, H. Insley, and H.F. McMurdie (1943) Relationship of the clay minerals halloysite and endellite. Amer. Mineral., 28, 1–18. (3) Brindley, G.W. and K. Robinson, (1948) X-ray studies of halloysite and metahalloysite. Mineral. Mag., 28, 392–428. (4) Bates, T.F., F.A. Hildebrand, and A. Swineford (1950) Morphology and structure of endellite and halloysite. Amer. Mineral., 35, 463–484. (5) Grim, R.E. (1953) Clay Mineralogy, McGraw Hill, 370. (6) Faust, G.T. (1955) The endellite-halloysite nomenclature. Amer. Mineral., 40, 1110–1118. (7) Deer, W.A., R.A. Howie, and J. Zussman (1963) Rock-forming minerals, v. 3, sheet silicates, 194–212. (8) Mitra, G.B. and S. Bhattacherjee (1975) The structure of halloysite. Acta Cryst., 31, 2851–2857. (9) Kohyama, N., K. Fukushima, and A. Fukami (1978) Observation of the hydrated form of tubular halloysite by an electron microscope equipped with an environmental cell. Clays and Clay Minerals, 26, 25–40.

Hancockite

$$(Ca, Pb, Sr)_2(Al, Fe^{3+})_3(SiO_4)(Si_2O_7)O(OH)$$

Crystal Data: Monoclinic. *Point Group:* 2/m. As very small lath-shaped crystals, showing {001}, {100}, {101}, {$\bar{1}$01}, and {$\bar{1}$11}, striated parallel to their length and with rounded faces; subhedral grains, sometimes hollow. In drusy cellular aggregates and compact masses.

Physical Properties: *Cleavage:* {001}, perfect. *Fracture:* Uneven. *Tenacity:* Brittle. Hardness = 6–7 D(meas.) = 4.03 D(calc.) = [4.03]

Optical Properties: Translucent. *Color:* Crystals yellowish brown, yellow-green; massive material dull brick-red, brownish red, or maroon. *Luster:* Vitreous.
Optical Class: Biaxial (–). *Pleochroism:* Strong; X = colorless, pale rose, greenish yellow; Y = pale brownish yellow, yellow; Z = pale rose, greenish yellow, green. *Dispersion:* r > v, perceptible. *Absorption:* Z > X. $\alpha = 1.788(3)$ $\beta = 1.81(1)$ $\gamma = 1.830(3)$ 2V(meas.) = ~50°

Cell Data: *Space Group:* $P2_1/m$. a = 8.958(20) b = 5.665(10) c = 10.304(20) $\beta = 114.4(4)°$ Z = 2

X-ray Powder Pattern: Franklin, New Jersey, USA. (ICDD 17-212).
2.91 (100), 3.49 (50), 2.60 (50), 2.81 (40), 2.71 (40), 2.18 (40), 1.90 (40)

Chemistry:

	(1)	(2)	(3)		(1)	(2)	(3)
SiO_2	30.99	29.1	27.48	PbO	18.53	26.3	32.44
TiO_2		0.10		MgO	0.52	trace	0.00
Al_2O_3	17.89	16.3	14.52	CaO	11.50	10.0	9.20
Fe_2O_3	12.33	14.5	13.09	SrO	3.89	3.6	
Mn_2O_3	1.38	2.7		BaO			0.32
MnO	2.12		0.19	H_2O^+	1.62	[1.37]	[2.66]
				Total	100.77	[103.9]	[100.00]

(1) Franklin, New Jersey, USA; corresponds to $(Ca_{1.17}Pb_{0.47}Sr_{0.21}Mn^{2+}_{0.17})_{\Sigma=2.02}(Al_{1.95}Fe_{0.88}$ $Mn^{3+}_{0.10}Mg_{0.07})_{\Sigma=3.00}(Si_{2.95}Al_{0.05})_{\Sigma=3.00}O_{12}OH$. (2) Do.; by electron microprobe, H_2O from theoretical $CaPbFeAl_2Si_3O_{12}(OH)$. (3) Jakobsberg, Sweden; by electron microprobe, H_2O by difference; corresponds to $(Ca_{1.07}Pb_{0.95}Mn_{0.02}Ba_{0.01})_{\Sigma=2.05}(Al_{1.87}Fe^{3+}_{1.08}Ti_{0.01})_{\Sigma=2.95}Si_{3.00}O_{12}OH$.

Mineral Group: Epidote group.

Occurrence: In a metamorphosed stratiform zinc deposit (Franklin, New Jersey, USA); in a metamorphosed manganese-iron orebody in skarns enclosed in dolomitic marble (Jakobsberg, Sweden).

Association: Andradite, franklinite, manganaxinite, clinohedrite, roeblingite, datolite, prehnite, willemite, barite, hendricksite, barian feldspar, phlogopite, lead, copper (Franklin, New Jersey, USA); melanotekite, hematite, garnet (Jakobsberg, Sweden).

Distribution: From Franklin, Sussex Co., New Jersey, USA. At Jakobsberg, Värmland, Sweden.

Name: For Elwood P. Hancock (1836–1916), of Burlington, New Jersey, USA, collector of minerals from Franklin.

Type Material: n.d.

References: (1) Dana, E.S. (1899) Dana's system of mineralogy, (6th edition), app. I, 32. (2) Penfield, S.L. and C.H. Warren (1899) Some new minerals from the zinc mines at Franklin, N.J., and note concerning the chemical composition of ganomalite. Amer. J. Sci., 8, 339–353. (3) Palache, C. (1935) The minerals of Franklin and Sterling Hill, Sussex County, New Jersey. U.S. Geol. Sur. Prof. Paper 180, 98. (4) Dollase, W.A. (1971) Refinement of the crystal structures of epidote, allanite and hancockite. Amer. Mineral., 56, 447–464. (5) Dunn, P.J. (1985) The lead silicates from Franklin, New Jersey: occurrence and composition. Mineral. Mag., 49, 721–727. (6) Holtstam, D. and J. Langhof (1994) Hancockite from Jakobsberg, Filipstad, Sweden: the second world occurrence. Mineral. Mag., 58, 172–174.

Crystal Data: Orthorhombic. *Point Group:* $2/m\ 2/m\ 2/m$. As tubular aggregates, to 2 mm; massive.

Physical Properties: *Cleavage:* Perfect on {010}, distinct on {100} and {001}. Hardness = 4.5   D(meas.) = 3.80   D(calc.) = 3.83

Optical Properties: Transparent to translucent. *Color:* Bright green. *Luster:* Vitreous. *Optical Class:* Biaxial (–). *Pleochroism:* X = colorless to very pale green; Y = colorless to light yellowish green; Z = bluish green. *Orientation:* X = a; Y = b; Z = c. *Dispersion:* $r < v$, very strong. $\alpha = 1.713(2)$ $\beta = 1.721(2)$ $\gamma = 1.734(2)$ 2V(meas.) = n.d.

Cell Data: *Space Group: Amam.*   a = 7.06   b = 14.64   c = 5.33   Z = 4

X-ray Powder Pattern: Synthetic.
3.20 (100), 2.88 (90), 3.65 (40), 2.65 (40), 2.12 (40), 2.04 (35), 7.30 (30)

Chemistry:

	(1)		(1)
SiO$_2$	38.38	CaO	1.27
TiO$_2$	0.06	SrO	27.08
VO$_2$	26.16	BaO	4.90
Al$_2$O$_3$	0.36	Na$_2$O	0.01
FeO	0.12	K$_2$O	0.04
MnO	0.19	H$_2$O$^+$	1.24
CuO	0.20	H$_2$O$^-$	0.20
PbO	0.02	Total	100.23

(1) Yamato mine, Japan; corresponds to $(Sr_{3.26}Ba_{0.40}Ca_{0.28}Mn_{0.03}Fe_{0.02}K_{0.01})_{\Sigma=4.00}$ $(V_{3.94}Al_{0.07}Ti_{0.01})_{\Sigma=4.02}(Si_{7.98}Al_{0.02})_{\Sigma=8.00}O_{28}$.

Occurrence: In coarse-grained rhodonite ore (Noda-Tamagawa mine, Japan); as veinlets cutting rhodonite-goldmanite ore (Yamato mine, Japan).

Association: Rhodonite, quartz.

Distribution: In Japan, in the Yamato mine and on Amamioshima Island, Kagoshima Prefecture; in the Noda-Tamagawa mine, Iwate Prefecture; and in the Matsuo mine, Kochi Prefecture. From the Gambatesa mine, near Chiavari, Val Graveglia, Liguria, Italy.

Name: For Zyunpei Harada (1898–), Emeritus Professor, Hokkaido University, Sapporo, Japan.

Type Material: National Science Museum, Tokyo, Japan, M15111.

References: (1) Watanabe, T., A. Kato, J. Ito, T. Yoshimura, H. Momoi, and K. Fukuda (1974) Haradaite, $Sr_2V_2(O_2)(Si_4O_{12})$, a new mineral from the Noda Tamagawa mine, Iwate Prefecture, and the Yamoto mine, Kagoshima Prefecture, Japan. IMA, 9th General Meeting, 97 (abs.). (2) (1975) Amer. Mineral., 60, 340 (abs. ref. 1). (3) Watanabe, T., A. Kato, J. Ito, T. Yoshimura, H. Momoi, and K. Fukuda (1974) Haradaite, $Sr_2V_2^{4+}[O_2Si_4O_{12}]$, from the Noda Tamagawa mine, Iwate Prefecture, and the Yamoto mine, Kagoshima Prefecture, Japan. Proc. Japan Acad., Ser. B, 58(2), 21–24 (in English). (4) (1975) Chem. Abs., 96, 146305 (abs. ref. 3). (5) Takéuchi, Y. and W. Joswig (1967) The structure of haradaite and a note on the Si:O bond lengths in silicates. Mineral. J. (Japan), 5, 98–123. (6) (1971) Amer. Mineral., 56, 1123 (abs. ref. 5). (7) Ito, J. (1965). Synthesis of vanadium silicates: haradaite, goldmanite and roscoelite. Mineral. J. (Japan), 4, 299–316.

Crystal Data: Tetragonal. *Point Group:* $\bar{4}2m$. In coarse, columnar masses; granular and as isolated grains.

Physical Properties: *Cleavage:* {001}, good; {100} and {110}, poor. *Tenacity:* Brittle. Hardness = 3–4 D(meas.) = 3.396–3.443 D(calc.) = 3.42 A dull violet fluorescence can be observed in some samples under UV.

Optical Properties: Translucent. *Color:* White, pinkish, light brown; colorless in thin section. *Luster:* Vitreous.
Optical Class: Uniaxial (–). $\omega = 1.669$ $\epsilon = 1.657$

Cell Data: *Space Group:* $P\bar{4}2_1m$. a = 7.8287(16) c = 5.0140(4) Z = 2

X-ray Powder Pattern: Franklin, New Jersey, USA. (ICDD 12-453).
2.868 (100), 3.085 (60), 3.711 (50), 5.018 (35), 1.761 (35), 2.473 (30), 2.037 (18)

Chemistry:

	(1)	(2)	(3)
SiO_2	38.10	36.80	38.31
Al_2O_3		0.94	
Fe_2O_3	0.57	0.20	
MnO	1.50	0.76	
ZnO	24.30	25.56	25.94
PbO		0.56	
MgO	1.62	0.39	
CaO	33.85	34.61	35.75
Na_2O		0.29	
LOI	0.52		
Total	100.46	100.11	100.00

(1) Franklin, New Jersey, USA. (2) Do.; by electron microprobe. (3) $Ca_2ZnSi_2O_7$.

Mineral Group: Melilite group.

Occurrence: In granular ore in a metamorphosed stratiform zinc deposit.

Association: Vesuvianite, apatite, franklinite, willemite, rhodonite, calcite, dolomite.

Distribution: From Franklin, Sussex Co., New Jersey, USA.

Name: For Hardyston Township in which Franklin is situated.

Type Material: Harvard University, Cambridge, Massachusetts, USA, 113594–113608.

References: (1) Dana, E.S. (1899) Dana's system of mineralogy, (6th edition), app. I, 32–33. (2) Palache, C. (1935) The minerals of Franklin and Sterling Hill, Sussex County, New Jersey. U.S. Geol. Sur. Prof. Paper 180, 93–94. (3) Louisnathan, S.J. (1969) Refinement of the crystal structure of hardystonite, $Ca_2ZnSi_2O_7$. Zeits. Krist., 130, 427–437.

Crystal Data: Hexagonal, pseudocubic. *Point Group:* $\bar{3}\,2/m$. As simple pseudo-octahedral crystals, to 3 mm. *Twinning:* Observed, with four rhombohedral individuals each ideally oriented ‖ to one of the pseudocubic 3-fold axes.

Physical Properties: Hardness = n.d. D(meas.) = 2.95–2.96   D(calc.) = [3.02]

Optical Properties: Semitransparent. *Color:* Colorless. *Luster:* Vitreous.
Optical Class: Uniaxial, but typically sensibly isotropic; sections may separate into four fields with weak birefringence in the diagonal position. $n = 1.649$–1.653

Cell Data: *Space Group:* $R\bar{3}m$.   a = 10.439   c = 51.303   Z = 3

X-ray Powder Pattern: Broadford area, Isle of Skye, Scotland.
2.61 (vs), 5.22 (s), 2.13 (s), 1.84 (s), 1.51 (s), 3.39 (m), 3.01 (m)

Chemistry:

	(1)	(2)	(3)
SiO_2	14.17	16.82	16.37
B_2O_3	7.77	7.0	7.11
Al_2O_3	2.84	3.13	3.47
Ce_2O_3		0.8	
Fe_2O_3	0.85	1.78	
FeO	0.46		
MnO	0.02	0.13	
MgO	11.15	10.01	10.98
CaO	46.23	44.13	45.85
SrO		0.2	
F		0.15	
Cl	1.36	0.06	
H_2O^+	0.81	1.25	1.23
H_2O^-	0.11		
CO_2	14.94	15.07	14.99
$-O = (F, Cl)_2$	0.31	0.07	
Total	100.40	100.46	100.00

(1) Camas Malag, Isle of Skye, Scotland. (2) Alban Hills, Italy; by electron microprobe, spectrophotometry, elemental analyzer, and mass spectrometer. (3) $Ca_{24}Mg_8Al_2$ $(SiO_4)_8(BO_3)_6(CO_3)_{10} \cdot 2H_2O$.

Occurrence: In skarns in dolomitic limestones at the contact with granite (Camas Malag, Scotland); in volcanic ejecta (Alban Hills, Italy).

Association: Calcite, monticellite, diopside, magnetite, vesuvianite, cuspidine, phlogopite.

Distribution: At Camas Malag, near Broadford, Isle of Skye, Scotland. On the Chersky Ridge, Tas-haiatah, northeastern Siberia, Russia. From the Alban Hills, Lazio, Italy.

Name: For Alfred Harker (1859–1939), British petrologist, Cambridge University, Cambridge, England.

Type Material: n.d.

References: (1) Tilley, C.E. (1951) The zoned contact-skarns of the Broadford area, Skye: a study of boron-fluorine metasomatism in dolomites. Mineral. Mag., 29, 621–667. (2) (1952) Amer. Mineral., 37, 359 (abs. ref. 1). (3) Ostrovskaya, I.V., N.N. Pertsev, and I.B. Nikitina (1966) Sakhaite, a new carbonate-borate of calcium and magnesium. Zap. Vses. Mineral. Obshch., 95, 193–202 (in Russian). (4) Giuseppetti, G., F. Mazzi, and C. Tadini (1977) The crystal structure of harkerite. Amer. Mineral., 62, 263–272. (5) Barbieri, M., C. Cozzupoli, M. Federico, M. Fornaseri, S. Merlino, P. Orlandi, and L. Tolomeo (1977) Harkerite from the Alban Hills, Italy. Lithos, 10, 133–141.

Harmotome

$$(Ba, K)_{1-2}(Si, Al)_8O_{16} \cdot 6H_2O$$

Crystal Data: Monoclinic. *Point Group:* $2/m$. Crystals elongated $\parallel$ [001] or tabular $\parallel$ {010}; typically striated on {010} $\parallel$ [001]. Commonly as interpenetrant twins, to 4 cm, alone, in groups, or radiating aggregates. *Twinning:* Repeated on {001}, {021}, {110}; three double twins may further twin to simulate a pseudotetragonal prism or dodecahedron.

Physical Properties: *Cleavage:* {010} distinct, {001} poor. *Fracture:* Uneven to subconchoidal. *Tenacity:* Brittle. Hardness = 4.5 D(meas.) = 2.41–2.47 D(calc.) = 2.448

Optical Properties: Transparent to translucent. *Color:* Colorless, white, gray, yellow, pink, brown; colorless in thin section. *Streak:* White. *Luster:* Vitreous.
Optical Class: Biaxial (+). *Orientation:* $Z = b$; $X \wedge a = 63°$–$67°$. *Dispersion:* Weak, crossed.
$\alpha = 1.503$–1.508 $\beta = 1.505$–1.509 $\gamma = 1.508$–1.514 2V(meas.) = $\sim 80°$

Cell Data: *Space Group:* $P2_1/m$. $a = 9.8688(21)$ $b = 14.1295(7)$ $c = 8.7092(9)$
$\beta = 124.74°$ $Z = 2$

X-ray Powder Pattern: Korsnäs mine, Finland.
6.38 (100), 3.13 (80), 2.670 (70), 4.08 (60), 3.24 (60), 3.17 (60), 2.730 (60)

Chemistry:

	(1)	(2)	(3)
SiO_2	48.49	47.52	49.80
Al_2O_3	16.35	16.94	14.09
BaO	20.08	20.25	21.18
Na_2O	trace	1.09	
K_2O	2.07	1.00	
H_2O	13.00	13.45	14.93
Total	99.99	100.25	100.00

(1) St. Andreasberg, Germany. (2) Strontian, Scotland. (3) $BaSi_6Al_2O_{16} \cdot 6H_2O$.

Mineral Group: Zeolite group.

Occurrence: Of hydrothermal origin, in cavities in basalts, phonolites, trachytes; in gneisses, and in some ore veins.

Association: Zeolites, calcite, leucite, hyalophane, strontianite, quartz, kaolinite, barite, pyrite, sphalerite, galena.

Distribution: Many occurrences; some for well-studied material follow. In Germany, from St. Andreasberg, Harz Mountains; at Idar-Oberstein, Rhineland-Palatinate; and on the Silberberg, near Bodenmais, Bavaria. At Příbram, Czech Republic. From Sarrabus, Sardinia, Italy. Fine crystals from Strontian, Argyllshire, Scotland. From Kongsberg, Norway. At Kupferberg-Rudelstadt, Silesia, Poland. In the Korsnäs lead mine, south of Vaasa, Finland. In Russia, large crystals at Bukan, in the Kotuy River basin, Taimyr. In the USA, from Glen Riddle, Delaware Co., Pennsylvania; at Sing Sing, near Ossining, Westchester Co., New York. In the Beaver mine, Thunder Bay district, Ontario, Canada. At Batopilas, Chihuahua, Mexico.

Name: From the Greek *harmos*, for *a joint*, and *tome*, for *a cutting*, as the twinned crystals may be divided.

References: (1) Dana, E.S. (1892) Dana's system of mineralogy, (6th edition), 581–583. (2) Deer, W.A., R.A. Howie, and J. Zussman (1963) Rock-forming minerals, v. 4, framework silicates, 386–400. (3) Sahama, T.G. and M. Lehtinen (1967) Harmotome from Korsnäs, Finland. Mineral. Mag., 36, 444–448. (4) Stuckenschmidt, E., H. Fuess, and Å. Kvick (1990) Investigation of the structure of harmotome by X-ray (293 K, 100 K) and neutron diffraction (15 K). Eur. J. Mineral., 861–874.

Crystal Data: Hexagonal. *Point Group:* $\bar{3}m$. In anhedral grains, to 2 mm; as rims on fluorapatite.

Physical Properties: *Cleavage:* Observed, very poor. *Fracture:* Conchoidal. *Tenacity:* Brittle. Hardness = < 5 D(meas.) = 4.02(3) D(calc.) = 4.01

Optical Properties: Translucent to transparent on thin edges. *Color:* Yellow-brown to orange-brown; pale yellow in thin section. *Streak:* Pale yellow. *Luster:* Vitreous. *Optical Class:* Uniaxial (–); slightly biaxial due to strain. *Pleochroism:* Slight, in shades of pale yellow. $\omega = 1.770(5)$ $\epsilon = 1.759(3)$ 2V(meas.) = $\sim 5°$

Cell Data: *Space Group:* $R\bar{3}m$. a = 6.248(1) c = 26.802(7) Z = 3

X-ray Powder Pattern: Arcedeckne Island, Canada.
3.119 (100), 2.558 (100), 2.689 (80), 2.505 (80), 1.560 (80), 5.00 (60), 1.903 (60)

Chemistry:

	(1)
SiO_2	15.99
FeO	52.27
MnO	0.11
MgO	3.89
CaO	7.65
P_2O_5	19.18
Total	99.09

(1) Arcedeckne Island, Canada; by electron microprobe, average of four analyses, Ti, Al, Cr, Na, K, Cl, F absent; corresponds to $Ca_{1.01}(Fe_{5.36}Mg_{0.71}Mn_{0.01})_{\Sigma=6.08}(Si_{0.98}O_4)_2(P_{0.99}O_4)_2$.

Occurrence: A minor but wide-spread constituent in a layered iron silicate-quartz-apatite body in gneisses, probably a shaly and phosphatic iron formation metamorphosed to the granulite facies.

Association: Fluorapatite, quartz, fayalite, ferrosilite, almandine, ilmenite, biotite, zircon, monazite-(Ce).

Distribution: From Arcedeckne Island, near the Boothia Peninsula, District of Franklin, Arctic Canada.

Name: To honor Dr. James Merritt Harrison (1915–1990), former Director of the Canadian Geological Survey.

Type Material: Canadian Geological Survey, Ottawa, 66402; Canadian Museum of Nature, Ottawa, Canada, 59685.

References: (1) Roberts, A.C., J.A.R. Stirling, J.D. Grice, T. Frisch, R.K. Herd, and J.L. Jambor (1993) Harrisonite, a new calcium iron silicate-phosphate from Arcedeckne Island, District of Franklin, Arctic Canada. Can. Mineral., 31, 775–780. (2) Grice, J.D. and A.C. Roberts (1993) Harrisonite, a well-ordered silico-phosphate with a layered crystal structure. Can. Mineral., 31, 781–785. (3) (1994) Amer. Mineral., 79, 1010–1011 (abs. refs. 1 and 2).

Crystal Data: Orthorhombic. *Point Group:* $2/m\ 2/m\ 2/m$. Crystals stout prismatic, to 2 cm, terminated by {110} and small {221}.

Physical Properties: *Fracture:* Small conchoidal to splintery. *Tenacity:* Brittle. Hardness = 5.5 D(meas.) = 3.16 D(calc.) = 3.19

Optical Properties: Transparent to translucent. *Color:* Colorless. *Luster:* Vitreous. *Optical Class:* Biaxial (+). *Orientation:* X = a; Y = b; Z = c. *Dispersion:* $r < v$, weak. $\alpha = 1.678$ $\beta = 1.68$ $\gamma = 1.683$ 2V(meas.) = 52°

Cell Data: *Space Group: Pnam.* a = 9.793(2) b = 13.636(3) c = 13.830(3) Z = 4

X-ray Powder Pattern: Harstig mine, Sweden.
2.695 (100), 2.817 (50), 2.788 (50), 2.268 (50), 4.35 (40), 3.222 (40), 3.545 (30)

Chemistry:

	(1)
SiO_2	40.00
MnO	7.05
BeO	11.49
MgO	0.94
CaO	37.82
F	0.15
H_2O	2.48
Total	99.93

(1) Harstig mine, Sweden; corresponds to $Ca_{5.88}(Mn_{0.86}Mg_{0.20})_{\Sigma=1.06}Be_{4.00}Si_{5.80}O_{21.34}(OH)_{2.40}$.

Occurrence: A rare mineral found in open fissures in a metamorphosed manganese deposit.

Association: Hausmannite, dolomite, calcite, rhodonite, andradite, manganoan humite, barite.

Distribution: At the Harstig mine, Pajsberg, near Persberg, Värmland, Sweden.

Name: For the Harstig mine, Pajsberg, Sweden, in which it was found.

Type Material: Swedish Natural History Museum, Stockholm, Sweden.

References: (1) Dana, E.S. (1892) Dana's system of mineralogy, (6th edition), 532–533. (2) Moore, P.B. (1968) Relations of the manganese-calcium silicates, gageite and harstigite. Amer. Mineral., 53, 309–315. (3) Moore, P.B. (1968) Relation of the manganese-calcium silicates, gageite and harstigite: a correction. Amer. Mineral., 53, 1418–1420. (4) Hesse, K.-F. and G. Stümpel (1986) Crystal structure of harstigite, $MnCa_6Be_4[SiO_4]_2[Si_2O_7]_2(OH)_2$. Zeits. Krist., 177, 143–148.

$NaCa_2[(Fe^{2+}, Mg)_4Fe^{3+}](Si_6Al_2)O_{22}(OH)_2$ Hastingsite

Crystal Data: Monoclinic. *Point Group: 2/m.* Prismatic crystals, to 4 cm.
Twinning: Simple or multiple twinning common ∥ {100}.

Physical Properties: *Cleavage:* Perfect on {110}, with intersections at 56° and 124°; partings on {001}, {100}. *Tenacity:* [Brittle.] Hardness = 5–6   D(meas.) = 3.35–3.5   D(calc.) = [3.42]

Optical Properties: Semitransparent. *Color:* Black, dark green, green-brown, yellow; dark green in thin section. *Luster:* [Vitreous.]
Optical Class: Biaxial (–). *Pleochroism:* X = yellow, greenish brown, yellowish green; Y = deep greenish blue, brownish green, very dark olive-green; Z = deep olive-green, smoky blue-green, very dark green. *Orientation:* Y = b; $Z \wedge c \simeq 12°$. *Dispersion:* r < v, moderate to strong.
$\alpha = 1.685–1.702$ $\beta = 1.71–1.729$ $\gamma = 1.71–1.728$ 2V(meas.) = 10°–45°

Cell Data: *Space Group:* C2/m. a = 9.8659(4)   b = 18.0139(8)   c = 5.3545(2)
$\beta = 105.082(1)°$ Z = 2

X-ray Powder Pattern: Hastings Co., Ontario, Canada. (ICDD 20-469).
8.43 (100), 3.13 (70), 2.706 (60), 3.39 (50), 3.28 (50), 2.594 (50), 2.559 (50)

Chemistry:

	(1)	(2)		(1)	(2)		(1)	(2)
SiO_2	33.50	37.68	MnO	0.26	0.66	F	0.25	
TiO_2	3.26	1.02	MgO	5.00	2.20	H_2O^+	1.37	
Al_2O_3	17.89	14.01	CaO	10.30	9.26	H_2O^-	0.04	
Fe_2O_3	6.65	7.97	Na_2O	3.14	3.31	P_2O_5	0.05	
FeO	16.49	20.63	K_2O	1.48	2.22	$-O = F_2$	0.10	
						Total	[99.58]	98.96

(1) Wolfe Belt, Kargus Ridge, Lyndoch, Ontario, Canada; original total given as 99.53%, corresponds to $(Ca_{1.74}Na_{0.95}K_{0.30})_{\Sigma=2.99}(Fe^{2+}_{2.17}Mg_{1.17}Fe^{3+}_{0.79}Al_{0.58}Ti_{0.39}Mn_{0.04})_{\Sigma=5.14}$ $(Si_{5.27}Al_{2.73})_{\Sigma=8.00}O_{22}(OH)_2$. (2) Hastings Co., Ontario, Canada; by electron microprobe, corresponds to $(Ca_{1.56}Na_{1.01}K_{0.45})_{\Sigma=3.02}(Fe^{2+}_{2.71}Fe^{3+}_{0.94}Al_{0.53}Mg_{0.52}Ti_{0.12}Mn_{0.09})_{\Sigma=4.91}$ $(Si_{5.93}Al_{2.07})_{\Sigma=8.00}O_{22}(OH)_2$.

Polymorphism & Series: Forms a series with magnesio-hastingsite.

Mineral Group: Amphibole (calcic) group: $Mg/(Mg + Fe^{2+}) < 0.30$; $Fe^{3+} > Al^{vi}$; $(Na + K)_A \geq 0.5$; $(Ca + Na)_B \geq 1.34$; $Na_B < 0.67$; Si < 6.25; Ti < 0.5.

Occurrence: In nepheline syenite and granite; in schists, gneisses, tactites, and amphibolites.

Association: Scapolite, apatite, magnetite, spinel (gneiss); garnet, epidote, hedenbergite, quartz (tactite).

Distribution: Analyzed material from many localities worldwide. In Canada, at Bancroft, and elsewhere in Hastings Co., Ontario; near Crescent Lake, Cassiar Mountains, Yukon Territory. In the USA, at Franklin, Sussex Co., New Jersey; Cornwall, Orange Co., New York; and Iron Hill, Gunnison Co., Colorado. From Koraput, Orissa, and elsewhere in India. In the Sampo mine, Okayama Prefecture, and the Obira mine, Bungo, Oita Prefecture, Japan. From Kakanui, New Zealand. In the Marangudzi ring complex, Zimbabwe. At Almunge, Sweden.

Name: For a locality in Hastings Co., Ontario, Canada.

Type Material: Royal Ontario Museum, Toronto, Canada, M13724.

References: (1) Dana, E.S. (1899) Dana's system of mineralogy, (6th edition), app. I, 3.
(2) Hawthorne, F.C. and H.D. Grundy (1977) The crystal chemistry of the amphiboles. III: Refinement of the crystal structure of a sub-silicic hastingsite. Mineral. Mag., 41, 43–50.
(3) Leake, B.E. (1968) A catalog of analyzed calciferous and subcalciferous amphiboles together with their nomenclature and associated minerals. Geol. Soc. Amer. Special Paper 98, 210 p. [analysis 936].

Crystal Data: n.d. *Point Group:* n.d. Crystals pseudohexagonal, to about 50 μ.

Physical Properties: Hardness = n.d. D(meas.) = n.d. D(calc.) = n.d.

Optical Properties: Transparent to translucent. *Color:* Light gray; colorless in thin section. *Optical Class:* Biaxial (–). n = n.d.; birefringence $\sim$0.006 2V(meas.) = Small.

Cell Data: *Space Group:* n.d. Z = n.d.

X-ray Powder Pattern: Hatrurim Formation, Israel; stronger lines overlapped by larnite. 3.03 (m), 1.767 (mw), 2.96 (vw), 1.759 (vw), 1.484 (vw), 1.452 (vw)

Chemistry:

	(1)
SiO_2	26.1
TiO_2	0.3
Al_2O_3	0.4
Fe_2O_3	0.2
MgO	trace
CaO	72.8
Total	99.8

(1) Hatrurim Formation, Israel; by electron microprobe, corresponds to $Ca_{2.97}Al_{0.01}Ti_{0.01}SiO_5$.

Occurrence: In a high-temperature contact metamorphic assemblage in a larnite-brownmillerite-mayenite rock, probably derived from fine-grained sediments.

Association: Nagelschmidtite, larnite, brownmillerite, mayenite.

Distribution: In the Hatrurim Formation, Israel.

Name: For the Hatrurim Formation in Israel.

Type Material: n.d.

References: (1) Gross, S. (1977) The mineralogy of the Hatrurim Formation, Israel. Geol. Sur. Israel Bull. 70, 35–36. (2) (1978) Amer. Mineral., 63, 425–427 (abs. ref. 1).

Crystal Data: Cubic. *Point Group:* $\bar{4}3m$. Crystals dodecahedra or pseudo-octahedra, to 3 cm; in rounded grains. *Twinning:* On {111}, common, rarely as penetration twins; also as polysynthetic or contact twins.

Physical Properties: *Cleavage:* {110}, distinct. *Fracture:* Uneven to conchoidal. *Tenacity:* Brittle. Hardness = 5.5–6   D(meas.) = 2.44–2.50   D(calc.) = n.d.   May show reddish orange to purplish pink fluorescence under LW UV.

Optical Properties: Transparent to translucent. *Color:* Bright blue to greenish blue; white or shades of black, gray, brown, green, yellow, red, may be patchy; colorless or pale blue in thin section. *Streak:* Slightly bluish to colorless. *Luster:* Vitreous to greasy. *Optical Class:* Isotropic; weakly birefringent when included. $n = 1.494–1.509$

Cell Data: *Space Group:* $P\bar{4}3n$.   a = 9.08–9.13   Z = 1

X-ray Powder Pattern: Niedermendig, Germany.
3.72 (100), 2.623 (25), 6.47 (16), 2.873 (14), 2.141 (14), 1.781 (10), 2.428 (8)

Chemistry:

	(1)	(2)		(1)	(2)
SiO_2	34.04	29.3	K_2O	5.44	3.71
Al_2O_3	28.27	29.0	Cl	0.76	
Fe_2O_3		0.07	H_2O	0.34	
FeO	0.69		CO_2	0.4	
MgO	0.48	0.15	SO_3	10.02	13.1
CaO	9.51	11.2	$-O = Cl_2$	0.17	
Na_2O	10.39	13.0	Total	100.17	99.53

(1) Monte Vulture, Italy; corresponds to $(Na_{3.55}Ca_{1.80}K_{1.22}Mg_{0.13})_{\Sigma=6.70}Al_{5.89}Fe_{0.11}Si_{6.00}O_{24}$ $[(SO_4)_{1.33}Cl_{0.22}]_{\Sigma=1.55}$. (2) Anguillara, Italy; by electron microprobe, corresponds to $(Na_{4.76}Ca_{2.26}$ $K_{0.90}Mg_{0.45})_{\Sigma=8.37}Al_{6.46}Fe_{0.01}Si_{5.53}O_{24}(SO_4)_{1.86}$.

Mineral Group: Sodalite group.

Occurrence: In phonolites and related leucite- or nepheline-rich igneous rocks; less commonly in nepheline-free extrusives.

Association: Nepheline, leucite, titanian andradite, melilite, augite, sanidine, biotite, phlogopite, apatite.

Distribution: In Italy, in Lazio, at many localities in the Alban Hills, as at Ariccia, Marino, Sacrofano, and Campagnaro; on Monte Somma, and near Melfi, on Monte Vulture, Campania; in the Pitigliano quarry, near Grosseto, Tuscany. From Mendig, Mayen, and elsewhere in the Eifel district, Germany. In the USA, from Winnett, Petroleum Co., Montana, and in the Edwards mine, St. Lawrence Co., New York. In the Niangniang Shan complex, Nanjing, Jiangsu Province, China. On the smaller island of Taiarupu, Tahiti. A few other localities are known.

Name: To honor Abbé René Just Haüy (1743–1822), French crystallographer and mineralogist.

References: (1) Dana, E.S. (1892) Dana's system of mineralogy, (6th edition), 431–432. (2) Deer, W.A., R.A. Howie, and J. Zussman (1963) Rock-forming minerals, v. 4, framework silicates, 289–302. (3) Taylor, D. (1967) The sodalite group of minerals. Contr. Mineral. Petrol., 16, 172–188. (4) Löhn, J. and H. Schulz (1968) Strukturverfeinerung am gestörten Hauyn, $(Na_5K_1Ca_2)Al_6Si_6O_{24}(SO_4)_{1.5}$. Neues Jahrb. Mineral., Abh., 109, 201–210 (in German with English abs.). (5) Burragato, F., A. Maras, and A. Rossi (1982) The sodalite group minerals in the volcanic areas of Latium. Neues Jahrb. Mineral., Monatsh., 433–445. (6) Hassan, I. and P. Buseck (1989) Cluster ordering and antiphase domain boundaries in hauyne. Can. Mineral., 27, 173–180.

Crystal Data: Monoclinic. *Point Group:* $2/m$. As thin laths, to 2 μm, and as aggregates of such laths.

Physical Properties: *Cleavage:* {001}, perfect. *Fracture:* Uneven. Hardness = 1–2 D(meas.) = ~2.3 D(calc.) = n.d. Swells on addition of H_2O. Positive identification of minerals in the smectite group may need data from DTA curves, dehydration curves, and X-ray powder patterns before and after treatment by heating and with organic liquids.

Optical Properties: Translucent, transparent in thin section. *Color:* White, cream, pale brown, mottled. *Luster:* Earthy to waxy, dull.
Optical Class: Biaxial (–). $\alpha = {\sim}1.49$ $\beta = 1.50$ $\gamma = 1.52$ 2V(meas.) = Small.

Cell Data: *Space Group:* $C2/m$. $a = 5.2$ $b = 9.16$ $c = 16.0$ $\beta = {\sim}\,99°$ Z = n.d.

X-ray Powder Pattern: Hector, California, USA; spacings variable by humidity, intensities variable by orientation.
4.58 (100), 1.53 (100), 15.8 (80), 2.66 (80), 1.32 (80), 1.30 (80), 2.48 (60)

Chemistry:

	(1)	(2)
SiO_2	53.68	53.95
TiO_2		trace
Al_2O_3	0.60	0.14
Fe_2O_3		0.03
MgO	25.34	25.89
CaO	0.52	0.16
Li_2O	1.12	1.22
Na_2O	3.00	3.04
K_2O	0.07	0.23
Cl	0.31	
H_2O^+	8.24	5.61
H_2O^-	7.28	9.29
Total	100.16	99.56

(1) Hector, California, USA; corresponds to $(Na_{0.42}Ca_{0.04}K_{0.01})_{\Sigma=0.47}(Mg_{2.73}Li_{0.33})_{\Sigma=3.06}$ $(Si_{3.89}Al_{0.05})_{\Sigma=3.94}O_{10}(OH)_2$. (2) Do.; corresponds to $(Na_{0.42}K_{0.02}Ca_{0.01})_{\Sigma=0.45}$ $(Mg_{2.78}Li_{0.36})_{\Sigma=3.14}(Si_{3.89}Al_{0.01})_{\Sigma=3.90}O_{10}(OH)_2 \bullet 0.35H_2O$.

Mineral Group: Smectite group.

Occurrence: In a bentonite deposit, altered from clinoptilolite derived from volcanic tuff and ash with a high glass content, related to hot spring activity (Hector, California, USA).

Association: Calcite, clinoptilolite (Hector, California, USA).

Distribution: In the USA, five km south of Hector, San Bernardino Co., California; in the Lyles deposit, 38 km northeast of Hillside, Yavapai Co., Arizona; and at Disaster Peak, in the Montana Mountains, near McDermitt, Humboldt Co., Nevada. From around Puy Chalard, Puy-de-Dôme, France. In the Balıkesir colemanite deposit, Balıkesir Province, Turkey.

Name: For the locality at Hector, California, USA.

Type Material: n.d.

References: (1) Foshag, W.F. and A.O. Woodford (1936) Bentonitic magnesian clay-mineral from California. Amer. Mineral., 21, 238–244. (2) Strese, H. and U. Hofmann (1941) Synthesis of magnesium silicate gels with two-dimensional regular structure. Zeit. anorginsche allgemeine Chemie, 247, 65–95. (3) (1944) Amer. Mineral., 29, 73 (abs. ref. 2). (4) Nagelschmidt, G. (1938) On the atomic arrangement and variability of the members of the montmorillonite group. Mineral. Mag., 25, 140–155. (5) Deer, W.A., R.A. Howie, and J. Zussman (1963) Rock-forming minerals, v. 3, sheet silicates, 226–245.

Crystal Data: Monoclinic. *Point Group:* $2/m$. As short prismatic crystals, to 5 cm. In columnar or acicular aggregates; granular, lamellar, massive. *Twinning:* Simple and polysynthetic twinning on {100} and {010}.

Physical Properties: *Cleavage:* Good on {110}, (110) $\wedge$ (1$\bar{1}$0) $\sim$87°; partings on {100} and {010}. *Fracture:* Uneven to conchoidal. *Tenacity:* Brittle. Hardness = 5.5–6.5 D(meas.) = 3.56   D(calc.) = [3.65]

Optical Properties: Transparent to opaque. *Color:* Black, dark green, green-brown; brownish green in thin section. *Luster:* Vitreous or dull.
Optical Class: Biaxial (+). *Pleochroism:* Weak; X = pale green, bluish green; Y = green, bluish green; Z = green, yellow-green. *Orientation:* $Y = b$; $Z \wedge c = 48°$; $Z \wedge a = 34°$.
Dispersion: $r > v$, strong. $\alpha = 1.732$ $\beta = 1.739$ $\gamma = 1.757$ 2V(meas.) = 63°

Cell Data: *Space Group:* $C2/c$.   a = 9.852(4)   b = 9.031(3)   c = 5.242(3) $\beta = 104.84(4)°$ Z = 4

X-ray Powder Pattern: Cordillera Blanca, Peru; magnesian. (ICDD 25-160).
2.995 (100), 2.528 (80), 2.970 (50), 1.638 (40), 1.629 (40), 3.26 (30), 2.890 (30)

Chemistry:

	(1)	(2)	(3)		(1)	(2)	(3)
SiO_2	48.00	48.50	48.44	MgO	2.12	0.83	
TiO_2		0.04		CaO	20.35	22.64	22.60
Al_2O_3	0.63	0.53		Na_2O	0.34		
Fe_2O_3	3.32	1.03		K_2O	0.18		
FeO	22.25	22.74	28.96	H_2O^-	1.72		
MnO	0.81	3.84		Total	99.72	100.15	100.00

(1) Tignitoio iron deposit, Elba, Italy; corresponds to $(Ca_{0.90}Mg_{0.13})_{\Sigma=1.03}(Fe^{2+}_{0.77}Fe^{3+}_{0.10}Mn_{0.03}$ $Na_{0.03}Al_{0.01}K_{0.01})_{\Sigma=0.95}(Si_{1.98}Al_{0.02})_{\Sigma=2.00}O_6$. (2) Dal'negorsk, Russia; corresponds to $Ca_{1.00}$ $(Fe^{2+}_{0.81}Mn_{0.13}Mg_{0.05}Al_{0.01})_{\Sigma=1.00}(Si_{1.99}Al_{0.01})_{\Sigma=2.00}O_6$. (3) $CaFeSi_2O_6$.

Polymorphism & Series: Forms two series, with diopside, and with johannsenite.

Mineral Group: Pyroxene group.

Occurrence: A common constituent of metamorphosed iron formations or other ferruginous siliceous sediments; common in Fe-Mn skarns. In alkalic granites, syenites, and in xenoliths in kimberlite.

Association: Grunerite (iron formations); arfvedsonite, quartz, fayalite (granites, syenites).

Distribution: A few localities for studied material include: in Sweden, at Nordmark, Värmland, and Yxsjö, Örebro. From Prägraten, Tirol, Austria. At Fürstenberg, Saxony, Germany. From Rio Marina, Elba, Italy. On Seriphos, Greece. In the USA, at Iron Hill, Gunnison Co., Colorado; fine crystals from the Laxey mine, South Mountain, Owyhee Co., Idaho; in the Pima district, Pima Co., and the Westinghouse mine, Santa Cruz Co., Arizona; at Hanover, Grant Co., New Mexico. In the Vesturhorn intrusion, southeast Iceland. Large crystals from Broken Hill, New South Wales, Australia. In the Obira mine, Bungo, Oita Prefecture, Japan. At Tirodi, Madhya Pradesh, and Kacharwali, Nagpur district, Maharashtra, India. Fine crystals from the Skardu area, Pakistan. At Dal'negorsk, Primorski Krai, Russia.

Name: To honor the Swedish chemist, M.A. Ludwig Hedenberg, who first described the species.

References: (1) Dana, E.S. (1892) Dana's system of mineralogy, (6th edition), 352–364, esp. 356. (2) Deer, W.A., R.A. Howie, and J. Zussman (1978) Rock-forming minerals, (2nd edition), v. 2A, single-chain silicates, 198–293. (3) Maslenikov, A.V. and V.N. Zaitsev (1978) Refinement of the structure of natural hedenbergite. Zap. Vses. Mineral. Obshch., 107, 113–115 (in Russian). (4) Phillips, W.R. and D.T. Griffen (1981) Optical mineralogy, 191–193.

Crystal Data: Monoclinic. *Point Group:* $2/m$. As laths and platy crystals, with dominant {100} and elongated along [001], up to 2 cm, and in groups.

Physical Properties: *Cleavage:* Perfect on {100}; poor parting on {0kl}. *Fracture:* Irregular. *Tenacity:* Brittle. Hardness = n.d. D(meas.) = 3.97–4.06, average 4.02 D(calc.) = 4.29

Optical Properties: Transparent to translucent. *Color:* Brownish or golden to yellow. *Streak:* Brownish yellow. *Luster:* Vitreous.
Optical Class: Biaxial (–). *Pleochroism:* X = light yellow-green; Y = dark golden yellow; Z = light yellow. *Orientation:* X = b; Y $\wedge$ c = 37.9°; Z $\wedge$ a = 15.9°. *Absorption:* Y > Z = X.
$\alpha = 1.814(1)$ $\beta = 1.846(1)$ $\gamma = 1.867(1)$ 2V(meas.) = 76.4° 2V(calc.) = 76.8°

Cell Data: *Space Group:* $P2_1/m$. a = 11.748(4) b = 13.768(5) c = 10.698(4) $\beta = 112.27(2)°$ Z = 8

X-ray Powder Pattern: Mbolwe Hill, Zambia.
2.726 (100), 5.472 (32), 3.241 (21), 3.215 (21), 3.455 (17), 2.180 (16), 3.669 (14)

Chemistry:

	(1)		(1)
SiO_2	23.52	BaO	30.20
TiO_2	13.27	Na_2O	0.06
Al_2O_3	0.37	K_2O	0.30
Nb_2O_5	1.4	F	3.3
FeO	11.29	Cl	< 0.1
MnO	14.12	H_2O	[1.86]
MgO	0.13	$-O = (F, Cl)_2$	1.39
CaO	0.00	Total	[98.43]

(1) Mbolwe Hill, Zambia; by electron microprobe, Nb by energy-dispersive analysis, H_2O from stoichiometry; corresponds to $(Ba_{1.04}K_{0.03}Mg_{0.02}Na_{0.01})_{\Sigma=1.10}(Mn_{1.05}Fe_{0.82})_{\Sigma=1.87}$ $(Ti_{0.87}Nb_{0.06})_{\Sigma=0.93}(Si_{2.06}Al_{0.04})_{\Sigma=2.10}O_8[(OH)_{1.09}F_{0.91}]_{\Sigma=2.00}$.

Polymorphism & Series: Forms a series with bafertisite.

Occurrence: In arfvedsonite veins in pegmatite cutting alkalic granite and syenite (Mbolwe Hill, Zambia).

Association: Manganoan arfvedsonite, albite, bastnäsite, chevkinite, manganoan ilmenite, apatite, aegirine (Mbolwe Hill, Zambia).

Distribution: On Mbolwe Hill, Mkushi River area, Central Province, Zambia. In the Dara-i-Pioz massif, Alai Range, Tien Shan, Tadzhikistan.

Name: To honor Dr. Bohuslav Hejtman, Emeritus Professor of Petrology, Charles University, Prague, Czech Republic.

Type Material: Charles University and the National Museum, Prague, Czech Republic.

References: (1) Vrána, S., M. Rieder, and M.E. Gunter (1992) Hejtmanite, a manganese-dominant analogue of bafertisite, a new mineral. Eur. J. Mineral., 4, 35–43. (2) (1992) Amer. Mineral., 77, 1306 (abs. ref. 1).

$(Ca, Y)_6(Al, Fe^{3+})Si_4B_4O_{20}(OH)_4$ **Hellandite**

Crystal Data: Monoclinic. *Point Group:* $2/m$. Crystals prismatic $\parallel$ [001], tabular $\parallel$ [010], to 30 cm. *Twinning:* Contact and polysynthetic, $\parallel$ {001} and {100}.

Physical Properties: *Cleavage:* Poor on {100}, {010}. Hardness = 4.5–6.5 D(meas.) = 2.95–3.63   D(calc.) = 3.161

Optical Properties: Semitransparent. *Color:* Nut-brown, brownish red; black, gray, green, yellow, cream. *Luster:* Vitreous to dull.
Optical Class: Biaxial (+). *Orientation:* X = b; $Z \wedge c = 44°$–$53°$. $\alpha = 1.652$–1.656
$\beta = 1.657$–1.712 $\gamma = 1.662$–1.668 2V(meas.) = 48°–86°

Cell Data: *Space Group:* $P2/a$.   a = 18.99(1)   b = 4.715(5)   c = 10.30(1) $\beta = 111.4(1)°$
Z = 4

X-ray Powder Pattern: Kragerö, Norway.
2.812 (100), 4.69 (80), 2.635 (80), 2.603 (80), 3.436 (70), 3.198 (70), 3.068 (70)

Chemistry:

	(1)	(2)	(3)
SiO_2	26.65	25.57	24.41
TiO_2	0.39	< 0.2	< 0.02
ThO_2	1.46		0.57
B_2O_3	10.5	[14.85]	8.47
Al_2O_3	2.58	3.04	4.59
Y_2O_3	21.68	18.07	26.53
RE_2O_3	15.63	16.0	19.19
Fe_2O_3	3.07	3.77	2.29
FeO	0.07		
MnO	0.41	0.70	0.58
MgO	0.60		
CaO	11.51	15.22	9.28
H_2O^+	3.75	[4.50]	5.3
H_2O^-	1.89		2.0
Total	100.19	[101.92]	103.23

(1) Kragerö, Norway. (2) Predazzo, Italy; total Fe as Fe_2O_3, B_2O_3 and H_2O from stoichiometry.
(3) Wakefield Lake, Canada; total Fe as Fe_2O_3.

Occurrence: In granite pegmatite (Kragerö, Norway; Wakefield Lake, Canada).

Association: Tourmaline, thorite, allanite, apatite, phenakite, zircon, titanite (Kragerö, Norway); quartz, chlorite, fergusonite, kainosite-(Y), tengerite, xenotime, wakefieldite, thorogummite (Wakefield Lake, Canada); microcline, albite, schorl, monazite, zircon (Crestmore, California, USA).

Distribution: From the Lindvikskollen pegmatite dike, near Kragerö, Norway. At Predazzo, Trentino-Alto Adige, and Lago Vico, Viterbo, Italy. In the Trimouns talc deposit, near Luzenac, Ariège, France. At Crestmore, Riverside Co., California, USA. In the Evans-Lou quarry, near Wakefield Lake, Quebec, Canada.

Name: For geologist Amund Theodor Helland (1846–1918), of Oslo, Norway.

Type Material: n.d.

References: (1) Dana, E.S. and W.E. Ford (1909) Dana's system of mineralogy, (6th edition), app. II, 50–51. (2) Oftedal, I. (1965) Über den Hellandit. Tschermaks Mineral. Petrog. Mitt., 10, 125–129 (in German with English abs.). (3) Hogarth, D.D., G.Y. Chao, and D.C. Harris (1972) New data on hellandite. Can. Mineral., 11, 760–776. (4) Mellini, M. and S. Merlino (1977) Hellandite: a new type of silicoborate chain. Amer. Mineral., 62, 89–99.

Crystal Data: Cubic. *Point Group:* $\bar{4}3m$. As tetrahedra, dodecahedra, and pseudo-octahedral crystals, up to 12 cm; as rounded grains.

Physical Properties: *Cleavage:* Distinct on {111}, {1$\bar{1}$1}. *Fracture:* Uneven to conchoidal. *Tenacity:* Brittle. Hardness = 6–6.5 VHN = 741–873 D(meas.) = 3.20–3.44 D(calc.) = [3.23] Pyroelectric.

Optical Properties: Translucent. *Color:* Yellow, brown, red, gray-yellow, yellow-green. *Streak:* White. *Luster:* Vitreous to resinous.
Optical Class: Isotropic. $n = 1.728$–1.749

Cell Data: *Space Group:* $P\bar{4}3n$. a = 8.24–8.29 Z = 2

X-ray Powder Pattern: Butte, Montana, USA.
3.382 (100), 1.955 (80), 2.215 (70), 2.618 (65), 1.692 (65), 1.510 (50), 1.466 (50)

Chemistry:

	(1)	(2)	(3)
SiO_2	31.06	29.62	32.47
Al_2O_3	0.30	0.03	
FeO	0.18	13.25	
MnO	52.48	24.34	51.12
ZnO	0.55	16.09	
BeO	[13.52]	[13.21]	13.52
CaO		0.07	
S	5.51	5.91	5.78
$-O = S$	[2.75]	[2.95]	2.89
Total	[100.85]	[99.57]	100.00

(1) East Moultan mine, Butte, Montana, USA; by electron microprobe; BeO calculated assuming $(Fe^{2+}, Mn, Zn, Mg, Ca):Be = 4:3$. (2) Pitkäranta, Finland; by electron microprobe, BeO calculated as in (1). (3) $Mn_4Be_3(SiO_4)_3S$.

Polymorphism & Series: Forms two series, with danalite, and with genthelvite.

Occurrence: May be locally abundant in granites, granite pegmatites, gneisses, and contact zones and skarns.

Association: Garnet, magnetite, fluorite, chlorite, diopside, vesuvianite.

Distribution: From Schwarzenberg and Breitenbrunn, Saxony, Germany. Exceptionally large crystals from Yxsjö, Örebro, Sweden. At Luppikko, Karelia, and near Miass, Ilmen Mountains, Southern Ural Mountains, Russia. From the Inyl'chek Mountains, Tien Shan, southeastern Kyrgyzstan. At Kara-Oba, Kazakhstan. From the Langesundsfjord, at Hørtekollen, and several other localities in Norway. In Finland, at Pitkäranta. In the USA, at Butte, Silver Bow Co., Montana; from Amelia, Amelia Co., Virginia; in the American Tunnel, Silverton, San Juan Co., Colorado; on the east fork of Benedict Creek, Sawtooth Mountains, Boise Co., Idaho; from the Clark vein, Rincon, San Diego Co., California; at Iron Mountain, Sierra Co., New Mexico. In the Hongo mine, Iwate Prefecture; the Oashi mine, Tochigi Prefecture; the Yagisawa mine, Nagano Prefecture; and several other places in Japan. A few additional localities are known.

Name: From the Greek for *sun*, in allusion to its yellow color.

Type Material: Mining Academy, Freiberg, Germany, 22941.

References: (1) Dana, E.S. (1892) Dana's system of mineralogy, (6th edition), 434–435. (2) Deer, W.A., R.A. Howie, and J. Zussman (1963) Rock-forming minerals, v. 4, framework silicates, 303–309. (3) Vlasov, K.A., Ed. (1966) Mineralogy of rare elements, v. II, 119–126. (4) Glass, J.J., R.H. Jahns, and R.E. Stevens (1944) Helvite and danalite from New Mexico and the helvite group. Amer. Mineral., 29, 163–191. (5) Dunn, P.J. (1976) Genthelvite and the helvine group. Mineral. Mag., 40, 627–636. (6) Hassan, I. and H.D. Grundy (1985) The crystal structures of helvite group minerals, $(Mn, Fe, Zn)_8(Be_6Si_6O_{24})S_2$. Amer. Mineral., 70, 186–192.

Crystal Data: Triclinic. *Point Group:* 1. As euhedral crystals exhibiting hemihedral symmetry, up to 1 cm. *Twinning:* Three laws recognized; by reflection on $(\overline{2}23)$, $(0\overline{1}2)$, and $(0\overline{1}0)$, with composition surface (010).

Physical Properties: Hardness = 3 D(meas.) = 6.42 D(calc.) = 6.50

Optical Properties: Translucent to transparent. *Color:* Bright orange, henna-brown, to almost black. *Streak:* Saffron-yellow. *Luster:* Vitreous.
Optical Class: Biaxial (–). *Pleochroism:* Feeble, yellow to orange. *Dispersion:* Noticeable, resembles horizontal dispersion. *Absorption:* $Z > Y > X$. $\alpha = 2.105(5)$ $\beta = 2.32(2)$
$\gamma = 2.65(2)$ 2V(meas.) = 92° 2V(calc.) = 88°

Cell Data: *Space Group:* P1. $a = 9.497(1)$ $b = 11.443(2)$ $c = 10.841(2)$ $\alpha = 120°30'$
$\beta = 92°06'$ $\gamma = 55°50'$ Z = 1

X-ray Powder Pattern: Florence Lead-Silver mine, Arizona, USA.
3.301 (100), 4.872 (90), 4.364 (80), 3.164 (80), 3.102 (80), 2.924 (55), 2.849 (45)

Chemistry:

	(1)	(2)
SiO_2	3.2	3.93
CrO_3	19.5	19.64
ZnO	[3.8]	2.66
PbO	70.5	73.05
F	5.1	1.24
H_2O	n.d.	
$-O = F_2$	2.1	0.52
Total	[100.0]	100.00

(1) Florence Lead-Silver mine, Arizona, USA; average of several analyses, corrected for $PbMoO_4$ and $PbCO_3$ impurities, ZnO originally given as 3.93%; recalculated to 100.0%.
(2) $Pb_{10}Zn(CrO_4)_6(SiO_4)_2F_2$.

Polymorphism & Series: Forms a series with iranite.

Occurrence: Formed through oxidation of galena, sphalerite, and pyrite, in the oxide zone of hydrothermal lead-bearing veins.

Association: Cerussite, phoenicochroite, vauquelinite, willemite, wulfenite, galena, sphalerite, pyrite, tennantite, chalcopyrite (Florence Lead-Silver mine, Arizona, USA).

Distribution: From the Florence Lead-Silver mine, Tortilla Mountains, Pinal Co., and the Pack Rat claim, the Moon Anchor mine, and the Potter-Cramer property, Maricopa Co., Arizona; also from near Boulder City, Clark Co., Nevada, USA. In the Seh-Changi mine, near Neyband, Khorassan, Iran.

Name: In allusion to its distinctive *hemihedral* morphology.

Type Material: The Natural History Museum, London, England, 1968,246; University of Arizona, Tucson, Arizona; National Museum of Natural History, Washington, D.C., USA, 141026.

References: (1) Williams, S.A. and J.W. Anthony (1970) Hemihedrite, a new mineral from Arizona. Amer. Mineral., 55, 1088–1102. (2) McLean, W.J. and J.W. Anthony (1970) The crystal structure of hemihedrite. Amer. Mineral., 55, 1103–1114. (3) Cesbron, F. and S.A. Williams (1980) Iranite–hemihedrite, bellite, phoenicochroite, vauquelinite et fornacite: synthèse et nouvelles donnèes. Bull. Minéral., 103, 469–477 (in French with English abs.).

Crystal Data: Orthorhombic. *Point Group:* $mm2$. Commonly crystallized, thin tabular $\perp$ {010}, striated $\parallel$ [001], to 10 cm; doubly-terminated crystals show hemimorphism. Typically in sheaflike or fan-shaped aggregates; stalactitic, mammillary, botryoidal; as compact fibrous and chalky coatings. *Twinning:* On {001}, rare, with antilogous poles in contact.

Physical Properties: *Cleavage:* Perfect on {110}, poor on {101}; {001}, rare. *Fracture:* Uneven to subconchoidal. *Tenacity:* Brittle. Hardness = 4.5–5 D(meas.) = 3.475 D(calc.) = 3.484 Strongly pyroelectric; may fluoresce bluish under SW UV.

Optical Properties: Translucent, transparent in small crystals. *Color:* Colorless, white; pale blue, pale green, gray, brown from impurities. *Streak:* White. *Luster:* Vitreous, subpearly, adamantine, rarely silky.
Optical Class: Biaxial (+). *Orientation:* $X = b$; $Y = a$; $Z = c$. *Dispersion:* $r > v$, strong. $\alpha = 1.614$ $\beta = 1.617$ $\gamma = 1.636$ 2V(meas.) = $46°$

Cell Data: *Space Group:* $Imm2$. a = 8.367(5) b = 10.730(6) c = 5.155(3) Z = 2

X-ray Powder Pattern: Sterling Hill, New Jersey, USA.
3.104 (100), 6.60 (86), 3.288 (75), 3.296 (73), 5.36 (55), 2.400 (54), 2.559 (51)

Chemistry:

	(1)	(2)
SiO_2	25.01	24.94
ZnO	67.42	67.58
H_2O	8.32	7.48
Total	100.75	100.00

(1) Pulaski Co., Virginia, USA. (2) $Zn_4Si_2O_7(OH)_2 \cdot H_2O$.

Occurrence: A secondary mineral typically found in the oxidized zone of zinc-bearing mineral deposits.

Association: Smithsonite, sphalerite, galena, cerussite, anglesite, calcite, aurichalcite, rosasite, hydrozincite, chrysocolla.

Distribution: Only a few localities for fine examples can be mentioned. From Băiţa (Rézbánya), Romania. At Bănská Štiavnica (Schemnitz), Slovakia. From Caldbeck Fells, Cumbria, England. At Moresnet, Belgium. In Germany, at Freiberg and Altenberg, Saxony. In the Sa Duchessa mine, Iglesias, Sardinia, Italy. Large crystals from Nerchinsk, Siberia, Russia. At Tchah Kuh, Esfahan, Iran. In the USA, at Franklin and Sterling Hill, Ogdensburg, Sussex Co., New Jersey; in Arizona, from Bisbee, Cochise Co., and in the 79 mine, Gila Co.; in Utah, at the Emma mine, Little Cottonwood Canyon, Salt Lake Co.; in the Ibex and Wolftone mines, Leadville, Lake Co., Colorado; in the Elkhorn mine, Elkhorn district, Jefferson Co., Montana. In Mexico, large crystals from Santa Eulalia, Chihuahua, and in the Mina Ojuela, Mapimi, Durango.

Name: In allusion to the mineral's *hemimorphic* morphology.

References: (1) Dana, E.S. (1892) Dana's system of mineralogy, (6th edition), 546–549. (2) Hill, R.J., G.V. Gibbs, and J.R. Craig (1977) A neutron-diffraction study of hemimorphite. Zeits. Krist., 146, 241–259. (3) (1953) NBS Circ. 539, 2, 62.

Crystal Data: Monoclinic. *Point Group:* 2/*m*. Euhedral crystals, to 14 cm, rare. Forms interlocking aggregates and anhedral plates, to over 30 cm, commonly bent, or showing other signs of mechanical deformation.

Physical Properties: *Cleavage:* {001}, perfect. Hardness = 2.5–3   D(meas.) = 2.86–3.43 D(calc.) = 3.30–3.37

Optical Properties: Translucent. *Color:* Coppery brown, bronze-brown, dark reddish brown to reddish black.
Optical Class: Biaxial (–). *Pleochroism:* X = pale yellow; Y = Z = light chestnut-brown. *Orientation:* X = b; Z ∧ c = 36°. *Dispersion:* r < v, slight. *Absorption:* Z = Y > X. α = 1.598–1.624   β = 1.658–1.686   γ = 1.660–1.697   2V(meas.) = 2°–8°

Cell Data: *Space Group:* C2/*m*. a = 5.340(2)   b = 9.254(2)   c = 10.235(3) β = 100.07(2)°   Z = 2

X-ray Powder Pattern: Franklin, New Jersey, USA; 1M.
10.20 (100), 3.398 (60), 5.094 (36), 2.546 (35), 1.696 (17), 2.652 (13), 1.554 (10)

Chemistry:

	(1)	(2)		(1)	(2)
SiO$_2$	31.58	31.9	BaO	0.65	0.3
TiO$_2$	0.32	0.35	Li$_2$O		0.04
Al$_2$O$_3$	13.72	13.6	Na$_2$O	0.24	0.1
Fe$_2$O$_3$	2.25	4.9	K$_2$O	7.91	8.6
FeO	0.34		F	0.45	
MnO	12.28	12.5	H$_2$O$^+$ + F		3.95
ZnO	22.97	19.8	H$_2$O$^+$	3.65	
MgO	3.69	2.7	H$_2$O$^-$		0.95
CaO		0.02	−O = F$_2$	0.19	
			Total	[99.86]	[99.71]

(1) Franklin, New Jersey, USA; original total given as 99.57%, corresponds to (K$_{0.85}$Na$_{0.04}$Ba$_{0.02}$)$_{\Sigma=0.91}$(Zn$_{1.43}$Mn$_{0.88}$Mg$_{0.46}$Fe$^{3+}_{0.14}$Ti$_{0.04}$Al$_{0.02}$Fe$^{2+}_{0.02}$)$_{\Sigma=3.00}$(Si$_{2.66}$Al$_{1.34}$)$_{\Sigma=4.00}$ O$_{10}$[(OH)$_{1.98}$F$_{0.02}$]$_{\Sigma=2.00}$. (2) Do.; original total given as 99.6%, corresponds to (K$_{0.94}$Na$_{0.02}$Li$_{0.02}$Ba$_{0.01}$)$_{\Sigma=0.99}$(Zn$_{1.24}$Mn$_{0.90}$Mg$_{0.34}$Fe$^{3+}_{0.32}$Al$_{0.08}$Ti$_{0.02}$)$_{\Sigma=2.90}$ (Si$_{2.72}$Al$_{1.28}$)$_{\Sigma=4.00}$O$_{10}$[(OH), F]$_{\Sigma=2.05}$.

Polymorphism & Series: 1M, 2M$_1$, 3A polytypes.

Mineral Group: Mica group.

Occurrence: Restricted to irregular lens- or sheetlike skarn bodies in a metamorphosed stratiform zinc deposit.

Association: Andradite, rhodonite, calcite, barium feldspars, franklinite, willemite, axinite, hancockite.

Distribution: From Franklin, Sussex Co., New Jersey, USA.

Name: For Dr. Sterling B. Hendricks (1902–), American crystallographer and chemist, a student of micas.

Type Material: Harvard University, Cambridge, Massachusetts, USA, 89818, 133712; The Natural History Museum, London, England, 1966,209.

References: (1) Frondel, C. and J. Ito (1966) Hendricksite, a new species of mica. Amer. Mineral., 51, 1107–1123. (2) Evans, B.W. and R.G.J. Strens (1966) Zinc mica from Franklin Furnace, New Jersey. Nature, 211, 619. (3) Robert, J.-L. and M. Gaspérin (1985) Crystal structure refinement of hendricksite, a Zn- and Mn-rich trioctahedral potassium mica: a contribution to the crystal chemistry of zinc-bearing minerals. Tschermaks Mineral. Petrog. Mitt., 34, 1–14.

Crystal Data: Orthorhombic. *Point Group:* $2/m\ 2/m\ 2/m$. Rarely in feltlike masses; as irregular aggregates, to 1 mm, imbedded in other minerals.

Physical Properties: Hardness = ~ 4 D(meas.) = n.d. D(calc.) = 3.68

Optical Properties: Translucent. *Color:* Yellow-brown; yellow-brown in thin section. *Luster:* Vitreous.
Optical Class: Biaxial. *Pleochroism:* Strong; from yellowish brown to dark red-brown. $n = > 1.82$. α = n.d. β = n.d. γ = n.d. 2V(meas.) = $63(1)°$

Cell Data: *Space Group: Cmcm.* a = 6.255(1) b = 9.034(2) c = 13.397(2) Z = 4

X-ray Powder Pattern: Wessels mine, South Africa; intensities calculated.
2.833 (100), 2.695 (98), 4.804 (86), 2.807 (82), 2.401 (68), 3.373 (66), 2.715 (58)

Chemistry:

	(1)
SiO_2	28.22
TiO_2	0.00
Al_2O_3	0.00
Fe_2O_3	0.53
Mn_2O_3	37.82
CaO	0.02
SrO	24.32
BaO	0.46
H_2O	[8.62]
Total	[99.99]

(1) Wessels mine, South Africa; by electron microprobe, average of 13 analyses, Li and F not detected by ion microprobe, H_2O from ideal stoichiometry; corresponds to $(Sr_{0.98}Ba_{0.01})_{\Sigma=0.99}$ $(Mn_{2.01}^{3+}Fe_{0.03}^{3+})_{\Sigma=2.04}Si_{1.97}O_7(OH)_2 \cdot H_2O$.

Occurrence: From a hand specimen, in veinlets of sérandite-pectolite cutting sugilite, probably of hydrothermal origin in a bedded manganese deposit.

Association: Sérandite-pectolite, sugilite, braunite, taikanite, kornite.

Distribution: In the Wessels mine, near Kuruman, Cape Province, South Africa.

Name: For Henno Martin, German geologist, who has worked on the Precambrian geology of the general area in which the mineral occurs.

Type Material: Natural History Museum, Bern, Switzerland, B5564.

References: (1) Armbruster, T., R. Oberhänsli, V. Bermanec, and R. Dixon (1993) Hennomartinite and kornite, two new Mn^{3+} rich silicates from the Wessels mine, Kalahari, South Africa. Schweiz. Mineral. Petrog. Mitt., 73, 349–355. (2) (1994) Amer. Mineral., 79, 763–764 (abs. ref. 1). (3) Armbruster, T., R. Oberhänsli, and V. Bermanec (1992) Crystal structure of $SrMn_2[Si_2O_7](OH)_2 \cdot H_2O$, a new mineral of the lawsonite type. Eur. J. Mineral., 4, 17–22.

Crystal Data: Tetragonal. *Point Group:* $4/m\ 2/m\ 2/m$. As crystals, to 3 mm, and as aggregates of small grains. *Twinning:* Common on {101}, sectored.

Physical Properties: *Fracture:* Conchoidal. Hardness = n.d. D(meas.) = 3.34(2) D(calc.) = 3.40

Optical Properties: Translucent. *Color:* Clove-brown to apricot-brown; lemon-yellow to pale yellow in thin section. *Streak:* White. *Luster:* Vitreous.
Optical Class: Uniaxial (+); may be anomalously biaxial. *Pleochroism:* Weak; O = very pale yellow; E = lemon-yellow. $\omega = 1.765(5)$ $\epsilon = 1.800(5)$ 2V(meas.) = Small.

Cell Data: *Space Group:* $I4_1/acd$. a = 12.39(1) c = 11.91(1) Z = 8

X-ray Powder Pattern: Tachgagalt mine, Morocco.
2.75 (vvs), 2.516 (vs), 4.37 (s), 3.09 (s), 2.98 (s), 2.684 (ms), 1.614 (ms)

Chemistry:

	(1)
SiO_2	24.65
Al_2O_3	5.95
Fe_2O_3	0.95
MnO	22.38
CaO	35.45
active O_2	2.56
H_2O^+	7.85
H_2O^-	0.08
Total	99.87

(1) Tachgagalt mine, Morocco; corresponds to $Ca_{2.97}(Mn^{3+}_{1.48}Al_{0.54}Fe_{0.06})_{\Sigma=2.08}$ $Si_{1.93}O_{7.90}(OH)_{4.10}$.

Occurrence: As small grains filling interstices between crystals of other minerals in a manganese ore deposit (Tachgagalt mine, Morocco).

Association: Marokite, hausmannite, gaudefroyite, calcite (Tachgagalt mine, Morocco); hausmannite, manganite, hematite, calcite, barite, andradite (N'Chwaning mine, South Africa).

Distribution: In the Tachgagalt manganese mine, Anti-Atlas Mountains, Morocco. From the N'Chwaning and Wessels mines, near Kuruman, Cape Province, South Africa.

Name: For Henri F.E. Termier (1897–), Professor of Geology at the Sorbonne, Paris, France.

Type Material: National School of Mines, Paris, France.

References: (1) Gaudefroy C., M. Orliac, F. Permingeat, and A. Parfenoff (1969) L'henritermiérite, une nouvelle espèce minérale. Bull. Soc. fr. Minéral., 92, 185–190 (in French with English abs.). (2) Aubry, A., Y. Dusausoy, A. Laffaille, and J. Protas (1969) Détermination et étude de la structure cristalline de l'henritermiérite, hydrogrenat de symétrie quadratique. Bull. Soc. fr. Minéral., 92, 126–133 (in French with English abs.). (3) (1969) Amer. Mineral., 54, 1739 (abs. refs. 1 and 2).

Crystal Data: Hexagonal. *Point Group:* $\bar{3}\ 2/m$. As small hexagonal plates or aggregates, composed of stacked individuals; rarely as minute spherules.

Physical Properties: *Cleavage:* Rarely on $\{10\bar{1}1\}$, distinct. *Fracture:* Uneven. *Tenacity:* Brittle. Hardness = 4–5 D(meas.) = 2.08–2.16 D(calc.) = [2.09]

Optical Properties: Transparent to translucent. *Color:* White. *Luster:* Vitreous. *Optical Class:* Uniaxial (+). $\omega = 1.471$–1.479 $\epsilon = 1.474$–1.481

Cell Data: *Space Group:* $R\bar{3}m$. $a = 13.799$ $c = 15.102$ $Z = 12$

X-ray Powder Pattern: Aci Reale, Sicily, Italy; very similar to chabazite.
2.930 (100), 4.32 (65), 9.36 (50), 5.03 (40), 2.897 (30), 3.88 (25), 6.89 (20)

Chemistry:

	(1)	(2)
SiO_2	47.39	42.45
Al_2O_3	20.90	22.06
Fe_2O_3		0.11
MgO		0.24
CaO	0.38	1.01
SrO		0.47
Na_2O	8.33	9.20
K_2O	4.39	4.70
H_2O^+		16.39
H_2O^-		3.34
H_2O	17.84	
Total	99.23	99.97

(1) Aci Castello, Sicily, Italy. (2) Aci Trezza, Sicily, Italy; corresponds to $(Na_{0.78}K_{0.26}$ $Ca_{0.05}Mg_{0.02}Sr_{0.01})_{\Sigma=1.12}Al_{1.13}Si_{1.85}O_6 \cdot 2.87H_2O$.

Mineral Group: Zeolite group.

Occurrence: In cavities in basalt. Also in bedded lake deposits altered from volcanic pyroclastic material.

Association: Phillipsite (Aci Castello, Sicily, Italy); analcime, chabazite, erionite, clinoptilolite, halite, thenardite (Bowie, Arizona, USA); calcite, stilbite, analcime (Horseshoe Dam, Arizona, USA).

Distribution: From Aci Castello and nearby on the flanks of Mt. Etna, Sicily, and on the Cyclopean Islands, Italy. In Scotland, near Lynedale, Isle of Skye. At Dalsnipa, on Sandoy, Faeroe Islands. In the Ilímaussaq intrusion, southern Greenland. At Richmond, Victoria, Australia. From the Tokatoka district, about 150 km north of Auckland, New Zealand. At Hayata, Saga Prefecture, Japan. In the USA, in the San Simon basin, 11 km northeast of Bowie, Cochise Co., and from the Horseshoe Dam area, Maricopa Co., Arizona.

Name: After Sir John Frederick William Herschel (1792–1871), British astronomer.

References: (1) Dana, E.S. (1892) Dana's system of mineralogy, (6th edition), 589–593 (with chabazite). (2) Mason, B. (1962) Herschelite, a valid species. Amer. Mineral., 47, 985–987. (3) Gude, A.J., 3rd and R.A. Sheppard (1966) Silica-rich chabazite from the Barstow Formation, San Bernardino County, southern California. Amer. Mineral., 51, 909–915. (4) Passaglia, E. (1970) The crystal chemistry of chabazites. Amer. Mineral., 55, 1278–1301. (5) Akizuki, M., H. Nishido, and M. Fujimoto (1989) Herschelite: morphology and growth sectors. Amer. Mineral., 74, 1337–1342.

Crystal Data: Monoclinic. *Point Group:* $2/m$. Crystals commonly tabular ‖ {010}, and elongated, widest at the center, hence called "coffin-shaped," to 12 cm; also granular to massive. *Twinning:* With {100} as twin and contact plane.

Physical Properties: *Cleavage:* {010}, perfect. *Fracture:* Subconchoidal to uneven. *Tenacity:* Brittle. Hardness = 3.5–4   D(meas.) = 2.10–2.20   D(calc.) = 2.17

Optical Properties: Transparent to translucent. *Color:* Colorless, white, gray, yellow, red, pink, orange, brown, black; colorless in thin section. *Streak:* White. *Luster:* Vitreous, pearly on {010}.
Optical Class: Biaxial (+). *Orientation:* $Z = b$; $X \wedge a = 0°–34°$; $Y \wedge c = 0°–32°$.
Dispersion: $r > v$, distinct, crossed. $\alpha = 1.491–1.505$ $\beta = 1.493–1.503$ $\gamma = 1.500–1.512$
2V(meas.) = 0°–55°   2V(calc.) = 34°

Cell Data: *Space Group:* $C2/m$.   a = 17.77(2)   b = 17.95(2)   c = 7.435(7)
$\beta = 116.46(5)°$ Z = 4

X-ray Powder Pattern: Giebelsbach, Switzerland.
3.917 (100), 2.959 (90), 8.84 (80), 7.80 (70), 5.096 (70), 3.420 (70), 2.805 (70)

Chemistry:

	(1)	(2)		(1)	(2)
SiO_2	60.06	57.17	BaO	0.75	0.31
Al_2O_3	13.65	17.03	Na_2O	1.07	1.39
Fe_2O_3	1.16	0.04	K_2O	0.66	0.73
MgO	0.38	0.02	H_2O^+	12.43	
CaO	4.44	7.13	H_2O^-	3.11	
SrO	1.35	0.19	H_2O		16.71
			Total	99.06	100.72

(1) Col Aut, Buffaure, Trentino–Alto Adige, Italy; corresponds to $(Ca_{0.56}Na_{0.24}K_{0.10}$ $Sr_{0.09}Mg_{0.07}Ba_{0.03})_{\Sigma=1.09}Al_{1.89}Fe_{0.10}Si_{7.05}O_{18} \cdot 6.08H_2O$. (2) Faeroe Islands; corresponds to $(Ca_{0.89}Na_{0.32}K_{0.11}Ba_{0.02}Sr_{0.01})_{\Sigma=1.35}Al_{2.34}Si_{6.68}O_{18} \cdot 6.50H_2O$.

Mineral Group: Zeolite group.

Occurrence: In cavities in basalts; in highly weathered andesites and diabases; as a devitrification product of volcanic glasses and tuffs.

Association: Zeolites, datolite, apophyllite, calcite.

Distribution: Many localities, even for fine specimens. Exceptional crystals from the Teigarhorn, Berufjord, Iceland. At Haldarsvik, on Streymoy, Faeroe Islands. Around Glasgow, Dumbartonshire, Scotland. At Giebelsbach, near Fiesch, Valais, Switzerland. In the USA, at Paterson, Passaic Co., and Bergen Hill, Hudson Co., New Jersey; at Goble, Columbia Co., and elsewhere in Oregon; at Skookumchuck Dam, near Bucoda, Thurston Co., Washington. From Cape Blomiden, Nova Scotia, Canada. From near Bento Gonçalves, Rio Grande do Sul, Brazil. Large crystals from the Nasik, Poona, and Bombay districts, Maharashtra, India. At Tambar Springs, near Gunnedah, New South Wales, Australia.

Name: For the English mineral collector and dealer John Henry Heuland (1778–1856).

References: (1) Dana, E.S. (1892) Dana's system of mineralogy, (6th edition), 573–576. (2) Deer, W.A., R.A. Howie, and J. Zussman (1963) Rock-forming minerals, v. 4, framework silicates, 377–384. (3) Alietti, A. (1972) Polymorphism and crystal chemistry of heulandites and clinoptilolites. Amer. Mineral., 57, 1448–1462. (4) Alberti, A. (1972) On the crystal structure of the zeolite heulandite. Tschermaks Mineral. Petrog. Mitt., 18, 129–146. (5) Hambley, T.W. and J.C. Taylor (1984) Neutron diffraction studies on natural heulandite and partially dehydrated heulandite. J. Solid State Chem., 54, 1–9. (6) Merkle, A.B. and M. Slaughter (1968) Determination and refinement of the structure of heulandite (1968) Amer. Mineral., 53, 1120–1138.

Crystal Data: Cubic. *Point Group:* $[4/m\ \bar{3}\ 2/m.]$ As shells (up to 60 μm thick) replacing andradite or other minerals.

Physical Properties: *Tenacity:* Very brittle. Hardness = 6 D(meas.) = 3.05–3.08 D(calc.) = [3.05–3.28]

Optical Properties: Semitransparent. *Color:* Colorless or pale yellow. *Optical Class:* Isotropic; may be anomalously biaxial. $n = 1.670$–1.677

Cell Data: *Space Group:* [$Ia3d$.] a = 12.02–12.22 Z = [8]

X-ray Powder Pattern: Mariánská Hora, Czech Republic.
2.736 (s), 3.045 (ms), 2.227 (m), 1.984 (m), 3.263 (mw), 1.634 (mw), 4.994 (w)

Chemistry:

	(1)	(2)
SiO_2	27.30	21.24
Al_2O_3	23.97	19.38
Fe_2O_3	0.30	4.08
MnO	0.19	
MgO	2.01	0.92
CaO	37.00	39.83
H_2O+	9.20	
H_2O-	0.20	
H_2O		[11.41]
SO_3		3.14
Total	100.17	[100.00]

(1) Bug River, Russia. (2) Mariánská Hora, Czech Republic; by electron microprobe, H_2O by difference; corresponds to $Ca_{3.00}(Al_{1.61}Fe_{0.22}Mg_{0.10})_{\Sigma=1.93}(Si_{1.50}S_{0.17})_{\Sigma=1.67}[O_{6.64}(OH)_{5.36}]_{\Sigma=12.00}$.

Polymorphism & Series: Forms a series with grossular and katoite.

Mineral Group: Garnet group.

Occurrence: In a calcareous marl (Mariánská Hora, Czech Republic); in zeolite-calcite rocks and low-grade metamorphosed marls (Hatrurim Formation, Israel).

Association: Quartz, tremolite, gypsum (Mariánská Hora, Czech Republic); calcite, aragonite, vaterite, portlandite, ettringite, tobermorite, jennite, afwillite (Hatrurim Formation, Israel).

Distribution: At Mariánská Hora (Marienberg), near Ústí nad Lábem (Aussig), Czech Republic. At Aubenas, Ardèche, France. In the Hatrurim Formation, Israel. From Nikortzminda, Caucasus Mountains, Georgia. Along the Bug River, locality not further specified, in Russia. At Crestmore, Riverside Co., California, USA. From Velardeña, Durango, Mexico. From the Tokatoka district, about 150 km north of Auckland, New Zealand.

Name: After Professor Josef Emanuel Hibsch of Tetschen, Czech Republic.

Type Material: Natural History Museum, Vienna, Austria, J4774.

References: (1) Dana, E.S. and W.E. Ford (1909) Dana's system of mineralogy, (6th edition), app. II, 52. (2) Nalivkina, É.B. (1960) Hibschite from the Bug River. Zap. Vses. Mineral. Obshch., 89, 714–718 (in Russian). (3) (1964) Mineral. Abs., 16, 454 (abs. ref. 2). (4) Gross, S. (1977) The mineralogy of the Hatrurim Formation, Israel. Geol. Sur. Israel Bull. 70, 32–35. (5) Passaglia, E. and R. Rinaldi (1984) Katoite, a new member of the $Ca_3Al_2(SiO_4)_3$ – $Ca_3Al_2(OH)_{12}$ series and a new nomenclature for the hydrogrossular group of minerals. Bull. Minéral., 107, 605–618. (6) Rinaldi, R. and E. Passaglia (1989) Hibschite topotype: crystal chemical characterization. Eur. J. Mineral., 1, 639–644.

Crystal Data: Hexagonal. *Point Group:* 32. As nearly equidimensional rhombohedral crystals, displaying {11$\bar{2}$0} and {01$\bar{1}$2}, to 4 mm. *Twinning:* Multiple twinning by rotation about [2$\bar{2}$01] and about [0001], common; typically multiply twinned after both laws, involving four, eight, or more individuals.

Physical Properties: *Fracture:* Conchoidal. *Tenacity:* Brittle. Hardness = > 4
D(meas.) = 2.724(4)   D(calc.) = 2.739

Optical Properties: Transparent to opaque. *Color:* Pale to dark brown, yellow, colorless, white, flesh-pink, rose-red. *Streak:* White. *Luster:* Vitreous to porcelaneous.
Optical Class: Uniaxial (–). ω = 1.609(1) ϵ = 1.596(1)

Cell Data: *Space Group:* R32.   a = 10.556(1)   c = 15.855(2)   Z = 6

X-ray Powder Pattern: Mont Saint-Hilaire, Canada.
5.28 (100), 6.00 (60), 3.168 (50), 3.046 (40), 1.759 (40), 2.994 (30), 2.639 (30)

Chemistry:

	(1)
SiO$_2$	42.08
TiO$_2$	0.04
ZrO$_2$	29.72
Al$_2$O$_3$	0.03
FeO	0.03
MnO	0.02
MgO	0.01
CaO	0.20
Na$_2$O	13.43
K$_2$O	0.52
H$_2$O	13.54
Total	99.62

(1) Mont Saint-Hilaire, Canada; by electron microprobe, H$_2$O by TGA; corresponding to (Na$_{1.85}$K$_{0.05}$Ca$_{0.02}$)$_{\Sigma=1.92}$Zr$_{1.03}$Si$_{2.99}$O$_{9.30}$·3.21H$_2$O.

Occurrence: In miarolitic cavities and altered pegmatite dikes in nepheline syenite in an intrusive alkalic gabbro-syenite complex (Mont Saint-Hilaire, Canada).

Association: Catapleiite, elpidite, gaidonnayite, aegirine, analcime, albite, microcline, chlorite, calcite, epididymite, natrolite, fluorite, pyrite, sphalerite, galena (Mont Saint-Hilaire, Canada).

Distribution: In Canada, at Mont Saint-Hilaire, Quebec. From Bratthagen, near Larvik, Norway. On Mt. Alluaiv, Lovozero massif, Kola Peninsula, Russia.

Name: For the locality at Mont Saint-Hilaire, Quebec, Canada.

Type Material: Canadian Museum of Nature, Ottawa; Royal Ontario Museum, Toronto, Canada, 34803.

References: (1) Chao, G.Y., D.H. Watkinson, and T.T. Chen (1974) Hilairite, Na$_2$ZrSi$_3$O$_9$·3H$_2$O, a new mineral from Mont St. Hilaire, Quebec. Can. Mineral., 12, 237–240. (2) Ilyushin, G.D, A.A. Voronkov, N.N. Nevskii, V.V. Ilyukhin, and N.V Belov (1981) Crystal structure of hilairite, Na$_2$ZrSiO$_3$O$_9$·3H$_2$O. Sov. Phys. Doklady Acad. Nauk SSSR, 260, 1118–1120 (in Russian).

Crystal Data: Monoclinic (?), pseudo-orthorhombic. *Point Group:* $2/m$. Acicular fibrous crystals, to 0.5 mm; radial spherulitic.

Physical Properties: *Cleavage:* {010}. Hardness = 5–6 D(meas.) = 2.66–2.69 D(calc.) = [2.65]

Optical Properties: Transparent to translucent. *Color:* White, colorless, pale greenish; colorless in thin section.
Optical Class: Biaxial (–). *Orientation:* $Z = c$. *Dispersion:* $r < v$, strong, producing characteristic abnormal blue interference colors. $\alpha = 1.605(5)$ $\beta = \sim1.61$ $\gamma = 1.612(3)$ 2V(meas.) = 60°–80°

Cell Data: *Space Group:* $P2_1/a$ (?). a = 16.60 b = 7.26 c = 11.85 $\beta = 90°$ Z = 12

X-ray Powder Pattern: Velardeña, Mexico.
2.92 (vvs), 4.76 (vs), 3.33 (vs), 3.02 (s), 2.82 (s), 2.76 (s), 2.37 (s)

Chemistry:

	(1)	(2)
SiO_2	32.59	31.58
TiO_2	0.02	
Al_2O_3	0.23	
Fe_2O_3	0.15	
MnO	0.01	
MgO	0.04	
CaO	57.76	58.95
Na_2O	0.03	
K_2O	0.05	
F	0.00	
H_2O	9.36	9.47
CO_2	0.00	
Total	100.24	100.00

(1) Velardeña, Mexico; total Fe as Fe_2O_3. (2) $Ca_2SiO_3(OH)_2$.

Occurrence: In limestone affected by high-grade contact metamorphism.

Association: Wollastonite, garnet (Velardeña, Mexico); calcite, vesuvianite, grossular (Carlingford, Ireland); foshagite (Crestmore, California, USA).

Distribution: In the Terneras mine, Velardeña, Durango, Mexico. In the USA, from Crestmore, Riverside Co., California. At Carlingford, Co. Louth, Ireland. In the Güneyce-İkizdere area, Trabzon Province, Turkey. From Mt. Chapchachi, north of the Caspian Sea, Kazakhstan. At Nugrah, Saudi Arabia. In the Hatrurim Formation, Israel. From Kushiro, Hiroshima Prefecture, and in the Mihari mine and at Fuka, near Bicchu, Okayama Prefecture, Japan.

Name: For William Francis Hillebrand (1853–1925), American geochemist and analytical chemist.

Type Material: National Museum of Natural History, Washington, D.C., USA, 86531.

References: (1) Dana, E.S. and W.E. Ford (1909) Dana's system of mineralogy, (6th edition), app. II, 52. (2) Heller, L. and H.F.W. Taylor (1956) Crystallographic data for the calcium silicates. H.M. Stationary Office, London, 59–62. (3) Mamedov, K.H. and N.V. Belov (1958) Crystal structure of hillebrandite. Doklady Acad. Nauk SSSR, 123, 741–743 (in Russian).

Crystal Data: Monoclinic. *Point Group:* n.d. As idiomorphic crystals, up to 5 mm.

Physical Properties: Hardness = n.d. D(meas.) = n.d. D(calc.) = [4.82]

Optical Properties: Semitransparent (?). *Color:* Light red-brown.
Optical Class: [Biaxial.] α = n.d. β = n.d. γ = n.d. 2V(meas.) = n.d.

Cell Data: *Space Group:* n.d. a = 9.996(10) b = 7.705(7) c = 4.792(4) β = 90.06(4)°
Z = [4]

X-ray Powder Pattern: Iwaguro Sekizai quarry, Japan.
3.159 (vs), 4.799 (s), 2.874 (s), 2.870 (s), 2.577 (s), 2.565 (s)

Chemistry:

	(1)
SiO_2	22.27
B_2O_3	trace
Y_2O_3	10.91
La_2O_3	3.40
Ce_2O_3	16.77
RE_2O_3	30.73
FeO	5.65
BeO	[9.27]
CaO	0.39
H_2O	[1.90]
Total	[101.29]

(1) Iwaguro Sekizai quarry, Japan; by electron microprobe, BeO calculated so Be:Si = 1:1; RE_2O_3 = Nd_2O_3 9.79%, Pr_2O_3 [3.5%], Sm_2O_3 4.70%, Eu_2O_3 trace, Gd_2O_3 4.18%, Tb_2O_3 [0.5%], Dy_2O_3 3.82%, Ho_2O_3 1.08%, Er_2O_3 1.84%, Tm_2O_3 trace, Yb_2O_3 1.02%, Lu_2O_3 [0.3%]; corresponds to $(Ce_{0.54}Y_{0.51}RE_{1.07})_{\Sigma=2.12}Fe_{0.41}Be_{1.96}Si_{1.96}O_{8.87}(OH)_{1.13}$.

Mineral Group: Gadolinite group.

Occurrence: In drusy pegmatite.

Association: Quartz, feldspar, mica, cassiterite, stokesite, fluorite, chlorite.

Distribution: In the Iwaguro Sekizai quarry, Tahara, Gifu Prefecture, Japan.

Name: For the predominance of *cerium* and relation to *hingganite*-(Y).

Type Material: n.d.

References: (1) Miyawaki, R., I. Nakai, K. Nagashima, A. Okamoto, and T. Isobe (1987) The first occurrences of hingganite, hellandite and wodginite in Japan. Kobutsugaku Zasshi, 18(1), 17–30 (in Japanese). (2) (1990) Amer. Mineral., 75, 432 (abs. ref. 1).

Crystal Data: Monoclinic. *Point Group:* $2/m$. Stout prismatic crystals, to 1.5 mm; commonly granular.

Physical Properties: Hardness = [5–5.5] VHN = 415–467 D(meas.) = 4.42–4.57 D(calc.) = 4.45

Optical Properties: Semitransparent. *Color:* Milky white, light yellow, light green; colorless in transmitted light. *Luster:* Vitreous.
Optical Class: Biaxial (+). *Orientation:* $Y = b$; $Z \wedge c = 6°–13°$; $Z \wedge a = 14°$.
Dispersion: $r < v$, strong. $\alpha = 1.744–1.748$ $\beta = 1.753–1.765$ $\gamma = 1.765–1.783$
2V(meas.) = 80°

Cell Data: *Space Group:* $P2_1/c$. $a = 4.790$ $b = 7.545$ $c = 9.989$ $\beta = 91°$ $Z = 4$

X-ray Powder Pattern: Heilungkiang Province, China.
3.148 (10), 2.685 (10), 2.569 (6), 4.790 (5), 3.462 (5), 1.985 (5), 1.880 (5)

Chemistry:

	(1)	(2)		(1)	(2)		(1)	(2)
SiO_2	25.20	26.43	Yb_2O_3		17.02	BeO	10.41	10.50
TiO_2	0.10		RE_2O_3		12.39	MgO	0.09	
Al_2O_3	1.70	0.10	Fe_2O_3	1.63	1.46	CaO	0.96	2.38
Y_2O_3	26.11	24.83	FeO	0.89		$(K, Na)_2O$	1.17	
Ce_2O_3	28.47		PbO	0.38	0.52	H_2O	2.94	[3.92]
						Total	100.05	[99.55]

(1) Greater Khingan Range, China; corresponds to $(Y_{0.33}Ce_{0.21}La_{0.16}Nd_{0.10}RE_{0.13})_{\Sigma=0.93}$
$[(K, Na)_{0.07}Fe^{3+}_{0.05}Ca_{0.04}Fe^{2+}_{0.03}]_{\Sigma=0.19}(Be_{1.00}Al_{0.08})_{\Sigma=1.08}Si_{1.02}O_{4.38}(OH)_{0.80}$. (2) Kola Peninsula, Russia; by electron microprobe, H_2O by difference, $RE_2O_3 = Tb_2O_3$ 0.33%, Dy_2O_3 2.13%, Ho_2O_3 0.21%, Er_2O_3 6.47%, Tm_2O_3 1.38%, Lu_2O_3 1.87%. (3) Heilungkiang Province, China, by XRF and wet chemical analysis, analytical results not available; stated to correspond to $(Y_{0.33}Ce_{0.19}Nd_{0.10}La_{0.05}RE_{0.20}Fe^{3+}_{0.05}Fe^{2+}_{0.03}Na_{0.03})_{\Sigma=0.98}(Be_{0.97}Al_{0.03})_{\Sigma=1.00}(Si_{0.98}Al_{0.05})_{\Sigma=1.03}$ $O_{4.00}[(OH)_{0.77}O_{0.23}]_{\Sigma=1.00}$.

Mineral Group: Gadolinite group.

Occurrence: In a RE, Be-bearing granophyre (Greater Khingan Range, China); in "amazonite"-rich pegmatite in a differentiated alkalic massif (Kola Peninsula, Russia).

Association: n.d.

Distribution: From an undefined locality in the Greater Khingan Range, Heilungkiang Province, China. In the Trimouns talc deposit, six km northeast of Luzenac, Ariège, France. From the [Keivy massif,] Kola Peninsula, Russia. In the Tahara area, Gifu Prefecture, Japan.

Name: Presumably for the occurrence in the Greater Khingan (Hinggan) Range, and predominance of *yttrium*.

Type Material: Geology Bureau, Chinese Academy of Geological Sciences, Beijing, China.

References: (1) Xiaoshi Ding, Ge Bai, Zhongxian Yuan, and Luren Sun (1981) Yttroceberysite, a new Ce-Be-rich silicate. Geol. Rev. China, 27, 459–465 (in Chinese with English abs.). (2) (1988) Amer. Mineral., 73, 442 (abs. ref. 1). (3) Xiaoshi Ding, Ge Bai, Zhongxian Yuan, and Jinding Liu (1984) Hingganite [(Y, Ce)BeSiO_5(OH)]: new data. Yanshi Kuangwu Ji Ceshi, 3(1), 46–48 (in Chinese). (4) (1987) Chem. Abs., 106, 70456 (abs. ref. 3). (5) Lulu Ximen and Peng Zhizlong (1985) Crystal structure of xinganite. Acta Mineral. Sinica, 5, 289–293 (in Chinese with English abs.). (6) (1988) Amer. Mineral., 73, 441–442 (abs. ref. 5). (7) (1988) Amer. Mineral., 73, 935 (errata). (8) Voloshin, A.V., Y.A. Pakhomovskii, Y.P. Men'shikov, A.S. Povarennykh, E.N. Matvinenko, and O.V. Yakubovich (1983) Hingganite-(Yb), a new mineral from amazonite pegmatites of the Kola Peninsula. Doklady Acad. Nauk SSSR, 270, 1188–1192 (in Russian). (9) (1984) Amer. Mineral., 69, 811 (abs. ref. 8).

Crystal Data: Monoclinic. *Point Group:* 2/m. As acicular crystals, to 2 mm, in spherical aggregates.

Physical Properties: Hardness = 6–7 D(meas.) = n.d. D(calc.) = 4.83

Optical Properties: Transparent. *Color:* Colorless. *Luster:* Vitreous.
Optical Class: Biaxial (+). *Orientation:* $X \wedge c = 20°$; $Z \wedge c = 23°$. $\alpha = 1.725$ $\beta = 1.738$
$\gamma = 1.760$ 2V(meas.) = 65°

Cell Data: *Space Group:* $P2_1/a$. a = 9.888(5) b = 7.607(3) c = 4.740(2) $\beta = 90.45(4)°$
Z = 4

X-ray Powder Pattern: Kola Peninsula, Russia.
3.13 (10), 2.85 (10), 2.572 (8), 2.542 (8), 1.977 (8), 6.07 (7)

Chemistry:

	(1)
SiO$_2$	22.11
Y$_2$O$_3$	8.56
Yb$_2$O$_3$	34.07
RE$_2$O$_3$	19.37
BeO	10.90
CaO	1.14
H$_2$O	[3.74]
Total	[99.89]

(1) Kola Peninsula, Russia; by electron microprobe, H$_2$O stated to be by difference; RE$_2$O$_3$ = Tb$_2$O$_3$ 0.05%, Dy$_2$O$_3$ 2.47%, Ho$_2$O$_3$ 1.03%, Er$_2$O$_3$ 8.22%, Tm$_2$O$_3$ 3.10%, Lu$_2$O$_3$ 4.50%; corresponds to $(Yb_{0.45}Y_{0.20}RE_{0.30}Ca_{0.05})_{\Sigma=1.00}Be_{1.13}Si_{0.96}O_{3.92}(OH)_{1.08}$.

Mineral Group: Gadolinite group.

Occurrence: Formed by very late-stage replacement reactions in "amazonite"-rich pegmatites.

Association: Plumbian microlite, fluorite, keiviite-(Yb), bastnäsite.

Distribution: From the [Keivy massif,] Kola Peninsula, Russia.

Name: For the predominance of *ytterbium* and its relation to *hingganite*-(Y).

Type Material: A.E. Fersman Mineralogical Museum, Academy of Sciences, Moscow, Russia.

References: (1) Voloshin, A.V., Y.A. Pakhomovskii, Y.P. Men'shikov, A.S. Povarennykh, E.N. Matvinenko, and O.V. Yakubovich (1983) Hingganite-(Yb), a new mineral from amazonite pegmatites of the Kola Peninsula. Doklady Acad. Nauk SSSR, 270, 1188–1192 (in Russian). (2) (1984) Amer. Mineral., 69, 811 (abs. ref. 1). (3) Yakubovich, O.V., E.N. Matvinenko, A.V. Voloshin, and M.A. Simonov (1983) The crystal structure of hingganite-(Yb), $(Y_{0.51}TR_{0.36}Ca_{0.13}) \cdot Fe_{0.065}Be[SiO_4](OH)$. Kristallografiya (Sov. Phys. Crystal.), 28, 457–460 (in Russian).

Hiortdahlite

$(\mathrm{Ca}, \mathrm{Na})_3(\mathrm{Zr}, \mathrm{Ti})\mathrm{Si}_2\mathrm{O}_7(\mathrm{O}, \mathrm{F})_2$

Crystal Data: Triclinic. *Point Group:* $\bar{1}$. Crystals tabular parallel to {100}, with {010} and {101}. *Twinning:* Polysynthetic twinning on {010} and {100}, common.

Physical Properties: *Cleavage:* Distinct on {110} and {1$\bar{1}$0}, the two intersecting at nearly 90°. *Tenacity:* Very brittle. Hardness = 5.5 D(meas.) = 3.25–3.31 D(calc.) = [3.21–3.24]

Optical Properties: Semitransparent. *Color:* Light yellow to honey-yellow, yellowish brown, green. *Luster:* Vitreous on crystal faces, greasy on fractures.
Optical Class: Biaxial (+). *Pleochroism:* X = nearly colorless; Y = bright yellow; Z = wine-yellow. *Dispersion:* $r < v$, strong. *Absorption:* $Z > Y > X$. $\alpha = 1.639$–1.658
$\beta = 1.643$–1.664 $\gamma = 1.646$–1.671 2V(meas.) = 80°–86°

Cell Data: *Space Group:* $P\bar{1}$. $a = 10.95(3)$ $b = 10.31(2)$ $c = 7.29(3)$ $\alpha = 90°19(10)'$
$\beta = 109°2(15)'$ $\gamma = 90°5(10)'$ $Z = [4]$, or *Space Group:* $P\bar{1}$. $a = 11.0149(9)$ $b = 10.9409(9)$
$c = 7.3534(3)$ $\alpha = 109.350(3)°$ $\beta = 109.879(4)°$ $\gamma = 83.434(4)°$ $Z = [4]$

X-ray Powder Pattern: Kipawa River, Canada.
2.87 (100), 3.00 (90), 3.28 (45), 1.70 (40), 1.84 (30), 2.03 (25), 1.80 (25)

Chemistry:

	(1)		(1)		(1)
SiO_2	32.20	$\mathrm{RE}_2\mathrm{O}_3$	2.08	SrO	0.03
TiO_2	0.30	$\mathrm{Fe}_2\mathrm{O}_3$	0.27	$\mathrm{Na}_2\mathrm{O}$	6.85
ZrO_2	18.76	$\mathrm{Nb}_2\mathrm{O}_5$	0.56	$\mathrm{K}_2\mathrm{O}$	0.02
HfO_2	0.27	$\mathrm{U}_3\mathrm{O}_8$	0.09	F	6.69
SnO_2	0.02	MnO	0.30	$\mathrm{H}_2\mathrm{O}^+$	0.42
$\mathrm{Al}_2\mathrm{O}_3$	0.07	MgO	0.09	$-\mathrm{O} = \mathrm{F}_2$	2.81
$\mathrm{Y}_2\mathrm{O}_3$	3.50	CaO	29.57	Total	99.28

(1) Kipawa River, Canada; corresponding to $(\mathrm{Ca}_{2.01}\mathrm{Na}_{0.85}\mathrm{Y}_{0.12}\mathrm{RE}_{0.04})_{\Sigma=3.02}$
$(\mathrm{Zr}_{0.58}\mathrm{Nb}_{0.02}\mathrm{Mn}_{0.02}\mathrm{Ti}_{0.01}\mathrm{Fe}_{0.01}\mathrm{Mg}_{0.01}\mathrm{Al}_{0.01})_{\Sigma=0.66}\mathrm{Si}_{2.05}\mathrm{O}_7[\mathrm{F}_{1.34}\mathrm{O}_{0.41}(\mathrm{OH})_{0.18}]_{\Sigma=1.93}$.

Polymorphism & Series: Topological variants I and II, both triclinic.

Occurrence: A rare mineral in alkalic rocks, their pegmatites, and metamorphosed equivalents; in miarolitic cavities in sanidinites.

Association: Feldspar, biotite, meliphanite, titanite (Langesundsfjord, Norway); aegirine, astrophyllite, nepheline (Korgeredaba massif, Russia).

Distribution: In Norway, on Mittel Arø, Lille Arø, and Stokkø Islands, in the Langesundsfjord. In Italy, at Monte Somma and Vesuvius, Campania. From Mayen, Eifel district, Germany. At the Kangerdlugssuaq Fjord, Greenland. In Russia, in the Korgeredaba massif, Sangilen, Tuva. From the [Sheffield Lake complex,] Kipawa River, Villedieu Township, Quebec, Canada. On the Los Islands, Guinea. From the Jingera complex, New South Wales, Australia.

Name: After Professor Thorstein Hallager Hiortdahl (1839–1925), mineralogist of Christiania (now Oslo), Norway.

References: (1) Dana, E.S. (1892) Dana's system of mineralogy, (6th edition), 377–378. (2) Vlasov, K.A., Ed. (1966) Mineralogy of rare elements, v. II, 379–381 (3) Aarden, H.M. and J. Gittins (1974) Hiortdahlite from Kipawa River, Villedieu Township, Temiscaming County, Quebec, Canada. Can. Mineral., 12, 241–247. (4) Eggleton, R.A., G.E. Halford, and S.D. Beams (1979) Hiortdahlite from Jingera, New South Wales. J. Geol. Soc. Aust., 26, 81–85. (5) Merlino, S. and N. Perchiazzi (1985) The crystal structure of hiortdahlite I. Tschermaks Mineral. Petrog. Mitt., 34, 297–310. (6) Merlino, S. and N. Perchiazzi (1987) The crystal structure of hiortdahlite II. Mineral. Petrol., 37, 25–35.

Crystal Data: Amorphous to poorly crystalline. *Point Group:* n.d. Commonly massive, compact; may be minutely spherical.

Physical Properties: *Fracture:* Conchoidal. *Tenacity:* Brittle. Hardness = 2.5–3 D(meas.) = 2.43–2.67   D(calc.) = n.d.

Optical Properties: Transparent to translucent. *Color:* Black, dark brown; dark green masses may turn brown in light; in thin section, greenish brown or golden yellow. *Streak:* Yellowish brown. *Luster:* Resinous, vitreous, greasy.
Optical Class: Isotropic; locally anisotropic. $n = 1.50$–1.66

Cell Data: *Space Group:* n.d.

X-ray Powder Pattern: Riddarhyttan, Sweden; easily confused with neotocite.
4.45 (s), 3.53 (s), 2.56 (s), 1.71 (s), 1.54 (s)

Chemistry:

	(1)	(2)
SiO_2	35.08	34.15
Al_2O_3	1.38	
Fe_2O_3	40.28	45.38
FeO	2.23	
MgO	0.35	
CaO	0.36	
H_2O	20.78	20.47
Total	100.46	100.00

(1) Riddarhyttan, Sweden. (2) $Fe_2Si_2O_5(OH)_4 \cdot 2H_2O$.

Occurrence: A secondary mineral, formed from the weathering, or late-stage deuteric or hydrothermal alteration, of iron-bearing silicates or sulfides; by late-stage hydrothermal activity during sulfide ore deposition.

Association: Olivine, pyroxene, pyrite, chalcopyrite, pyrrhotite.

Distribution: In small amounts, easily overlooked, from many localities worldwide. Some for described material are: at Riddarhyttan, Västmanland, and Långban, Värmland, Sweden. From Fagul Cetatii, Balan, Romania. At Salberg, Norway. Found near Helsingfors, Finland. From Llallagua, Bolivia. In the USA, in the Hibbing district, St. Louis Co., Minnesota; in Arizona, at the Castle Dome mine, Gila Co., and on the Mildren and Steppe claims, Cababi district, Pima Co.; at the Gap Nickel mine, Lancaster Co., Pennsylvania. In Canada, from the Wilcox mine, Parry Sound, Ontario; at the Tetrault mines, near Montauban-les-mines, Quebec; and from Goldfields, Saskatchewan. In the Kawayama mine, Yamaguchi Prefecture; the Sano mine, Wakayama Prefecture; the Suzuyama mine, Kagoshima Prefecture; and other localities in Japan.

Name: For the Swedish chemist and mineralogist, Wilhelm Hisinger (1765–1852).

References: (1) Dana, E.S. (1892) Dana's system of mineralogy, (6th edition), 702–703. (2) Whelan, J.A. and S.S. Goldich (1961) New data for hisingerite and neotocite. Amer. Mineral., 46, 1412–1423. (3) Eggleton, R.A., J.H. Pennington, R.S. Freeman, and I.M. Threadgold (1983) Structural aspects of the hisingerite-neotocite series. Clay Minerals, 18, 21–31.

Crystal Data: Monoclinic. *Point Group:* $2/m$. As euhedral crystals, stout prismatic and terminated by steep pyramids, also dominated by pyramidal forms, to 2 cm. Granular to massive in veinlets.

Physical Properties: *Cleavage:* Perfect on {001}. Hardness = < 5 D(meas.) = 4.06–4.08 D(calc.) = 4.07 Fluoresces weak pinkish red in LW UV.

Optical Properties: Transparent to translucent. *Color:* Bright pink or red to reddish brown and orange, rarely yellow; black from tarnish or inclusions.
Optical Class: Biaxial (–). *Pleochroism:* In shades of lavender and pale purple. *Orientation:* $Y = b$; $Z \wedge c = 38°$. *Dispersion:* $r > v$, strong. $\alpha = 1.720–1.724$ $\beta = 1.741–1.742$ $\gamma = 1.746$ 2V(meas.) = 50°–60°

Cell Data: *Space Group:* $P2_1/a$. $a = 8.171(2)$ $b = 5.316(2)$ $c = 11.761(2)$ $\beta = 95°15(5)'$ $Z = 4$

X-ray Powder Pattern: Franklin, New Jersey, USA.
2.864 (100), 2.957 (90), 1.547 (85), 2.567 (65), 1.412 (65), 2.115 (60), 1.594 (60)

Chemistry:

	(1)	(2)	(3)
SiO_2	19.86	19.6	19.27
FeO		0.2	
MnO	20.68	22.7	22.75
ZnO	52.93	50.9	52.20
MgO	0.04	0.4	
CaO	0.93	0.3	
H_2O^+	5.77	5.76	5.78
Total	100.21	99.86	100.00

(1) Franklin, New Jersey, USA; average of three analyses. (2) Do.; by electron microprobe, average of eight samples, H_2O by the Penfield method. (3) $MnZn_2SiO_4(OH)_2$.

Occurrence: As seams in massive willemite-franklinite granular ore in a metamorphosed stratiform zinc deposit (Franklin, New Jersey, USA).

Association: Barite, willemite, franklinite, tephroite, pyrochroite, calcite, manganoan garnet, copper.

Distribution: From Franklin and Sterling Hill, Ogdensburg, Sussex Co., New Jersey, USA.

Name: For H.H. Hodgkinson, Assistant Underground Superintendent of the Franklin mine, who discovered the mineral.

Type Material: Harvard University, Cambridge, Massachusetts, 89875, 89880, 89881; National Museum of Natural History, Washington, D.C., USA, 87231.

References: (1) Palache, C. and W.T. Schaller (1913) Hodgkinsonite, a new mineral from Franklin Furnace, New Jersey. J. Wash. Acad. Sci., 3(19), 474–478. (2) Palache, C. (1935) The minerals of Franklin and Sterling Hill, Sussex County, New Jersey. U.S. Geol. Sur. Prof. Paper 180, 108–111. (3) Roberts, W.M.B. and F.M. Quodling (1962) X-ray, optical, and morphological observations on hodgkinsonite from Franklin Furnace. Mineral. Mag., 33, 343–346. (4) Rentzeperis, P.J. (1963) The crystal structure of hodgkinsonite, $Zn_2Mn[(OH)_2SiO_4]$. Zeits. Krist., 119, 117–138. (5) Dunn, P.J. and R.C. Bostwick (1982) Hodgkinsonite from Franklin and Sterling Hill, New Jersey: a review. Mineral. Record, 13, 229–232.

Crystal Data: Triclinic, pseudomonoclinic. *Point Group:* $\bar{1}$. As prismatic crystals, to 4 cm, with well-developed striations ‖ elongation; rarely in radiating groups. *Twinning:* Polysynthetic, ubiquitous, twin axis ⊥ pseudomonoclinic (010).

Physical Properties: *Cleavage:* Two good, intersecting at ∼55°; a parting ⊥ elongation. *Fracture:* Uneven. *Tenacity:* Brittle. Hardness = 5.5 D(meas.) = 3.85–3.88 D(calc.) = 3.92–3.98

Optical Properties: Opaque; transparent to translucent only in ultrathin sections. *Color:* Black. *Streak:* Dark green. *Luster:* Subadamantine to nonmetallic. *Optical Class:* Biaxial (–) (?). *Pleochroism:* Very strong; X = bronze; Z = green. $\alpha = 1.78~(\alpha')$ $\beta = $ n.d. $\gamma = 1.82~(\gamma')$ 2V(meas.) = Large.

Cell Data: *Space Group:* $P\bar{1}$. $a = 10.317(1)$ $b = 10.724(1)$ $c = 8.855(1)$ $\alpha = 105.77(1)°$ $\beta = 96.21(1)°$ $\gamma = 124.77(1)°$ Z = 2

X-ray Powder Pattern: Near Mo i Rana, Norway.
2.529 (100), 8.48 (90), 2.098 (63), 2.925 (59), 2.676 (48), 2.075 (47), 3.125 (46)

Chemistry:

	(1)	(2)		(1)	(2)
SiO_2	31.60	30.09	MgO	0.42	2.74
TiO_2	2.77	6.02	CaO	10.44	13.38
SnO_2	0.53		Na_2O	1.52	1.35
Al_2O_3	2.64	3.55	K_2O	0.00	0.30
Fe_2O_3	[19.03]	11.12	H_2O^+		0.35
FeO	[28.06]	26.91	H_2O^-		0.07
MnO	0.27	1.26	F	0.00	
BeO	2.65	2.32	LOI		0.22
			Total	[99.93]	[99.68]

(1) Near Mo i Rana, Norway; by electron microprobe and ICP, average of four samples, $Fe^{2+}:Fe^{3+}$ by wet chemical analysis and Mössbauer spectroscopy; corresponding to $(Ca_{1.63}Na_{0.43})_{\Sigma=2.06}$ $(Fe^{2+}_{3.42}Fe^{3+}_{2.08}Ti_{0.30}Mg_{0.09}Mn_{0.03}Sn_{0.03})_{\Sigma=5.95}(Si_{4.60}Be_{0.92}Al_{0.45})_{\Sigma=5.97}O_{20}$. (2) Ilmen Mountains, Russia; original total given as 100.28%; corresponding to $(Ca_{2.09}Na_{0.38}K_{0.03})_{\Sigma=2.50}(Fe^{2+}_{3.29}Fe^{3+}_{1.22}$ $Ti_{0.66}Mg_{0.30}Mn_{0.08})_{\Sigma=5.55}(Si_{4.39}Be_{0.81}Al_{0.61})_{\Sigma=5.81}O_{20}$.

Mineral Group: Aenigmatite group.

Occurrence: As a late-stage metamorphic mineral in peraluminous granitic gneisses and mafic pegmatites associated with a beryllium deposit (near Mo i Rana, Norway).

Association: Quartz, albite, microcline, biotite, phenakite, zircon, fluorite, calcite, many minor beryllium and sulfide minerals (near Mo i Rana, Norway).

Distribution: Found about 16 km northwest of Mo i Rana, Norway. In the Ilmen Mountains, Southern Ural Mountains, Russia.

Name: For Høgtuva Mountain, near the type locality in Norway.

Type Material: University of Oslo, Oslo, Norway; National Museum of Natural History, Washington, D.C., USA.

References: (1) Grauch, R.I., I. Lindahl, H.T. Evans, Jr., D.M. Burt, J.J. Fitzpatrick, E.E. Foord, P.-R. Graff, and J. Hysingjord (1994) Høgtuvaite, a new beryllian member of the aenigmatite group from Norway, with new X-ray data on aenigmatite. Can. Mineral., 32, 439–448. (2) Yakubovich, O.V., Y.A. Malinovskii, and O.V. Polyakov (1990) Crystal structure of makarochkinite [høgtuvaite]. Kristallografiya (Sov. Phys. Crystal.), 35, 1388–1394 (in Russian). (3) (1992) Amer. Mineral., 77, 448 (abs. ref. 2).

Holdenite

$(Mn^{2+}, Mg)_6Zn_3(AsO_4)_2(SiO_4)(OH)_8$

Crystal Data: Orthorhombic. *Point Group:* $2/m\ 2/m\ 2/m$. As complex crystals tabular on {100}, with {110}, {7.16.2}, {211}, {131}, to 1 cm. In aggregates of highly irregular microcrystals; fibrous, fine-grained, massive.

Physical Properties: *Cleavage:* {010}, poor. *Fracture:* Subconchoidal. Hardness = 4 D(meas.) = 4.11(1)   D(calc.) = 4.11

Optical Properties: Transparent to translucent. *Color:* Pink, orange, yellowish to deep red; pink in thin section. *Luster:* Vitreous.
Optical Class: Biaxial (+). *Orientation:* X = c; Y = b; Z = a. *Dispersion:* $r > v$, distinct.
$\alpha = 1.769$ $\beta = 1.770$ $\gamma = 1.785$ 2V(meas.) = 30°20′   2V(calc.) = 28°58′

Cell Data: *Space Group:* $Abma$. a = 11.99(1)   b = 31.46(4)   c = 8.697(6)   Z = 8

X-ray Powder Pattern: Franklin, New Jersey, USA.
2.838 (10), 3.582 (8), 2.583 (8), 1.531 (8), 2.464 (7), 3.406 (6), 5.740 (5)

Chemistry:

	(1)	(2)	(3)
SiO_2	6.4	5.9	5.92
As_2O_5	21.4	22.3	22.65
MnO	38.1	39.1	38.09
ZnO	23.9	23.4	24.06
MgO	3.5	2.5	2.18
H_2O	[7.1]	[7.1]	7.10
Total	[100.4]	[100.3]	100.00

(1) Franklin, New Jersey, USA; by electron microprobe, H_2O from theoretical composition.
(2) Sterling Hill, New Jersey, USA; by electron microprobe, H_2O from theoretical composition.
(3) $(Mn_{5.45}Mg_{0.55})_{\Sigma=6.00}Zn_3(AsO_4)_2(SiO_4)(OH)_8$.

Occurrence: A secondary mineral in veinlets, on slip surfaces, and as interstitial fillings within a metamorphosed stratiform zinc deposit (Franklin, New Jersey, USA).

Association: Franklinite, willemite, pyrochroite, barite, kolicite, sussexite, kraisslite, zincite, sphalerite, galena, calcite, rhodochrosite.

Distribution: From Franklin and Sterling Hill, Ogdensburg, Sussex Co., New Jersey, USA.

Name: For the mining engineer Albert Fairchild Holden (1866–1913), of Salt Lake City, Utah, USA, in whose collection the mineral was first noticed.

Type Material: Harvard University, Cambridge, Massachusetts, 89996; National Museum of Natural History, Washington, D.C., USA, 95434, 162601.

References: (1) Palache, C. and E.V. Shannon (1927) Holdenite, a new arsenate of manganese and zinc, from Franklin, New Jersey. Amer. Mineral., 12, 144–148. (2) Palache, C., H. Berman, and C. Frondel (1951) Dana's system of mineralogy, (7th edition), v. II, 775–777. (3) Moore, P.B. and T. Araki (1977) Holdenite, a novel cubic close-packed structure. Amer. Mineral., 62, 513–521. (4) Dunn, P.J. (1981) Holdenite from Sterling Hill and new chemical data. Mineral. Record, 12, 373–375.

Crystal Data: Orthorhombic. *Point Group:* $2/m\ 2/m\ 2/m$. As prismatic or acicular crystals, to 10 cm, typically striated on {210}. In columnar sheaflike aggregates, massive.

Physical Properties: *Cleavage:* Perfect on {210}, intersecting at $\sim54°$ and $\sim126°$; partings on {001}, {112}, and {113}. *Tenacity:* Brittle. Hardness = 5–6   D(meas.) = 2.95–3.13 D(calc.) = 3.09

Optical Properties: Transparent to translucent. *Color:* Black, dark violet to light sky-blue; pale yellow to violet in thin section. *Streak:* White, with sky-blue tinge. *Luster:* Vitreous. *Optical Class:* Biaxial (–). *Pleochroism:* Strong in shades of violet when dark-colored. *Orientation:* $X = a$; $Y = b$; $Z \wedge c = 0°{-}4°$. *Dispersion:* $r > v$, weak. *Absorption:* $Z > Y > X$. $\alpha = 1.622{-}1.642$ $\beta = 1.642{-}1.660$ $\gamma = 1.646{-}1.666$ 2V(meas.) = $45°{-}52°$ 2V(calc.) = $78(8)°$

Cell Data: *Space Group:* $Pnma$.   a = 18.30   b = 17.69   c = 5.30   Z = 4

X-ray Powder Pattern: Barraute, Canada.
8.107 (100), 3.00 (90), 4.43 (70), 3.34 (60), 3.61 (50), 2.538 (50), 2.797 (40)

Chemistry:

	(1)	(2)		(1)	(2)		(1)	(2)
SiO_2	59.58	59.73	MnO	0.41	0.20	K_2O	0.27	0.15
TiO_2		0.17	MgO	11.66	10.16	F	0.21	0.24
Al_2O_3	7.19	11.21	CaO	0.06	0.56	H_2O^+	2.23	2.08
Fe_2O_3	9.35	2.97	Li_2O	3.54	3.56	H_2O^-	0.03	0.02
FeO	4.88	8.92	Na_2O	0.50	0.18	$-\text{O} = \text{F}_2$	0.10	0.10
						Total	[99.81]	100.05

(1) Utö, Sweden; original total given as 99.82%, corresponding to $(\text{Li}_{1.90}\text{Na}_{0.13}\text{K}_{0.05}\text{Ca}_{0.01})_{\Sigma=2.09}$ $(\text{Mg}_{2.32}\text{Fe}^{2+}_{0.54}\text{Mn}_{0.05})_{\Sigma=2.91}(\text{Al}_{1.08}\text{Fe}^{3+}_{0.94})_{\Sigma=2.02}(\text{Si}_{7.95}\text{Al}_{0.05})_{\Sigma=8.00}\text{O}_{22}[(\text{OH})_{1.99}\text{F}_{0.09}]_{\Sigma=2.08}$.
(2) Barraute, Canada; corresponding to $(\text{Li}_{1.90}\text{Ca}_{0.08}\text{Na}_{0.05}\text{K}_{0.03})_{\Sigma=2.06}(\text{Mg}_{2.01}\text{Fe}^{2+}_{0.99}\text{Mn}_{0.02})_{\Sigma=3.02}$ $(\text{Al}_{1.67}\text{Fe}^{3+}_{0.30}\text{Ti}_{0.02})_{\Sigma=1.99}(\text{Si}_{7.92}\text{Al}_{0.08})_{\Sigma=8.00}\text{O}_{22}[(\text{OH})_{1.84}\text{F}_{0.10}]_{\Sigma=1.94}$.

Polymorphism & Series: Dimorphous with clinoholmquistite; forms a series with ferro-holmquistite and magnesio-holmquistite.

Mineral Group: Amphibole (Fe–Mn–Mg) group: $0.1 \le \text{Mg}/(\text{Mg} + \text{Fe}^{2+}) \le 0.89$; $(\text{Ca} + \text{Na})_\text{B} < 1.34$; Li ≥ 1.0.

Occurrence: As metasomatic replacements near the outer margins of lithium-rich pegmatites.

Association: Quartz, clinoholmquistite, tourmaline, spodumene, plagioclase, biotite, clinozoisite, "hornblende."

Distribution: At Utö, Sweden. From Brandrücken, Austria. In the USA, from Hiddenite, Alexander Co., and the Foote mine, Kings Mountain, Cleveland Co., North Carolina; in the Harding pegmatite, Dixon, Taos Co., New Mexico; and at Keystone, Pennington Co., South Dakota. At Barraute, and near St. Benoit, Quebec, and in the Tanco pegmatite, Bernic Lake, Manitoba, Canada. In the Benson pegmatite mine, Mtoko-Fungwe area, Zimbabwe. From the Vredefort Ring crater, Orange Free State, South Africa. At Greenbushes, Western Australia.

Name: To honor the Swedish petrologist, Per Johan Holmquist (1866–1946), of Stockholm.

Type Material: n.d.

References: (1) Osann, A. (1913) Über Holmquistit, einen Lithionglaucophan von der Insel Utö. Sitzungsber. Heidelberg. Akad. Wiss., Abt. A., Abh. 23. (2) Deer, W.A., R.A. Howie, and J. Zussman (1963) Rock-forming minerals, v. 2, chain silicates, 230–233. (3) Vlasov, K.A., Ed. (1966) Mineralogy of rare elements, v. II, 13–17. (4) Nickel, E.H., B.C. Karpoff, J.A. Maxwell, and J.F. Rowland (1960) Holmquistite from Barraute, Quebec. Can. Mineral., 6, 504–512. (5) Whittaker, E.J.W. (1969) The structure of the orthorhombic amphibole holmquistite. Acta Cryst., 25, 394–397. (6) Walter, F., E.M. Walitzi, and K. Mereiter (1989) Verfeinerung der Kristallstruktur von Holmquistit vom Brandrücken/Weinebene, Koralpe, Österreich. Zeits. Krist., 188, 95–101 (in German with English abs.).

Holtite

$$Al_6(Al, Ta)(BO_3)[(Si, Sb, As)O_4]_3(O, OH)_3$$

Crystal Data: Orthorhombic. *Point Group:* $2/m \ 2/m \ 2/m$. In blocky crystals, to 2 cm; as pseudohexagonal needles, elongated along [100], typically S-shaped; finely crystalline. *Twinning:* Multiple on {011}.

Physical Properties: *Cleavage:* Good on {001}, may be parting. Hardness = 8.5 D(meas.) = 3.60–3.90 D(calc.) = n.d. Fluoresces dull orange under SW and bright yellow under SW UV.

Optical Properties: Semitransparent. *Color:* Light buff to cream-buff or deep olive-buff, brown, russet. *Streak:* Light buff. *Luster:* Resinous, vitreous, dull when weathered. *Optical Class:* Biaxial (−). *Pleochroism:* Various shades of yellow to colorless. *Orientation:* $X = a$. *Dispersion:* $r < v$. $\alpha = 1.705$–1.746 $\beta = 1.728$–1.759 $\gamma = 1.730$–1.761 2V(meas.) = 20°–55°

Cell Data: *Space Group:* $Pnma$. $a = 4.6914(5)$ $b = 11.896(2)$ $c = 20.383(4)$ $Z = 4$

X-ray Powder Pattern: Greenbushes, Western Australia.
10.28 (100), 2.94 (40), 5.89 (34), 5.93 (32), 5.08 (32), 2.338 (26), 5.12 (24)

Chemistry:

	(1)	(2)		(1)	(2)
SiO_2	20.30	26.74	BeO	0.05	
TiO_2	0.09		MgO		0.10
B_2O_3	1.82	4.67	CaO		0.10
Al_2O_3	46.43	43.44	Li_2O		0.59
Fe_2O_3	0.27	0.20	Na_2O		0.35
Sb_2O_3	13.89		K_2O		0.45
Sb_2O_5	4.61	6.49	Rb_2O		0.11
As_2O_5		2.92	Cs_2O		0.57
Nb_2O_5	0.76	0.15	H_2O^+	0.38	1.13
Ta_2O_5	11.24	11.70	H_2O^-	0.08	
MnO	0.05				
			Total	99.97	99.71

(1) Greenbushes, Western Australia; B_2O_3 thought to be low; corresponding to $Al_{24.50}Ta_{1.36}$ $Nb_{0.16}Fe^{3+}_{0.10}Be_{0.05}Ti_{0.03}Mn_{0.02}B_{1.40}Si_{9.03}Sb^{3+}_{2.56}Sb^{5+}_{0.76}O_{66.62}(OH)_{1.13}$. (2) Kola Peninsula, Russia; corresponding to $Al_{23.23}Ta_{1.49}Nb_{0.03}Fe_{0.09}Mg_{0.06}Ca_{0.06}Ti_{0.03}B_{3.84}Si_{10.64}Sb^{5+}_{1.15}As^{5+}_{0.72}$ $O_{69.57}(OH)_{2.09}$.

Occurrence: As coatings on stibiotantalite and replacing tantalite in a pegmatite (Greenbushes, Western Australia); in pegmatites cutting amphibolites (Kola Peninsula, Russia).

Association: Stibiotantalite, tantalite, tourmaline, microlite, quartz (Greenbushes, Western Australia); stibiotantalite, microlite (Kola Peninsula, Russia).

Distribution: At Bunbury Gully, Greenbushes, Western Australia. From the Voronya massif, Kola Peninsula, Russia.

Name: For Harold Edward Holt (1908–1967), Prime Minister of Australia (1966–1967).

Type Material: Government Chemical Laboratories, Perth, Australia, MDC550.

References: (1) Pryce, M.W. (1971) Holtite: a new mineral allied to dumortierite. Mineral. Mag., 38, 21–25. (2) (1972) Amer. Mineral., 57, 1556 (abs. ref. 1). (3) Voloshin, A.V., V.V. Gordienko, E.M. Gel'man, M.L. Zorina, N.A. Elina, E.A. Kul'chitskaya, Y.P. Men'shikov, L.I. Polezhaeva, R.I. Ryzhova, P.B. Sokolov, and G.I. Utochkina (1977) Holtite (first find in the USSR) and its interaction with other tantalum minerals in rare metal pegmatites. Zap. Vses. Mineral. Obshch., 106, 337–347 (in Russian). (4) (1979) Mineral. Abs., 30, 62 (abs. ref. 3). (5) Hoskins, B.F., W.G. Mumme, and M.W. Pryce (1989) Holtite, $(Si_{2.25}Sb_{0.75})$ $B[Al_6(Al_{0.43}Ta_{0.27}\square_{0.30})O_{15}(O, OH)_{2.25}]$: crystal structure and crystal chemistry. Mineral. Mag., 53, 457–463.

Crystal Data: Monoclinic. *Point Group:* $2/m$. Crystals commonly tabular on {001}, also with {100} prominent; pseudo-octahedral by development of {110} and {012} or {$\bar{1}11$}, to 5 cm. *Twinning:* On {001} and {100}; also on {034} to produce cruciform twins.

Physical Properties: *Cleavage:* Indistinct. *Fracture:* Subconchoidal. *Tenacity:* Brittle. Hardness = 5 D(meas.) = 3.34–3.38 D(calc.) = 3.451

Optical Properties: Opaque, translucent only in thin splinters. *Color:* Black to blackish brown. *Streak:* Grayish. *Luster:* Resinous to vitreous. *Optical Class:* Biaxial (+). *Pleochroism:* X = bluish green; Y = deep brownish red to brownish gray; Z = smoky gray or brownish yellow. *Orientation:* $Z = b$; $Y \wedge c = -1°$. *Dispersion:* $r > v$, distinct, with strong horizontal dispersion. *Absorption:* $Y > X > Z$. $\alpha = 1.715$ $\beta = 1.725$ $\gamma = 1.738$ 2V(meas.) = 80°

Cell Data: *Space Group:* $P2_1/a$. a = 9.786(2) b = 7.621(2) c = 4.776(1) $\beta = 90.61(2)°$ Z = 2

X-ray Powder Pattern: Store-Arø Island, Langesundsfjord, Norway. (ICDD 17-211). 3.10 (100), 2.52 (100), 2.83 (90), 2.97 (70), 2.18 (70), 2.23 (60), 1.86 (50)

Chemistry:

	(1)	(2)	(3)
SiO_2	31.87	31.58	30.01
B_2O_3	[18.08]	[21.46]	18.62
Al_2O_3	1.50		
Fe_2O_3	2.15		
FeO	16.25	17.03	19.22
MnO		0.50	
MgO	0.52		
CaO	27.28	29.43	32.15
Na_2O	1.50		
LOI	0.85		
Total	[100.00]	[100.00]	100.00

(1) Langesundsfjord, Norway; B_2O_3 by difference. (2) Do.; by electron microprobe, B_2O_3 by difference. (3) $Ca_2FeB_2Si_2O_{10}$.

Mineral Group: Gadolinite group.

Occurrence: In pegmatite.

Association: Allanite, meliphanite, titanite, zircon, aegirine, löllingite, astrophyllite, melanocerite, nordenskiöldine, wöhlerite, hiortdahlite, molybdenite.

Distribution: On Stokkø, Store-Arø, and Øvre-Arø Islands, in the Langesundsfjord, Norway.

Name: From the Greek for *to occur together*, in allusion to its association with meliphanite and "erdmannite" (allanite).

References: (1) Dana, E.S. (1892) Dana's system of mineralogy, (6th edition), 505–507. (2) Miyawaki, R., I. Nakai, and K. Nagashima (1985) Structure of homilite, $Ca_{2.00}(Fe_{0.90}Mn_{0.03})B_{2.00}Si_{2.00}O_{9.86}(OH)_{0.14}$. Acta Cryst., C41, 13–15.

$$Na(Fe^{2+}, Mn)_{10}(Fe^{3+}, Al)_2Si_{12}O_{31}(OH)_{13}$$

Crystal Data: Triclinic. *Point Group:* $\bar{1}$ or 1. As bladed crystals, to 1 cm; in plumose aggregates and rosettes.

Physical Properties: *Cleavage:* Good on {010}, fair on {100}, poor on {2$\bar{1}$0}. Hardness = n.d. D(meas.) = 3.378 D(calc.) = [3.34]

Optical Properties: Transparent to translucent. *Color:* Dark green to black. *Luster:* Greasy. *Optical Class:* Biaxial (–). *Pleochroism:* Marked; X = golden; Y = dark lilac gray; Z = green. *Dispersion:* $r < v$, strong. $\alpha = 1.701$ $\beta = 1.720$ $\gamma = 1.734$ 2V(meas.) = 65°

Cell Data: *Space Group:* $P\bar{1}$ or $P1$. $a = 10.170(4)$ $b = 9.774(4)$ $c = 9.589(4)$ $\alpha = 91.22(5)°$ $\beta = 70.76(5)°$ $\gamma = 108.09(5)°$ $Z = 1$

X-ray Powder Pattern: Laytonville, California, USA. (ICDD 19-571). 9.18 (100), 7.91 (80), 3.25 (65), 2.62 (60), 2.68 (45), 2.78 (40), 3.06 (35)

Chemistry: (1) Laytonville district, California, USA; analysis not given, stated to correspond to $(Na_{1.03}Ca_{0.02})_{\Sigma=1.05}(Fe^{2+}_{6.41}Mn_{2.98}Mg_{0.45})_{\Sigma=9.84}(Fe^{3+}_{1.57}Al_{0.62})_{\Sigma=2.19}$ $(Si_{11.96}Ti_{0.04})_{\Sigma=12.00}[O_{31.31}(OH)_{12.69}]_{\Sigma=44.00}$.

Occurrence: An essential mineral in some of the metamorphosed shales, siliceous ironstones, and impure limestones of the Franciscan Formation (Laytonville district, California, USA).

Association: Deerite, zussmanite, stilpnomelane, spessartine, riebeckite, quartz, aegirine, grunerite, aragonite, manganoan siderite, ferroan kutnohorite (Laytonville district, California, USA).

Distribution: In the USA, in California, from the Laytonville quarry, and at Covelo, Mendocino Co.; at Ward Creek, Sonoma Co.; in Panoche Pass, San Benito Co.; at Pacheco Pass, Santa Clara and Merced Cos.; and in the Powers quarry, Coos Co., Oregon. From Brezovica, Yugoslavia. In the Tanemaya mine, Kumamoto Prefecture, Japan.

Name: For Professor Robert Andrew Howie (1923–), British mineralogist and petrologist, London University, London, England.

Type Material: National Museum of Natural History, Washington, D.C., USA, 109453, 144184; The Natural History Museum, London, England, 1964,544.

References: (1) Agrell, S.O., M.G. Bown, and D. McKie (1965) Deerite, howieite and zussmanite, three new minerals from the Franciscan of the Laytonville District, Mendocino Co., California. MSA meeting, Bozeman, Montana, July 26–31, 1964. Amer. Mineral., 50, 278 (abs.). (2) Wenk, H.R. (1974) Howieite, a new type of chain silicate. Amer. Mineral., 59, 86–97. (3) Muir Wood, R. (1979) The iron-rich blueschist facies minerals: 2. Howieite. Mineral. Mag., 43, 363–370.

Crystal Data: Monoclinic. *Point Group:* 2/*m*. Crystals tabular, flattened on {100}, with {100} dominant and terminated by {011} and {001}, to 1 mm. Most commonly nodular. When massive, chalky to porcelaneous.

Physical Properties: *Fracture:* For porcelaneous types, nearly even and smooth. Hardness = 3.5   D(meas.) = 2.53–2.59   D(calc.) = 2.61

Optical Properties: Translucent in thin splinters. *Color:* White; in thin section, colorless. *Luster:* Subvitreous, glimmering.
Optical Class: Biaxial (–). *Orientation:* X = *b*; Z ∧ *c* ≃ 51°.   α = 1.583–1.586
β = 1.596–1.598   γ = 1.605   2V(meas.) = 73°

Cell Data: *Space Group:* $P2_1/c$.   a = 12.820(3)   b = 9.351(1)   c = 8.608(2)
β = 104.84(2)°   Z = 4

X-ray Powder Pattern: Sterling Borax mine, California, USA.
6.2 (100), 3.10 (90), 3.90 (80), 2.04 (70), 2.07 (50), 1.794 (50), 12.4 (40)

Chemistry:

	(1)	(2)	(3)
SiO$_2$	15.33	15.50	15.35
B$_2$O$_3$	44.52	44.38	44.48
CaO	27.94	28.45	28.66
Na$_2$O	0.53		
K$_2$O	0.13		
H$_2$O	11.55	11.58	11.51
rem.		0.09	
Total	[100.00]	100.00	100.00

(1) Windsor, Canada; recalculated to 100.00% after deduction of 4.32% gypsum. (2) Daggett, California, USA; remainder is Na$_2$O and MgO. (3) Ca$_2$B$_5$SiO$_9$(OH)$_5$.

Occurrence: In borate deposits.

Association: Colemanite, ulexite, bakerite.

Distribution: In Canada, near Windsor, and at Wentworth, Nova Scotia. In the USA, in California, in the Sterling Borax mine, Tick Canyon, near Lang, Los Angeles Co.; near Calico, San Bernardino Co.; in Gower Gulch, near Ryan, Inyo Co.; at the Russell and other borate mines north of Lockwood Valley, Ventura Co.; and in the Kramer deposit, Kern Co. From Magdalena, Sonora, Mexico. At Susurluk, Turkey. From Baljevac na Ibru, near Kraljevo, and near Bela Stena, Yugoslavia.

Name: For Henry How (1828–1879), Canadian chemist, geologist, and mineralogist, University of King's College, Windsor, Nova Scotia, Canada, who first described the species.

References: (1) Palache, C., H. Berman, and C. Frondel (1951) Dana's system of mineralogy, (7th edition), v. II, 362–363. (2) Murdoch, J. (1957) Crystallography and X-ray measurements of howlite from California. Amer. Mineral., 42, 521–524. (3) Griffen, D.T. (1988) Howlite, Ca$_2$SiB$_5$O$_9$(OH)$_5$: structure refinement and hydrogen bonding. Amer. Mineral., 73, 1138–1144.

Crystal Data: Cubic. *Point Group:* 432. Crystals modified by the dodecahedron, to 9 mm; granular, massive.

Physical Properties: *Tenacity:* Brittle. Hardness = 6.5   VHN = 896   D(meas.) = 2.97–3.00   D(calc.) = 2.944

Optical Properties: Transparent to translucent. *Color:* Milk-white to colorless.
Luster: Vitreous.
Optical Class: Isotropic. $n = 1.613$

Cell Data: *Space Group:* $I4_132$. a = 12.897(4)   Z = 8

X-ray Powder Pattern: Hunan Province, China.
2.746 (100), 2.209 (100), 2.090 (90), 1.753 (70), 3.443 (60), 1.691 (52), 1.228 (45)

Chemistry:

	(1)	(2)
SiO_2	35.66	36.64
Al_2O_3	0.50	
Fe_2O_3	0.22	0.06
BeO	15.78	16.30
MgO	0.18	0.17
CaO	34.60	35.18
Li_2O	5.85	5.60
Na_2O	0.13	0.03
K_2O	0.06	0.03
F	7.81	7.27
LOI	1.28	
$-O = F_2$	3.2	3.06
Total	98.87	98.22

(1) Hunan Province, China; corresponds to $Ca_{3.07}Li_{1.95}Be_{3.13}(SiO_4)_3F_{2.00}$. (2) Do.; corresponds to $Ca_{3.08}Li_{1.85}Be_{3.20}(SiO_4)_3F_{2.06}$.

Occurrence: In phlogopite veins in a fluorine-rich metamorphosed Devonian limestone intruded by a beryllium-bearing granite.

Association: Fluorite, zinnwaldite, chrysoberyl, taaffeite, liberite.

Distribution: On Hsianghua Ridge, Linwu Co., Hunan Province, China.

Name: For the place of discovery; the word means *fragrant flower* in Chinese.

Type Material: n.d.

References: (1) Wen-Hui Huang, Shao-Hua Tu, K'ung-Hai Wang, Chun-Lin Chao, and Cheng-Chih Yu (1958) Hsiang-hua-shih [hsianghualite], a new beryllium mineral. Ti-chih-yueh-k'an, 7, 35 (in Chinese). (2) (1959) Amer. Mineral., 44, 1327–1328 (abs. ref. 1). (3) Beus, A.A. (1960) Geochemistry of beryllium and genetic types of beryllium deposits., 69–71. (4) (1961) Amer. Mineral., 46, 244 (abs. ref. 3). (5) Vlasov, K.A., Ed. (1966) Mineralogy of rare elements, v. II, 127–129.

Crystal Data: Orthorhombic. *Point Group:* $2/m \; 2/m \; 2/m$. Crystals typically highly modified, to 1 cm; granular.

Physical Properties: *Cleavage:* {100}, poor. *Fracture:* Uneven to subconchoidal. *Tenacity:* Brittle. Hardness = 6   D(meas.) = 3.20–3.32   D(calc.) = 3.201

Optical Properties: Transparent to translucent. *Color:* White, yellow, dark orange, brown; colorless to yellow-brown in thin section. *Luster:* Vitreous.
Optical Class: Biaxial (+). *Pleochroism:* X = very pale yellow to dark yellow; Y = Z = colorless, pale yellow. *Orientation:* X = a; Y = c; Z = b.   *Dispersion:* r > v. $\alpha = 1.607–1.643$
$\beta = 1.619–1.655$ $\gamma = 1.639–1.675$ 2V(meas.) = 65°–84°

Cell Data: *Space Group:* Pmcn.   a = 20.8526(4)   b = 4.7408(1)   c = 10.2580(2)   Z = 4

X-ray Powder Pattern: Synthetic $\text{Mg}_7(\text{SiO}_4)_3\text{F}$.
2.256 (100), 2.438 (70), 1.4786 (70), 1.7387 (65), 3.64 (50), 2.691 (50), 2.572 (40)

Chemistry:

	(1)	(2)
SiO_2	35.79	36.43
TiO_2	2.00	0.10
Al_2O_3	0.79	
Fe_2O_3	0.33	
FeO	3.31	5.03
MnO	0.84	0.65
MgO	54.51	53.84
CaO	0.00	0.01
F	2.77	4.07
H_2O^+	0.91	[1.69]
H_2O^-	0.00	
$-O = F_2$	1.17	[1.71]
Total	100.08	[100.11]

(1) Lohja, Finland; corresponds to $(\text{Mg}_{6.75}\text{Fe}^{2+}_{0.23}\text{Ti}_{0.12}\text{Al}_{0.08}\text{Mn}_{0.06}\text{Fe}^{3+}_{0.02})_{\Sigma=7.26}(\text{Si}_{0.99}\text{O}_4)_3$ $[\text{F}_{0.73}(\text{OH})_{0.50}]_{\Sigma=1.23}$. (2) Sillböle, Finland; by electron microprobe, H_2O calculated from stoichiometry; corresponds to $(\text{Mg}_{6.60}\text{Fe}_{0.35}\text{Mn}_{0.04})_{\Sigma=6.99}(\text{SiO}_4)_3[\text{F}_{1.06}(\text{OH})_{0.93}\text{O}_{0.01}]_{\Sigma=2.00}$.

Mineral Group: Humite group.

Occurrence: Typically in contact metamorphic zones in limestones and dolostones associated with felsic, or more rarely, alkalic plutonic rocks, especially where metasomatism has introduced Fe, B, and F.

Association: Grossular, wollastonite, forsterite, monticellite, cuspidine, fluoborite, ludwigite, spinel, brucite, calcite, dolomite, serpentine, diopside, corundum, phlogopite, pyrrhotite.

Distribution: At Monte Somma and Vesuvius, Campania, Italy. At Lohja, Sillböle, and Hermala, Finland. In the Norberg area, and at the Ladu mine, Persberg, Värmland, Sweden. From Sorfinnset, Glomfjord, Norway. In the Tilly Foster mine, Brewster, Putnam Co., New York, and at Franklin, Sussex Co., New Jersey, USA. From Llanos de Juanar, Málaga Province, Spain. At Anzahamazonono, Madagascar.

Name: After Sir Abraham Hume (1749–1838), English connoisseur and collector of works of art, gems, and minerals.

References: (1) Dana, E.S. (1892) Dana's system of mineralogy, (6th edition), 535–536. (2) Deer, W.A., R.A. Howie, and J. Zussman (1982) Rock-forming minerals, (2nd edition), v. 1A, orthosilicates, 380–417. (3) Ribbe, P.H. and G.V. Gibbs (1971) Crystal structures of the humite minerals: III. Mg/Fe ordering in humite and its relation to other ferromagnesian silicates. Amer. Mineral., 56, 1155–1173. (4) (1962) NBS Mono. 25, 30.

Crystal Data: Monoclinic. *Point Group:* $2/m$. As anhedral grains, may be flattened on $\{100\}$, to 0.2 mm.

Physical Properties: *Cleavage:* Distinct on $\{001\}$; another poor cleavage is of unspecified orientation. *Fracture:* Conchoidal. Hardness = [4.5] VHN = 318 (150 g load). D(meas.) = 7.1 D(calc.) = 7.18 Fluoresces dull white with a pink tinge under SW UV; radioactive.

Optical Properties: Transparent to translucent. *Color:* Colorless to very pale cream. *Optical Class:* Biaxial (+). *Orientation:* $Y = b$; $Z \simeq c$. *Dispersion:* $r < v$, moderate. $\alpha = 1.898(3)$ $\beta = [1.900]$ $\gamma = 1.922$ 2V(meas.) = 25(1)°

Cell Data: *Space Group:* $P2_1/n$. $a = 6.80(3)$ $b = 6.96(3)$ $c = 6.54(3)$ $\beta = 104°55(10)'$ $Z = 4$

X-ray Powder Pattern: Gillespie's Beach, New Zealand.
3.09 (100), 2.89 (90), 4.23 (75), 3.29 (75), 4.71 (60), 4.08 (50), 3.53 (50)

Chemistry:

	(1)	(2)	(3)
SiO_2	19.7	12.8	18.54
ThO_2	76.6	69.9	81.46
UO_2	0.0	1.0	
RE_2O_3	2.6	4.5	
Fe_2O_3	1.2		
FeO		< 0.09	
MnO	trace		
CaO	0.0	2.5	
MgO	0.0		
F		1.0	
OH		[1.8]	
H_2O	0.0	[0.0]	
P_2O_5	trace	7.1	
Total	100.1	[100.6]	100.00

(1) Gillespie's Beach, New Zealand. (2) Bogatynia area, Poland; by electron microprobe, $\text{RE}_2\text{O}_3 = \text{La}_2\text{O}_3$ 0.9%, Ce_2O_3 2.7%, $\text{Pr}_2\text{O}_3 \leq 0.15\%$, Nd_2O_3 0.9%; corresponds to $\text{Th}_{0.78}\text{Ca}_{0.13}\text{RE}_{0.09}\text{U}_{0.01}[(\text{Si}_{0.63}\text{P}_{0.22})_{\Sigma=0.85}\text{O}_4](\text{OH})_{0.31}\text{F}_{0.16}$. (3) ThSiO_4.

Polymorphism & Series: Dimorphous with thorite.

Mineral Group: Monazite group.

Occurrence: In beach sands (New Zealand).

Association: Scheelite, titanian cassiterite, uranothorite, ilmenite, gold (New Zealand); cheralite, thorogummite, thorian ningyoite (Bogatynia area, Poland).

Distribution: From Harihari, Saltwater Creek, Okirito, Five Mile Beach, Bruce Bay, north and south of the mouth of the Waikukupa River, and Gillespie's Beach, South Westland, New Zealand. From the Bogatynia area, Lower Silesia, Poland. On Mont Most, Great St. Bernard, Val d'Aosta, Italy. In the Holiday mine, Hawthorne, Mineral Co., Nevada, USA.

Name: For Colin Osborne Hutton (1910–1971), New Zealand-American mineralogist of Stanford University, Palo Alto, California, USA.

Type Material: The Natural History Museum, London, England, 1950,340; National School of Mines, Paris, France; National Museum of Natural History, Washington, D.C., USA, 106265.

References: (1) Pabst, A. and C.O. Hutton (1951) Huttonite, a new monoclinic thorium silicate. Amer. Mineral., 36, 60–69. (2) Taylor, M. and R.C. Ewing (1978) The crystal structures of the ThSiO_4 polymorphs: huttonite and thorite. Acta Cryst., 34, 1074–1079. (3) Kucha, H. (1980) Continuity in the monazite-huttonite series. Mineral. Mag., 43, 1031–1034.

Crystal Data: Monoclinic. *Point Group:* $2/m$. Crystals similar to adularia or orthoclase, to 20 cm; also granular or massive. *Twinning:* Commonly simple twins according to the Carlsbad, Manebach, or Baveno laws.

Physical Properties: *Cleavage:* Perfect on {001}, good on {010}, intersecting at 90°. *Fracture:* Conchoidal. *Tenacity:* Brittle. Hardness = 6–6.5   D(meas.) = 2.58–2.82 D(calc.) = [2.88]

Optical Properties: Transparent to translucent. *Color:* Colorless, white, also flesh-red; colorless in thin section. *Luster:* Vitreous. *Optical Class:* Biaxial (–). *Orientation:* $Z = b$; $Y \wedge c = -25°$ to $-45°$; $X \wedge a = 1°$ to $-19°$. *Dispersion:* $r > v$, weak. $\alpha = 1.520$–1.542 $\beta = 1.524$–1.545 $\gamma = 1.526$–1.547 2V(meas.) = 48°–79°

Cell Data: *Space Group:* $C2/m$. $a = 8.557$ $b = 13.040$ $c = 7.200$ $\beta = 115.69°$ $Z = 4$

X-ray Powder Pattern: Busovača, Bosnia-Herzegovina. (ICDD 19-2). 3.24 (100), 3.31 (90), 3.00 (70), 3.46 (50), 3.78 (45), 2.57 (45b), 2.91 (40)

Chemistry:

	(1)	(2)		(1)	(2)
SiO$_2$	59.85	49.54	CaO	0.86	0.19
TiO$_2$		trace	BaO	6.92	19.01
Al$_2$O$_3$	20.64	23.14	Na$_2$O	0.93	1.65
Fe$_2$O$_3$		0.11	K$_2$O	11.07	6.37
MgO	0.21	0.04	H$_2$O	0.24	
			Total	100.72	100.05

(1) Slyudyanka, Russia; corresponding to $(K_{0.67}Ba_{0.13}Na_{0.09}Ca_{0.04}Mg_{0.01})_{\Sigma=0.94}$ $(Si_{2.85}Al_{1.16})_{\Sigma=4.01}O_8$. (2) Busovača, Bosnia-Herzegovina; by electron microprobe, corresponding to $(K_{0.43}Ba_{0.39}Na_{0.17}Ca_{0.01})_{\Sigma=1.00}(Si_{2.59}Al_{1.42})_{\Sigma=4.01}O_8$.

Polymorphism & Series: Intermediate member of the series orthoclase-celsian.

Mineral Group: Feldspar group.

Occurrence: Typically associated with metamorphosed manganiferous rocks or in mineral deposits rich in manganese.

Association: Manganiferous epidote, rhodonite, rhodochrosite, spessartine, manganiferous tremolite, plagioclase, analcime.

Distribution: From the Lengenbach quarry, Binntal, Valais, Switzerland. At Jakobsberg, Långban, and in the Harstig mine, Pajsberg, near Persberg, Värmland, and in the Sjö mine, near Grythyttan, Örebro, Sweden. From Slyudyanka, near Lake Baikal, Siberia, Russia. Large crystals from Zagradski Creek, near Busovača, Bosnia-Herzegovina. At Aberfeldy, Scotland. In the USA, from Franklin, Sussex Co., New Jersey, and Johnsburg, Warren Co., New York. At Nisikkatch Lake, Saskatchewan, Canada. From Piggery Creek, Broken Hill, New South Wales, Australia. In the Kaso mine, Tochigi Prefecture, and at Minakami, Gumma Prefecture, Japan. From Otjosondu, Namibia. A few other localities are known.

Name: From the Greek for *glass* and *to appear*, in reference to its transparency in crystals.

References: (1) Dana, E.S. (1892) Dana's system of mineralogy, (6th edition), 321–322. (2) Deer, W.A., R.A. Howie, and J. Zussman (1963) Rock-forming minerals, v. 4, framework silicates, 166–178. (3) De Pieri, R., S. Quareni, and K.M. Hall (1977) Refinement of the structures of low and high hyalophanes. Acta Cryst., 33, 3073–3076. (4) Phillips, W.R. and D.T. Griffen (1981) Optical mineralogy, 360–363.

Hyalotekite $(Ba, Pb, Ca)_6(B, Si, Al)_2(Si, Be)_{10}O_{28}(F, Cl)$

Crystal Data: Triclinic. *Point Group:* $\bar{1}$. Coarsely crystalline, massive.

Physical Properties: *Cleavage:* Easy in two directions at $\sim 90°$; indistinct in a third direction in the same zone as the other two. *Tenacity:* Brittle. Hardness = 5–5.5 D(meas.) = 3.81–3.82 D(calc.) = 3.83 Fluoresces blue, light brownish orange, or yellow in SW UV; bright yellow or blue cathodoluminescence.

Optical Properties: Transparent in very thin plates. *Color:* White to pearly gray. *Luster:* Vitreous to greasy.
Optical Class: Biaxial (+). *Dispersion:* $r < v$, weak to strong. $\alpha = 1.646$–1.656 $\beta = 1.649$–1.660 $\gamma = 1.659$–1.671 2V(meas.) = 57° 60.5° 2V(calc.) = 55.4°–62.5°

Cell Data: *Space Group:* $I\bar{1}$. $a = 11.310(2)$ $b = 10.955(2)$ $c = 10.317(3)$ $\alpha = 90.43(2)°$ $\beta = 90.02(2)°$ $\gamma = 90.16(2)°$ $Z = 2$

X-ray Powder Pattern: Långban, Sweden. (ICDD 19-572).
3.45 (100), 3.53 (80), 2.94 (80), 3.81 (70), 2.297 (65), 2.143 (65), 7.7 (60)

Chemistry:

	(1)	(2)		(1)	(2)
SiO_2	39.47	40.74	SrO		0.07
B_2O_3	3.73	4.46	BaO	20.08	27.30
Al_2O_3	0.18	≤ 0.03	Na_2O	0.17	0.77
Fe_2O_3	0.06	0.05	K_2O	0.89	0.74
MnO	0.29	0.00	Rb_2O		0.01
CuO	0.09		F	0.99	0.77
PbO	25.11	17.46	Cl	0.06	
BeO	0.75	0.57	LOI	0.59	
MgO	0.09	< 0.01	$-O = F_2$	[0.43]	0.32
CaO	7.82	6.81	Total	[99.94]	99.43

(1) Långban, Sweden; original total given as 100.37%, corresponds to $(Ca_{2.07}Ba_{1.94}Pb_{1.67}$ $K_{0.28}Na_{0.08})_{\Sigma=6.04}(B_{1.60}Si_{0.19}Al_{0.05}Mn_{0.06}Mg_{0.03}Cu_{0.02}Fe_{0.01})_{\Sigma=1.96}(Si_{9.56}Be_{0.44})_{\Sigma=10.00}O_{28}$ $(F_{0.78}Cl_{0.02})_{\Sigma=0.80}$. (2) Dara-i-Pioz, Tadzhikistan; by electron and ion microprobe, corresponds to $(Ba_{2.58}Ca_{1.76}Pb_{1.13}Na_{0.36}K_{0.23}Sr_{0.01})_{\Sigma=6.07}(B_{1.85}Si_{0.14}Mg_{0.03}Fe_{0.01})_{\Sigma=2.03}$ $(Si_{9.67}Be_{0.33})_{\Sigma=10.00}O_{28.5}F_{0.58}$.

Occurrence: Sparingly in feldspar in a metamorphosed manganese deposit (Långban, Sweden); in reedmergnerite pegmatite in an alkaline massif (Dara-i-Pioz, Tadzhikistan).

Association: Quartz, calcite, hematite, barylite, barite, manganoan pectolite, ferrian potassic feldspar, calderitic andradite, melanotekite, barian hedyphane, plumboan taramellite, rhodonite, manganoan aegirine (Långban, Sweden); feldspars, reedmergnerite, aegirine, pyrochlore, eudialyte, polylithionite (Dara-i-Pioz, Tadzhikistan).

Distribution: At Långban, Värmland, Sweden. From the Dara-i-Pioz massif, Alai Range, Tien Shan, Tadzhikistan.

Name: From the Greek for *glass* and *to melt*, in allusion to its easy fusibility.

References: (1) Dana, E.S. (1892) Dana's system of mineralogy, (6th edition), 422. (2) Moore, P.B., T. Araki, and S. Ghose (1982) Hyalotekite, a complex lead borosilicate: its crystal structure and the lone-pair effect of Pb(II). Amer. Mineral., 67, 1012–1020. (3) Grew, E.S., M.G. Yates, D.I. Belakovskiy [Belakovskii], R.C. Rouse, S.-C. Su, and N. Marquez (1994) Hyalotekite from reedmergnerite-bearing peralkaline pegmatite, Dara-i-Pioz, Tajikistan, and from Mn skarn, Långban, Sweden: a new look at an old mineral. Mineral. Mag., 58, 285–297.

$(H_3O, K, Ca)_3(Fe^{2+}, Mn)_{5-6}Ti_2Si_8(O, OH)_{31}$ Hydroastrophyllite

Crystal Data: Triclinic. *Point Group:* n.d.   In aggregates of blocky crystals.

Physical Properties: *Cleavage:* Two sets.   Hardness = n.d.   D(meas.) = 3.151
D(calc.) = [2.82]

Optical Properties: Semitransparent. *Color:* Dark brown.
Optical Class: Biaxial (–). *Pleochroism:* X = bright yellow; Y = orange-yellow; Z = dull yellow.
Absorption: Z > Y > X. $\alpha = 1.660$ $\beta = 1.720$ $\gamma = 1.728$ 2V(meas.) = 40°

Cell Data: *Space Group:* n.d.   a = 11.86   b = 11.98   c = 5.42 $\alpha = 103°25'$ $\beta = 95°9'$
$\gamma = 112°12'$ Z = [1]

X-ray Powder Pattern: Sichuan Province, China.
3.51 (10), 10.55 (9), 2.643 (8), 2.118 (6), 1.768 (6), 2.584 (5)

Chemistry:

	(1)		(1)
SiO_2	25.72	MgO	0.51
TiO_2	9.63	CaO	2.40
MnO_2	3.37	BaO	0.32
Al_2O_3	3.92	Na_2O	0.53
Fe_2O_3	24.24	K_2O	1.28
Nb_2O_5	5.01	F	1.83
Ta_2O_5	0.82	H_2O^+	6.65
FeO	0.05	H_2O^-	5.74
MnO	8.05	$-O = F_2$	0.77
		Total	99.30

(1) Sichuan Province, China; corresponds to $[(H_3O)_{1.96}Ca_{0.51}K_{0.31}Na_{0.20}Ba_{0.02}]_{\Sigma=3.00}$
$(Fe^{3+}_{3.48}Mn^{2+}_{1.30}Mn^{4+}_{0.44}Mg_{0.14})_{\Sigma=5.36}(Ti_{1.38}Nb_{0.43}Ta_{0.04})_{\Sigma=1.85}(Si_{4.92}Al_{0.88})_{\Sigma=5.80}$
$[O_{20.01}(OH)_{9.89}F_{1.10}]_{\Sigma=31.00}$.

Mineral Group: Astrophyllite group.

Occurrence: A product of weathering of an alkalic pegmatite.

Association: n.d.

Distribution: From an undefined locality in Sichuan Province, China.

Name: Presumably for the high water content, and its relation to *astrophyllite*.

Type Material: n.d.

References: (1) Hubei [Hupei] Geological College (1974) The crystal chemistry of astrophyllite group minerals. Sci. Geol. Sin., 1, 18–33 (in Chinese). (2) (1975) Amer. Mineral., 60, 736–737 (abs. ref. 1).

Hydrobiotite

$$[K(Mg, Fe^{2+})_3(Al, Fe^{3+})Si_3O_{10}(OH, F)_2] \cdot$$
$$[(Mg, Fe^{2+}, Al)_3(Si, Al)_4O_{10}(OH)_2 \cdot 4H_2O]$$

Crystal Data: [Monoclinic.] (by analogy to biotite and vermiculite). *Point Group:* [2/m.] In cleavage plates and flakes.

Physical Properties: *Cleavage:* [{001}, perfect.] *Tenacity:* [Brittle to flexible, elastic.] Hardness = [~2] D(meas.) = 2.49–2.64 D(calc.) = n.d.

Optical Properties: Semitransparent. *Color:* Blackish, brownish; golden yellow, pinkish. *Luster:* [Dull.]
Optical Class: Biaxial (–). *Pleochroism:* X = light brown; $Y = Z$ = brown. *Absorption:* $X > Y$ = Z. α = n.d. β = 1.560–1.562; 1.575 γ = 1.565–1.567 2V(meas.) = 10°–13°

Cell Data: *Space Group:* [C2/m.] a = n.d. b = n.d. c = 24.5–25.5 β = n.d. Z = n.d.

X-ray Powder Pattern: Northeastern Transvaal, South Africa.
12.23 (60), 3.493 (50), 4.909 (30), 2.725 (20), 2.041 (16) 3.071 (15), 24.3 (7)

Chemistry:

	(1)	(2)		(1)	(2)
SiO_2	35.60		SrO	0.01	
TiO_2	1.13		BaO	0.17	
Al_2O_3	11.85		Na_2O	0.16	
Fe_2O_3	10.28		K_2O	3.17	
Cr_2O_3	0.03		Rb_2O	0.01	
FeO	0.81		F	0.21	
MnO	0.08		H_2O^+	7.56	
MgO	20.17		H_2O^-	7.20	
CaO	1.44		P_2O_5	0.07	
			$-O = F_2$	0.09	
			Total	99.86	

(1) Libby. Montana; corresponds to $(Mg_{2.38}Fe^{3+}_{0.61}K_{0.32}Ca_{0.12}Ti_{0.07}Fe^{2+}_{0.05}Na_{0.03}Ba_{0.01})_{\Sigma=3.59}$ $(Si_{2.82}Al_{1.18})_{\Sigma=4.00}O_{10}[(OH)_{1.95}F_{0.05}]_{\Sigma=2.00} \cdot 3.01H_2O$.

Polymorphism & Series: A 1:1 regular interstratification of biotite and vermiculite.

Mineral Group: Mica group.

Occurrence: As an alteration product of other micas.

Association: Vermiculite, biotite, apatite, zircon.

Distribution: Perhaps at Křemže, Czech Republic. A few other localities for well-characterized material include: in the USA, in the Rainy Creek complex, near Libby, Lincoln Co., Montana, and from the Enoree area, Spartanburg Co., South Carolina. At Palabora, Transvaal, South Africa.

Name: For its resemblance to *biotite*, but hydrated.

References: (1) Dana, E.S. (1892) Dana's system of mineralogy, (6th edition), 632 [biotite], 664 [vermiculite]. (2) Gruner, J.W. (1934) The structures of vermiculites and their collapse by dehydration. Amer. Mineral., 19, 557–575. (3) Boettcher, A.L. (1966) Vermiculite, hydrobiotite, and biotite in the Rainy Creek igneous complex near Libby, Montana. Clay Minerals, 6, 283–296. (4) Brindley, G.W., P.E. Zalba, and C.M. Bethke (1983) Hydrobiotite, a regular 1:1 interstratification of biotite and vermiculite layers. Amer. Mineral., 68, 420–425.

Crystal Data: Orthorhombic.   *Point Group: mm*2.   Habit not stated.

Physical Properties: *Cleavage:* {010}, very perfect; {100} and {001}, imperfect. Hardness = ~4   D(meas.) = 2.168   D(calc.) = 2.22

Optical Properties: Semitransparent.   *Color:* Grayish white.   *Luster:* Vitreous. *Optical Class:* Biaxial. α = 1.503 β = n.d. γ = 1.518   2V(meas.) = n.d.

Cell Data: *Space Group: Pnm*2$_1$.   a = 6.6483   b = 23.8462   c = 7.0727   Z = 2

X-ray Powder Pattern: Khibiny massif, Russia.
2.923 (100), 3.069 (75), 2.800 (55), 3.319 (43), 6.79 (38)

Chemistry:

	(1)	(2)		(1)	(2)
SiO$_2$	55.53	55.56	Na$_2$O	0.22	
TiO$_2$	0.01		K$_2$O	6.18	6.22
Al$_2$O$_3$	8.46	6.74	F	0.00	
Fe$_2$O$_3$	0.65		Cl	0.15	
MnO	0.18		H$_2$O$^+$	9.62	
MgO	0.21		H$_2$O$^-$	5.58	
CaO	12.72	14.82	H$_2$O		16.66
SrO	0.22		$-$O = Cl$_2$	0.09	
			Total	99.64	100.00

(1) Mt. Rasvumchorr, Khibiny massif, Russia. (2) KCa$_2$AlSi$_7$O$_{17}$(OH)$_2$·6H$_2$O.

Occurrence: A secondary alteration product of delhayelite from ijolite-urtite pegmatites in a differentiated alkalic massif.

Association: Delhayelite.

Distribution: On Mts. Yukspor and Rasvumchorr, in the Khibiny massif, Kola Peninsula, Russia.

Name: For its relation to *delhayelite*, and the water in its composition.

Type Material: n.d.

References: (1) Dorfman, M.D. and M.I. Chiragov (1979) Hydrodelhayelite, a product of supergene alteration of delhayelite. New data on minerals of the USSR, 28, 172–175. (2) (1980) Mineral. Mag., 31, 496 (abs. ref. 1). (3) (1987) Amer. Mineral., 72, 1024 (abs. ref. 1). (4) Chiragov, M.I. and M.G. Dorfman (1981) Crystal chemistry of minerals in the delhayelite group. Doklady Acad. Nauk SSSR, 260, 458–461 (in Russian).

Crystal Data: Tetragonal. *Point Group:* $4/m \, 2/m \, 2/m$. Crystals are equant or tabular on {001}, to 6 cm, with {100}, {001}, and {111}; some crystals striated on {100} ∥ [001]. *Twinning:* Multiply twinned.

Physical Properties: *Cleavage:* Perfect on {001}. *Fracture:* [Uneven] (by analogy with fluorapophyllite). *Tenacity:* [Brittle.] Hardness = 4.5–5 D(meas.) = 2.37 D(calc.) = 2.36

Optical Properties: Transparent to translucent, opaque. *Color:* Colorless to white; may be pink, light green, or light yellow. *Streak:* White. *Luster:* Vitreous on {100} and pearly on {001}.
Optical Class: Uniaxial (+). $\omega = 1.536$–1.542 $\epsilon = 1.537$–1.543

Cell Data: *Space Group:* $P4/mnc$. a = 8.978(3) c = 15.83(1) Z = 2

X-ray Powder Pattern: Jefferson, North Carolina, USA.
3.965 (100), 2.990 (67), 1.588 (26), 4.554 (23), 2.494 (22), 7.90 (13), 7.82 (13)

Chemistry:

	(1)	(2)	(3)
SiO_2	51.93	52.59	53.10
Al_2O_3	0.30	0.12	
CaO	25.78	25.44	24.78
K_2O	4.76	5.00	5.20
Na_2O	0.09	0.00	
H_2O	16.76	16.89	16.92
Total	99.62	100.04	100.00

(1) Jefferson, North Carolina, USA; by electron microprobe, H_2O by the Penfield method.
(2) Kimberley, South Africa; by electron microprobe, H_2O by the Penfield method.
(3) $KCa_4Si_8O_{20}(OH) \cdot 8H_2O$.

Polymorphism & Series: Forms a series with fluorapophyllite.

Occurrence: Typically a secondary mineral in amygdules or druses in basalts.

Association: Zeolites, fluorapophyllite, datolite, pectolite, calcite.

Distribution: In the USA, in the Ore Knob mine, Jefferson, Ashe Co., and the Foote mine, Kings Mountain, Cleveland Co., North Carolina; at Great Notch and Paterson, Passaic Co., and Franklin, Sussex Co., New Jersey; from the Fairfax quarry, Centreville, Fairfax Co., Virginia. From the Valenciana and La Luz mines, Guanajuato, Mexico. In India, around Bombay, Poona, and Nasik, Maharashtra. From Kimberley, Cape Province, South Africa. In Norway, from the Mofjellet mine, Mo i Rana. These occurrences have been authenticated, which requires chemical analysis; there are undoubtedly additional ones.

Name: For the predominance of *hydroxyl* and relation to other *apophyllite* species.

Type Material: The Natural History Museum, London, England, 1977,58; Geological Survey of Canada, Ottawa, 14099; Royal Ontario Museum, Toronto, Canada; Harvard University, Cambridge, Massachusetts; National Museum of Natural History, Washington, D.C., USA, 115268.

References: (1) Dana, E.S. (1892) Dana's system of mineralogy, (6th edition), 566–569. (2) Dunn, P.J., R.C. Rouse, and J.A. Norberg (1978) Hydroxyapophyllite, a new mineral, and a redefinition of the apophyllite group. I. Description, occurrences, and nomenclature. Together with: Rouse, R.C, D.R. Peacor, and P.J. Dunn: II. Crystal structure. Amer. Mineral., 63, 196–202. (3) Dunn, P.J. and W.E. Wilson (1978) Nomenclature revisions in the apophyllite group: hydroxyapophyllite, apophyllite, fluorapophyllite. Mineral. Record, 9, 95–98.

Crystal Data: Hexagonal. *Point Group:* 6. Massive aggregates, with grains to 1.5 cm.

Physical Properties: *Cleavage:* Perfect on $\{10\bar{1}0\}$. *Fracture:* Uneven. *Tenacity:* Brittle. Hardness = 6   D(meas.) = 2.32(2)   D(calc.) = 2.26

Optical Properties: Transparent. *Color:* Light blue; colorless in thin section. *Streak:* White. *Luster:* Vitreous.
Optical Class: Uniaxial (+). $\omega = 1.494(2)$ $\epsilon = 1.501(2)$

Cell Data: *Space Group:* $P6_3$. $a = 12.740(3)$ $c = 5.182(2)$ Z = 1

X-ray Powder Pattern: Lovozero massif, Russia; resembles cancrinite.
3.26 (100), 3.68 (70), 4.70 (60), 2.756 (50), 2.433 (30), 6.43 (25), 4.17 (20)

Chemistry:

	(1)
SiO_2	36.32
Al_2O_3	31.15
Fe_2O_3	0.18
MnO	0.03
MgO	0.11
CaO	0.92
Na_2O	23.43
K_2O	0.45
H_2O	5.41
CO_2	1.59
Total	99.59

(1) Lovozero massif, Russia, corresponding to $(Na_{7.46}Ca_{0.16}K_{0.10}Mg_{0.03}Fe^{3+}_{0.02})_{\Sigma=7.77}$ $(Al_{6.03}Si_{5.97})_{\Sigma=12.00}O_{24}[(OH)_{1.23}(CO_3)_{0.36}]_{\Sigma=1.59} \cdot 2.35H_2O$.

Mineral Group: Cancrinite group.

Occurrence: In veins in ultra-agpaitic pegmatites in a differentiated alkalic massif.

Association: Natrolite, steenstrupine, vuonnemite, epistolite, mountainite, ilmajokite, nastrophite.

Distribution: On Mt. Karnasurt, Lovozero massif, Kola Peninsula, Russia.

Name: For predominant *hydroxyl* in its composition, and its relation to *cancrinite*.

Type Material: Vernadskii Geological Museum, Moscow; A.E. Fersman Mineralogical Museum, Academy of Sciences, Moscow, Russia.

References: (1) Khomyakov, A.P., T.N. Nadezhina, R.K. Rastsvetaeva, and E.A. Pobedimskaya (1992) Hydroxycancrinite $Na_8[Al_6Si_6O_{24}](OH)_2 \cdot 2H_2O$ – a new mineral. Zap. Vses. Mineral. Obshch., 121(1), 100–105 (in Russian). (2) (1993) Amer. Mineral., 78, 1315 (abs. ref. 1). (3) (1994) Mineral. Abs., 45, 111 (abs. ref. 1). (4) Nadezhina, T.N., R.K. Rastsvetaeva, E.A. Pobedimskaya, and A.P. Khomyakov (1991) Crystal structure of natural hydroxyl-containing cancrinite. Kristallografiya (Sov. Phys. Crystal.), 36, 591–595 (in Russian). (5) Hassan, I. and H.D. Grundy (1991) The crystal structure of basic cancrinite, ideally $Na_8[Al_6Si_6O_{24}](OH)_2 \cdot 3H_2O$. Can. Mineral., 29, 377–383.

Crystal Data: Monoclinic, pseudohexagonal. *Point Group:* $2/m$. As cleavable aggregates, to 2 cm.

Physical Properties: Hardness = 4.5 D(meas.) = 3.01(8) D(calc.) = ~3.080

Optical Properties: Translucent. *Color:* Pale purplish. *Luster:* Vitreous. *Optical Class:* Uniaxial (–). $\omega = 1.654(1)$ $\epsilon = 1.650(1)$

Cell Data: *Space Group:* $P2_1/m$, pseudo- $P6_3/m$. $a = 9.476(2)$ $b = 9.508(2)$ $c = 6.919(1)$ $\beta = 119.53(2)°$ Z = [2]

X-ray Powder Pattern: Chichibu mine, Japan.
2.839 (100), 2.739 (60), 2.655 (45), 2.801 (44), 1.853 (43), 3.462 (40), 1.484 (20)

Chemistry:

	(1)		(1)
SiO_2	17.30	P_2O_5	0.66
Al_2O_3	trace	F	0.28
Fe_2O_3	0.21	Cl	0.91
MnO	0.04	H_2O^+	2.04
MgO	trace	H_2O^-	0.72
CaO	54.51	CO_2	1.65
SrO	0.28	SO_3	21.56
Na_2O	0.34	$-O = (F, Cl)_2$	0.32
K_2O	0.07	Total	100.25

(1) Chichibu mine, Japan; corresponds to $(Ca_{9.66}Na_{0.11}Sr_{0.03}Fe_{0.03}K_{0.02})_{\Sigma=9.85}$ $[(SiO_4)_{2.86}(SO_4)_{2.68}(CO_3OH)_{0.37}(PO_4)_{0.09}]_{\Sigma=6.00}[(OH)_{1.88}Cl_{0.26}F_{0.15}]_{\Sigma=2.29}$.

Polymorphism & Series: Forms a series with fluorellestadite.

Mineral Group: Ellestadite group.

Occurrence: In pre-ore contact metamorphic skarns.

Association: Diopside, wollastonite, xanthophyllite, vesuvianite, calcite.

Distribution: In the Chichibu mine, Saitama Prefecture, Japan.

Name: For *hydroxyl* in the chemical composition and analogy to *ellestadite*.

Type Material: National Science Museum, Tokyo, Japan, M15761.

References: (1) Harada, K., K. Nagashima, K. Nakao, and A. Kato (1971) Hydroxylellestadite, a new apatite from Chichibu mine, Saitama Prefecture, Japan. Amer. Mineral., 56, 1507–1518. (2) Sudarsanan, K. (1980) Structure of hydroxylellestadite. Acta Cryst., 36, 1636–1639. (3) Rouse, R.C. and P.J. Dunn (1982) A contribution to the crystal chemistry of ellestadite and the silicate sulfate apatites. Amer. Mineral., 67, 90–96.

Crystal Data: Triclinic. *Point Group:* $\bar{1}$. As masses up to 3 cm.

Physical Properties: *Cleavage:* Distinct on {011}. Hardness = 5.5–6 D(meas.) = 4.47 D(calc.) = 4.91

Optical Properties: Semitransparent. *Color:* Buff-tan, light purplish gray; in thin section, colorless. *Streak:* White. *Luster:* Vitreous to resinous.
Optical Class: Biaxial (–). $\alpha = 1.75$–1.79 $\beta = 1.82$–1.83 $\gamma = 1.83$ 2V(meas.) = 31°
2V(calc.) = 31.5°

Cell Data: *Space Group:* $P\bar{1}$. a = 6.573(1) b = 6.651(1) c = 6.454(1) $\alpha = 116.44°$
$\beta = 92.34°$ $\gamma = 95.63°$ Z = 2

X-ray Powder Pattern: Bokan Mountain, Alaska, USA.
2.881 (100), 2.954 (80), 2.784 (40), 3.023 (36), 2.841 (35), 2.705 (29), 3.20 (27)

Chemistry:

	(1)
SiO$_2$	19.0
TiO$_2$	0.05
Y$_2$O$_3$	45.7
RE$_2$O$_3$	22.8
Fe$_2$O$_3$	0.4
CaO	0.1
MnO	0.01
H$_2$O$^+$	0.39
H$_2$O$^-$	0.10
CO$_2$	11.3
P$_2$O$_5$	< 0.2
Total	99.85

(1) Bokan Mountain, Alaska, USA; primarily by emission spectrographic analysis; corresponds to $(Y_{1.44}RE_{0.49}Fe^{3+}_{0.20}Ca_{0.02})_{\Sigma=2.15}Si_{1.12}O_4[(CO_3)_{0.91}(H_2O)_{0.14}]_{\Sigma=1.05}$.

Occurrence: In quartz-microcline pegmatite (Fusamata, Japan); in veins in peralkalic riebeckite-aegirine granite (Bokan Mountain, Alaska, USA).

Association: Biotite, monazite, fergusonite, uraninite, thalenite (Fusamata, Japan); quartz, albite, siderite, pyrite, fluorite, sphalerite, galena, zircon (Bokan Mountain, Alaska, USA).

Distribution: In Japan, at Fusamata and Suishoyama, Fukushima Prefecture, and Soraku, Kyoto Prefecture. On the I and L No. 4 claim, Bokan Mountain, Prince of Wales Island, Alaska, USA. From the Trimouns talc deposit, six km northeast of Luzenac, Ariège, France.

Name: For Takeo Iimori and Satoyasu Iimori (1885–?), Japanese mineralogists, and *yttrium* in the composition.

Type Material: National Science Museum, Tokyo, Japan, M16288; National Museum of Natural History, Washington, D.C., USA, 120635.

References: (1) Kato, A. and K. Nagashima (1970) In: Introduction to Japanese minerals, 39, 85–86. (2) (1973) Amer. Mineral., 58, 140 (abs. ref. 1). (3) Foord, E.E., M.H. Staatz, and N.M. Conklin (1984) New data for iimoriite. Amer. Mineral., 69, 196–199.

Ilímaussite-(Ce) $\qquad$ $Ba_2Na_4CeFe^{3+}Nb_2Si_8O_{28} \cdot 5H_2O$

Crystal Data: Hexagonal. *Point Group:* $6/m\ 2/m\ 2/m$, $\bar{6}m2$, or $6mm$. As lamellar aggregates up to 15 mm. *Twinning:* Polysynthetic twinning may be observed.

Physical Properties: *Fracture:* Conchoidal. Hardness = ~4 D(meas.) = 3.6 D(calc.) = 3.7

Optical Properties: Transparent to translucent. *Color:* Brownish yellow. *Luster:* Resinous. *Optical Class:* Uniaxial (+). $\omega = 1.689$ $\epsilon = 1.695$

Cell Data: *Space Group:* $P6_3/mcm$, $P\bar{6}c2$, or $P6_3cm$. $a = 10.80(4)$ $c = 20.31(7)$ $Z = 3$

X-ray Powder Pattern: Ilímaussaq intrusion, Greenland.
2.67 (10), 3.25 (6), 3.12 (5), 2.98 (4), 2.24 (3), 2.50 (2), 2.022 (2b)

Chemistry:

	(1)	(2)
SiO_2	31.28	31.81
TiO_2	1.64	
Ce_2O_3		10.86
RE_2O_3	10.60	
Fe_2O_3	3.18	5.29
Nb_2O_5	13.20	17.59
BaO	23.62	20.29
Na_2O	7.00	8.20
K_2O	3.80	
H_2O		5.96
LOI	6.43	
Total	100.75	100.00

(1) Ilímaussaq intrusion, Greenland; alkalis by flame photometry, RE in proportions $Ce_{55}La_{23.6}Nd_{14.7}Pr_{6.7}$. (2) $Ba_2Na_4CeFeNb_2Si_8O_{28} \cdot 5H_2O$.

Occurrence: In a hydrothermal ussingite-analcime vein cutting sodalite syenite in an alkalic massif.

Association: Chkalovite, epistolite.

Distribution: At Nákâlâq, in the Ilímaussaq intrusion, southern Greenland.

Name: For the Ilímaussaq intrusion, in which it was discovered.

Type Material: n.d.

References: (1) Semenov, E.I., M.E. Kazakova, and V.J. Bukin (1968) Ilimaussite, a new rare-earth-niobium-barium silicate from Ilímaussaq, South Greenland. Medd. Grønland, 181(7), 3–7. (2) (1969) Amer. Mineral., 54, 992–993 (abs. ref. 1).

Crystal Data: Monoclinic. *Point Group:* $2/m$ or m. As bunches of crystals to 2 mm, granular deposits, and crusts.

Physical Properties: *Cleavage:* Perfect on rhombic prism and pinacoid, intersecting at 72°. *Tenacity:* Brittle. Hardness = 1   D(meas.) = 2.20(2)   D(calc.) = n.d.   Water-soluble.

Optical Properties: Transparent to translucent. *Color:* Bright yellow. *Luster:* Vitreous. *Optical Class:* Biaxial (+). $\alpha = 1.573$ $\beta = 1.576$ $\gamma = 1.579$ 2V(meas.) = 90°

Cell Data: *Space Group:* $C2/c$ or Cc.   a = 39.80(4)   b = n.d.   c = 29.83(5) $\beta = 96°38'$ Z = n.d.

X-ray Powder Pattern: Lovozero massif, Russia.
11.5 (10), 4.3 (10), 2.44 (10), 10.2 (9), 3.1 (9), 10.9 (7), 3.7 (7)

Chemistry:

	(1)		(1)
SiO_2	35.55	CaO	0.04
TiO_2	16.95	SrO	0.02
ZrO_2	0.03	BaO	2.65
Al_2O_3	0.28	Na_2O	12.40
RE_2O_3	4.93	K_2O	0.40
Fe_2O_3	0.07	H_2O^+	16.78
Nb_2O_5	0.02	H_2O^-	7.76
Ta_2O_5	0.01	CO_2	1.46
		Total	99.36

(1) Lovozero massif, Russia; $RE_2O_3 = La_2O_3$ 26.8%, Ce_2O_3 51.2%, Pr_2O_3 3.6%, Nd_2O_3 17.8%, Sm_2O_3 17.8%; after deduction of CO_2 as nahcolite, corresponds to $(Na_{8.8}RE_{0.7}Ba_{0.5})_{\Sigma=10.0}$ $Ti_5(Si_{13.9}Al_{0.1})_{\Sigma=14.0}O_{22}(OH)_{44} \cdot nH_2O$.

Occurrence: On the walls of cavities in the central natrolite zone of pegmatites in a differentiated alkalic massif.

Association: Sphalerite, halite, mountainite, nahcolite, aegirine.

Distribution: In the Lovozero massif, near the valley of the Ilmajok River, Kola Peninsula, Russia.

Name: For the locality near the Ilmajok River, Kola Peninsula, Russia.

Type Material: Geology Museum, Kola Branch, Academy of Sciences, Russia; National School of Mines, Paris, France.

References: (1) Bussen, I.V., L.F. Gannibal, E.A. Goiko, A.N. Mer'kov, and A.P. Nedorezova (1972) Ilmajokite, a new mineral from the Lovozero Tundra. Zap. Vses. Mineral. Obshch., 101, 75–79 (in Russian). (2) (1973) Amer. Mineral., 58, 139–140 (abs. ref. 1). (3) Goiko, E.A., I.V. Bussen, L.F. Gannibal, and E.A. Lipatova (1974) Ilmajokite. Uch. Zap. Leningr. Gos. Univ., Ser. Biol. Nauk 278, 174–181 (in Russian). (4) (1976) Chem. Abs., 84, 7517 (abs. ref. 3).

Crystal Data: Orthorhombic or monoclinic. *Point Group:* $2/m\ 2/m\ 2/m$ or $2/m$. Crystals prismatic, striated $\parallel \{001\}$, to 25 cm; columnar, radiating, or compact massive.

Physical Properties: *Cleavage:* Distinct on $\{001\}$ and $\{010\}$. *Fracture:* Uneven. *Tenacity:* Brittle. Hardness = 5.5–6 D(meas.) = 3.99–4.05 D(calc.) = 4.064

Optical Properties: Opaque, translucent in fine fragments. *Color:* Iron-black or dark grayish black. *Streak:* Black, inclining to green or brown. *Luster:* Submetallic.
Optical Class: Biaxial (+). *Pleochroism:* In thin section, intense; X = dark green; Y = yellow-brown to dark brown; Z = dark brown; in reflected light, light gray to bluish gray, pinkish red to violet. *Orientation:* $X = c$; $Y = b$; $Z = a$. *Dispersion:* $r < v$, strong.
Absorption: $X > Y > Z$. $\alpha = 1.727$ $\beta = 1.870$ $\gamma = 1.883$ 2V(meas.) = 20°–30°
R_1–R_2: (400) 8.7–10.4, (420) 8.6–10.3, (440) 8.5–10.2, (460) 8.2–10.1, (480) 7.9–10.0, (500) 7.6–9.9, (520) 7.3–9.8, (540) 7.0–9.8, (560) 6.7–9.8, (580) 6.3–9.8, (600) 5.8–9.8, (620) 5.5–9.8, (640) 5.2–9.8, (660) 5.0–9.8, (680) 5.1–9.7, (700) 5.4–9.6 *Anisotropism:* Very strong.

Cell Data: *Space Group:* $Pbnm$. a = 8.800 b = 13.019 c = 5.852 Z = 4, or
Space Group: $P2_1/a$. a = 13.0103(5) b = 8.8039(4) c = 5.8517(3) $\beta = 90.209(5)°$ Z = 4

X-ray Powder Pattern: Gruba, Oberhalbstein Valley, Graubünden, Switzerland; monoclinic.
2.676 (100), 2.849 (93), 2.840 (93), 7.305 (70), 2.865 (70), 2.721 (70), 2.714 (70)

Chemistry:

	(1)	(2)
SiO_2	29.30	29.40
Fe_2O_3	20.30	19.53
FeO	33.50	35.15
MnO	1.97	
CaO	13.71	13.72
H_2O	1.90	2.20
Total	100.68	100.00

(1) Ilímaussaq intrusion, Greenland. (2) $CaFe_2^{2+}Fe^{3+}OSi_2O_7(OH)$.

Polymorphism & Series: Orthorhombic and monoclinic forms are known.

Occurrence: In contact metasomatic rocks and iron deposits; in metamorphosed limestones and dolostones with other calc-silicate minerals.

Association: Quartz, magnetite, hedenbergite, sphalerite, fluorite.

Distribution: Many localities, even for fine material. From Elba, at Rio Marina and Capo Calamita; on Mt. Mulatto, near Predazzo, Trentino-Alto Adige, Italy. Very large crystals from Seriphos, Cyclades Islands, Greece. In Germany, at St. Andreasberg, Harz Mountains; near Herborn, Hesse; and Schneeberg, Saxony. In Norway, at Fossum, near Skeen. At Thyrill, Iceland. From the Kangerdluarssuk Plateau, in the Ilímaussaq intrusion, southern Greenland. In the Kamioka mine, Gifu Prefecture, and the Obira mine, Bungo, Oita Prefecture, Japan. Fine crystals at Dal'negorsk, Primorski Krai, Russia. In the USA, splendid crystals from the Laxey mine, South Mountain, Owyhee Co., Idaho; from Middlemarch Pass, Dragoon Mountains, Cochise Co., Arizona; on Copper Mountain, Prince of Wales Island, Alaska.

Name: From the Latin name, *Ilva*, for the Island of Elba, on which it was first found.

References: (1) Dana, E.S. (1892) Dana's system of mineralogy, (6th edition), 541–542.
(2) Dietrich, V. (1972) Ilvait, Ferroantigorit und Greealith als Begleiter oxidisch-sulfidischer Vererzungen in den Oberhalbsteiner Serpentiniten. Schweiz. Mineral. Petrog. Mitt., 52, 57–74 (in German with English abs.). (3) Ghose, S., A.W. Hewat, and M. Marezio (1984) A neutron powder diffraction study of the crystal and magnetic structures of ilvaite from 305 K to 5 K – a mixed valence iron silicate with an electronic transition. Phys. Chem. Minerals, 11, 67–74.
(4) Finger, L.W. and R.M. Hazen (1987) Crystal structure of monoclinic ilvaite and the nature of the monoclinic-orthorhombic transition at high pressure. Zeits. Krist., 179, 415–430.

Crystal Data: Orthorhombic. *Point Group:* $2/m \ 2/m \ 2/m$. Anhedral grains, to 3 mm; as rims around eudialyte.

Physical Properties: Hardness = $\sim$4   D(meas.) = 2.93   D(calc.) = 2.92

Optical Properties: Semitransparent. *Color:* Honey-yellow. *Luster:* Vitreous.
Optical Class: Biaxial (+). $\alpha = 1.605$ $\beta = 1.608$ $\gamma = 1.612$ 2V(meas.) = 75°

Cell Data: *Space Group: Pmnn.*   a = 10.331(1)   b = 10.546(1)   c = 7.426(4)   Z = 1

X-ray Powder Pattern: Khibiny massif, Russia.
2.63 (100), 1.853 (70), 3.33 (60), 3.73 (50), 1.520 (50)

Chemistry:

	(1)	(2)
SiO_2	50.95	50.75
TiO_2	0.95	
ZrO_2	1.43	
Fe_2O_3	5.47	11.24
MnO	2.30	
MgO	0.26	
CaO	12.00	11.84
Na_2O	26.66	26.17
Total	100.02	100.00

(1) Khibiny massif, Russia; by electron microprobe, total Fe as Fe_2O_3. (2) $Na_{12}Ca_3Fe_2Si_{12}O_{36}$.

Mineral Group: Lovozerite group.

Occurrence: In apatite-bearing alkalic pegmatitic rocks in a differentiated alkalic massif.

Association: Eudialyte, aegirine, orthoclase, alkalic amphibole, pectolite.

Distribution: From a drill core in the Khibiny massif, Vourennem River area, near Lake Imandra, Kola Peninsula, Russia.

Name: For Lake Imandra on the Kola Peninsula, Russia.

Type Material: A.E. Fersman Mineralogical Museum, Academy of Sciences, Moscow, Russia; The Natural History Museum, London, England, 1994,10.

References: (1) Khomyakov, A.P., N.M. Chernitsova, S.M. Sandomirskaya, and G.L. Vasil'eva (1979) Imandrite, a new mineral of the lovozerite family. Mineral. Zhurnal, 1(1), 89–93 (in Russian). (2) (1980) Amer. Mineral., 65, 810 (abs. ref. 1). (3) (1980) Mineral. Mag., 31, 496 (abs. ref. 1). (4) Chernitsova, N.M., Z.V. Pudovkina, A.A. Voronkov, V.V. Ilyukhin, and Y.A. Pyatenko (1980) Imandrite $Na_{12}Ca_3Fe_2[Si_6O_{18}]_2$ as a representative of a new branch in the lovozerite structural family. Doklady Acad. Nauk SSSR, 252, 618–621 (in Russian). (5) (1980) Chem. Abs., 93, 141277 (abs. ref. 4).

Crystal Data: n.d. *Point Group:* n.d.   Conchoidal to earthy; as microscopic threadlike particles, and bundles of fine tubes, each about 20 Å in diameter.

Physical Properties: *Fracture:* Conchoidal, earthy. *Tenacity:* Brittle.   Hardness = 2–3 D(meas.) = 2.70   D(calc.) = 2.70

Optical Properties: Transparent to translucent. *Color:* White, blue, green, brown, black. *Luster:* Vitreous, resinous, waxy.
Optical Class: Isotropic. $n = 1.47–1.51$

Cell Data: *Space Group:* n.d.   c = 8.4; 5.1 $\perp$ c   Z = n.d.

X-ray Powder Pattern: Uemura, Japan; by electron diffraction.
21.0 (100b), 4.12 (100), 1.40 (100), 11.7 (80b), 7.8 (80b), 3.75 (80b), 2.32 (80b)

Chemistry: An analysis of natural material does not appear to be available.

Occurrence: In soils derived from volcanic ash.

Association: Allophane, quartz, cristobalite, gibbsite, vermiculite, limonite.

Distribution: Probably quite widespread in volcanic-ash-derived soils. In the Misutsuchi bed, Iijima, Nagano Prefecture; the Kanumatsuchi bed, Kanuma, Tochigi Prefecture; and from Uemura, Kumamoto Prefecture, Japan.

Name: For the name, *Imogo*, of the brownish yellow volcanic ash soil of Japan in which it occurs.

Type Material: n.d.

References: (1) Yoshinaga, N. and S. Aomine (1962) Allophane in some Ando soils. Soil Sci. and Plant Nutrition (Japan), 8, 6–13. (2) (1963) Amer. Mineral., 48, 434 (abs. ref. 1). (3) Russell, J.D, W.J. McHardy, and A.R. Fraser (1969) Imogolite: a unique aluminosilicate. Clay Minerals, 8, 87–99. (4) Cradwick, P.D.G., V.C. Farmer, J.D. Russell, C.R. Masson, K. Wada, and N. Yoshinaga (1972) Imogolite, a hydrated aluminum silicate of tubular structure. Nature, Phys. Sci., 240, 187–189. (5) Wada, S.I. and K. Wada (1977) Density and structure of allophane. Clay Minerals, 12, 289–298. (6) Goodman, B.A., J.D. Russell, B. Montez, E. Oldfield, and R.J. Kirkpatrick (1985) Structural studies of imogolite and allophanes by aluminum-27 and silicon-29 nuclear magnetic resonance spectroscopy. Phys. Chem. Minerals, 12, 342–346. (7) Bayliss, P. (1987) Mineral nomenclature: imogolite. Mineral. Mag., 51, 327. (8) (1988) Amer. Mineral., 73, 198 (abs. ref. 7).

Crystal Data: Hexagonal. *Point Group:* $6/m \, 2/m \, 2/m$. As small hexagonal grains and clusters of spiral-shaped grains, some with tentacles or multiple terminations.

Physical Properties: Hardness = 7–7.5   D(meas.) = 2.512 (synthetic $\mathrm{Mg}_2\mathrm{Al}_4\mathrm{Si}_5\mathrm{O}_{18}$). D(calc.) = 2.59

Optical Properties: Transparent. *Color:* Colorless in thin section. *Luster:* Vitreous. *Optical Class:* Uniaxial (–). $\epsilon = 1.532$–1.535 $\omega = 1.537$–1.539

Cell Data: *Space Group:* $P6/mcc$.   a = 9.800(3)   c = 9.345(3)   Z = 2

X-ray Powder Pattern: Synthetic $\mathrm{Mg}_2\mathrm{Al}_4\mathrm{Si}_5\mathrm{O}_{18}$. 8.48 (100), 3.027 (85), 3.138 (65), 3.379 (55), 4.094 (50), 4.89 (30), 1.6882 (30)

Chemistry:

	(1)	(2)
SiO_2	48.6	47.33
Al_2O_3	34.7	32.96
FeO	7.8	12.30
MnO		0.35
MgO	9.3	5.64
Na_2O		0.36
Total	100.4	98.94

(1) Bokaro coalfield, India; by electron microprobe, corresponding to $(\mathrm{Mg}_{1.40}\mathrm{Fe}_{0.66})_{\Sigma=2.06}$ $\mathrm{Al}_{4.11}\mathrm{Si}_{4.89}\mathrm{O}_{18}$. (2) Unazuki, Japan; by electron microprobe, corresponding to $(\mathrm{Fe}_{1.07}\mathrm{Mg}_{0.88}$ $\mathrm{Na}_{0.07}\mathrm{Mn}_{0.03})_{\Sigma=2.05}\mathrm{Al}_{4.06}\mathrm{Si}_{4.95}\mathrm{O}_{18}$.

Polymorphism & Series: Dimorphous with cordierite.

Occurrence: Formed by fusion and recrystallization of sedimentary rocks as a result of the burning of underlying coal seams (Bokaro coalfield, India); in cordierite veins in a polymetamorphosed pelitic rock (Unazuki, Japan).

Association: Enstatite, magnetite, labradorite, corundum, glass (Bokaro coalfield, India); cordierite, andalusite, sillimanite, biotite, quartz (Unazuki, Japan).

Distribution: In the Bokaro coal seam, southwest of Hazaribagh, Bihar, India. From the Unazuki area, Toyama Prefecture, Japan. At Pyramid Lake, Washoe Co., Nevada, USA. From the Bellerberg volcano, two km north of Mayen, Eifel district, Germany.

Name: For the country of first occurrence, India.

Type Material: n.d.

References: (1) Miyashiro, A. and T. Iiyama (1954) A preliminary note on a new mineral, indialite, polymorphic with cordierite. Proc. Japan Acad., 30, 746–751. (2) (1955) Amer. Mineral., 40, 787 (abs. ref. 1). (3) Miyashiro, A., T. Iiyama, M. Yamasaki, and T. Miyashiro (1955) The polymorphism of cordierite and indialite. Amer. J. Sci., 253, 185–208. (4) Meagher, E.P. and G.V. Gibbs (1977) The polymorphism of cordierite: II. The crystal structure of indialite. Can. Mineral., 15, 43–49. (5) Venkatesh, V. (1952) Development and growth of cordierite in para-lavas. Amer. Mineral., 37, 831–847. (6) Daniels, P. (1990) What is the true space group of high-cordierite? Zeits. Krist., 190, 271–276. (7) (1961) NBS Mono. 25, 29.

Crystal Data: Triclinic. *Point Group:* $\overline{1}$. Crystals slender laths, may be chisel-shaped, to 7 cm, typically forming radiating fan-shaped clusters; fibrous, in veinlets, and massive.

Physical Properties: *Cleavage:* Perfect on {010}, good on {100}. *Fracture:* Uneven. *Tenacity:* Brittle. Hardness = 5.5–6 D(meas.) = 3.03–3.04 D(calc.) = 3.03

Optical Properties: Semitransparent. *Color:* Rose-red, pink, orange-pink, orange-red-brown; changes to ginger-brown on exposure. *Streak:* White. *Luster:* Vitreous to silky. *Optical Class:* Biaxial (–). *Orientation:* $X \wedge c = 74°$; $Y \wedge c = 32°$; $Z \wedge c = 62°$. *Dispersion:* $r > v$, weak, distinct. $\alpha = 1.6178$–1.6183 $\beta = 1.6384$–1.6390 $\gamma = 1.6519$–1.6526 2V(meas.) = 74°–77°

Cell Data: *Space Group:* $P\overline{1}$. a = 8.889(2) b = 9.247(2) c = 11.975(3) $\alpha = 88.15(2)°$ $\beta = 132.07(2)°$ $\gamma = 96.64(2)°$ Z = 1

X-ray Powder Pattern: Broken Hill, Australia.
9.16 (100), 2.92 (80), 2.84 (80), 2.73 (70), 2.19 (60), 4.59 (50), 4.01 (50)

Chemistry:

	(1)	(2)	(3)
SiO_2	45.67	45.00	45.60
Al_2O_3		0.32	
FeO	0.92	1.73	
MnO	35.10	37.48	37.69
MgO	0.86	0.47	
CaO	9.33	6.78	8.51
K_2O		0.01	
H_2O	8.66	8.27	8.20
Total	100.54	[100.06]	100.00

(1) Quinault, Washington, USA. (2) Equity mine, Colorado, USA; original total given as 100.10%. (3) $Ca_2Mn_7Si_{10}O_{28}(OH)_2 \cdot 5H_2O$.

Occurrence: A late-stage hydrothermal mineral in manganese deposits.

Association: Rhodochrosite, bementite, hausmannite (Hale Creek mine, California, USA); datolite, pectolite, apophyllite, ruizite, orientite, quartz (Wessels mine, South Africa).

Distribution: In Germany, from Nanzenbach, near Dillenburg, Hesse. At Långban and in the Harstig mine, Pajsberg, near Persberg, Värmland, Sweden. From Săcărâmb (Nagyág), Romania. At Banská Štiavnica (Schemnitz), Slovakia. From the Wessels and N'Chwaning mines, near Kuruman, Cape Province, South Africa. In the USA, from Quinault, Grays Harbor Co., and in the Crescent mine, Olympic Peninsula, Clallam Co., Washington; in the Equity mine, near Creede, Mineral Co., Colorado; at the Hale Creek mine, near Red River, Trinity Co., and several other places in California. At Broken Hill, New South Wales, Australia. In Japan, at the Yugashima, Seikoshi, and Kawazu mines, Shizuoka Prefecture; the Innai mine, Akita Prefecture; the Kocho mine, Kochi Prefecture; and the Todoroki mine, Hokkaido.

Name: From the Greek for *flesh fibers*, in allusion to its color and structure.

References: (1) Dana, E.S. (1892) Dana's system of mineralogy, (6th edition), 564. (2) Richmond, W.E. (1942) Inesite, $Mn_7Ca_2Si_{10}O_{28}(OH)_2 \cdot 5H_2O$. Amer. Mineral., 27, 563–568. (3) Ryall, W.R. and I.M. Threadgold (1968) Inesite from the Broken Hill lode, New South Wales, Australia. Amer. Mineral., 53, 1614–1634. (4) Che'ng Wan and S. Ghose (1978) Inesite, a hydrated calcium manganese silicate with five-tetrahedral-repeat double chains. Amer. Mineral., 63, 563–571. (5) Van Loenen, R.E. (1980) Inesite, a new U.S. occurrence near Creede, Mineral County, Colorado. Mineral. Record, 11, 35–36.

Crystal Data: Triclinic. *Point Group:* 1. As plates, to 1 cm, rarely with faces, with {001} and {010} the most common; in radiating groups. *Twinning:* Polysynthetic Manebach twinning.

Physical Properties: *Cleavage:* Perfect on {010}, {110}, {1$\bar{1}$0}, good on {001}. *Tenacity:* Brittle. Hardness = 4.75 VHN = 435 D(meas.) = 3.96 D(calc.) = [3.81] Piezoelectric; slightly electromagnetic.

Optical Properties: Transparent to translucent. *Color:* Pale yellow to brown. *Luster:* Vitreous on cleavages, slightly resinous on fractures. *Optical Class:* Biaxial (+). *Pleochroism:* X = Y = light yellow; Z = pale brownish yellow. *Orientation:* Z $\simeq$ a. *Dispersion:* $r > v$, strong; sections cut $\perp$ {001} show anomalous bluish interference color. $\alpha = 1.726(1)$ $\beta = 1.737(1)$ $\gamma = 1.766(1)$ 2V(meas.) = 82(2)°

Cell Data: *Space Group:* $P1$. $a = 14.76$ $b = 7.14$ $c = 5.38$ $\alpha = 90°$ $\beta = 95°$ $\gamma = 99°$ Z = 1

X-ray Powder Pattern: Inagli massif, Russia.
3.92 (10), 3.04 (6), 2.95 (6), 1.964 (6), 1.845 (6), 1.735 (6), 6.31 (5)

Chemistry:

	(1)		(1)
SiO_2	18.78	BaO	44.16
TiO_2	18.50	Na_2O	5.63
Al_2O_3	0.23	K_2O	0.72
Fe_2O_3	0.66	F	0.40
FeO	0.57	H_2O^+	0.88
MnO	1.04	H_2O^-	0.09
MgO	0.83	SO_3	7.19
CaO	0.72	$-O = F_2$	0.17
		Total	100.23

(1) Inagli massif, Russia; corresponds to $(Na_{2.33}Mg_{0.27}Ca_{0.16}Fe^{2+}_{0.11}Fe^{3+}_{0.10})_{\Sigma=2.97}$ $(Ba_{3.69}K_{0.20}Mn_{0.19})_{\Sigma=4.08}Ti_{2.96}(Si_2O_7)_2(S_{0.58}O_4)_2[O_{2.75}(OH)_{1.25}F_{0.27}]_{\Sigma=4.27}$.

Occurrence: In miarolitic cavities of aegirine-eckermannite-microcline pegmatites in dunites; in pulaskite and shonkinite.

Association: Natrolite, albite, lorenzenite, batisite.

Distribution: In the Inagli massif, 30 km west of Aldan, and the Yakokutsk massif, near Schelochnoy Spring, Yakutia, Russia.

Name: From the Yakut name, *Inneli*, for the Inagli River, Yakutia, Russia.

Type Material: Institute of Mineralogy and Geochemistry of Rare Elements, Moscow; A.E. Fersman Mineralogical Museum, Academy of Sciences, Moscow, Russia; National Museum of Natural History, Washington, D.C., USA, 143822.

References: (1) Kravchenko, S.M., E.V. Vlasova, M.E. Kazakova, V.V. Ilokhin, and K.K. Abrashev (1961) Innelite, a new barium silicate. Doklady Acad. Nauk SSSR, 141, 1198–1199 (in Russian). (2) (1962) Amer. Mineral., 47, 805–806 (abs. ref. 1). (3) Chernov, A.N., V.V. Ilyukhin, B.A. Maksimov, and N.V. Belov (1971) Crystal structure of innelite, $Na_2Ba_3(Ba, K, Mn)(Ca, Na)Ti(TiO_2)_2[Si_2O_7]_2(SO_4)_2$. Kristallografiya (Sov. Phys. Crystal.), 16, 87–92 (in Russian).

Crystal Data: Triclinic. *Point Group:* 1 (probable). As equant to flattened euhedral crystals, to 1 mm, showing prominent $\{\overline{1}00\}$, $\{010\}$, $\{011\}$, and also $\{\overline{1}0\overline{1}\}$, $\{00\overline{1}\}$, $\{10\overline{2}\}$. *Twinning:* Common on $(12\overline{1})$.

Physical Properties: Hardness = n.d. D(meas.) = n.d. D(calc.) = 5.8

Optical Properties: Semitransparent. *Color:* Brown to orange. *Streak:* Saffron-yellow. *Luster:* Resinous.
Optical Class: Biaxial. *Pleochroism:* On (100), brownish orange $\parallel$ elongation; yellow-orange $\perp$ elongation. *Orientation:* Extinction 5° $\wedge$ elongation. $\alpha = 2.25$–2.30 β = n.d. $\gamma = 2.40$–2.50 2V(meas.) = Very large.

Cell Data: *Space Group:* $P1$ (probable). a = 9.57 b = 11.42 c = 10.84 $\alpha = 120°23'$ $\beta = 92°27'$ $\gamma = 56°6'$ Z = [1]

X-ray Powder Pattern: Sébarz mine, Iran.
3.60 (10), 3.49 (10), 3.28 (10), 3.18 (10), 3.08 (10), 4.84 (8), 4.42 (8)

Chemistry:

	(1)	(2)	(3)	(4)
SiO_2			3.9	3.94
CrO_3	28.8		20.1	19.65
CuO		1.85	2.8	2.60
ZnO		0.58	2.4	
PbO	66.2		72.7	73.09
F				1.24
H_2O^+				
$-O = F_2$				0.52
Total			101.9	100.00

(1) Sébarz mine, Iran; by electron microprobe, partial analysis. (2) Seh-Changi mine, Iran; partial analysis. (3) Tchah Khuni mine, Iran; by electron microprobe. (4) $Pb_{10}Cu(CrO_4)_6$ $(SiO_4)_2F_2$.

Polymorphism & Series: Forms a series with hemihedrite.

Occurrence: In the oxidized portions of lead-bearing hydrothermal veins.

Association: Dioptase, fornacite, wulfenite, mimetite, cerussite, diaboleite.

Distribution: In Iran, from the Sébarz and Tchah Khuni mines, Anarak district; in the Seh-Changi mine, near Neyband, Khorassan. From the Mammoth-St. Anthony mine, Tiger, Pinal Co., Arizona, USA.

Name: For the country of first occurrence, Iran.

Type Material: University of P. and M. Curie, Paris; National School of Mines, Paris, France; The Natural History Museum, London, England, 1969,52.

References: (1) Bariand, P. and P. Herpin (1963) Une nouvelle espèce minérale: l'iranite, chromate hydraté de plomb. Bull. Soc. fr. Minéral., 86, 133–135 (in French). (2) (1963) Amer. Mineral., 48, 1417 (abs. ref. 1). (3) Adib, D., J. Ottemann, and B. Nuber (1972) Further data on khuniite [iranite-hemihedrite] from the Tschah Khuni mine, Anarak, Iran. Neues Jahrb. Mineral., Monatsh., 328–335. (4) Williams, S.A. (1974) The naturally occurring chromates of lead. Bull. British Museum (Nat. Hist.) Mineral. 2, 377–419. (5) (1976) Amer. Mineral., 61, 186 (abs. ref. 4). (6) Bariand, P. and J.F. Poullen (1980) Rare chromates from Seh-Changi, Iran. Mineral. Record, 11, 293–297. (7) Cesbron, F. and S.A. Williams (1980) Iranite-hémihédrite, bellite, phoenicochroite, vauquelinite et fornacite: synthèse et nouvelles données. Bull. Minéral., 103, 469–477.

Crystal Data: Tetragonal. *Point Group:* $4/m \ 2/m \ 2/m$. Anhedral, massive.

Physical Properties: *Cleavage:* Three orthogonal, two good and one poor. *Fracture:* Uneven. Hardness = 4.5 VHN = 245–314 D(meas.) = [3.27] (slightly corrected for impurities). D(calc.) = 3.28

Optical Properties: Semitransparent. *Color:* Pale greenish yellow; in thin section, colorless. *Streak:* Whitish. *Luster:* Dull to pearly. *Optical Class:* Uniaxial (–); anomalously biaxial. *Orientation:* Extinction angles to 7°. $\omega = 1.590$ $\epsilon = 1.585$

Cell Data: *Space Group:* [$P4/mcc$] (by analogy to ekanite). $a = 7.61(1)$ $c = 14.72(2)$ Z = 1

X-ray Powder Pattern: Shakhi-Rash Mountain, Iraq. 5.28 (100), 3.31 (100), 2.64 (100), 7.36 (80), 3.38 (80), 3.40 (60), 2.17 (40)

Chemistry:

	(1)		(1)
SiO_2	51.7	CaO	12.00
ZrO_2	0.17	Na_2O	0.27
ThO_2	9.54	K_2O	2.76
UO_2	0.65	F	0.07
Al_2O_3	0.77	H_2O^+	3.51
RE_2O_3	15.06	H_2O^-	0.90
Fe_2O_3	0.22	CO_2	1.00
CuO	0.07	P_2O_5	0.01
PbO	0.35	S	0.14
MgO	0.02	$-O = F_2$	0.03
		Total	99.18

(1) Shakhi-Rash Mountain, Iraq; Zr, Th, RE by XRF, U by delayed neutron activation; $RE_2O_3 = La_2O_3$ 6.78%, Ce_2O_3 6.44%, Pr_2O_3 0.44%, Nd_2O_3 0.88%, Sm_2O_3 0.17%, Gd_2O_3 0.10%, Yt_2O_3 [sic] 0.25%; after deduction of calcite 2.27% and pyrite 0.26%, corresponds to $K_{1.07}[(La, Ce)_{1.33}Th_{0.66}]_{\Sigma=1.99}[Ca_{3.49}(La, Ce)_{0.35}Na_{0.16}]_{\Sigma=4.00}$ $(Si_{15.69}Al_{0.27})_{\Sigma=15.96}(O_{39.93}F_{0.07})_{\Sigma=40.00}$.

Occurrence: In granite in contact with dolomitic marble containing olivine and diopside.

Association: n.d.

Distribution: At Shakhi-Rash Mountain, Hero Town, Qala-Diza, Iraq.

Name: For the country of origin, *Iraq*, and *lanthanum* in the composition.

Type Material: The Natural History Museum, London, England, 1973,481; National School of Mines, Paris, France.

References: (1) Livingstone, A., D. Atkin, D. Hutchison, and H.M. Al-Hermezi (1976) Iraqite, a new rare-earth mineral of the ekanite group. Mineral. Mag., 40, 441–445. (2) (1976) Amer. Mineral., 61, 1054 (abs. ref. 1).

Crystal Data: Monoclinic. *Point Group:* $2/m$. Crystals rare, prismatic with $\{110\}$, $\{111\}$, and $\{100\}$ as dominant forms, showing cleavage faces to 1 cm. Commonly massive, or fibrous, granular, compact. *Twinning:* Single and lamellar twinning on $\{100\}$ and $\{001\}$.

Physical Properties: *Cleavage:* Good on $\{110\}$, $(110) \wedge (1\bar{1}0) \sim 87°$. *Fracture:* Splintery. *Tenacity:* Very tough when massive. Hardness = 6–7 D(meas.) = 3.24–3.43 D(calc.) = 3.330

Optical Properties: Translucent. *Color:* Apple-green, emerald-green, bluish green, leek-green, greenish white, white, may show green spots, rarely blue or violet; colorless in thin section. *Streak:* White. *Luster:* Subvitreous, pearly on cleavages.
Optical Class: Biaxial (+). *Orientation:* $Y = b$; $Z \wedge c = 32°–55°$. *Dispersion:* $r > v$, moderate to strong. $\alpha = 1.640–1.681$ $\beta = 1.645–1.684$ $\gamma = 1.652–1.692$ 2V(meas.) = $60°–96°$

Cell Data: *Space Group:* $C2/c$. $a = 9.418$ $b = 8.562$ $c = 5.219$ $\beta = 107.58°$ Z = 4

X-ray Powder Pattern: Clear Creek, California, USA.
2.831 (100), 2.922 (75), 4.29 (45), 3.10 (30), 2.069 (30), 2.417 (25), 2.490 (20)

Chemistry:

	(1)	(2)		(1)	(2)
SiO_2	59.06	61.66	MgO	0.17	0.98
TiO_2	0.08	0.05	CaO	0.35	1.38
Al_2O_3	24.62	21.81	Na_2O	14.95	12.27
Fe_2O_3	0.41	0.32	K_2O	0.01	0.57
FeO	0.18	0.24	H_2O^+	0.07	0.44
MnO	0.03	0.05	H_2O^-	0.03	0.10
			Total	99.96	99.87

(1) New Idria district, California, USA; corresponds to $(Na_{0.98}Ca_{0.01})_{\Sigma=0.99}$ $(Al_{0.98}Fe^{3+}_{0.01}Mg_{0.01})_{\Sigma=1.00}Si_{2.00}O_6$. (2) Cloverdale, California, USA; corresponds to $(Na_{0.80}Ca_{0.05}K_{0.02})_{\Sigma=0.87}(Al_{0.87}Mg_{0.05}Fe^{3+}_{0.01}Fe^{2+}_{0.01})_{\Sigma=0.94}Si_{2.08}O_6$.

Mineral Group: Pyroxene group.

Occurrence: In high-pressure metamorphic rocks of the glaucophane facies; a component of eclogite.

Association: Albite, quartz, muscovite, omphacite, glaucophane, calcite, aragonite, analcime, zeolites.

Distribution: In small amounts at many places; the origin of many specimens is obscure. Some well-known localities are: around Tawmaw, Myitkyina-Mogaung district, Kachin State, northern Myanmar (Burma). In the Ohmi area, along the Hashidate and Kotake Rivers, Niigata Prefecture, and at Shibukawa, Gumma Prefecture, Japan. Around Lake Baikal, Siberia, Russia. In the USA, in California, as crystals in boulders along the Russian River, about 3.5 km north of the Sonoma-Mendocino Co. line, near Cloverdale; at Clear Creek, near New Idria, San Benito Co.; and at several places in Mendocino Co. At Manzanal, in the Motagua Valley, near the Sierra de las Minas, Guatemala.

Name: From *jade*, a material commonly containing jadeite, in turn from the Spanish *Piedra de yjada*, or *colic stone*, for its supposedly curative powers for nephritic colic.

References: (1) Dana, E.S. (1892) Dana's system of mineralogy, (6th edition), 369–371. (2) Deer, W.A., R.A. Howie, and J. Zussman (1978) Rock-forming minerals, (2nd edition), v. 2A, single-chain silicates, 461–481. (3) Wolfe, C.W. (1955) Crystallography of jadeite crystals from near Cloverdale, California. Amer. Mineral., 40, 248–260. (4) Prewitt, C.T. and C.W. Burnham (1966) The crystal structure of jadeite, $NaAlSi_2O_6$. Amer. Mineral., 51, 956–975. (5) Coleman, R.G. and J.R. Clark (1968) Pyroxenes in the blueschist facies of California. Amer. J. Sci., 266, 43–59.

Crystal Data: Hexagonal. *Point Group:* $\bar{3}$. As euhedral to subhedral crystals elongated ‖ [0001], to 0.4 mm, showing hexagonal cross sections, and with {0001} and {10$\bar{1}$0}.

Physical Properties: *Cleavage:* {10$\bar{1}$0}, imperfect, suspected. *Fracture:* Conchoidal. *Tenacity:* Brittle. Hardness = n.d. D(meas.) = 2.65(3) D(calc.) = 2.58(2)

Optical Properties: Transparent. *Color:* Colorless. *Streak:* White. *Luster:* Vitreous. *Optical Class:* Uniaxial (+). $\omega = 1.596(2)$ $\epsilon = 1.604(2)$

Cell Data: *Space Group:* [$P\bar{3}$] (by analogy with synthetic Ca$_6$Si$_2$O$_7$(OH)$_6$). $a = 10.026(5)$ $c = 7.482(4)$ Z = 2

X-ray Powder Pattern: Kombat mine, Namibia.
8.66 (100), 2.996 (90), 2.833 (90), 2.887 (70), 3.279 (50), 2.466 (30), 2.083 (30)

Chemistry:

	(1)
SiO$_2$	23.96
CaO	64.98
H$_2$O	[11.06]
Total	[100.00]

(1) Kombat mine, Namibia; by electron microprobe, average of three analyses, includes Mn and Cl < 0.1%, H$_2$O by difference; corresponds to Ca$_{5.86}$Si$_{2.02}$O$_{13}$H$_{6.21}$.

Occurrence: Closely associated with or enclosed by defernite within low-grade metamorphic rocks.

Association: Defernite, hausmannite, apatite, brucite, hillebrandite, vesuvianite, glaucochroite, galena, copper.

Distribution: At the Kombat mine, 49 km south of Tsumeb, Namibia.

Name: In honor of Professor Howard Jaffe of the University of Massachusetts, Amherst, Massachusetts, USA.

Type Material: Museum of Natural History, Geneva, Switzerland, 500/32; National Museum of Natural History, Washington, D.C., USA, 163802.

References: (1) Sarp, H. and D.R. Peacor (1989) Jaffeite, a new hydrated calcium silicate from the Kombat mine, Namibia. Amer. Mineral., 74, 1203–1206.

Crystal Data: Hexagonal. *Point Group:* $\bar{6}m2$. Fine-grained micaceous aggregates of plates.

Physical Properties: *Cleavage:* Perfect on {0001}. *Tenacity:* Plates are flexible.
Hardness = 3   D(meas.) = 5.43   D(calc.) = n.d.

Optical Properties: Semitransparent. *Color:* Yellow-green. *Streak:* Yellow.
Luster: Vitreous, shining on cleavages.
Optical Class: Uniaxial (–). $n = \sim 2.0$; birefringence = 0.025.

Cell Data: *Space Group:* $P\bar{6}2c$. $a = 8.528(8)$ $c = 33.33(3)$ $Z = [6]$

X-ray Powder Pattern: Långban, Sweden.
3.40 (100), 2.80 (80), 4.16 (50), 2.99 (50), 2.50 (40), 2.60 (30), 2.47 (30)

Chemistry:

	(1)		(1)
SiO_2	22.35	MgO	0.60
TiO_2	0.10	CaO	0.65
Al_2O_3	0.50	Na_2O	0.61
Fe_2O_3	7.00	K_2O	0.37
FeO	0.00	Cl	3.25
MnO	0.88	H_2O^+	0.17
PbO	64.26	H_2O^-	0.19
BeO	0.12	$-O = Cl_2$	0.73
		Total	100.32

(1) Långban, Sweden; corresponds to $(Pb_{2.43}Na_{0.17}Ca_{0.10}K_{0.07})_{\Sigma=2.77}(Fe^{3+}_{0.74}Mg_{0.12}$
$Mn_{0.10}Ti_{0.01})_{\Sigma=0.97}(Si_{3.14}Al_{0.08}Be_{0.04})_{\Sigma=3.26}O_{9.98}[Cl_{0.77}(OH)_{0.16}]_{\Sigma=0.93}$.

Occurrence: In hematite ore.

Association: Melanotekite, quartz.

Distribution: At Långban, Värmland, Sweden.

Name: For John B. Jago Trelawney (1909–), mineral collector of Palo Alto, California, USA, and benefactor of mineral collections in the USA.

Type Material: n.d.

References: (1) Blix, R., O. Gabrielson, and F.E. Wickman (1957) Jagoite, a new lead-silicate mineral from Långban in Sweden. Arkiv Mineral. Geol., 2, 315–317. (2) (1958) Amer. Mineral., 43, 387 (abs. ref. 1). (3) Mellini, M. and S. Merlino (1981) The crystal structure of jagoite. Amer. Mineral., 66, 852–858.

Crystal Data: Monoclinic. *Point Group:* $2/m$. As sprays of prismatic crystals, to 1.5 cm, slightly bent, flattened on {010} and striated || [001]; in lamellar aggregates; as subhedral grains. *Twinning:* Rarely on {100}.

Physical Properties: *Cleavage:* Distinct on {010}. *Tenacity:* Very brittle. Hardness = 5 D(meas.) = 3.60(5)   D(calc.) = 3.71

Optical Properties: Semitransparent. *Color:* Reddish brown. *Streak:* Light brown. *Luster:* Vitreous.
Optical Class: Biaxial (+). *Pleochroism:* Weak; X = nearly colorless; Y = beige. *Orientation:* $Z = b$; $X \wedge c = 12°–18°$. $\alpha = 1.770(4)$ $\beta = 1.828(4)$ $\gamma = [1.910]$ 2V(meas.) = 70°–90°

Cell Data: *Space Group:* $P2_1/n$.   a = 10.668(2)   b = 9.787(4)   c = 13.931(3) $\beta = 107.82(2)°$ Z = 4

X-ray Powder Pattern: Gjerdingen, Norway.
2.839 (100), 2.833 (90), 2.782 (90), 3.202 (60), 1.744 (50), 3.920 (40), 2.742 (40)

Chemistry:

	(1)
SiO_2	29.79
TiO_2	15.18
ZrO_2	5.98
Nb_2O_5	5.00
Ta_2O_5	0.26
FeO	5.52
MnO	22.26
CaO	1.45
Na_2O	11.02
K_2O	0.19
F	2.69
H_2O	[1.63]
$-O = F_2$	1.13
Total	[99.84]

(1) Gjerdingen, Norway; by electron microprobe, average of three analyses, H_2O calculated; corresponds to $(Na_{2.75}Ca_{0.20}K_{0.03})_{\Sigma=2.98}(Mn_{2.43}Fe_{0.60})_{\Sigma=3.03}(Ti_{1.32}Zr_{0.38}Nb_{0.29}Ta_{0.01})_{\Sigma=2.00}$ $(Si_{3.84}Ti_{0.15})_{\Sigma=3.99}O_{15}[(OH)_{1.40}F_{1.10}O_{0.50}]_{\Sigma=3.00}$.

Occurrence: In both the groundmass and miarolitic cavities in sodium-rich granite.

Association: Pyrophanite, elpidite, monazite, dalyite, kupletskite.

Distribution: From Gjerdingen, 30 km north of Oslo, Norway.

Name: Honors Jan Haug, an amateur mineralogist who first observed the mineral.

Type Material: University of Oslo, Oslo, Norway.

References: (1) Raade, G. and M.H. Mladeck (1983) Janhaugite, $Na_3Mn_3Ti_2Si_4O_{15}$ $(OH, F, O)_3$, a new mineral from Norway. Amer. Mineral., 68, 1216–1219. (2) Annehed, H., L. Fälth, and G. Raade (1985) The crystal structure of janhaugite, a sorosilicate of the cuspidine family. Neues Jahrb. Mineral., Monatsh., 7–18.

Jasmundite

$Ca_{11}(SiO_4)_4O_2S$

Crystal Data: Tetragonal. *Point Group:* $\bar{4}2m$. As equant crystals, with forms {110}, {101}, {100} and {001}; in irregular grains up to several mm.

Physical Properties: *Fracture:* Conchoidal. Hardness = ~5 D(meas.) = 3.03 D(calc.) = 3.23

Optical Properties: Semitransparent. *Color:* Dark brown, greenish brown, brownish green; light brown in thin section. *Streak:* White. *Luster:* Resinous.
Optical Class: Uniaxial (+). $\omega = 1.715$ $\epsilon = 1.728$

Cell Data: *Space Group:* $I\bar{4}m2$. a = 10.461(1) c = 8.813(1) Z = 2

X-ray Powder Pattern: Bellerberg volcano, Germany.
2.832 (> 100), 3.242 (42), 2.615 (35), 1.849 (34), 1.5491 (21), 2.756 (17), 1.920 (16)

Chemistry:

	(1)	(2)
SiO_2	27.3	27.52
Al_2O_3	0.4	
FeO	0.8	
MgO	0.9	
CaO	67.6	70.64
S^{2-}	2.7	3.67
$-O = S$	1.4	1.83
Total	98.3	100.00

(1) Bellerberg volcano, Germany; by electron microprobe, average of seven analyses; corresponds to $(Ca_{10.6}Mg_{0.2}Fe_{0.1}Al_{0.05})_{\Sigma=10.95}Si_{4.0}O_{18.25}S_{0.75}$. (2) $Ca_{11}(SiO_4)_4O_2S$.

Occurrence: In metamorphosed limestone inclusions in basalt.

Association: Mayenite, brownmillerite, larnite, portlandite, ettringite, calcite, vaterite, tobermorite, thaumasite.

Distribution: From the Bellerberg volcano, two km north of Mayen, Eifel district, Germany.

Name: Honors Professor Karl Jasmund, retired Director of the Mineralogical-Petrographic Institute, University of Cologne, Cologne, Germany.

Type Material: Mineralogical-Petrographic Institute, University of Cologne, Cologne, Germany; Department of Chemistry, University of Aberdeen, Aberdeen, Scotland.

References: (1) Dent Glasser, L.S. and C.K. Lee (1981) The structure of jasmundite, $Ca_{22}(SiO_4)_8O_4S_2$. Acta Cryst., 37, 803–806. (2) Hentschel, G., L.S. Dent Glasser, and C.K. Lee (1983) Jasmundite, $Ca_{22}(SiO_4)_8O_4S_2$, a new mineral. Neues Jahrb. Mineral., Monatsh., 337–342. (3) (1984) Amer. Mineral., 69, 566–567 (abs. refs. 1 and 2).

Crystal Data: Orthorhombic. *Point Group:* 222. As thin micaceous pseudotetragonal plates, to 1.2 mm, composed of {001} and {110}. *Twinning:* On {100}.

Physical Properties: *Cleavage:* Perfect on {001} and {110}. *Tenacity:* Brittle. Hardness = ∼5   D(meas.) = 2.99(2)   D(calc.) = 2.98

Optical Properties: Transparent. *Color:* Colorless, but commonly with a brown coating. *Optical Class:* Biaxial (–). *Orientation:* X = c; Y = a; Z = b. $\alpha = 1.625(2)$ $\beta = 1.641(2)$ $\gamma = 1.643(2)$ 2V(meas.) = 40(2)°   2V(calc.) = 39°

Cell Data: *Space Group:* $C222_1$.   a = 14.90(1)   b = 14.90(1)   c = 40.41(8)   Z = 64

X-ray Powder Pattern: Jeffrey mine, Canada.
2.774 (100), 2.993 (90), 2.541 (60), 1.755 (50), 5.00 (40), 2.360 (40), 2.229 (40)

Chemistry:

	(1)
SiO_2	46.7
Al_2O_3	2.8
BeO	8.1
CaO	37.4
Na_2O	2.3
H_2O	1.8
Total	99.1

(1) Jeffrey mine, Canada; by electron microprobe, average of nine analyses, Be by AA, H_2O by TGA; corresponds to $(Ca_{1.69}Na_{0.19})_{\Sigma=1.88}(Be_{0.82}Al_{0.14})_{\Sigma=0.96}Si_{1.97}[O_{6.49}(OH)_{0.51}]_{\Sigma=7.00}$.

Polymorphism & Series: Dimorphous with gugiaite.

Occurrence: In a cavity in a highly calcium-metasomatized granite dike.

Association: Grossular.

Distribution: In the Jeffrey mine, Asbestos, Quebec, Canada.

Name: For the type locality, the Jeffrey mine, Quebec, Canada.

Type Material: Canadian Museum of Nature, Ottawa, Canada, 48740.

References: (1) Grice, J.D. and G.W. Robinson (1984) Jeffreyite, $(Ca, Na)_2(Be, Al)$ $Si_2(O, OH)_7$, a new mineral species and its relation to the melilite group. Can. Mineral., 22, 443–446. (2) (1985) Amer. Mineral., 70, 872 (abs. ref. 1).

Crystal Data: Triclinic. *Point Group:* $\bar{1}$ or 1. As blade-shaped crystals, elongated along [010], to 6 mm, or as fibrous aggregates.

Physical Properties: *Cleavage:* Distinct on {001}. Hardness = n.d. D(meas.) = 2.32–2.33 D(calc.) = [2.34]

Optical Properties: Transparent to translucent. *Color:* White; colorless in thin section. *Luster:* Vitreous.
Optical Class: Biaxial (–). *Orientation:* $X \perp$ {001} cleavage; $Y \wedge b = 35°$–$40°$.
$\alpha = 1.548$–1.552 $\beta = 1.562$–1.564 $\gamma = 1.570$–1.571 2V(meas.) = 74°

Cell Data: *Space Group:* $P\bar{1}$ or $P1$.   a = 10.56   b = 7.25   c = 10.81 $\alpha = 99°42'$
$\beta = 97°42'$ $\gamma = 110°4'$ Z = 1

X-ray Powder Pattern: Crestmore, California, USA.
10.5 (vvs), 2.92 (vs), 3.04 (s), 2.83 (s), 2.66 (s), 6.46 (ms), 3.47 (ms)

Chemistry:

	(1)	(2)		(1)	(2)
SiO_2	34.2	33.20	Na_2O		0.00
TiO_2		0.02	K_2O		0.01
Al_2O_3		0.09	H_2O^+		18.53
FeO		0.01	H_2O^-		1.69
MnO		0.02	H_2O	19.2	
MgO		0.03	P_2O_5		0.04
CaO	46.6	46.85	Total	100.0	100.49

(1) Crestmore, California, USA; by electron microprobe; H_2O by TGA; corresponds to $Ca_{8.76}H_{2.00}$ $Si_{6.00}O_{17.76}(OH)_{8.00} \cdot 6.24H_2O$. (2) Fuka, Japan; by electron microprobe, H_2O by gravimetry; corresponds to $Ca_{9.00}H_{2.00}(Si_{5.95}Al_{0.02})_{\Sigma=5.97}O_{17.93}(OH)_{8.00} \cdot 6.07H_2O$.

Occurrence: A late-stage mineral, partially filling open spaces and in veins in fractured skarns.

Association: Tobermorite, scawtite, calcite (Crestmore, California, USA); afwillite, oyelite, spurrite (Fuka, Japan).

Distribution: From Crestmore, Riverside Co., California, USA. At Campomorto, Montalto di Castro, Lazio, Italy. In Germany, from the Bellerberg volcano, two km north of Mayen, Eifel district, and fine crystals from the Zeilberg, near Maroldsweisach, Bavaria. In the Hatrurim Formation, Israel. In the Wessels mine, near Kuruman, Cape Province, South Africa. In Japan, from Fuka, near Bicchu, Okayama Prefecture.

Name: For Colonel Clarence M. Jenni (1896–1974?), Director of the Geological Museum, University of Missouri, Rollo, Missouri, USA, the mineral's discoverer.

Type Material: National Museum of Natural History, Washington, D.C., USA, 119007.

References: (1) Carpenter, A.B., R.A. Chalmers, J.A. Gard, K. Speakman, and H.F.W. Taylor (1966) Jennite, a new mineral. Amer. Mineral., 51, 56–74. (2) Gard, J.A., H.F.W. Taylor, G. Cliff, and G.W. Lorimer (1977) A reexamination of jennite. Amer. Mineral., 62, 365–368. (3) Kusachi, I., C. Henmi, and K. Henmi (1989) Afwillite and jennite from Fuka, Okayama Province, Japan. Mineral. J. (Japan), 14, 279–292.

Crystal Data: Orthorhombic. *Point Group:* $mm2$ (probable), or $2/m \; 2/m \; 2/m$. As interlocking anhedral crystals, to 2 mm.

Physical Properties: *Cleavage:* Imperfect on {001}. Hardness = ~5.5   D(meas.) = 4.00(2) D(calc.) = 4.045

Optical Properties: Transparent and translucent lamellae alternating ∥ {001}.
Color: Violet-pink, with a brownish tinge; light pink in thin section. *Streak:* Light pink.
Luster: Vitreous.
Optical Class: Biaxial (–). *Orientation:* $X = b$; $Y = c$; $Z = a$. *Dispersion:* $r > v$, moderate.
$\alpha = 1.772(4)$ $\beta = 1.783(4)$ $\gamma = 1.789(4)$ 2V(meas.) = 72°

Cell Data: *Space Group:* $Pbn2_1$ (probable), or *Pbnm*.   a = 4.875(2)   b = 10.709(6) c = 28.18(2)   Z = 4

X-ray Powder Pattern: Franklin, New Jersey, USA.
2.557 (100), 1.806 (100), 2.869 (78), 2.752 (49), 2.702 (46), 2.362 (39), 2.661 (34)

Chemistry:

	(1)
SiO_2	27.1
FeO	0.3
MnO	64.1
ZnO	3.9
MgO	1.4
CaO	0.4
F	0.0
H_2O	2.13
Total	99.3

(1) Franklin, New Jersey, USA; by electron microprobe, H_2O by the Penfield method; corresponds to $(Mn_{7.86}Zn_{0.59}Mg_{0.24}Ca_{0.16}Fe_{0.14})_{\Sigma=8.99}(SiO_4)_4(OH)_2$.

Polymorphism & Series: Dimorphous with sonolite.

Mineral Group: Leucophoenicite group.

Occurrence: In a metamorphosed stratiform Zn-Mn deposit.

Association: Franklinite, willemite, zincite, sonolite, leucophoenicite, tephroite.

Distribution: From Franklin, Sussex Co., New Jersey, USA.

Name: In honor of Professor Gerald V. Gibbs, Virginia Polytechnic Institute and State University, Blacksburg, Virginia, USA.

Type Material: Harvard University, Cambridge, Massachusetts, 125030; National Museum of Natural History, Washington, D.C., USA, C3209, R18772, 149037; The Natural History Museum, London, England1983,236.

References: (1) Dunn, P.J., D.R. Peacor, W.B. Simmons, and E.J. Essene (1984) Jerrygibbsite, a new polymorph of $Mn_9(SiO_4)_4(OH)_2$ from Franklin, New Jersey, with new data on leucophoenicite. Amer. Mineral., 69, 546–552. (2) Kato, T., Y. Ito, and N. Hashimoto (1989) The crystal structures of sonolite and jerrygibbsite. Neues Jahrb. Mineral., Monatsh., 410–430, 1989.

Jervisite $(Na, Ca, Fe^{2+})(Sc, Mg, Fe^{2+})Si_2O_6$

Crystal Data: Monoclinic. *Point Group:* $2/m$. As sprays of elongated platy crystals, < 1 mm.

Physical Properties: *Cleavage:* Perfect on {110}. Hardness = n.d. D(meas.) = n.d. D(calc.) = [3.31]

Optical Properties: Semitransparent. *Color:* Light green. *Luster:* Vitreous. *Optical Class:* Biaxial. $\alpha = 1.663$ β = n.d. $\gamma = 1.684$ 2V(meas.) = n.d.

Cell Data: *Space Group:* $C2/c$. a = 9.853(11) b = 9.042(10) c = 5.312(7) $\beta = 106°37(7)'$ Z = 4

X-ray Powder Pattern: Baveno, Italy.
3.038 (s), 2.979 (m), 2.543 (m), 1.647 (m), 6.51 (w), 4.51 (w), 3.389 (w)

Chemistry:

	(1)	(2)
SiO_2	50.42	54.59
TiO_2	0.55	
Al_2O_3	0.42	
Sc_2O_3	18.48	31.33
FeO	8.59	
MnO	0.44	
MgO	2.80	
CaO	7.25	
Na_2O	5.55	14.08
Total	94.50	100.00

(1) Baveno, Italy; by electron microprobe, average of multiple analyses; corresponds to $(Na_{0.43}Ca_{0.31}Fe^{2+}_{0.14}Mn_{0.01})_{\Sigma=0.89}(Sc_{0.66}Mg_{0.17}Fe^{2+}_{0.15}Al_{0.02}Ti_{0.02})_{\Sigma=1.02}Si_2O_{6.02}$. (2) $NaScSi_2O_6$.

Mineral Group: Pyroxene group.

Occurrence: In miarolitic cavities in granite.

Association: Cascandite, quartz, orthoclase, albite.

Distribution: In the Diverio quarry, Mt. Mottarone, near Baveno, Piedmont, Italy.

Name: Honors William P. Jervis, Curator of the Museo Industriale Italiano di Torino, Torino, Italy.

Type Material: Municipal Museum of Natural History, Milan, Italy, 23270.

References: (1) Mellini, M., S. Merlino, P. Orlandi, and R. Rinaldi (1982) Cascandite and jervisite, two new scandium silicates from Baveno, Italy. Amer. Mineral., 67, 599–603. (2) Hawthorne, F.C. and H.D. Grundy (1973) Refinement of the crystal structure of $NaScSi_2O_6$. Acta Cryst., 29, 2615–2616.

Crystal Data: Orthorhombic. *Point Group:* $2/m\ 2/m\ 2/m$. As radiating sprays of crystals, to 5 cm; as fibrous intergrowths parallel to {010} in anthophyllite and cummingtonite.

Physical Properties: *Cleavage:* Perfect on {210}, intersecting at 38° and 142°; breakage on {100} and {010} may be partings. Hardness = n.d. D(meas.) = n.d. D(calc.) = [3.02]

Optical Properties: Transparent. *Color:* Colorless to very light pinkish brown; colorless in thin section.
Optical Class: Biaxial (–). *Orientation:* $X = a$; $Y = b$; $Z = c$. *Dispersion:* $r > v$, weak.
$\alpha = 1.605(5)$ $\beta = 1.626(5)$ $\gamma = 1.633(5)$ 2V(meas.) = 62(2)°

Cell Data: *Space Group:* $Pbca$. $a = 18.6263(3)$ $b = 27.2303(6)$ $c = 5.2970(3)$ $Z = 8$

X-ray Powder Pattern: Calculated. (ICDD 31-638).
8.812 (100), 13.6 (57), 3.092 (53), 2.601 (42), 3.250 (38), 2.547 (28), 3.814 (27)

Chemistry:

	(1)
SiO_2	57.78
Al_2O_3	0.29
FeO	12.22
MnO	0.72
MgO	25.14
CaO	0.38
Na_2O	0.12
H_2O	[2.92]
Total	[99.57]

(1) Chester, Vermont, USA; by electron microprobe; H_2O assuming (OH) sites filled by $(OH)^{1-}$.

Polymorphism & Series: Dimorphous with clinojimthompsonite.

Occurrence: In the black wallrock between chlorite and actinolite zones of a metamorphosed ultramafic body.

Association: Chesterite, clinojimthompsonite, anthophyllite, cummingtonite, talc.

Distribution: In the Carleton talc quarry, near Chester, Windsor Co., Vermont, USA.

Name: For Professor James Burleigh Thompson, Jr. (1921–), eminent petrologist of Harvard University, Cambridge, Massachusetts, USA.

Type Material: Royal Ontario Museum, Toronto, Canada, M36083; Harvard University, Cambridge, Massachusetts; National Museum of Natural History, Washington, D.C., USA, 145689.

References: (1) Veblen, D.R. and C.W. Burnham (1978) New biopyriboles from Chester, Vermont: I. Descriptive mineralogy. Amer. Mineral., 63, 1000–1009. (2) Veblen, D.R. and C.W. Burnham (1978) New biopyriboles from Chester, Vermont: II. The crystal chemistry of jimthompsonite, clinojimthompsonite, chesterite, and the amphibole-mica reaction. Amer. Mineral., 63, 1053–1073. (3) Veblen, D.R. and P.R. Buseck (1979) Chain-width order and disorder in biopyriboles. Amer. Mineral., 64, 687–700.

Jinshajiangite

$$Na_2KBaCa(Fe^{2+}, Mn^{2+})_8Ti_4Si_8O_{32}(O, F, H_2O)_6$$

Crystal Data: Monoclinic. *Point Group:* $2/m$, m, or 2. As tabular crystals, up to 2 cm.

Physical Properties: *Cleavage:* Perfect on $\{010\}$ and $\{100\}$. *Fracture:* Uneven. Hardness = n.d. VHN = 430 D(meas.) = 3.61 D(calc.) = 3.56

Optical Properties: Semitransparent. *Color:* Blackish red, brownish red, golden red. *Streak:* Light yellow. *Luster:* Vitreous.
Optical Class: Biaxial (+). *Pleochroism:* Strong; X = light golden yellow; Y = brownish yellow; Z = brownish red. *Orientation:* $X \wedge c = 13°$. *Dispersion:* $r < v$. *Absorption:* $X = Y > Z$.
$\alpha = 1.729$ $\beta = 1.802$ $\gamma = 1.852$ 2V(meas.) = 72° 2V(calc.) = 76°

Cell Data: *Space Group:* $C2/m$, Cm, or $C2$. a = 10.732 b = 13.847 c = 20.817 $\beta = 95°3'$ Z = 2

X-ray Powder Pattern: Near the Jinshajiang River, China.
3.44 (10), 3.15 (8), 2.570 (8), 10.2 (7), 2.85 (7), 2.63 (7), 1.715 (5b)

Chemistry:

	(1)		(1)
SiO_2	27.10	MgO	0.28
TiO_2	15.90	CaO	2.94
$(Zr, Hf)O_2$	0.70	SrO	0.08
Al_2O_3	0.36	BaO	9.80
RE_2O_3	0.30	Na_2O	3.15
Fe_2O_3	1.64	K_2O	2.31
Nb_2O_5	1.03	F	2.66
Ta_2O_5	0.07	H_2O^+	0.33
FeO	19.07	H_2O^-	0.36
MnO	12.93	$-O = F_2$	1.12
		Total	99.89

(1) Near the Jinshajiang River, China; corresponds to $Na_{1.81}K_{0.87}Ba_{1.14}Ca_{0.93}RE_{0.10}Sr_{0.01}$ $(Fe^{2+}_{4.73}Mn_{3.25}Mg_{0.12})_{\Sigma=8.10}(Ti_{3.55}Fe^{3+}_{0.37}Nb_{0.14}Zr_{0.10})_{\Sigma=4.16}(Si_{8.04}Al_{0.12})_{\Sigma=8.16}O_{32}[O_{2.83}F_{2.49}$ $(H_2O)_{0.56}(OH)_{0.12}]_{\Sigma=6.00}$.

Occurrence: In an arfvedsonite dike in alkalic syenites.

Association: Albite, arfvedsonite, aegirine, pyrochlore, monazite, chevkinite.

Distribution: Found near the Jinshajiang River, western Sichuan Province, China.

Name: For the Jinshajiang River, China.

Type Material: Institute of Geochemistry, Academy Sinica, Guiyang, Ghizhou Province, China.

References: (1) Hong Wenxing and Fu Pingqiu (1982) Jinshajiangite, a new Ba-Mn-Fe-Ti-bearing silicate mineral. Geochemistry (China), 1, 458–464 (in English). (2) (1984) Amer. Mineral., 69, 567 (abs. ref. 1). (3) Chao, G.Y. (1991) Perraultite, a new hydrous Na-K-Ba-Mn-Ti-Nb silicate species from Mont Saint-Hilaire, Quebec. Can. Mineral., 29, 355–358.

Crystal Data: Monoclinic, pseudo-orthorhombic. *Point Group:* 2. As equant or tabular crystals, flattened $\perp$ [001], up to 1.2 cm; intimately intergrown with orthojoaquinite-(Ce). *Twinning:* On {001}, polysynthetic, common.

Physical Properties: *Cleavage:* {001}, good. Hardness = 5.5   D(meas.) = 3.89–3.98 D(calc.) = [3.93]

Optical Properties: Transparent to translucent. *Color:* Honey-yellow to brown. *Luster:* Vitreous.
Optical Class: Biaxial (+). *Pleochroism:* Weak; $X = Y$ = colorless; Z = pale yellow. *Orientation:* $X = a$; $Y = b$; $Z = c$. *Dispersion:* $r < v$, perceptible. *Absorption:* $Z > Y > X$. $\alpha = 1.748$–1.753 $\beta = 1.767$ $\gamma = 1.822$–1.823 2V(meas.) = 30°–55°

Cell Data: *Space Group:* $C2$. $a = 10.516(3)$ $b = 9.686(3)$ $c = 11.833(4)$ $\beta = 109.67(3)°$ Z = 2

X-ray Powder Pattern: San Benito Co., California, USA.
2.943 (100), 4.43 (95), 2.890 (85), 3.29 (60), 2.606 (60), 3.05 (40), 2.978 (40)

Chemistry:

	(1)		(1)
SiO_2	34.97	MgO	0.05
TiO_2	11.83	CaO	0.21
ThO_2	0.27	SrO	3.20
Y_2O_3	0.70	BaO	22.44
RE_2O_3	18.46	Na_2O	1.87
FeO	4.09	K_2O	0.03
MnO	0.00	H_2O	[1.88]
		Total	[100.00]

(1) San Benito Co., California, USA; by electron microprobe, average of six points on five grains, intergrown with orthojoaquinite-(Ce) of presumably nearly identical composition; RE_2O_3 = La_2O_3 2.14%, Ce_2O_3 10.69%, Pr_2O_3 1.25%, Nd_2O_3 3.21%, Sm_2O_3 0.70%, Gd_2O_3 0.26%, Dy_2O_3 0.21%, Er_2O_3 0.00%, H_2O by difference; corresponds to $Ba_{2.01}Ca_{0.05}Mg_{0.02}Na_{0.83}K_{0.01}$ $(Ce_{0.90}RE_{0.72}Sr_{0.42})_{\Sigma=2.04}Fe_{0.78}Ti_{2.04}Th_{0.02}Si_{8.00}O_{24.68}(OH)_{3.32}$.

Polymorphism & Series: Dimorphous with orthojoaquinite-(Ce).

Mineral Group: Joaquinite group.

Occurrence: In a natrolite vein cutting a glaucophane schist inclusion in a serpentinite body (San Benito Co., California, USA); in fenitized gneisses and alkalic syenites (Seal Lake, Canada).

Association: Orthojoaquinite-(Ce), benitoite, neptunite, natrolite (San Benito Co., California, USA); aegirine, barylite, eudidymite, neptunite (Seal Lake, Canada).

Distribution: At the Gem mine and to its north, on Santa Rita peak; at Mina Numero Uno and on the Victor claim, San Benito Co., California, USA. In Canada, at Seal Lake, Labrador, Newfoundland, and Mont Saint-Hilaire, Quebec. Along the Narssaq river, near Kvanefjeld, in the Ilímaussaq intrusion, southern Greenland.

Name: For Joaquin Ridge, near the original locality at the Gem mine, California, USA.

Type Material: Harvard University, Cambridge, Massachusetts, USA, 90840.

References: (1) Louderback, G.D. (1909) Benitoite, its paragenesis and mode of occurrence. Univ. of Calif., Bull. Dept. of Geol., 5, 331–380. (2) Palache, C. and W.F. Foshag (1932) The chemical nature of joaquinite. Amer. Mineral., 17, 308–312. (3) Laird, J. and A.L. Albee (1972) Chemical composition and physical, optical, and structural properties of benitoite, neptunite, and joaquinite. Amer. Mineral., 57, 85–102. (4) Dowty, E. (1975) Crystal structure of joaquinite. Amer. Mineral., 60, 872–878. (5) Wise, W.S. (1982) Strontiojoaquinite and bario-orthojoaquinite: two new members of the joaquinite group. Amer. Mineral., 67, 809–816.

Crystal Data: Monoclinic.   *Point Group: 2/m.*   As prismatic, crudely doubly terminated, highly-modified crystals, elongated along [001] and flattened on {100}, up to 1 cm. Principal forms are {$\bar{1}$10}, {100}, {010}, {011}, {$\bar{1}$12}, and {$\bar{1}$13}.

Physical Properties: *Cleavage:* Perfect on {110}.   Hardness = 5.5   D(meas.) = 3.83(1) D(calc.) = 3.91

Optical Properties: Opaque to transparent in thin fragments.   *Color:* Black; olive-brown in thin fragments.   *Streak:* Pale brown.   *Luster:* Subadamantine.
Optical Class: Biaxial (+).   *Pleochroism:* $X = Z =$ olive; $Y =$ brown with olive tint.
Absorption: $Y > X = Z$. $\alpha = 1.747(5)$ $\beta = 1.765(5)$ $\gamma = 1.78(1)$ 2V(meas.) = 60°–70°

Cell Data: *Space Group: P2/a.* $a = 9.915(2)$ $b = 17.951(4)$ $c = 5.243(1)$
$\beta = 105.95(2)°$ Z = 2

X-ray Powder Pattern: Långban, Sweden.
3.33 (10), 2.564 (6), 2.530 (6), 2.740 (5.5), 3.70 (5), 2.903 (5), 2.676 (5)

Chemistry:

	(1)		(1)
SiO_2	35.4	CaO	9.9
Al_2O_3	0.5	BaO	0.3
Fe_2O_3	12.3	Na_2O	0.5
FeO	2.6	K_2O	0.0
MnO	2.3	H_2O^+	[1.5]
PbO	20.7	F	0.3
BeO	[4.2]	$-O = F_2$	0.1
MgO	9.6	Total	[100.0]

(1) Långban, Sweden; by electron microprobe, average of four analyses, presence of Be confirmed by ion microprobe; Fe^{2+}:Fe^{3+} calculated from charge balance, BeO from stoichiometry, H_2O by difference; corresponds to $(Pb_{0.98}Ba_{0.02})_{\Sigma=1.00}(Ca_{1.86}Na_{0.16})_{\Sigma=2.02}(Mg_{2.52}Fe^{3+}_{1.63}Fe^{2+}_{0.38}Mn_{0.34}$ $Al_{0.10})_{\Sigma=4.97}(Si_{6.22}Be_{1.78})_{\Sigma=8.00}O_{22}[(OH)_{1.81}F_{0.19}]_{\Sigma=2.00}$.

Mineral Group: Amphibole (calcic) group: $(Na + K)_A < 0.5$; $Na_B < 0.67$; $(Ca + Na)_B \geq 1.34$.

Occurrence: A rare mineral in a metamorphosed manganese orebody.

Association: Hematite, magnetite, manganoan aegirine, quartz, barite, calcite.

Distribution: At Långban, Värmland, Sweden.

Name: For Professor Joseph Victor Smith (1928–), English-American mineralogist and petrologist of the University of Chicago, Chicago, Illinois, USA.

Type Material: National Museum of Natural History, Washington, D.C., USA, 120064, 162612.

References: (1) Moore, P.B. (1968) Joesmithite, a new amphibole-like mineral from Långban. Arkiv Mineral. Geol., 4, 487–492. (2) Moore, P.B. (1968) The crystal structure of joesmithite: a preliminary note. Mineral. Mag., 36, 876–879. (3) (1969) Amer. Mineral., 54, 577–578 (abs. ref. 1 and 2). (4) Moore, P.B. (1969) Joesmithite: a novel amphibole crystal chemistry. Min. Soc. Am. Spec. Pap. 2, 111–115. (5) Moore, P.B. (1988) The joesmithite enigma: note on the $6s^2$ Pb^{2+} lone pair. Amer. Mineral., 73, 843–844. (6) Moore, P.B., A.M. Davis, D.G. Van Derveer, and P.K. Sen Gupta (1993) Joesmithite, a plumbous amphibole revisited and comments on bond valences. Mineral. Petrol., 48, 97–113.

Crystal Data: Monoclinic. *Point Group:* $2/m$. As prismatic crystals, to 10 cm. In columnar, radiating, and spherulitic aggregates of fibers and prisms. *Twinning:* Simple and lamellar twinning common on {100}.

Physical Properties: *Cleavage:* Good on {110}, (110) $\wedge$ (1$\bar{1}$0) $\sim$87°; partings on {100}, {001}, and {010}. *Fracture:* Uneven to conchoidal. *Tenacity:* Brittle. Hardness = 6 D(meas.) = 3.27–3.54 D(calc.) = [3.52]

Optical Properties: Translucent to opaque. *Color:* Clove-brown, gray, colorless, blue, green; colorless in thin section.
Optical Class: Biaxial (+). *Orientation:* $Y = b$; $Z \wedge c = -46°$ to $-55°$. *Dispersion:* $r < v$ or $r > v$, weak to moderate. $\alpha = 1.699$–1.710 $\beta = 1.710$–1.719 $\gamma = 1.725$–1.738 2V(meas.) = 58°–72°

Cell Data: *Space Group:* $C2/c$. a = 9.978(9) b = 9.156(9) c = 5.293(5) $\beta = 105°29(2)'$ Z = 4

X-ray Powder Pattern: Borieva deposits, Erzebezirk Maden, Bulgaria. (ICDD 18-299). 3.02 (100), 2.547 (80), 2.600 (60), 2.564 (50), 2.243 (50), 1.633 (50), 6.58 (40)

Chemistry:

	(1)	(2)		(1)	(2)
SiO_2	47.62	48.81	CaO	22.18	21.87
TiO_2		0.01	Na_2O		0.07
Al_2O_3	0.91	0.74	K_2O		0.02
Fe_2O_3	0.04	0.79	H_2O^+	0.40	0.32
FeO	0.70	1.54	H_2O^-	0.09	0.35
MnO	27.47	22.58	CO_2	0.24	
MgO	0.53	2.29	P_2O_5		0.01
			Total	100.18	99.40

(1) Tetela de Ocampo, Mexico; corresponds to $(Ca_{0.97}Mg_{0.03})_{\Sigma=1.00}(Mn^{2+}_{0.96}Fe^{2+}_{0.02}Al_{0.02})_{\Sigma=1.00}$ $(Si_{1.97}Al_{0.03})_{\Sigma=2.00}O_6$. (2) Aravaipa district, Arizona, USA; corresponds to $(Ca_{0.96}Na_{0.01})_{\Sigma=0.97}$ $(Mn^{2+}_{0.78}Mg_{0.14}Fe^{2+}_{0.05}Al_{0.03}Fe^{3+}_{0.02})_{\Sigma=1.02}(Si_{1.99}Al_{0.01})_{\Sigma=2.00}O_6$.

Polymorphism & Series: Forms two series, with diopside, and with hedenbergite.

Mineral Group: Pyroxene group.

Occurrence: In metasomatized limestones and manganiferous skarns; in quartz or calcite veins cutting rhyolite.

Association: Rhodonite, manganese oxides.

Distribution: From Tetela de Ocampo, Puebla, and Pachuca, Hidalgo, Mexico. In the USA, at Franklin, Sussex Co., New Jersey; from the Bohemia district, Lane Co., Oregon; at the Aravaipa district, Graham Co., Arizona; and in the Empire Zinc Co. mine, Hanover, Grant Co., New Mexico. From Schio, Vicenza, and Campiglia, Tuscany, Italy. In the Akatani mine, Niigata Prefecture; the Nakatatsu mine, Fukui Prefecture; the Onagusa mine, Katsuyam, Okayama Prefecture; and elsewhere in Japan. A number of other localities are known.

Name: To honor Professor Albert Johannsen (1871–1962), petrologist, University of Chicago, Chicago, Illinois, USA.

Type Material: National Museum of Natural History, Washington, D.C., USA, R3118, 97484.

References: (1) Schaller, W.T. (1938) Johannsenite, a new manganese pyroxene. Amer. Mineral., 23, 575–582. (2) Deer, W.A., R.A. Howie, and J. Zussman (1978) Rock-forming minerals, (2nd edition), v. 2A, single-chain silicates, 415–422. (3) Freed, R.L. and D.R. Peacor (1967) Refinement of the crystal structure of johannsenite. Amer. Mineral., 52, 709–720.

Crystal Data: Triclinic. *Point Group:* 1 or $\bar{1}$. As fibrous aggregates, may be plumose, to 4.5 cm; individuals are elongated along [001]. *Twinning:* An indeterminate twinninglike intergrowth was observed.

Physical Properties: *Cleavage:* Good on {100}, poor on {010}. *Tenacity:* Friable in aggregate. Hardness = n.d. D(meas.) = 3.48(4) D(calc.) = 3.51

Optical Properties: Transparent. *Color:* Light yellowish brown; colorless in transmitted light. *Streak:* Light brownish yellow. *Luster:* Vitreous.
Optical Class: Biaxial (–). *Dispersion:* $r > v$, distinct. $\alpha = 1.6742(4)$ $\beta = 1.6968(3)$ $\gamma = 1.6999(3)$ 2V(meas.) = 41.9(2)° 2V(calc.) = 40.9°

Cell Data: *Space Group:* $P1$ or $P\bar{1}$. a = 10.44(2) b = 11.064(6) c = 9.62(1) $\alpha = 107.43(7)°$ $\beta = 82.7(1)°$ $\gamma = 111.6(1)°$ Z = 1

X-ray Powder Pattern: Kombat mine, Namibia.
2.676 (100), 9.8 (60), 3.23 (60), 5.99 (40), 3.38 (40), 2.479 (40), 1.539 (40b)

Chemistry:

	(1)	(2)
SiO_2	35.5	34.68
As_2O_5	10.6	11.05
FeO	0.1	
MnO	40.7	40.94
MgO	8.2	7.75
Na_2O	3.1	2.98
H_2O	2.6	2.60
Total	100.8	100.00

(1) Kombat mine, Namibia; by electron microprobe, total As as As_2O_5, total Mn as MnO, H_2O by the Penfield method. (2) $Na_2Mg_4Mn_{12}As_2Si_{12}O_{43}(OH)_6$.

Occurrence: In low-temperature hydrothermal veins cutting metamorphosed Fe-Mn ores.

Association: Kentrolite, rhodonite, richterite, barite, calcite.

Distribution: From the Kombat mine, 49 km south of Tsumeb, Namibia.

Name: Honors John Innes, mineralogist of the Tsumeb Corporation, for his contributions to the mineralogy of the Tsumeb and Kombat mines.

Type Material: National Museum of Natural History, Washington, D.C., USA, 163207.

References: (1) Dunn, P.J., D.R. Peacor, S.-C. Su, J.A. Nelen, and O. von Knorring (1986) Johninnesite, a new sodium manganese arsenosilicate from the Kombat Mine, Namibia. Mineral. Mag., 50, 667–670. (2) (1988) Amer. Mineral., 73, 928 (abs. ref. 1).

Crystal Data: Orthorhombic. *Point Group:* 222. Crystals are slender blades, with prismatic {210}, {310}, and terminated by the dome {101}. Commonly as rosettes of many such crystals, to 3 mm.

Physical Properties: *Cleavage:* Distinct on {010}. *Fracture:* Irregular. Hardness = 3–4 D(meas.) = 3.25(1)   D(calc.) = 3.239   Fluoresces dull orange under SW UV.

Optical Properties: Transparent to translucent. *Color:* Colorless.
Optical Class: Biaxial (+). *Orientation:* $X = b$; $Y = a$; $Z = c$. $\alpha = 1.641(1)$ $\beta = 1.660(1)$ $\gamma = 1.682(1)$ 2V(meas.) = 76°–78°

Cell Data: *Space Group:* $B22_12$.   a = 13.730(5)   b = 25.904(5)   c = 10.608(3)   Z = 4

X-ray Powder Pattern: Gem mine, California, USA.
12.95 (100), 3.031 (45), 2.652 (30), 2.228 (20), 3.008 (18), 2.073 (18), 2.596 (17)

Chemistry:

	(1)		(1)
SiO_2	32.54	MgO	0.111
TiO_2	17.31	CaO	0.045
B_2O_3	0.008	SrO	0.051
Al_2O_3	5.52	BaO	34.88
Fe_2O_3	0.16	Na_2O	0.67
Nb_2O_5	0.022	K_2O	2.70
MnO	0.035	H_2O	5.9
		Total	99.952

(1) Gem mine, California, USA; weighted average of 20 electron microprobe and ion probe analyses; corresponds to $(Ba_{4.19}Mg_{0.05}Fe_{0.04}Mn_{0.01}Ca_{0.01}Sr_{0.01})_{\Sigma=4.31}(K_{1.06}Na_{0.40})_{\Sigma=1.46}$ $Ti_{3.99}Al_{1.99}Si_{9.97}O_{36} \cdot 6H_2O$.

Occurrence: In fractures and cavities of brecciated greenstone enclosed in blueschist.

Association: Neptunite, benitoite, joaquinite, natrolite.

Distribution: In the USA, at the Gem mine, San Benito Co., California.

Name: For Francis Tucker Jones (1905–1993), Research Chemical Microscopist of Berkeley, California, USA, who discovered the mineral.

Type Material: University of California, Santa Barbara, California, 7325; American Museum of Natural History, New York, New York, USA.

References: (1) Wise, W.S., A. Pabst, and J.R. Hinthorne (1977) Jonesite, a new mineral from the Benitoite Gem mine, San Benito County, California. Mineral. Record, 8, 453–456.

Crystal Data: Orthorhombic (?). *Point Group:* n.d. As slender fibers in sheaflike aggregates, to about 1 mm; massive.

Physical Properties: Hardness = 5.5 D(meas.) = 3.01–3.3 D(calc.) = n.d.

Optical Properties: Semitransparent. *Color:* Nearly white to pistachio-green. *Luster:* Dull. *Optical Class:* Biaxial (+). $\alpha = 1.640(3)$ $\beta =$ n.d. $\gamma = 1.647(3)$ 2V(meas.) = 50° (?)

Cell Data: *Space Group:* n.d. Z = n.d.

X-ray Powder Pattern: Kuznetsk, Kazakhstan.
3.27 (100), 2.98 (100), 1.93 (100), 2.90 (90), 1.97 (80), 3.58 (50), 3.41 (50)

Chemistry:

	(1)	(2)	(3)
SiO_2	42.05	37.72	42.45
TiO_2	0.00	0.00	
Al_2O_3	5.19	10.79	6.55
Fe_2O_3	3.26	5.18	
FeO		1.08	
MnO	0.09	0.16	
MgO	9.52	4.99	10.35
CaO	34.68	34.0	36.02
Na_2O	1.06	1.01	
K_2O	0.14		
H_2O	4.45	0.00	4.63
P_2O_5		0.03	
LOI		5.87	
Total	100.44	100.83	100.00

(1) Iron Hill, Colorado, USA. (2) Kuznetsk, Kazakhstan. (3) $Ca_{10}Mg_4Al_2Si_{11}O_{39} \cdot 4H_2O$.

Occurrence: Formed by hydrothermal alteration of a coarse-grained melilite rock (uncompahgrite) previously subjected to deuteric alteration (Iron Hill, Colorado, USA); in skarns at a contact zone with carbonates in a gabbro-urtite massif (Kuznetsk, Kazakhstan).

Association: Perovskite, phlogopite, magnetite, titaniferous garnet, melilite, diopside, vesuvianite, brugnatellite, hastingsite, tremolite, aegirine, cebollite (Iron Hill, Colorado, USA); melilite, pyroxene, wollastonite, vesuvianite, cebollite, calcite (Kuznetsk, Kazakhstan).

Distribution: In the Iron Hill carbonatite, Powderhorn-Cibolla district, Gunnison Co., Colorado, USA. From the Kiya-Shaltyrsk massif, Kuznetsk, Ala-Tau Range, Tien Shan, Kazakhstan.

Name: For the San Juan Mountains, Colorado, USA, where the mineral was discovered.

Type Material: Harvard University, Cambridge, Massachusetts; National Museum of Natural History, Washington, D.C., USA, 105990.

References: (1) Larsen, E.S. and E.A. Goranson (1932) The deuteric and later alterations of the uncompahgrite of Iron Hill, Colorado. Amer. Mineral., 17, 343–356. (2) Rodygina, V. (1971) First find of juanite and cebollite in the Kuznetsk Ala-Tau. Geol. Geofiz., 12, 62–70. (3) (1972) Chem. Abs., 76, 156622 (abs. ref. 2).

$Ca_2Fe^{2+}(Fe^{3+}, Al)_2(SiO_4)(Si_2O_7)(OH)_2 \cdot H_2O$ **Julgoldite-(Fe^{2+})**

Crystal Data: Monoclinic. *Point Group:* $2/m$. Crystals flat prismatic to bladed, to 2 mm, elongated along [010] and flattened ∥ {100}. In fan-shaped, plumose crystal groups; granular. *Twinning:* Twin plane {001}, typically repeated, common.

Physical Properties: *Cleavage:* Perfect on {100} and {001}. *Tenacity:* Brittle. Hardness = 4.5   D(meas.) = 3.58–3.60   D(calc.) = 3.56

Optical Properties: Transparent. *Color:* Deep black, greenish black to green in small fragments; in thin section, brilliant interference colors in greens or blues. *Streak:* Greenish olive with a bluish tinge. *Luster:* Nearly submetallic.
Optical Class: Biaxial (–). *Pleochroism:* Strong; X = pale brown; Y = pale brownish green; Z = deep emerald-green. *Orientation:* $Y = b$. *Absorption:* $Z \gg Y > X$. $\alpha = 1.776(4)$ $\beta = 1.814(4)$ $\gamma = 1.836(4)$ 2V(meas.) = 50°–70°   2V(calc.) = 73°

Cell Data: *Space Group:* $A2/m$.   a = 8.922(4)   b = 6.081(3)   c = 19.432(9) $\beta = 97.60(6)°$ Z = 4

X-ray Powder Pattern: Scotland.
2.958 (100), 2.780 (80), 2.574 (80), 1.519 (80), 4.817 (70), 3.859 (70), 2.501 (60)

Chemistry:

	(1)	(2)		(1)	(2)
SiO$_2$	34.0	32.04	MnO	0.2	
TiO$_2$	0.1		MgO	0.2	0.23
Al$_2$O$_3$	1.3	0.68	CaO	22.0	19.93
Fe$_2$O$_3$	29.6	30.28	BaO	0.01	
FeO	8.7	9.5	H$_2$O	4.69	[7.34]
			Total	100.8	[100.00]

(1) Långban, Sweden; by emission spectroscopy, corresponds to $(Ca_{8.8}Mn_{0.1})_{\Sigma=8.9}(Fe^{2+}_{2.7}Fe^{3+}_{1.2}$ $Mg_{0.1})_{\Sigma=4.0}(Fe^{3+}_{7.2}Al_{0.6})_{\Sigma=7.8}Si_{12.7}O_{44.7}(OH)_{11.7}$. (2) Auchinstarry quarry, Scotland; by electron microprobe, $Fe^{2+}:Fe^{3+}$ by Mössbauer spectroscopy, H_2O by difference; corresponds to $Ca_{8.01}Fe^{3+}_{8.55}$ $Al_{0.30}Fe^{2+}_{3.00}Mg_{0.13}Si_{12.03}O_{48.47} \cdot (H_2O)_{9.18}$.

Polymorphism & Series: Forms two series, with pumpellyite-(Fe^{2+}), and with pumpellyite-(Mg).

Mineral Group: Pumpellyite group.

Occurrence: In hematite-magnetite ore (Långban, Sweden); in quartz-diabase (Scotland).

Association: Apophyllite, barite, hematite, magnetite, ilvaite, calcite, quartz, chlorite, prehnite, epistilbite, stilbite, pectolite, laumontite, babingtonite, titanite.

Distribution: At Långban, Värmland, Sweden. Between Tafjord and Fjøra, Sunnmøre district, Norway. In the Ratho quarry, near Edinburgh, and the Auchinstarry quarry, Kilsyth, Scotland. In Germany, from the Wolfmühl quarry, Waldgrehweiler, Rhineland-Palatinate. From Hale Creek, Trinity Co., California, and in the Clark mine, Copper Harbor, Keweenaw Co., Michigan, USA. At Sawda, near Jalgoan, and Bombay, Maharashtra, India.

Name: For Professor Julian Royce Goldsmith (1918–), mineralogist and geochemist, University of Chicago, Chicago, Illinois, USA, and its *ferrous iron* content.

Type Material: Swedish Natural History Museum, Stockholm, Sweden; Harvard University, Cambridge, Massachusetts; National Museum of Natural History, Washington, D.C., USA, 137014.

References: (1) Moore, P.B. (1971) Julgoldite, the $Fe^{+2}-Fe^{+3}$ dominant pumpellyite; a new mineral from Långban, Sweden. Lithos, 4, 93–99. (2) (1971) Amer. Mineral., 56, 2157–2158 (abs. ref. 1). (3) Allmann, R. and G. Donnay (1973) The crystal structure of julgoldite. Mineral. Mag., 39, 271–281. (4) Passaglia, E. and G. Gottardi (1973) Crystal chemistry and nomenclature of pumpellyites and julgoldites. Can. Mineral., 12, 219–223. (5) Livingstone, A. (1976) Julgoldite, new data and occurrences; a second recording. Mineral. Mag., 40, 761–763.

Crystal Data: Orthorhombic. *Point Group:* $mm2$. Only as crystals, to 5 mm, with excellent faces and good evidence of hemihedrism; as sprays of individuals.

Physical Properties: *Cleavage:* Good on {100}; poor on {010} and {011}. *Tenacity:* Brittle to semi-sectile as a result of alteration. Hardness = 4.5 D(meas.) = 3.5(1) D(calc.) = 3.516 Strongly pyroelectric.

Optical Properties: Transparent to translucent. *Color:* Colorless, milk-white, or colored in various tints as a result of alteration. *Luster:* Vitreous.
Optical Class: Biaxial (+). *Orientation:* $X = b$; $Y = a$; $Z = c$. *Dispersion:* $r < v$, very weak.
$\alpha = 1.656$ $\beta = 1.664$ $\gamma = 1.672$ 2V(meas.) = n.d.   2V(calc.) = 86°

Cell Data: *Space Group:* $Ama2$. a = 12.510(7)   b = 6.318(3)   c = 8.561(6)   Z = 4

X-ray Powder Pattern: Christmas, Arizona, USA.
3.528 (100), 2.816 (100), 2.540 (100), 2.352 (70), 1.540 (60), 4.703 (50), 2.521 (50)

Chemistry:

	(1)	(2)
SiO_2	31.0	33.65
ZnO	44.8	45.59
CaO	15.5	15.71
H_2O	5.8	5.05
Total	97.1	100.00

(1) Christmas, Arizona, USA. (2) $CaZn_2Si_2O_7 \cdot H_2O$.

Occurrence: In a retrogressively altered tactite zone, closely related to the breakdown of sphalerite in the ores.

Association: Kinoite, apophyllite, calcite, xonotlite, smectite.

Distribution: In the USA, from the Christmas copper mine, Gila Co., Arizona.

Name: For Dr. Jun Ito (1926–1978), Japanese-American mineral chemist, Harvard University, Cambridge, Massachusetts, USA.

Type Material: University of Arizona, Tucson, Arizona; Harvard University, Cambridge, Massachusetts, 119097; National Museum of Natural History, Washington, D.C., USA, 136688; University of Paris, Paris; National School of Mines, Paris, France; The Natural History Museum, London, England, 1980,535.

References: (1) Williams, S.A. (1976) Junitoite, a new hydrated calcium zinc silicate from Christmas, Arizona. Amer. Mineral., 61, 1255–1258. (2) Hamilton, R.D. and J.J. Finney (1985) The structure of junitoite, $CaZn_2Si_2O_7 \cdot H_2O$. Mineral. Mag., 49, 91–95.

$NaCa_2[(Mg, Fe^{2+})_4Ti](Si_6Al_2)O_{22}(OH)_2$ **Kaersutite**

Crystal Data: Monoclinic. *Point Group:* $2/m$. Commonly as well-formed phenocrysts with rhombic basal sections; prismatic, to 10 cm; as granular aggregates. *Twinning:* Simple or multiple twinning ∥ {100}.

Physical Properties: *Cleavage:* Perfect on {110}, intersecting at 56° and 124°; partings on {100}, {001}. *Tenacity:* Brittle. Hardness = 5–6   D(meas.) = 3.2–3.28   D(calc.) = [3.11]

Optical Properties: Semitransparent. *Color:* Dark brown to black, typically zoned; yellow-brown, green-brown, or red-brown in thin section. *Luster:* Vitreous. *Optical Class:* Biaxial (–). *Pleochroism:* Strong; X = yellow, yellow-brown; Y = red, red-brown; Z = deep brown, dark red-brown. *Orientation:* Y = b; Z ∧ c = 0°–19°. *Dispersion:* $r > v$. *Absorption:* Z ≥ Y > X. α = 1.670–1.689 β = 1.690–1.741 γ = 1.700–1.772 2V(meas.) = 66°–82°

Cell Data: *Space Group:* $C2/m$. a = 9.8903(3)   b = 18.0596(5)   c = 5.3152(2) β = 105.4(1)°   Z = 2

X-ray Powder Pattern: Boulder Dam area, Arizona, USA. (ICDD 17-478). 2.693 (100), 3.11 (80), 8.38 (65), 3.36 (65), 2.548 (65), 2.589 (55), 1.439 (55)

Chemistry:

	(1)	(2)		(1)	(2)
SiO_2	39.50	39.88	MgO	12.90	14.10
TiO_2	10.33	4.69	CaO	10.91	13.05
Al_2O_3	11.12	14.25	Na_2O	3.82	1.86
Fe_2O_3	0.06	9.58	K_2O	1.43	1.80
FeO	9.44	0.10	F		0.14
MnO	0.10	0.10	H_2O^+	0.59	0.53
			Total	100.20	100.08

(1) Qaersut, Greenland; corresponds to $(Na_{1.11}K_{0.28})_{\Sigma=1.39}Ca_{1.76}(Mg_{2.89}Fe^{2+}_{1.19}Ti_{1.17}$ $Fe^{3+}_{0.01}Mn_{0.01})_{\Sigma=5.27}(Si_{5.93}Al_{1.97})_{\Sigma=7.90}O_{22}(OH)_{0.59}$. (2) Vlčí Hora, Czech Republic; by microanalysis, corresponds to $(Na_{0.53}K_{0.41})_{\Sigma=0.94}Ca_{2.06}(Mg_{3.10}Fe^{3+}_{1.06}Ti_{0.52}Al_{0.34}$ $Fe^{2+}_{0.01}Mn_{0.01})_{\Sigma=5.04}(Si_{5.87}Al_{2.13})_{\Sigma=8.00}O_{22}[(OH)_{1.94}F_{0.06}]_{\Sigma=2.00}$.

Polymorphism & Series: Forms a series with ferro-kaersutite.

Mineral Group: Amphibole (calcic) group: $Mg/(Mg + Fe^{2+}) \geq 0.50$; $Na_B < 0.67$; $(Ca + Na)_B \geq 1.34$; Si < 6.5; Ti ≥ 0.5.

Occurrence: Common as phenocrysts in alkalic volcanic rocks; in gabbroic and peridotitic nodules in alkalic basalts; in syenites, monzonites, carbonatite tuffs, and alkalic gabbros.

Association: Titanian augite, rhönite, olivine, ilmenite, spinel, plagioclase, titanian pargasite.

Distribution: In Greenland, from Østerfjeld, near Qaersut, at Nûgssuaq, and elsewhere in the Kangerdlugssuaq Fjord and Skaergaard areas. On Linosa, Pelagian Islands, south of Sicily, Italy. From Vlčí Hora, Czech Republic. From near Boulder Dam, Mohave Co., and near San Carlos, Gila Co., Arizona, USA. At Mont Saint-Hilaire, Quebec, Canada. From Chikaishi, Oki Island, Shimane Prefecture, and at Mushozu and Numazu, Iki Island, Nagasaki Prefecture, Japan. At Kakanui, New Zealand. A number of other localities are known.

Name: For the occurrence near Qaersut (formerly Kaersut), Greenland.

References: (1) Dana, E.S. (1892) Dana's system of mineralogy, (6th edition), 386, 392. (2) Deer, W.A., R.A. Howie, and J. Zussman (1963) Rock-forming minerals, v. 2, chain silicates, 321–327. (3) Pechar, F., H. Fuess, and W. Joswig (1989) Refinement of the crystal structure of kaersutite (Vlčí Hora, Bohemia) from neutron diffraction. Neues Jahrb. Mineral., Monatsh., 137–143.

Crystal Data: Orthorhombic. *Point Group:* $2/m\ 2/m\ 2/m$. Crystals short, equant, to slender, prismatic, elongated along [001], up to 2 cm; most crystals display a granular core. Also as sheaflike aggregates of crystals.

Physical Properties: *Cleavage:* Good on {110}. *Fracture:* Subconchoidal. *Tenacity:* Brittle. Hardness = 5–6 D(meas.) = 3.52 D(calc.) = 3.542

Optical Properties: Transparent to translucent. *Color:* Colorless, white, straw-yellow, yellow-brown, chestnut-brown, may be tipped with rose or pink. *Luster:* Vitreous to resinous. *Optical Class:* Biaxial (–). *Orientation:* $X = c$; $Y = b$; $Z = a$. $\alpha = 1.662$–1.665 $\beta = 1.682$–1.689 $\gamma = 1.687$–1.691 2V(meas.) = 40° 2V(calc.) = 49.5°

Cell Data: *Space Group:* $Pmnb$. a = 13.011(1) b = 14.310(1) c = 6.757(1) Z = 4

X-ray Powder Pattern: Bancroft, Canada.
6.52 (100), 2.764 (100), 3.29 (80), 3.19 (75), 3.45 (70), 2.170 (70), 1.929 (60)

Chemistry:

	(1)	(2)		(1)	(2)
SiO_2	35.24	34.55	CaO	16.78	16.69
TiO_2	0.02		Na_2O	0.04	
ThO_2	0.03		K_2O	0.01	
Al_2O_3	1.20	1.02	H_2O^+	2.59	[2.68]
Y_2O_3	25.27	25.63	H_2O^-	0.06	
RE_2O_3	12.62	12.65	CO_2	4.60	6.55
Fe_2O_3	0.57	0.23	FeS_2	0.98	
			Total	100.01	[100.00]

(1) Bancroft, Canada; RE and Th by XRF, Fe reported as Fe_2O_3 above that required to form pyrite with S 0.52%; $RE_2O_3 = Ce_2O_3$ 0.38%, Nd_2O_3 0.19%, Sm_2O_3 0.27%, Gd_2O_3 2.44%, Dy_2O_3 3.10%, Er_2O_3 3.53%, Yb_2O_3 2.71%; corresponds to $Ca_{2.02}(Y_{1.50}RE_{0.50})_{\Sigma=2.00}$ $(Si_{4.02}Al_{0.08})_{\Sigma=4.10}O_{12}(CO_3)_{0.70} \cdot 1.93H_2O$. (2) Baveno, Italy; Ca, Y, RE by gamma-ray spectrometry, Si, Al, Fe by AA, CO_2 by gas volumetric analysis, H_2O by difference; $RE_2O_3 = La_2O_3$ 0.24%, Ce_2O_3 0.39%, Pr_2O_3 0.09%, Nd_2O_3 0.20%, Sm_2O_3 0.29%, Eu_2O_3 0.08%, Tb_2O_3 0.68%, Ho_2O_3 0.22%, Tm_2O_3 0.20%, Lu_2O_3 0.21%, Gd_2O_3 1.89%, Dy_2O_3 3.08%, Er_2O_3 2.65%, Yb_2O_3 2.71%; corresponds to $Ca_2(Y_{1.53}RE_{0.45}Fe_{0.02})_{\Sigma=2.00}Si_{3.87}O_{12}(CO_3) \cdot H_2O$.

Occurrence: In vugs in pegmatites in granites and alkalic complexes.

Association: Chamosite, pyrite, quartz, calcite, sphalerite, fluorite, uraninite, zircon, molybdenite (Bancroft, Canada); diopside, magnetite, clinochlore, apatite (Nordmark, Sweden).

Distribution: At Igeltjern, on Hitterö Island, Flekkefjord, Norway. In the Ko mine, Nordmark, Värmland, Sweden. In Switzerland, at Grubhorn, Baltschiedertal, Valais; near Guttannen, Uri; Piz Gannaretsch, Val Curnera, in the Gotthard freeway tunnel, Ticino; at Tavetsch, Graubünden; and elsewhere. From Hopffeldboden, in the Obersulzbachtal, Salzburg, Austria. At Baveno, Piedmont, Italy. In the USA, from near Cotopaxi, Fremont Co., Colorado; in Washington Pass, Okanogan Co., Washington; and east of Porthill, Boundary Co., Idaho. In Canada, from North Burgess Township, Lanark Co., and from the Bicroft mine, near Bancroft, Ontario; in the Evans-Lou quarry, near Hull, and at Mont Saint-Hilaire, Quebec; and in a commercial deposit at the Strange Lake complex, northern Quebec and Labrador, Newfoundland.

Name: From the Greek for *unusual*, for its rarity and exotic composition.

References: (1) Dana, E.S. (1892) Dana's system of mineralogy, (6th edition), 698. (2) Vlasov, K.A., Ed. (1966) Mineralogy of rare elements, v. II, 246–247. (3) Pouliot, G., J.A. Maxwell, and S.C. Robinson (1964) Cenosite from Bancroft, Ontario. Can. Mineral., 8, 1–10. (4) Giuseppetti, G., C. Tadini, and M. Oddone (1989) Cenosite-(Y) from Baveno, Novara (Italy): crystal structure. Neues Jahrb. Mineral., Monatsh., 153–164.

Crystal Data: Tetragonal. *Point Group:* $\bar{4}2m$. As grains up to 2 mm.

Physical Properties: *Cleavage:* Perfect on {110}. Hardness = ~6 VHN = 733–897, 838 average. D(meas.) = 2.5 D(calc.) = 2.48

Optical Properties: Translucent to transparent. *Color:* Colorless with a slight brownish rose tint. *Luster:* Vitreous to pearly on cleavages.
Optical Class: Uniaxial (+). $\omega = 1.525$ $\epsilon = 1.525$; birefringence < 0.001.

Cell Data: *Space Group:* $P\bar{4}2_1c$. a = 9.851(5) c = 13.060(5) Z = 2

X-ray Powder Pattern: Mt. Rasvumchorr, Russia.
3.08 (10), 2.79 (9), 3.44 (8), 2.94 (8), 3.26 (5), 2.24 (5), 2.08 (5)

Chemistry:

	(1)
SiO$_2$	38.54
B$_2$O$_3$	4.98
Al$_2$O$_3$	21.18
K$_2$O	30.06
Cl	3.71
H$_2$O	[3.81]
$-$O = Cl$_2$	0.84
Total	[101.44]

(1) Mt. Rasvumchorr, Russia; by electron microprobe, average of three analyses, H$_2$O from stoichiometry; corresponds to K$_{6.04}$B$_{1.35}$(Al$_{3.93}$Si$_{6.07}$)$_{\Sigma=10.00}$O$_{22.58}$(OH)$_4$Cl$_{0.99}$.

Occurrence: Forms rims around lovozerite in segregations of pectolite, in rischorrite pegmatite from a differentiated alkalic massif.

Association: Pectolite, lovozerite.

Distribution: On Mt. Rasvumchorr, Khibiny massif, Kola Peninsula, Russia.

Name: For potassium, Kalium, ALuminum, BORate, and SIlicon in the composition.

Type Material: A.E. Fersman Mineralogical Museum, Academy of Sciences, Moscow, Russia.

References: (1) Khomyakov, A.P., S.M. Sandomirskaya, and Y.A. Malinovskii (1980) Kalborsite, K$_6$BAl$_4$Si$_6$O$_{20}$(OH)$_4$Cl, a new mineral. Doklady Acad. Nauk SSSR, 252, 1465–1468 (in Russian). (2) Malinovskii, Y.A. and N.V. Belov (1980) Crystal structure of kalborsite. Doklady Acad. Nauk SSSR, 252, 611–615 (in Russian). (3) (1981) Amer. Mineral., 66, 879 (abs. refs. 1 and 2).

Kaliophilite

KAlSiO$_4$

Crystal Data: Hexagonal. *Point Group:* 622. In bundles of slender acicular to threadlike crystals.

Physical Properties: *Cleavage:* {0001}, perfect. *Tenacity:* Brittle. Hardness = 6
D(meas.) = 2.493–2.649   D(calc.) = [2.65]

Optical Properties: Transparent. *Color:* Colorless. *Luster:* Silky, brilliant.
Optical Class: Uniaxial (–). $\omega = 1.536$ $\epsilon = 1.531$

Cell Data: *Space Group:* $P6_322$.   a = 26.930   c = 8.522   Z = 54

X-ray Powder Pattern: Vesuvius, Italy.
3.09 (100), 2.593 (30), 2.131 (25), 4.26 (15), 3.76 (12), 3.02 (12), 2.814 (12D)

Chemistry:

	(1)	(2)	(3)
SiO$_2$	39.2	38.87	37.99
TiO$_2$		0.00	
Al$_2$O$_3$	33.36	31.89	32.23
FeO		0.07	
MnO		0.00	
CaO	0.47	0.05	
Na$_2$O	2.88	0.77	
K$_2$O	24.13	28.76	29.78
Total	100.04	100.41	100.00

(1) Monte Somma, Italy. (2) Do.; by electron microprobe. (3) KAlSiO$_4$.

Polymorphism & Series: Polymorphous with kalsilite, panunzite, and trikalsilite.

Occurrence: In blocks of biotite-pyroxenite volcanic ejecta.

Association: Biotite, pyroxene, augite, melilite, calcite (Monte Somma, Italy); clinopyroxene, garnet, leucite, haüyne, latiumite (Albano, Italy).

Distribution: In Italy, from Monte Somma and Vesuvius, Campania; at Albano and Marino, Colle Cimino, near Rome, Lazio.

Name: From the Greek for potassium, *kalium*, and for *friend*, recognizing potassium's presence.

References: (1) Dana, E.S. (1892) Dana's system of mineralogy, (6th edition), 427.
(2) Bannister, M.A. (1931) A chemical, optical, and X-ray study of nepheline and kaliophilite. Mineral. Mag., 22, 569–608. (3) Smith, J.V. and O.F. Tuttle (1957) The nepheline–kalsilite system: I. X-ray data for the crystalline phases. Amer. J. Sci., 255, 282–305. (4) Deer, W.A., R.A. Howie, and J. Zussman (1963) Rock-forming minerals, v. 4, framework silicates, 231–270. (5) Abbott, R.N., Jr. (1984) KAlSiO$_4$ stuffed derivatives of tridymite: phase relationships. Amer. Mineral., 69, 449–457.

Crystal Data: Orthorhombic, probably, twinned to a pseudohexagonal aspect. *Point Group:* 622, apparent. As anhedral embedded grains and as patchy to fine intergrowths with nepheline or potassic feldspar; massive. *Twinning:* Ubiquitous on a microscopic scale, on $\{10\bar{1}0\}$, $\{33\bar{6}5\}$, and $\{11\bar{2}2\}$.

Physical Properties: *Cleavage:* Poor on $\{10\bar{1}0\}$, $\{0001\}$. *Tenacity:* Brittle. Hardness = 6 D(meas.) = 2.59–2.62 D(calc.) = [2.62]

Optical Properties: Transparent to translucent. *Color:* Colorless, white, gray; in thin section, colorless. *Luster:* Vitreous to greasy.
Optical Class: Uniaxial (–). $\omega = 1.535$–1.554 $\epsilon = 1.530$–1.539

Cell Data: *Space Group:* $P6_322$ apparent. a = 5.16 c = 8.69 Z = 2

X-ray Powder Pattern: Synthetic.
3.118 (100), 2.579 (50), 3.973 (45), 2.175 (17), 2.472 (15), 4.351 (12), 2.432 (10)

Chemistry:

	(1)	(2)
SiO$_2$	39.6	37.99
Al$_2$O$_3$	21.3	32.23
Fe$_2$O$_3$	5.9	
MgO	3.7	
CaO	5.0	
Na$_2$O	1.6	
K$_2$O	20.1	29.78
Total	97.2	100.00

(1) Mafuru crater, Uganda. (2) KAlSiO$_4$.

Polymorphism & Series: Polymorphous with kaliophilite, panunzite, and trikalsilite.

Occurrence: In the groundmass of some potassium-rich and silica-deficient lavas and tuffs; rare in syenites; may be formed by the breakdown of silica-rich leucite in alkalic ultramafic rocks; from a granulite-facies emery deposit.

Association: Olivine, melilite, clinopyroxene, phlogopite, nepheline, leucite.

Distribution: In the Mafuru and Katunga craters, Uganda. From Zaire, in the Baruta crater, Nyiragongo area, Kivu Province. From Italy, at Monte Somma and Vesuvius, Campania; Ariccia, near Rome, Lazio; and in the San Venanzo and Pian di Celle volcanos, Umbria. At Ödersdorf, Eifel district, Germany. In the Batbjerg alkalic ultramafic intrusion, Kangerdlugssuaq Glacier, Greenland. From the Khibiny massif, Kola Peninsula; the Murun massif, southwest of Olekminsk, Yakutia; and in the Tazheran alkalic intrusive, Baikal, eastern Siberia, Russia. From the Punalar district, Kerala, India. In the Colima graben, about 50 km south-southwest of Guadalajara, Jalisco, Mexico. A few other localities are known.

Name: For potassium, Kalium, ALuminum, and SILicate in the composition.

Type Material: The Natural History Museum, London, England, 1942,42; Harvard University, Cambridge, Massachusetts, 101998; National Museum of Natural History, Washington, D.C., USA, 105856.

References: (1) Bannister, F.A. and M.H. Hey (1942) Kalsilite, a polymorph of KAlSiO$_4$ from Uganda. Mineral. Mag., 26, 218–224. (2) Smith, J.V. and O.F. Tuttle (1957) The nepheline-kalsilite system. I. X-ray data for the crystalline phases. Amer. J. Sci., 255, 282–305. (3) Deer, W.A., R.A. Howie, and J. Zussman (1963) Rock-forming minerals, v. 4, framework silicates, 231–269. (4) Perrotta, A.J. and J.V. Smith (1965) The crystal structure of kalsilite, KAlSiO$_4$. Mineral. Mag., 35, 588–595. (5) Capobianco, C. and M. Carpenter (1989) Thermally induced changes in kalsilite (KAlSiO$_4$). Amer. Mineral., 74, 797–811.

Crystal Data: Tetragonal. *Point Group:* n.d. As grains up to about 0.1 mm.

Physical Properties: Hardness = n.d. D(meas.) = n.d. D(calc.) = [2.82]

Optical Properties: Transparent. *Color:* Colorless.
Optical Class: Isotropic, nearly. $n = 1.629$

Cell Data: *Space Group:* Body-centered cell. $a = 8.850$ $c = 8.770$ $Z = 4$

X-ray Powder Pattern: Kamaishi mine, Japan.
3.607 (100), 2.799 (85), 2.777 (40), 2.547 (40), 1.557 (35), 2.085 (30)

Chemistry:

	(1)
SiO_2	20.03
Al_2O_3	34.15
FeO	0.21
MgO	0.02
CaO	37.42
H_2O^+	6.1
H_2O^-	0.2
Total	98.13

(1) Kamaishi mine, Japan; by electron microprobe, H_2O by wet chemical analysis; corresponds to $(Ca_{1.99}Fe_{0.01})_{\Sigma=2.00}Al_{2.00}Si_{0.99}O_{5.98}(OH)_{2.02}$.

Polymorphism & Series: Dimorphous with bicchulite.

Occurrence: In vesuvianite skarn in a marble, apparently altering from vesuvianite.

Association: Vesuvianite, perovskite, calcite, magnetite, chalcopyrite.

Distribution: At the Kamaishi mine, Iwate Prefecture, Japan.

Name: For the Kamaishi mine, Japan.

Type Material: National Science Museum, Tokyo, Japan, M23560.

References: (1) Uchida, E. and J.T. Iiyama (1981) On kamaishilite, $Ca_2Al_2SiO_6(OH)_2$, a new mineral (tetragonal), dimorphous with bicchulite, from the Kamaishi mine, Japan. Proc. Japan Acad., 57B, 239–243 (in English). (2) (1982) Amer. Mineral., 67, 855 (abs. ref. 1).

Crystal Data: Orthorhombic. *Point Group:* $2/m\ 2/m\ 2/m$. Crystals platy with {010} dominant; in spherulites, to 2 mm, isolated and as aggregations.

Physical Properties: *Cleavage:* Perfect on {010}, good on {100}. Hardness = $\sim$4 D(meas.) = 1.926(4)   D(calc.) = 1.933

Optical Properties: Translucent to transparent. *Color:* White to brown. *Luster:* Silky. *Optical Class:* Biaxial (–). *Orientation:* $X = b$; $Y = a$; $Z = c$. *Dispersion:* $r > v$. $\alpha = 1.451(2)$ $\beta = 1.470(2)$ $\gamma = 1.478(2)$ 2V(meas.) = 46(2)°

Cell Data: *Space Group:* $Pnmb$.   a = 7.282(2)   b = 20.507(5)   c = 4.956(1)   Z = 4

X-ray Powder Pattern: Andjia, Chad.
10.33 (10), 4.014 (10), 3.435 (9), 2.480 (8), 3.162 (7), 3.093 (7), 2.386 (6)

Chemistry:

	(1)	(2)
SiO$_2$	56.68	57.5
Al$_2$O$_3$	0.40	
CaO	0.20	0.5
Na$_2$O	13.25	13.2
K$_2$O	0.10	1.2
H$_2$O($> 280°$)	4.25	
H$_2$O($170° - 280°$)	8.45	
H$_2$O($< 170°$)	16.30	
H$_2$O		27.5
Total	99.63	99.9

(1) Andjia, Chad; corresponding to $(Na_{0.93}K_{0.07})_{\Sigma=1.00}H_{1.03}(Si_{2.06}Al_{0.01})_{\Sigma=2.07}O_{4.13}$ $(OH)_{2.04}\cdot1.97H_2O$. (2) Lake Bogoria, Kenya; by AA, H$_2$O by TGA.

Occurrence: In evaporites in an interdunary depression (Andjia, Chad).

Association: Gaylussite, trona (Andjia, Chad); magadiite, analcime, mordenite (Lake Bogoria, Kenya).

Distribution: In the Kanem region, near Andjia, Chad. From Lake Bogoria, Rift Valley, Kenya.

Name: After Kanem, a region on the northeastern edge of Lake Chad, Chad, where the mineral was discovered.

Type Material: National School of Mines, Paris, France; National Museum of Natural History, Washington, D.C., USA, 123219, 142501.

References: (1) Johan, Z. and G.F. Maglione (1972) La kanemite, nouveau silicate de sodium hydraté de néoformation. Bull. Soc. fr. Minéral., 95, 371–382 (in French with English abs.). (2) (1974) Amer. Mineral., 59, 210 (abs. ref. 1). (3) Perinet, G., J.-J. Tiercelin, and C.E. Barton (1982) Présence de kanemite dans les sédiments récents du lac Bogoria, Rift Gregory, Kenya. Bull. Minéral., 105, 633–639 (in French with English abs.). (4) Beneke, K. and G. Lagaly (1977) Kanemite – innercrystalline reactivity and relations to other sodium silicates. Amer. Mineral., 62, 763–771.

Kanoite $(Mn^{2+}, Mg)_2Si_2O_6$

Crystal Data: Monoclinic. *Point Group:* $2/m$, (probable). As grains up to 0.1 mm.
Twinning: Polysynthetic on {100}, common.

Physical Properties: *Cleavage:* Perfect on {110}, (110) ∧ (1$\bar{1}$0) ~88°. Hardness = 6
D(meas.) = 3.66   D(calc.) = 3.60

Optical Properties: Semitransparent. *Color:* Light pinkish brown; colorless in thin section.
Streak: White. *Luster:* Vitreous.
Optical Class: Biaxial (+). *Orientation:* $Y = b$; $Z \wedge c = 42°$. $\alpha = 1.715(2)$ $\beta = 1.717(2)$
$\gamma = 1.728(2)$ 2V(meas.) = 40°–42°

Cell Data: *Space Group:* $P2_1/c$ (probable).   a = 9.739   b = 8.939   c = 5.260
$\beta = 108.56°$ Z = 4

X-ray Powder Pattern: Tatehira, Japan.
3.211 (100), 3.021 (90), 2.910 (90), 2.921 (80), 2.493 (40), 1.627 (40), 2.573 (30)

Chemistry:

	(1)
SiO_2	50.20
Al_2O_3	0.04
Fe_2O_3	0.39
FeO	2.64
MnO	31.19
MgO	15.08
CaO	[0.61]
Na_2O	0.03
K_2O	0.03
Total	[100.21]

(1) Tatehira, Japan; by electron microprobe, average of three analyses; Fe_2O_3, FeO, Na_2O,
and K_2O by wet chemical analysis; original CaO 0.57% and original total given as 100.17%;
corresponds to $(Mn^{2+}_{1.04}Mg_{0.88}Fe^{2+}_{0.09}Ca_{0.02}Fe^{3+}_{0.01})_{\Sigma=2.04}Si_{1.97}O_6$.

Polymorphism & Series: Dimorphous with donpeacorite.

Mineral Group: Pyroxene group.

Occurrence: In a seam cutting a pyroxmangite-cummingtonite metamorphic rock.

Association: Spessartine, manganoan cummingtonite, pyroxmangite.

Distribution: From near Tatehira, Oshima Peninsula, Hokkaido, Japan.

Name: To honor Dr. Hiroshi Kano, Professor of Petrology, Akita University, Akita, Japan.

Type Material: Shimane University, Matuse; National Science Museum, Tokyo, Japan,
M21331.

References: (1) Kobayashi, H. (1977) Kanoite, $(Mn^{2+}, Mg)_2[Si_2O_6]$, a new clinopyroxene in the
metamorphic rock from Tatehira, Oshima Peninsula, Hokkaido, Japan. J. Geol. Soc. Japan, 83,
537–542. (2) (1978) Amer. Mineral., 63, 598 (abs. ref. 1).

Crystal Data: Orthorhombic. *Point Group:* $2/m\ 2/m\ 2/m$. As anhedral nodular porphyroblasts, to 1.2 cm.

Physical Properties: *Cleavage:* Poor on {110}. Hardness = 6.5   VHN = 906–1017 (100 g load). D(meas.) = n.d. D(calc.) = 3.395

Optical Properties: Transparent to translucent. *Color:* Greenish black. *Streak:* Gray-green. *Luster:* Vitreous.
Optical Class: Biaxial (+). *Pleochroism:* Strong; X = yellow-green; Y = bluish green; Z = deep golden yellow. *Orientation:* $X = a$; $Y = b$; $Z = c$. $\alpha = 1.702$ $\beta = 1.730$ $\gamma = 1.823$ 2V(meas.) = 53(3)°

Cell Data: *Space Group:* $Pnnm$.   a = 7.959(2)   b = 8.047(2)   c = 5.616(1)   Z = 4

X-ray Powder Pattern: Kanona, Zambia.
5.669 (100), 2.827 (94), 3.577 (90), 2.517 (90), 2.212 (83), 4.590 (75), 2.299 (69)

Chemistry:

	(1)	(2)
SiO_2	32.2	32.48
TiO_2	0.01	
Al_2O_3	33.9	31.60
Fe_2O_3	0.66	2.57
Mn_2O_3	32.2	33.48
CuO	0.01	
ZnO	0.13	
PbO	0.01	
MgO	0.04	
CaO	0.01	
BaO	0.04	
Total	99.21	100.13

(1) Kanona, Zambia; by electron microprobe; corresponds to $(Mn^{3+}_{0.76}Al_{0.23}Fe^{3+}_{0.02})_{\Sigma=1.01}$ $Al_{1.00}Si_{0.99}O_5$. (2) Salmchâteau, Belgium; by electron microprobe, corresponds to $(Mn^{3+}_{0.79}Al_{0.15}$ $Fe^{3+}_{0.06})_{\Sigma=1.00}Al_{1.00}Si_{1.00}O_5$.

Polymorphism & Series: Forms a series with andalusite.

Occurrence: Formed under low-grade metamorphic conditions, in a gahnite schist (Kanona, Zambia); in a schist (Salmchâteau, Belgium).

Association: Gahnite, magnesian chlorite, coronadite, braunite (Kanona, Zambia); quartz, muscovite, paragonite, chlorite, hematite, braunite, rutile, apatite (Salmchâteau, Belgium).

Distribution: From 13 km north of Kanona, Serenje, Zambia. On the Salm River, one km north of Salmchâteau, near Ottré, Ardennes Mountains, Belgium.

Name: For Kanona, near the locality in Zambia.

Type Material: Charles University, Prague, Czech Republic, 17352; National Museum of Natural History, Washington, D.C., USA, 144523.

References: (1) Vrána, S., M. Rieder, and J. Podlaha (1978) Kanonaite, $(Mn^{3+}_{0.76}Al_{0.23}Fe^{3+}_{0.02})^{[6]}$ $Al^{[5]}[O|SiO_4]$, a new mineral isotypic with andalusite. Contr. Mineral. Petrol., 66, 325–332. (2) (1979) Amer. Mineral., 64, 655 (abs. ref. 1). (3) Kramm, U. (1979) Kanonaite-rich viridines from the Venn-Stavelot massif, Belgian Ardennes. Contr. Mineral. Petrol., 69, 387–395. (4) Weiss, Z., S.W. Bailey, and M. Rieder (1981) Refinement of the crystal structure of kanonaite, $(Mn^{3+}, Al)^{[6]}(Al, Mn^{3+})^{[5]}O[SiO_4]$. Amer. Mineral., 66, 561–567. (5) Gunter, M. and F.D. Bloss (1982) Andalusite-kanonaite series: lattice and optical parameters. Amer. Mineral., 67, 1218–1228.

Crystal Data: Triclinic. *Point Group:* 1. Rarely as crystals, thin platy or stacked, to 2 mm. More commonly as microscopic pseudohexagonal plates and clusters of plates, aggregated into compact, claylike masses.

Physical Properties: *Cleavage:* Perfect on {001}. *Tenacity:* Flexible but inelastic. Hardness = 2–2.5 D(meas.) = 2.61–2.68 D(calc.) = 2.63

Optical Properties: Transparent to translucent as single crystals. *Color:* White to tan, may be variously colored by impurities. *Luster:* Pearly to dull earthy.
Optical Class: Biaxial (–). *Orientation:* $X \wedge c = -13°$ to $-10°$; $Y \wedge a = 1°–4°$.
Dispersion: $r > v$, weak. $\alpha = 1.553–1.565$ $\beta = 1.559–1.569$ $\gamma = 1.560–1.570$
2V(meas.) = 24°–50°

Cell Data: *Space Group:* $P1$. a = 5.15 b = 8.95 c = 7.39 $\alpha = 91.8°$
$\beta = 104.5°–105.0°$ $\gamma = 90°$ Z = [2]

X-ray Powder Pattern: Scalby, Yorkshire, England (1A).
7.16 (vvs), 3.573 (vvs), 4.336 (vs), 2.491 (s), 2.289 (s), 2.558 (ms), 2.379 (ms)

Chemistry:

	(1)
SiO_2	45.80
Al_2O_3	39.55
Fe_2O_3	0.57
FeO	0.18
MgO	0.14
CaO	0.41
K_2O	0.03
H_2O^+	13.92
H_2O^-	0.17
Total	100.77

(1) Mikawo mine, Niigata Prefecture, Japan; corresponds to $(Al_{2.00}Fe^{3+}_{0.02}Mg_{0.01}Ca_{0.02})_{\Sigma=2.05}$ $Si_2O_5(OH)_{3.99}$.

Polymorphism & Series: Dickite, halloysite, and nacrite are polymorphs.

Mineral Group: Kaolinite-serpentine group.

Occurrence: Replaces other aluminosilicate minerals during hydrothermal alteration and weathering. A common constituent of the clay-size fraction of sediments, where it may be formed by direct precipitation.

Association: Quartz, feldspar, muscovite.

Distribution: Pure material from many localities, including: at Kauling, Kiangsi Province, China. In numerous china-clay pits in Cornwall and Devon, England. At Limoges, Haute-Vienne, France. Near Dresden, Kemmlitz, and Zettlitz, Saxony, and elsewhere in Germany. Large deposits in the Donets basin, Ukraine. In the USA, at Macon, Bibb Co., Georgia; at the Dixie Clay Company mine, and in the Lamar Pit, near Bath, Aikin Co., South Carolina; near Webster, Jackson Co., North Carolina; near Murfreesboro, Pike Co., and at Greenwood, Sebastian Co., Arkansas; from Mesa Alta, Rio Arriba Co., New Mexico. At Huberdeau, Quebec, and near Walton, Nova Scotia, Canada.

Name: From a corruption of the Chinese *Kauling*, *high ridge*, for a Chinese occurrence.

References: (1) Dana, E.S. (1892) Dana's system of mineralogy, (6th edition), 684–687. (2) Deer, W.A., R.A. Howie, and J. Zussman (1963) Rock-forming minerals, v. 3, sheet silicates, 194–212. (3) Brindley, G.W. and K. Robinson (1946) The structure of kaolinite. Mineral. Mag., 27, 242–253. (4) Goodyear, J. and W.J. Duffin (1961) An X-ray examination of an exceptionally well crystallized kaolinite. Mineral. Mag., 32, 902–907. (5) Young, R.A. and A.W. Hewat (1988) Verification of the triclinic crystal structure of kaolinite. Clays and Clay Minerals, 36, 225–232.

Crystal Data: Amorphous; hexagonal (?) *Point Group:* n.d. Platy crystals up to 1 cm; as aggregates and individual grains.

Physical Properties: *Cleavage:* Good in one direction, imperfect in another.
Tenacity: Brittle. Hardness = 2 D(meas.) = 2.89–2.95 D(calc.) = n.d.

Optical Properties: Semitransparent. *Color:* Honey-yellow when fresh to pale yellow when altered; in thin section, pale yellow. *Streak:* Yellow. *Luster:* Greasy.
Optical Class: Uniaxial (–); may be anomalously slightly biaxial. $\omega = 1.617$ $\epsilon = 1.595$

Cell Data: *Space Group:* n.d. Z = n.d.

X-ray Powder Pattern: Mt. Karnasurt, Russia; X-ray amorphous; on heating to 900°, it yields a pattern close to that of monazite.
3.10 (7), 2.88 (7), 3.29 (6), 3.49 (5), 1.723 (5), 2.44 (4), 2.14 (4)

Chemistry:

	(1)	(2)		(1)	(2)
SiO$_2$	20.47	24.91	CaO	1.66	3.20
TiO$_2$	12.14	12.33	BaO	0.68	
ZrO$_2$		1.20	Na$_2$O	0.91	
ThO$_2$	6.22	6.04	K$_2$O	0.76	
Al$_2$O$_3$	7.46	5.52	F	0.91	
La$_2$O$_3$		8.55	H$_2$O$^+$	7.22	7.29
Ce$_2$O$_3$	17.58	8.11	H$_2$O$^-$	10.84	12.29
Fe$_2$O$_3$	1.10	1.07	P$_2$O$_5$	3.29	6.81
Nb$_2$O$_5$	8.20	2.20	$-$O = F$_2$	0.38	
MgO	0.91	0.60	Total	99.97	100.12

(1) Mt. Karnasurt, Russia; corresponds to $(Ce_{0.54}Na_{0.15}Ca_{0.14}Th_{0.12}K_{0.08}Ba_{0.02})_{\Sigma=1.05}$ $(Ti_{0.76}Nb_{0.31})_{\Sigma=1.07}(Al_{0.73}Mg_{0.11}Fe^{3+}_{0.07})_{\Sigma=0.91}(Si_{1.70}P_{0.23})_{\Sigma=1.93}O_7[(OH)_{3.46}F_{0.23}]_{\Sigma=3.69}\cdot$3H$_2$O.
(2) Mt. Punkaruaiv, Russia.

Occurrence: In the intermediate replacement zone, composed largely of microcline, of a zoned pegmatite stock in a differentiated alkalic massif (Mt. Karnasurt, Russia).

Association: Manganoan pectolite, natrolite, epididymite, polylithionite, ussingite.

Distribution: On Mts. Karnasurt and Punkaruaiv, Lovozero massif, Kola Peninsula, Russia.

Name: For the locality, Mt. Karnasurt, Kola Peninsula, Russia, and its high *cerium* content.

Type Material: National Museum of Natural History, Washington, D.C., USA, 143823.

References: (1) Kuz'menko, M.V. and S.I. Kozhanov (1959) The new mineral karnasurtite. Trudy Inst. Mineral., Geokhim. i Kristallokhim. Redkikh Elementov, 2, 95–98 (in Russian). (2) Vlasov, K.A., M.V. Kuz'menko, and E.M. Es'kova (1959) The Lovozero alkaline massif. Akad. Nauk SSSR, 423–425 (in Russian). (3) (1960) Amer. Mineral., 45, 1133–1134 (abs. refs. 1 and 2).

Crystal Data: Monoclinic (?). *Point Group:* n.d. As minute plates and monoclinic prisms, to 0.8 mm; appears to be cryptocrystalline.

Physical Properties: Hardness = 2.5–3 D(meas.) = 2.53–2.63 D(calc.) = n.d.

Optical Properties: Semitransparent. *Color:* Colorless, light blue to deep greenish blue. *Luster:* Dull to weakly greasy.
Optical Class: Biaxial (–). *Orientation:* $X \wedge c = 0°–12°$. $\alpha = 1.553–1.570$ $\beta = $ n.d.
$\gamma = 1.569–1.594$ 2V(meas.) = n.d.

Cell Data: *Space Group:* n.d. Z = n.d.

X-ray Powder Pattern: Nizhni Tagil massif, Russia.
11., 7.71, 4.76, 3.75, 1.555 [strongest lines]

Chemistry:

	(1)
SiO_2	47.55
Al_2O_3	0.48
NiO	21.12
CuO	0.01
MgO	17.56
CaO	0.80
H_2O^+	6.50
H_2O^-	3.50
LOI	2.30
Total	99.82

(1) Nizhni Tagil massif, Russia; corresponds to $(Mg_{1.30}Ni_{0.70})_{\Sigma=2.00}Si_2O_5(OH)_2$.

Occurrence: As veinlets in "kerolitized" serpentinite.

Association: n.d.

Distribution: In the Nizhni Tagil massif, Ural Mountains, Russia.

Name: For Alexander Petrovich Karpinsky (1846–1936), Russian geologist and President of the Russian Academy of Sciences.

Type Material: n.d.

References: (1) Rukavishnikova, I.A. (1956) Some magnesium-nickel hydrous silicates of the Nizhne-Tagilsk serpentine massif. Kora Vyvetrivaniya [The crust of weathering], 2, 124–178 (in Russian). (2) (1957) Amer. Mineral., 42, 584 (abs. ref. 1).

Crystal Data: Monoclinic. *Point Group:* $2/m$. Crystals stout prismatic, flattened on {001} and lathlike by extension along [010], to several mm. As groups of divergent crystals, rosettes, and radial fibrous aggregates; in dense gumlike crusts and compact, colloform masses.

Physical Properties: *Cleavage:* Perfect on {001}, indistinct on {100} and {010}. *Tenacity:* Brittle. Hardness = 4–5 D(meas.) = 5.83–6.5 D(calc.) = 6.256 Radioactive.

Optical Properties: Transparent to translucent, opaque. *Color:* Ocher-yellow to brownish yellow; amber-brown when transparent; rarely lemon-yellow to green or reddish orange. *Streak:* Pale yellow-brown. *Luster:* Subadamantine to greasy, dull to earthy when massive. *Optical Class:* Biaxial (+). *Pleochroism:* Weak, rarely; $X = Y =$ very pale yellow; $Z =$ colorless to slightly grayish. *Orientation:* $X = b$; $Z \simeq c$. $\alpha = 1.877$–1.900 $\beta = 1.880$–1.910 $\gamma = 1.935$–1.970 2V(meas.) = 43°

Cell Data: *Space Group:* $P2_1/a$. a = 13.24 b = 6.94 c = 6.70 $\beta = 104°20'$ Z = 4

X-ray Powder Pattern: Kasolo, Zaire.
3.26 (10), 2.93 (9), 4.19 (8b), 3.53 (7), 6.61 (6), 3.07 (5), 1.962 (5)

Chemistry:

	(1)	(2)
SiO_2	9.42	10.23
UO_3	49.28	48.70
Fe_2O_3	0.41	
PbO	36.2	38.00
MgO	0.03	
CaO	0.06	
H_2O	3.59	3.07
CO_2	0.85	
Total	99.84	100.00

(1) Kasolo, Zaire. (2) $Pb(UO_2)SiO_4 \cdot H_2O$.

Occurrence: An oxidization product of uraninite.

Association: Uraninite, torbernite, curite, dewindtite (Kasolo, Zaire); rutherfordine, sklodowskite, curite (Nabarlek, Australia).

Distribution: In the Shinkolobwe mine, Kasolo, and the Musonoi mine, Kolwezi, Shaba Province, Zaire. From Mounana, Gabon. In Germany, at Wölsendorf, Bavaria. In France, at Kersegalec, Lignol, Morbihan; Grury, Saône-et-Loire; and at Reliez and Bigay, Lachaux, Puy-de-Dôme. In the Nabarlek uranium deposit, Northern Territory, and at Wodgina, Pilbara district, Western Australia. In the USA, in the Ruggles pegmatite, near Grafton Center, Grafton Co., New Hampshire; from the Green Monster mine, Goodsprings district, Clark Co., and the 4-D mine, Marietta, Mineral Co., Nevada; and in the Ross-Adams mine, Bokan Mountain, Prince of Wales Island, Alaska. In Canada, at Great Bear Lake, Northwest Territories, and at Lake Athabasca, Saskatchewan, from the Nicholson mines. In the Santo Domingo mine, Santa Eulalia, Chihuahua, Mexico. A few other minor localities are known.

Name: For the occurrence at Kasolo, Zaire.

Type Material: n.d.

References: (1) Schoep, A. (1921) Sur la kasolite, nouveau minéral radioactif. Compt. Rendus Acad. Sci. Paris, 173, 1476–1477 (in French). (2) (1922) Amer. Mineral., 7, 128–129 (abs. ref. 1). (3) Frondel, C. (1958) Systematic mineralogy of uranium and thorium. U.S. Geol. Sur. Bull. 1064, 315–319. (4) Stohl, F.V. and D.K. Smith (1981) The crystal chemistry of the uranyl silicate minerals. Amer. Mineral., 66, 610–625.

Crystal Data: Cubic. *Point Group:* $4/m\,\bar{3}\,2/m$. Octahedra crystals, typically rounded, to 0.3 mm, in thin crusts of columnar aggregates.

Physical Properties: Hardness = n.d. D(meas.) = n.d. D(calc.) = 2.76

Optical Properties: Transparent to translucent. *Color:* Milky white; in transmitted light, colorless.
Optical Class: Isotropic; may be weakly birefringent. $n = 1.632(1)$

Cell Data: *Space Group: Ia3d.* a = 12.358(2) Z = 8

X-ray Powder Pattern: Campomorto quarry, Italy.
2.763 (100), 2.257 (58), 2.004 (58), 3.089 (50), 5.046 (37), 1.6507 (37), 3.303 (32)

Chemistry:

	(1)
SiO_2	10.58
Al_2O_3	24.01
MgO	0.07
CaO	42.27
H_2O	20.8
SO_3	2.27
Total	[100.00]

(1) Campomorto quarry, Italy; by electron microprobe, H_2O by TGA, recalculated slightly to 100.00%; corresponds to $Ca_{2.96}(Al_{1.85}Mg_{0.01})_{\Sigma=1.86}(Si_{0.69}S_{0.11})_{\Sigma=0.80}[(OH)_{9.07}O_{2.93}]_{\Sigma=12.00}$.

Polymorphism & Series: Forms a series with grossular and hibschite.

Mineral Group: Garnet group.

Occurrence: A hydrothermal mineral in cavities in a phonolitic lava flow that erupted through an argillaceous marl.

Association: Tobermorite, afwillite, gehlenite, hydrocalumite, "opal," portlandite, apophyllite, cordierite, jennite, strätlingite, chabazite, gismondine, phillipsite, vertumnite, ettringite, garnet, wollastonite, gypsum, calcite, quartz, hematite.

Distribution: In the Campomorto quarry, near Montalto di Castro, Lazio, Italy.

Name: In honor of Akira Kato, mineralogist of the National Science Museum, Tokyo, Japan.

Type Material: Municipal Museum of Natural History, Milan, Italy; National Museum of Natural History, Washington, D.C., USA, 163797.

References: (1) Passaglia, E. and R. Rinaldi (1984) Katoite, a new member of the $Ca_3Al_2(SiO_4)_3 - Ca_3Al_2(OH)_{12}$ series and a new nomenclature for the hydrogrossular group of minerals. Bull. Minéral., 107, 605–618. (2) (1985) Amer. Mineral., 70, 873 (abs. ref. 1).

$Na[NaCa][(Fe^{2+}, Mg)_4Al](Si_7Al)O_{22}(OH)_2$ **Katophorite**

Crystal Data: Monoclinic. *Point Group:* $2/m$. Prismatic, granular, also as fibrous aggregates or dendrites and skeletal crystals. Commonly rimming other minerals. *Twinning:* ∥ {100}.

Physical Properties: *Cleavage:* Perfect on {110}, intersecting at ~56° and ~124°; parting on {010}. *Tenacity:* Brittle. Hardness = 5–6 D(meas.) = 3.2–3.5 D(calc.) = [3.31]

Optical Properties: Transparent to translucent. *Color:* Black, dark green-black, bluish black; reddish yellow, bluish green in thin section. *Luster:* Vitreous.
Optical Class: Biaxial (–). *Pleochroism:* Strong; reddish yellow, reddish brown, and dark green to black. *Orientation:* $Z = b$. *Dispersion:* $r > v$. *Absorption:* $Z > Y > X$. $\alpha = 1.639–1.681$ $\beta = 1.658–1.688$ $\gamma = 1.600–1.690$ 2V(meas.) = ~0°–50°

Cell Data: *Space Group:* $C2/m$. $a = 10.019(2)$ $b = 18.036(7)$ $c = 5.286(3)$ $\beta = 104.98(3)°$ $Z = 2$

X-ray Powder Pattern: n.d.

Chemistry:

	(1)	(2)		(1)	(2)
SiO₂	48.04	45.98	MnO	1.11	1.12
TiO₂	2.09	1.92	MgO	6.42	5.21
Al₂O₃	3.86	2.79	CaO	8.08	5.76
Cr₂O₃	0.09		Na₂O	4.18	4.85
FeO	25.58	28.97	K₂O	1.34	1.23
			Total	100.79	97.83

(1) Baie-des-Moutons complex, Canada; by electron microprobe, corresponds to $(Na_{0.54}K_{0.26})_{\Sigma=0.80}(Ca_{1.31}Na_{0.69})_{\Sigma=2.00}(Fe^{2+}_{3.25}Mg_{1.45}Ti_{0.23}Mn_{0.15})_{\Sigma=5.08}(Si_{7.29}Al_{0.69}$ $Ti_{0.01}Cr_{0.01})_{\Sigma=8.00}O_{22}(OH)_2$. (2) Rallier-du-Baty Peninsula, Kerguelen Island; by electron microprobe, corresponds to $(Na_{0.80}K_{0.25})_{\Sigma=1.05}(Ca_{0.98}Na_{0.70}Fe^{2+}_{0.32})_{\Sigma=2.00}(Fe^{2+}_{3.54}Mg_{1.24}$ $Mn_{0.15}Ti_{0.07})_{\Sigma=5.00}(Si_{7.32}Al_{0.52}Ti_{0.16})_{\Sigma=8.00}O_{22}(OH)_2$.

Polymorphism & Series: Forms a series with magnesio-katophorite.

Mineral Group: Amphibole (sodic-calcic) group: $Mg/(Mg + Fe^{2+}) < 0.5$; $(Na + K)_A \geq 0.5$; $0.67 \leq Na_B \leq 1.33$; $(Ca + Na)_B \geq 1.34$; $6.5 \leq Si \leq 7.49$.

Occurrence: In alkalic volcanic and plutonic igneous rocks; in blueschist facies jadeitites.

Association: Arfvedsonite, aegirine, nepheline, pyroxenes, eckermannite, chromite.

Distribution: May occur in the Oslo (Christiania) district, Norway, from where it was originally described. On the Rallier-du-Baty Peninsula, Kerguelen Island, in the south Indian Ocean. In the Baie-des-Moutons complex, La Tabatière, Quebec, Canada. At Tawmaw, Kachin State, northern Myanmar (Burma).

Name: From the Greek for *a carrying down*, in allusion to its volcanic origin.

Type Material: n.d.

References: (1) Dana, E.S. (1899) Dana's system of mineralogy, (6th edition), app. I, 14 [catophorite]. (2) Deer, W.A., R.A. Howie, and J. Zussman (1963) Rock-forming minerals, v. 2, chain silicates, 359–363. (3) Giret, A., B. Bonin, and J.-M. Leger (1980) Amphibole compositional trends in oversaturated and undersaturated alkaline plutonic ring complexes. Can. Mineral., 18, 481–495. (4) Lalonde, A.E. and R.F. Martin (1983) The Baie-des-Moutons syenitic complex, La Tabatière, Québec, II. The ferromagnesian minerals. Can. Mineral., 21, 81–91. (5) Hawthorne, F.C. (1983) The crystal chemistry of the amphiboles. Can. Mineral., 21, 173–480, esp. 363. (6) Mével, C. and J.-R. Kiénast (1986) Jadeite-kosmochlor solid solution and chromian sodic amphiboles in jadeitites and associated rocks from Tawmaw (Burma), Bull. Minéral., 109, 617–633. (7) Phillips, W.R. and D.T. Griffen (1981) Optical mineralogy, 241–242.

Katoptrite

$$(Mn^{2+}, Mg)_{13}(Al, Fe^{3+})_4 Sb_2^{5+} Si_2 O_{28}$$

Crystal Data: Monoclinic. *Point Group:* $2/m$. Crystals commonly tabular on {010}; elongated along [100], [101], or equant. As minute grains or anhedral masses, to 1 cm.

Physical Properties: *Cleavage:* Perfect micaceous on {001}. *Tenacity:* Brittle. Hardness = 5.5 D(meas.) = 4.56 D(calc.) = 4.65

Optical Properties: Opaque, translucent in thin flakes. *Color:* Iron-black to jet-black; fire-red in thin section. *Luster:* Metallic, brilliant.
Optical Class: Biaxial (–). *Pleochroism:* Strong; red-brown to orange. *Orientation:*
$X \simeq \perp$ {001}; $Y = b$; $Z \wedge a \simeq -3°$. *Dispersion:* $r > v$, strong, inclined. $\alpha = 1.92$ $\beta = 1.95$ $\gamma = 1.95$ 2V(meas.) = Small.

Cell Data: *Space Group:* $C2/m$. $a = 5.617(2)$ $b = 23.02(2)$ $c = 9.079(6)$
$\beta = 101°23(3)'$ Z = 2

X-ray Powder Pattern: Brattfors mine, Sweden.
2.957 (100), 8.88 (65), 2.600 (50), 4.43 (45), 2.488 (45), 2.810 (40), 1.944 (35)

Chemistry:

	(1)	(2)
SiO_2	7.75	8.0
Al_2O_3	9.50	10.9
Fe_2O_3	3.58	
Sb_2O_5	20.76	22.4
FeO	2.44	2.5
MnO	52.61	54.3
ZnO		0.0
MgO	3.06	3.7
CaO	0.58	< 0.05
H_2O	0.11	
Total	100.39	101.8

(1) Långban, Sweden; average of four analyses. (2) Do.; by electron microprobe; corresponds to $(Mn_{11.7}^{2+}Mg_{1.4})_{\Sigma=13.1}(Al_{3.3}Fe_{0.5})_{\Sigma=3.8}Sb_{2.1}^{5+}Si_{2.0}O_{28}$.

Occurrence: In metamorphosed limestone (Brattfors mine, Sweden).

Association: Magnetite, garnet, manganosite (Brattfors mine, Sweden).

Distribution: In Sweden, in the Brattfors, Moss, and Jakobsberg mines, near Nordmark, and at Långban, Värmland; in the Sjö mine, near Grythyttan, Örebro.

Name: From the Greek for *mirror*, for the perfectly reflecting cleavage surfaces.

References: (1) Palache, C., H. Berman, and C. Frondel (1951) Dana's system of mineralogy, (7th edition), v. II, 1029–1030. (2) Moore, P.B. (1966) Catoptrite [katoptrite] and yeatmanite – stuffed pyrochroite structures? Amer. Mineral., 51, 1494–1500. (3) Moore, P.B., T. Araki, and G.D. Brunton (1976) Catoptrite [katoptrite], $(Mn_5^{2+}Sb_2^{5+})^{VI}(Mn_8^{2+}Al_4Si_2)^{IV}O_{28}$, a novel close-packed oxide sheet structure. Neues Jahrb. Mineral., Abh., 127, 47–61.

Crystal Data: Hexagonal. *Point Group:* $\bar{3}\,2/m$. Crystals, to 2 mm, show the forms $\{11\bar{2}1\}$ and $\{11\bar{2}4\}$; as disseminated grains. *Twinning:* Twinning axis normal to $\{11\bar{2}4\}$; simple, complex, and polysynthetic twins.

Physical Properties: *Fracture:* Uneven to subconchoidal. Hardness = 4   D(meas.) = 2.84–2.90   D(calc.) = 2.97

Optical Properties: Semitransparent. *Color:* Yellow; pale yellow in thin section. *Luster:* Vitreous to greasy.
Optical Class: Uniaxial (–). $\omega = 1.648$–1.650 $\epsilon = 1.625$–1.638

Cell Data: *Space Group:* $R\bar{3}m$. a = 10.174   c = 13.053   Z = 3

X-ray Powder Pattern: Mt. Karnasurt, Russia.
2.60 (10), 2.52 (8), 1.816 (8), 3.60 (7), 1.480 (7), 3.28 (6), 3.17 (6)

Chemistry:

	(1)
SiO_2	52.44
TiO_2	7.62
Al_2O_3	0.84
Fe_2O_3	2.30
Nb_2O_5	1.40
MnO	4.40
MgO	0.10
CaO	0.40
Na_2O	25.50
K_2O	0.47
H_2O^+	3.66
H_2O^-	0.14
P_2O_5	0.70
Total	99.97

(1) Mt. Karnasurt, Russia; corresponds to $(Na_{5.51}K_{0.07}Ca_{0.05})_{\Sigma=5.63}Mn^{2+}_{0.42}(Ti_{0.64}Fe_{0.20}$ $Nb_{0.07}Mg_{0.02}Al_{0.02})_{\Sigma=0.95}(Si_{5.84}Al_{0.09}P_{0.07})_{\Sigma=6.00}O_{18}$.

Mineral Group: Lovozerite group.

Occurrence: In sodalite-rich syenites in an alkalic massif (Lovozero massif, Russia).

Association: Nordite, belovite, vuonnemite, alkalic feldspars, ussingite (Lovozero massif, Russia); villiaumite, delhayelite, lamprophyllite, lomonosovite, feldspars, nepheline, aegirine (Khibiny massif, Russia).

Distribution: On Mts. Karnasurt and Alluaiv, Lovozero massif, and in the Khibiny massif, Kola Peninsula, Russia.

Name: For Maria Efimovna Kazakova, analytical chemist who provided the mineral's analysis.

Type Material: A.E. Fersman Mineralogical Museum, Academy of Sciences, Moscow, Russia.

References: (1) Khomyakov, A.P., E.I. Semenov, E.M. Es'kova, and A.A. Voronkov (1974) Kazakovite – a new mineral of the lovozerite group. Zap. Vses. Mineral. Obshch., 103, 342–345 (in Russian). (2) (1975) Amer. Mineral., 60, 161–162 (abs. ref. 1). (3) (1975) Mineral. Abs., 26, 126 (abs. ref. 1). (4) Povarennykh, A.S. and Y.P. Men'shikov (1977) Kazakovite from the Khibiny Mountains and IR spectra of some titano- and zirconosilicates. Geol. Zh., 37, 115–120 (in Russian). (5) (1977) Chem. Abs., 87, 41912 (abs. ref. 4). (6) Voronkov, A.A., Z.V. Pudovkina, V.A. Blinov, V.V. Ilyukhin, and Y.A. Pyatenko (1979) Crystal structure of kazakovite $Na_6Mn\{Ti[Si_6O_{18}]\}$. Doklady Acad. Nauk SSSR, 245, 106–109 (in Russian).

Crystal Data: Monoclinic, pseudohexagonal. *Point Group:* $2/m$, 2, or m. As pseudohexagonal plates with dominant $\{100\}$, to 0.3 mm, forming spherical aggregates.

Physical Properties: *Cleavage:* $\{100\}$, perfect. *Tenacity:* Extremely flexible. Hardness = n.d. D(meas.) = ~ 4.5 D(calc.) = 4.76

Optical Properties: Transparent to translucent. *Color:* Colorless to white. *Luster:* Vitreous. *Optical Class:* Biaxial ($-$). $n = 1.81 \parallel \{100\}$. 2V(meas.) = n.d.

Cell Data: *Space Group:* $A2/m$, $A2$, or Am. a = 21.04(1) b = 15.55(1) c = 8.986(6) $\beta = 91.0(1)°$ Z = 3

X-ray Powder Pattern: Tsumeb, Namibia.
21.0 (100), 2.591 (90), 3.82 (80), 7.01 (50), 3.006 (50), 2.339 (50), 3.74 (40)

Chemistry:

	(1)
SiO_2	17.0
Al_2O_3	7.1
FeO	trace
MnO	0.0
CuO	1.5
ZnO	trace
PbO	60.9
CaO	0.0
H_2O	2.6
CO_2	7.0
SO_3	5.5
Total	101.6

(1) Tsumeb, Namibia; by electron microprobe, CO_2 by a C analyzer, H_2O and SO_3 by TGA and EGA; corresponding to $Pb_{7.55}Cu_{0.52}Al_{3.85}Si_{7.83}O_{19.22}(SO_4)_{1.90}(CO_3)_{4.40}(OH)_{7.98}$.

Occurrence: In a deep oxidation zone in a polymetallic mineral deposit.

Association: Quartz, galena, mimetite, hematite, leadhillite, anglesite, fleischerite, melanotekite, alamosite.

Distribution: From Tsumeb, Namibia.

Name: For Friedrich Wilhelm Kegel, Director of mining operations (1922–1938) at Tsumeb, Namibia.

Type Material: National Museum of Natural History, Washington, D.C., USA, 134514, 147460.

References: (1) Medenbach, O. and K. Schmetzer (1975) Kegelite – ein neues Bleisilikat von Tsumeb. Naturwiss., 62, 137 (in German with English abs.). (2) (1976) Amer. Mineral., 61, 175–176 (abs. ref. 1). (3) Medenbach, O. and K. Schmetzer (1976) Kegelite – ein neues Bleisilikat von Tsumeb. Neues Jahrb. Mineral., Monatsh., 110–114 (in German). (4) (1977) Amer. Mineral., 62, 175 (abs. ref. 3). (5) Dunn, P.J., R.S.W. Braithwaite, A.C. Roberts, and R.A. Ramik (1990) Kegelite from Tsumeb, Namibia: a redefinition. Amer. Mineral., 75, 702–704. (6) Braithwaite, R.S.W. (1991) Kegelite: infrared spectroscopy and a structural hypothesis. Mineral. Mag., 55, 127–134.

Crystal Data: Monoclinic.   *Point Group:* $2/m$.   As prismatic crystals, up to 1 mm.

Physical Properties: *Fracture:* Uneven.   Hardness = 4–5   D(meas.) = 4.45   D(calc.) = 4.48 Yellow-green cathodoluminescence.

Optical Properties: Semitransparent.   *Color:* Colorless to white.   *Streak:* White.
Luster: Vitreous.
Optical Class: Biaxial (–).   *Orientation:* $Z = b$; $X \wedge c = 4°$; $Y \wedge a = 7°$.   *Dispersion:* $r < v$.
$\alpha = 1.713(1)$ $\beta = 1.748(1)$ $\gamma = 1.758(1)$ 2V(meas.) = 56(2)°   2V(calc.) = 55°

Cell Data: *Space Group:* $[C2/m]$ (by analogy to keiviite-(Yb)).   a = 6.845(5)   b = 8.960(5) c = 4.734(3) $\beta = 101.65(5)°$ Z = 2

X-ray Powder Pattern: Kola Peninsula, Russia; nearly identical to keiviite-(Yb).
3.23 (10), 4.65 (9), 3.04 (8), 2.280 (7), 5.40 (5), 2.729 (5), 2.692 (5)

Chemistry:

	(1)	(2)
SiO_2	30.85	30.49
Y_2O_3	40.86	45.86
Gd_2O_3	0.08	0.62
Tb_2O_3	0.10	0.00
Dy_2O_3	1.53	2.92
Ho_2O_3	0.85	1.27
Er_2O_3	6.50	4.40
Tm_2O_3	1.99	1.11
Yb_2O_3	15.36	11.81
Lu_2O_3	2.61	1.05
CaO	0.00	0.30
Total	100.73	99.83

(1–2) Kola Peninsula, Russia; by electron microprobe; when averaged with another intermediate analysis, corresponds to $(Y_{1.55}Yb_{0.20}Er_{0.10}Dy_{0.08}Ca_{0.04}Tm_{0.02}Lu_{0.02}Ho_{0.02}Gd_{0.01})_{\Sigma=2.04}Si_{1.98}O_7$.

Polymorphism & Series: Forms a series with keiviite-(Yb).

Occurrence: In fissures in quartz and fluorite in "amazonite" pegmatites.

Association: Keiviite-(Yb), thalenite, xenotime, bastnäsite, kuliokite-(Y), quartz, fluorite.

Distribution: From the [Keivy massif,] Kola Peninsula, Russia.

Name: For the relation to *keiviite*-(Yb) and the *yttrium* content.

Type Material: A.E. Fersman Mineralogical Museum, Academy of Sciences, Moscow, Russia.

References: (1) Voloshin, A.V., Y.A. Pakhomovskii, and F.N. Tyusheva (1985) Keiviite-(Y) – a new yttrian diorthosilicate, and thalenite from amazonite pegmatites of the Kola Peninsula. Diortho- and triorthosilicates of yttrium. Mineral. Zhurnal, 7(6), 79–94 (in Russian with English abs.). (2) (1988) Amer. Mineral., 73, 191–192 (abs. ref. 1).

Crystal Data: Monoclinic. *Point Group:* $2/m$. As elongated platy and prismatic crystals, to 0.8 mm. *Twinning:* Polysynthetic, common.

Physical Properties: *Cleavage:* Perfect on {110}, imperfect on {001}. Hardness = n.d. D(meas.) = 5.95 D(calc.) = 5.99 Faint green cathodoluminescence.

Optical Properties: Transparent. *Color:* Colorless. *Luster:* Vitreous.
Optical Class: Biaxial (–). *Orientation:* $Z = b$; $X \wedge c = 3°$–$5°$; $Y \wedge a = 7°$–$8°$.
Dispersion: $r < v$, strong. $\alpha = 1.723$ $\beta = 1.758$ $\gamma = 1.768$ 2V(meas.) = 58°

Cell Data: *Space Group:* $C2/m$. a = 6.840(2) b = 8.916(4) c = 4.745(1) $\beta = 102.11(3)°$ Z = 2

X-ray Powder Pattern: Kola Peninsula, Russia; pattern nearly identical to keiviite-(Y).
3.24 (10), 3.20 (10), 3.03 (9), 4.64 (8), 2.720 (7), 2.674 (7), 2.262 (7)

Chemistry:

	(1)	(2)
SiO_2	23.47	26.71
Y_2O_3	1.02	15.42
Gd_2O_3	0.06	0.15
Tb_2O_3	0.04	0.04
Dy_2O_3	1.23	3.86
Ho_2O_3	0.65	1.23
Er_2O_3	6.24	9.85
Tm_2O_3	3.10	3.19
Yb_2O_3	55.06	34.57
Lu_2O_3	8.97	5.22
FeO	0.00	0.09
CaO	0.03	0.07
Total	99.87	100.40

(1–2) Kola Peninsula, Russia; by electron microprobe; when averaged with three other intermediate analyses, corresponds to $(Yb_{1.43}Lu_{0.23}Er_{0.17}Tm_{0.08}Y_{0.05}Dy_{0.03}Ho_{0.02})_{\Sigma=2.01}Si_{1.99}O_7$.

Polymorphism & Series: Forms a series with keiviite-(Y).

Occurrence: In microcline-bearing pegmatites; two generations of the mineral are present.

Association: Fluorite, bastnäsite, hingganite, wulfenite.

Distribution: From the [Keivy massif,] Kola Peninsula, Russia.

Name: For Keivy on the Kola Peninsula, Russia, and *ytterbium* in its composition.

Type Material: A.E. Fersman Mineralogical Museum, Academy of Sciences, Moscow, Russia.

References: (1) Voloshin, A.V., Y.A. Pakhomovskii, and F.N. Tyusheva (1983) Keiviite $Yb_2Si_2O_7$, a new ytterbium silicate from amazonitic pegmatites of the Kola Peninsula. Mineral. Zhurnal, 5(5), 94–99 (in Russian with English abs.). (2) (1984) Amer. Mineral., 69, 1191 (abs. ref. 1).

Crystal Data: Triclinic. *Point Group:* $\bar{1}$. As irregular grains, to 4 mm, or granular aggregates. *Twinning:* Very fine polysynthetic twinning seen under the microscope.

Physical Properties: *Cleavage:* Two poor, intersecting $\sim 90°$. *Fracture:* Irregular. *Tenacity:* Very brittle. Hardness = 3.8–4.3   VHN = 157–231   D(meas.) = 3.22–3.30 D(calc.) = 3.26

Optical Properties: Translucent, transparent in thin fragments. *Color:* White; colorless in thin section. *Luster:* Vitreous to greasy.
Optical Class: Biaxial (–). $\alpha = 1.670$ β = n.d. $\gamma = 1.710$ 2V(meas.) = 78°

Cell Data: *Space Group:* $P\bar{1}$. a = 9.0(1)   b = 5.34(2)   c = 6.96(3) $\alpha = 92(1)°$ $\beta = 116(1)°$ $\gamma = 88(1)°$ Z = 2

X-ray Powder Pattern: Lovozero massif, Russia.
3.97 (10), 4.11 (7), 1.542 (7), 1.097 (6), 1.013 (6), 2.95 (5), 2.66 (5)

Chemistry:

	(1)	(2)
SiO_2	39.39	39.69
TiO_2	0.60	0.32
ZrO_2	40.35	39.53
$Fe_2O_3 + FeO$	0.31	0.26
CaO		1.37
Na_2O	16.03	18.35
K_2O	0.94	trace
H_2O^+	0.95	
H_2O^-	0.35	
Total	98.92	99.52

(1) Lovozero massif, Russia; average of three partial analyses. (2) Do.; corresponds to $(Na_{1.79}Ca_{0.07})_{\Sigma=1.86}(Zr_{0.97}Ti_{0.01}Fe_{0.01})_{\Sigma=0.99}Si_2O_{6.94}$.

Occurrence: A primary mineral in foyaites composed of partly albitized microcline, nepheline, sodalite, aegirine, and alkali amphibole, in a differentiated alkalic massif (Lovozero massif, Russia).

Association: Eudialyte, lorenzenite (Lovozero massif, Russia).

Distribution: In the vicinity of the Tavaiok and Angoundasiok Rivers and on Mt. Alluaiv, Lovozero massif, and at Tachtarvumchorr, Khibiny massif, Kola Peninsula, Russia. From Lågendalen, near Larvik, Norway.

Name: For Russian mathematician Mstislav Vsevolodovich Keldysh (1911–), President of the Academy of Sciences, Russia.

Type Material: n.d.

References: (1) Gerasimovskii, V.I. (1962) Keldyshite, a new mineral. Doklady Acad. Nauk SSSR, 142, 916–918 (in Russian). (2) (1962) Amer. Mineral., 47, 1216 (abs. ref. 1). (3) Khomyakov, A.P., M.E. Kazakova, and A.A. Voronkov (1969) New data on keldyshite. Doklady Acad. Nauk SSSR, 189, 166–168 (in Russian). (4) (1970) Amer. Mineral., 55, 1072–1073 (abs. ref. 3). (5) (1972) Amer. Mineral., 57, 1317 (corr. ref. 4). (6) Khalikov, A.D., A.P. Khomyakov, and S.A. Makhmudov (1978) Crystal structure of keldyshite. Doklady Acad. Nauk SSSR, 238, 573–575 (in Russian). (7) Vlasov, K.A., Ed. (1966) Mineralogy of rare elements, v. II, 390–391.

Crystal Data: Hexagonal. *Point Group:* 6. As irregular to platy tablets and laths, to 1 mm.

Physical Properties: *Cleavage:* Perfect on {0001}. Hardness = n.d. D(meas.) = 3.07 D(calc.) = 3.11

Optical Properties: Transparent. *Color:* Yellow; light yellow in thin section. *Optical Class:* Biaxial (−). *Pleochroism:* X = colorless to greenish yellow; Y = Z = pale yellow to reddish brown. *Orientation:* X ⊥ (0001). *Dispersion:* $r > v$, moderate. $\alpha = 1.639(1)$ $\beta = 1.646(1)$ $\gamma = 1.646(1)$ 2V(meas.) = 30°–16°

Cell Data: *Space Group:* $P6_3$. a = 5.438(8) c = 14.04(1) Z = [4]

X-ray Powder Pattern: Bald Knob, North Carolina, USA. 3.51 (100), 7.0 (90), 2.53 (60), 2.67 (40), 2.34 (40), 1.954 (30), 1.570 (30)

Chemistry:

	(1)
SiO_2	17.60
TiO_2	< 0.05
Al_2O_3	28.55
Fe_2O_3	2.18
MnO	38.84
MgO	2.97
H_2O	[10.82]
Total	[100.96]

(1) Bald Knob, North Carolina, USA; by electron microprobe, H_2O calculated from stoichiometry; corresponds to $(Mn_{1.82}Al_{0.84}Mg_{0.24}Fe_{0.09})_{\Sigma=2.99}(Al_{1.02}Si_{0.98})_{\Sigma=2.00}O_5(OH)_4$.

Polymorphism & Series: two-layer hexagonal and six-layer rhombohedral polytypes noted.

Mineral Group: Kaolinite-serpentine group.

Occurrence: In banded Ca-Mn carbonate-rich rocks.

Association: Kutnohorite, galaxite, jacobsite, manganoan chlorite, alleghanyite, sonolite.

Distribution: In the Bald Knob mine, near Sparta, Alleghany Co., North Carolina, USA.

Name: To honor Professor William Crowe Kelly (1929–), geologist, of the University of Michigan, Ann Arbor, Michigan, USA.

Type Material: Department of Geological Sciences, University of Michigan, Ann Arbor, Michigan; National Museum of Natural History, Washington, D.C., USA, 127101, 137140, 137141.

References: (1) Peacor, D.R., E.J. Essene, W.B. Simmons, Jr., and W.C. Bigelow (1974) Kellyite, a new Mn-Al member of the serpentine group from Bald Knob, North Carolina, and new data on grovesite. Amer. Mineral., 59, 1153–1156.

Crystal Data: Orthorhombic. *Point Group:* $2/m \, 2/m \, 2/m$. As tiny, short prismatic crystals with {010}, {110}, and {111} common; as sheaflike groupings and massive.

Physical Properties: *Cleavage:* Distinct on {110}. *Fracture:* Uneven. *Tenacity:* Brittle. Hardness = 5 D(meas.) = 6.19 D(calc.) = [6.29]

Optical Properties: Semitransparent. *Color:* Dark reddish brown, with deep red internal reflections; tarnishes black on the surface. *Streak:* Yellowish brown. *Luster:* Vitreous to submetallic, greasy, dull.
Optical Class: Biaxial (+). *Pleochroism:* Strong; X = bright yellowish pink; Y = reddish brown; Z = deep brownish red. *Orientation:* $X = a$. *Dispersion:* $r < v$, strong. *Absorption:* $Z > Y > X$. $\alpha = 2.10$ $\beta = 2.20$ $\gamma = 2.31$ 2V(meas.) = 88°
R_1–R_2: (400) 14.7–18.7, (420) 14.4–18.1, (440) 14.1–17.5, (460) 13.9–16.9, (480) 13.6–16.4, (500) 13.4–16.0, (520) 13.2–15.7, (540) 13.0–15.4, (560) 12.9–15.2, (580) 12.8–15.0, (600) 12.6–14.9, (620) 12.5–14.8, (640) 12.5–14.7, (660) 12.4–14.5, (680) 12.4–14.4, (700) 12.3–14.3

Cell Data: *Space Group: Pbcn.* $a = 6.961(2)$ $b = 11.018(3)$ $c = 9.964(5)$ $Z = 4$

X-ray Powder Pattern: Långban, Sweden.
2.90 (100), 2.86 (100), 2.74 (100), 3.71 (80), 3.51 (80), 3.24 (80), 2.84 (80)

Chemistry:

	(1)	(2)	(3)		(1)	(2)	(3)
SiO_2	15.95	16.45	16.59	Mn_2O_3	22.26	13.55	21.79
TiO_2		0.38		ZnO		0.05	
Al_2O_3		0.30		PbO	59.79	59.59	61.62
Fe_2O_3		6.62		MgO		0.05	
				Total	98.00	96.99	100.00

(1) "Chile." (2) Långban, Sweden; by electron microprobe, corresponding to $Pb_{1.98}(Mn_{1.28}Fe_{0.61} Al_{0.04}Ti_{0.04}Mg_{0.01})_{\Sigma=1.98}Si_{2.02}O_9$. (3) $Pb_2Mn_2O_2Si_2O_7$.

Polymorphism & Series: Forms a series with melanotekite.

Occurrence: In a cavity in calcite in a metamorphosed stratiform zinc deposit (Franklin, New Jersey, USA); in veinlets cutting franklinite ore in bedded manganese deposits (Ushkatyn deposits, Kazakhstan).

Association: Quartz, barite, apatite ("Chile"); braunite, richterite, barite (Långban, Sweden); calcite, willemite, franklinite (Franklin, New Jersey, USA); garnet, tephroite, franklinite, braunite (Ushkatyn deposits, Kazakhstan).

Distribution: The unknown type locality is "southern Chile." From Långban; Jakobsberg; the Harstig mine, Pajsberg, near Persberg; and the Klintgruvan mine, near Kryibo, Värmland, Sweden. From Bona de Padru, near Ozieri, Sardinia, Italy. At Higher Pitts Farm, Priddy, Somerset, England. In the Kombat mine, 49 km south of Tsumeb, Namibia. In South Africa, in the Wessels mine, near Kuruman, Cape Province. From the Ushkatyn deposits, Atasui area, Kazakhstan. At Franklin, Sussex Co., New Jersey, USA.

Name: From the Greek for *spike* or *thorn*, for its prismatic habit.

References: (1) Dana, E.S. (1892) Dana's system of mineralogy, (6th edition), 544, 1039. (2) Gabrielson, O. (1961) The crystal structures of kentrolite and melanotekite. Arkiv Mineral. Geol., 3, 141–151. (3) Glasser, F.P. (1967) New data on kentrolite and melanotekite: ternary phase relations in the system $PbO-Fe_2O_3-SiO_2$. Amer. Mineral., 52, 1085–1093. (4) Moore, P.B., P.K. Sen Gupta, J. Shen, and E.O. Schlemper (1991) The kentrolite-melanotekite series, $4Pb_2(Mn, Fe)_2^{3+}O_2[Si_2O_7]$: chemical crystallographic relations, lone-pair splitting, and cation relation to $8URe_2$. Amer. Mineral., 76, 1389–1399.

Crystal Data: Monoclinic, probable. *Point Group:* n.d. As nodular concretions, commonly with chert cores.

Physical Properties: Hardness = n.d. D(meas.) = n.d. D(calc.) = [2.33]

Optical Properties: Transparent to translucent. *Color:* [White.]
Optical Class: n.d. $n = \sim 1.48$

Cell Data: *Space Group:* n.d. $a = 7.79(5)$ $b = 19.72(5)$ $c = 7.30(5)$ $\beta = 95°54(5)'$
$Z = 1$

X-ray Powder Pattern: Lake Magadi, Kenya; pattern varies with interlayer $Na:H_2O$.
19.68 (100), 3.428 (85), 3.198 (55), 9.925 (50), 3.320 (45), 4.965 (35), 4.689 (28)

Chemistry:

	(1)	(2)
SiO_2	83.50	84.52
TiO_2	0.02	
Al_2O_3	0.22	
Fe_2O_3	0.09	
MnO	trace	
MgO	0.04	
CaO	0.11	
Na_2O	3.96	3.96
K_2O	0.04	
H_2O^+	4.90	
H_2O^-	7.10	
H_2O		11.52
Total	99.98	100.00

(1) Lake Magadi, Kenya. (2) $Na_2Si_{22}O_{41}(OH)_8 \cdot 6H_2O$.

Occurrence: As concretions, with chert cores, imbedded in magadiite; perhaps as a reaction product between the two (Lake Magadi, Kenya); in altered volcanic rocks (Trinity Co., California, USA).

Association: Magadiite, quartz.

Distribution: From Lake Magadi, Rift Valley, Kenya. In Niger, at Kafra. In the USA, from five km north of Trinity Lake, Trinity Co., California.

Name: For the locality in Kenya.

Type Material: National School of Mines, Paris, France; National Museum of Natural History, Washington, D.C., USA, 121336, 121337.

References: (1) Eugster, H.P. (1967) Hydrous sodium silicates from Lake Magadi, Kenya: precursors of bedded chert. Science, 157, 1177–1180. (2) (1968) Amer. Mineral., 53, 510 (abs. ref. 1). (3) McAtee, J.L., Jr., R. House, and H.P. Eugster (1968) Magadiite from Trinity County, California. Amer. Mineral., 53, 2061–2069. (4) Beneke, K. and G. Lagley (1983) Kenyaite – synthesis and properties. Amer. Mineral., 68, 818–826.

Crystal Data: Monoclinic, pseudohexagonal. *Point Group:* 2/m. As irregular grains, to 3 mm, in small ovoids zoned with eudialyte.

Physical Properties: Hardness = 4.5–5.5   D(meas.) = 3.40   D(calc.) = 3.33

Optical Properties: Translucent. *Color:* Pale yellowish to white. *Luster:* Dull or greasy. *Optical Class:* Biaxial (–). $\alpha = 1.665$ $\beta = 1.715$ $\gamma = 1.715$ 2V(meas.) = 6°–16°

Cell Data: *Space Group:* B2/m (synthetic). $a = 19.188$ $b = 14.072$ $c = 11.075$ $\gamma = 117°04'$ Z = 16

X-ray Powder Pattern: Khibiny massif, Russia.
2.76 (100), 2.95 (70), 1.630 (65), 2.133 (50), 1.381 (50), 1.252 (45), 1.595 (40)

Chemistry:

	(1)
SiO$_2$	33.8
TiO$_2$	0.6
ZrO$_2$	37.8
CaO	trace
K$_2$O	27.0
Total	99.2

(1) Khibiny massif, Russia; by electron microprobe, corresponds to $K_{1.98}(Zr_{1.06}Ti_{0.03})_{\Sigma=1.09}$ $Si_{1.94}O_{7.05}$.

Occurrence: In aegirine-rich metasomatic rocks in a differentiated alkalic massif.

Association: Eudialyte, zircon.

Distribution: From the Khibiny massif, in the Hakman Valley, Kola Peninsula, Russia.

Name: For the locality in the Khibiny massif, Kola Peninsula, Russia.

Type Material: Institute of Mineralogy and Geochemistry of Rare Elements, Moscow; A.E. Fersman Mineralogical Museum, Academy of Sciences, Moscow, Russia; National School of Mines, Paris, France; The Natural History Museum, London, England, 1994,11.

References: (1) Khomyakov, A.P. and A.A. Voronkov (1973) New zirconium silicates in the Lovozero and Khibina massifs. Trudy Mineral. Muzeya Akad. Nauk SSSR, 22, 215–217 (in Russian). (2) (1974) Amer. Mineral., 59, 1140 (abs. ref. 1). (3) Khomyakov, A.P., A.A. Voronkov, S.I. Lebedeva, V.P. Bykov, and K.V. Yurkiva (1974) Khibinskite, K$_2$ZrSi$_2$O$_7$, a new mineral. Zap. Vses. Mineral. Obshch., 103, 110–116 (in Russian). (4) (1975) Amer. Mineral., 60, 340 (abs. ref. 3). (5) Nosyrev, N.A., E.N. Treushnikov, A.A. Voronkov, V.V. Ilyukhin, R.M. Ganiev, and N.V. Belov (1976) Crystal structure of synthetic khibinskite. Doklady Acad. Nauk SSSR, 231, 1351–1353 (in Russian). (6) (1977) Chem. Abs., 86, 64045 (abs. ref. 5).

Crystal Data: Orthorhombic. *Point Group:* $2/m\ 2/m\ 2/m$ or $mm2$. No crystal forms observed as typically a replacement of rankinite; massive, to 2 mm.

Physical Properties: Hardness = n.d. D(meas.) = 2.992(2) D(calc.) = [3.00]

Optical Properties: Semitransparent. *Color:* Colorless; in thin section, interference colors are typically weak ultra-blue or ultra-brown.
Optical Class: Biaxial (–). *Dispersion:* $r > v$, distinct to strong. $\alpha = 1.646–1.647$ $\beta = 1.648$ $\gamma = 1.650$ 2V(meas.) = $46°–60°$

Cell Data: *Space Group: Imam* or *Ima2.* a = 11.42(5) b = 5.09(5) c = 21.95(5) Z = 8

X-ray Powder Pattern: Kilchoan, Scotland.
2.89 (s1), 3.07 (s2), 2.68 (s3), 3.56 (ms), 2.36 (ms), 1.964 (ms), 5.17 (m)

Chemistry: (1) Kilchoan, Scotland; an analysis, not given, from which CO_2 was deducted as spurrite, yielded $(Ca_{2.93}Fe^{3+}_{0.01})_{\Sigma=2.94}(Si_{1.96}Al_{0.04})_{\Sigma=2.00}[O_{6.94}(OH)_{0.06}]_{\Sigma=7.00}$. (2) Fuka, Japan; an analysis, not available, yielded $(Ca_{3.13}Fe^{2+}_{0.01}Na_{0.01}K_{0.01})_{\Sigma=3.16}(Si_{1.91}Al_{0.02})_{\Sigma=1.93}O_7$.

Polymorphism & Series: Dimorphous with rankinite.

Occurrence: In limestones thermally metamorphosed by invasion of gabbro; thought to have formed during retrograde metamorphism following decarbonation of limestone and thus under low pressure of CO_2 (Kilchoan, Scotland; Carlingford, Ireland).

Association: Rankinite, spurrite, tilleyite, melilite, cuspidine, grossular, wollastonite, vesuvianite (Kilchoan, Scotland); rankinite, larnite, spurrite (Golden Gully, New Zealand).

Distribution: From near Kilchoan, Ardnamurchan, Argyllshire, Scotland. At Carlingford, Co. Louth, Ireland. In New Zealand, at Golden Gully, Tokatoka district, about 150 km north of Auckland. In Japan, from Fuka, near Bicchu, Okayama Prefecture. In the Ozerskii massif, near Lake Baikal, eastern Siberia, Russia.

Name: For Kilchoan, Scotland, where it was first found.

Type Material: National Museum of Natural History, Washington, D.C., USA, 119424, 119425.

References: (1) Agrell, S.O. and P. Gay (1961) Kilchoanite, a polymorph of rankinite. Nature, 189, 743. (2) (1961) Amer. Mineral., 46, 1203 (abs. ref. 1). (3) (1962) Amer. Mineral., 47, 420 (corr. ref. 1). (4) Taylor, H.F. (1971) The crystal structure of kilchoanite, $Ca_6(SiO_4)(Si_3O_{10})$, with some comments on related phases. Mineral. Mag., 38, 26–31. (5) Deer, W.A., R.A. Howie, and J. Zussman (1986) Rock-forming minerals, (2nd edition), v. 1B, disilicates and ring silicates, 272–277 [rankinite].

Crystal Data: Monoclinic. *Point Group: 2/m.* Crystals are well-formed, elongated along [010], to 2 mm. *Twinning:* Complex penetration twinning, the composition plane of which includes [010]; they display characteristic rhombic cross sections showing crossed "bow tie" twins under crossed nicols.

Physical Properties: *Cleavage:* Perfect on {100}; poor on {010} and {001}. Hardness = n.d. D(meas.) = n.d.   D(calc.) = 2.94

Optical Properties: Translucent. *Color:* [White]; colorless in thin section. *Optical Class:* Biaxial (–). *Orientation:* $Y = b$; $Z \wedge c \simeq 16°$. $\alpha = 1.634–1.635$ $\beta = 1.646$ $\gamma = 1.642–1.648$ 2V(meas.) = 26(2)°   2V(calc.) = 38°

Cell Data: *Space Group: P2$_1$/m.* $a = 6.807$ $b = 15.459$ $c = 6.811$ $\beta = 97.76°$ Z = 4

X-ray Powder Pattern: Killala Bay, Ireland.
2.824 (100), 3.03 (80), 2.724 (60), 2.275 (45), 2.224 (45), 1.413 (40), 1.688 (35)

Chemistry:

	(1)	(2)
SiO$_2$	39.8	39.22
CaO	57.0	54.90
H$_2$O	[3.2]	5.88
Total	[100.0]	100.00

(1) Killala Bay, Ireland; by electron microprobe, H$_2$O by difference, MgO, FeO, and Al$_2$O$_3$ each < 0.1%. (2) Ca$_3$Si$_2$O$_7$·H$_2$O.

Occurrence: A secondary mineral in cavities and veins in hydrothermally altered and thermally metamorphosed limestones.

Association: Calcite, afwillite, spurrite, wollastonite (Killala Bay, Ireland); larnite, magnetite, perovskite, spinel, spurrite (Carneal, Ireland).

Distribution: From near Inishcrone, along the east shore of Killala Bay, Co. Sligo, and at Carneal, Co. Antrim, Ireland. In Turkey, from the Güneyce-İkizdere region, Trabzon Province.

Name: For the locality at Killala Bay, Ireland.

Type Material: Ulster Museum, Belfast, Ireland; The Natural History Museum, London, England, 1973,484; National Museum of Natural History, Washington, D.C., USA, 128672.

References: (1) Nawaz, R. (1974) Killalaite, a new mineral from Co. Sligo, Ireland. Mineral. Mag., 39, 544–548. (2) (1974) Amer. Mineral., 59, 1331 (abs. ref. 1). (3) Nawaz, R. (1977) A second occurrence of killalaite. Mineral. Mag., 41, 546–548. (4) Taylor, H.F.W. (1977) The crystal structure of killalaite. Mineral. Mag., 41, 363–369. (5) Sarp, H., J. Deferne, and E. Sarman (1982) Second occurrence of killalaite in a skarn from the Guneyce-Ikizdere region (eastern Pontids, Turkey). Arch. Sci., 35(3), 275–278. (6) (1983) Chem. Abs., 98, 188 (abs. ref. 5).

Crystal Data: Cubic. *Point Group:* $4/m \, \bar{3} \, 2/m$. As dodecahedra modified by trapezohedra, up to 5 mm; also in rounded grains.

Physical Properties: Hardness = ~7 D(meas.) = 4.0 D(calc.) = 4.03

Optical Properties: Transparent. *Color:* Dark brown; light brown in thin section. *Streak:* Light brown. *Luster:* Vitreous. *Optical Class:* Isotropic. $n = 1.94(1)$

Cell Data: *Space Group: Ia3d.* a = 12.46 Z = 8

X-ray Powder Pattern: Magnet Cove, Arkansas, USA. 1.667 (10), 2.539 (9), 2.79 (8), 3.12 (6), 4.42 (4), 1.728 (4), 1.395 (4)

Chemistry:

	(1)	(2)
SiO_2	9.6	16.63
TiO_2	5.6	6.95
ZrO_2	29.9	27.38
Al_2O_3	11.0	9.37
Fe_2O_3	13.4	
Nb_2O_5	1.0	
FeO	0.8	6.43
MgO	0.5	3.32
CaO	29.8	30.25
Total	101.6	100.33

(1) Magnet Cove, Arkansas, USA; corresponding to $(Ca_{3.11}Fe^{2+}_{0.07}Mg_{0.07})_{\Sigma=3.25}$ $(Zr_{1.42}Ti_{0.40}Nb_{0.05})_{\Sigma=1.87}(Al_{1.26}Fe^{3+}_{0.98}Si_{0.94})_{\Sigma=3.18}O_{12}$. (2) Stromboli, Italy; by electron microprobe, average of three analyses; corresponding to $(Ca_{2.94}Mg_{0.06})_{\Sigma=3.00}$ $(Zr_{1.21}Ti_{0.47}Mg_{0.32})_{\Sigma=2.00}(Si_{1.51}Al_{1.00}Fe^{3+}_{0.49})_{\Sigma=3.00}O_{12}$.

Mineral Group: Garnet group.

Occurrence: In a carbonatite associated with ijolite (Magnet Cove, Arkansas, USA); in a shoshonitic basaltic lava (Stromboli, Italy).

Association: Apatite, calcite, monticellite, magnetite, perovskite, anhydrite, vesuvianite, biotite, pyrite (Magnet Cove, Arkansas, USA).

Distribution: In the Kimzey quarry, Magnet Cove, Hot Springs Co., Arkansas, USA. On Stromboli, Lipari Islands, Italy.

Name: For Joe Kimzey, the mineral's discoverer, and the Kimzey family, of Magnet Cove, Arkansas, USA.

Type Material: National Museum of Natural History, Washington, D.C., USA, 117124.

References: (1) Milton, C., B.L. Ingram, and L.V. Blade (1961) Kimzeyite, a zirconium garnet from Magnet Cove, Arkansas. Amer. Mineral., 46, 533–548. (2) Munno, R., G. Rossi, and C. Tadini (1980) Crystal chemistry of kimzeyite from Stromboli, Aeolian Islands, Italy. Amer. Mineral., 65, 188–191.

Crystal Data: Monoclinic. *Point Group:* $2/m$. As well-formed crystals, to 1.5 mm, tabular on [100], somewhat elongated along [001]; also in veinlets, massive.

Physical Properties: *Cleavage:* Excellent on {010}; distinct on {100} and {001}. Hardness = 4–5   D(meas.) = 3.16(3)   D(calc.) = [3.20]

Optical Properties: Transparent to translucent. *Color:* Deep azurite-blue. *Optical Class:* Biaxial (–). *Pleochroism:* Strong; X = pale greenish blue; Y = blue; Z = deep blue. *Orientation:* $X = b$; $Z \wedge c \simeq 0°$. *Dispersion:* $r < v$, distinct. *Absorption:* $Z > Y > X$. $\alpha = 1.638–1.640$ $\beta = 1.663–1.665$ $\gamma = 1.676–1.680$ 2V(meas.) = 68°   2V(calc.) = 64°–80°

Cell Data: *Space Group:* $P2_1/m$.   a = 6.991(2)   b = 12.884(3)   c = 5.655(2) $\beta = 96°11(2)'$ Z = 2

X-ray Powder Pattern: Santa Rita Mountains, Arizona, USA.
4.72 (100), 3.052 (81), 6.44 (74), 2.116 (41), 3.138 (30), 2.315 (30), 3.951 (26)

Chemistry:

	(1)	(2)
SiO_2	35.90	36.97
CuO	31.10	32.63
MgO	0.15	
CaO	23.55	23.01
H_2O	8.16	7.39
Total	98.86	100.00

(1) Santa Rita Mountains, Arizona, USA. (2) $Ca_2Cu_2Si_3O_8(OH)_4$.

Occurrence: In vugs and veinlets in skarn (Santa Rita Mountains, Arizona, USA); in amygdules in basaltic lava flows (Calumet, Michigan, USA).

Association: Apophyllite, copper (Santa Rita Mountains, Arizona, USA); quartz, calcite, copper, silver, epidote, pumpellyite, chlorite (Calumet, Michigan, USA).

Distribution: In the USA, in Arizona, between Helvetia and Rosemont, Santa Rita Mountains, Pima Co., and in the Christmas copper mine, Gila Co.; in the Bawana mine, about six km northwest of Milford, Beaver Co., Utah; and in Michigan, in the Laurium mine, Calumet, Houghton Co.

Name: For Fr. Eusebio Francisco Kino (1645–1711), Jesuit pioneer of the Sonora-Arizona-California frontier.

Type Material: National Museum of Natural History, Washington, D.C., USA, 122395.

References: (1) Anthony, J.W. and R.B. Laughon (1970) Kinoite, a new hydrous copper calcium silicate mineral from Arizona. Amer. Mineral., 55, 709–715. (2) Laughon, R.B. (1971) The crystal structure of kinoite. Amer. Mineral., 56, 193–200. (3) Bauer, W.H. (1971) The prediction of bond length variations in silicon-oxygen bonds. Amer. Mineral., 56, 1573–1599. (4) Ruotsala, A.P. and M.L. Wilson (1977) Kinoite from Calumet, Michigan. Amer. Mineral., 62, 1032–1033.

Crystal Data: Monoclinic. *Point Group:* $2/m$. Forms small scales, < 1 mm.

Physical Properties: *Cleavage:* $\{001\}$, perfect. *Tenacity:* Brittle. Hardness = 2.5–3
D(meas.) = 3.30   D(calc.) = 3.33

Optical Properties: Semitransparent. *Color:* Yellow-brown to colorless; light yellow to colorless in thin section. *Luster:* Vitreous.
Optical Class: Biaxial (–). *Pleochroism:* X = very light yellow to light yellow; $Y = Z$ = light yellow with brownish tinge. *Absorption:* $Y \simeq Z > X$. $\alpha = 1.619$ $\beta = 1.628$–1.633 $\gamma = 1.635$
2V(meas.) = 23°

Cell Data: *Space Group:* $C2/m$. a = 5.345(3)   b = 9.250(4)   c = 10.256(8)
$\beta = 99.99(6)°$ Z = 2

X-ray Powder Pattern: Noda-Tamagawa mine, Japan.
3.37 (100), 2.52 (55), 2.020 (55), 5.05 (50), 10.1 (45), 1.684 (15), 3.16 (5)

Chemistry:

	(1)	(2)		(1)	(2)
SiO_2	24.58	23.43	BaO	17.85	27.60
TiO_2	0.16		Na_2O	0.68	0.11
Al_2O_3	22.06	19.25	K_2O	3.30	0.24
Fe_2O_3	0.71	1.87	F	0.21	
Mn_2O_3	3.24		H_2O^+	2.90	
FeO	0.04		H_2O^-	0.20	
MnO	7.38	2.62	H_2O		3.50
MgO	16.60	21.95	$-O = F_2$	0.09	
CaO	0.05	0.05	Total	99.87	100.62

(1) Noda-Tamagawa mine, Japan; corresponds to $(Ba_{0.58}K_{0.35}Na_{0.11}Ca_{0.01})_{\Sigma=1.05}(Mg_{2.06}Mn^{2+}_{0.52}$ $Al_{0.22}Mn^{3+}_{0.21}Fe^{3+}_{0.04}Ti_{0.01})_{\Sigma=3.06}Si_{2.05}Al_{1.94}O_{10}[(OH)_{1.62}O_{0.33}F_{0.06}]_{\Sigma=2.01}$. (2) Netra, India; by electron microprobe, total Fe as Fe_2O_3; corresponding to $(Ba_{0.93}K_{0.03}Na_{0.02}Ca_{0.01})_{\Sigma=0.99}$ $(Mg_{2.80}Mn_{0.19}Fe_{0.08})_{\Sigma=3.07}Si_{2.01}(Al_{1.94}Fe_{0.05})_{\Sigma=1.99}O_{10}(OH)_2$.

Polymorphism & Series: 1M, $2M_1$ polytypes.

Mineral Group: Mica group.

Occurrence: In hausmannite-tephroite ore (Noda-Tamagawa mine, Japan); in manganese-rich rocks invaded by silicic pegmatite and carbonate veins (Netra, India).

Association: Hausmannite, tephroite, celsian, quartz, spessartine, rhodonite, chalcopyrite, pyrrhotite, rhodochrosite, hübnerite, sonolite (Noda-Tamagawa mine, Japan); braunite, hausmannite, bixbyite, alkalic feldspar, hematite, calcite, dolomite, quartz (Netra, India).

Distribution: In the Noda-Tamagawa mine, Iwate Prefecture, and at Hokkejino, Kyoto Prefecture, Japan. From Netra, Balaghat district, Madhya Pradesh, India. On Trumbull Peak, Mariposa Co., California, USA.

Name: For Dr. Kameki Kinoshita (1896–1974), investigator of ore deposits in Japan.

Type Material: National Science Museum, Tokyo, Japan, M19511; National School of Mines, Paris, France.

References: (1) Yoshii, M., K. Maeda, T. Kato, T. Watanabe, S. Yui, A. Kato, and K. Nagashima (1973) Kinoshitalite, a new mineral from the Noda-Tamagawa mine, Iwate Prefecture. Chigaku Kenkyu (Geosci. Mag.), 24, 181–190 (in Japanese). (2) (1975) Amer. Mineral., 60, 486–487 (abs. ref. 1). (3) Kato, T., Y. Miúra, M. Yoshii, and K. Maeda (1979) The crystal structures of 1M-kinoshitalite, a new barium brittle mica and 1M-manganese trioctahedral micas. Mineral. J. (Japan), 9, 392–408. (4) Dasgupta, S., S. Chakraborti, P. Sengupta, P.K. Bhattacharya, H. Banerjee, and M. Fukuoka (1989) Compositional characteristics of kinoshitalite from the Sausar group, India. Amer. Mineral., 74, 200–202.

Crystal Data: Orthorhombic. *Point Group:* $2/m \, 2/m \, 2/m$. In crystals, to 0.5 mm; as skeletal rims on other minerals; crystalline massive.

Physical Properties: Hardness = n.d. D(meas.) = 3.434 D(calc.) = 3.596

Optical Properties: Transparent. *Color:* Pale green; colorless in thin section.
Optical Class: Biaxial (–). *Orientation:* $X = b$; $Y = c$; $Z = a$. $\alpha = 1.660$–1.689 $\beta = 1.720$
$\gamma = 1.694$–1.728 2V(meas.) = 51(1)°   2V(calc.) = 53°–61°

Cell Data: *Space Group:* [*Pbnm*] (by analogy to the olivine group). $a = 4.859$ $b = 11.132$
$c = 6.420$ $Z = 4$

X-ray Powder Pattern: Mt. Shaheru, Zaire.
2.949 (100), 2.680 (85), 2.604 (80), 3.658 (70), 1.830 (60), 2.414 (40), 5.569 (35)

Chemistry:

	(1)	(2)
SiO$_2$	32.71	31.96
TiO$_2$	0.23	
Al$_2$O$_3$	0.26	
Fe$_2$O$_3$	0.66	
FeO	29.34	38.21
MnO	1.65	
MgO	4.95	
CaO	29.30	29.83
Na$_2$O	0.34	
K$_2$O	0.36	
H$_2$O$^+$	0.25	
H$_2$O$^-$	0.06	
P$_2$O$_5$	0.07	
Total	100.18	100.00

(1) Mt. Shaheru, Zaire. (2) CaFeSiO$_4$.

Polymorphism & Series: Forms a series with monticellite.

Occurrence: In melilite-nephelinite lava (Mt. Shaheru, Zaire); in calcareous skarn (Tazheran intrusive, Russia).

Association: Melilite, nepheline, clinopyroxene, kalsilite, götzenite, combeite, sodalite, magnetite, perovskite, apatite, "hornblende," biotite (Mt. Shaheru, Zaire); titanian augite, wollastonite, melilite, garnet, calcite, cuspidine, diopside, perovskite, troilite, graphite (Tazheran intrusive, Russia).

Distribution: On Mt. Shaheru, the extinct southern cone of Mt. Nyiragongo, Kivu Province, Zaire. In the Wessels mine, near Kuruman, Cape Province, South Africa. From the Tazheran alkalic intrusive, Baikal, eastern Siberia, Russia. In the Angra dos Reis meteorite.

Name: For Dr. Egon Kirschstein, German geologist, a pioneer in the geological exploration of Kivu Province, Zaire.

Type Material: Harvard University, Cambridge, Massachusetts, USA.

References: (1) Sahama, T.G. and K. Hytönen (1957) Kirschsteinite, a natural analogue to synthetic iron monticellite, from the Belgian Congo. Mineral. Mag., 31, 698–699. (2) (1958) Amer. Mineral., 43, 790 (abs. ref. 1). (3) Konev, A.A., Z.F. Ushchapovskaya, and V.S. Lebedeva (1970) First find of magnesian kirschsteinite in the USSR. Doklady Acad. Nauk SSSR, 190, 932–935 (in Russian).

Kittatinnyite

$$Ca_4Mn_2^{2+}Mn_4^{3+}Si_4O_{16}(OH)_8 \cdot 18H_2O$$

Crystal Data: Hexagonal. *Point Group:* $6/m\ 2/m\ 2/m$, $6mm$, or $\bar{6}m2$. As very thin, distorted crystals, to 0.2 mm, flattened on {0001}, and in subparallel aggregates.

Physical Properties: *Cleavage:* {0001}, perfect. *Tenacity:* Brittle. Hardness = ~4 D(meas.) = 2.61 D(calc.) = 2.62

Optical Properties: Translucent to opaque. *Color:* Bright to golden yellow; in thin section, medium yellow. *Streak:* Light yellow. *Luster:* Vitreous.
Optical Class: Uniaxial (–). *Pleochroism:* Very weak in medium yellow. *Absorption:* $E \geq O$.
$\omega = 1.727(3)$ $\epsilon = $ n.d.

Cell Data: *Space Group:* $P6_3/mmc$, $P6_3mc$, or $P\bar{6}2c$.   a = 6.498(4)   c = 22.78(2)   Z = 1

X-ray Powder Pattern: Franklin, New Jersey, USA.
11.2 (100), 5.61 (60), 2.733 (60), 2.822 (50), 2.525 (40), 2.279 (40), 5.03 (30)

Chemistry:

	(1)	(2)
SiO_2	18.2	18.23
Mn_2O_3	23.6	23.95
As_2O_5	0.8	
FeO	0.0	
MnO	10.6	10.76
CuO	0.0	
ZnO	0.3	
MgO	0.0	
CaO	17.3	17.01
H_2O	[29.2]	30.05
Total	[100.0]	100.00

(1) Franklin, New Jersey, USA; by electron microprobe, $Mn^{2+}:Mn^{3+}$ from charge balance, H_2O by difference. (2) $Ca_4Mn_2^{2+}Mn_4^{3+}Si_4O_{16}(OH)_8 \cdot 18H_2O$.

Occurrence: A secondary mineral coating fractures in a metamorphosed stratiform zinc deposit.

Association: Bostwickite, calcite, franklinite, fluorite.

Distribution: From Franklin, Sussex Co., New Jersey, USA.

Name: Derived from *kittatinny*, which in the language of the Algonquin Indians meant *endless hills*, in allusion to the topography of the Franklin area.

Type Material: Harvard University, Cambridge, Massachusetts, 113514; National Museum of Natural History, Washington, D.C., USA, C4222.

References: (1) Dunn, P.J. and D.R. Peacor (1983) Kittatinnyite and wallkilldellite, silicate/arsenate analogues containing calcium and manganese, from Franklin and Sterling Hill, New Jersey. Amer. Mineral., 68, 1029–1032.

Crystal Data: Cubic. *Point Group:* $4/m\,\bar{3}\,2/m$. Massive and as minute grains.

Physical Properties: Hardness = ~7   D(meas.) = 3.756   D(calc.) = 3.852

Optical Properties: Semitransparent. *Color:* Bluish green, greenish. *Luster:* Vitreous. *Optical Class:* Isotropic. $n = 1.803$

Cell Data: *Space Group: Ia3d.* a = 11.65   Z = 8

X-ray Powder Pattern: Kao kimberlite pipe, Lesotho.
2.92 (100), 2.61 (100), 2.382 (80), 1.560 (80), 2.489 (60), 2.288 (60), 1.894 (60)

Chemistry:

	(1)	(2)
SiO_2	39.92	39.78
TiO_2	0.11	
Al_2O_3	9.74	
Cr_2O_3	17.47	33.54
Fe_2O_3	1.20	
FeO	6.53	
MnO	0.60	
MgO	16.97	26.68
CaO	8.14	
Total	100.68	100.00

(1) Kao kimberlite pipe, Lesotho; by electron microprobe, corresponding to $(Mg_{1.90}Ca_{0.66}$ $Fe^{2+}_{0.41}Mn_{0.17})_{\Sigma=3.14}(Cr_{1.04}Al_{0.86}Fe^{3+}_{0.07})_{\Sigma=1.97}Si_{3.01}O_{12}$. (2) $Mg_3Cr_2(SiO_4)_3$.

Polymorphism & Series: Forms a series with pyrope.

Mineral Group: Garnet group.

Occurrence: A rare component of ultramafic kimberlite nodules.

Association: Olivine, enstatite, chromian diopside, chromian pyrope, chromian spinel, ilmenite, perovskite, zircon, diamond, omphacite, rutile, carbonates, micas.

Distribution: In the Kao kimberlite pipe, Lesotho.

Name: For Oleg von Knorring, Department of Earth Sciences, Leeds University, Leeds, England.

Type Material: n.d.

References: (1) Nixon, P.H. and G. Hornung (1968) A new chromium garnet end member, knorringite, from kimberlite. Amer. Mineral., 53, 1833–1840. (2) Nixon, P.H., O. von Knorring, and J.M. Rooke (1963) Kimberlites and associated inclusions of Basutoland: a mineralogical and geochemical study. Amer. Mineral., 48, 1090–1132.

Koashvite $Na_6(Ca, Mn)(Fe^{3+}, Ti)Si_6O_{18} \cdot H_2O$

Crystal Data: Orthorhombic. *Point Group:* $2/m \ 2/m \ 2/m$. Habit not stated.

Physical Properties: *Fracture:* Conchoidal. Hardness = 6   VHN = 680–740 (20 g load). D(meas.) = 2.98–3.02   D(calc.) = 3.069

Optical Properties: Transparent. *Color:* Pale yellow. *Luster:* Vitreous. *Optical Class:* Biaxial (–). *Dispersion:* $r > v$, weak. $\alpha = 1.637$ $\beta = 1.643$ $\gamma = 1.648$ 2V(meas.) = 83°

Cell Data: *Space Group: Pmnb.*   a = 10.179(1)   b = 20.899(2)   c = 7.335(1)   Z = 4

X-ray Powder Pattern: Mt. Koashva, Russia. 2.581 (100), 1.820 (70), 3.66 (50), 3.28 (50), 1.504 (50), 2.620 (40), 1.476 (40)

Chemistry:

	(1)
SiO_2	51.00
TiO_2	5.07
Fe_2O_3	5.42
MnO	4.20
MgO	trace
CaO	6.00
Na_2O	25.60
K_2O	0.38
H_2O	2.27
Total	99.94

(1) Mt. Koashva, Russia; corresponds to $(Na_{5.88}K_{0.06})_{\Sigma=5.94}(Ca_{0.76}Mn_{0.42})_{\Sigma=1.18}$ $(Fe^{3+}_{0.48}Ti_{0.45})_{\Sigma=0.93}Si_{6.05}O_{17.88} \cdot 0.90H_2O$.

Mineral Group: Lovozerite group.

Occurrence: As veinlets replacing lomonosovite in an ultra-agpaitic pegmatite in an alkalic massif.

Association: Lomonosovite, pectolite, villiaumite, natrophosphate.

Distribution: On Mt. Koashva, Khibiny massif, Kola Peninsula, Russia.

Name: For the locality, Mt. Koashva, Kola Peninsula, Russia.

Type Material: A.E. Fersman Mineralogical Museum, Academy of Sciences, Moscow, Russia; National School of Mines, Paris, France.

References: (1) Kapustin, Y.L., Z.V. Pudovkina, A.V. Bykova, and G.V. Lyubomilova (1974) Koashvite, a new mineral. Zap. Vses. Mineral. Obshch., 103, 559–566 (in Russian). (2) (1975) Amer. Mineral., 60, 487 (abs. ref. 1). (3) Chernitsova, N.M., Z.A. Pudovkina, A.A. Voronkov, and Y.A. Pyatenko (1980) Crystal structure of koashvite $Na_6(Ca, Mn)_{1+0.5x}(Fe^{3+}_xTi_{1-x})[Si_6O_{18}]$. Mineral. Zhurnal, 2(5), 40–44 (in Russian). (4) (1981) Chem. Abs., 94, 39866 (abs. ref. 3).

Crystal Data: Orthorhombic. *Point Group:* $2/m\ 2/m\ 2/m$. As crystals, to 0.5 mm, of pinacoidal habit, predominantly tabular; as fractured grains.

Physical Properties: *Fracture:* Even. *Tenacity:* Brittle. Hardness = ~4.5 D(meas.) = 4.17(2)   D(calc.) = 4.20

Optical Properties: Semitransparent. *Color:* Bright yellowish orange. *Streak:* Light orange. *Luster:* Vitreous.
Optical Class: Biaxial (–). *Pleochroism:* Strong; X = colorless or pale yellow; Y = yellowish orange; Z = light yellow. *Orientation:* $X = b$; $Y = c$; $Z = a$. *Dispersion:* $r < v$, strong. *Absorption:* $Z = Y > X$. $\alpha = 1.779(2)$ $\beta = 1.786(2)$ $\gamma = 1.790(2)$ 2V(meas.) = 78(2)° 2V(calc.) = 74°

Cell Data: *Space Group:* $Cmca$.   a = 18.59(3)   b = 8.789(5)   c = 12.04(1)   Z = 4

X-ray Powder Pattern: Sterling Hill, New Jersey, USA.
2.970 (100), 1.540 (70), 3.58 (60), 2.608 (50), 2.815 (40), 2.479 (40), 2.342 (40)

Chemistry:

	(1)
SiO_2	10.4
As_2O_5	18.9
FeO	0.4
MnO	39.2
ZnO	26.0
MgO	0.8
H_2O	5.5
Total	101.2

(1) Sterling Hill, New Jersey, USA; by electron microprobe, H_2O by TGA; corresponds to $(Mn_{6.70}Mg_{0.24}Fe_{0.06})_{\Sigma=7.00}Zn_{3.87}(AsO_4)_2(Si_{1.05}O_4)_2(OH)_8$.

Occurrence: Incrusting willemite-franklinite ore from metamorphosed stratiform zinc orebodies.

Association: Willemite, franklinite, sonolite, friedelite, holdenite, calcite.

Distribution: At Sterling Hill, Ogdensburg, and at Franklin, Sussex Co., New Jersey, USA.

Name: For John Kolic, of Rockaway, New Jersey, USA, who first discovered the mineral.

Type Material: Harvard University, Cambridge, Massachusetts; National Museum of Natural History, Washington, D.C., USA, 143765.

References: (1) Dunn, P.J., D.R. Peacor, and B.D. Sturman (1979) Kolicite, a new manganese zinc silicate arsenate from Sterling Hill, Ogdensburg, New Jersey. Amer. Mineral., 64, 708–712. (2) Peacor, D.R. (1980) The crystal structure of kolicite, $Mn_7(OH)_4[As_2Zn_4Si_2O_{16}(OH)_4]$. Amer. Mineral., 65, 483–487.

Komarovite

$(Ca, Mn)Nb_2(Si_2O_7)(O, F)_3 \cdot 3.5H_2O$

Crystal Data: Orthorhombic. *Point Group:* n.d. Platy, foliated, massive.

Physical Properties: *Cleavage:* Fair on {001}. Hardness = 1.5–2 D(meas.) = 3.0 D(calc.) = 2.96

Optical Properties: Transparent to translucent. *Color:* Pale rose. *Streak:* White. *Luster:* Dull.
Optical Class: Biaxial (+). *Orientation:* X = a; Y = c; Z = b. $\alpha = 1.750(2)$ $\beta = 1.766(2)$ $\gamma = 1.85(2)$ 2V(meas.) = 48°

Cell Data: *Space Group:* n.d. a = 21.30(7) b = 14.00(5) c = 17.19(7) Z = 18

X-ray Powder Pattern: Mt. Karnasurt, Russia.
3.16 (100), 12.2 (70), 1.783 (45), 3.118 (42), 6.35 (35), 2.740 (35), 2.715 (35)

Chemistry:

	(1)
SiO_2	23.50
TiO_2	2.50
Al_2O_3	1.00
Fe_2O_3	1.50
Nb_2O_5	47.00
MnO	5.00
CaO	4.70
Na_2O	0.85
K_2O	0.30
F	1.21
H_2O	12.00
$-O = F_2$	0.51
Total	99.05

(1) Mt. Karnasurt, Russia; corresponding to $(Ca_{0.43}Mn_{0.36}Na_{0.14}K_{0.03})_{\Sigma=0.96}(Nb_{1.82}Ti_{0.16}Al_{0.10}Fe_{0.10})_{\Sigma=2.18}Si_2O_9(O_{0.88}F_{0.33})_{\Sigma=1.21} \cdot 3.5H_2O$.

Occurrence: With late albite and redeposited fine-grained natrolite in alkalic rocks in a differentiated alkalic massif.

Association: Natrolite, albite.

Distribution: On Mt. Karnasurt, Lovozero massif, Kola Peninsula, Russia.

Name: For the Russian cosmonaut, Vladimir M. Komarov (1927–1967).

Type Material: A.E. Fersman Mineralogical Museum, Academy of Sciences, Moscow, Russia.

References: (1) Portnov, A.M., G.K. Krivokoneva, and T.I. Stolyarova (1971) Komarovite, a new niobosilicate of calcium and manganese. Zap. Vses. Mineral. Obshch., 100, 599–602 (in Russian). (2) (1972) Amer. Mineral., 57, 1315–1316 (abs. ref. 1).

Crystal Data: Hexagonal. *Point Group:* 32. As equant crystals terminated by trigonal pyramids, to 5 mm.

Physical Properties: *Tenacity:* Brittle. Hardness = 3–4   D(meas.) = 3.31(5) D(calc.) = 3.31   Light blue cathodoluminescence.

Optical Properties: Semitransparent. *Color:* Brown. *Streak:* Light brown. *Luster:* Vitreous.
Optical Class: Uniaxial (–). $\omega = 1.671(1)$ $\epsilon = 1.644(1)$

Cell Data: *Space Group:* $R32$.   a = 10.526(6)   c = 15.736(9)   Z = 6

X-ray Powder Pattern: Vuoriyärvi complex, Russia.
5.23 (100), 2.96 (90), 3.59 (80), 3.02 (80), 2.57 (60), 2.106 (60)

Chemistry:

	(1)
SiO$_2$	34.44
ZrO$_2$	24.94
HfO$_2$	0.46
FeO	0.33
MnO	0.00
CaO	0.08
BaO	28.19
K$_2$O	0.13
H$_2$O	[11.43]
Total	[100.00]

(1) Vuoriyärvi complex, Russia; by electron microprobe, H$_2$O by difference (H$_2$O 10% to 11.5% by coulometric analysis); corresponds to $(Ba_{0.95}Fe_{0.02}Ca_{0.01}K_{0.01})_{\Sigma=0.99}(Zr_{1.04}Hf_{0.01})_{\Sigma=1.05}$ $Si_{2.95}O_9$·3.08H$_2$O.

Occurrence: Apparently altering from catapleiite, in dolomitic veinlets cutting metasomatically altered pyroxenites, in a carbonatite.

Association: Dolomite, strontianite, phlogopite, barite, georgechaoite, pyrite.

Distribution: In the Vuoriyärvi carbonatite complex, Kola Peninsula, Russia.

Name: To honor Russian mineralogist and crystallographer A.I. Komkov (1926–1987).

Type Material: A.E. Fersman Mineralogical Museum, Academy of Sciences, Moscow, Russia.

References: (1) Voloshin, A.V., Y.A. Pakhomovskii, Y.P. Men'shikov, Y.V. Sokolova, and Y.K. Yegorov-Tismenko (1990) Komkovite – a new hydrous barium zirconosilicate from the carbonatites of Vuoriyarvi (Kola Peninsula). Mineral. Zhurnal, 12(3), 69–73 (in Russian).
(2) (1992) Amer. Mineral., 77, 207–208 (abs. ref. 1). (3) Sokolova, E.V., A.V. Araktscheeva, and A.V. Voloshin (1991) Crystal structure of komkovite. Doklady Acad. Nauk SSSR, 320, 1384–1388 (in Russian). (4) (1993) Amer. Mineral., 78, 454 (abs. ref. 3).

Kornerupine

$(Mg, Fe^{2+})_4Al_6(Si, Al, B)_5O_{21}(OH)$

Crystal Data: Orthorhombic. *Point Group:* $2/m \, 2/m \, 2/m$. Crystals prismatic, to over 20 cm, showing principally $\{110\}$, $\{100\}$, and $\{010\}$, rarely terminated; as radiating aggregates.

Physical Properties: *Cleavage:* Good on $\{110\}$. Hardness = 6–7 D(meas.) = 3.29–3.35 D(calc.) = 3.288

Optical Properties: Transparent to translucent and opaque. *Color:* Colorless, white, blue, green, dark green, greenish yellow, yellow-brown, black. *Luster:* Vitreous.
Optical Class: Biaxial (–); may be pseudouniaxial (–). *Pleochroism:* X = colorless to green; Y = colorless, pale brownish yellow, pale yellowish green; Z = pale brownish green, green, light amber. *Orientation:* X = c; Y = a; Z = b. *Dispersion:* $r < v$ or $r > v$. $\alpha = 1.660$–1.671 $\beta = 1.673$–1.683 $\gamma = 1.674$–1.684 2V(meas.) = 3°–48°

Cell Data: *Space Group:* $Cmcm$. $a = 16.041(3)$ $b = 13.746(2)$ $c = 6.715(2)$ $Z = 4$

X-ray Powder Pattern: Mautia Hill, Tanzania.
2.639 (100), 3.03 (80), 3.37 (60), 2.118 (60), 1.503 (40), 2.096 (30), 1.685 (30)

Chemistry:

	(1)		(1)
SiO_2	29.80	MgO	18.96
TiO_2	0.06	CaO	0.07
B_2O_3	1.44	Li_2O	0.08
Al_2O_3	45.85	Na_2O	0.04
FeO	1.60	F	0.01
MnO	0.05	H_2O	[1.21]
		Total	[99.17]

(1) Fiskenæsset, Greenland; by electron microprobe, Li, B, F by ion microprobe, H_2O from stoichiometry; corresponds to $(Mg_{3.48}Fe_{0.16}Li_{0.04}Ca_{0.01}Na_{0.01})_{\Sigma=3.70}(Al_{5.64}Ti_{0.01})_{\Sigma=5.65}$ $(Si_{3.67}Al_{1.02}B_{0.31})_{\Sigma=5.00}O_{21}[(OH)_{0.99}F_{0.01}]_{\Sigma=1.00}$.

Occurrence: In boron-rich volcanic and sedimentary rocks subjected to amphibolite to granulite facies metamorphism; in metamorphosed anorthosite complexes.

Association: Sapphirine, cordierite, spinel, corundum, tourmaline, grandidierite, dumortierite, kyanite, sillimanite, andalusite, biotite, phlogopite, magnetite, ilmenite, hematite, rutile.

Distribution: Some localities for well-studied material follow. In Greenland, at Fiskenæsset. At Bjordam, near Kragerö, Norway. In Germany, from Waldheim, Saxony. From Mautia Hill, Tanzania. On the Bok se Puts Farm, Namaqualand, Cape Province, South Africa. Gem crystals from Itrongahy, near Betroka, and elsewhere in Madagascar. From gem gravels of the Matale and Ratnapura districts, Sri Lanka. In the Harts and Strangways Ranges, Northern Territory, Australia. At Lac Ste-Marie, Quebec, Canada.

Type Material: University of Copenhagen, Copenhagen, Denmark, 1883.754.

Name: For the Danish geologist, Andreas Nikolaus Kornerup (1857–1883).

References: (1) Dana, E.S. (1892) Dana's system of mineralogy, (6th edition), 560–561. (2) McKie, D. (1965) The magnesium aluminium borosilicates: kornerupine and grandidierite. Mineral. Mag., 34, 346–357. (3) Grew, E.S., R.K. Herd, and N. Marquez (1987) Boron-bearing kornerupine from Fiskenæsset, West Greenland: a re-examination of specimens from the type locality. Mineral. Mag., 51, 695–708. (4) Moore, P.B., P.K. Sen Gupta, and E.O. Schlemper (1989) Kornerupine: chemical crystallography, comparative crystallography, and its cation relation to olivine and to Ni_2In intermetallic. Amer. Mineral., 74, 642–655. (5) Grew, E.S., J.V. Chernosky, G. Werding, K. Abraham, N. Marquez, and J.R. Hinthorne (1990) Chemistry of kornerupine and associated minerals, a wet chemical, ion microprobe, and X-ray study emphasizing Li, Be, B and F contents. J. Petrol., 31, 1025–1070.

$KNa_2(Mg, Mn^{3+}, Fe^{3+}, Li)_5Si_8O_{22}(OH)_2$ **Kornite**

Crystal Data: Monoclinic. *Point Group:* $[2/m.]$ In bundles of fibers, elongated along [001], commonly bent, to 200 μm; rimming hennomartinite.

Physical Properties: *Cleavage:* Parallel to [001]. *Tenacity:* Brittle. Hardness = [5–6] D(meas.) = n.d. D(calc.) = 3.15

Optical Properties: Transparent. *Color:* Dark red to brownish lilac. *Luster:* Vitreous. *Optical Class:* Biaxial (–). *Pleochroism:* X = pink; Y = dark red; Z = red-orange. *Orientation:* Z = b; Y ∧ c = 60°–65°. $\alpha = 1.654(4)$ $\beta = [1.675]$ $\gamma = 1.696(4)$ 2V(meas.) = 88°–92°

Cell Data: *Space Group:* $P2_1/m$ or $P2_1/a$. a = 9.94(1) b = 17.80(2) c = 5.302(4) $\beta = 105.5(2)°$ Z = 2

X-ray Powder Pattern: Wessels mine, South Africa; assembled from single crystal patterns. 3.257 (s), 3.132 (s), 2.812 (s), 2.553 (s), 8.890 (m), 8.427 (m), 5.077 (m)

Chemistry:

	(1)
SiO_2	56.06
Al_2O_3	0.00
Fe_2O_3	4.93
Mn_2O_3	13.17
MgO	10.03
CaO	0.00
Li_2O	1.96
Na_2O	7.61
K_2O	3.56
F	n.d.
H_2O	[2.68]
Total	[100.00]

(1) Wessels mine, South Africa; by electron microprobe, Li by ion microprobe, average of seven analyses, H_2O by difference; corresponds to $(K_{0.65}Na_{0.31})_{\Sigma=0.96}(Na_{1.79}Li_{0.21})_{\Sigma=2.00}(Mg_{2.12}Mn^{3+}_{1.43}Li_{0.91}Fe^{3+}_{0.52})_{\Sigma=4.98}Si_{8.00}O_{22}(OH)_2$.

Mineral Group: Amphibole (alkali) group: $Na_B \geq 1.34$; $Li_C \geq 0.5$; $Mn^{3+} > Fe^{3+}$.

Occurrence: From a hand specimen, in veinlets of sérandite-pectolite cutting sugilite, probably of hydrothermal origin in a bedded manganese deposit.

Association: Sérandite-pectolite, sugilite, braunite, taikanite, hennomartinite.

Distribution: From the Wessels mine, near Kuruman, Cape Province, South Africa.

Name: For Hermann Korn (?–1946), German geologist, professionally associated with Henno Martin, for whom an associated mineral is named.

Type Material: Natural History Museum, Bern, Switzerland, B5564.

References: (1) Armbruster, T., R. Oberhänsli, V. Bermanec, and R. Dixon (1993) Hennomartinite and kornite, two new Mn^{3+} rich silicates from the Wessels mine, Kalahari, South Africa. Schweiz. Mineral. Petrog. Mitt., 73, 349–355. (2) (1994) Amer. Mineral., 79, 763–764 (abs. ref. 1).

Crystal Data: Monoclinic. *Point Group:* $2/m$. Short prismatic crystals, to 2 mm; in fibrous and polycrystalline aggregates. *Twinning:* Simple, lamellar on {100}, {001}.

Physical Properties: *Cleavage:* Good on {110}, (110) $\wedge$ ($1\bar{1}0$) $\sim 87°$; parting on {001}. Hardness = ~ 6 D(meas.) = 3.51–3.60   D(calc.) = 3.60

Optical Properties: Semitransparent. *Color:* Emerald-green; green or yellow in thin section. *Optical Class:* Biaxial (–). *Pleochroism:* Strong; X = yellowish green; Y = blue-green, grass-green; Z = emerald-green. *Orientation:* $Y = b$; $Z \wedge a = 8°$–$22°$. *Dispersion:* $r > v$, moderate to strong. $\alpha = 1.740$–1.766 $\beta = 1.756$–1.778 $\gamma = 1.745$–1.781 2V(meas.) = $6°$–$70°$

Cell Data: *Space Group:* $C2/c$.   a = 9.550   b = 8.712   c = 5.273 $\beta = 107.44°$ Z = 4

X-ray Powder Pattern: Synthetic.
2.956 (100), 2.867 (100), 2.5166 (90), 2.4476 (70), 2.0985 (70), 1.5957 (70), 1.3802 (60)

Chemistry:

	(1)	(2)		(1)	(2)
SiO_2	55.5	54.81	MgO	0.8	0.49
TiO_2		0.03	CaO	1.7	0.54
Al_2O_3		4.02	Na_2O	11.6	12.94
Fe_2O_3	0.2	3.61	K_2O		0.02
Cr_2O_3	30.6	23.67	P_2O_5		0.09
MnO		0.02	Total	100.4	[100.24]

(1) Coahuila meteorite; by electron microprobe, corresponds to $(Na_{0.83}Ca_{0.07})_{\Sigma=0.90}$ $(Cr_{0.90}Mg_{0.04})_{\Sigma=0.94}Si_{2.06}O_6$. (2) Myanmar; by electron microprobe, average of seven analyses, average sum originally given as 100.19%; corresponds to $(Na_{0.93}Ca_{0.02})_{\Sigma=0.95}(Cr_{0.69}Al_{0.17}$ $Fe_{0.10}Mg_{0.03})_{\Sigma=0.99}Si_{2.02}O_6$.

Mineral Group: Pyroxene group.

Occurrence: A major constituent of some jadeitites; an accessory constituent of some iron meteorites.

Association: "Cliftonite" [graphite], chromian diopside, troilite (Toluca); daubréelite (Coahuila); krinovite, roedderite, high albite, richterite, chromite (Canyon Diablo); jadeite, chromite, chlorite (Myanmar).

Distribution: In the Toluca, Coahuila, Hex River Mountains, and Canyon Diablo meteorites. Around Tawmaw and other towns, Myitkyina-Mogaung district, Kachin State, Myanmar (Burma). At Mocchie, Susa, Piedmont, Italy. From Williams Creek, Mendocino Co., California, USA.

Name: From the German *kosmisch*, for *cosmic*, in allusion to its meteoritic occurrence, and the Greek *chlor*, for *green*.

Type Material: National Museum of Natural History, Washington, D.C., USA, 81869, 81870; The Natural History Museum, London, England, 81869–81870.

References: (1) Dana, E.S. (1899) Dana's system of mineralogy, (6th edition), app. I, 20. (2) Deer, W.A., R.A. Howie, and J. Zussman (1978) Rock-forming minerals, (2nd edition), v. 2A, single-chain silicates, 520–525. (3) Frondel, C. and C. Klein, Jr. (1965) Ureyite [kosmochlor], $NaCr_2Si_2O_6$, a new meteoritic pyroxene. Science, 149, 742–744. (4) (1965) Amer. Mineral., 50, 2096 (abs. ref. 3). (5) Clark, J.R., D.E. Appleman, and J.J. Papike (1969) Crystal-chemical characterization of clinopyroxenes based on eight new structure refinements. MSA Spec. Paper 2, 31–50. (6) Yang, C.M.O. (1984) A terrestrial source of ureyite [kosmochlor]. Amer. Mineral., 69, 1180–1183. (7) Harlow, G.E. and E.P. Olds (1987) Observations on terrestrial ureyite [kosmochlor] and ureyitic pyroxene. Amer. Mineral., 72, 126–136.

Crystal Data: Monoclinic. *Point Group:* 2/m. As crystals elongated ∥ [001], showing {001}, {010}, {100}, and {011}. *Twinning:* On {100}.

Physical Properties: *Cleavage:* {110}, perfect. Hardness = ~5 VHN = 428–435 D(meas.) = 2.74 D(calc.) = 2.79

Optical Properties: Transparent. *Color:* Colorless. *Luster:* Vitreous.
Optical Class: Biaxial (+). *Orientation:* X = b; Y ∧ c = 45°. *Dispersion:* r < v, weak.
$\alpha = 1.595(2)$ $\beta = 1.598(2)$ $\gamma = 1.610(2)$ 2V(meas.) = 48°

Cell Data: *Space Group:* $P2_1/a$. a = 13.171(4)   b = 11.717(4)   c = 6.565(2) $\beta = 105.26°$
Z = 2

X-ray Powder Pattern: Khibiny massif, Russia.
3.087 (100), 5.60 (60), 3.336 (53), 2.802 (53), 6.42 (47), 5.86 (31), 5.24 (31)

Chemistry:

	(1)	(2)
SiO_2	42.01	43.37
TiO_2	2.06	
ZrO_2	23.90	29.64
HfO_2	0.61	
Fe_2O_3	0.02	
CaO	0.00	
Na_2O	0.00	
K_2O	22.14	22.66
H_2O		4.33
Total	90.74	100.00

(1) Khibiny massif, Russia; by electron microprobe. Presence of H_2O is shown by strong IR absorption bands; if 1.00 H_2O is assumed to satisfy structural requirements and measured density, note the analysis will then sum to only ~95% however. (2)$K_2ZrSi_3O_9 \cdot H_2O$.

Polymorphism & Series: Dimorphous with umbite.

Occurrence: In alkalic pegmatite in a differentiated alkalic massif.

Association: Umbite, wadeite, eudialyte, potassian feldspar, aegirine.

Distribution: In the valley of the Vuonnemiok River, Khibiny massif, Kola Peninsula, Russia.

Name: For Ekaterina E. Kostyleva-Labuntsova (1894–1974), Soviet mineralogist.

Type Material: Geology Museum, Kola Branch, Academy of Sciences, Apatity; A.E. Fersman Mineralogical Museum, Academy of Sciences, Moscow, Russia.

References: (1) Khomyakov, A.P., A.A. Voronkov, L.I Polezhaeva, and N.N. Smol'yaninova (1983) Kostylevite, $K_4Zr_2Si_6O_{18} \cdot 2H_2O$, a new mineral. Zap. Vses. Mineral. Obshch., 112, 469–474 (in Russian). (2) Ilyushin, G.D., A.P. Khomyakov, N.V. Shumyatskaya, A.A. Voronkov, N.N. Nevskii, V.V. Ilyukhin, and N.V. Belov (1981) Crystal structure of a new natural zirconosilicate, $K_4Zr_2Si_6O_{18} \cdot 2H_2O$. Doklady Acad. Nauk SSSR, 256, 860–863 (in Russian). (3) (1984) Amer. Mineral., 69, 812 (abs. refs. 1 and 2).

Kôzulite

$NaNa_2[Mn_4^{2+}(Fe^{3+}, Al)]Si_8O_{22}(OH)_2$

Crystal Data: Monoclinic. *Point Group:* $2/m$. As prismatic crystals, to 3.5 mm, in banded aggregates.

Physical Properties: *Cleavage:* Perfect on $\{110\}$. *Tenacity:* [Brittle.] Hardness = 5 D(meas.) = 3.30   D(calc.) = 3.36

Optical Properties: Semitransparent. *Color:* Reddish black to black. *Streak:* Light purplish brown. *Luster:* Vitreous.
Optical Class: Biaxial (–). *Pleochroism:* Strong; X = yellow-brown; Y = reddish brown; Z = dark brown. *Orientation:* $Y = b$; $X \wedge c = 25°$. *Dispersion:* $r > v$, weak. *Absorption:* $Z > Y > X$. $\alpha = 1.685$ $\beta = 1.717$ $\gamma = 1.720$ 2V(meas.) = $34°$–$36°$

Cell Data: *Space Group:* $C2/m$.   a = 9.914   b = 18.111   c = 5.308 $\beta = 104.50°$ Z = 2

X-ray Powder Pattern: Tanohata mine, Japan.
8.51 (100), 3.15 (65), 2.827 (30), 3.30 (17), 4.53 (10), 3.40 (10), 2.722 (10)

Chemistry:

	(1)	(2)		(1)	(2)
SiO_2	51.38	49.24	MgO	2.71	0.03
TiO_2		0.00	CaO	1.12	0.11
Al_2O_3	1.69	0.02	Na_2O	8.41	8.91
Fe_2O_3	2.85		K_2O	1.36	0.03
Cr_2O_3		0.03	F	0.08	
FeO		0.06	H_2O^+	2.10	
MnO	27.96	37.06	H_2O^-	0.06	
ZnO	0.03		$-O = F_2$	0.03	
			Total	99.72	95.48

(1) Tanohata mine, Japan; corresponding to $(Na_{2.54}K_{0.27}Ca_{0.19})_{\Sigma=3.00}(Mn_{3.69}Mg_{0.63}Fe_{0.33}^{3+}Al_{0.31})_{\Sigma=4.96}Si_{8.00}O_{21.78}[(OH), F]_{2.22}$. (2) Woods mine, Australia; by electron microprobe.

Mineral Group: Amphibole (alkali) group: $Mn^{2+}/(Mg + Fe^{2+} + Mn^{2+}) > 0.33$; $Fe^{3+}/(Fe^{3+} + Al^{vi}) \geq 0.5$; $(Na + K)_A \geq 0.5$; $Na_B \geq 1.34$; $Mn_C \geq 2.5$.

Occurrence: In metamorphosed manganese-rich sediments along the contact of a granodiorite intrusion (Tanohata mine, Japan).

Association: Braunite, rhodonite, manganiferous alkalic pyroxene, quartz (Tanohata mine, Japan).

Distribution: In the Tanohata and Noda-Tamagawa mines, Iwate Prefecture, Japan. From the Woods mine, 30 km north-northeast of Tamworth, New South Wales, Australia.

Name: To honor Shukusuke Kôzu (1880–1955), Professor at Tohoku University, Sendai, Japan.

Type Material: Tohoku University, Sendai, Japan.

References: (1) Nambu, M., K. Tanida, and T. Kitamura (1969) Kôzulite, a new alkali amphibole from Tanohata mine, Iwate Prefecture, Japan. J. Japan. Assoc. Mineral. Petrol. Econ. Geol., 62, 311–328. (2) (1970) Amer. Mineral., 55, 1815 (abs. ref. 1). (3) Kitamura, M. and N. Morimoto (1972) Crystal structure of kôzulite and tetrahedral Al in amphiboles. Acta Cryst., A28, S71. (4) Kawachi, Y. and D.S. Coombs (1993) Namansilite, $NaMn^{3+}Si_2O_6$: a widespread clinopyroxene? Mineral. Mag., 57, 533–538.

$(Mn^{2+}, Mg)_{24}Zn_3Fe^{3+}$
$(As^{3+}O_3)_2(As^{5+}O_4)_3(SiO_4)_6(OH)_{18}$ Kraisslite

Crystal Data: Hexagonal. *Point Group:* $6/m\ 2/m\ 2/m$. As thin, curved, compact foliated scales and coarse plates, to 1 cm; in sprays; in veinlets and lenses.

Physical Properties: *Cleavage:* Perfect on {0001}. *Tenacity:* Brittle. Hardness = 3–4
D(meas.) = 3.876   D(calc.) = 3.918

Optical Properties: Translucent. *Color:* Deep coppery brown; rich brown in thin section. *Streak:* Golden brown. *Luster:* Submetallic.
Optical Class: Uniaxial (+). $\omega = 1.805(2)$ ϵ = n.d. ; weak birefringence.

Cell Data: *Space Group:* $P6_322$.   a = 8.22(1)   c = 43.88(5)   Z = 2

X-ray Powder Pattern: Sterling Hill, New Jersey, USA.
2.74 (100), 2.437 (55), 2.194 (55), 4.385 (45), 3.651 (35), 3.13 (20), 1.219 (7)

Chemistry:

	(1)	(2)
SiO_2	13.8	12.9
Al_2O_3	0.21	0.2
Fe_2O_3		2.0
As_2O_3	0.87	6.69
As_2O_5	17.7	10.35
FeO	1.92	
MnO	51.6	52.0
ZnO	8.47	8.6
MgO	2.53	2.6
H_2O	3.68	[3.68]
Total	100.78	[99.02]

(1) Sterling Hill, New Jersey, USA. (2) Do.; by electron microprobe and other chemical tests; average of five analyses, excepting As_2O_3 and As_2O_5 which are averages of three; H_2O from (1); corresponds to $(Mn_{21.96}Mg_{1.90})_{\Sigma=23.86}Zn_{3.16}Fe^{3+}_{0.74}(As^{3+}O_3)_{2.02}(As^{5+}O_4)_{2.70}(SiO_4)_{6.44}(OH)_{18}$.

Occurrence: In the zincite zone, as films and lenses in fractures, probably of secondary origin, in a metamorphosed stratiform zinc deposit.

Association: Zincite, willemite, franklinite, calcite, pyrochroite, barite, sphalerite, rhodochrosite, adelite, holdenite.

Distribution: In the Sterling Hill mine, Ogdensburg, Sussex Co., New Jersey, USA.

Name: For Frederick Kraissl, Jr. (1899–1986) and Alice L. Kraissl (1905–1986), of Hackensack, New Jersey, USA, American amateur mineralogists who specialized in Franklin and Sterling Hill minerals.

Type Material: National Museum of Natural History, Washington, D.C., USA, 137017, 137018.

References: (1) Moore, P.B. and J. Ito (1978) Kraisslite, a new platy arsenosilicate from Sterling Hill, New Jersey. Amer. Mineral., 63, 938–940. (2) Dunn, P.J. and J.A. Nelen (1980) Kraisslite and mcgovernite: new chemical data. Amer. Mineral., 65, 957–960.

Crystal Data: Monoclinic. *Point Group:* $2/m$. As equant to short prismatic grains, subhedral to anhedral, to 6 mm.

Physical Properties: *Cleavage:* Perfect on {010} and {001} intersecting at 90°, with a third poor cleavage at a high angle to the other two. Hardness = ∼4 D(meas.) = 3.14(2) for impure material. D(calc.) = 3.10

Optical Properties: Transparent to translucent. *Color:* White to colorless; colorless in thin section. *Streak:* White. *Luster:* Subvitreous, pearly on cleavages.
Optical Class: Biaxial (–). *Orientation:* $X = b$; $Y \wedge a = 6°$; $Z \wedge c = 10.5°$. *Dispersion:* $r > v$, distinct. $\alpha = 1.574(2)$ $\beta = 1.587(2)$ $\gamma = 1.599(2)$ 2V(meas.) = 88(5)° 2V(calc.) = 87°42′

Cell Data: *Space Group:* $P2_1/a$. $a = 8.460(5)$ $b = 10.622(6)$ $c = 7.837(4)$ $\beta = 94°32(8)′$ $Z = 4$

X-ray Powder Pattern: Rush Creek area, California, USA.
3.84 (100), 6.36 (45), 5.34 (45), 3.01 (40), 6.7 (30), 3.94 (30), 3.66 (30)

Chemistry:

	(1)
SiO_2	35.2
TiO_2	0.01
Al_2O_3	0.05
FeO	< 0.01
MnO	< 0.002
MgO	< 0.05
CaO	0.11
SrO	< 0.05
BaO	47.7
K_2O	< 0.1
LOI	16.7
Total	[100.0]

(1) Rush Creek area, California, USA; by D-C arc spectrography, loss on ignition taken as H_2O; stated to be recalculated to 100.0%; corresponds to $(Ba_{1.03}K_{0.01}Ca_{0.01})_{\Sigma=1.05}Si_{1.95}O_{3.95}$ $(OH)_2 \cdot 2.08H_2O$.

Occurrence: In veins cutting sanbornite-quartz-bearing gneissic metamorphic rocks.

Association: Macdonaldite, "opal," witherite, sanbornite, quartz (Rush Creek area, California, USA).

Distribution: In the Rush Creek area, Fresno Co., and Chickencoop Canyon, Tulare Co., California, USA.

Name: For Konrad Bates Krauskopf (1910–), Professor of Geochemistry, Stanford University, Palo Alto, California, USA.

Type Material: n.d.

References: (1) Alfors, J.T., M.C. Stinson, R.A. Matthews, and A. Pabst (1965) Seven new barium minerals from eastern Fresno County, California. Amer. Mineral., 50, 314–340. (2) Coda, A., A. Dal Negro, and G. Rossi (1967) The crystal structure of krauskopfite. Atti Rend. Accad. Lincei, 42(6), 859–873. (3) (1968) Chem. Abs., 69, 46894 (abs. ref. 2).

Crystal Data: Triclinic, pseudomonoclinic.   *Point Group:* $\bar{1}$.   As disseminated subhedral grains, to 200 μm.   *Twinning:* Multiple twinning common.

Physical Properties: Hardness = 5.5–7   D(meas.) = 3.38   D(calc.) = 3.44

Optical Properties: Semitransparent.   *Color:* Deep emerald-green.
Optical Class: Biaxial (+).   *Pleochroism:* Intense; X = yellow-green; Y = blue-green; Z = greenish black, may be an anomalous dark reddish brown. *Orientation:* X = b. $\alpha = 1.712(2)$
$\beta = 1.725(2)$ $\gamma = 1.760(5)$ 2V(meas.) = 61(2)°   2V(calc.) = 64°

Cell Data: *Space Group:* $P\bar{1}$.   a = 10.238(4)   b = 10.642(4)   c = 8.780(3) $\alpha = 105.15(3)$°
$\beta = 96.50(4)$° $\gamma = 125.15(3)$°   Z = 2

X-ray Powder Pattern: Canyon Diablo meteorite.
2.501 (100), 2.655 (90), 2.893 (80), 2.080 (70), 7.92 (60), 3.639 (60), 3.104 (60)

Chemistry:

	(1)	(2)
SiO_2	48.1	49.00
TiO_2	0.5	
Al_2O_3	0.6	
Cr_2O_3	19.1	20.66
FeO	1.8	
MnO	0.1	
MgO	19.7	21.91
CaO	0.1	
Na_2O	9.1	8.43
K_2O	0.0	
Total	99.1	100.00

(1) Canyon Diablo meteorite; by electron microprobe. (2) $Na_2Mg_4Cr_2Si_6O_{20}$.

Mineral Group: Aenigmatite group.

Occurrence: Disseminated within graphite nodules in some octahedrite iron meteorites.

Association: Graphite, roedderite, high albite, richterite, forsterite, kosmochlor, chromite.

Distribution: In the Canyon Diablo, Wichita County, and Youndegin meteorites.

Name: For Evgeny Leonidovich Krinov, noted Russian investigator of meteorites.

Type Material: n.d.

References: (1) Olsen, E. and L. Fuchs (1968) Krinovite, $NaMg_2CrSi_3O_{10}$: a new meteorite mineral. Science, 161, 786–787. (2) (1969) Amer. Mineral., 54, 578 (abs. ref. 1). (3) Bonaccorsi, E., S. Merlino, and M. Pasero (1989) The crystal structure of the meteoritic mineral krinovite, $NaMg_2CrSi_3O_{10}$. Zeits. Krist., 187, 133–138.

Crystal Data: Orthorhombic. *Point Group:* $2/m \, 2/m \, 2/m$. Crystals prismatic along [001], flattened on {100}, up to 7 mm, in fan-shaped aggregates replacing lamprophyllite.

Physical Properties: *Cleavage:* Parting on {010}. *Fracture:* Splintery, uneven. *Tenacity:* Elastic. Hardness = 5.5–6 VHN = 517–571, average 544 (20 g load). D(meas.) = 2.90 D(calc.) = 2.95 Yellow-green cathodoluminescence.

Optical Properties: Transparent. *Color:* White to colorless, rarely with a silver tint; colorless in transmitted light. *Streak:* White. *Luster:* Vitreous. *Optical Class:* Biaxial (–). *Orientation:* X = a; Y = b; Z = c. $\alpha = 1.676(3)$ $\beta = 1.746(3)$ $\gamma = 1.795(3)$ 2V(meas.) = n.d. 2V(calc.) = 77°

Cell Data: *Space Group:* Pccn. a = 28.889(4) b = 8.604(4) c = 5.215(3) Z = 2

X-ray Powder Pattern: Khibiny massif, Russia.
3.009 (100), 14.49 (90), 4.815 (80), 3.722 (65), 6.42 (60)

Chemistry:

	(1)
SiO_2	42.66
TiO_2	27.03
Nb_2O_5	0.83
FeO	0.23
MnO	0.61
ZnO	5.99
Na_2O	16.27
F	0.20
H_2O	[6.26]
$-O = F_2$	[0.08]
Total	[100.00]

(1) Khibiny massif, Russia; by electron microprobe, contains F 0.20%, H_2O by difference; corresponding to $Na_{5.96}(Zn_{0.84}Mn_{0.10})_{\Sigma=0.94}(Ti_{3.85}Nb_{0.07}Fe_{0.04})_{\Sigma=3.96}Si_{8.08}O_{28} \cdot 4.03H_2O$.

Occurrence: In pegmatite veins cutting ijolite-urtite rocks in a differentiated alkalic massif.

Association: Lamprophyllite, arfvedsonite, microcline, nepheline, natrolite.

Distribution: At the Kukisvumchorr apatite deposit, Khibiny massif, Kola Peninsula, Russia.

Name: For the Kukisvumchorr deposit, Kola Peninsula, Russia, in which it occurs.

Type Material: Leningrad Mining Institute, Leningrad; A.E. Fersman Mineralogical Museum, Academy of Sciences, Moscow, Russia.

References: (1) Yakovenchuk, V.N., Y.A. Pakhomovskii, and A.N. Bogdanova (1991) Kukisvumite – a new mineral from the alkaline pegmatites of the Khibiny massif (Kola Peninsula). Mineral. Zhurnal, 13(2), 63–67 (in Russian). (2) (1992) Amer. Mineral., 77, 1116 (abs. ref. 1).

Crystal Data: Triclinic. *Point Group:* 1. As platy crystals up to 0.5 mm.

Physical Properties: *Cleavage:* Poor on {010}. Hardness = 4–5 D(meas.) = 4.3(5) D(calc.) = 4.26 Yellow-green cathodoluminescence.

Optical Properties: Transparent. *Color:* Colorless. *Streak:* White. *Luster:* Adamantine. *Optical Class:* Biaxial (–). *Orientation:* $Z = c$; $X \wedge a = 7°$; $Y \wedge b = 28°$. *Dispersion:* $r > v$. $\alpha = 1.656(1)$ $\beta = 1.700(1)$ $\gamma = 1.703(1)$ 2V(meas.) = 19(1)° 2V(calc.) = 29°

Cell Data: *Space Group:* $P1$. $a = 8.606(6)$ $b = 8.672(8)$ $c = 4.317(3)$ $\alpha = 102.79(6)°$ $\beta = 97.94(5)°$ $\gamma = 116.66(6)°$ $Z = 1$

X-ray Powder Pattern: Kola Peninsula, Russia.
2.793 (10b), 3.710 (9), 3.490 (9), 2.459 (8b), 1.702 (8b), 2.144 (7), 3.310 (6)

Chemistry:

	(1)
SiO_2	17.75
Al_2O_3	7.36
Y_2O_3	56.40
Gd_2O_3	0.39
Dy_2O_3	1.38
Ho_2O_3	0.19
Er_2O_3	2.24
Tm_2O_3	0.10
Yb_2O_3	2.71
Lu_2O_3	0.11
F	13.45
H_2O	[3.58]
$-O = F_2$	5.66
Total	[100.00]

(1) Kola Peninsula, Russia; by electron microprobe, H_2O by difference; corresponding to $(Y_{3.58}RE_{0.27})_{\Sigma=3.85}Al_{1.04}Si_{2.12}O_{7.60}(OH)_{2.88}F_{5.08}$.

Occurrence: As inclusions in fluorite or on fissures in microcline pegmatites.

Association: Thalenite, xenotime, kainosite, bastnäsite, keiviite-(Y), keiviite-(Yb), quartz, fluorite.

Distribution: From the [Keivy massif,] Kola Peninsula, Russia.

Name: For the Kuliok River, Kola Peninsula, Russia, and *yttrium* in its composition.

Type Material: Mining Institute, St. Petersburg; A.E. Fersman Mineralogical Museum, Academy of Sciences, Moscow, Russia.

References: (1) Voloshin, A.V., Y.A. Pakhomovskii, F.N. Tyusheva, Y.V. Sokolova, and Y.K. Yegorov-Tismenko (1986) Kuliokite-(Y) – a new yttrium–aluminum fluoride–silicate from amazonite pegmatites of the Kola Peninsula. Mineral. Zhurnal, 8(2), 94–99 (in Russian with English abs.). (2) (1988) Amer. Mineral., 73, 192 (abs. ref. 1).

Crystal Data: Monoclinic. *Point Group:* n.d. As platy to subhedral crystals, to 2 mm, and in aggregates.

Physical Properties: *Cleavage:* Perfect on {001}. Hardness = ~2 D(meas.) = n.d. D(calc.) = 2.70

Optical Properties: Transparent. *Color:* Colorless. *Streak:* White. *Luster:* Pearly on the cleavage.
Optical Class: Biaxial (–). *Orientation:* $X = c$; $Y = a$; $Z = b$. *Dispersion:* $r < v$. $\alpha = 1.552$
$\beta = 1.5605$ $\gamma = 1.5610$ 2V(meas.) = 24(3)° 2V(calc.) = 27°

Cell Data: *Space Group:* n.d. $a = 5.319(1)$ $b = 9.195(2)$ $c = 23.897(10)$ $\beta = 97°1(2)'$
$Z = 2$

X-ray Powder Pattern: Derrag, Algeria.
7.90 (10), 11.9 (8), 3.38 (8), 4.74 (6), 2.55 (6), 2.96 (5), 2.46 (5)

Chemistry:

	(1)
SiO_2	40.53
Al_2O_3	12.64
MgO	33.19
CaO	0.06
Na_2O	1.20
K_2O	0.07
H_2O	[9.24]
Total	[96.93]

(1) Derrag, Algeria; by electron microprobe, H_2O from theoretical formula; corresponding to $(Na_{0.38}K_{0.01})_{\Sigma=0.39}Mg_{8.02}Al_{0.99}(Si_{6.57}Al_{1.43})_{\Sigma=8.00}O_{20}(OH)_{10}$.

Polymorphism & Series: A 1:1 regular interstratification of trioctahedral chlorite and talc.

Occurrence: In dolostone, metamorphosed below 400 °C, with evaporites.

Association: Talc, chlorite, phlogopite, tourmaline.

Distribution: From El Mourdur Hill, near Derrag, 35 km west of Ksar El Boukhari, Tell Atlas Mountains, Algeria.

Name: For Dr. Holger Kulke, geologist of Essen, Germany, who provided the original specimen.

Type Material: Ruhr University, Bochum, Germany; National Museum of Natural History, Washington, D.C., USA, 147361.

References: (1) Abraham, K., W. Schreyer, O. Medenbach, and W. Gebert (1980) Kulkeite, ein geordnetes 1:1 Mixed-Layer-Mineral zwischen Klinochlor und Talk. Fortschr. Mineral., Beiheft 1, 58, 4–5 (in German). (2) (1981) Amer. Mineral., 66, 218 (abs. ref. 1). (3) Schreyer, W., O. Medenbach, K. Abraham, W. Gebert, and W.F. Müller (1982) Kulkeite, a new metamorphic phyllosilicate mineral: ordered 1:1 chlorite/talc mixed layer. Contr. Mineral. Petrol., 80, 103–109.

$(K, Na)_3(Mn, Fe^{2+})_7(Ti, Nb)_2Si_8O_{24}(O, OH)_7$ **Kupletskite**

Crystal Data: Monoclinic, or triclinic, pseudomonoclinic. *Point Group:* n.d. In thin plates and lamellar masses, to 5 cm; as radial masses.

Physical Properties: *Cleavage:* Perfect on {100}. Hardness = 3–4   D(meas.) = 3.20–3.36 D(calc.) = [3.05]

Optical Properties: Semitransparent. *Color:* Bright straw-yellow when fresh, dark brown to black when altered to manganese oxides. *Streak:* Brown. *Luster:* Vitreous. *Optical Class:* Biaxial (–). *Pleochroism:* X = yellow-orange; Z = cinnamon-brown. *Dispersion:* $r > v$, strong. $\alpha = [1.656]–1.660$ $\beta = 1.699–1.702$ $\gamma = 1.731–1.734$ 2V(meas.) = 79°–85°

Cell Data: *Space Group:* n.d.   a = 13.14   b = 12.82   c = 5.42 $\alpha = 93°94'$ $\beta = 101°40'$ $\gamma = 113°38'$ Z = 1

X-ray Powder Pattern: Lovozero massif, Russia; nearly identical with astrophyllite. 3.505 (8), 2.642 (8), 2.573 (4), 2.099 (3), 1.732 (3), 3.249 (1), 2.998 (1)

Chemistry:

	(1)	(2)		(1)	(2)
SiO_2	32.60	33.54	CaO	3.60	1.45
TiO_2	12.04	10.64	BaO		0.32
ZrO_2	1.19	0.00	Na_2O	2.14	2.14
Al_2O_3	1.68	1.00	K_2O	4.38	5.63
Fe_2O_3	5.44	5.35	F	1.22	
Nb_2O_5	0.66	2.48	H_2O^+	3.83	3.90
FeO		7.80	H_2O^-	1.08	0.80
MnO	27.65	23.60	$-O = F_2$	0.51	
MgO	2.98	1.63	Total	99.98	100.28

(1) Mt. Kuivchorr, Lovozero massif, Russia; total Fe as Fe_2O_3. (2) Mt. Nepkhe, Lovozero massif, Russia.

Polymorphism & Series: Forms two series, with astrophyllite, and with cesium kupletskite.

Mineral Group: Astrophyllite group.

Occurrence: In the natrolite-rich central part and in the wall rocks of nepheline syenite pegmatites in a differentiated alkalic massif (Lovozero massif, Russia).

Association: Manganoan pectolite, neptunite, microcline, eudialyte, nepheline, aegirine (Lovozero massif, Russia); aegirine, analcime, gonnardite, apophyllite, eggletonite, pyrophanite (Little Rock, Arkansas, USA).

Distribution: In Russia, in the Lovozero massif, Kola Peninsula, and some other less-well-defined localities. From the Oktyabr massif, eastern Azov area, Ukraine. In the Ilímaussaq intrusion and the Werner Berge complex, Greenland. At Gjerdingen, north of Oslo, Norway. In the 3M quarry, Little Rock, Pulaski Co., Arkansas, USA. At Mont Saint-Hilaire, Quebec, Canada.

Name: For Boris Mikhailovich Kupletsky (1894–?) and Elsa Maximilianovna Bohnshtedt-Kupletskaya (?–1974), Russian geologists.

Type Material: n.d.

References: (1) Semenov, E.I. (1956) Kupletskite – a new mineral of the astrophyllite group. Doklady Acad. Nauk SSSR, 108, 933–936 (in Russian). (2) (1957) Amer. Mineral., 42, 118–119 (abs. ref. 1). (3) Vlasov, K.A., M.V. Kuz'menko, and E.M. Es'kova (1966) The Lovozero alkaline massif. Akad. Nauk SSSR, 370–373 (in English). (4) Val'ter, A.A, G.K. Eremenko, and T.A. Lysenko (1965) Kupletskite from the Azov area alkaline rocks. Mineralog. Sb., L'vovsk. Gos. Univ., 19(2), 248–252 (in Russian). (5) (1966) Chem. Abs., 65, 495 (abs. ref. 4). (6) Chih-Chung P'eng and Che-Sheng Ma (1964) Crystal structure of triclinic manganastrophyllite. Sci. Sinica, 13(7), 1180–1183 (in Russian). (7) (1966) Chem. Abs., 64, 7460 (abs. ref. 6).

Kurumsakite

$$(\mathrm{Zn, Ni, Cu})_8 \mathrm{Al}_8 \mathrm{V}_2^{5+} \mathrm{Si}_5 \mathrm{O}_{35} \cdot 27\mathrm{H}_2\mathrm{O}(?)$$

Crystal Data: Orthorhombic (?). *Point Group:* n.d. As radiating to finely felted fibers.

Physical Properties: Hardness = n.d. D(meas.) = 4.03 D(calc.) = n.d.

Optical Properties: Transparent to translucent. *Color:* Yellowish green to bright yellow. *Luster:* Vitreous to silky.
Optical Class: Biaxial (+). $\alpha = 1.616$ $\beta =$ n.d. $\gamma = 1.622$–1.623 2V(meas.) = $\sim$35°

Cell Data: *Space Group:* n.d. Z = n.d.

X-ray Powder Pattern: Kurumsak, Kazakhstan. (ICDD 29-571, corrected lines).
1.53 (100), 3.91 (75), 2.61 (50), 2.28 (38), 4.91 (25), 1.42 (25), 2.42 (13)

Chemistry:

	(1)
SiO_2	13.82
Al_2O_3	20.51
Fe_2O_3	2.15
V_2O_5	8.50
NiO	7.33
CuO	3.05
ZnO	17.55
MgO	0.92
CaO	1.24
H_2O	23.25
SO_3	1.15
Total	99.47

(1) Kurumsak, Kazakhstan.

Occurrence: In bituminous schists in the walls of cavities and open fissures.

Association: n.d.

Distribution: From Kurumsak, near Dzhambul, Kara-Tau Mountains, Kazakhstan.

Name: For the locality at Kurumsak, Kazakhstan.

Type Material: n.d.

References: (1) Ankinovich, E.A. (1954) [title unknown] Izv. Akad. Nauk Kazakhstan SSR, 134, Ser. Geol. 19, 116. (2) Bohnshtedt-Kupletskaya, E.N. (1955) Zap. Vses. Mineral. Obshch., 84, 343–344 (abs. ref. 1, in Russian). (3) (1957) Mineral. Abs., 13, 207 (abs. ref. 1). (4) (1957) Amer. Mineral., 42, 583–584 (abs. ref. 2).

Crystal Data: Orthorhombic. *Point Group:* $2/m\ 2/m\ 2/m$. Anhedral plates, flattened on {010}; as aggregates, to 15 mm.

Physical Properties: *Cleavage:* Good on {010}; poor on {101}, intersecting at 83°. Hardness = 5.5–6 D(meas.) = 2.55 D(calc.) = 2.53

Optical Properties: Transparent. *Color:* Pink with a violet hue. *Luster:* Vitreous, pearly on {010} cleavage.
Optical Class: Biaxial (+). *Orientation:* $X = a$; $Y = c$; $Z = b$. $\alpha = 1.522(1)$ $\beta = 1.522(1)$ $\gamma = 1.543(1)$ 2V(meas.) = 0°–9°

Cell Data: *Space Group:* $Pcab$. $a = 10.213(2)$ $b = 15.878(4)$ $c = 9.058(1)$ $Z = 4$

X-ray Powder Pattern: Ilímaussaq intrusion, Greenland.
3.117 (10), 4.355 (7), 3.880 (7), 4.447 (6), 3.388 (6), 3.313 (5), 2.576 (5)

Chemistry:

	(1)
SiO_2	65.00
TiO_2	< 0.02
Al_2O_3	< 0.02
Y_2O_3	0.40
FeO	0.05
MnO	1.59
MgO	< 0.02
CaO	8.49
Na_2O	22.06
K_2O	< 0.02
Total	97.59

(1) Ilímaussaq intrusion, Greenland; by electron microprobe, average of 16 analyses on two grains; TGA gave 4% weight loss at 600 °C; corresponds to $Na_{3.96}(Ca_{0.84}Mn_{0.12}Y_{0.02})_{\Sigma=0.98}$ $Si_{6.01}O_{14}(OH)_2$.

Occurrence: As patches, streaks, and veinlets cutting arfvedsonite-bearing nepheline syenite; probably of late-stage hydrothermal origin.

Association: Villiaumite, analcime.

Distribution: On the Kvanefjeld Plateau, in the Ilímaussaq intrusion, southern Greenland.

Name: For the type locality, which is Danish for *the mountain where the angelica grows.*

Type Material: University of Copenhagen, Copenhagen, Denmark, 1983.136, GGU 47887; Harvard University, Cambridge, Massachusetts, 127475; National Museum of Natural History, Washington, D.C., USA, 162234.

References: (1) Petersen, O.V. and O. Johnsen (1984) Kvanefjeldite, a new mineral species from the Ilímaussaq alkaline complex, southwest Greenland. Can. Mineral., 22, 465–467. (2) Johnsen, O., E.S. Leonardsen, L. Fälth, and H. Annehed (1983) Crystal structure of kvanefjeldite: the introduction of $^2_\infty[Si_3O_7OH]$ layers with eight-membered rings. Neues Jahrb. Mineral., Monatsh., 505–512. (3) (1985) Amer. Mineral., 70, 873 (abs. refs. 1 and 2).

Crystal Data: Triclinic. *Point Group:* $\bar{1}$. Crystals bladed and tabular on {100}, elongated ‖ to [001], to 0.5 m; typically bent or twisted. *Twinning:* Lamellar on {100}, twin axis ⊥ {100} or ‖ [010] or [001], common.

Physical Properties: *Cleavage:* Perfect on {100}, good on {010}, (100) ∧ (010) = 79°; parting on {001}. *Fracture:* Splintery. Hardness = 5.5 ‖ [001], 7 ‖ [100]. D(meas.) = 3.53–3.65 D(calc.) = [3.67]

Optical Properties: Transparent to translucent. *Color:* Blue, white, rarely green, gray, yellow, pink, black, can be zoned; colorless to pale blue in thin section. *Luster:* Vitreous to pearly. *Optical Class:* Biaxial (–). *Pleochroism:* Weak; in thick sections, X = colorless; Y = violet-blue; Z = cobalt blue. *Orientation:* $X \simeq$ ⊥ {100} on {100}; $X' \wedge a = 0°$–25° on {001}; $Z' \wedge c = 27°$–32° on {100}; $Z' \wedge c = 5°$–8° on {010}. *Dispersion:* $r > v$, weak. $\alpha = 1.710$–1.718 $\beta = 1.719$–1.724 $\gamma = 1.724$–1.734 2V(meas.) = 78°–83°

Cell Data: *Space Group:* $P\bar{1}$. a = 7.1262(12) b = 7.8520(10) c = 5.5724(10) $\alpha = 89.99(2)°$ $\beta = 101.11(2)°$ $\gamma = 106.03(1)°$ Z = 4

X-ray Powder Pattern: Zillertal, Austria. (ICDD 11-46). 3.18 (100), 1.377 (75), 3.35 (65), 1.962 (55), 1.935 (50), 1.930 (50), 2.520 (30)

Chemistry:

	(1)	(2)		(1)	(2)
SiO$_2$	36.68	37.08	MnO	0.00	
TiO$_2$	0.00		MgO	0.02	
Al$_2$O$_3$	63.28	62.92	CaO	0.01	
Cr$_2$O$_3$	0.06		Na$_2$O	0.00	
FeO	0.37		K$_2$O	0.01	
			Total	100.43	100.00

(1) Elovyi, Russia; by electron microprobe. (2) Al$_2$SiO$_5$.

Polymorphism & Series: Trimorphous with andalusite and sillimanite.

Occurrence: In gneisses, schists, included pegmatites, and quartz veins, from moderately high-pressure regional metamorphism of principally pelitic rocks; detrital in sedimentary rocks.

Association: Staurolite, andalusite, sillimanite, talc, "hornblende," gedrite, mullite, corundum.

Distribution: Widely distributed, even in good crystals. From Mt. Greiner, Zillertal, Tirol, Austria. At Alpe Sponda, Pizzo Forno, and at Alpe Campolungo, Tessin, Switzerland. In the Pfitschtal, Trentino-Alto Adige, Italy. From Röros, Norway. Around Yekaterinburg (Sverdlovsk), Ural Mountains; at Elovyi, Klavoloke, Karelia, Russia. From Sultan Hamud, Machakos district, Kenya. Large crystals from a number of localities in Minas Gerais, Brazil, as at Barro Prêto, São José do Jacuri. In the USA, at Lyme, Grafton Co., New Hampshire; at Judd's Bridge, Litchfield Co., Connecticut; at Darby, Delaware Co., Pennsylvania; from near Bakersville, Mitchell Co., and near Burnsville, Yancy Co., North Carolina.

Name: From the Greek for *blue*, in allusion to its common dark blue color.

Type Material: Mining Academy, Freiberg, Germany, 22491.

References: (1) Dana, E.S. (1892) Dana's system of mineralogy, (6th edition), 500–501 [cyanite]. (2) Deer, W.A., R.A. Howie, and J. Zussman (1982) Rock-forming minerals, (2nd edition), v. 1A, orthosilicates, 780–800. (3) Parkin, K.M., B.M. Loeffler, and R.G. Burns (1977) Mössbauer spectra of kyanite, aquamarine, and cordierite showing intervalence charge transfer. Phys. Chem. Minerals, 1, 301–311. (4) Winter, J.K. and S. Ghose (1979) Thermal expansion and high-temperature crystal chemistry of the Al$_2$SiO$_5$ polymorphs. Amer. Mineral., 64, 573–586.

Crystal Data: Triclinic. *Point Group:* $\overline{1}$. Crystals typically thin, tabular along [010], and rhombic in section, to 4 cm. Cleavable, granular, or massive. *Twinning:* Common according to Albite, Pericline, Carlsbad, Baveno, or Manebach laws.

Physical Properties: *Cleavage:* Perfect on {001}, less perfect on {010}, intersecting at 90°; distinct on {110}. *Fracture:* Uneven to conchoidal. *Tenacity:* Brittle. Hardness = 6–6.5 D(meas.) = 2.68–2.72 D(calc.) = 2.685–2.710

Optical Properties: Translucent to transparent. *Color:* Gray, brown, greenish, blue, yellow, colorless; commonly exhibits iridescence, especially on {010}; colorless in thin section. *Streak:* White. *Luster:* Vitreous; pearly on cleavages.
Optical Class: Biaxial (+). *Dispersion:* $r > v$, weak. $\alpha = 1.555$–1.565 $\beta = 1.558$–1.569 $\gamma = 1.563$–1.573 2V(meas.) = 78°–87° (low); 80° to −81° (high).

Cell Data: *Space Group:* $C\overline{1}$ (low). $a = 8.1648$ $b = 12.8585$ $c = 7.0973$ $\alpha = 90°34'$ $\beta = 116°06'$ $\gamma = 89°47'$ Z = 4

X-ray Powder Pattern: Synthetic $Na_{0.40}Ca_{0.60}Al_{1.60}Si_{2.40}O_8$ (high).
3.20 (100), 4.03 (80), 3.75 (80), 3.17 (80), 3.63 (70), 2.94 (70), 2.51 (70)

Chemistry:

	(1)	(2)	(3)		(1)	(2)	(3)
SiO_2	52.97	55.59	50.54	Na_2O	3.97	5.73	3.40
Al_2O_3	29.41	28.30	31.70	K_2O	0.26		
Fe_2O_3	0.76			H_2O	0.22		
CaO	12.59	10.38	14.36	Total	100.18	100.00	100.00

(1) San Marcos Mountains, California, USA. (2) $Na_{0.50}Ca_{0.50}Al_{1.50}Si_{2.50}O_8$.
(3) $Na_{0.30}Ca_{0.70}Al_{1.70}Si_{2.30}O_8$.

Polymorphism & Series: High- and low-temperature structural varieties are recognized.

Mineral Group: Feldspar group, plagioclase series.

Occurrence: In mafic igneous rocks and anorthosites. Rarer in metamorphic rocks, as amphibolites. As detrital grains in sedimentary rocks.

Association: Olivine, pyroxenes, amphiboles, magnetite.

Distribution: Widespread. From Ford Harbour, Pauls Island, Labrador, Newfoundland; at Lake St. John, Quebec; and elsewhere in Canada. In the USA, especially in northern New York, forming the Adirondack Mountains; in the Ouachita Mountains of Arkansas; crystals from Sagebrush Flat, about 37 km north of Plush, Lake Co., Oregon; atop the San Marcos Mountains, San Diego Co., and in the western San Gabriel Mountains, Los Angeles Co., California. Abundant gem crystals in the Pinacate volcanic field, Sonora, Mexico. At Vesuvius, Campania, and on Mt. Etna, Sicily, Italy. From Ylämaa, near Lappeenranta, Finland. In the Langesundsfjord-Larvik-Tvedalen area, Norway. On Surtsey Island, south of Iceland.

Name: For the type locality on Pauls Island, off the coast of Labrador, Canada.

References: (1) Dana, E.S. (1892) Dana's system of mineralogy, (6th edition), 325–327, 334–337. (2) Deer, W.A., R.A. Howie, and J. Zussman (1963) Rock-forming minerals, v. 4, framework silicates, 94–165. (3) Phillips, W.R. and D.T. Griffen (1981) Optical mineralogy, 352–360. (4) Goodyear, J. and W.J. Duffin (1954) The identification and determination of plagioclase feldspars by the X-ray powder method. Mineral. Mag., 30, 306–326. (5) Stewart, D.B, G.W. Walker, T.L. Wright, and J.J. Fahey (1966) Physical properties of calcic labradorite from Lake County, Oregon. Amer. Mineral., 51, 177–197. (6) Wenk, H.-R., W. Joswig, T. Tagai, M. Korekawa, and B.K. Smith (1980) The average structure of An 62-66 labradorite. Amer. Mineral., 65, 81–95.

Labuntsovite

$(K, Na, Ba)_8(Ti, Nb)_9(Si_4O_{12})_4(O, OH)_{10} \cdot nH_2O$

Crystal Data: Monoclinic. *Point Group:* $2/m$. As simple prismatic crystals, to several cm, with $\{001\}$, $\{010\}$, $\{110\}$, $\{011\}$, and $\{111\}$; in radiating clusters.

Physical Properties: *Cleavage:* Perfect on $\{100\}$. *Fracture:* Subconchoidal. *Tenacity:* Brittle. Hardness = ~6 D(meas.) = 2.89–3.02 D(calc.) = [3.00]

Optical Properties: Transparent to translucent. *Color:* Rose to brownish yellow. *Streak:* White. *Luster:* Greasy. *Optical Class:* Biaxial (+). *Pleochroism:* X = yellowish; Z = yellowish brown. *Orientation:* $X = b$; $Y = c$. *Absorption:* $Z > Y > X$. $\alpha = 1.684$–1.694 $\beta = 1.693$–1.702 $\gamma = 1.792$–1.814 2V(meas.) = $41°$

Cell Data: *Space Group:* $I2/m$. $a = 14.18$ $b = 15.48$ $c = 13.70$ $\gamma = 117°$ $Z = [2]$

X-ray Powder Pattern: Lovozero massif, Russia.
3.15 (10), 2.56 (9), 1.543 (9), 1.413 (9), 3.09 (8), 1.677 (8), 2.47 (7)

Chemistry:

	(1)	(2)		(1)	(2)
SiO_2	39.59	39.53	MgO	0.42	1.86
TiO_2	25.49	25.56	CaO	1.19	0.02
Al_2O_3	1.30	0.42	BaO	[8.61]	9.25
Fe_2O_3	1.56	0.42	Na_2O	3.18	5.66
Nb_2O_5	1.45	0.25	K_2O	7.23	7.16
MnO	2.34		H_2O	7.91	9.25
			Total	[100.27]	99.38

(1) Lovozero massif, Russia; BaO originally given as 6.61%; corresponds to $(K_{3.44}Na_{2.32}Ba_{1.28}Mn_{0.72}Ca_{0.48})_{\Sigma=8.24}(Ti_{7.12}Fe^{3+}_{0.40}Nb_{0.24}Mg_{0.24})_{\Sigma=8.00}(Si_{14.64}Al_{0.56})_{\Sigma=15.20}[O_{46.32}(OH)_{9.68}]_{\Sigma=56.00} \cdot 4.8H_2O$. (2) Green River Formation, Wyoming, USA.

Occurrence: A hydrothermal alteration product of murmanite in alkalic pegmatites in a differentiated alkalic massif (Kola Peninsula, Russia); in carbonatite (Kovdor massif, Russia).

Association: Albite, natrolite, aegirine, nepheline, microcline, eudialyte, lorenzenite, murmanite, anatase (Kola Peninsula, Russia); elpidite (Mont Saint-Hilaire, Canada).

Distribution: In the Lovozero, Khibiny, and Kovdor massifs and the Vuoriyärvi carbonatite complex, Kola Peninsula, and the Murun massif, southwest of Olekminsk, Yakutia, Russia. From the Green River Formation, in the Trona mine, Sweetwater Co., Wyoming, USA. At Mont Saint-Hilaire, Quebec, Canada.

Name: For Aleksander Nikolaevich Labuntsov and Ekaterina Eutikhieva Labuntsov-Kostyleva, Russian mineralogists.

Type Material: National School of Mines, Paris, France; National Museum of Natural History, Washington, D.C., USA, 113995.

References: (1) Semenov, E.I. and T.A. Burova (1955) On the new mineral labuntsovite and on the so-called titanoelpidite. Doklady Acad. Nauk SSSR, 101, 1113–1116 (in Russian). (2) (1956) Amer. Mineral., 41, 163 (abs. ref. 1). (3) Golovastikov, N.I. (1973) Crystal structure of the alkali titanosilicate labuntsovite. Kristallografiya (Sov. Phys. Crystal.), 18, 950–955 (in Russian). (4) Vlasov, K.A., M.V. Kuz'menko, and E.M. Es'kova (1966) The Lovozero alkaline massif. Akad. Nauk SSSR, 354–356 (in English). (5) Kapustin, Y.L. (1971) Mineralogy of carbonatites. Nauka, Moscow, 187–189 (in English). (6) Mandarino, J.A. and V. Anderson (1989) Monteregian Treasures. Cambridge Univ. Press, 120.

Crystal Data: Monoclinic, pseudo-orthorhombic. *Point Group:* $2/m$. Crystals subhedral, thick tabular to short prismatic, to 1 mm; as anhedral grains.

Physical Properties: *Cleavage:* Perfect on {100} and {010}, two less perfect $\sim\perp$ {100} and intersecting at 60°. Hardness = 5.5–6.5 VHN = 617–833 D(meas.) = 3.967 D(calc.) = 4.11 Moderately magnetic and electromagnetic.

Optical Properties: Opaque, translucent in very thin section. *Color:* Black; gray in reflected light. *Streak:* Pale brown, blackish brown. *Luster:* Submetallic to metallic. *Pleochroism:* Gray to grayish black. *Anisotropism:* Weak.

Cell Data: *Space Group:* $P2_1/b$. $a = 4.805(2)$ $b = 10.189(9)$ $c = 17.403(9)$ $\alpha = 91.0(2)°$ $Z = 12$

X-ray Powder Pattern: Little Lai-He Village, China.
3.488 (10b), 2.521 (10), 2.774 (9), 1.745 (8), 2.405 (7), 2.246 (7), 2.175 (6)

Chemistry:

	(1)
SiO_2	31.07
Fe_2O_3	44.24
FeO	23.64
MgO	0.87
CaO	0.21
Total	100.03

(1) Little Lai-He Village, China; corresponds to $(Fe_{1.18}^{2+}Mg_{0.08})_{\Sigma=1.26}Fe_{2.00}^{3+}(Si_{0.93}O_4)_2$.

Polymorphism & Series: 3M, 2M polytypes.

Occurrence: In a Precambrian metamorphic iron deposit, probably derived by oxidation of fayalite (Little Lai-He Village, China).

Association: Quartz, "hypersthene," magnetite, "hornblende," augite, plagioclase (Little Lai-He, China).

Distribution: At Little Lai-He village, Liaoning Province, China. From Kamitaga, Shizuoka Prefecture, and Yugawara, Kanagawa Prefecture, Japan. In the Cherkassk massif, Kuraminsk, Siberia, Russia.

Name: For the occurrence at Little Lai-He Village, China.

Type Material: n.d.

References: (1) Laihunite Research Group, Guiyang Institute of Geochemistry, Academia Sinica and Geological Team 101, Liaoning Metallurgical and Geological Prospecting Company (1976) Laihunite, a new iron silicate mineral. Geochimica, 2, 95–103 (in Chinese with English abs.). (2) X-ray Laboratory, Guiyang Institute of Geochemistry, Academia Sinica (1976) The crystal structure of laihunite. Geochimica, 2, 104–105 (in Chinese with English abs.). (3) (1977) Amer. Mineral., 62, 1058 (abs. refs. 1 and 2). (4) Ferrifayalite Research Group, Department of Geology, Peking University and Institute of Geology and Mineral Resources, Chinese Academy of Geological Sciences (1976) Ferrifayalite and its crystal structure. Acta Geologica Sinica, 2, 161–175 (in Chinese with English abs.). (5) (1978) Amer. Mineral., 63, 424–425 (abs. ref. 4). (6) Shen, B., O. Tamada, M. Kitamura, and N. Morimoto (1986) Superstructure of laihunite-3M ($\square_{0.40}Fe_{0.80}^{2+}Fe_{0.80}^{3+}SiO_4$). Amer. Mineral., 71, 1455–1460.

Lamprophyllite $(Na, Mn, K)_3(Sr, Ba)_2Ti_3O_2(Si_2O_7)_2(O, OH, F)_2$

Crystal Data: Monoclinic. *Point Group:* $2/m$. Crystals flattened on $\{100\}$, elongated along $[001]$, rarely terminated, to 20 cm; as stellate clusters and sheaflike aggregates. *Twinning:* Common on $\{100\}$, rarely polysynthetic.

Physical Properties: *Cleavage:* Perfect micaceous on $\{100\}$; poor on $\{010\}$. *Fracture:* Uneven. *Tenacity:* Brittle. Hardness = 2–3 D(meas.) = 3.44–3.53 D(calc.) = [3.48]

Optical Properties: Opaque, transparent to translucent in thin section. *Color:* Dark brown, brownish yellow, straw-yellow, reddish brown with a golden tint. *Streak:* Pale yellow, may have an orange tint. *Luster:* Submetallic on cleavages.
Optical Class: Biaxial (+). *Pleochroism:* X = yellow; Y = straw-yellow; Z = orange or straw-yellow. *Orientation:* $X \perp b$; $Z \wedge c \simeq 4°–8°$. *Dispersion:* $r > v$, strong. $\alpha = 1.735–1.749$ $\beta = 1.747–1.754$ $\gamma = 1.770–1.781$ 2V(meas.) = 21°–41°

Cell Data: *Space Group:* $C2/m$. $a = 19.431(3)$ $b = 7.086(1)$ $c = 5.392(1)$ $\beta = 96.75(5)°$ $Z = 2$

X-ray Powder Pattern: Kola Peninsula, Russia. (ICDD 17-751).
2.773 (100), 3.43 (55), 2.130 (45), 1.477 (45), 3.73 (40), 3.27 (40), 2.874 (40)

Chemistry:

	(1)	(2)		(1)	(2)
SiO_2	30.40	30.93	SrO	14.58	14.12
TiO_2	27.48	27.14	BaO	1.25	3.49
Al_2O_3		0.56	Na_2O	12.35	10.64
Fe_2O_3	3.87	1.36	K_2O	2.31	1.26
$(Nb, Ta)_2O_5$		0.16	F	1.82	1.10
FeO	1.97	2.31	H_2O^+		0.56
MnO	2.33	3.46	H_2O^-		0.16
MgO	0.25	0.40	LOI	0.60	
CaO	1.41	2.48	$-O = F_2$	0.77	0.46
			Total	99.85	99.67

(1) Kukisvumchorr apatite deposit, Khibiny massif, Russia. (2) Mt. Karnasurt, Lovozero massif, Russia.

Occurrence: In all rock types and most pegmatites of a differentiated alkalic massif (Lovozero massif, Russia).

Association: Microcline, nepheline, aegirine, lorenzenite, eudialyte, lomonosovite, murmanite, villiaumite (Lovozero massif, Russia); microcline, nepheline, catapleiite, aegirine, calcite, fluorite, analcime, pectolite (Pilansberg, South Africa).

Distribution: In Russia, in the Khibiny and Lovozero massifs, Kola Peninsula; in the Konder, Inagli, and Yllymakh massifs, Yakutia. In the Gardiner complex, beyond the head of Kangerdlugssuaq Fjord, Greenland. From the Langesundsfjord, Norway. At Pilansberg, Transvaal, South Africa. In the Rocky Boy stock, Bear Paw Mountains, Hill Co., Montana, USA. At Mont Saint-Hilaire, Quebec, Canada. From Pedro Balao, Pocos dos Caldas, Minas Gerais, and Serrote, Cascata, São Paulo, Brazil.

Name: From the Greek for *shining* and *leaf*, for its lustrous cleavage.

References: (1) Dana, E.S. (1899) Dana's system of mineralogy, (6th edition), app. I, 40. (2) Vlasov, K.A., Ed. (1966) Mineralogy of rare elements, v. II, 207–209. (3) Vlasov, K.A., M.V. Kuz'menko, and E.M. Es'kova (1966) The Lovozero alkaline massif. Akad. Nauk SSSR, 361–370 (in English). (4) Saf'yanov, Y.N., N.O. Vasil'eva, V.P. Golovachev, E.A. Kuz'min, and N.V. Belov (1983) Crystal structure of lamprophyllite. Doklady Acad. Nauk SSSR, 269, 117–120 (in Russian).